Electrophoresis '82
Advanced Methods
Biochemical and Clinical Applications

Electrophoresis '82

Advanced Methods
Biochemical and Clinical Applications

Proceedings of the International
Conference on Electrophoresis
Athens, Greece, April 21–24, 1982

Editor
D. Stathakos

Walter de Gruyter · Berlin · New York 1983

Editor:

Dimitri Stathakos, Dr. rer. nat.
Nuclear Research Center Demokritos
Department of Biology
Aghia Paraskevi
Athens
Greece

CIP-Kurztitelaufnahme der Deutschen Bibliothek

Electrophoresis ...: advanced methods, biochem. and clin. applications;
proceedings of the Internat. Conference on Electrophoresis. –
Berlin; New York: de Gruyter
NE: International Conference on Electrophoresis
4. 1982. Athens, Greece, April 21 – 24, 1982. – 1983.
ISBN 3-11-008791-X

Library of Congress Cataloging in Publication Data

International Conference on Electrophoresis (4th: 1982: Athens, Greece)
Electrophoresis '82
Bibliography: p.
Includes index.
1. Electrophoresis-Congresses. 2. Biological chemistry-Technique-Congresses.
3. Chemistry, Clinical-Technique-Congresses. I. Stathakos, D. (Dimitri), 1931 –
II. Title.
QP519.9.E434157 1982 574.19'285 83-7199
ISBN 3-11-008791-X

Printing: Gerike GmbH, Berlin. – Binding: Lüderitz & Bauer Buchgewerbe GmbH, Berlin.
Printed in Germany.

PREFACE

The International Conference "Electrophoresis '82" was held at the School of Natural Sciences in Athens on April 21-24, 1982, under the auspices of the Ministry of Culture and Sciences and the Greek Atomic Energy Commission. It was the fourth of a projected series of international meetings organized with the objective of stimulating communication of ideas, information exchange and the advancement of knowledge in all areas of electrophoresis.

This Conference, held in conjunction with the Annual Meeting of the Electrophoresis Society, was attended by about 340 participants from 26 countries, representing a broad spectrum of scientific disciplines and research fields and included most of the recognized leaders in the field.

In an experimental approach, toward a present-day *agora*, we aimed at maximal level of communication in theory and ideas, laboratory expertise, new results, layman's applications, as well as speculation in future trends. Accordingly, leading companies in the field were asked to hold up-to-date "laboratory workshops" during the Conference. There, in the actual milieu of the laboratory, the participants could work *in situ* on the latest methodology. This integration of lectures, poster sessions, laboratories and round-tables within the same building on the slope of Mount Hymettus created a lively community and led to a very successful meeting.

This volume contains a large portion of the papers presented at "Electrophoresis '82". They are compiled in the following sections: I. Plenary Lecture. II. Theory and Methods. III. Two-Dimensional Separations. IV. Biological and Biomedical Applications. V. Round-Table Discussions. As it was already felt during the sessions and became fully evident now, very few manuscripts belong solely where they are listed; many expand thematically beyond one or even two sections. This high sophistication in a field of composite methodology, which characterized the whole meeting, was most vividly reflected in the final round-table discussions. These are also included here, thanks to the indefatigable efforts of the Chairmen who succeeded, each in his own manner, in conveying on paper the exciting uniqueness of each session. The editor contributed only his persistence.

This volume could not have been produced without the cooperation of all contributors in the preparation of their manuscripts and I would like to express my thanks to them. I also greatly appreciate the efforts of the staff of Walter de Gruyter towards the publication of this volume.

Athens Dimitri Stathakos

ACKNOWLEDGEMENTS

I am indebted to the members of the Organizing Committee of "Electrophoresis '82", Drs. C. Coutsogeorgopoulos, E. Frangoulis and C. Sekeris for their wholehearted and valuable collaboration. I would also like to thank the staff of the Biochemistry Department, School of Natural Sciences, and in particular Mr. L. Pastelakos of the Nuclear Research Center "Demokritos", for their splendid help in running the whole meeting. Furthermore, I wish to thank FMC-Marine Colloids, LKB Produkter, Pharmacia Fine Chemicals and Serva Feinbiochemica for their contribution through their excellent "laboratory workshops", and to assure them that, to my knowledge, all participants and especially the students share this statement.

Financial aid for the Conference was provided by the Greek Ministry of Culture and Sciences, the National Tourist Organization of Greece, Olympic Airways, AC&R Advertising and the Electrophoresis Society.

CONTENTS

SECTION V: ROUND-TABLE DISCUSSIONS

SECTION I PLENARY LECTURE

EVOLUTION OF IDEAS IN ELECTROPHORETIC DEVELOPMENTS - Selected Highlights

Alexander Kolin
Molecular Biology Institute
University of California, Los Angeles

To begin with, I wish to thank Dr. Stathakos and the Organizing Committee of the International Electrophoresis Congress for their kind invitation to deliver this lecture in Athens, the cradle of western science and civilization. It is particularly gratifying to do so under the auspices of the research center that bears the name of the father of the atomic concept.

The content of this lecture is limited and determined to a great extent by shortage of time. It is impossible to do justice to developments of the past two centuries in about an hour. Consequently, instruments and techniques will not be described and we shall limit ourselves to methodological highlights. Their choice is obviously personal and to some extent arbitrary. Inclusion or exclusion of a method or an investigator's name is not a reflection on their deemed importance but, rather, on the limitation of time. Thus, for instance, continuous flow methods will not be discussed, although I have devoted much time and effort to methods of this type. Finally, the amount of time devoted to a subject will not be proportional to its importance. Significant and generally well known topics will get only brief treatment. Unfamiliar ones will be discussed more fully.

Antiquity

Appropriately for this location, we shall begin in ancient

Electrophoresis '82

Greece. Thales of Miletus (about 600 B.C.), who is generally considered to be the father of Greek science, described a curious transformation in amber ("electron" in Greek) when it was rubbed. It exhibited the ability to attract small light objects, like a piece of straw, and then to repel them after contact. Thales had the means at his disposal to discover electrophoresis in air. He could have observed dust particles in dark-field illumination by viewing sideways a beam of light penetrating into a dark temple through a crack in the wall. Approaching an electrified piece of amber to the light beam, he could have observed attraction of oppositely charged dust particles toward it and repulsion of particles of like charge from it. But there is no record of such an observation in antiquity.

The Discoveries of Reuss

Despite the potential for discovery of electrophoresis in air more than half a millennium B.C., it was not until 1809 that electrophoresis in liquids was discovered. Why did this discovery have to wait for two and a half thousand years? --The answer is very simple. There is essentially no energy required to maintain an electrical field in air by means of an object carrying an electrostatic charge. In a conductive liquid, however, an electrical current must flow in accordance with Ohm's law in a region where an electric field is maintained. A source of energy is needed to maintain such a current. Such energy sources did not exist prior to the discoveries of Galvani and Volta. The Voltaic battery (1800) provided the first electrochemical energy source for sustaining electrical currents in conductive solutions. It is therefore no mere coincidence that electrophoresis was discovered by Reuss in 1809, nine years after Volta's historic publication (1). To appreciate the state of knowledge at the time of

Reuss, one should keep in mind that the year 1800 was twenty-eight years ahead of the discovery of Ohm's law and thirty-two years before Faraday's discovery of his laws of electrolysis. Oersted's discovery of electromagnetism was nineteen years in the future so that electromagnetic current meters were not available. The intensity of electric currents was semi-quantitatively estimated by observing the rate of evolution of bubbles in an acid solution.

In 1800, the nature of the electric current was still mysterious. One spoke of an "electrical effluvium." As soon as voltaic batteries made maintenance of electric currents possible, they became a subject of intense investigation and electrical conduction through all kinds of materials was subjected to scrutiny. Professor F. F. Reuss of the University of Moscow was studying electrical conductivity of moist soil in a garden on the shores of the Moskva river. His electric battery was a voltaic pile. It was a multiple sandwich of ninety-two silver rubel coins separated from ninety-two Zink discs by layers of acid-soaked cloth. He estimated the current intensity by the rate of evolution of bubbles in an acid solution traversed by the current. He was surprised to note that the current was not greatly affected by the distance between his electrodes in the soil. He decided to repeat his experiments under better controlled conditions in the laboratory.

The simple apparatus of Reuss was a V-tube shown in Fig. 1. Instead of soil it contained quartz sand Q at the bottom. The effect of the electric current that he observed was a rise of liquid in the negative electrode leg and a drop in liquid level in the positive electrode leg. There was no effect in the absence of the sand. In this first experiment Reuss discovered electroosmosis and reported it on April 15, 1908 in a lecture entitled: "On a New Effect of Galvanic Electricity."

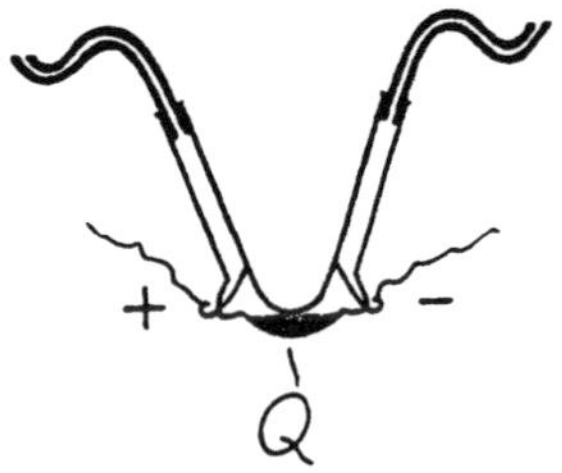

<u>Fig. 1</u> *Reuss' apparatus in which he observed electroosmosis. Q: quartz sand. (From H. A. Abramson, <u>Electrokinetic Phenomena</u>. New York: Chemical Catalogue Company, 1934.)*

In a second experiment, he changed the texture of his soil. He replaced the coarse sand with very small clay particles. Fig. 2 illustrates the scheme of this experiment. He inserted two vertical glass tubes into a slab of wet clay and filled the tube with a conductive aqueous solution. As the current passed between the electrodes, clay particles migrated upward in the positive leg, but not in the negative leg. Reuss, thus discovered electrophoresis in his second experiment.

<u>Fig. 2</u> *Set-up in which Reuss observed electrophoresis. C: moist clay slab.* T_1, T_2 *glass tubes.*

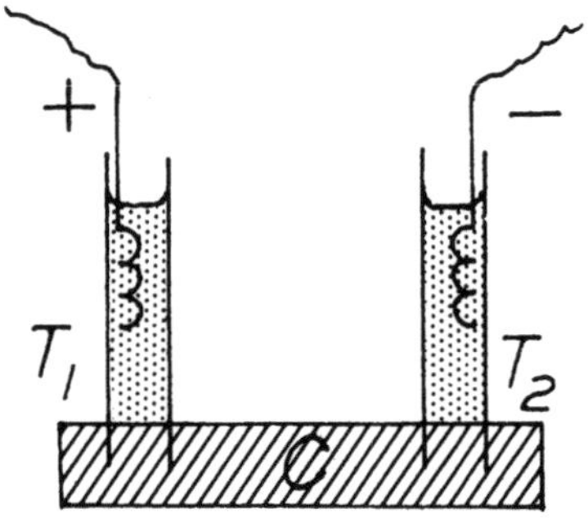

Prelude to Contemporary Developments

The early pioneers in the study of electrophoresis and electroosmosis were not analytical chemists but, rather, pure physicists. It is surprising that the unspectacular effects described by Reuss should have attracted the attention of some of the foremost physicists. Most notable among them was Helmholtz, the codiscoverer of the law of conservation of energy. He was followed by other first-rate physicists like Quincke, Smoluchowski, Lodge and later by Nobel Laureates like Stern, Perrin, Debye and Onsager. The early physicists may have been attracted by the insight that electrification of matter by contact is related to the structure of matter. The contact between a liquid and an immersed solid is much more reproducible and well defined than the contact between two solids used to separate electric charges by friction. Indeed, the work of Helmholtz and Smoluchowski led to a neat mathematical formulation of a common theory for electrophoresis and electroosmosis on the basis of the notion of an interfacial electric double layer (2, 3). This notion gradually evolved into a more sophisticated theory of the diffuse ionic atmosphere in the subsequent treatments of Guye, Chapman and Stern (4, 5, 6). And the Debye and Hückel theory of counter-ion atmospheres and ionic migration can be considered as an extension of the theories originally designed for plane surfaces to a spherical configuration (7).

The physicists of the past century were interested in studies of electrolytic conduction (let us call it "ionics") for similar reasons as contemporary solid state physicists study electronics. Such studies resulted in revolutionary changes in our notions. Thus, for instance, Helmholtz was able to infer the quantum nature of the electric charge from Faraday's laws of electrolysis (8). Arrhenius' theory of ionic dissociation was so controversial (revolutionary) that he barely squeezed through with the lowest passing grade in his doctoral examination with a thesis for which he later received a Nobel Prize. His theory assumed that supposedly immutable atoms radically change their chemical properties by acquisition of an electric charge in becoming ions.

The development of theories went hand in hand with experimental studies, notably of the electrical conductivities of ionic solutions. Such measurements revealed lower conductivities for strong electrolytes than would be expected from complete dissociation. This discrepancy was resolved by Debye's and Hückel's theory on the basis of the interaction between a migrating ion and the cloud of counter-ions. Most notable among the pioneers in the study of electrolyte conductance was Friedrich Kohlrausch. The laws which he discovered for concentration shifts and boundary migration in electrolyte columns with concentration gradients and abrupt changes in electrolyte composition have provided a foundation for some of the most important recent developments in electrophoretic methodology (9).

Sharp boundary discontinuities in electrolyte columns were of importance and interest in connection with new methods for studies of ion mobilities which were being developed by Oliver Lodge (10). In 1886 Lodge was interested in direct observation and measurement of migration rates of ions in solution. He proposed the moving boundary method which was perfected for analytical use by Tiselius nearly half a century later. Lodge also suggested zone electrophoresis in gels. The ideas of Lodge present an interesting example of a phenomenon that we shall encounter on more than one occasion in the field of electrophoresis. We may call it "hybernation" of ideas which often lie dormant in the literature until they are retrieved or reinvented when a practical need for them arises. Another important instance of "hybernation" is the invention of paper chromatography by Tswett in 1903 (11) which was resurected by

Consden, Gordon and Martin in 1944. One cannot help wondering how many more important ideas may be lying buried in the literature.

Tiselius Initiates Analytical Use of Electrophoresis

Analytical chemists showed no interest in electrophoresis for 120 years. The ground for analytical uses was prepared by physicists (notably Lodge (10)) who showed how electrophoretic mobilities could be measured by the moving boundary method. In 1930 (12) this method was transformed into a powerful analytical tool by Tiselius. At that time fractionation of serum proteins by ultracentrifugation revealed only two components, the albumin and the globulin fractions. By means of electrophoresis, Tiselius was able to resolve the globulin fraction into several components (Fig. 3). We shall not discuss his

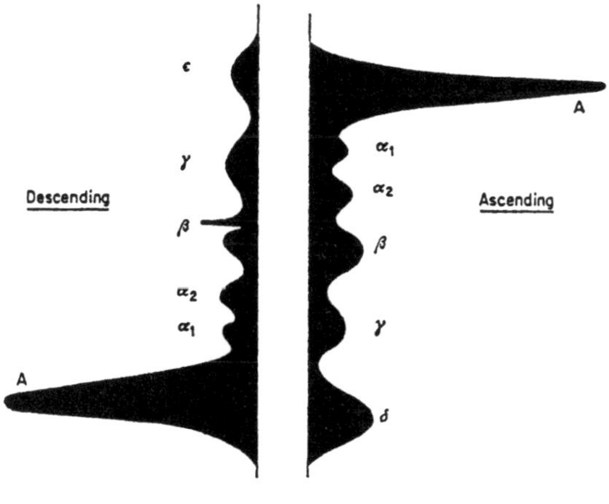

Fig. 3 *Descending and ascending electrophoretic separation patterns obtained for human serum proteins with a Tiselius apparatus. A: albumin;* α_1, α_2, β *and* γ*: globulins;* δ *and* ϵ*: salt boundaries. (From D. J. Shaw: Electrophoresis. New York: Academic Press, 1969.)*

well-known apparatus and will limit ourselves to the enumerations of the main innovations which Tiselius introduced in it: I) A method for generation of sharp boundaries. II) Suppression of thermal convection by carrying out electrophoresis at the density maximum of the solution near 4°C. III) An optical method for visualization of boundaries between colorless components.

Tiselius' apparatus was complex and costly. It represented a great leap forward in biochemical analysis. It had, however, serious limitations. Its usefulness was great in providing information about the number of fractions in protein mixtures and in characterizing the components by their electrophoretic mobilities. Its preparative capabilities were, however, very limited. Only a fraction of the fastest component and a portion of the slowest one could be isolated in pure form. The growing need for preparative isolation of all components of a mixture, as well as a desire for simplification of the apparatus, provided the stimulus for the development of "new" zonal methods. Actually these methods represent a revival of old ideas of Lodge (1886) (10) and of Tswett (1903) (11).

Evolution of Zonal Methods

The evolution of zonal methods from the Tiselius U-tube scheme is an almost self-evident logical step. If we imagine the U-tube straightened out, then the protein mixture in the lower portion of the U-tube will become a straight zone which we can make as narrow as we please. In the course of electrophoresis the protein components of this zone will get far enough apart to exhibit complete separation without overlap. The narrower the original zone, the less time will be required for a complete separation of the ingredients. We shall classify the zonal separation procedures in three types of methods. I) Non-focusing methods, II) focusing methods and III) stacking methods. The following survey of zonal separation methods is intended to be illustrative rather than exhaustive.

Zonal Non-focusing Methods

DENSITY GRADIENT ELECTROPHORESIS

The classical moving boundary electrophoresis (12) was performed in liquid columns. Zone electrophoresis can also be performed in liquid columns stabilized by density gradients (13, 14, 15). A sucrose concentration gradient in a vertical liquid column will be accompanied by a concomitant density gradient (and viscosity gradient). A downward increase in density will stabilize the column against thermal convection and will permit to establish concomitant gradients of other parameters, such as pH and electrical conductivity (13, 14). Withdrawal of liquid from a level in the density gradient generates a density jump ("density shelf"). Reinjection of a sample whose density lies within the range of the density jump produces a stabilized narrow sample zone. The electrophoretic column becomes a race-track and electrophoresis separates the original sample into migrating zones of different electrophoretic mobilities.

HJERTÉN'S ROTATING TUBE METHOD

Density gradient electrophoresis has certain drawbacks. Because of the electrophoretic mobility-and electrical conductivity gradients that are concomitant with the density gradient, it is not suitable for analytical characterization of the electrophoretic components by measurement of absolute values of their electrophoretic mobilities. Such measurements would require an homogeneous electrophoretic column. An ingenious solution to this problem was provided by Hjertén in which thermal stability is obtained without recourse to a density gradient (16). Gravitationally induced movements, like sedimentation of denser particles or thermal convection, are inhibited by rotation of a fluid about a horizontal axis (17). In Hjertén's "free zone electrophoresis," a horizontal buffer-filled tube rotates slowly (about one revolution per second) about a horizontal axis. The solution in the tube is homogeneous (i.e., devoid of concentration gradients). Fig. 4 shows Hjerten's apparatus in which four dye stuffs have been separated into four distinct zones.

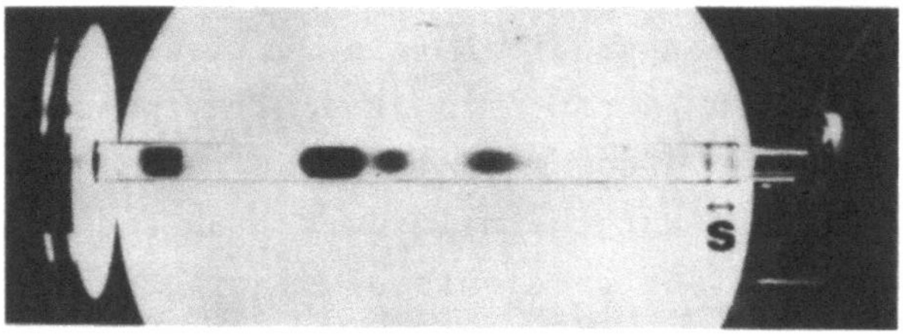

Fig. 4 *Hjertén's rotating tube "free zone electrophoresis" apparatus separation pattern obtained in 2 hours from a sample introduced at S containing: naphtholgreen, phenol red, bromo thymol blue and methyl orange. (From: S. Hjertén in Protids of the Biological Fluids. 7, 28, 1959.)*

The mixture is introduced at the marked location (S) by means of a fine plastic tube and the separated fractions can be withdrawn in similar fashion. This method is capable of separating molecules as well as cells and subcellular particles and it provides an elegant means for absolute determination of electrophoretic mobilities.

STABILIZATION BY ELECTROMAGNETIC ROTATION

Hjerten's rotating tube apparatus involves exacting mechanical precision work. This factor, as well as its effectiveness, stimulated attempts to achieve fluid rotation by non-mechanical means. These attempts failed in a horizontal tube but succeeded in the annular space (about 1.5 mm in width) between two horizontal plastic cylinders of different diameters (18). The buffer solution in the annulus is traversed by the axial electrophoretic current. By penetration of the annulus with a radial magnetic field, electromagnetic forces are engendered at right angles to both the electric and the magnetic field; i.e., these forces are tangential and the fluid is set into a circulating (rotational) motion about the horizontal cylinder axis. When a liquid sample containing electrically neutral particles is injected as a pulse centrally between the plastic cylinders into the annular interspace, the particles are aligned in a thin short streak and go into a circular orbit. If the particles are electrically charged, they exhibit simultaneously an axial electrophoretic migration, so that their actual path becomes helical. With each revolution two particles of different electrophoretic mobilities will gain an increment S in separation, so that their separation after n revolutions will be n·s. Clearly visible separations can be achieved in less than five seconds. The separated zones can be intercepted and retrieved by a collector at the end of the migration path. Fig. 5 illustrates the scheme of the electromagnetic endless belt electrophoresis apparatus as applied to zonal separation. (The main purpose of this principle is continuous-flow separation which will not be discussed (19)). It should be mentioned in this context that zonal separations can also be achieved by pulsed injection into a fluid curtain electrophoresis apparatus (20).

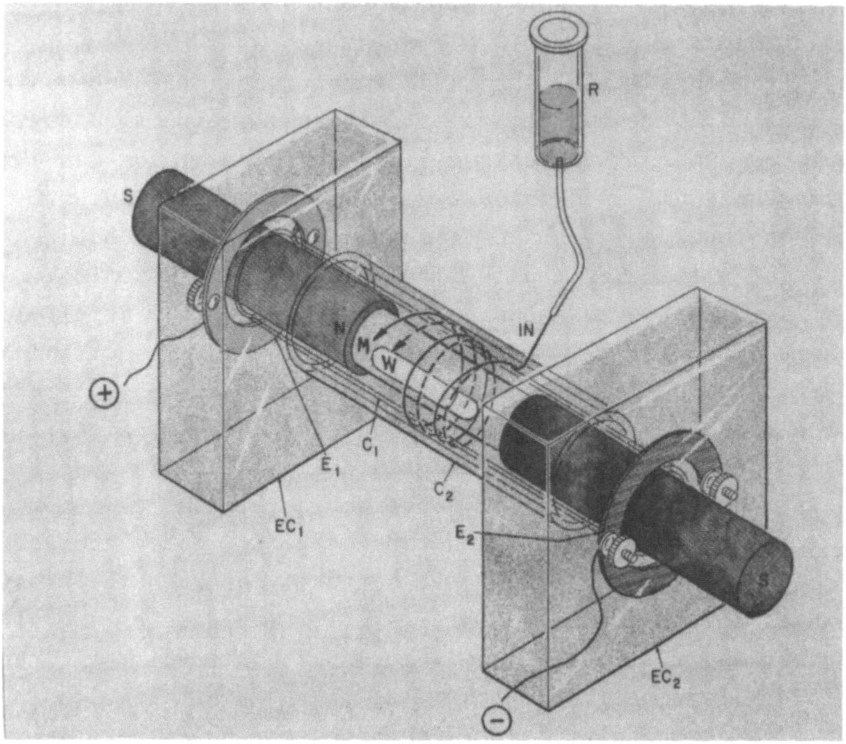

Fig. 5 *Scheme of rotational stabilization in a fluid endless belt. N-S : Bar magnets; m : soft-iron cylinder;* E_1, E_2 *: electrodes;* EC_1, EC_2 *: electrode compartments;* C_1 *: inner lucite cylinder;* C_2 *: outer lucite cylinder; IN : injector; R : reservoir for mixture to be analyzed.*

ANTICONVECTION STABILIZATION BY FLUID ENTRAPMENT IN THE MESHWORK OF A MATRIX - Filter Paper and Gel Electrophoresis

These methods, which came into general use only relatively recently, have their roots in the distant past. Deposition of mixtures of pigments to be separated in form of a narrow zone on filter paper was initiated by Tswett in 1903 in connection with the method of chromatography which he invented (11). This idea was transferred to electrophoresis in 1937 by König in Brazil who used it for fractionation of snake venoms (21). Köing's use of filter paper electrophoresis actually preceded by several years the revival of filter paper chromatography by Consden, Gordon and Martin (22). The successes obtained in these uses of sample zones on filter paper stimulated an avalanche of applications of filter paper zone electrophoresis beginning about 1950.

The advantages of simplicity of technique and completeness of zonal separations on filter paper were gained at the expense of emergence of new artifacts. Pronounced electroosmosis and streaming caused by evaporation of water, as well as non-uniformity of buffer concentration and temperature along the paper strip made filter paper electrophoresis unsuitable for measurement of electrophoretic mobilities for characterization of components. Adsorptive interaction between filter paper and some of the components caused not only retardation and possibly immobilization of some of the zones but also "tailing" which broadens the zones and degrades the separation pattern. This led to replacement of paper by other materials, such as cellulose acetate in which some of these problems are less severe. The use of a thin supporting matrix introduced, however, one advantage: The possibility of cooling the paper strip. This permitted the use of high potential gradients with resulting gain in speed of the separation process ("high voltage electrophoresis") (23).

The search for materials in which the above-mentioned artifacts were less pronounced than in paper and in thin plastic films led to the discovery of many different types of gel with suitable properties, most notably polyacrylamide gel. But this again was more a revival than an innovation. Oliver Lodge

(10) originated this approach to electrophoresis in 1886 but analytical chemistry did not adopt it until recent times. The transition from a paper— or thin-film matrix to a gel had the important advantage of adding a significant third dimension which permitted an increase in the sample quantity for preparative separations. But the gel layer can also be made very thin so as to permit cooling and application of high potential gradients to achieve rapid separations. The filter paper- and gel methods have one factor in common. Both inhibit thermal convection in the liquid buffer by viscous shearing stress in the interstices of the fiber—or molecular network of the matrix in which the liquid is entrapped. There is, however, one property of gels which can be utilized concurrently with electrophoresis which makes them superior to other matrices; it is the possibility of superimposing a molecular sieving effect upon electrophoresis, thus adding the capability of resolution on the basis of differences in molecular dimensions. This type of interaction will be considered later in connection with other methodological aspects.

Zonal Focusing Methods

ISOELECTRIC FOCUSING

It would be logical to surmise that electrophoretic focusing methods resulted from the quest of analytical biochemists for an increase in resolving power of zone electrophoresis. But this is not the case. The concept was originated by a physicist and was described in 1954 as a physical effect in the Journal of Chemical Physics (13) with indications of the analytical potentialities of an electrophoretic method based on this phenomenon. In fact, analytical biochemists came very close to this concept in designing "electrodialysis apparatus" with compartments separated by membranes and containing buffers of different pH (24, 25). But, in spite of the time-lapse between 1929 (resp. 1941) and 1954 they did not make the essential transition from a pH step-function to a continuous pH gradient. This conceptual transition was apparently not self-evident since even the prolific innovator Tiselius did not make it. The conceptual step required was analogous to the transition from algebra to infinitesimal calculus.

The consideration that led to the conception of isoelectric focusing was an imaginary experiment ("Gedanken experiment") in which the titration of a solution containing one single protein molecule was visualized. A positive protein ion in an originally acidic solution was seen to acquire a negative charge after passing through the neutral state at the isoelectric point as the pH was gradually changing from acidic to basic. In a "thought-experiment" it was possible to accomplish this titration (i.e., change in the pH of the environment of the molecule) without modifying the buffer solution. The buffer column was imagined to have a continuously monotonically changing pH. The pH of the surroundings of the protein molecule could be changed by mechanically pulling it through the pH gradient in the buffer column (13).

Two additional steps in this train of thought led to the concept of "isoelectric condensation" or "isoelectric focusing." First, the entire column was populated with protein molecules of the same species. It was at once clear that there was variation in the charge of the protein ions in the pH gradient. The ampholyte had to have opposite charges on either side of the isoelectric zone. In the second step the mechanical propulsion of the ions was replaced by electrical transport, i.e., by electrophoresis. Here, an unusual situation presented itself. Ions of the same substance were moving in opposite directions in the same electric field, depending on their location in the electrophoretic column.

There had to be three related effects: I) "Isoelectric condensation" in a current flowing in the direction from low pH to high pH. In this case, the ampholyte ions had to be swept toward the isoelectric zone where they would loose their charge and accumulate. II) "Isoelectric evacuation" in an oppositely directed current. Under these conditions, the oppositely charged ions located on either side of the isoelectric zone would move away from this zone, creating a "protein ion vacuum" on each side of it. Protein molecules removed from the isoelectric zone by diffusion or stirring would acquire an electric charge and would be swept away by the electric field. III) The isoelectric condensation zones had to exhibit electrochemical stability. Molecules removed mechanically or thermally from such a zone in a current directed toward increasing pH values would acquire a charge in the pH gradient of such a sign as to be returned to the isoelectric zone.

The isoelectric focusing effect was presented as a basis for a new analytical approach in which the parameter of electrophoretic mobility would be replaced by the isoelectric point as a separation and characterization parameter. The predicted effects were demonstrated in a simple U-tube cell shown in Fig. 6 (13). In order to produce the anticipated effects, some additional new ideas had to be introduced: 1) stabilization of the electrophoretic column against hydrostatic imbalance and thermal convection by density gradients, 2) coupling pH and/or other physical parameters with a concomitant density gradient, and 3) making the electrical conductivity in the sample volume much lower than in the adjacent buffer solution so that most of the voltage applied to the cell electrodes appeared as a voltage drop across the sample zone producing a high electric field intensity and rapid electrophoretic migration.

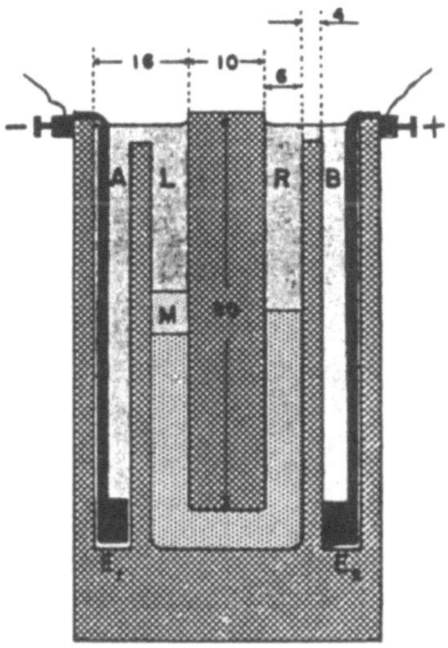

Fig. 6 *Cell which was used to demonstrate isoelectric focusing and velocity gradient focusing. E_1, E_2: electrodes; A, B: electrode compartments; L, R: left and right legs of U-tube; M: middle layer containing sample. (From Ref. 13.)*

M is the sample volume ("M-layer") in Fig. 6. Its density is intermediate between the densities of the acidic buffer below and the basic buffer above it. Channels A and B harbor the electrodes E_1, E_2. A pH gradient is established in the sample volume M by diffusion and/or by stirring its boundaries. Fig. 7 shows simultaneous demonstration of isoelectric "focusing" and "evacuation" effects in the cell of Fig. 6. In this case (Fig. A) two M layers containing a hemoglobin solution are symmetrically placed in the left and right legs L, R of the U-tube section. The current is flowing, as indicated, by the polarity of the electrodes in Fig. 6, from high to low pH in the R leg and in the direction of increasing pH in the L leg. As seen in Fig. 7B, there is a clear-cut condensation ("focusing") of hemoglobin in the left leg and an evacuation zone in the right leg after passage of a current of 9 mA for 40 seconds. Fig. 8 (26) shows an "isoelectric spectrum" (i.e., sorting pattern) produced within four minutes in the same apparatus in an M-layer containing four proteins. The three lower zones have been obtained by isoelectric focusing, the top zone by non-isoelectric focusing to be discussed below.

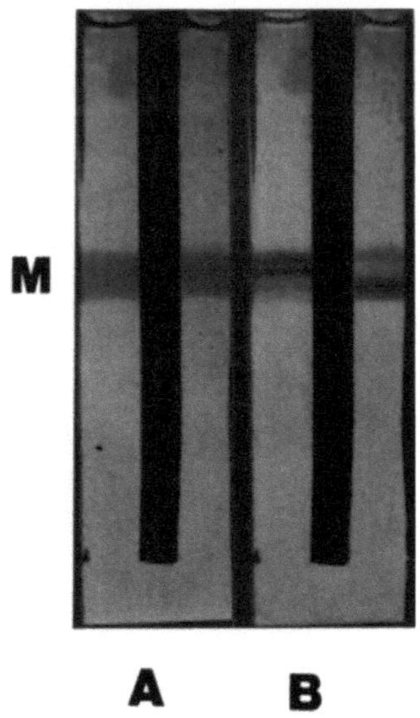

Fig. 7 *Demonstration of isoelectric focusing and evacuation of hemoglobin in the cell shown in Fig. 6 in a pH gradient (pH 2.6-9.6) within 40 sec. A: U-tube with hemoglobin belts M in both legs before passage of current. (The polarity is indicated in Fig. 6.) B: Focusing of hemoglobin by a current flowing in the direction of increasing pH in left leg of U-tube and evacuation of hemoglobin in right leg by a current flowing in the direction of decreasing pH. (From Ref. 13.)*

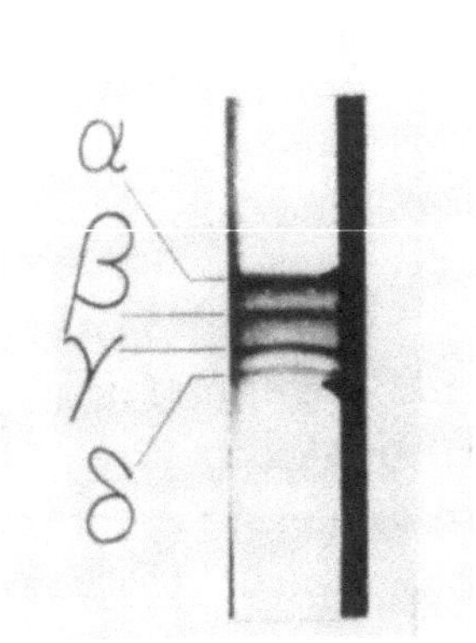

Fig. 8 *Isoelectric focusing of a mixture of proteins obtained in 4 min. The top componenet (α), cytochrome c, is focused non-isoelectrically in a conductivity gradient. The lower ones, hemoglobin (β), catalase (γ), and Azocoll (δ), are focused isoelectrically in pH gradients ranging from pH 4.8 to 7.7. (From Ref. 26.)*

This method gave impetus to several developments in electrophoresis: 1) density gradient electrophoresis, 2) electrophoresis in discontinuous multiphasic buffer systems (e.g., disc electrophoresis), 3) non-isolectric focusing, and 4) isoelectric focusing. As an approach to isoelectric focusing (27), it represented just a first step. It was an ephemeral link in an evolutionary chain. The prepared pH gradients on which it relied proved short-lived under passage of current. The survival of the method as a practical analytical tool required stable pH gradients that would show no objectionable drift in analytical runs extending over many hours. The achievement of pH gradients of adequate stability and durability was a decisive contribution of Svensson (Rilbe) and Vesterberg (28, 29, 30, 31).

In 1961 Svensson (28) introduced his so-called "natural pH gradients." He extended Tiselius' analysis of steady-state electrolysis (25) and showed, at first theoretically, that a stable stationary pH gradient could be established and maintained by prolonged (e.g., a day or longer) passage of a current through an electrolyte column containing a mixture of low molecular weight ampholytes ("carrier ampholytes") of closely spaced isoelectric points. Each carrier ampholyte migrates until it reaches a pH corresponding to its pI and becomes stationary. Eventually the zones of isoelectric carrier ampholytes form a continuous chain as the electric field ties them to their isoelectric zones and diffusion causes them to broaden and penetrate the neighboring ampholyte zones. Proteins to be subjected to isoelectric analysis present in small concentration ("trace ampholytes") will migrate in this stable pH gradient and condense or "focus" in very sharp thin zones at locations where the pH equals their isoelectric pH. The synthesis of appropriate carrier ampholytes was crucial to the success of this technique. This was accomplished by Vesterberg (31) who synthesized aminocarboxylic acids of a wide and quasi-continuous spectrum of pI values which have been made commercially available by LKB under the trade name of "ampholines." Fig. 9 shows Svensson's "natural pH gradient" column apparatus displaying an isoelectric separation pattern. Resolution of

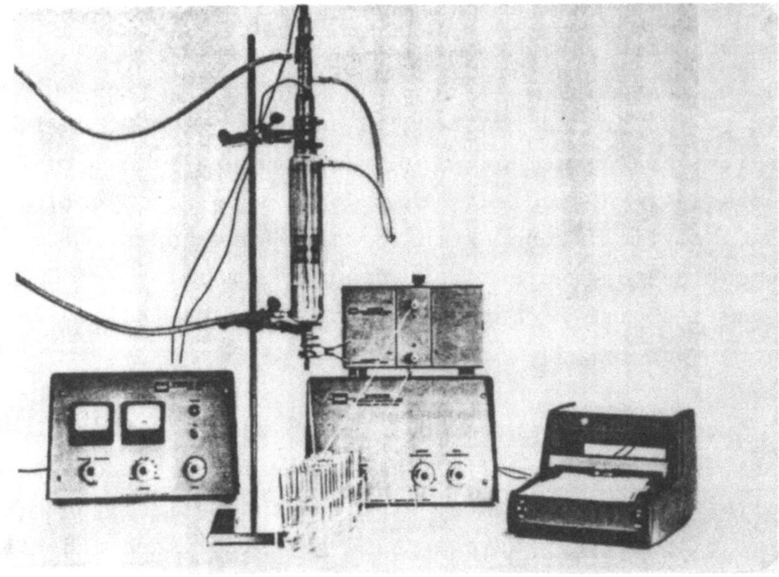

Fig. 9 *Svensson's isoelectric focusing equipment consisting of electrofocusing column (LKB 8100-10) (110 ml), power supply (LKB 3371), UV monitor (LKB 8300), recorder. (From H. Haglund, Meth. Biochem. Analysis. 19, 1-104, 1970.)*

componenets differing as little as 0.01 pH unit in their pI values is currently possible. (A more extensive review of the genesis and early evolution of isoelectric focusing can be found elsewhere (32, 33)).

A drawback of this powerful technique is the contamination of the separated fractions by the "ampholines" and by sucrose which have to be removed. Ideas to avoid "ampholines" by generating pH gradients by means of temperature gradients have been suggested but not worked out (34, 35).

NON-ISOELECTRIC FOCUSING: VELOCITY GRADIENT FOCUSING

Isoelectric focusing is a special case of a general principle which we shall call "velocity gradient focusing." Let us consider a zone of width w in an electrophretic column. This zone width will be preserved if the rear of the zone migrates with the same velocity as the front of the zone. Let us assume now that the electrical conductivity σ of the buffer in the column increases in the direction of migration. Since by Ohm's law the electric feild intensity is E = J/σ (where J is the current density), the electric field intensity will have to decrease in the direction of the zone migration. As a result, the ions forming the front of the zone will move more slowly than the rearguard ions and the zone will become progressively narrower i.e., we will observe non-isoelectric focusing (14). The final width of the zone will be determined by a balance between the tendency of the zone to diminish in width in the electrophoretic velocity gradient and the tendency of diffusion to broaden the zone. The gradient that determines the ion velocity need not be shallow. Zone narrowing will occur equally well when the zone migrates across a boundary at which the ion velocity changes abruptly. Such non-isoelectric focusing was first demonstrated in the same cell in which the first isoelectric focusing experiments were performed (13, 14) (Fig. 6). In Fig. 8 the top component has been focused non-isoelectrically because of the sharp increase in buffer conductivity across the upper boundary of the M layer. Fig. 10 shows non-

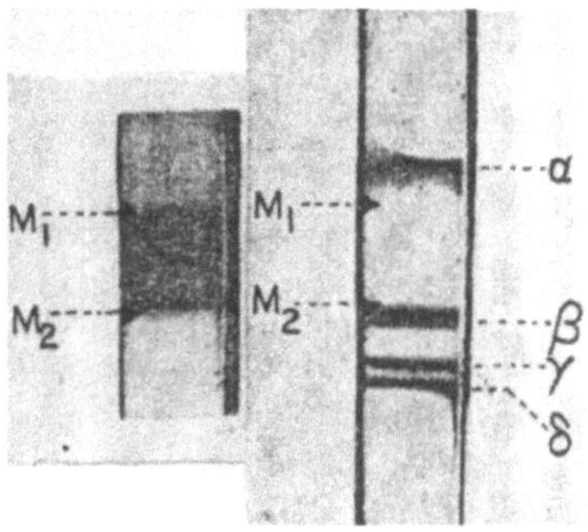

Fig. 10 *Velocity gradient focusing in the cell shown in Fig. 6. Separation of a mixture of dyes (time, 10 minutes). α: basic fuchsin, β: chrome hematoxylin, γ: brom thymol blue, δ: brom phenol blue. The original M layer is shown on the left of the separation pattern.*

isoelectric focusing of dyes which migrate upward (α) and downward (β, γ, δ) from a sample zone of low electrical conductivity (between the marks M_1, M_2) into buffer solutions above M_1 and below M_2 of higher electrical conductivity. (Such zone narrowing has later been employed in other methods, such as "disc electrophoresis" (29).

Another means of achieving velocity gradient focusing is to use a molecular sieving effect. If the zone of ions must traverse a boundary between two gels of different porosity which are arranged so that the front of the zone migrates from a wide-pore gel into a small-pore gel, the front of the zone will be retarded by molecular sieving while the rear boundary is still migrating almost unimpeded with greater speed causing a narrowing of the zone. This effect is used along with other focusing means in "disc electrophoresis" (29) where zones can be compressed to the thickness of a few micra. Electrophoresis of ion zones in gentle gel-pore gradients also results in narrowing of migrating zones (37).

Finally, the ion velocity gradient can be created by means of a pH gradient in the electrophoretic column. Due to the pH-dependence of the ion mobility, the ions would move with different velocities at different points in the column even if the electric field is uniform throughout the column (13). Thus, if the pH gradient diminishes the ion mobility in the direction of migration, the rear of the zone will travel faster than the front and the zone will be narrowed (i.e., "focused") progressively. This method of modification of ion velocities has also been used in "disc electrophoresis" (see below). The converse will, of course, happen if the zone migrates in the direction of increasing ion mobility in the pH gradient. It will become progressively wider.

We can now easily see that isoelectric focusing is a special case of zonal velocity-gradient focusing where the ion mobility is controlled by a varying pH in the migration column.

We obtain isoelectric focusing if the pH gradient encompasses the isoelectric pH of the migrating protein and if the direction of migration is toward the isoelectric point. In this case the ion velocity relative to the pH distribution function is decelerated, reaching zero at the isoelectric point and the front of the protein zone moves more slowly than its rear boundary so that maximal condensation (or "focusing") is achieved when all protein ions have lost their charge in the isoelectric layer. In the special case where the isoelectric point lies inside the protein zone, its boundaries migrate toward each other if the current flows in the direction of increasing pH. We obtain again a focusing effect. If, on the other hand, the current flows in the direction of decreasing pH values, the velocities of the two zone boundaries are still opposite to each other, but they are directed away from the zone center and we obtain zone broadening ("defocusing"). The special features of isoelectric focusing, when viewed as a velocity-gradient focusing effect, are thus the existence of a plane of zero ampholyte velocity relative to the buffer and the condition that the zero mobility in this plane is brought about by the local pH value.

Stacking Methods

ISOTACHOPHORESIS

Ordinary zone electrophoresis does not have a clear-cut completion point beyond the stage where the original mixed-sample zone has been split into several perceptibly separated zones of different electrophoretic mobilities. The separation between the various components can, in principle, be increased indefinitely by continued electromigration which proceeds with different velocities for the individual components. Isoelectric focusing analysis, on the other hand, can be considered completed when the components of different pI values have become stationary at their isoelectric points in the pH gradient. The separation method to be considered now sorts the components of a mixture according to differences in electrophoretic mobilities and arranges them as contiguous zones in the sequence of their mobilities. Unlike isoelectric focusing, however, these zones do not become stationary and, in contrast to zone electrophoresis, these zones do not separate from each other in the course of further electromigration because they continue to move indefinitely with the same speed. (Hence the name "isotachophoresis," derived from the Greek words "isos" for equal and "tachos" for speed.) Because of the resemblance of the contiguous separated zones to a stack of coins, we have classified isotachophoresis as a "stacking method."

The recent emergence and cultivation of isotachophoresis (ITP) is another example of "hybernation" of a scientific idea. It was first proposed by Kendall in 1923 (32) for electrochemical separation of isotopes. Its theoretical foundation is based on Kohlrausch's publication of 1897 (9). Kendall did not succeed in separating isotopes, but later others did so on the basis of his idea (39). (Kendall's principle is encountered in the literature under a variety of names, such as isotachophoresis, displacement electrophoresis, cons electrophoresis, transphoresis, etc.) The key to ITP is the following insight which was pointed out by Kohlrausch. Let us assume that we have an electrolyte column in which a sharp boundary exists between an electrolyte P^+L^- and P^+T^-. (We assume, for simplicity, all ions to be monovalent.) The positive ion species is continuous and the boundary is a discontinuity in the negative ion composition. Let us further assume the mobility of the L^- ion (leading ion) to be greater than that of the T^- ion (terminating ion) ($\mu_L > \mu_T$). When a current flows through this column, the boundary between the L^- and T^- ions migrates with the speed of the leading ions. This implies that the terminating ions migrate at the boundary with the same speed in spite of their lower mobility. This can happen only if the electric field intensity on the T-side of the boundary is higher than on the L-side. Kohlrausch showed that this inequality in field intensity is caused by a readjustment of the ion concentration which alters the electrical conductivity in the column in such a way that a certain function of the ion mobilities and concentrations (the Kohlrausch "regulating function") has the same value on both sides of the boundary. This statement is valid not only for the case of one single boundary but is applicable to a column containing an arbitrary number of boundaries between ions of equal sign having a common counter-ion.

Fig. 11 illustrates the ion configuration in isotachophoresis. The long electrolyte column terminates in large reservoirs with electrodes at both ends which are not shown. The common positive ions P^+ are not indicated in the figure which shows only the negative ion species that migrate to the left. The L^- ions have a higher mobility than the T^- ions. Let us assume at first that $(ABC)^-$ is a species of monovalent ions of an intermediate mobility ($\mu_L > \mu_{(ABC)} > \mu_T$) (Fig. 11a). Then, changes in concentration and in potential gradients will occur at each boundary of the zone, as mentioned above, to insure equality of the Kohlrausch regulating function across the boundaries. The concentration changes will lead to a change in the width of the zone. After establishment of a steady state, both boundaries will migrate with equal speed to the left and the electric field intensities in the column will change abruptly across the boundaries which are demarcations of discontinuous changes in ion composition.

Let us now assume, that the ABC zone actually contains three different ion species A^-, B^-, C^- whose mobilities lie between those of the L^- and the T^- ions. Sharp boundaries will then develop between the individual ion species which will migrate with equal speed in the final steady state. The distances between the boundaries (i.e., the widths of the zones A, B, C)

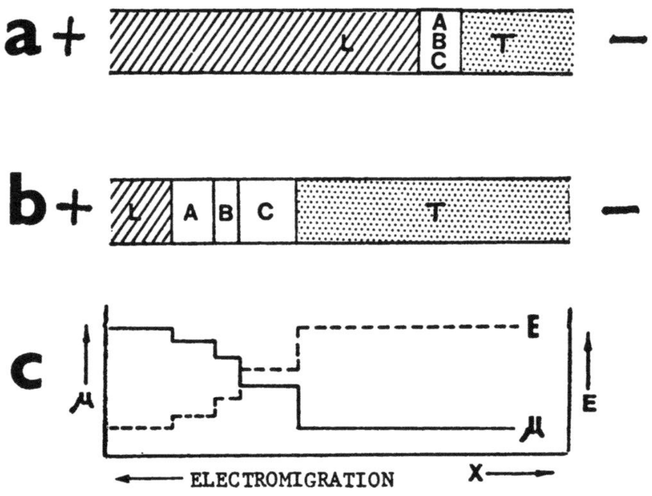

Fig. 11 *Illustration of isotachophoretic stacking. The current polarity is indicated in a and b. (a) The zone containing the mixture of anions to be separated is sandwiched between the leading anions L and terminating anions T. (b) Stack of contiguous zones of ionic species A, B, C transported to the left by isotachophoresis. (c) Correlation between the electric field intensity E and electophoretic mobility μ in the migration column. (Redrawn from M. Bier and T. Algyer in Electrokinetic Separation Methods, P. G. Righetti et al, Eds. (New York: Elsevier, 1979.)*

will change until the concentrations satisfying the Kohlrausch condition have been established (Fig. 11b). The sequence of the ions in these zones is such that their mobilities increase from T to L. Once the species A, B, C have been separated, their boundaries migrate with equal speed without further separation. Fig. 11c shows how the variations in the electric field intensity E within the various zones are coordinated with variations in the ion mobilities μ.

Isotachophoresis has been ingeniously and very effectively utilized, along with other principles, in the method of "disc electrophoresis" (36). Ornstein presents a calculation based on a modified Kohlrausch equation to show how a 1cm wide protein zone sandwiched between ions of glycinate and chlorine can be reduced to a thickness of 25 micra and concentrated by a factor of 380. In addition to isotachophoresis, disc electrophoresis utilizes principles of velocity gradient focusing (14, 26) and discontinuous buffer systems (13, 14, 26). The scheme of disc electrophoresis shown in Fig. 12 has some common

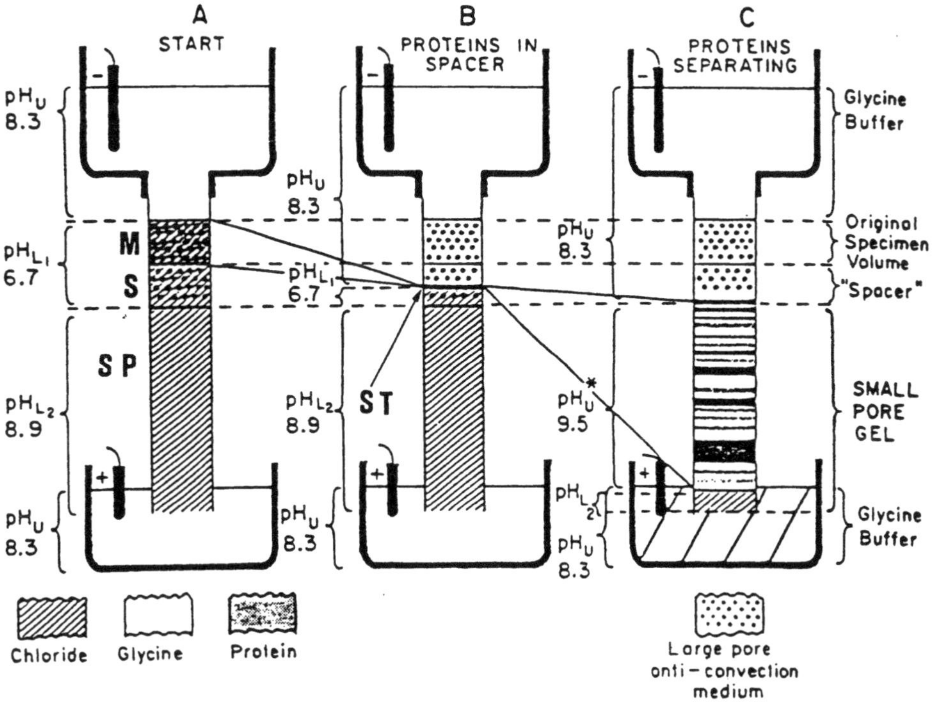

Fig. 12 *Scheme of Disc Electrophoresis showing 3 stages of the process.* **A:** *Initial stage. The sample is in zone M.* **B:** *The components of the sample have migrated into the "spacer gel" and are condensed by isotachophoresis into a narrow stack ST.* **C:** *The chloride ions have left the small-pore gel and separation proceeds in it by normal zone electrophoresis combined with molecular sieving. (Slightly modified from Ref. 36.)*

features with the scheme of velocity gradient electrophoresis shown in Fig. 6 (14, 16). An important difference is the use of a polyacrylamide gel for inhibition of thermal convection in disc electrophoresis instead of density gradients used in the scheme of Fig. 6. The second important difference is the use of molecular sieving in a small-pore gel in disc electrophoresis to achieve velocity gradient focusing. The process of disc electrophoresis is illustrated in Fig. 12. The sample is contained in the large-pore gel section M. The sample ions are sandwiched between the low-mobility "terminating" glycine ions and the high-mobility "leading" chlorine ions so that the migration of the sample ions results in an isotachophoretic sorting or stacking process. The original sample zone M is thus compressed into a thin stack ST of contiguous separated zones by isotachophoresis. In Fig. 12c the chlorine ions are shown to have left the small-pore gel where the pH has risen from 8.9 to 9.5. Transition of the zones from the large-pore gel into the small-pore gel leads to further zone narrowing due to velocity gradient focusing and from then on, the electromigration through the small-pore gel is governed, in addition, by molecular sieving. Thus, the contiguous zones are separated from each other in accordance with the differences in molecular shape and dimensions of the ions in the zones.

Quest for Higher Resolution

TWO-DIMENSIONAL METHODS

Even with modern high-resolution separation methods, one comes across closely-spaced components in electropherograms that are difficult to separate. To get such narrow multipletts further apart, a practice that had been used in chromatography since 1948 may have provided the stimulus for introduction of two-dimensional techniques into electrophoresis (40). Durrum (41) appears to have been the first to suggest a two-step process in which zonal filter paper electrophoresis is carried out first in one direction and then in the transverse direction after drying the square sheet and replacing the previous buffer by a new solution of different composition and, if useful, of different pH. The advent of new materials (such as various gels) and new electrophoretic techniques opened up new, almost unlimited possibilities of two-dimensional combinations of consecutive transverse separation processes with the consequence of greatly enhanced resolution. The separation criteria of the two transverse procedures must be different.

In view of the large and growing number of possible method combinations, which is augmented by the desirability to specify which method was used as the first step, it will be very helpful to represent two-dimensional methods by a diagram (Fig. 13). This diagram may also have a heuristic value by helping to conceive new combination methods. In analogy to the complex-number plane in mathematics, we define a method-plane in which the primed numbers along the vertical axis, as well as the unprimed ones along the horizontal axis are each correlated with a method. The unprimed numbers (horizontal axis) designate the first-step method and the primed ones the second step method. (In some instances, like electromagnetophoresis combined with sedimentation, the two processes can occur simultaneously.)

Thus, a point with integral coordinates in the method plane represents a combination method. Points along the 45° line represent cases where the same method is used in both transverse directions (but with different parameters). For instance, the point (1, 1') represents two-step chromatography where two-different elution fluids are used in two mutually perpendicular directions. If a combination method AB can be carried out with either the method A or the method B forming the first step, we will obtain two points symmetrical with respect to the 45° axis representing the combination method. In view of the large number of possible combination methods, which is practically limitless considering the rapid evolution of new techniques, we cannot review them all in the limited scope of this lecture.

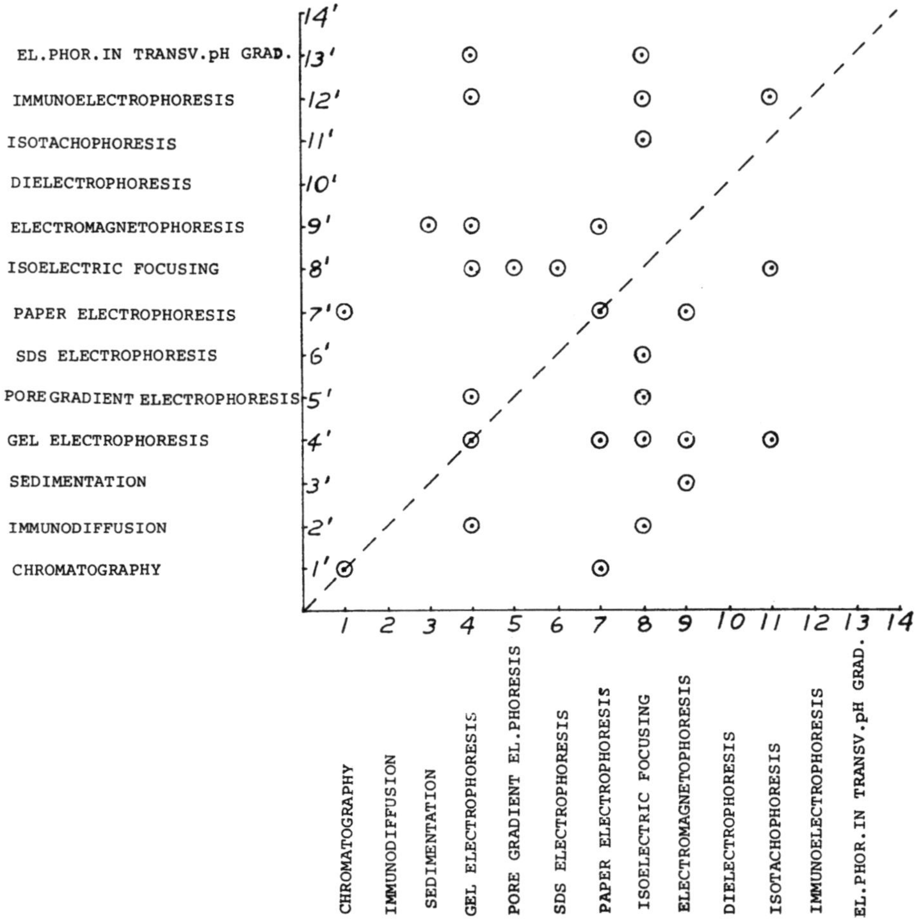

Fig. 13 *Coordinate representation of two-dimensional methods. The methods listed along the abscissa represent the first step of a two-dimensional procedure, the methods listed along the ordinate represent the second step.*

It will suffice to make a few comments on less well-known hybrid methods represented in Fig. 13. This figure is not intended to be comprehensive and the arrangement of methods along the coordinate axis is rather arbitrary. The methods will be designated below by the coordinates of the points representing them in Fig. 13.

Examples

(1, 7') - (1', 7). This is the well-known technique of "fingerprinting" (42). It proved particularly effective in resolving mixtures of many amino acid and polypeptide components, that could not be adequately separated by a one-dimensional process, into a two-dimensional map of clearly separated fractions. It exemplifies the basic advantage of two-dimensional methods.

(4-2'). Classical immuno-electrophoresis after Grabar and Williams (43).

(4-12') (44). The first step consists of isoelectric focusing of the antigen mixture in a polyacrylamide gel. The second step subjects the separation pattern to transverse electrophoresis into an antiserum-containing agarose gel where precipitation zones of antigen-antibody complexes are formed according to the diffusion-free immunoelectrophoresis method of Laurel (45).

(9-7') - (7-9'). Simultaneous application of transverse electric and magnetic fields imposes simultaneous electrophoresis and electromagnetophoresis upon changed particles in a quasi-two-dimensional matrix of filter paper or thin gel layer (46, 47).

(8-13'). Titration curves. This is a particularly interesting and beautiful procedure in which the amphoteric ions delineate their own pH-mobility curve (37, 48). Fig. 14 shows two examples of this process in A and B. To begin with, a stable pH gradient is generated in the ampholine-containing gel by prolonged passage of current in the horizontal direction, as indicated by the double arrow (IEF) in Fig. 14B. Then, the

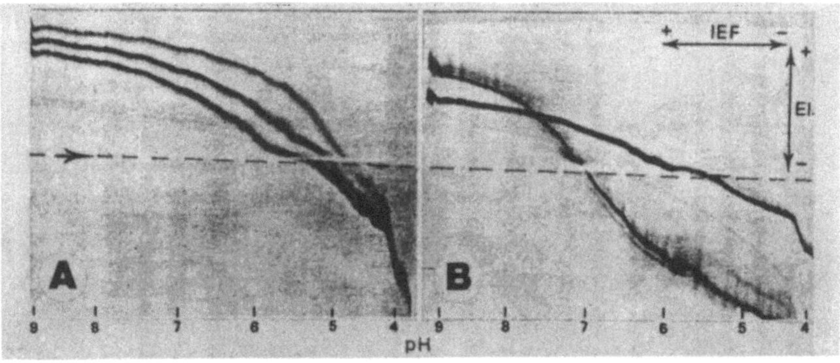

Fig. 14 *Vertical electrophoresis across a horizontal pH gradient. The sample is placed after the first step (establishment of the horizontal pH gradient) into a linear horizontal trough in the gel (dashed line). Electromigration along the vertical axis in the second step distorts the zone as shown, delineating the pH-mobility curve (titration curve). The intersection of this curve with the dashed line marks the isoelectric point of the component. The two electropherograms represent titration curves of Hp-Hb complexes in a 3:1 molar excess of Hp (A) and in a 3:1 molar excess of Hb (B). The three curves in A are: free, excess Hp (pI 4.5), Hp-Hb, 1:0.5 molar, complex (pI 5.0) and Hp-Hb, 1:1 molar, complex (pI 5.5). The two curves in B are: Hp-Hb, 1:1 molar, complex (pI 5.5) and excess free Hb (pI 7.0). The two arrows and positive and negative symbols represent the direction and polarity of isoelectric focusing (IEF) and electrophoresis (El.) The arrowhead indicates the sample application zone (i.e, the zero-mobility or isoelectric plane) (slightly modified from Ref. 37).*

electrode axis is turned 90° (as indicated by the double arrow (EL) in Fig. 14B) so that a current can be passed in the vertical direction at right angles to the established pH gradient. (The electrofocused ampholines are, of course, not affected by this current since they are located at their isolation points in the pH gradient). A long trough is now cut into the gel about centrally in the direction of the pH gradient as indicated by the arrow and dashed line in Figs. 14A and B. The sample is placed into this trough and the gel is subjected to a vertical electric field. If we consider one particular protein in the sample, it will not move at the intersection of the dashed line with the vertical line which passes through the pI value of the protein on the horizontal pH scale. On both sides of this intersection point, the protein ions will carry opposite charges and, hence, move in opposite directions (upward on the left and downward on the right side of this point in Fig. 14). Since the rate of migration depends on pH, the vertical distance traversed by the ions will vary from point to point of the original horizontal zone which is thus deformed from a straight line into a curve depicting the pH-mobility function of the given ion species. This method provides a characterization which is superior to the mere specification of the isoelectric pH or of the ion mobility at a given pH value.

The basic principle of this method was conceived by Michl in 1952 (49) before the advent of isoelectric focusing (13) in 1954! But, using ordinary buffers, he was unable to obtain good results. The successful revival of this idea (48) became possible after invention of the ampholines (31). This method was further developed by Righetti and Gianazza (37).

It is safe to conclude that two-dimensional techniques will enjoy increasing use and development because of the obvious advantage of wider distribution of separated fractions in two dimensions than can be accomplished in a one-dimensional process. It is not unlikely that further improvements will be achieved in the future by development of three-dimensional methods.

Quest for New Separation and Characterization Parameters

MOLECULAR DIMENSIONS (DETERMINATION BY SDS-PAGE)

In the methods that we surveyed, separation and characterization of the molecular species were based on ion mobilities and isoelectric points. Is electrophoretic methodology necessarily limited to these parameters? --Developments have shown that there is no such restriction. One of the main parameters of chemistry, the molecular weight, can be linked with the electrophoretic process. If we had a means of imparting the same electric charge density to a variety of molecules (nucleic acids, proteins) of similar conformation (e.g., random coil) but different in size, their migration through a gel (e.g., polyacrylamide gel) of suitable porosity could provide information about their molecular dimensions. In this case, the electrophoretic mobility of the molecular species under study in the gel can be compared to mobilities of standard molecules (ions) of known molecular weight (MW). This would lead to a determination of relative molecular size from which the (MW) could be inferred. To achieve uniform electrical charge densities, the unknown as well as the known standard molecules are treated (by "cooking") with sodium dodecyl sulfate (SDS) (50). Electrophoresis is performed in a crosslinked polyacrylamide gel of suitable concentration.

MOLECULAR DIMENSIONS (DETERMINATION BY ELECTROPHORESIS IN GEL-PORE-GRADIENT)

Another approach to (MW) determination is based on electrophoresis in a polyacrylamide gel of graded porosity. If a zone of a molecular species migrates in such a gel parallel to the pore-size gradient in the direction of diminishing pore-size, it will become progressively narrower (due to greater retardation of the front of the zone in the more concentrated gel matrix as compared to the rear boundary (velocity gradient focusing (14)). If the original zone contains a mixture of proteins, they split into progressively narrowing zones as they travel. Studies on mixtures of known molecular weight have established (51) that the relative migration speeds of the proteins (based on the slowest component serving as a standard) become constant after all components have passed beyond a certain gel concentration ("pore limit"). The plot of the logarithm of the molecular weight versus the logarithm of the relative migration speed becomes linear. Thus, if a mixture of proteins of unknown molecular weights with two components of known (MW) is examined, the line fixed in the log log plot by the two known components permits the determination of the (MW) of the unknown fractions (51).

IMMUNO-ELECTROPHORESIS (Immunochemical Specificity as Analytical Parameter)

Many proteins are hard to differentiate because of closeness of their electrophoretic mobilities, isoelectric points and/or enzymatic or staining reactions. This is especially true of proteins derived from genetically related plants and animals. The idea of Grabar and Williams (43) made such a differentiation possible with immunochemical specificity. Their method is a two-step process where electrophoresis is followed by transverse immunodiffusion. The end result is a two-dimensional pattern which enables one to identify antigens and/or antibodies.

It goes beyond the scope of this lecture to review all the variants of this method. Fig. 15 illustrates one typical use.

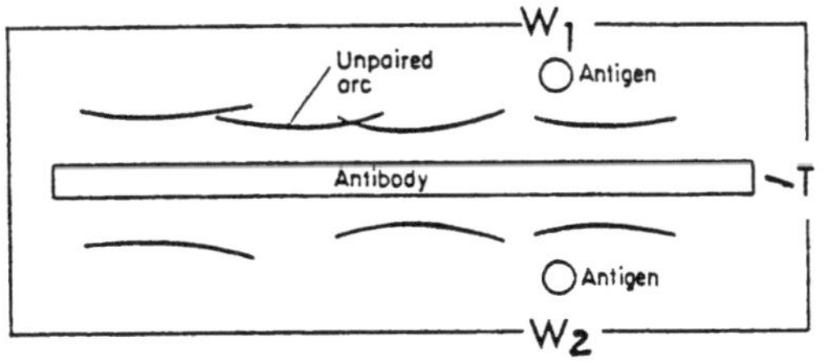

Fig. 15 Schematic representation of immunoelectrophoretic comparison of antigen mixtures. (Slightly modified from D. J. Shaw, Electrophoresis. New York: Academic Press, 1969.)

Let us assume that a mixture of unknown antigens is placed in the circular well W_1 of the gel. Electrophoresis of the antigen ions toward the left has resolved this sample into four invisible circular protein zones. The problem of visualizing them is analogous to the development of a latent photographic image. The idea of Grabar and Williams was to accomplish it by "development" with a mixture of antibodies so that an immunoprecipitate would reveal and identify an antigen present in the mixture. But this is not a trivial experimental problem since the formation of the precipitate requires a certain antigen/antibody concentration ratio. This difficulty was overcome by the ingenious idea of creating concentration gradients of the reactants by diffusion. Thus, there will be arc-shaped loci within the diffusion field where the concentration ratio will be optimal for the precipitation reaction.

After termination of the first step of electrophoresis in Fig. 15, a mixture of antibodies is placed in a long trough T in the gel. The macromolecules of the antigens and antibodies are now allowed to diffuse at the low rates determined by their diffusion coefficients for a long period of time (e.g., 2 days). The radial diffusion patterns of the circular antigen zones and the linear diffusion front which proceeds from the antibody trough penetrate each other and characteristic precipitation arcs are formed at proper locations within the concentration gradients of the diffusion fields for those antigens whose antibodies are contained in the trough. By concurrent electrophoresis of a mixture of known antigens placed in the circular well W_2, one obtains a second pattern of arcs below the trough T. Thus, the paired ones in the two patterns permit identification of antigens present in the unknown mixture and the presence of an unpaired arc in the unknown sample reveals an antigen without a counterpart in the known antigen mixture. This is a highly simplified description of a powerful method that is much more complex in actual practice. The most time-consuming part of the method is the preparation of the antiserum which may require many weeks.

The loss of time due to the slow diffusion process has been eliminated by development of new immunoelectrophoretic methods in which the diffusion step is eliminated. This is achieved by incorporating the antibodies in the gel under electrochemical conditions which immobilize them in an electric field while the antigens migrate and form precipitate arcs with the antibodies. Such methods permit a quantitative estimation of the amounts of antigens in the sample (e.g., Laurel's method (45)).

DIELECTROPHORESIS (Charge-Independent Migration)

Dielectrophoresis is the oldest known electrical phenomenon (except for lightning). It is the effect through which electrical attraction has been discovered in ancient times. A small electrified object produces a diverging electric field at a distance. A distant object subjected to this field is polarized, i.e., becomes an electric dipole with zero net charge. As early as in 1600 Gilbert observed the deformation of a hanging drop of water exposed to an electric field. A, say, positively charged object induces a negative charge on the proximal side of the sphere facing it and a positive charge on the distal side. The attraction of the proximal opposite charge on the sphere and the simultaneous repulsion of the like distal charge distort the hanging drop. Suppose now that the drop is detached. The greater force of attraction on the proximal charge will surpass the repulsion force exerted on the distal charge and the drop will move toward the electrified object under the predominant action of the attractive force. The direction of this force depends, however, on the relative values of the dielectric constants of the drop (or other polarizable object) and the surrounding medium. The force exerted by the source of a radial electric field upon a dielectric particle will be attractive if it is embedded in a medium of lower dielectric constant.

It will be repulsive if the dielectric constant of the medium (ε') surpasses that of the particle (ε'') and will be zero if the two dielectric constants are equal. This interaction is described by the equation (52):

$$\vec{F} = 2\pi a^3 \varepsilon' \left[\frac{\varepsilon'' - \varepsilon'}{\varepsilon'' + 2\varepsilon'}\right] \vec{\nabla}|E|^2. \qquad (1)$$

($\vec{F}$ is the force, a the radius of the spherical particle, and E the magnitude of the electrical field intensity. The force upon a conductive sphere in a conductive medium is given by a more complex equation and will not be considered here (53)).

Thus, attractive or repulsive forces can be exerted upon suspended particles in a non-uniform electric field depending on the relative magnitudes of the dielectric constants of the particle and the medium. The nature of the force (attraction or repulsion) is not affected by the direction of the electric field. Thus, alternating fields can be used.

Practical analytical use of this effect in cell biology is made difficult by the required non-uniformity of the electric field. Particles located in different portions of the non-uniform field experience forces of different magnitude. This creates a problem in design of a separation method of good resolving power and throughput. In addition, non-uniform heating of the fluid in the non-uniform electric field engenders thermal convection.

ELECTROMAGNETOPHORESIS (Charge-Independent Migration)

There is a growing interest in separation of cells, cell organelles and viruses in connection with research in cell biology and molecular genetics. Electrophoresis is capable of accomplishing such separations not only zonally but on a continuous-flow basis as well (18, 19, 20). However, electrophoresis is not an ideal method for cell separations. Good resolution and fast separations require a high electric field intensity. Avoidance of overheating the buffer and the cells requires a low current density. This leads to the use of low ionic strength buffers in cell electrophoresis which subjects the cells to the hazards of clumping and deterioration during the separation process. Ideally, one would wish to maintain the cells in their native body fluid that would have a much higher ionic strength than the commonly used buffers. The solution of this problem may be offered by a relatively recently discovered electrokinetic phenomenon, electromagnetophoresis (EMP) (54, 55, 56), which requires for optimum performance media of high ionic strength and would operate well in an electrolyte such as blood serum. The reason for the preference of a high ionic strength medium in EMP is that the particle velocity in EMP is proportional to the current density whereas in electrophoresis, it is proportional to the voltage gradient.

Electromagnetophoresis does not require the particles to be electrically charged and their migration direction does not coincide with the electric field. The effect requires, in addition to an electric current, the presence of a magnetic field. Only the magnetic field component perpendicular to the current contributes to the force exerted upon the particle and the migration is perpendicular to both the electric and the magnetic field. Reversal of either field reverses the direction of the particle migration. Simultaneous reversal of both fields does not reverse the direction of particle migation. Thus, unidirectional EMP can be obtained with an alternating electric current, provided it is accompanied by a perpendicular alternating magnetic field of equal frequency and phase.

The force $\vec{F}$ and the migration velocity $\vec{v}$ for a spherical particle of radius a and electrical conductivity σ' suspended in a fluid of electrical conductivity σ'' which is traversed by an electric current of current density $\vec{J}$ and a magnetic field of intensity $\vec{B}$ is given by equations (2) and (3):

$$\vec{F} = 2\pi a^3 \left(\frac{\sigma'-\sigma''}{\sigma'+2\sigma''}\right) [\vec{B} \times \vec{J}] . \qquad (2)$$

$$\vec{v} = \frac{a^2}{3\eta} \left(\frac{\sigma'-\sigma''}{\sigma'+2\sigma''}\right) [\vec{B} \times \vec{J}] . \qquad (3)$$

As we see, the particle velocity does not depend on the particle charge, but rather on the electrical conductivities σ' and σ'' of the particle and the suspension medium. If the particle moves to the right for $\sigma' > \sigma''$, it will move to the left for $\sigma' < \sigma''$ and will remain at rest for $\sigma' = \sigma''$. The particle speed increases with the square of its radius. The force upon a non-spherical particle depends on its shape and orientation relative to the electric and magnetic fields (54). In addition, to the electrical conductivity of a particle, its dielectric constant and/or its magnetic susceptibility can be used as a separation parameter in addition to particle size and shape (55, 57).

Fig. 16 shows an arrangment in which the EMP effect can be demonstrated. The Helmholtz coils C_1, C_2 generate a uniform magnetic field $\vec{B}$ in the central region of their interspace where a plastic cell with electrodes E_1, E_2 is placed. They form the top and bottom surfaces of the cell which contains an electrolyte, such as physiological saline or blood serum, with suspended particles. $\vec{J}$ is the current density vector and $\vec{F}$ indicates the direction of the force exerted upon the current-carrying electrolyte as well as on suspended particles whose conductivity exceeds that of the electrolyte. Particles of lesser conductivity than the electrolyte's experience a force in the opposite direction.

The force experienced by a conductive particle consists of two components. The first one is an electromagnetic force exerted by the perpendicular magnetic field upon the moving ions of the current passing through the particle. Its nature is the same

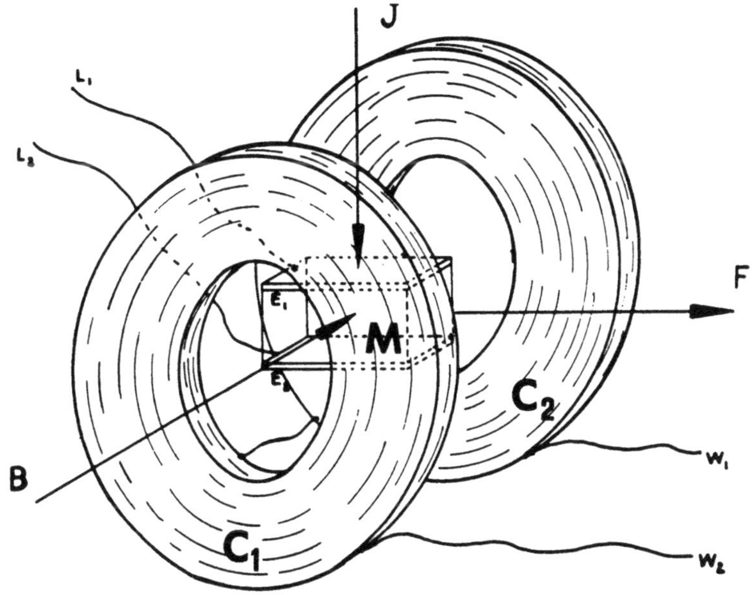

Fig. 16 *Scheme of electromagnetophoresis apparatus. M is the migration cell with electrodes E_1, E_2. C_1, C_2 is a pair of Helmholtz coils generating a uniform magnetic field within the cell M. $\vec{B}$: magnetic field vector. $\vec{J}$: electric current density vector. $\vec{F}$: electromagnetic force vector. W_1, W_2: magnet coil lead wires. L_1, L_2: electrode lead wires. (From Ref. 19.)*

as that of the force exerted upon a wire in the rotor of an electric motor in its magnetic field. The second component is a hydrostatic force exerted upon the particle surface by the electrolyte surrounding the particle. It is due to a hydrostatic pressure gradient which is set up in the electrolyte by the interaction of the ion current traversing it with the transverse magnetic field. This latter force component is very similar to the force of buoyancy exerted upon a floating object, whereas the first force component is roughly analogous to the gravitational force. The resultant force on the particle is the sum of these two opposing force components. A perfectly non-conducting particle experiences only the force of "electromagnetic buoyancy." The principle of Archimedes which is valid in the gravitational field can be qualitatively extended to electromagnetophoresis by assigning to the electrical conductivity in the case of EMP the same role as is played by the specific gravity in the gravitational case.

Having seen the lapse of time between the discoveries of chromatography (1903) and electrophoresis (1809) and the advent of their practical analytical uses, we should not be surprised that electromagnetophoresis has not yet reached this stage. The growing interest in electrokinetic separation of biological particles and the availability of superconducting magnets of field intensisites two orders of magnitude above those used in the initial demonstration of electromagnetophoresis will undoubtedly have a catalytic effect upon further developments in this field.

It is reasonable to expect that being a physical effect like electrophoresis, EMP will give rise to a comparably large variety of methods and techniques as we now see in the numerous implementations of electrophoresis. It has been shown, for instance, theoretically (55, 57) that the electrical conductivity is not the only parameter according to which particles can be sorted by EMP. The dielectric constant and the magnetic susceptibility can also be used as separation parameters. A further example will show the variety of ways in which electromagnetophoresis can be implemented. It can, for instance be used to achieve electrokinetic centrifugation. Fig. 17 shows an arrangement using the simplest conceivable electromagnet, namely, a wire w carring an electric current I. The magnetic field it generates is depicted by circular field lines H. The electrolyte-filled cell C containing suspended particles surrounds this wire and carries a vertical current of current density J between its electrodes E_1, E_2. This field configuration of circular magnetic field lines with an axial electric cell-current parallel to the wire generates a force

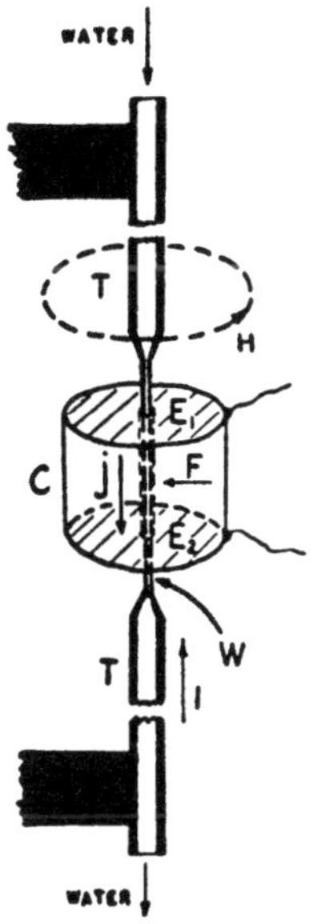

Fig. 17 *Scheme of non-rotational electrokinetic centrifuge based on electromagnetophoresis. T_1: watercooled copper tube carrying the current I which generates the magnetic field $\vec{H}$ within the cylindrical migration cell C with electrodes E_1, E_2. W is the narrowed portion of the copper tube that runs along the central axis of the cell C. $\vec{F}$ electromagnetic force (centripetal or centrifugal) exerted upon particles suspended in the cell C. (From Ref. 55.)*

upon suspended particles that is perpendicular to the electric and magnetic field lines. This force is thus radial and is directed toward the wire (centripetal force) or away from it (centrifugal force) depending on the phase relationship between the alternating current and alternating magnetic field, which optimally, should be either in phase or phase opposition. We have thus in this EMP variant an electrokinetic centrifuge without rotational motion.

We can now remove needless restraints from our imagination and remember that a plane electromagnetic wave (such as a radiowave) has an electric field vector and a magnetic one perpendicular to the electric field and in phase with it. This is the configuration for generation of the EMP effect and the force should be perpendicular to these two vectors, i.e., it should point toward the source of the wave or away from it. Going one step further, we allow the frequency ofthe electromagnetic wave to reach the domain of visible light. A particle in the path of such a wave should exhibit "photophoresis" in the direction of the light ray. This prediction of electromagnetophoresis at optical frequencies reminds us, of course, of the well-known effect of "light pressure." In fact, it was this mode of propulsion by light pressure of microscopic organisms through interstellar space that Arrhenius assumed in his hypothesis of panspermia to explain the origine of life on earth. Thus, if Arrhenius' intuition is correct, electromagnetophoresis may play a role in disseminating life in the universe by intergalactic photophoretic transport between distant inhabited and inhabitable worlds.

EPILOGUE

We can confidently close this lecture with the conclusion that discoveries of new effects and inventions of new instruments will continue to provide new building blocks to the ingenious analytical chemist to create new electrokinetic separation methods. But what we have shown in this lecture also suggests that potentially fruitful ideas, discoveries and inventions may be lying buried in the massive literature until some perceptive scientist will disturb their winter sleep or until they are resurrected through an independent act of creative imagination.

References

1. Reuss, F. F.: Mémoirs de la Société Impériale de Naturalistes de Moscow, 2, 327, (1809).
2. Helmholtz, H.: Wied. Ann. 7, 337, (1879).
3. Smoluchowski, M.: Krak. Anz. 182 (1903).
4. Guye, G.: J. Phys. Radium 9, 457 (1910).
5. Chapman, D. L.: Phil. Mag. 6, 25, 457 (1913).

6. Stern, O.: Z. Electrochem. 30, 508 (1924).

7. Debye, P., Hückel, E.: Phys. Z. 24, 185 (1923).

8. Helmholtz, H. von: Vorträge und Reden, vol. 2, 272 (Vieweg und Sohn, Braunschweig 1903).

9. Kohlrausch, F.: Ann. d. Physik und chemie 62, 209 (1897).

10. Lodge, O.: Brit. Ass. 389 (1886).

11. Tswett, M.: Trav. Soc. Nat. Varsovie, 14, #6 (1903).

12. Tiselius, A.: The Moving Boundary Method of Studying the Electrophoresis of Proteins. Thesis, Uppsala 1930.

13. Kolin, A.: J. Chem. Phys. 22, 1628 (1954).

14. Kolin, A.: J. Chem. Phys. 23, 407 (1955).

15. Svensson, H., Valmet, E.: Science Tools 2, 11 (1955).

16. Hjertén, S.: Free Zone Electrophoresis. Protids of the Biological Fluids 7, 28 (1959).

17. Kolin, A.: J. Appl. Physics 25, 1442-43 (1954).

18. Kolin, A.: Proc. Nat. Ac. Sci. 46, 509 (1960).

19. Kolin, A.: J. Chromatog. 159, 147 (1978).

20. Hanning, K., Wirth, M., Zeiller, K.: Z. Physiol. Chem. 356, 1209 (1975).

21. König, P.: Actas e trabalhos congr. Sul-Amer. Chim. 3° Congr., Rio de Janeiro 2, 334 (1937).

22. Consden, R., Gordon, A. H., Martin, A.J.P.: Biochem J. 38, 224 (1944).

23. Michl, H.: Hochspannungs Elektrophorese, Thieme, Stuttgart 1962.

24. Williams, R. R., Waterman R. E.: Proc. Soc. Exper. Biol. and Med. 27, 56 (1929).

25. Tiselius, A.: Svensk. Kem. tidskr. 58, 305 (1941).

26. Kolin, A.: Proc. Nat. Ac. Sci. 41, 101 (1955).

27. Electrochemical Sorting, Scientific American, p. 60, Feb. (1955).

28. Svensson, H.: Acta chem. scandinav. 15, 325 (1961).

29. Svensson, H.: Acta chem. scandinav. 16, 456 (1962).

30. Svensson, H.: Arch. Biochem. Suppl. I, 132 (1962).

31. Vesterberg, O.: Acta chem. scandinav. 23, 2653 (1969).

32. Kolin, A.: in Isoelectric Focusing, N. Catsimpoolas, Ed., Academic Press, New York 1976, 1-10.

33. Kolin, A.: in Electrofocusing and Isotachophoresis, B. J. Radola and D. Graesslin, Eds., Walter de Gruyter, Berlin · New York 1977.

34. Luner, S. J., Kolin, A.: Proc. Nat. Ac. Sci. 66, 898 (1970).

35. Lundahl, P., Hjertén, S.: Ann. N.Y. Ac. Sci. 209, 94 (1973).

36. Davis, B. J., Ornstein, L.: Ann. N.Y. Ac. Sci. 121, 321 404 (1964).

37. Gianazza, E., Righetti, P. G.: in Electrokinetic Separation Methods, P. G. Righetti, C. J. van Oss, J. W. Vanderhoff, Eds., Elsevier, Amsterdam 1979, 293.

38. Kendall, J., Crittenden, E. D.: Proc. Nat. Ac. Sci. 9, 75 (1923).

39. Konstantinov, B. P., Fiks, V. B.: Russ. J. Phys. Chem. 38, 895 (1964).

40. Williams, R. J., Kirby, H.: Science 107, 481 (1948).

41. Durrum, E. L.: J. Coll. Sci. 6, 274 (1951).

42. Ingram, V. M.: Biochim. Biophys. Acta 28, 539 (1958).

43. Grabar, P., Williams, C. A.: Biochim. Biophys. Acta. 17, 67 (1955).

44. Rotbøl, L.: Clin. Chim. Acta 29, 101 (1970).

45. Laurel, L. B.: Anal. Biochem. 10, 358 (1965).

46. Kowalczyk, J., Pompowski, T.: Wiad. Chem. 15, 107 (1961).

47. Kowalczyk, J.: Zesz. Nauk. Politech. Gdansk, Chem. 15, 1, (1966).

48. Rosengren, A., Bjelqvist, B., Gasparic, V.: in Electrofocusing and Isotachophoresis, B. J. Radola and D. Graesslin, Eds., Walter de Gruyter, Berlin, 1977.

49. Michl, H.: Monat. Chemie 83, 210 (1952).

50. Stegmann, H.: in Electrokinetic Separation Methods, P. G. Righetti, C. J. van Oss, J. W. Vanderhoff, Eds., Elsevier, New York 1979, 313.

51. Slater, G. G.: Analyt. Chem. 41, 1039 (1969).

52. Pohl, H. A.: in Electrostatics and its Applications, A. D. Moore, Ed., John Wiley and Sons, New York 1973, 336.

53. Pohl, H. A., Crane, J. S.: in Electrostatics and its Applications, A. D. Moore, Ed., John Wiley and Sons, New York 1973, 363.

54. Kolin, A.: Science 117, 134 (1953).

55. Kolin, A.: in Medical Physics, O. Glasser, Ed., Yearbook Publ. Comp., Chicago 1960, 268.

56. Kolin, A.: in Electrophoresis 1981, R. C. Allen and P. Arnaud, Eds., Walter de Gruyter, Berlin, 1981.

57. Leenov, D., Kolin, A: J. Chem. Phys. 22, 683 (1954).

SECTION II THEORY AND METHODS

A UNIFIED MATHEMATICAL THEORY OF ELECTROPHORETIC PROCESSES

M. Bier, O.A. Palusinski, R.A. Mosher, A. Graham
Biophysics Technology Laboratory, University of Arizona
Tucson, AZ 85721

and D.A. Saville
Department of Chemical Engineering, Princeton University
Princeton, NJ 08540

Introduction

The last thirty years have seen the proliferation of a rather bewildering number of electrophoretic instruments and methods. Most of these remained confined to the laboratory of the developer, but some gained rapid and widespread acceptance. As a result of the diversity of existing methods, their rational classification is rather difficult and somewhat subject to the personal bias or interest of the individual reviewer. Nevertheless, most authors have emphasized four classical modes of electrophoresis: zone electrophoresis (ZE), moving boundary electrophoresis (MBE), isotachophoresis (ITP) and isoelectric focusing (IEF). The reason for this emphasis is that elements of one or another of these four modes are usually recognizable in most electrophoretic techniques.

Past theoretical treatments of electrophoretic processes have been usually confined to a single mode, such as the reviews of Longsworth on MBE (1), the work of Martin, Everaerts, and others (2-4) on ITP, and the papers of Svensson, Almgren, Cann and others on the theory of IEF (5-7). This compartmentalization of the theory of electrophoresis is not only unneces-

Electrophoresis '82

sary but counterproductive, as it sets artificial barriers to the understanding of the underlying processes. The object of the present paper is to show that each of these electrophoretic modes is based on the same general principles and can collectively be described in terms of a single set of equations. The mathematical model so derived is an amplification of concepts first utilized in our model of the steady state in IEF (8) and has been previously described in general terms (9).

For a qualitative understanding of the characteristic features of the four modes the pioneering work of Kohlrausch in 1897 (10) and the still earlier work of Hittorf (11) is essential and is here briefly summarized. Kohlrausch was the first to demonstrate that the passage of an electric current induces no concentration changes in a system with uniform electrolyte distribution. He has further shown that any concentration boundary, i.e. boundary resulting from dilution of the electrolytes will not migrate under the influence of the electric field but will gradually broaden due to diffusion. Such 'stationary' boundaries are commonly observed in MBE there being a slight change in electrolyte concentration at the protein-buffer interface. It is also observed in ITP as a result of the concentration readjustment of samples and terminators to conform to the Kohlrausch regulatory function.

Kohlrausch has also considered species discontinuities, where one of the ionic constituents is replaced by another species across a boundary. This boundary may migrate in two different modes: if the leading ion has a net mobility higher than that of the trailing ion, a steady state, governed by the well-known Kohlrausch regulatory function, will be reached (3). If there are several such boundaries in the system, they will all migrate at the same velocity. Moreover, the structure of the boundaries will remain constant, once steady state is reached, and they will not broaden as a result of diffusion. This is the essential feature of ITP. Alternatively, if the trailing ions have the higher net mobility, no steady state will be reached, each boundary migrating at its own characteristic velocity and broadening as a result of diffusion.

Finally, electrodes are present in all forms of electrophoresis. Assuming that the solutes undergo no electrochemical reactions at the electrodes, cations will accumulate at the cathode and anions at the anode, as first studied by Hittorf (11). This is one of the methods whereby a natural pH gradient can be established in IEF, if no, or minimal electrode buffer volumes are utilized. Most other electrophoretic systems comprise large electrode buffer reservoirs, thereby avoiding the necessity of considering these electrolytic changes.

The above considerations encompass the essential features of the various modes in electrophoresis, which are seen to differ only in terms of initial conditions (i.e., presence of concentration gradients and/or species discontinuities) and boundary conditions (i.e., the presence or absence of large electrode buffer volumes). The mathematical model presented in the following pages is applicable irrespective of the imposed initial or boundary conditions.

Model construction

The model was designed to predict the evolution of the four classical electrophoretic modes as a function of time. The modeled system is one-dimensional, and the effects of electroosmosis, temperature gradients, or any bulk flows of liquid are neglected. It was further stipulated that the only electrolytes in the system are monovalent buffers or biprotic ampholytes, such as simple amino acids. The model is based on equations describing the components' dissociation equilibria, the mass transport due to electromigration and diffusion, electroneutrality, and the conservation of mass and charge. The resulting model consists of a system of coupled partial differential and non-linear algebraic equations.

The model is far too complex for an analytical solution and a program was developed for its numerical solution. The equations were put in dimensionless form to provide proper physical scaling (9). Discretization of the spatial derivatives at a set of grid points generated a set of ordinary differential equations with time as the independent variable. The resulting equations were integrated numerically using DARE-P software (12) and the calculations were carried out with a CDC-Cyber-175 processor.

Simulation studies

Our first objective was to verify the versatility of our model to simulate the typical features of ZE, MBE and ITP. A three component system, comprising cacodylate, tris hydroxymethylaminomethane (TRIS), and histidine, i.e., a weak acid, a weak base and an ampholyte, was sufficient for this purpose. The initial conditions for ZE and MBE were similar: histidine was assumed to be the sample, migrating in a buffer of TRIS-cacodylate. For the simulation of ITP, cacodylate was assumed to be the leading ion, histidine the terminator, and TRIS the common counterion. In all three cases, unrestricted permeability of column ends to all ions was specified as the boundary condition. This is the situation prevailing in experimental runs, where large electrode buffer volumes are generally utilized. All simulations were run at constant current, and the model was able to predict the characteristic behavior of each mode, such as Kohlrausch adjustment of terminator concentration, the diffusional spreading of stationary boundaries, etc. Unfortunately, space does not permit us to describe in detail all the simulations performed.

Because of our prior studies with the simulation of the steady state in IEF (8), we were more interested in the simulation of transient states in focusing. Uniform distribution of the same three compounds was stipulated as the initial

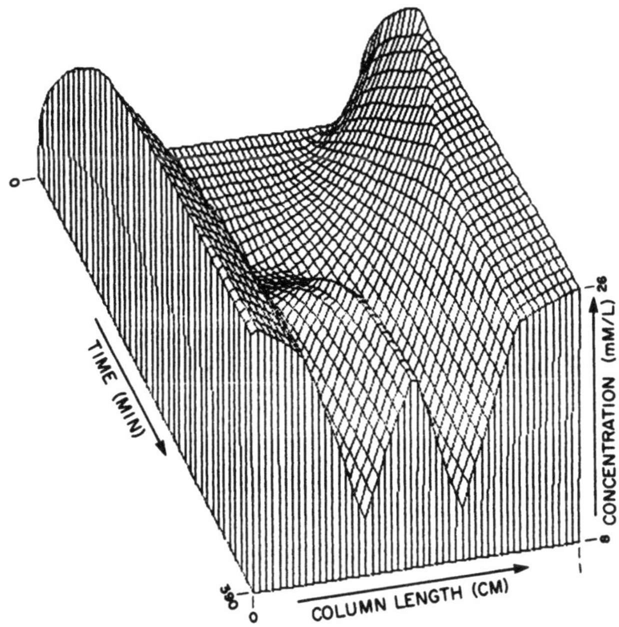

Fig. 1. Time dependent evolution of concentration profiles resulting from isoelectric focusing at constant current. Only the predominant component along the column axis is shown, cacodylate focusing at the left, histidine in the middle and TRIS at the right end of the simulated column.

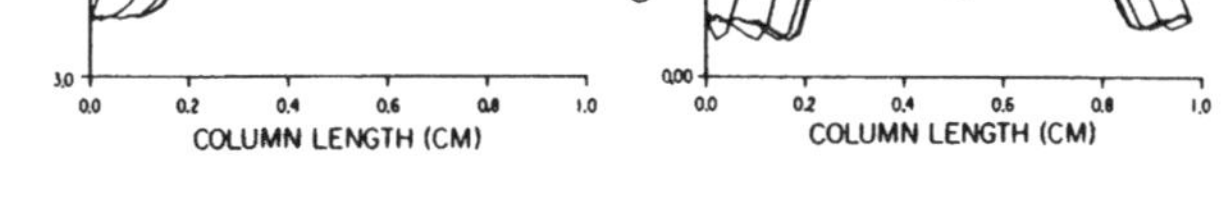

Fig. 2. Plots of pH and conductivity corresponding to the simulation presented in Fig. 1. The horizontal lines in both plots represent the initial condition in the column, as a uniform distribution of all components was specified. The other lines represent the plots at 30, 90, 150, 210, 270, 330 and 390 minutes of focusing.

condition. As the isoelectric point of histidine (pI = 7.60) is intermediate to the pK values of cacodylate (pK = 6.21) and TRIS (pK = 8.30), histidine was expected to focus towards the center of the column. IEF, however, requires a different set of boundary conditions than the other three modes: the ends of the column were assumed to be permeable to hydrogen and hydroxyl ions only. These conditions prevail in experimental flat bed IEF with electrodes in close contact with the gel surface.

Again, a constant current was assumed. The time-dependent progress of focusing is reproduced in the three-dimensional plot in Fig. 1. It can be seen that the focusing starts with the migration of cacodylate and TRIS to the anode and the cathode, resp., followed by the later focusing of histidine in the center of the column. The concomitant changes in pH and conductivity are shown in Fig. 2. The horizontal lines in both plots represent the initial conditions, as uniform initial concentration of all components was assumed. The remaining lines represent the simulation at 30, 90, 150, 210, 270, 330 and 390 minutes of constant current. The pH plots show that the pH gradient is generated at the electrodes and slowly propagate towards the center. The conductivity plots show the development of three 'conductivity gaps' corresponding to the accumulation of relatively undissociated acid and base at the electrodes, and of isoelectric histidine in the center of the column. Most gratifying, however, was to find that the predicted profiles of concentration, pH and conductivity, derived from the transient model, matched those previously derived from our older steady state model. This verifies the numerical integrity of the two programs.

Discussion

Our mathematical model was found useful for computer simulation of the four classical modes of electrophoresis, i.e., ZE, MBE, ITP and IEF. It not only correctly predicts the characteristic features of each mode, but gives details of concentration, pH and conductivity profiles not easily amenable to direct experimental measurement. While the present program is limited to simple chemical systems, ongoing work is directed to its expansion to more complex mixtures. The practical applications of such a model are somewhat limited by the paucity of reliable data on components dissociation constants and mobilities. It is well known, in particular, that anionic and cationic ampholyte subspecies differ in their mobilities, but exact data are virtually unavailable.

We wish to propose that our model provides the basis for a rational classification of all electrophoretic techniques, and is not limited to ZE, MBE, ITP and IEF. These four classical modes differ only in terms of the initial distribution of components along the column axis and ionic permeabilities of column ends, i.e. the previously described Kohlrausch-Hittorf constraints. These two conditions define an 'electrophoretic plane', in which the classical modes are but idealized points within a virtual infinity of possible conditions. To model the manifold electrophoretic techniques in actual usage, such as immunoelectrophoresis or density gradient gel electrophoresis, other constraints have to be incorporated. Some of these constraints are presented schematically in Fig. 3. Accordingly, a technique is characterized by its location within the illustrated multidimensional 'electrophoretic space'. Our model can easily be adapted to encompass those additional non Kohlrausch-Hittorf constraints which involve only changes of transport characteristics of the molecules. For instance, to model molecular sieving in porosity gradients, the mobility of the components would have

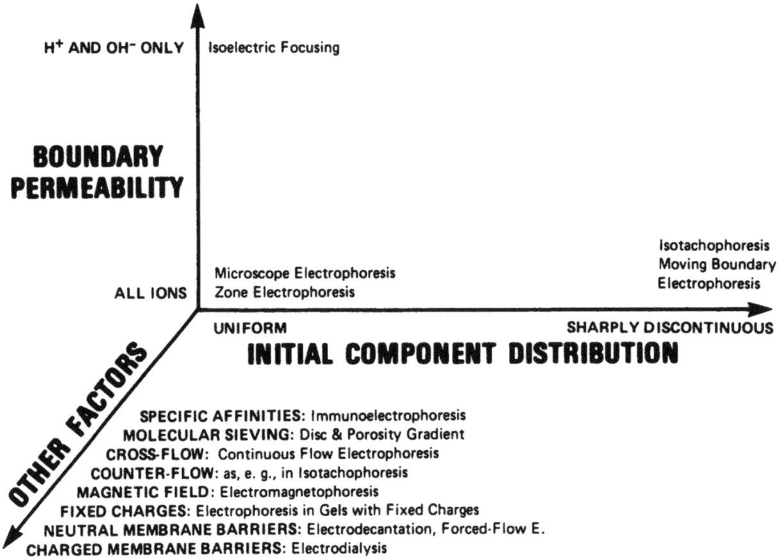

Fig. 3. A graphic representation of the suggested classification of electrophoretic processes. For an explanation see text.

to be made a function of their location along the column axis. We have already modeled ITP in gels with immobilized counterions, as well as electrodialysis, i.e. the transport of ions across a membrane incorporating fixed charges. In both instances, it was sufficient to assign a zero mobility to the stationary species. On the other hand, some of the other constraints, such as incorporation of cross-flow fluid motion (13), as in continuous flow electrophoresis, would be far more difficult. Thus, we foresee that our model may be most useful in evaluating the possibility of developing new separation techniques involving conditions not yet tried in the laboratory, or even of conditions which may not be easily obtainable in practice. An example of such conditions would be the modification of boundary conditions, rendering the ends of the column permeable to selected ions only.

Acknowledgements

Supported in part by the NASA Grante NSG-7333, the NASA Contract NAS8-32950 and the NSF Grant CPE 8103079.

References

1. L.G. Longsworth in Electrophoresis, Vol. I, M. Bier, Ed. (Academic Press, New York 1959) pp. 91-136, 137-177.

2. A.J.P. Martin and F.M. Everaerts, Anal. Chim. Acta, 38, 233 (1967).

3. F.M. Everaerts, J.L. Beckers and T.P.E.M. Verheggen, Isotachophoresis (Elsevier, Amsterdam 1979).

4. R.J. Routs, Dissertation (University of Technology, Eindhoven, 1971); G.T. Moore, J. Chromatogr. 106, 1 (1975). See also recent work of Dr. E. Schumacher at the University of Bern: P. Ryser, Dissertation (1976), W. Thormann, Dissertation (1982) and P. Radi, Thesis (1982).

5. H. Svensson, Acta Chem. Scand. 15, 325 (1961); H. Rilbe (formerly H. Svensson) in Isoelectric Focusing, N. Catsimpoolas, Ed. (Academic Press, New York, NY 1976) pp. 13-52.

6. M. Almgren, Chemica Scripta 1, 69 (1971).

7. J.R. Cann and D.I. Stimpson, Biophys. Chem. 7, 103 (1977); J.R. Cann and K.S. Gardiner, ibid. 10, 203 (1979).

8. O.A. Palusinski, T.T. Allgyer, R.A. Mosher, M. Bier and D.A. Saville, Biophys. Chem. 13, 193 (1981); M. Bier, R.A. Mosher and O.A. Palusinski, J. Chromatogr. 211, 313 (1981).

9. O.A. Palusinski, M. Bier and D.A. Saville, Biophys. Chem. 14, 389 (1981).

10. F. Kohlrausch, Ann. Physik. 62, 209 (1897).

11. J.W. Hittorf, Poggendorff's Ann. 89, 177 (1853); 98, 1 (1856); 103, 1 (1858); 106, 337 (1859).

12. G.A. Korn and J.C. Wait, Digital Continuous System Simulation (Prentice Hall, Englewood Cliffs, N.J., 1978).

13. D.A. Saville in Physicochemical Hydrodynamics, D.B. Spalding, Ed. (Adv. Publications, London 1978) pp. 893-912.

ISOELECTRIC FOCUSING IN IMMOBILIZED pH GRADIENTS. I: PRINCIPLE AND METHODOLOGY

Bengt Bjellqvist, Kristina Ek

LKB Produkter AB, Box 305, S-161 26 Bromma, Sweden

Pier Giorgio Righetti, Elisabetta Gianazza

Department of Biochemistry, University of Milano, Via Celoria 2, Milano 20133, Italy

Angelika Görg and Wilhelm Postel

Lehrstuhl für Allgemeine Lebensmitteltechnologie der Technischen Universität, München, D-8050 Freising-Weihenstephan, G.F.R.

Introduction

In 1966 LKB introduced the first commercial ampholyte mixture, based on Rilbe's isoelectric focusing (IEF) theory (1). Since then the technique of IEF in carrier ampholyte generated pH gradients has developed to one of the most commonly used analytical and preparative methods in work with amphoteric molecules such as proteins and peptides. The resolution given by analytical IEF is amongst the highest available from present biochemical separation techniques.

Sixteen years of experience with IEF show that, notwithstanding the extreme success of the method, the carrier ampholyte based technique has certain inherent limitations and problems. The problem most frequently discussed in the literature is the so called "cathodic drift" (2) or "plateau phenomenon" (3-5), that is to say the continuous slow change of what was expected to be a stable pH gradient with time. This causes problems mainly when narrow pH gradients are used, especially at high pH values, where the drift is most pronounced. In experiments using a matrix for convectional stabilization, a large part of these phenomena could be ascribed to electroendosmosis caused by minor amounts of charges bound to the matrix (6), but there

Electrophoresis '82

also exist good reasons to believe that it is partly connected with more basic features of the carrier ampholyte pH gradient. Therefore IEF, as it is commonly used today, is a non-equilibrium method. One of the consequences of this is that the pH gradient obtained in an experiment depends on the salt content of the sample. To avoid displacements of the pH gradient in the anodic or cathodic directions, either the salt content or the sample volume must be kept small. In flat bed IEF runs, with a number of parallel sample tracks separated by sample-free zones, the effect of high salt concentrations, as a consequence of diffusional mass transport perpendicular to the focusing direction, can be seen as distorted zones (7). Conventional IEF can give curved zones also for a number of other reasons, such as skewed end-electrolyte distribution, uneven cooling and bad contact between the platinum wires and the electrode strips in flat-bed techniques. In no case introduced disturbances are eliminated as a result of focusing and, if large amounts of persulphate have been used for polymerization, disturbances even seem to be amplified (6).

Another inherent weakness of the carrier ampholyte concept is that the pH gradient is generated with the aid of a large number of amphoteric compounds. With techniques available today, it could neither be guaranteed that these compounds are present in equal amounts nor that essential properties, such as buffering capacities, are identical. There exists overwhelming evidence for the fact that a large number of the individual species in commercial carrier ampholytes are focused as sharp zones (8). As a consequence of this, not only the buffering capacity but also the ionic strength, which directly influences pK- and pI-values, might vary in a dramatic manner along the pH gradient. The schlieren pattern (9) resulting in polyacrylamide gels containing focused carrier ampholytes is a clear indication of the fact that the conditions in the pH gradient vary.

From what has been described above, present day IEF techniques show certain limitations and problems; therefore, there is a need for a complementary technique solving some of the above problems. Different possibilities to create stable pH gradients without carrier ampholytes have also been tried (10-12) but no work, except the LKB patent by Gasparic *et al.* (13), has until now been published on what might seem as the most obvious possibility, namely to bind the buffering groups generating the pH gradient to the matrix used for conventional stabilization. There is probably a number of reasons for this. Thus, it might seem unrealistic that sufficient conductivity would result in a gradient of this type; moreover, bound charges could result in electroendo-osmosis and the ion-exchange character of the matrix might lead to binding of proteins. On the contrary, we have found that the demand on electric conductivity is relativey small and that neither electroendo-osmosis nor binding creates any problem, and thus any matrix suitable for generating the necessary conventional stabilization in IEF could be modified to generate an immobilized pH gradient suitable for IEF.

The present paper will describe the production and use of immobilized pH gradients made with the aid of Immobilines^{Tm}. These are a number of buffering acryloil monomers developed by LKB Produkter AB for the purpose of immobilizing pH gradients in polyacrylamide gels. This work as well as the following ones (14, 15) will show that the use of Immobiline^{Tm} based pH gradients not only solves the problems associated with pH drift and distorted protein bands, but also allows isoelectric focusing under controlled conditions with increased resolution and loading capacity compared to what can be achieved with carrier ampholyte generated pH gradients.

Materials and Methods

Apparatus. The IEF experiments were carrier out using the LKB Ultrophor apparatus together with the LKB 2197 constant power supply and for cooling the LKB 2209 Multitemp. For pH measurements, a LKB 2117-111 surface pH electrode was used, together with a Beckmann model 3500 digital pH meter.

Chemicals. Immobilines^{Tm}, acrylamide, N,N'-methylene bisacrylamide, ammonium persulphate, TEMED, Repelsilane and Coomassie Brilliant Blue R-250 were all from LKB Produkter AB. Sulfosalicylic acid and trichloroacetic acid were from Merck AG. All the standard proteins used were purchased from Sigma. All other chemicals were of analytical grade.

Gel casting. The gel casting has been made according to Görg *et al.* (16) using the LKB 2117-901 Gradient Gel kit. The gels were 250x110x0.5 mm and, for sample application, they contained slots of the size 5x2x0.2 mm. Before use, the U-frame with the slots was coated with Repelsilane to avoid that the gel could stick to the glass surface when opening the cassette. Acrylamide and N,N'-methylene bisacrylamide have been used as separate stock solutions with the concentrations 29.1% (w/v) and 0.9% (w/v), respectively. The chambers of the gradient mixer were filled with 7 ml of each solution containing acrylamide and Bis corresponding to a 5%T, 3%C gel and Immobilines^{Tm} in concentrations adjusted to give the desired pH gradient. The acidic solution was always chosen as heavy solution and contained 20% (v/v) glycerol. The gel solutions were not degassed as the gradient mixer was used with both chambers open. The catalysts (5 µl TEMED and 5 µl of 40% persulphate per chamber) were added directly to the gradient mixer immediately before filling the gel in the cassette. The gels were polymerized for one hour at 50°C (17). After removing the gels from the cassette they were placed in one liter distilled water for half an hour whereafter excess water was allowed to drain from the vertically standing gels for another half hour at room temperature. The drying step is essential as a gel containing too large amounts of water will sweat in the focusing step giving water droplets on its surface. These droplets result in blue spots when the gel is stained as described below.

Running conditions. In most experiments the electrodes have been applied directly on the gel surface, alternatively 10^{-2}M H_3PO_4 and NaOH could be used as electrode solutions. The samples were applied without pre-running, with the power set at 10 W, the voltage at 2500 V and the current at 25 mA. Focusing was continued overnight at 10°C for any given pH range selected. With pH values higher than 7, two strips of absorbent cloth 200x20x5 mm were soaked in 1M NaOH and placed inside the safety lid of the 2217 Ultrophor chamber in order to minimize the influence of CO_2. The gels were fixed, stained and destained as with conventional carrier ampholytes PAG-plates using Coomassie Brilliant Blue R-250 (18).

Results

1) General physico chemical properties of Immobiline^{Tm}

The Immobilines^{Tm} are acrylamide derivatives with the general structure (where R contains either a carboxyl or a tertiary amino group):

TABLE I

PROPERTIES OF IMMOBILINES™

Apparent pK-values[1] $I = 10^{-2}$

	H_2O		Polyacrylamide gel[2] T=5% C=3%		Polyacrylamide gel[2] T=5% C=3% Glycerol 25%-w/v		Physical state at room temperature
	10°C	25°C	10°C	25°C	10°C	25°C	
Acids with carboxyl as buffering group							
Immobiline pK 3,6	3,57	3,58	-	-	3,68±0,02	3,75±0,02	solid
Immobiline pK 4,4	4,39	4,39	4,30±0,02	4,36±0,02	4,40±0,03	4,47±0,03	solid
Immobiline pK 4,6	4,60	4,61	4,51±0,02	4,61±0,02	4,61±0,02	4,71±0,03	solid
Bases with tertiary amine as buffering group							
Immobiline pK 6,2	6,41	6,23	6,21±0,05	6,15±0,03	6,32±0,08	6,24±0,07	solid
Immobiline pK 7,0	7,12	6,97	7,06±0,07	6,96±0,05	7,08±0,07	6,95±0,06	solid
Immobiline pK 8,5	8,96	8,53	8,50±0,06	8,38±0,06	8,66±0,09	8,45±0,07	liquid
Immobiline pK 9,3	9,64	9,28	9,59±0,08	9,31±0,07	9,57±0,06	9,30±0,05	liquid

[1] pK-values measured with glass surface electrode given without any corrections.

[2] Mean-values of 10 determinations. Due to the slow response of the electrode the pK-values for the amines are uncertain.

$$CH_2 = CH - \underset{\underset{O}{\|}}{C} - \underset{\underset{H}{|}}{N} - R$$

Provided that the correct polymerization conditions are chosen, these monomers are efficiently incorporated in the gel. As a result of their presence, the gel (in the absence of other compounds taking part in protolytic equilibria) will have a pH defined by the concentrations and dissociations constants on the Immobilines^{Tm}. The gel will also have a conductivity not only due to H^+ and OH^- ions, but also connected to the amount of incorporated Immobilines^{Tm} and their buffering capacity.

Table I lists the relevant physico chemical data of Immobiline^{Tm}: at present three acids and four bases are available, with pK's spanning the pH range 3.6 to 9.3. Immobiline^{Tm} pK 9.3 has been used as counter ion for determination of the pK values of the acids; conversely, Immobiline^{Tm} pK 3.6 has been used for the pK determination of the bases. pK measurements in gels have been performed by including 10^{-2}M buffering Immobiline^{Tm} and $5x10^{-3}$M Immobiline counter ion in the solution used to generate a gel of the size 240x110x1 mm. For catalyst removal, the gel was electrolyzed on the Ultrophor overnight and then the pH was measured. All pH measurements were made with a glass surface electrode calibrated with NBS standard buffers. The distance between the double bond and the group taking part in the protolytic equilibrium has in all cases been chosen long enough so that the influence of the double bond on the dissociation constant can be neglected. As a result, the pK difference between the Immobiline^{Tm} monomer and the buffering group fixed in the immobilized pH gradient when used for IEF is mainly due to media effects and temperature variations.

2) Band eveness

As can be seen from Fig. 1, the present technique gives straight iso-pH lines covering the whole width of the gel. In order to obtain this type of result it is essential that the pH gradient is approximately linear and the buffering power even. The result shown in Fig. 1 is also typical in that no distorted protein bands result when using Immobiline^{Tm} gels. This will be true also when the sample contains large amounts of salt. If the salt content in the samples is high and variable, protein bands from samples with high conductivity will spread out over an appreciably wider part of an Im-

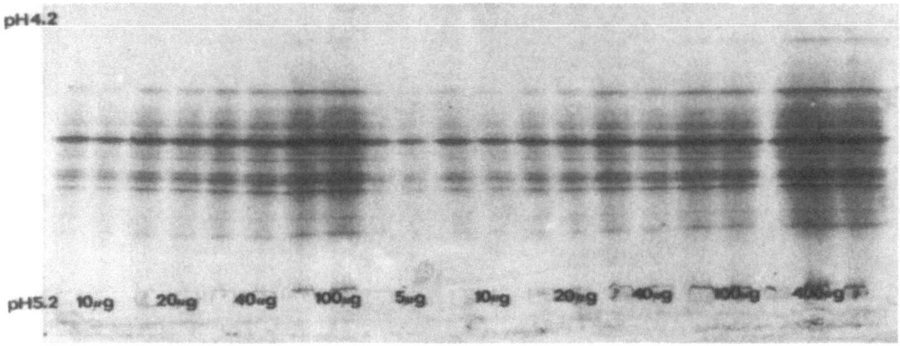

Fig. 1. Varying amounts of ovalbumin focused in an Immobiline^{Tm} gradient pH 4.2-5.2. Immobiline^{Tm} concentrations used for gradient mixing: acidic solution: $5x10^{-3}$M Immobiline^{Tm} pK 4.6 and $1.4x10^{-3}$M Immobiline ^{Tm} pK 9.3; basic solution: $5x10^{-3}$M Immobiline^{Tm} pK 4.6 and $4x10^{-3}$M Immobiline^{Tm} pK 9.3. Running conditions: overnight at 250 V/cm and 10°C.

mobiline^{Tm} gel than what corresponds to the width of the application zone. Thus, when running this type of samples it is important to use individual gel strips for each sample track.

3) Resolution

The band width and resolution obtainable in Immobiline^{Tm} pH gradients are comparable to conventional carrier ampholyte pH gradients. Thus narrower pH gradients decrease the band sharpness, but the resolution is increased. Fig. 2 shows the results of electrofocusing ovalbumin in a carrier ampholyte pH gradient with an approximate slope of 0.2 pH units/cm and in Immobiline pH gradients with the following slopes: 0.1, 0.02 and 0.01 pH units/cm, respectively. In the two last experiments, part of the gels were cut off and stained after 16 hours in order to localise the position of the main band. After 18 hours the distance between the electrodes was shortened to 5 and 2.5 cm, respectively, to allow delivering 500 V/cm to one gel and 1000 V/cm to the other. The experiment was then continued for 3 more hours. In going from 0.1 pH units/cm to 0.02 pH units/cm, the ovalbumin bands are resolved into doublets. With the slope 0.01 pH units/cm, the distance between the bands in the doublets is approximately 2 mm, which should correspond to a pI difference of 0.002 pH units. From this we estimate that it should be

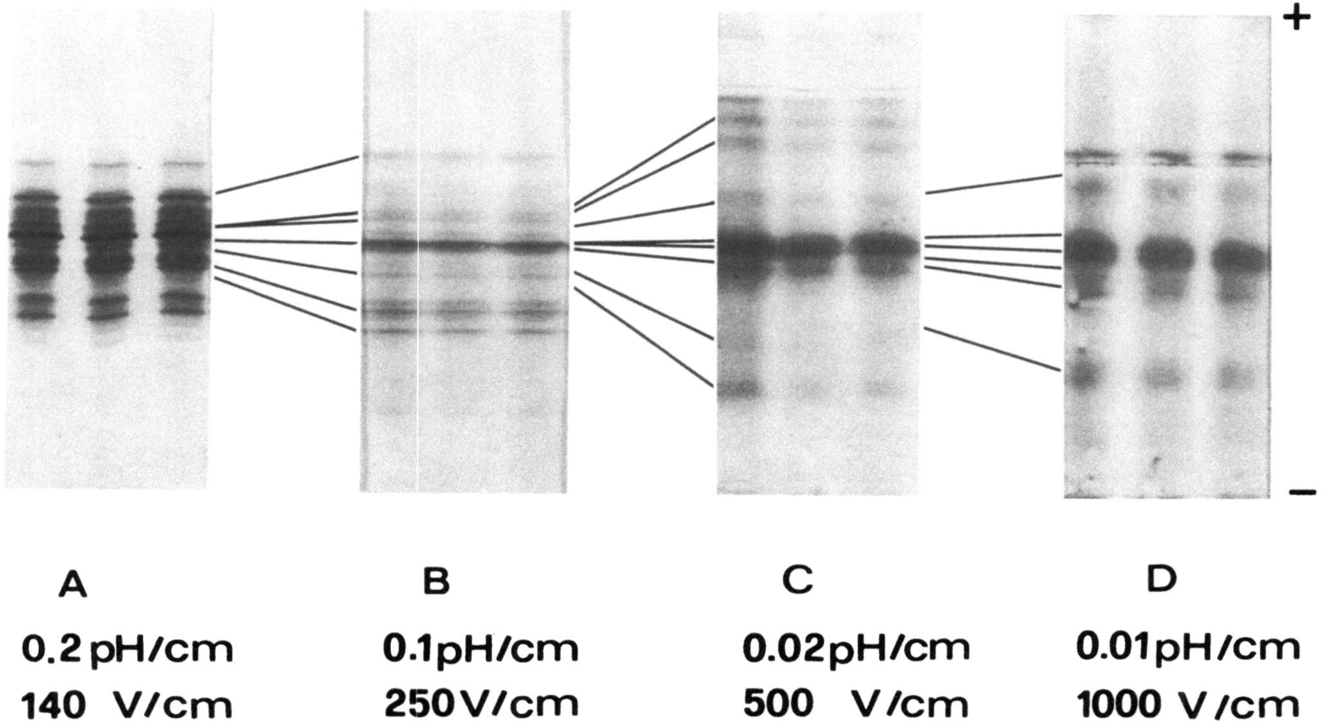

Fig. 2. Ovalbumin focused on a narrow Ampholine pH gradient (A) and on Immobiline^{Tm} gradients with varying pH slopes (B-D). Strips B-D contain $5x10^{-3}$M Immobiline^{Tm} pK 4.6 titrated with Immobiline^{Tm} pK 9.3 to the respective pH slopes. Ovalbumin load in the sample tracks (from left to right): 40, 20 and 20 µg.

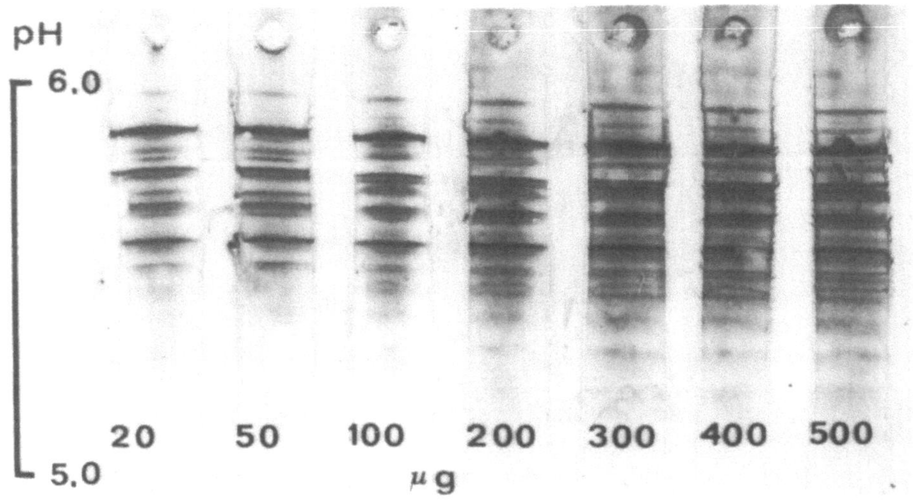

Fig. 3. Varying amounts of carbonic anhydrase focused on Immobiline^{Tm} gels in the pH range 4.3-6.8. Acidic end concentrations: $10x10^{-3}$M Immobiline pK 4.3 and $5x10^{-3}$M Immobiline^{Tm} pK 6.2; basic end concentrations: $3x10^{-3}$M Immobiline^{Tm} pK 4.3 and $10x10^{-3}$M Immobiline pK 6.8.

possible to resolve bands differing by as little as 0.001 pH units. From the early work of Vesterberg and Rilbe (19) the resolution limit for carrier ampholyte gradients has often been given as 0.02 pH units. In reality even slightly higher resolution is possible as long as the pH drift and conductivity gaps do not interfere if narrow pH ranges and high field strengths are used. Allen *et al.* (20) have as an example claimed the resolution of α_1-antitrypsin bands to be of the order of only 0.005 pH units in a narrow Ampholine pH gradient.

4) Loading capacity

Fig. 3 shows the effect of loading increasing amounts of carbonic anhydrase on 110 mm long, 10 mm wide and 0.5 mm thick Immobiline^{Tm} pH gradient strips. Up to 500 μg can be loaded on a gel strip of this size, which is appreciably more than what can be loaded on a carrier ampholyte gel. The low current in Immobiline^{Tm} pH gradients makes it possible to increase the gel thickness 5-10 times without problems with the cooling capacity. As also narrower pH ranges increase the loading capacity, it should be possible to handle more than 100 mg in one run in a gel with an area corresponding to

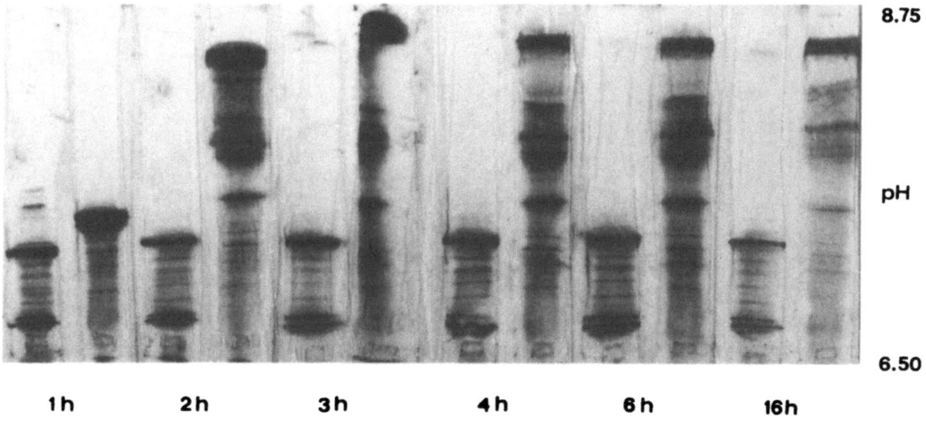

Fig. 4. Horse heart and sperm whale myoglobins, in alternate tracks, focused for varying times on an Immobiline^{Tm} gel, pH 6.5-8.75. The gradient was created as follows: a mixture of 2.5×10^{-3}M Immobiline^{Tm} pK 6.2, 5×10^{-3} M Immobiline^{Tm} pK 7.0 and 5×10^{-3}M Immobiline^{Tm} pK 8.5 was titrated with Immobiline pK 3.6 to the stated pH interval (pH 6.5 in the dense and pH 8.75 in the light gradient solutions).

an Ampholine PAG plate (110 x 265 mm): thus Immobiline^{Tm} pH gradients are clearly of interest in preparative runs. In this connection Immobiline^{Tm} gels are also advantageous because they do not contain any other extractable components than the focused proteins. Moreover the band position can be easily localized after focusing without any staining as ridges appear in the gel at the positions of focused proteins when large protein loads are used. It is thus very easy to cut out the parts of the gel containing the protein components of interest.

5) Behavior of proteins. pH gradient stability

Due to the low conductivity of Immobiline^{Tm} pH gradients, the sample components will show a different behavior than on a carrier ampholyte pH gradient. If the sample is introduced in the pH gradient after the unbound ions have been electrolysed away, the conductivity of the sample will be high compared with the background conductivity and the proteins will stay at the application area for a long period of time, with an increased risk of precipitates. It is thus normally preferable to apply the sample without pre-running the gel, so that it can move together with the salt background present in the gel plate. When coloured proteins are used, they can be seen as sharp bands moving with the refractile line marking the rear border of an ionic compound moving towards an electrode. The protein band stops as it reaches the pH zone in the gel where it is isoelectric, becoming rather diffuse and then sharpening again as the focusing is continued.

The time needed by a protein to reach its pI position depends on a number of factors: slope of its titration curve around its pI, slope of the immobilized pH gradient, voltage gradient applied, medium viscosity, etc. Once a protein has reached its equilibrium position in an Immobiline^{Tm} gradient, however, it will stay in this position indefinitely. Fig. 4 shows the behavior of sperm whale and horse heart myoglobins in a pH 6.5 to 8.75 Immobiline^{Tm} gradient. The proteins are focused in four hours and after 16 hours at 250 V/cm they are still in the same position. Under these conditions most proteins,in conventional carrier ampholyte systems,would have drifted out in the cathodic compartment (21).

Discussion

A comparison of IEF in Immobiline^{Tm} pH gradients and in carrier ampholyte generated pH gradients shows a number of advantages for the Immobiline^{Tm} gel:

a) true equilibrium method - no drift;
b) higher resolution;
c) higher loading capacity;
d) better control of form, width, ionic strength and buffering capacity;
e) possibility of generating extremely shallow pH gradients;
f) easier separation of buffering species from proteins in preparative runs;
g) insensitivity to salts and buffers in the sample.

IEF with good quality carrier ampholytes is still an excellent analytical method which in most cases gives results which correspond to the demands of the user. Immobiline^{Tm} pH gradients require longer focusing time and, although it is easy to cast a gradient gel, it is still easier to cast a carrier ampholyte containing gel or use a ready-made PAG plate. Thus, for most routine uses, a carrier ampholyte based pH gradient is still the natural choice and Immobiline^{Tm} should be used mainly when its advantages are needed. IEF in ultra-narrow pH gradients when very high resolution is required is an obvious case when Immobiline^{Tm} pH gradients ought to be used. This type of application is further exemplified and discussed in the second and third articles of this series (14, 15).

In order to detect the presence of minor components in a sample, it is often necessary to heavily overload major components, which disturb the IEF process in carrier ampholyte pH gradients. The high loading capacity of Immobiline^{Tm} pH gradients makes them the natural choice also for this type of application. The situation described frequently appears in two-dimensional applications using isoelectric focusing as first dimension. The other advantages with Immobiline^{Tm} pH gradients, such as insensitivity to disturbances and ease of control of form and width of pH gradients, are also of major importance to 2-D techniques and thus Immobiline pH gradients should be preferred as first dimension in 2-D runs. This important application is exemplified and described in detail in the third article of this series (15).

In the not too uncommon case when the samples contain very large amounts of salt, it should be more convenient to use Immobiline^{Tm} pH gradient gels than to first remove the salt before applying the samples to a conventional isoelectric focusing gel. Due to the fact that none of the naturally occurring amino acids have side chains with groups buffering in the pH range 7.5-9.5, many proteins isoelectric in this range will approach their pI relatively slowly. This phenomenon can be directly visualized with the aid of electrophoretically run titration curves on prefocused Ampholine pH gradients (22). Carrier ampholyte pH gradients in the pH range 7.5-9.5 are characterized by a fairly large pH drift, especially in the absence of urea (23). The combination of the slower focusing and the pH drift can result in less sharp bands and bad focusing also in pH gradients as wide as two pH units. In this case Immobiline^{Tm} gradients represent a valid alternative.

Another field in which the use of Immobiline^{Tm} should be of great interest is in preparative isoelectric focusing. The high load capacity, the ease with which a narrow, specially designed pH gradient can be generated and the fact that the problem of separating the focused proteins from the carrier ampholytes is avoided, clearly ought to make the Immobiline^{Tm} pH gradient attractive for preparative work.

To state that Immobiline^{Tm} pH gradients revolution the art of isoelectric focusing is perhaps too strong, but quite clearly it is a technique with large future possibilities. Immobiline[Tm] pH gradients are complementary to carrier ampholyte gradients and will certainly increase the usefulness of IEF as an analytical and preparative tool in biochemistry.

Acknowledgements

We thank Prof. H. Rilbe (formerly Svensson) for helpful suggestions and criticism during the lay out of this manuscript. P.G.R. and E.G. are supported by grants from Consiglio Nazionale delle Ricerche (CNR) and Ministero della Pubblica Istruzione (MPI, Roma).

References

1. Svensson, H.:Acta Chem. Scand. 15, 325-341 (1961)
2. Righetti, P.G., Drysdale, J.W.: Ann. N. Y. Acad. Sci. 209, 163-186 (1973)
3. Chrambach, A., Doerr, P., Finlayson, G.R., Miles; L.E.M., Sherins, R., Rodbard, D.: Ann. N. Y. Acad. Sci. 209, 44-69 (1973)
4. Fawcett, J.S.: In Progress in Isoelectric Focusing and Isotachophoresis, Righetti, P.G., ed., North Holland/American Elsevier, Amsterdam, pp. 25-37 (1975)
5. Rilbe, H.: In Electrofocusing and Isotachophoresis, Radola, B.J., Graesslin, D., eds., de Gruyter, Berlin, pp. 35-50 (1977)
6. Righetti, P.G., Macelloni, C.: J. Biochem. Biophys. Methods 6, 1-15 (1982)
7. Jonsson, M.: Electrophoresis 1, 141-149 (1980)
8. Righetti, P.G.: In Electrokinetic Separation Methods, Righetti, P.G., Van Oss, C.J., Vanderhoff, J.W., eds., Elsevier/North Holland, Amsterdam, pp. 389-441 (1979)
9. Edwards, J.J., Andersson, N.G.: Electrophoresis 2, 161-168 (1981)
10. Rilbe, H.: J. Chromatogr. 159, 193-205 (1978)
11. Martin, A.J.P., Hampson, F.J.: J. Chromatogr. 159, 101-110 (1978)
12. Troitski, G.V., Savialov, P.V., Kirjukhin, I.F., Abramov, V.M., Agitsky, G.J. (1975) Biochim. Biophys. Acta 400, 24-3L (1975)
13. Gasparic, V., Bjellqvist, B., Rosengren, A.: Swedish patent 75 140 49-1 (1975); US patent 4,130,470 (1978); German patent 2 656 162 (1981)

14. Righetti, P.G., Gianazza, E., Bjellqvist, B., Ek, K., Görg, A., Westermeier, R.: In Electrophoresis '82, Stathakos, D., ed., de Gruyter, Berlin, pp. 75-82 (1983)

15. Görg, A., Postel, W., Westermeier, R., Bjellqvist, B., Ek, K., Gianazza, E., Righetti, P.G.: In Electrophoresis '82, Stathakos, D., ed., de Gruyter, Berlin, pp. 353-361 (1983)

16. Görg, A., Postel, W., Westermeier, R., Gianazza, E., Righetti, P.G.: J. Biochem. Biophys. Methods 3, 273-284 (1980)

17. Gelfi, C., Righetti, P.G.: Electrophoresis 2,220-228 (1981)

18. Winter, A., Ek, K., Andersson, U.B.: LKB Application Note No. 250 (1977)

19. Vesterberg, O., Svensson, H.: Acta Chem. Scand. 20, 820-834 (1966)

20. Allen, R.C., Hasley, R.A., Talamo, R.C.: Amer. J. Clin. Pathol. 62, 732-739 (1974)

21. Davies, H.: In Isoelectric Focusing, Arbuthnott, J.P., Beeley, J.A., eds., Butterworths, London, pp. 97-113 (1975)

22. Ek, K.: LKB Application Note No. 319 (1980)

23. Gianazza, E., Astorri, C., Righetti, P.G.: J. Chromatogr. 171, 161-169 (1979)

ISOELECTRIC FOCUSING IN IMMOBILIZED pH GRADIENTS. II: APPLICATION TO HEMOGLOBIN ANALYSIS.

Pier Giorgio Righetti, Elisabetta Gianazza

Department of Biochemistry, University of Milano, Via Celoria 2, Milano 20133, Italy

Bengt Bjellqvist, Kristina Ek

LKB Produkter AB, Box 305, Bromma, Sweden

Angelika Görg and Rainer Westermeier

Lehrstuhl für Allgemeine Lebensmitteltechnologie der Technischen Universität, München, D-8050 Freising-Weihenstephan, G.F.R.

Introduction

Isoelectric focusing (IEF) of hemoglobins (Hb) was first described in 1971 by Drysdale *et al.* (1), utilizing gel cylinders. Subsequently, Jeppsson and Berglund (2) introduced thin layer separations (TLIEF), which are today, by far, the most widely used methods. Rosa's group in Créteil has extensively used TLIEF for mass screening of Hb's (3, 4). These authors have constructed a map of more than 80 human Hb variants occupying nearly 50 distinct pI positions. As seen in this map, most Hb's focus within less than 1 pH unit, from pI *ca.* 6.6 up to pI 7.5. Three clusters can be distinguished in this map: a group of "C-like" variants, a second of "S-like" and a third of "A-like". While the "C-like" zone is the least crowded, and allows resolution of most mutants by IEF, the "S-like" and "A-like" regions are overcrowded, so that several genetic variants co-focus within a single band of apparently identical pI. Clearly, an increment of resolution is needed in these two regions, for unambiguous determination of Hb mutants by IEF. The "S- and A-like" regions are very important, since they contain several common pathological Hb's. For instance Hb S is the most widely distributed abnormal Hb in the world and its homozygous S/S state, which is responsible for sickle-cell anemia, could be confused with other, less critical

Electrophoresis '82

disorders (e.g. double heterozygotes for Hb S/D or Hb S/P or Hb S/G) due to mutant Hb's D, P and G, which co-focus in the same band or in its immediate surrounding.

In addition to the unambiguous detection of new genetic variants, IEF has been applied also to the quantitation of some minor Hb components. In man it is especially important to measure three of these minor Hb's, since changes in their levels can play a role in the diagnosis of some disease. For example, the HbA_2 ($\alpha_2\delta_2$) level is about twice as high in the blood of carriers of the β-thalassemia trait as in the blood of non carriers (5).Hb F ($\alpha_2\gamma_2$), which is the major component in fetal blood and at birth, is normally present at trace levels in the adult but may be elevated in genetically determined or acquired disorders, including thalassemia, hereditary persistence of HbF and leukemias (6). The third minor component of importance in phatology is HbA_{1c}, which is a glycosylated HbA whose level is elevated in diabetes mellitus (7). Precise quantitation of HbA_{1c} is of great interest since it may provide the best means of assessing accurately the degree of hyperglycemia on a long-term basis (8). Accurate determination of these minor Hb's, as well as of several genetic mutants, is often a difficult proposition, since they exhibit minute pI differences from major Hb bands commonly present in the blood of healthy individuals. In the present article we report how, by a proper use of immobilized pH gradients (Immobiline^{Tm}), it is possible to fully resolve any of the minor Hb's or genetic mutants commonly found during Hb screening.

Materials and Methods

All experiments were run in polyacrylamide slabs supported by silanized glass plates (9) (gel composition: 5%T, 4%C_{Bis}; size: 125x125x0.5mm). For carrier ampholyte IEF, the gels were made to contain LKB Ampholine in the pH ranges 6-8 (2%), 7-9 (0.5%) and 3.5-10 (0.2%); in some experiments, 0.2 M β-alanine (β-Ala) and 6-amino caproic acid (6-ACA) were added as pH gradient modifiers (10). The runs lasted 90-120 min when delivering 5W-1500V (maximum). For immobilized pH gradients, two solutions were linearly mixed, one made to contain 20 mM Immobiline^{Tm} pK 3.6, 27 mM Immobiline pK 7.0, 18 mM Immobiline^{Tm} pK 6.2 and 20% glycerol, pH 6.5; the other 20 mM Immo-

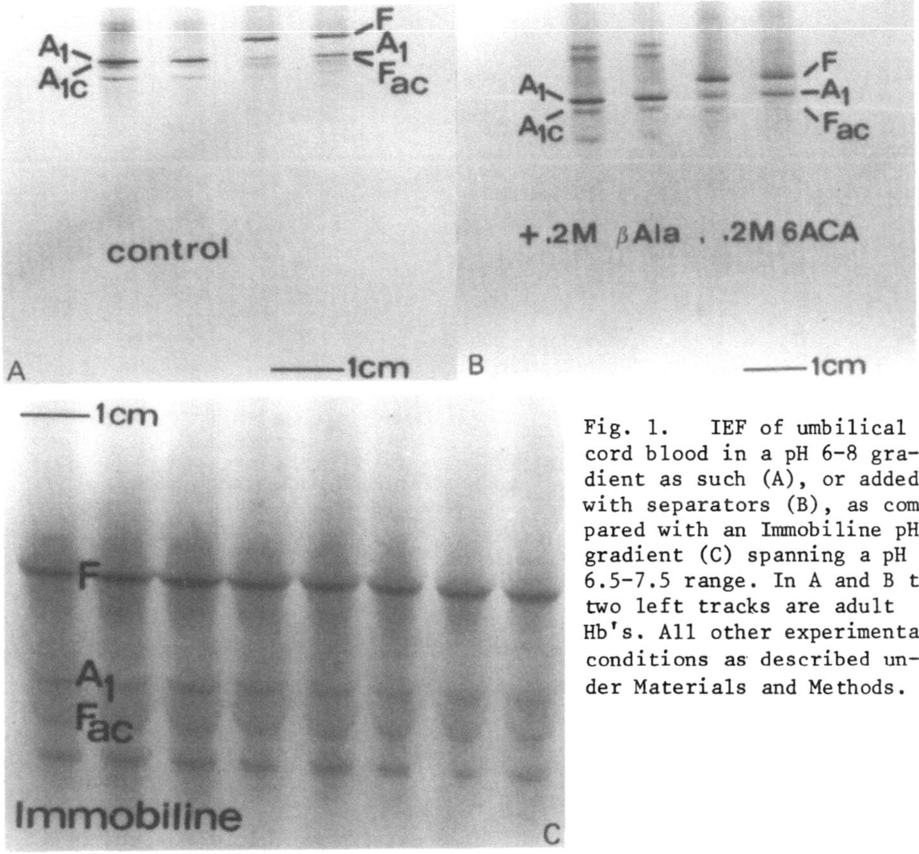

Fig. 1. IEF of umbilical cord blood in a pH 6-8 gradient as such (A), or added with separators (B), as compared with an Immobiline pH gradient (C) spanning a pH 6.5-7.5 range. In A and B the two left tracks are adult Hb's. All other experimental conditions as described under Materials and Methods.

biline^{Tm} pK 3.6, 27 mM Immobiline ^{Tm} pK 7.0 and 20 mM Immobiline^{Tm}pK 8.5, pH 7.5. The gels were run overnigth at 2000 V. To prevent Hb oxidation, the surface between the electrodes was covered with a sheet of polyester. The sample load was 5 µl of 1% dilution of hemolysate. The gels were stained by the fast stain method (11). Hb A_2 and C were separated in 1% pH 6-8, 1% pH 7-9 and 0.2% pH 3.5-10 LKB Ampholine; for improved resolution, 25 mM His was added. The immobilized pH gradient used was as follows: the dense and light gradient solutions were made 10 mM in Immobiline^{Tm} pK 8.5 and then titrated to pH 7.7 and pH 8.2, respectively, with Immobiline^{Tm} pK 3.6.

Results

Fig. 1 shows the separation of umbilical cord blood Hb components (Hb F, Hb

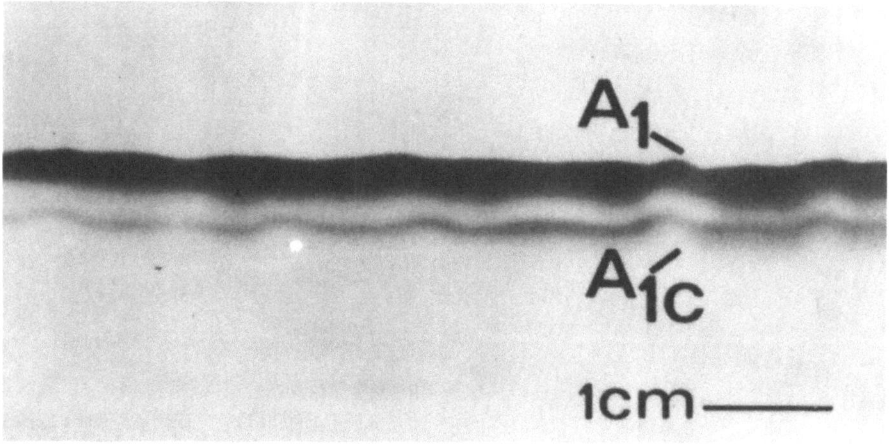

Fig. 2. IEF in an Immobiline™ pH gradient (nominal range pH 6.5-7.5) of adult lysates from diabetic patients. The run lasted overnight at 200 V/cm. About 300 µg Hb were applied per track, resulting in lateral diffusion of Hb in a continuous line. The cathode was uppermost.

A and Hb F_{ac}) in a standard pH 6-8 gradient (Fig. 1A), in an Ampholine gradient containing an equimolar mixture of "separators" (β-Ala and 6-ACA) (Fig. 1B) and in an immobilized pH gradient covering a nominal pH range 6.5 to 7.5 (Fig. 1C). Analysis of cord blood components has been suggested by Cossu *et al.* (12) for thalassemia screening in neonatal units. In conventional pH gradients, the distance between Hb A and Hb F_{ac} is too short (less than 1 mm) to allow for a proper densitometric evaluation of these two components. In "separator" IEF, Hb A is almost equidistant between Hb F and Hb F_{ac}, with ample resolution among the three components. When using immobilized pH gradients, a further increment of resolution (by a factor of *ca.* three) is obtained. The pH gradient is so shallow that minor components (such as the more alkaline Hb A_2 and the more acidic Hb Bart) practically condense against the cathodic and anodic filter paper strips, respectively.

In adult blood, an even more difficult separation problem is presented by Hb A and its glycosilated derivative, Hb A_{1c}. In commercial pH 6-8 gradients, the separation between these two components is almost nihil (Fig. 1A). Somewhat better results are obtained in "separator" IEF (Fig. 1B): the distance between these two species is slightly more than 1 mm, which might be sufficient for most densitometric scanners, provided carefully control-

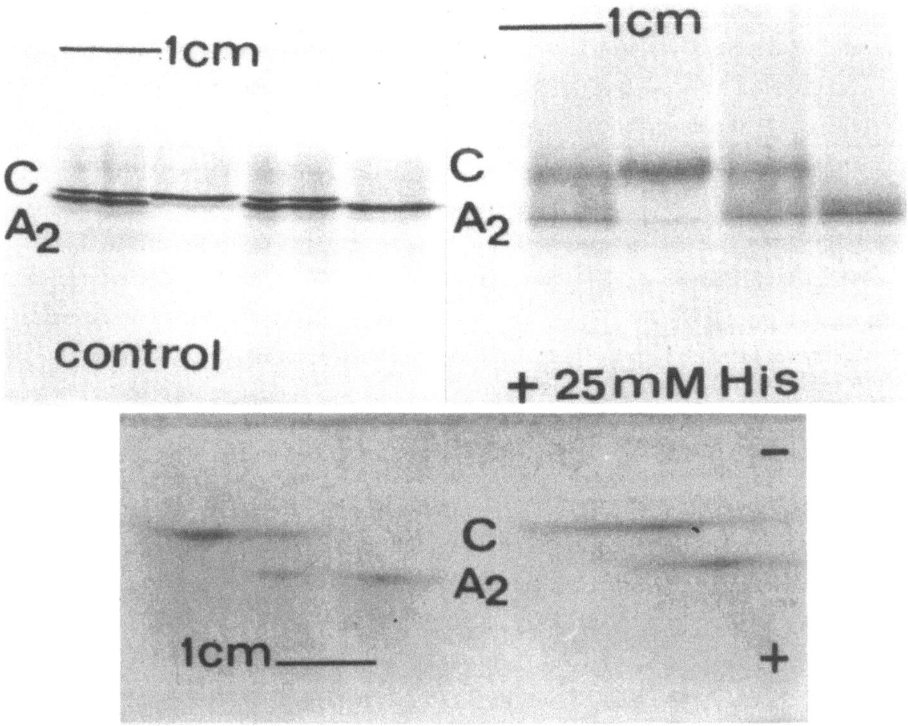

Fig. 3. IEF of Hb's C and A_2 in a pH 6-9 gradient as such (upper left), or added with 25 mM histidine (upper right), as compared with an Immobiline pH gradient (lower panel) spanning a nominal pH 7.7 to 8.2 range. Hb C was from a homozygous patient while Hb A_2 was purified chromatographically from adult blood. In all three experiments the cathode was uppermost.

led experimental conditions are adopted (e.g. overloading, which is often necessary for detection of small amounts of Hb A_{1c}, often obscures the separation). A quite remarkable separation is obtained by using immobilized pH gradients (Fig. 2): the distance between the center of each band is now 5 mm, with 3 mm of empty space between the outer edges of these two Hb zones. The separation is even more remarkable if one considers that the sample load here was considerably higher than in conventional pH gradients (by a factor of ten) so that the minor component, Hb A_{1c}, could be seen focused in the gel prior to staining.

As another example of a difficult separation problem, we have attempted to resolve Hb A_2 from the mutant Hb C. In the early days of IEF, it was thought that these two Hb's had identical pI's (pI 7.40 at 24°C). According to Drysdale *et al.* (1), in fact, when "Hb C was added to a normal hemolysate in an amount equivalent to the known concentration of hemoglobin A_2 (2.5%), no separation of the two hemoglobins was achieved". Indeed, a proper use of thin layer IEF shows that these two species are barely separated (Fig. 3, upper left). If the carrier ampholyte pH gradient is admixed with the "separator" His, which flattens the pH gradient around pH 7.6, a considerably improved separation is obtained (Fig. 3, upper right), which is matched, and slightly incremented, by immobilized pH gradients (Fig. 3, lower panel). In this last case, the Hb zones are much sharper than the bands obtained with the separator His, as clearly seen in the pictures.

Discussion

We have presented here some difficult separation problems which are of considerable importance in routine clinical analysis. In the case of β-thalassemia, Cossu *et al.* (12) have proposed IEF of intact Hb tetramers in cord blood in order to substitute the β/γ ratio obtained upon IEF of globin chains (13) with the Hb A/Hb F_{ac} (or Hb F/Hb A) ratios obtained by IEF of native Hb's from umbilical blood at delivery. In the case of diabetes, we have seen that quantitation of Hb A_{1c} is a useful parameter for assessing hyperglycemia over a long period of time (8). In 1978, no less than four articles appeared in the press describing the separation and quantitation by IEF of Hb A and Hb A_{1c} (10, 14-16). Since the two species had minute pI differences, it was necessary to resort to the use of separators, such as β-Ala or the dipeptide His-Gly. Even so, the two bands are often separated by barely 1 mm, which makes quantitation by common densitometry a very difficult task. As a third, complex separation problem we have shown the resolution between Hb A_2 and Hb C. Correct quantitation of Hb A_2 is fundamental for an independent evaluation of thalassemia conditions, in doubtful cases. One of the standard methods used is microchromatography in carboxymethyl cellulose (17). In carriers of Hb C, this determination of Hb A_2 could not be performed by IEF up to now.

In all the cases examined, we have seen that immobilized pH gradients fully achieve resolution, even in the most difficult separation problems. It might be argued that satisfactory resolution can be obtained also by a proper use of "separators", and this is certainly so. However, this requires the use of a specific "separator" tailored to each separation problem, and this compound might not always be available. On the other hand, Immobiline^{Tm} can be easily used to cast a gel bearing a given pH gradient just suited for the desired separation problem. While gelling an immobilized pH gradient is certainly more cumbersome than casting a gel containing ordinary carrier ampholytes, it should not be a deterrent in any biochemical laboratory, where techniques for production of linear pH gradients are part of the common knowledge of the research staff.

In addition, as stressed in the first article of this series (18), several advantages of this newly discovered technique make it appetible for routine biochemical and clinical separation problems: the gradients are indefinitely stable (i.e. drift-free), are insensitive to salts and ionic species in the sample, have a high buffering power and an extremely low conductivity and are compatible with very high sample loads. As a consequence of this, extremely high voltages can be applied to sharpen up the focused sample zones: e.g., by exploiting the movable electrodes of the Ultrophor chamber, we have used voltages as high as 1000 V/cm with no undesirable joule effects.

At the moment, we have some problems with the use of Immobiline^{Tm} in hemoglobin separation: the movement of Hb to its pI position is quite slow (presumably due to some ion-exchange process between the highly charged Hb molecule and the fixed charges in the matrix) so that often overnight runs are required. During this time, Hb could auto-oxidize, generating partially or fully oxidized species which might complicate the separation. In order to minimize this phenomenon, we use CO-saturated hemoglobins and cover the gel surface with a polyester foil, so as to minimize O_2 adsorption.

Ackowledgements

We thank Prof. M. Perrella and Dr. G. Cossu for the generous gift of Hb's. Supported in part by grants from Consiglio Nazionale delle Ricerche (CNR) and Ministero della Pubblica Istruzione (MPI), Roma.

References

1. Drysdale, J.W., Righetti, P.G., Bunn, H.F.: Biochim. Biophys. Acta 229, 42-53 (1971)

2. Jeppsson, J.O., Berglund, S.: Clin. Chim. Acta 40, 153-158 (1972)

3. Monte, M., Beuzard, Y., Rosa, J.: Amer. J. Clin. Pathol. 66, 753-760, (1976)

4. Basset, P., Beuzard, Y., Garel, M.C., Rosa, J.: Blood 51, 971-980 (1978 (1978)

5. Kunkel, H., Ceppellini, R., Müller-Eberhard, U., Wolf, J.: J. Clin. Invest. 36, 1615-1620 (1957)

6. Weatherall, D.J., Pembrey, M.E., Pritchard, J.: Clin. Haematol. 3, 467-472 (1974)

7. Rahbar, S.: Clin. Chim. Acta 22, 296-301 (1968)

8. Gabbay, K.H.: New England J. Med. 295, 443-447 (1976)

9. Bianchi Bosisio, A., Loehrlein, C., Snyder, R.S., Righetti, P.G.: J. Chromatogr. 189, 317-330 (1980)

10. Beccaria, L., Chiumello, G., Gianazza, E., Luppis, B., Righetti, P.G.: Am. J. Hematol. 4, 367-374 (1978)

11. Righetti, P.G., Chillemi, F.: J. Chromatogr. 157, 243-251 (1978)

12. Cossu, G., Manca, M., Gavina, P.M., Bullitta, R., Bianchi Bossio, A, Gianazza, E., Righetti, P.G.: Am. J. Hematol. (1982) in press

13. Righetti, P.G., Gianazza, E., Gianni, A.M., Comi, P., Giglioni, B., Ottolenghi, S., Secchi, C., Rossi-Bernardi, L.: J. Biochem. Biophys. Methods 1, 45-57 (1979)

14. Schoos, R., Schoos-Barbette, S., Lambotte, C.: Clin. Chim. Acta 86, 61-70 (1978)

15. Spicer, K.M., Allen, R.G., Buse, M.G.: Diabetes 27, 284-390 (1978)

16. Jeppsson, J.O., Franzén, B., Nilsson, K.O.: Sci Tools 25, 69-73 (1978)

17. Efremov, G.D., Huisman, T.H.J., Bowman, K., Wrightstone, R.N., Schroeder, W.A.: J. Lab. Clin. Med. 83, 657-664 (1974)

18. Bjellqvist, B., Ek, K., Righetti, P.G., Gianazza, E., Görg, A., Postel, W.: In Electrophoresis '82, Stathakos, D., ed., de Gruyter, Berlin, pp. 61-74 (1983)

COMPUTER SIMULATION OF MODEL ISOELECTRIC FOCUSING EXPERIMENTS

P. Just Svendsen and C. Schafer-Nielsen
The Protein Laboratory, University of Copenhagen
34 Sigurdsgade, 2200 Copenhagen N, Denmark.

Introduction

We have applied our unifying model for steady-state electrophoresis systems (1) to describe and simulate the development of an isoelectric focusing experiment. The model has been extended to include multivalent amphoteric constituents (2). We treat such model experiments by dividing the progress into different phases 1: The initial situation, where all constituents are distributed homogeneously throughout the separation path. 2: The transient state (3), which we define as a series of intermediate and local steady-state situations, that changes with time. 3: A focused phase, where the constituents have concentrated at - or close to - their isoelectric point, and 4: An ultimate situation where the constituents will migrate isotachophoretically in water - the decay (4). We assume that the separation path is open to the anolyte and catholyte. Since a detailed description of our model for multicomponent steady-state electrophoresis systems is given elsewhere (1), and as a full report on the application of this model to isoelectric focusing will be published shortly we refer to these papers for details.

Below we give some examples of our simulated experiments. The influence of the hydrogen ion and the hydroxyl ion is taken into account, but in these examples the diffusional effects on the migrating boundaries are neglected. This will be treated elsewhere (2,5,6). Electroendosmosis effects due to anti-convection media are neglected, i.e. we assume the experiments to be performed in pure water without secondary effects from stabilizing media.

Electrophoresis '82

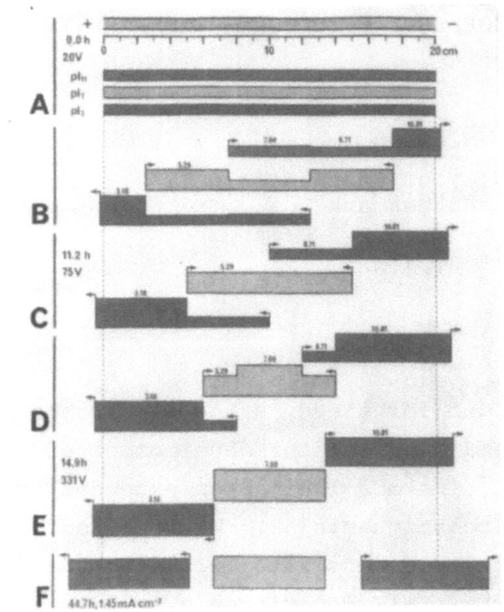

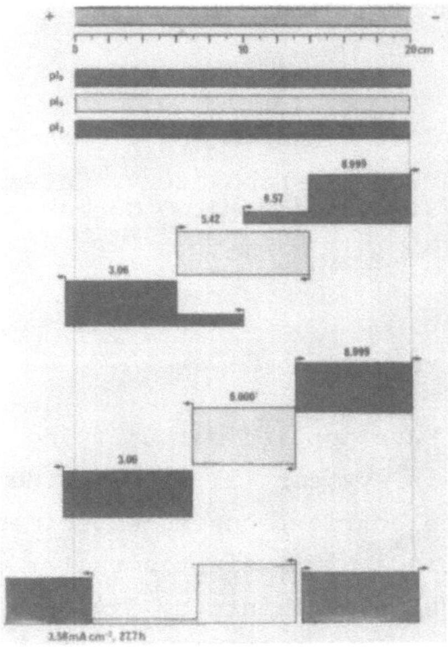

Figure 1. Figure 2.
Phases in the development of isoelectric focusing experiments. Anode to the left.

Examples on simulations

Let us consider a mixture of 3 model ampholytes mixed homogeneously in water. We assume these to exhibit the same characteristics with regard to ΔpK and mobility. In this example the difference between the pK-values is 4-pH units, it is 1,1 ampholytes and the mobility is assumed to be equal for both the cationic species and the anionic species. We have assigned a mobility value of $25\times10^{-5} cm^2 sec^{-1} V^{-1}$, and pI-values of 3, 7 and 11 respectively. The concentration of each ampholyte is 0.03 M, and the pH is therefore 7.0. At this pH the pI_{11} constituent is positively charged, the pI_3 constituent is negatively charged and the pI_7 constituent neutral. In fig. 1A the starting situation is indicated. The upper part of fig. 1A indicates a separation tube and is underlined with a scale.

The three shaded rectangles in the lower part of fig. 1A indicates the concentration distribution of the constituents at the initial stage before the current is switched on. We start the experiment with 1 V cm^{-1} and 1.45 $mAcm^{-2}$, and the pI_{11} constituent will migrate towards the cathode while the pI_3 constituent will migrate towards the anode, i.e. boundaries defined by these constituents will migrate cathodically and anodically respectively, see fig. 1B. The pI_{11} boundary will leave behind a zone containing the pI_3 -and pI_7 constituents, where the pH has decreased and where the relative concentrations of these constituents have changed as indicated in fig. 1B. The pI_3 constituent boundary will leave behind a zone containing the pI_{11} -and pI_7 constituents and where the pH has increased. The changes in the concentrations are shown in fig. 1B. However, as seen in fig. 1B, two boundaries defined by the pI_7 constituent has formed, and both of these boundaries migrate towards the center of the separation tube, leaving behind zones of pI_3 constituent (the anodic part of the system) and pI_{11} constituent (the cathodic part of the system) respectively. If the experiment is performed with constant current the situation shown in fig. 1C has been reached after approximately 11 hours, and the total voltage over a 20 cm separation tube has increased from 20 to 75 V. At this point the pI_3 and pI_{11} constituent boundaries meet or cross over one another. As these constituents continue towards the anode and cathode they both leave behind a zone containing only pI_7 constituent. After approximately 15 hours the situation shown in fig. 1E is obtained, and this is where the system is focused, i.e. all constituents have concentrated into zones containing only one ampholyte in water. Since we include the hydrogen ion and the hydroxyl ion in the calculations only the zone containing the pI_7 constituent has a pH that corresponds to the pI of focused constituents. As electroneutrality must be maintained, focused constituents with pI-values <7 and >7 will allways be negatively or positively charged respectively to an extent which is a function of decreasing and increasing pI-values respectively. There is of course nothing new in this statement, but we find this to cause the pH-drift in isoelectric focusing experiments and the effect appears from fig. 1B - 1F. The vertical dotted lines indicate the edges of the separation tube, and the migration of constituents with high and low pI-values is seen to occur already when the experiment is started, mostly pronounced at the cathodic end of the system. This is due to the conductivity difference between the hydroxyl ion and hydrogen ion. As mentioned above, we assume that isotachophoresis is the continuation of an experiment that goes beyond the situation indicated in fig. 1E, and this is indicated in fig. 1F. Since we have neglected diffusional effects in this example and since we have chosen one of the constituents to have a pI value of 7.0, this means that the pI_7 constituent is left in the center of the separation tube and that the pI_3 and pI_{11} constituents after a certain time would have migrated out of the separation tube.

Fig. 2 shows a simulation with a 3 component system, where we have omitted to include a constituent with a neutral pI-value. In this example we have simulated to focus 3 constituents with pI-values 3, 6, and 9. At the same time these constituents are assumed to be 3,3-ampholytes (more details will be given in ref. 2). As seen the constituents focus after a certain time, but in this case, going beyond the situation where the constituents are optimally focused, the pI_6 constituent migrates isotachophoretically out of the system following the pI_3 constituent boundary, but in a very low concentration, fig. 2. The pI_9 zone is only drifting very slowly to the cathodic end of the system.

Fig. 3 shows a simulation of isotachophoresis, where the pH of the leading ion (pK = 5, mobility $40\times10^{-5} cm^2 sec^{-1} V^{-1}$) electrolyte is decreasing until and including the case, where the concentration of counterion is zero, i.e. isotachophoresis in water. The pI-range analyzed is from 4 to 10. It can be seen, that ampholytes with a pI-value below 7.0 migrates isotachophoretically behind the leading ion, with decreasing concentration as the pI-value approaches 7. Since we do include the hydrogen ion (and hydroxyl ion) the hydrogen ion acts in this case as the counter ion. This simulation also applies if one assumed a reversed system, i.e. the leading ion would be a base. This is confirming the situations shown in fig. 1 and fig. 2 when the experiments shown in these figures goes beyond the fully focused phase.

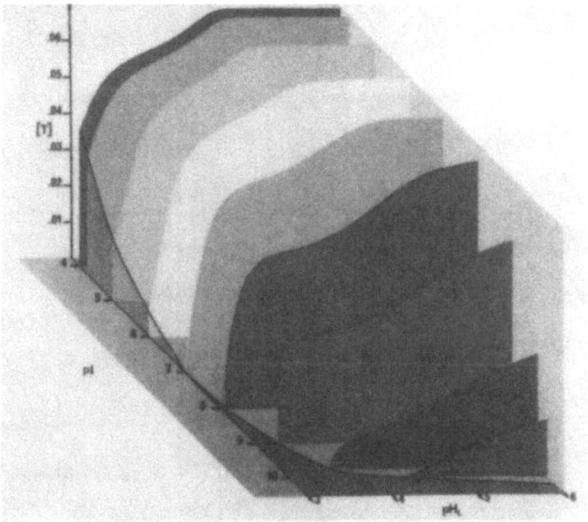

Figure 3. Isotachophoresis
pH in leading electrolyte decreases.

Fig. 4 shows the situation when a 9 component system is simulated. The pI-range is from 3 to 11, and the constituents are 3,3-ampholytes. The simulation illustrates the situation at the point, where the pI_3 constituent boundary crosses the pI_{11} constituent boundary. Fig. 4A shows the concentration distribution of each constituent, and it is seen that even though the number of ampholytes has been increased, the pI_{11} -and pI_3 constituents tends to be lost. Fig. 4B shows the pH profile, a, and the potential gradient, b, over the separation path. The dotted line indicates the ideal end situation of an isoelectric focusing experiment with ampholytes covering a pI-range from 3 to 11. Already at this point the pH-gradient is close to an acceptable shape, and we assume that the stable phase of an isoelectric focusing experiment exists from a situation starting shortly after the one shown and lasting until a situation present in the system will be more or less equivalent to an isotachophoresis system, which we suggest is the situation during the fatal decay.

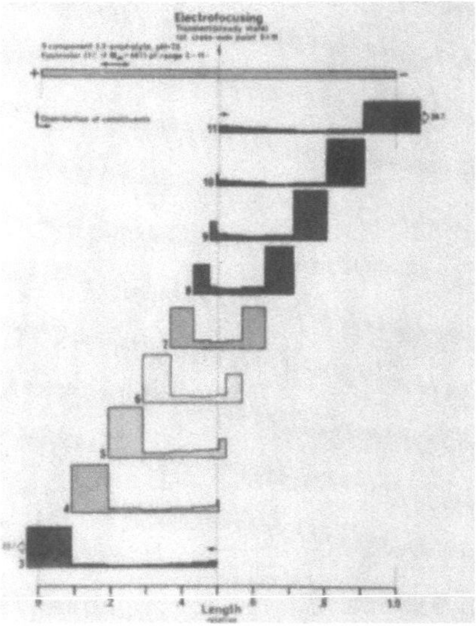

Figure 4A. Transient state of a 9 component simulated experiment. Distribution of the constituents

The above is only a brief presentation of some of our results. We have developed a computer program (UNIGEN, Fortran IV, Digital PDP-1103,(RT-11)) for simulating multicomponent electrophoresis systems. A report on the application of this program to a multitude of electrophoresis systems is in preparation.

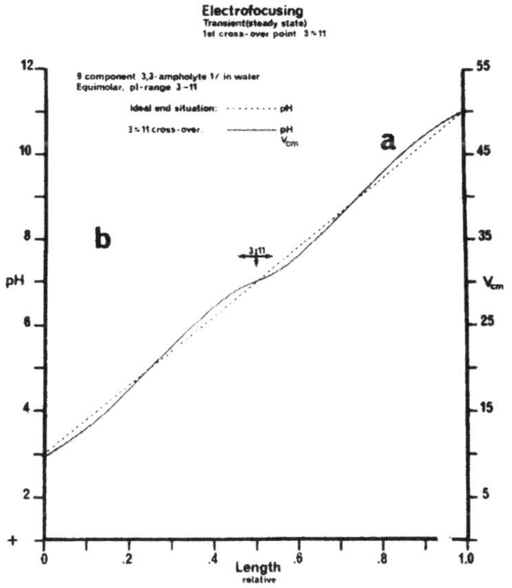

Figure 4B. Transient state
of a 9 component simulated experiment.
a. pH gradient. b. Potential gradient.

References

1. Schafer-Nielsen, C., Svendsen, P. J.: Anal. Biochem. 114, 244-262 (1981)
2. Svendsen, P. J., Schafer-Nielsen, C.: Manuscript in preparation.
3. Catsimpoolas, N.: in Progress in Isoelectric Focusing and Isotachophoresis (Righetti, P. G., ed.), 77-92, North-Holland Publishing Company, Amsterdam 1975
4. Baumann, G., Chrambach, A.: in Progress in Isoelectric Focusing and Isotachophoresis (Righetti, P. G., ed), 13-23, North-Holland Publishing Company, Amsterdam 1975
5. Schafer-Nielsen, C., Svendsen, P. J.: Electrophoresis '82, Proceedings, Walter de Gruyter, Berlin - New York

QUANTITATIVE MODELS FOR THE MIGRATION OF ELECTROLYTE GRADIENTS IN AN ELECTRIC FIELD.

Claus Schafer-Nielsen and Per Just Svendsen
The Protein Laboratory, University of Copenhagen, Sigurdsgade 34, DK-2200 Copenhagen N, Denmark.

Introduction

The design of electrophoresis systems other than simple isotachophoresis systems has for a number of years been hampered by the lack of generally applicable models describing the electrophoretic migration and relative concentrations of the ionized components defining such systems. Only recently have models been worked out that are useful in quantitative studies of steady-state electrophoresis systems with multi-component electrolyte phases (e.g. isoelectric focusing systems with polyampholytes),(1,2), or in simple electrolyte gradients with controlled boundary conditions (3). The aim of this paper is to give a brief outline of different approaches that can be employed in quantitative descriptions of the migration of electrolyte gradients in an electric field, with special emphasis on models that are suitable for computerization.

Results

The general equation describing mass conservation in gradients of migrating particles is:

$$\frac{dx}{dt} = \frac{dJ}{dC}, \qquad (1)$$

where J is the particle flux and C is the concentration of particles in position x at time t. Together with the law of electroneutrality and Ohm's law eqn. (1) forms the basis for all

Electrophoresis '82

quantitative models describing the ionic composition of steady-state elctrophoresis systems on the macroscopic level. Therefore, the point in the development of such models is not to derive principally new equations, but to express eqn. (1) in a form that is useful for actual calculations. dx/dt describes the migration velocity of the gradient profile (rather than of the particles themselves) and in steady-state electrophoresis systems with discontinuous electrolyte phases:

$$\frac{dx}{dt} = v_b = \frac{\Delta J}{\Delta C} = \frac{V_1 \times nm(I)_1 \times [I]_1 - V_2 \times nm(I)_2 \times [I]_2}{[I]_1 - [I]_2}, \quad (2)$$

where v_b is the (constant) velocity of the boundary between two neighbouring electrolyte phases, V is the voltage gradient, nm(I) is the electrophoretic net mobility of constituent I with concentration $[I]$. Eqn. (2) is the general mass balance in discontinuous steady-state electrophoresis systems in which the charged components are present on both sides of a given boundary (1). However, the equation has been derived under the assumption that diffusion can be disregarded as a factor in the calculation of particle flux. Insertion of Fick' s expression of the diffusion flux ($J_d = D_I(dI/dx)$) in the two terms in the nominator of eqn. (2) and rearranging gives:

$$(V_1 \times nm(I)_1 - v_b)[I]_1 - D_I\frac{dI}{dx_1} = (V_2 \times nm(I)_2 - v_b)[I]_2 - D_I\frac{dI}{dx_2}$$

where I_1 and I_2 is the same ion at two different points x_1 and x_2 on the concentration profile between the two electrolyte phases. At some distance from the "center" of the boundary, $dI/dx \approx 0$, and eqn. (2) can therefore be used to calculate v_b if the composition of the two zones is known at "infinite" distance from the boundary. Having calculated v_b, the field gradient and the net mobility in one of these zones the values can inserted on either side of the equation above to give a constant value K_I :

$$(V_x \times nm(I)_x - v_b)[I]_x - D_I\frac{dI}{dx} = K_I \quad (3)$$

In eqn.(3), K_I is a constant that is characteristic of the ion in the particular system at the defined current density. In isotachophoresis systems the leading and the terminating ions migrate with the same velocity as the boundary, so in these systems $K_I = 0$ for these ions. A similar conclusion was reached by MacInnes and Longsworth in their determination of transference numbers by the moving boundary method (4). In all other systems (1) and for the counterions in isotachophoresis systems $K_I \neq 0$. The K_I-value can be physically interpreted as the ion flux that is measured by an observer that moves in the same direction and with the same velocity as the boundary.
Actual calculations using eqn.(3) can (as will be detailled elsewhere) be performed by assuming linearity of the concentration profiles over small Δ I-values for one of the ions followed by calculation of the Δ I-values of the other ions over identical Δ x-distances. By repetition of this procedure and summation of the calculated Δ x-values, one can draw up a graph showing the concentration gradients of the individual ions across the boundary. The accuracy of this proces depends solely on the number of Δ I-intervals used for the calculation. We have designed a computerprogram that calculates the boundary conditions and works out the individual profiles in the boundary under the assumption that the diffusion coefficients are constant throughout the systems . The diffusion coefficients can be defined serarately for each ion, e.g. related to the mobility values by the Nernst equation $D_I = (RT/F) \times m_I$, where m is the mobility.

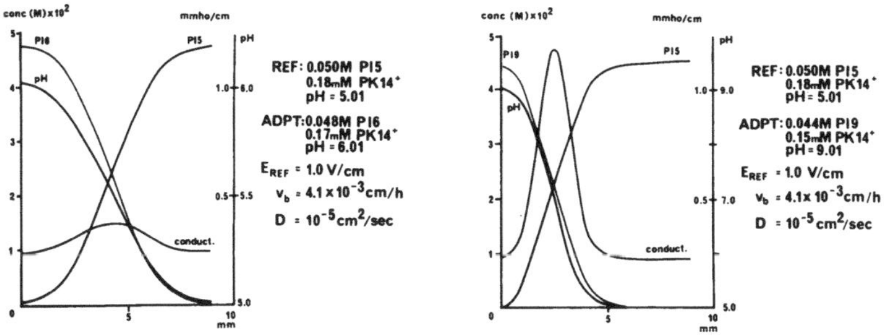

FIG.1

Fig. 1A and B shows examples of such profiles in isotachophoresis systems where a 1,1-ampholyte is the leading ion (pI = 5.0, pK1 = 4.0, pK2 = 6.0) in a concentration of 0.05 M at pH = 5.01 with a strong base (pK_b = 14.0) as counterion in a concentration of 0.18 mM. The voltage gradient in the leading electrolyte was set to 1.0 V/cm. In Fig. 1A, the terminating ion is a 1,1-ampholyte (pI = 6.0, pK1 = 5.0, pK2 = 7.0) and in Fig.1B the terminating ion is a 1,1-ampholyte (pI = 9.0, pK1 = 8.0, pK2 = 10.0). The mobility values of all species were set to 30 x $10^{-5} cm^2$/Vsec and the diffusion coefficients to $10^{-5} cm^2$/sec. The pH-value in the terminating electrolytes were calculated to be 6.0094 (Fig.1A) and 9.0088 (Fig. 1 B) at infinite distance from the boundary, i.e. in all zones the pH-value deviated with a maximum of 0.01 pH- unit from the isoelectric point of the ampholyte defining the zones. For all practical purposes, then, the systems may be considered as isoelectric focusing systems involving two ampholytes. The presence of the small amount of common counterion (not drawn in in Fig. 1A and B, since its concentration is too low) is necessary to obtain the situation in Fig. 1B. This is due to the fact, that when dissolved in pure water an ampholyte- with pI above 7.0 will carry a slight positive net-charge, whereas ampholytes with pI below 7.0 will carry a slight negative net charge (2), so that the two ampholytes in Fig. 1B would migrate in opposite directions thereby preventing attainment of a steady-state boundary. In the cases shown, a negatively charged counterion could equally well have been chosen, in which case the boundary would have migrated towards the cathode instead of towards the anode as in Fig.1. With the chosen voltage (1.0 V/cm in the leading electrolyte) the migration velocity of the two boundaries is about 41 microns per hour. At this voltage the major part of the electrolyte gradients forming the boundary extends over a distance of 5 - 10 mm which is far too short for the gradients to be of value in analytical separations. This is especially so since the applied voltage is at least one order of magnitude lower than what is normally used to obtain sharp bands in isoelectric focusing systems, and since (as can be deduced from eqn. (3)) the width of the gradients is inversely proportional to the current density and thus to the voltage gradient in the leading electrolyte. A notable feature in Fig. 1 is the local maximum in conductivity in the middle region of the boundary gradients. This maximum is most pronounced in Fig. 1 B, where the difference between the pI-values of the two ampholytes is greatest. Local maxima like these will also exist in isoelectric focusing systems with multiple ampholytes. However, their importance in practical runs is unclear, since the maxima tend to be less pronounced if the neighbouring ampholytes possess pK-values in between the two pI-values.

Since the profiles in Fig. 1 are steady-state gradients, any system made by mixing of two different ampholytes to give a pH-gradient extending over a given distance will end up forming a boundary, and in most cases the boundary will extend over the rather limited space indicated in Fig. 1.

The reverse proces, i.e. expansion of a pH-gradient can be obtained by use of the telescope effect (1) in systems with many ions with different pK-values. Theoretically, all these ions may be present in suitable ratios at any point in such systems, but in the practical case the ions + counterions are placed only in the separation vessel and in the anode reservoir (anodically migrating pH-gradients). The cathode reservoir may contain the solvent only or counterion(s) with or without a final terminating ion.

Fig. 2 shows the result of calculations with eqn. (3) in a system where 4 1,1-ampholytes (pI = 5,6,7 and 8 with delta pI-pK values as in Fig. 1) each in a concentration of 0.025 M forms the leading electrolyte together with a strong monovalent acid (pK_a = 1.0) at pH = 4.0. The mobility values and diffusion coefficients for the ampholytes were as in fig. 1. The strong acid were assigned a mobility value of 70 x 10^{-5} cm^2/V x sec. The field gradient in the initial mixture was set to 1.0 V/cm. In this run the ampholytes are left behind the the strong acid

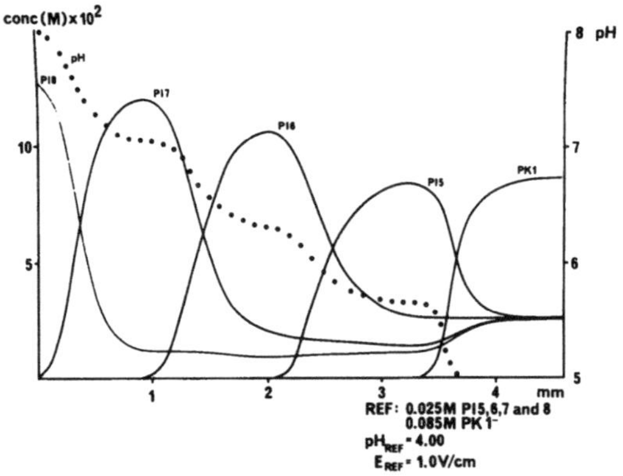

FIG. 2

and form zones that each are dominated by one of the ampholytes. The pH-value in each of these zones approaches the pI-value of the dominating ampholyte and thus a pH gradient is formed. Each zone is in steady state with respect to composition and increases in length as the proces is continued so that the maximum points on the concentration profiles (which are the steady-state concentrations for each ampholyte at "infinite" distance from the boundary) becomes plateaus of maximum concentration. The proces which resembles closely the transient state in an isoelectric focusing system (2) can be continued for any length of time and thus provides a means for generation of stable albeit expanding pH-gradients. Since the electrophoretic migration velocities of the individual boundaries are different (from left to right in Fig.2: 0.50/1.04/1.63 and 2.52 cm/h)the acidic part of the pH-gradient will expand faster than the basic part. Due to the continuous expansion of the zones, a smooth pH-gradient will only be obtainable if the number of different ampholytes is large, so that the space dominated by a given ampholyte is small compared to the length of the gradient in its entirety. In practice such a situation is only obtainable if the commercially available, but electrochemically ill-defined polyampholyte mixtures are used.

An alternative to the self-forming polyampholyte gradients described above may be found in preformed gradients made by variation of the relative concentrations of a few well characterized acids and bases to give a pH-gradient in the separation vessel. To get a basic idea of the mechanisms behind electrophoretic migration of such gradients , consider eqn.(1) which states, that the migration velocity of a given point on the profile equals the change in flux relative to the change in concentration of the ion defining the profile. If the flux does not change with concentration the velocity of the profile equals zero. This is very nearly the case in electrolyte gradients made by simple dilution of a given electrolyte, since in such gradients a decrease in concentration is parallelled by an increase in voltage and vice versa, so that the flux is maintained at a constant level throughout the system. Of course, diffusion processes and non-linearity between conductivity and concentration will result in deviations from this rule, but essentially one must expect very slow electrophoretic migration of such gradients relative to the migration velocity of the ions from which they are formed. A similar situation is found in two-component systems made of a weak monovalent acid and a weak monovalent base mixed in different relative concentrations to obtain a pH-gradient. Here, electroneutrality requires that changes in the relative concentrations are paralelled by inverse changes in the ion-fractions and thereby of the net mobilities of the components. The net result is therefore nearly the same as when two-component systems of strong electrolytes are considered, the essential difference being that the unequal concentrations of protons and hydroxylions must be taken into account outside the near neutral pH-range (i.e. the systems will in effect become 3-component systems at high or low pH-values).

To obtain significant electrophoretic migration of electrolyte gradients it is therefore necessary to use more than two ions or to use two ions of which at least one has more than one pK-value. As will be detailled elsewhere, eqn.(1) can be solved explicitly for monovalent 3-component systems under the simplifying assumptions that diffusion processes and the presence of

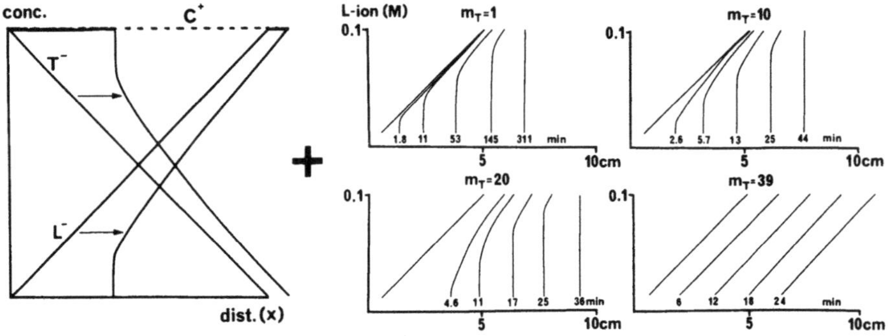

FIG. 3.

protons and hydroxyl-ions can be neglected. Fig. 3 (left) shows an electrolyte gradient in which the concentration of two strong monovalent acids (L^- and T^-) vary in inverse proportion so that the concentration of counterion (C^+) is constant throughout the gradient. As seen, all possible relative concentrations of the two acids are found in such a system. The solution of eqn. (1) gives us a means for calculation of the migration velocity of any point on this gradient at a given current density and therefore in any system consisting of the three components. If we assume L^- to possess a higher electrophoretic mobility than T^-, the gradients will migrate towards the anode in the way depicted in Fig.3 (left). The gradient is seen to migrate fastest in its cathodic end so that eventually it will collapse to give a dimensionless boundary between the L^-- and the T^--ions in the final (isotachophoretic) steady state. Due to the lower mobility of the T^--ions compared to the L^--ions, the conductivity increases linearly from left to right in the gradient in the starting situation. Now, consider a situation in which such a conductivity gradient is brougth to migrate in a way such that the collapse occurs over a distance that is great compared to the length of the separation vessel. This situation would resemble the parallel displacement in the conductivity gradient in isotachophoresis systems with multiple spacers (5) and would be attractive due to the known resolving power of such systems. The 4 graphs

in Fig. 3 (right) shows the calculated position of the L^--profile at different times after the onset of electrophoresis in 4 different systems. The initial setup of the systems were as in Fig. 3 (left), and the only difference between the 4 systems is the electrophoretic mobility of the T^--ion. The mobility of the L^--ion and the C^+-ion was set to 40 and 25 ($x10^{-5}cm^2/Vx$ sec) respectively and the voltage in the extreme anodical end of the gradients to 10 V/cm. As seen, the collapse of the gradients occurs over a comparatively small distance in the systems where the mobility of the T^--ion is low (the mobility of the T^--ions are indicated on the figure). This is unfortunate, since these systems contain the steepest conductivity gradients and therefore are the most interesting for separation purposes. The almost perfect parallel displacement obtained in the system with $m_T = 39 \times 10^{-5} cm^2/V \times$ sec has little interest in this context, since the conductivity gradient is virtually nonexistent. A number of similar analyses of 3-component systems of the "diagonal" type has left us with little hope, that they will be of much value in practical separations of macromolecules.
However, the "diagonal" systems only represent a subclass of the continuous electrolyte gradients, and their sole virtue seems to be, that they can be described with closed-form mathematics.
Electrolyte gradients involving more than three ions in varying relative concentrations will generally be too complex to be described by formula on a closed form. As a general approach to such gradients, we have therefore developed a simple model that in principle can be used to study the electrophoretic migration of any n-component electrolyte gradient whose initial composition and spatial distribution has been defined by the experimenter. Fig. 4 (upper left) shows a graphical representation of a two-component gradient of arbitrary length. This length has by vertical lines been subdivided into 20 intervals of length Δx. When the electric field is turned on, electrophoretic migration will result in transfer of ions from one subinterval to

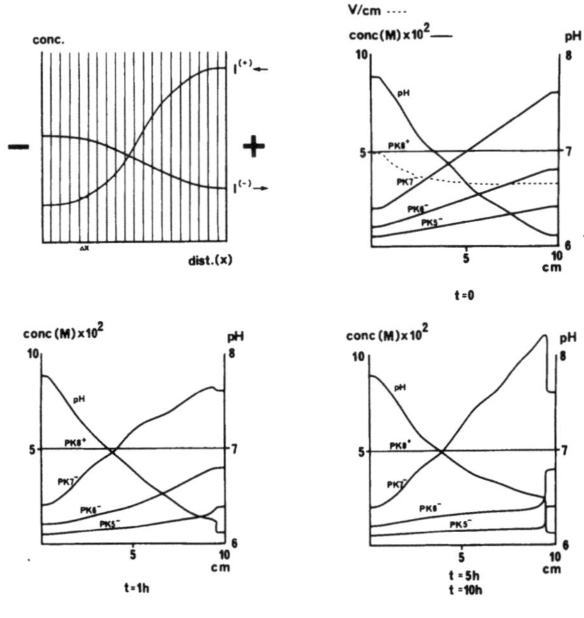

FIG. 4.

the next, and if the influx and efflux of a given ion in a given subinterval is not equal, the concentration of the ion in that subinterval will change. By considering each subinterval as a compartment in which instantaneous mixing of the constituents occur so as to give a uniform concentration of each component throughout the compartment one can describe the migration of the total gradient in terms of time-dependent changes in the concentrations in the individual compartments. If the influx into a compartment is assumed to be constant in a time interval Δt and the efflux is assumed to be proportional to the concentration in the compartment, the following equation describes the concentration $C_{2(t_o+\Delta t)}$ in the compartment at the end of the time-interval t:

$$C_{2(t_o+\Delta t)} = \frac{v_1}{v_2} C_1 + \left(C_{2_{t_o}} - \frac{v_1}{v_2} C_1\right) \times e^{-v_2 \frac{\Delta t}{\Delta x}}, \quad (4)$$

where v_1 is the migration velocity of the ion in the donor-compartment. Further assumptions in the derivation of this equation are, that the migration velocities in the two compartments remains constant during the time interval considered, and that diffusional flux can be disregarded. The mean concentration $\overline{C}_2$ in compartments 2 over the time interval Δt (to be used as constant value during calculation of $C_{3(t_o + \Delta t)}$ in which compartment 2 is the donor compartment) is defined by:

$$v_2\overline{C}_2\Delta t = v_2\int_0^{\Delta t} C_{2_t}\,dt$$

giving:

$$\overline{C}_2 = \frac{v_1}{v_2}C_1 + \frac{\Delta x}{v_2\Delta t}\left(C_{2_{t_o}} - \frac{v_1}{v_2}C_1\right)\left(1 - e^{-v_2\frac{\Delta t}{2\Delta x}}\right) \qquad (5)$$

As will be shown elsewhere the mean concentration can easily be employed in a correction for diffusional flux as well. The 3 remaining graphs in Fig. 4 shows the result of an analysis performed with these equations on a system consisting of 3 monovalent acids (pK_a = 5,6 and 7) and one monovalent base (pK_b=8) initially distributed in linear gradients over a distance of 10 cm. Mobility values were as in Fig.1. The calculations were performed over 20 Δx-intervals (Δx= 0.5 cm) with Δt= 300 sec. The concentration of components in compartment 1 and 20 were kept constant to simulate a constant composition of the anolyte and catholyte. After each time interval the new pH and voltage gradient in each compartment were calculated together with the new net mobilities of the components. This proces was repeated up to the times (1,5 and 10 hours) indicated in Fig. 4. With this particular system, a steady state was reached after about 5 hours after which time no significant changes were observed in the gradients. The final system still contains most of the initial pH-gradient, and it may in fact be considered as a stable system for focusing of proteins in the pI-range from 6.5 to 7.5. This is admittedly a narrow range, but the system was chosen at random to illustrate the principle rather than to demonstrate any particular separation features. A more detailled analysis of the possibilities for maintenance or controlled migration of pH- and conductivity gradients in continuous systems is out of the scope of the present paper, but will be published elsewhere. We wish, however, at the present stage to draw attention to this type of models, since they may represent the most realistic way of handling complex electrophoresis systems on the quantitative level. Such systems include the steady-state electrophoresis systems for which models have already been worked out and in addition any random system whose initial spatial composition has been defined by the investigator. The compartment models can be used to study the electrophoretic migration of sample components who initially are applied in one end of the separation vessel and to study the effect of controlled (e.g. time dependent) changes of the composition of the anolyte and catholyte during a run. The latter feature makes it possible to describe multi-component systems of the rheo-electrolysis type (3) and through introduction of ion-impermeable barriers to describe closed systems (6) in which the electrodes define the ends of the separation vessel.

References

1. Schafer-Nielsen, C. and Svendsen, P.J.: Analytical Biochemistry, 114, 244-262 (1981).

2. Svendsen, P.J. and Schafer-Nielsen, C.: These proceedings.

3. Rilbe, H.: Electrophoresis, 2, 261-267, (1981).

4. MacInnes, D.A. and Longsworth, L.G.: Chemical Reviews, 11, 171-230, (1932).

5. Svendsen, P.J.: in "Electrophoresis - A Survey of Techniques and Applications", Deyl, Z. (ed.), J. Chromat. Lib. 18,(1979).

6. Bier, M., Mosher, R.A. and Palusinski,O.A.: J. Chromatography, 211, 313-335, (1981).

APPLICABILITY OF THE CORRELATION LOG SIZE - SQUARE ROOT OF MIGRATION DISTANCE TO SDS ELECTROPHORESIS IN LINEAR POLYACRYLAMIDE GRADIENT GELS

Gunter M. Rothe

Institut für Allgemeine Botanik der Johannes Gutenberg-Universität Mainz, Saarstrasse 21
D-6500 Mainz, Germany

Introduction

While investigating the migration of non-denatured proteins in linear polyacrylamide (PAA) gradient gels we found a linear correlationship between the logarithm of the size of proteins (log mol mass (MM) or log Stokes'radius) and the square root of their migration distance ($\sqrt{D}$) (1). This correlation was found to be unaffected by the time of electrophoresis, but the constants of the corresponding regression lines such as slope and intercept changed as a function of time (1). In this paper it is demonstrated that the log MM - $\sqrt{D}$ relationship can be applied equally well to the conditions of sodium dodecylsulfate (SDS) electrophoresis, provided that linear PAA gradient gels are used.

Results

Shapiro, Viñuela and Maizel (2) and Weber and Osborne (3) first demonstrated the reliability of SDS polyacrylamide gel electrophoresis in the determination of the size of denatured proteins and protein subunits. The size range of this method is limited to MMs from 1.5×10^4 to 9×10^4, whereas SDS electrophoresis in linear PAA gradient gels permits the estimation of the size of denatured proteins over a range from 10^4-10^6 (4-6).

Electrophoresis '82

As Lambin (5) has shown a log MM - log D plot can be used to calculate the mol mass of a SDS-denatured protein when migrating in linear PAA gradient gels. This relationship, however, as pointed out by Lambin himself (7), cannot be used for the size estimation of non-denatured proteins migrating in PAA gradient gels. Under non-denaturing conditions the intrinsic charge of proteins hardly influences their migration velocity in PAA gradient gels, provided that pH > pI (8), which means that the position of a protein in this type of gel is governed by its apparent size (1,8-11). If this is true, a generally applicable relationship between protein size and protein migration should hold equally well for both non-denaturing and denaturing conditions, since under the influence of SDS most proteins, except for the histones (12), assume a spherical shape, a constant charge, and a constant density per unit mass (12-15).

To check whether the log MM - $\sqrt{D}$ relationship holds equally well for PAA gradient gel electrophoresis under denaturing conditions we started to re-evaluate electrophoretic data published on SDS polyacrylamide gradient gel electrophoresis (PAGGE) (6,8). Lasky (8) published electropherograms which show the migration distances of 11 different calibration proteins. We re-determined the D-values using the photographs and set up a log MM - $\sqrt{D}$ plot (Fig.1). The line calculated by fitting the obtained data to the least-squares method for linear regression analysis served for the estimation of the MMs to be expected. These values and the MMs as given in the literature are summarized in Table 1. At an average the MMs calculated deviate from the literature values by ±10% (Table 1). These relatively strong deviations have obviously 2 reasons: 1. Lasky prepared linear PAA gradients with an inversely increasing amount of N,N′-Methylene diacrylamide (BIS), and 2. thyroglobulin tends to dissociate into subunits (Fig.1.). On the other hand, an average deviation of ±10% almost equals the value given by Lasky, and therefore it is no objection against an

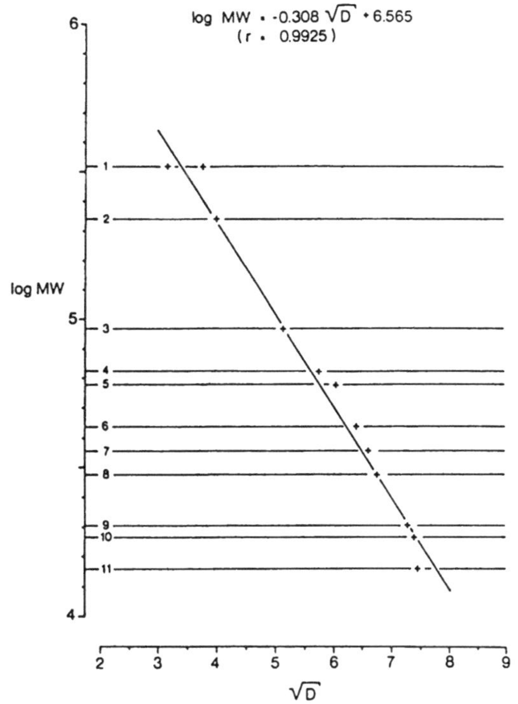

Fig.1: Re-evaluation of electropherograms as published by Lasky (8) according to the log MM-$\sqrt{D}$ relationship. Numbers indicate the log MM of the following proteins: 1.thyroglobulin, 2.ferritin, 3.phosphorylase b, 4.bovine serum albumin, 5.catalase, 6.ovalbumin, 7. lactate dehydrogenase, 8.carbonic anhydrase, 9. soybean trypsin inhibitor, 10.ferritin subunit, 11.α-lactalbumin. Deviations of calculated to expected MMs are summarized in Table 1.

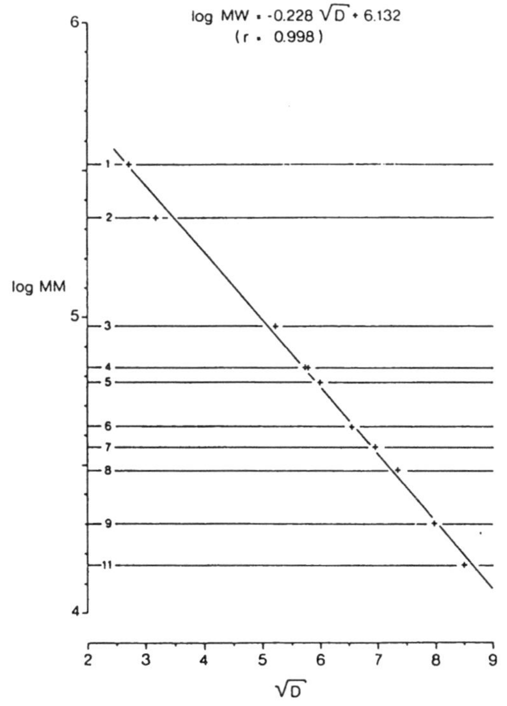

Fig.2: Re-evaluation of electropherograms as published by Poduslo and Rodbard (6) according to the log MM-$\sqrt{D}$ relationship. Numbers indicate the same proteins as given in Fig.1.

Table 1. SDS-PAGGE: % deviation of the calculated MMs of calibration proteins from the MMs expected

I Migration distances taken from an electropherogram published by Lasky (8)
II Migration distances taken from an electropherogram published by Poduslo and Rodbard (6)
I and II: calculation of MMs according to the equation $\log MM = a\sqrt{D} + b$
III % deviation as given by Poduslo and Rodbard (6)

Protein	Expected mol mass	I %	II %	III %
Thyroglobulin	330,000	+27 -17	- 2.7	- 5.3
Ferritin	220,000	+ 3	+13.3	+13.2
Phosphorylase b	94,000	+ 5	- 6.9	- 4.3
Bovine Serum Albumin	67,000	- 8 + 5	- 1.2	+ 1.3
Catalase	60,000	-14	- 3.2	- 4.3
Ovalbumin	43,000	-11	+ 1.9	+ 1.2
Lactate Dehydrogenase	36,000	- 5	- 2.9	- 4.4
Carbonic Anhydrase	30,000	+ 2	- 4.9	- 4.4
Soybean Trypsin Inhibitor	20,100	+ 3	+ 4.0	+ 2.0
Ferritin	18,500	+ 4	+ 8.2	+ 7.9
α-Lactalbumin	14,400	+24	+ 8.2	+ 7.9
average % deviation		±10	± 6.4	± 4.8

application of the log MM - $\sqrt{D}$ relationship.

As can be taken from Fig.2 and Table 1 the re-evaluation of electropherograms published by Poduslo and Rodbard (6) strongly supports this view. The correlation coefficient (r) calculated for the 10 different calibration proteins is in this case practically 1, and the average % deviation of the calculated MMs from the expected ones is minimized to 6%. Most probably the better correlation results from a constant ratio of BIS throughout the gradient and the excellent linearity of the gradients (6). Poduslo and Rodbard used a log MM - log D relationship for the calculation of the MMs and found them to deviate from the expected ones by ±5% (Table 1). Thus, the values calculated by these authors, Lambin (5), and Lasky (8) are in accordance with ours which we estimated by the log MM - $\sqrt{D}$ relationship.

Discussion

Lasky (8) studied the migration of SDS-denatured proteins in 3-30% T-gradients with an inversely increasing amount of BIS, while Poduslo and Rodbard (6) used gel gradients from 7-25% T with a constant amount of BIS. Nevertheless, the linear relationship between log MM and $\sqrt{D}$ can be applied demonstrating its independency on the range of % T of the gradient. The constants of the regression lines, i.e. slope and intercept, however, change. Since in both laboratories different types of buffer systems have been used it can be concluded that the linear relationship is uninfluenced by the buffer system in which linear PAGGE is performed.

To date at least 9 different mathematical procedures were suggested to determine the size of proteins and protein subunits in PAGGE and SDS-PAGGE (1,4-6,8), but none fits equally well to both electrophoretical systems. The log MM - $\sqrt{D}$ relationship, however, is the only one described so far which can be applied to both PAGGE and SDS-PAGGE indicating a basic physico-chemical coherence.

References

1. Rothe,G.M., Purkhanbaba,H.: Electrophoresis 3, 33-42 (1982)
2. Shapiro,A.L., Viñuela,E., Maizel,J.V.: Biochem.Biophys.Res. Comm. 28, 815-820 (1967)
3. Weber,K., Osborne,M.: J.Biol.Chem. 244, 4406-4412 (1967)
4. Lambin,P., Rochu,D., Fine,J.M.: Anal.Biochem. 74, 567-575 (1976)
5. Lambin,P.: Anal.Biochem. 85, 114-125 (1978)
6. Poduslo,J.F., Rodbard,D.: Anal.Biochem. 101, 394-406 (1980)
7. Lambin,P., Fine,J.M.: Anal.Biochem. 98, 160-168 (1979)
8. Lasky,M. in Electrophoresis ´78 (Catsimpoolas,N., ed.) Elsevier North Holland, Inc., Amsterdam,New York, 195-210 (1978)
9. Margolis,J.,Kendrick,K.G.: Biochem.Biophys.Res.Comm. 27, 68-73 (1967)
10. Slater,G.C.: Anal.Biochem. 41, 1039-1041 (1969)
11. Anderson,L.O., Borg,H., Mikaelsson,M.: FEBS-Letters 20, 199-202 (1972)
12. Panyim,S., Chalkley,R.: J.Biol.Chem. 246, 7557-7560 (1971)
13. Pitt-Rivers,R., Impiombato,F.S.A.: Biochem.J. 109, 825-830 (1968)
14. Reynolds,J.A., Tanford,C.: Proc.Natl.Acad.Sci.U.S. 66, 1002-1007 (1970)
15. Frank,R.N., Rodbard,D.: Arch.Biochem.Biophys. 171, 1-13 (1973)

VOLTAGE REGULATION ACROSS MULTIPLE GEL TUBES IN ELECTROFOCUSING, USING WEAKLY ACIDIC AND BASIC ANOLYTES AND CATHOLYTES.

Sarah Ben-Or

Endocrinology and Reproduction Research Branch, National Institute of Child Health and Human Development,

James V. Sullivan*

*Biomedical Engineering and Instrumentation Branch, Division of Research Services,

Andreas Chrambach

Endocrinology and Reproduction Research Branch, National Institute of Child Health and Human Development,

National Institutes of Health, Bethesda MD 20205

Abstract

Platinum electrode manifolds were designed and constructed which allow for voltage regulation across multiple gel tubes in electrofocusing with weakly acidic and basic anolytes and catholytes. This allows both for positioning individual electrode termini close to both ends of individual gels; and their connection in parallel to the power supply which thus provides the identical level of regulated voltage to each gel. Application of the device allowed for enhanced pH gradient stability and for electrofocusing under controlled conditions of an active glucocorticoid receptor to obtain its isoelectric pH over an extended electrofocusing time.

Electrophoresis '82

Introduction

Voltage regulation poses a problem in gel electrofocusing when 1) efficient temperature control and the need to follow the progress of electrofocusing as a function of time necessitate the use of gel tube apparatus (1,2); 2) the electrofocusing of a slowly migrating protein requires an enhanced pH gradient stability over an electrofocusing time which is sufficient for the attainment of its isoelectric position on the gel. Under these conditions, weakly acidic and basic anolytes and catholytes serve to provide enhanced pH gradient stability (3,4). The substantial voltage drop across weakly acidic and basic electrolyte reservoirs of large dimensions (required for approximate constancy of electrolyte properties during electrofocusing) makes it undesirable to use the conventional mode of regulating voltage across electrodes positioned at the top and bottom of electrolyte reservoirs, since the voltage across the system under these conditions is significantly higher than that across the gel. Furthermore, the voltage across the gel is variable and unknown, although the voltage across the system is regulated. To measure the voltage across the gel at the exclusion of that across the anolyte and catholyte, an electrode device was designed which allowed one both to measure and to regulate the voltage across a single gel tube [Fig.1 of (5)]. However, when this device is used to measure the voltage across one out of multiple identical gels, equal voltage across all gels is obtained, but that voltage varies with time, while the voltage across the system is regulated. Periodic voltage adjustments become necessary to keep the voltage across the gels approximately constant with time. Alternatively, when the device is connected to the power supply and voltage is regulated across a single gel, the voltage across all other gels will be variable for geometric reasons, since the distance from the electrodes across highly resistant electrolyte to all other tubes is increased and variable from tube to tube. To overcome these problems, electrode manifolds were designed which provided electrodes connected in parallel and identically positioned close to the ends of each gel tube.

Design

Electrode manifolds were designed as integral parts of a multi-modular Pyrex gel electrophoresis apparatus [Fig.1 of (1)] and for its application to gel tubes of both 6 and 18 mm I.D. This apparatus responds to the 3 requirements listed in the Introduction by providing efficient cooling, ease of sampling through periodic tube withdrawal, and large electrolyte capacity. The cooling efficiency of the apparatus rests on the good heat transfer characteristics of an all-Pyrex construction and allows for operation at -4°C when gel and electrolytes are made in 20% glycerol. The apparatus is applicable to "non-restrictive", mechanically labile gel media since the gel tube is hydrostatically equilibrated.

PROTEIN	IEF TECHNIQUE
VERY LARGE	SEPHADEX
TEMPERATURE SENSITIVE	GEL TUBE
LOW MOBILITY	STABLE pH GRADIENTS WEAKLY ACIDIC-BASIC ANOLYTE-CATHOLYTE

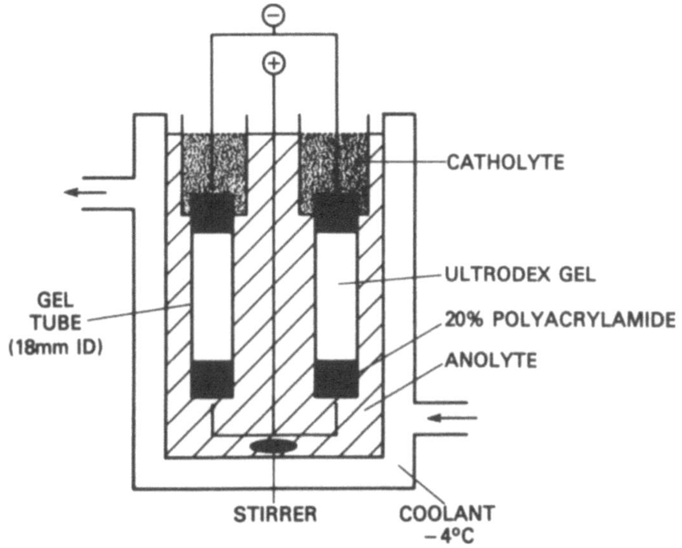

Fig.1: Representative set of conditions, under which voltage control across multiple electrofocusing gels by means of electrode manifolds becomes necessary (top panel). Gel electrofocusing apparatus applied under those conditions (bottom panel) to the electrofocusing of glucocorticoid receptor (6).

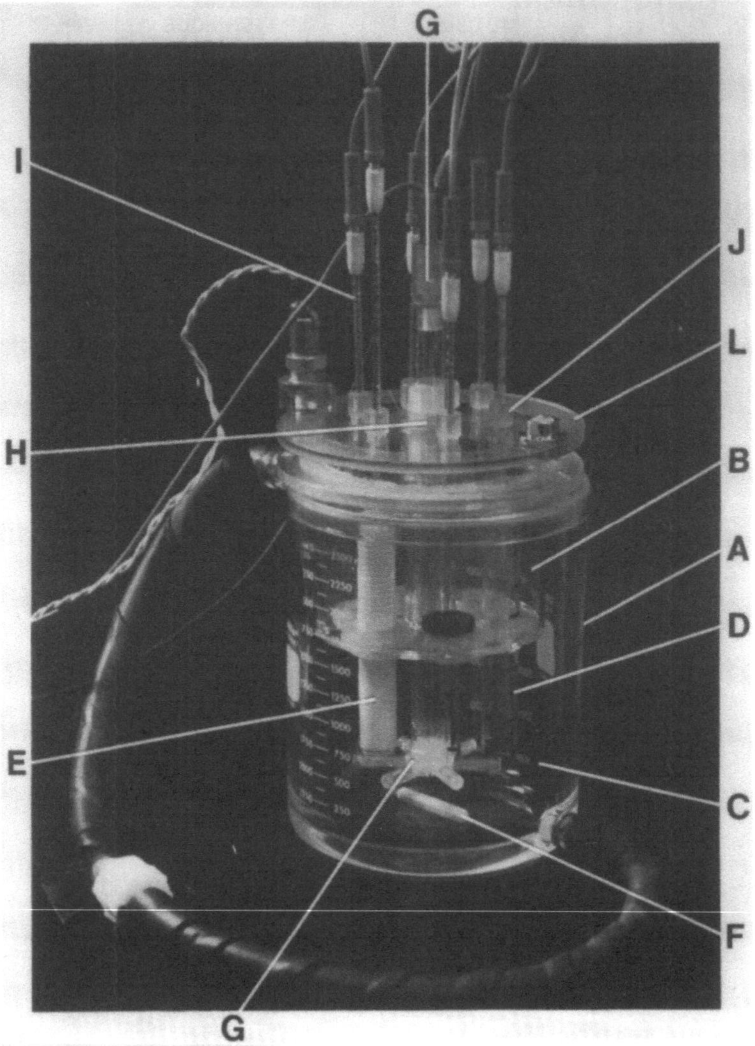

Fig.2: Assembled gel electrophoresis apparatus with electrode manifolds. A) Gel tube apparatus [Fig.1 of (1)]; B) Catholyte reservoir, 500 ml; C) Anolyte reservoir, 2,500 ml; D) Gel tube; E) Stopper and tube extrusion device; F) Magnetic stirrer; G) Electrode manifold (anode); H) Manifold holder and positioning device; I) Electrode array (cathode) for individual gel tubes; J) Electrode holders for I).

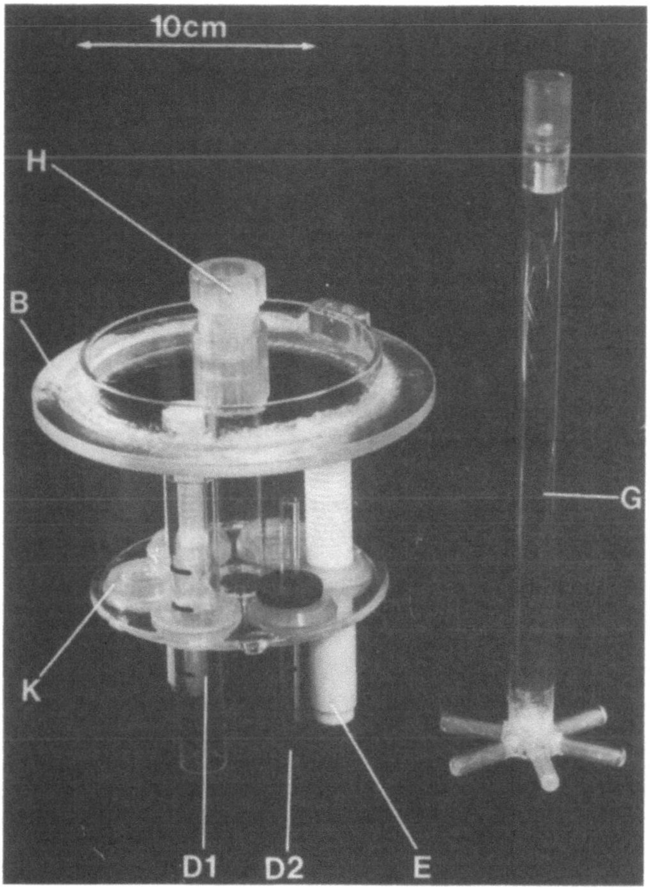

Fig.3: Upper electrolyte reservoir and electrode manifold. D1) 18 mm ID gel tube; D2) 6 mm ID gel tube. Other symbols as in Fig.2.

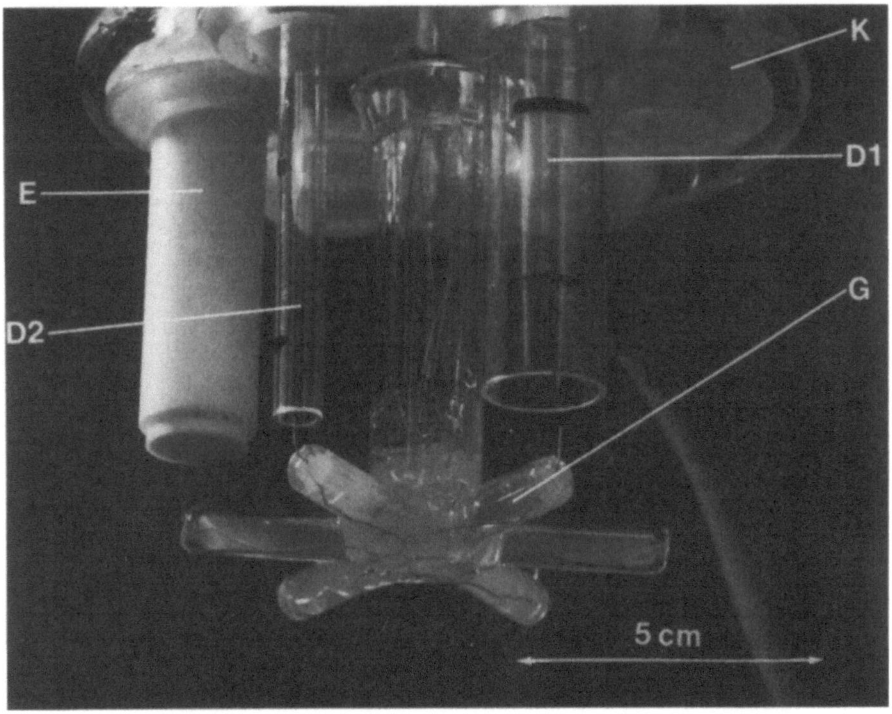

Fig.4: Positioning of electrode manifold in relation to the gel tubes. K) Silicone rubber grommet. Other symbols as in Fig.2.

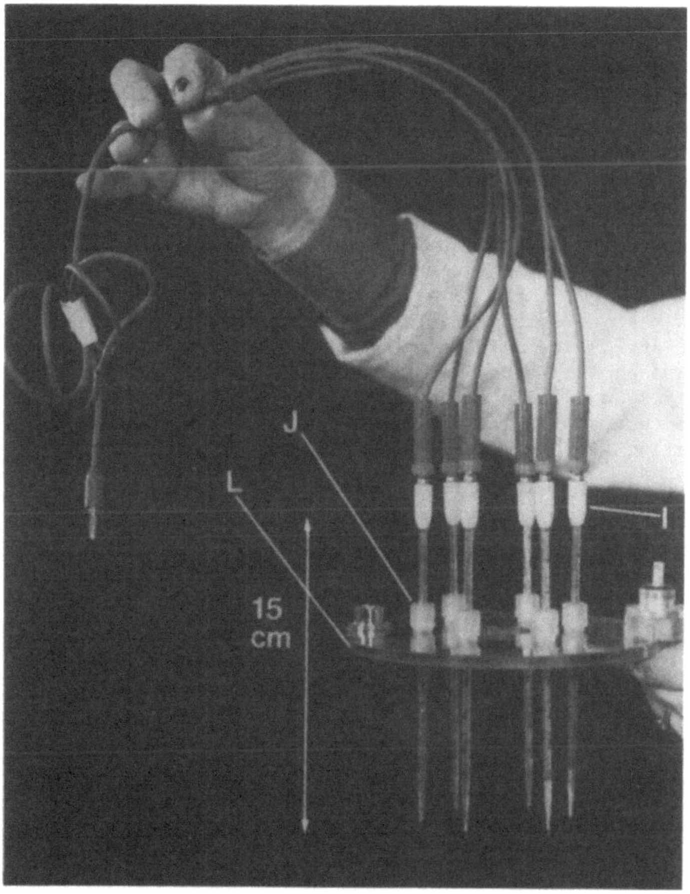

Fig.5: Upper electrode array connecting electrodes for individual tubes in parallel to a common power supply. L) Electrode top. Other symbols as in Fig.2.

Fig.1 illustrates its application to a large molecular weight, very heat labile glucocorticoid receptor (GR). In the application, the apparatus was used with Sephadex G-75 (Ultrodex, LKB) gel (6) (Fig.1). The assembly of the manifolds in the gel electrophoresis apparatus is shown in Fig.2.

A) Lower electrode manifold: The lower electrode consists of an insulated central strand of 6 electrode wires which are radially distributed at the bottom of the manifold to juxtapose and approximate the lower end of the gel tubes (Figs.3 and 4, Part G). The 6 strands of electrode wires are combined into a single receptacle on top. The electrode (Part G) is positioned centrally in the apparatus (Fig.2), proceeds through and is positioned vertically by a threaded sleeve holder (Figs.2 and 3, Part H) attached to the center of the upper electrolyte reservoir. The electrode is positioned at a sufficient height to allow for central magnetic stirring (Fig.1).

B) Upper electrode array: The upper electrode provided for each gel consists of an insulated linear Pt wire proceeding through the apparatus top (Figs.2 and 5) and positioned vertically by a threaded sleeve holder (Fig.2 and 5, Part J). Six such electrodes can be individually adjusted in height so as to approximate the top surface of the gels. The electrode leads are combined into a single cord which connects them in parallel to the power supply (Fig.5).

Construction

All construction details are available upon request in form of blueprints (No. 71-73-103, sheat no. 7).
1) Upper electrolyte reservoir (Figs. 2 and 3, Part B): The upper electrolyte reservoir was constructed from Pyrex, with a polycarbonate collar, as previously described (1), except that the central electrode (Part G) is held in position by a threaded collet made of polycarbonate. The nut holding the collet is bonded to the inner Pyrex tube in Part B by means of RTV-118 (General Electrics). The grommets were cast of silicone rubber (Selastic S-2000 Dow Corning) (Part K). Teflon stoppers (Part E) serve both to seal the grommets and to push 18 mm ID tubes through the grommet without loss of upper electrolyte. For the latter purpose, the Teflon stopper is reduced to 18 mm diameter at the bottom. Grommets for 18 mm and 6 mm ID gel tubes are exchangeable.
2) Lower electrode manifold (Figs.2-4, Part G): A 3/4 inch OD Pyrex heavy-wall tube was fused to 6 short pieces of tubing (8 mm OD, heavy wall) at one end. The short pieces were constricted at the distal end to approximately 1.5 mm ID. The fusion of the short pieces simultaneously closes the bottom of the central tube but leaves the short pieces open. Pt wire of 0.025 inch OD was threaded through each of the short

pieces of tubing and soldered to a single Cu wire connected to a male banana plug at the top of the part. The banana plug is held in place at the top by a polycarbonate threaded sleeve bonded to the tube by epoxy adhesive (Helix epoxy R-313, Applied Plastics). The Pt wire has to be initially at a sufficient length to allow one to perform the soldering outside of the 3/4 inch tube; thereafter, the Pt wire is pulled back through the short pieces of 8 mm tube and cut to the appropriate length. The appropriate length of wire must be sufficient to allow one to bend the wire vertically at the end of the 8 mm tube, with 2 mm of wire protruding vertically above the OD of the tube. The holes in the 8 mm tubes through which the wires protrude are sealed with RTV-118 which is injected with a syringe into the holes. To provide mechanical strength, the entire 8 mm tubes are filled with RTV.

3) Upper electrode array (Figs.2 and 5, Part L): The electrode top of the gel electrophoresis apparatus (1) was modified for application to multiple electrode wires for individual gel tubes of the type previously described (5) (Part I).

The individual upper electrodes were constructed by inserting a 3/4 inch length of 0.025 inch OD Pt wire into a length of 19 gauge type 304 stainless steel hypodermic tubing. The tubing was crimped so that 3/8 inch of the Pt wire protruded. The hypodermic tube was inserted into a hole drilled into a banana plug and was soldered. The length of the tube was such that it ended inside the pipet at 9/32 inches from its mouth, after the tube was cemented into the pipet (using epoxy adhesive or RTV-118). The Pt wire then protrudes 2.5-3 mm from the mouth of the pipet. The male banana plug endings of the electrodes are slightly flattened in a vise so as to reduce the force required to connect or disconnect the wire connections ending in the female banana plugs.

The individual electrodes are held in place by electrode holders (Part J). The threaded polycarbonate holders are drilled for a free fit with the individual electrodes. The holders are threaded with 1/8 inch pipe thread, and slotted so that when they are screwed into the electrode top (Part L) they will enclose the pipets tightly to hold them securely.

Electrode top (Part L) was constructed of polycarbonate as described (1) except that the male safety interlock was reversed so as to prevent interference with the positioning of the individual electrodes by the electrode holders (Part J). A portion of the interlock overhangs the perimeter of the electrode top. The electrode top (Part L) is drilled and tapped to accept electrode holders (Part J). The wire connections between the individual electrodes (Part I) and the power supply are soldered together to provide a single power inlet wired in parallel (Fig.5).

Performance

The performance of the electrode manifolds to provide constant and uniform voltage to multiple electrofocusing gels was tested by appli-

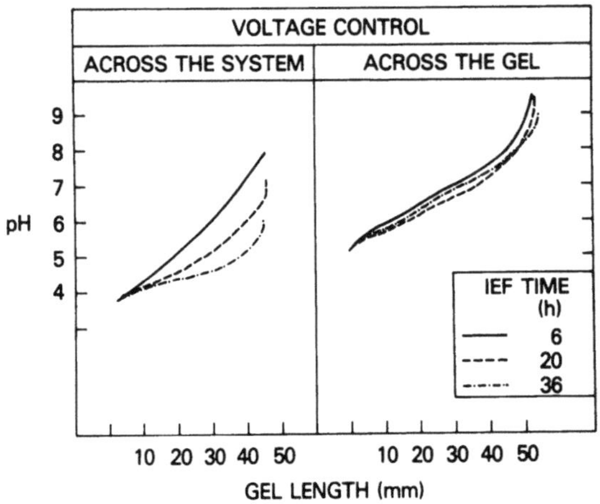

Fig.6: Improved stability of pH gradients obtained under conditions of voltage regulation through electrode manifolds. Presumably, the improvement is due to the avoidance of voltage surges between periodic voltage adjustments to a constant value of 20 V/cm of gel. The increased voltage between adjustments accelerates pH gradient decay. Anolyte: 0.1 M TES; Catholyte: 0.1 M lysine. Ultrodex gel cylinders, 18 mm diameter, 1% Ampholine (pI-range 5-10) (6).

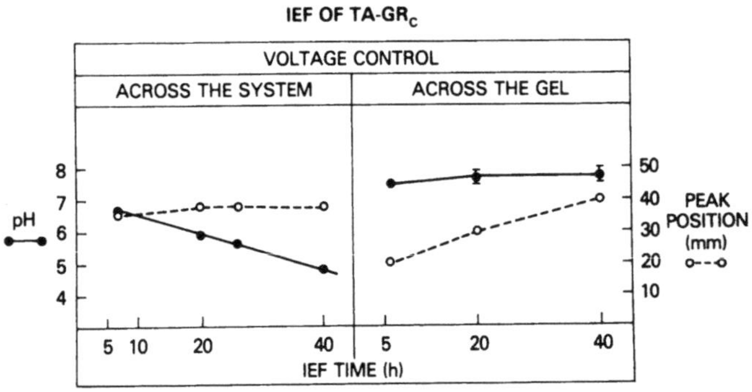

Fig.7: Prevention of irreversible protein precipitation during electrofocusing by voltage control, using electrode manifolds. Glucocorticoid receptor (TA-GR) was subjected to electrofocusing on Ultrodex gel tubes as described (6). When the voltage is regulated in all gel tubes by use of electrode manifolds, a constant pI' is reached and maintained, while the protein position on the gel shifts due to the changing pH gradient (cathodic drift). When the voltage is allowed to fluctuate between periodic voltage adjustments, the protein precipitates irreversibly and occupies an unvarying position on the gel, while the pH at that position changes with the cathodic drift, giving the erroneous impression of a protein at the transient state of its electrofocusing. Conditions as in Fig.6.

cation to a relatively slowly migrating, large (aggregated) and highly temperature sensitive protein, a glucocorticoid receptor. In view of the size of the protein and the tendency of the protein to aggregate at low ionic strength at the high protein concentrations at the surface of polyacrylamide gels (7), gel electrophoresis was carried out on Ultrodex using gel tubes (6).

The enhanced pH gradient stability achieved under conditions of effective voltage control of multiple electrofocusing gels is illustrated in Fig.6. When the voltage is applied conventionally, i.e. across the weakly acidic and basic electrolytes, monitored across one gel by use of the measuring device depicted in Fig.1 of (5) and adjusted periodically according to that measurement, voltage varied between periodic adjustments by as much as a factor of 4. The increased and variable voltage across the gels under those conditions (Fig.6, left panel) lead to a greater degree of pH gradient instability [since the rate of pH gradient decay is proportional to voltage (8)] than observed under conditions of rigorous voltage control achieved by use of the electrode manifolds (Fig.6, right panel).

The increased and variable voltage across the gels obtained under conventional conditions led to precipitation of the protein as evidenced by the constancy of its position on the gel when the pH gradient changed as a function of electrofocusing time (Fig.7, left panel). The problem of protein precipitation, presumably due to either the more rapidly progressing acidification of the gel from the anodic end, concomitant with the cathodic drift (8), and/or due to Joule heating, could be prevented by means of rigorous voltage control across the gels by use of the electrode manifolds (Fig.7, right panel).

Discussion

Rigorous voltage control across the gel is no problem in electrofocusing between highly conducting electrolytes as commonly applied in gel electrofocusing. It becomes a problem when the instability of pH gradients formed between strongly acidic and basic anolyte and catholyte relative to the rate of migration of proteins toward their pI positions forces one to employ weak acids and bases as anolyte and catholyte (3,4). The rate of approach of proteins toward their pI positions may be slow compared to the rate of decay of the pH gradient due to the low charge density of the protein, or it may relate to a very large particle size compared to the porosity of the gel medium (gel restrictiveness). Furthermore, the gradual loss of net charge upon the approach toward the isoelectric position leading to an asymptotically slowing migration rate [Fig.7 of (5)] is an inherent characteristic of electrofocusing systems. Moreover, the low ionic strength and high protein concentrations within zones promote protein aggregation [Fig.4 of (9)] which again diminishes the protein migration rates.

On a horizontal gel slab, the voltage control across the gel even with weakly acidic and basic reservoir electrolytes would not be difficult (assuming large enough electrolyte reservoirs), since electrodes merely would need to be placed across the ends of gel. However, in the study for which we applied the method (Ben-Or, S. and Chrambach, A., in preparation), the need to run gels at -4°C and the extreme temperature sensitivity of the protein made it imperative to work with gels entirely surrounded by thermostated liquid across heat-conductive (thin glass) surfaces, i.e. in gel tubes. Furthermore, for preparative purposes, a method of gel electrofocusing in wide-diameter preparative tubes was needed, with obvious lowering of the maximally tolerable voltage per gel surface area and increased need for operation at a voltage concisely controlled at the maximum tolerable value. The electrode manifold described here allows for such operation with multiple gels. Provision of a maximally tolerable controlled voltage has lead to quantitative recoveries of receptor protein activity from gel electrofocusing on multiple Ultrodex gels of 6 and 18 mm diameter (6).

References

1) Chrambach, A., Rodbard, D. In Gel Electrophoresis of Proteins: A Practical Approach (Hames, B.D. and Rickwood D., eds.) Information Retrieval Ltd. Press, London and Washington DC, pp. 93-143 (1981).

2) An der Lan, B., Chrambach, A. In Gel Electrophoresis of Proteins: A Practical Approach (Hames, B. D. and Rickwood D., eds.) Information Retrieval Ltd. Press, London and Washington DC, pp. 157-187 (1981).

3) Nguyen, N. Y., Chrambach, A. Anal. Biochem., 79, 462-469 (1977).

4) An der Lan, B., Chrambach, A. Electrophoresis 1, 23-27 (1980).

5) Chrambach, A., Hjelmeland, L. M., Nguyen, N. Y., An der Lan, B. (1980) In Electrophoresis '79 (Radola, B. J., ed.) Walter de Gruyter, Berlin-New York NY, pp. 3-22.

6) An der Lan, B., Ben-Or, S., Allenmark, S., Sullivan, J. V., Fitze, P. and Jackiw, A. Electrophoresis, submitted (1982).

7) Ben-Or, S. and Chrambach, A. Archives Biochem. Biophys. 206, 318-330 (1981).

8) Baumann, G., Chrambach, A. In Progress in Isoelectric Focusing and Isotachophoresis (Righetti, P. G. ed.) Elsevier, Excerpta Medica, North Holland, Assoc. Sci. Publ., Amsterdam, The Netherlands, pp.13-23 (1975).

9) Hjelmeland, L. M., Chrambach, A. Electrophoresis 2, 1-11. 1-11 (1981).

A GENERAL METHOD FOR THE STUDY OF SOME PROPERTIES OF POLYMERIC PROTEINS

Dominique de Vienne

Laboratoire d'Amélioration des Plantes, Bât 360,
Université de Paris-Sud, 91405 ORSAY CEDEX 05 FRANCE

Introduction

Most cellular proteins are oligomers, often composed of an even number of subunits. For example, in humans, more than three quarters of the enzymes are not monomers.

For about fifteen years, various techniques, especially electrophoresis, have been used to reveal genetic polymorphism. In particular, multiple forms of oligomeric proteins have been shown to result from multi-allelism. In the case of enzymes, some authors have compared these different forms, in order to find out : - if the different subunits have identical biochemical properties.

- if the effects of subunits in a heteromeric protein are non-additive (i.e. if there is molecular heterosis).

In this paper we give a general presentation of this kind of data, and propose statistical tests to answer these two questions.

Method

A polymeric protein composed of n subunits, which are taken among s different possible types, can exist in $N = (n+s-1)!\,/\,(s-1)!\,n!$ forms. When a parameter is measured for each of these N forms, the results can be represented by a set of N points which define a hypersurface in a space of s dimensions: one dimension corresponds to the variable Y associated to the parameter studied, and each of the s-1 others corresponds to the number r_i of times each subunit is included in the protein ($1 \leqslant i \leqslant s-1$, and $\forall i$, $0 \leqslant r_i \leqslant n$). The number of sth subunits is determined by difference from the sum of the

Electrophoresis '82

s-1 other ones.

If this supersurface is a hyperplane, there is no molecular heterosis. If it is "horizontal", the subunits are equivalent for the parameter measured.

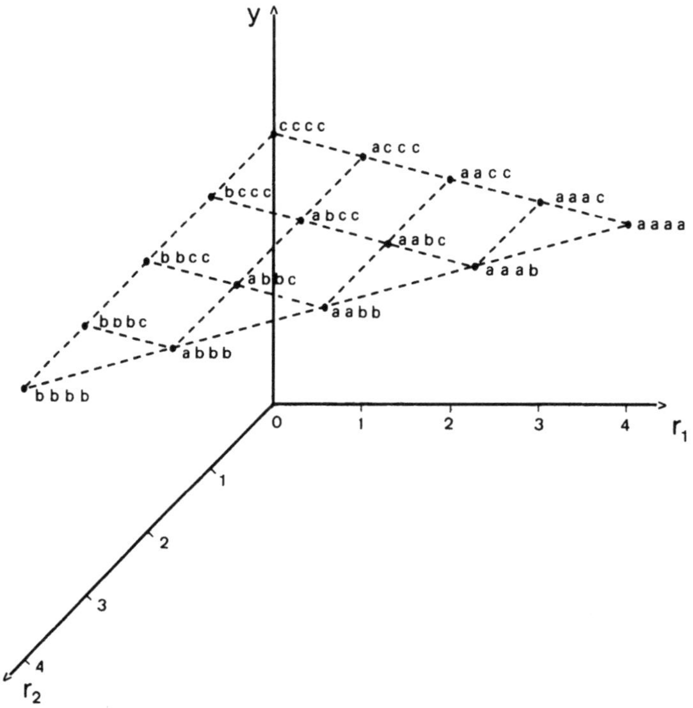

Figure 1 - Example of graph obtained when n = 4 and s = 3. This surface is plane and oblique (a_1 = 1 ; a_2 = 1,5 ; a_3 = 1,5. The subunits are respectively called a, b and c).

To test statistically the deviation from the plane and the orientation of the surface, independant repetitions of each measurement are necessary ; let n_j be the number of repetitions of the measure of the jth form. Let a_i the effect of subunit i on the parameter, which is measured on the oligomer. Under the hypothesis of additivity of these effects, the model is :

$$Y_{jk} = \sum_{i=1}^{i=s-1} (a_i - a_s) r_{ij} + n a_s + E_{jk} \qquad (1)$$

$$\text{where } 1 \leqslant j \leqslant N, \; 1 \leqslant k \leqslant n_j \text{ and } \forall j, \sum_{i=1}^{i=s} r_{ij} = n$$

The coefficients $(a_i - a_s)$ can be estimated : the vector

$$\hat{\underline{a}} = \begin{pmatrix} \widehat{a_1 - a_s} \\ \vdots \\ \widehat{a_{s-1} - a_s} \end{pmatrix}$$

is the solution of the matricial equation :

$$\underline{r}\ \hat{\underline{a}} = \underline{ry}$$

where $\underline{r}$ = matrice of the sums of the squares of deviates and of the products of deviates of the parameters r_i, and $\underline{ry}$: vector of the sums of the products of deviates of the variable Y and of r_i. The matrix $\underline{r}$ has a simple expression : it is possible to demonstrate that, for each i, the elements of the diagonal are :

$$\sum_{j=1}^{j=N} (r_{ij} - r_{i.})^2 = (s-1)(n+s)!/\ s(n-1)!(s+1)!$$

and for each i and each i', the other elements are

$$\sum_{j=1}^{j=N} (r_{ij} - r_{i.})(r_{i'j} - r_{i'.}) = -(n+s)!/\ s(n-1)!(s+1)!$$

The general term of the vector $\underline{ry}$ is

$$\sum_{j=1}^{j=N} (y_{j.} - y_{..})(r_{ij} - r_{i.}) = \sum_{j=1}^{j=N} y_{j.}r_{ij} - Ny_{..}r_{i.}$$

It is easy to demonstrate that for each i, $r_{i.} = n/s$

The coefficient a_s will be estimated as :

$$\hat{a}_s = y_{..}/n - \sum_{i=1}^{i=s-1} (\widehat{a_i - a_s})/s \qquad \text{and } \forall\, i \neq s,\ \hat{a}_i = (\widehat{a_i - a_s}) + \hat{a}_s$$

Before testing, the following quantities must be calculated

$$SR = \sum_{i=1}^{i=s-1} (\widehat{a_i - a_s}) \sum_{j=1}^{j=N} n_j(y_{j.} - y_{..})(r_{ij} - r_{i.}) =$$

$$\sum_{i=1}^{i=s-1} (\widehat{a_i - a_s}) \left[\sum_{j=1}^{j=N} n_j y_{j.} r_{ij} - Ny_{..} n/s \right]$$

$$SC = \sum_{j=1}^{j=N} n_j (y_{j.} - y_{..})^2 = \sum_{j=1}^{j=N} n_j y_{j.}^2 - \mathcal{N} y_{..}^2$$

$$SE = \sum_{j=1}^{j=N} \sum_{k=1}^{k=n_j} (y_{jk} - y_{j.})^2 = \sum_{j=1}^{j=N} \left[\sum_{k=1}^{k=n_j} y_{jk}^2 - n_j y_{j.}^2 \right]$$

$$ST = \sum_{j=1}^{j=N} \sum_{k=1}^{k=n_j} (y_{jk} - y_{..})^2 = \sum_{j=1}^{j=N} \sum_{k=1}^{k=n_j} y_{jk}^2 - \mathcal{N} y_{..}^2$$

where $\mathcal{N} = \sum_{j=1}^{j=N} n_j$

Thus, the table of analysis of variance for the model with equation *(1)* is:

	S. S.	d. F.	M. S.	
Deviation from linear model	SC - SR	N - s	$\frac{SC - SR}{N - s}$	*(2)*
Effects a_i	SR	s - 1	$\frac{SR}{s - 1}$	*(3)*
Residual	SE	$\mathcal{N}$ - N	$\frac{SE}{\mathcal{N} - N}$	*(4)*
Total	ST	$\mathcal{N}$ - 1	$\frac{ST}{\mathcal{N} - 1}$	

Then, it is possible to statistically answer the two questions posed at the beginning of this paper :

- the deviation from the linear model (molecular heterosis) is tested by the ratio : *(2) / (4)*, which is compared, at α level, to the value of $F(N - s, \mathcal{N} - N ; \alpha)$.
- the equality or inequality of the coefficients a_i is tested by the ratio *(3) / (4)*, which is compared, at α level, to the value of $F(s-1, \mathcal{N}-N ; \alpha)$.

Example : consider an enzyme constituted by the random association of s subunits. Let a'_1 ...., a'_s be the real activities of the s subunits (in the polymers). The variable Y is here associated to the relative activities of the isoenzymes ; these activities can be measured as follows. From electrophoresis under non-denaturing conditions, the activity of each isoenzyme (unweighted by its relative quantity) can be determined, expressed by the following relation :

$$(n!/\prod_{i=1}^{i=s} r_{ij}!)\prod_{i=1}^{i=s} p_i^{r_{ij}} \sum_{i=1}^{i=s} r_{ij}a'_i \Big/ \sum_{j=1}^{j=N}\left[(n!/\prod_{i=1}^{i=s} r_{ij}!)\prod_{i=1}^{i=s} p_i^{r_{ij}} \sum_{i=1}^{i=s} r_{ij}a'_i\right]$$

where p_i is the relative concentration of the subunit i. A second analysis - a second dimension of immunoelectrophoresis or the electrophoresis of the purified enzyme followed by non-specific coloration - will give the quantities of each isoenzyme, as given by the multinomial distribution :

$$(n!/\prod_{i=1}^{i=s} r_{ij}!)\prod_{i=1}^{i=s} p_i^{r_{ij}}$$

The ratio, for each isoenzyme, of the two values obtained, is a measure of the relative activity of this isoenzyme.
The abovementioned statistical tests can then be applied. The method functions even if the association of subunits is not random, since it is based on the ratio.

Conclusion

The proposed method is very general. Firstly, numerous biochemical properties can be studied, in particular in enzymes : specific activity, heat sensitivity, optimal pH, Km, etc ... (1, 2). Moreover it could be interesting to follow the deformation of the hypersurface of a parameter under various experimental conditions, especially in the case of allosteric enzymes (3). Properties less directly connected to the enzyme, such as the flux of its metabolic pathway, can also be studied (4).

Above all, this method is not limited to the study of polymeric proteins : it is possible to test one-locus heterosis and to compare the effects of alleles, by making n equal to the ploidy level and s equal to the number of alleles at the locus. In this case, the abovementioned formulae work for any measurable variable measured in all genotypic combinations. The study of the obtained hypersurfaces could be of great interest in population genetics.

References

1. Scandalios, J.G., Liu, E.H., Campeau, M.A. : Arch. Biochem. Bioph. 153, 695- 705 (1972).
2. Berger, E. : Amer. Natur., 110 (975), 823-839 (1976).
3. Fincham, J.R.S. : Heredity, 28, 387-391 (1972)
4. Kascser, H., Burns, J.A. : Genetics, 97, 639-666 (1981).

HIGHLY CROSS-LINKED GRADIENT GELS FOR DETERMINING MOLECULAR WEIGHT WITH NATIVE PROTEINS OF ALL SIZES

C.W. Wrigley, W.P. Campbell
CSIRO Wheat Research Unit, P.O. North Ryde N.S.W. 2113, Australia

and J. Margolis
Gradient Laboratories Pty Ltd, P.O. Box 65, Pyrmont Australia

Introduction

Polyacrylamide has proved to be much more versatile as an electrophoresis medium than alternatives, such as starch and agarose gels, because its formulation can be varied to provide a very wide range of sieving properties, whilst also providing a transparent, charge-free gel with good mechanical strength. At low concentrations, clear polyacrylamide accommodates proteins of a million daltons and larger.

Polyacrylamide gels of even greater porosity (1) can be made by increasing the percentage (%C) of crosslinking agent (N,N'-methylene bisacrylamide, bis) up to 50% (2), whilst maintaining a low overall concentration of acrylamide and bis (%T, nomenclature of Hjerten (3)). Polyacrylamide, highly crosslinked at low or moderate %T, is opaque. Nevertheless it is still a useful medium when molecular sieving needs to be minimized or when proteins in the multimillion-dalton range are to be fractionated.

On the other hand, an increase in total gel concentration (generally at constant %C) has been repeatedly used to restrict effective 'pore size'. By increasing the frequency of

Electrophoresis '82

polyacrylamide chains in the electrophoretic medium, protein mobilities are reduced due to frictional retardation and molecules are thus sorted according to size. The progressive restriction of protein migration has been demonstrated in polyacrylamide gels of up to 40 and even 50% T (at 5% C) (4), but there are practical limitations to further increases in %T.

At any given level of %T, the sieving properties of a gel may be enhanced by increasing the level of crosslinking, though this approach has limitations. Some degree of crosslinking is needed for a gel to form: without bis, polyacrylamide is merely a viscous solution. Presumably the progressive introduction of crosslinks between gel strands increases their rigidity thereby offering more resistance to the passage of macromolecules. However, with further increases in bis, there comes a point when no more crosslinks can be incorporated between chains with normal spacing, and gel opacity results. Beyond this 'critical point' (5), further increases in %C lead to reduced sieving properties (6-8).

Uniform Small-Pore Gels

In the study to be described, we have pursued the possibility of raising both %T and %C within the limitations described above, with the aim of obtaining an acrylamide-bis gel suitable for molecular sieving of very small proteins. This possibility has been explored previously (4,8), but not to its fullest extent. Margolis and Wrigley (8) reported that higher degrees of crosslinking could be usefully incorporated as total gel concentrations rose from 10% to 27%. Our further study showed that at 40% T, nearly 14% C is possible without exceeding the critical point. Gel formulations in which %C exceeded the critical point were opaque with poor mechanical properties. These gels were polymerized at 25°C. A slight shift in the position of the curve would be expected for gels polymerized at other temperatures (7).

A gel containing 40% acrylamide-plus-bis and 12.5% C was chosen as a practical formulation for testing the maximum sieving properties of bis-linked polyacrylamide. Its effective pore size was such that most proteins were excluded from it, but polypeptides from proteolytic digestion entered the gel and were fractionated as discreet zones. Small proteins (10 to 20 kilo daltons) entered the gel but had a very low mobility, indicating approach-to-pore-limit conditions.

Homogeneous gels with high levels of both %T and %C (if below the critical point) can thus be seen to provide a medium suitable for molecular sieving of polypeptides and nucleic acids in a range (below 10,000 daltons) not normally exploited for gel electrophoresis. Such formulations may also form a useful basis for extending gel filtration in polyacrylamide-gel beads to the separation of quite small molecules.

Gradient Gels with Small Pore Limit

Our main objective in developing a very small pored gel was to use it as the limit concentration of a gradient gel, thereby ensuring the retention of small proteins. The gradient gel concept (9-12) has proved to be a very popular means of providing the most suitable gel concentration for fractionating each of a mixture of proteins of different sizes. Furthermore, gradient gels are becoming the normal medium for the second dimension in high-resolution fractionation of heterogeneous samples (13).

Due to the progressive increase in sieving properties in a gradient gel, the migration of proteins progressively decreases, zones are sharpened, and the components are arranged

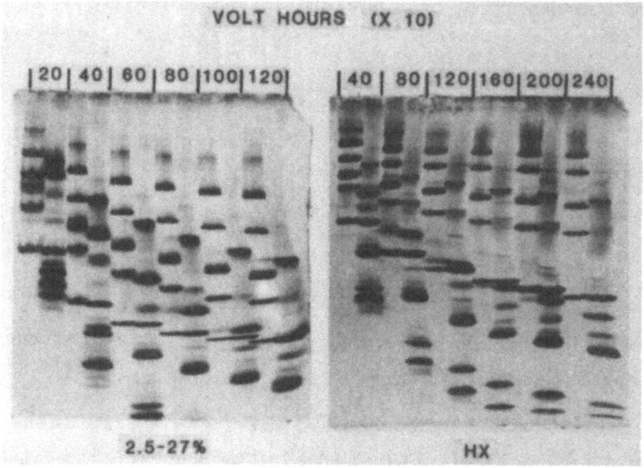

Figure 1. Electrophoresis of native proteins of known molecular weights in two types of gradient gel - 2.5-27%T and a highly crosslinked (HX) gel (3 to 40% T with 12.5% C at upper limit). Both gels were manufactured by Gradient Laboratories Pty Ltd, Pyrmont 2009, Australia. The buffer for prerun and electrophoresis was 1% Tris, 0.5% boric acid, 0.11% Na EDTA, pH 8.3. Each of two proteins were applied at hourly intervals. The 2.5-27% gel was run at 200 volts, the HX gel at 400 volts. The temperature of both was maintained at 25°C. Proteins applied (and size in kilodaltons) were, for the left track of each pair, thyroglobulin (670), ferritin (440), catalase (232), beef heart lactate dehydrogenase (140) and bovine serum albumin (67); for each right track, bovine serum albumin (67), ovalbumin (43), soybean trypsin inhibitor (20) and bovine alpha-lactalbumin (14). Gels were stained in Coomassie brilliant blue (14).

in the gel according to molecular size whether sodium dodecyl sulfate (SDS) is used or not. Although the risk of losing zones of many proteins from the gel is greatly reduced, this risk has remained for proteins smaller than about 40,000 daltons, especially if electrophoresis is prolonged to permit size estimation (8,12).

Figure 1 shows that this problem can be overcome by using 40% T and 12.5% C as the limit in the gradient gel formulation. Alpha-lactalbumin (14,000 daltons) and soybean trypsin inhibitor (21,000 daltons) have been retained on the highly crosslinked (HX) gradient gel even after six hours' electrophoresis at 400 volts. By comparison, both these proteins were lost from a conventional 2.5 to 27% gel after one third of this time. On the other hand, the resolution of large proteins such as thyroglobulin is still obtained on the HX gel since %C is maintained at 4% for its upper third. HX gradient gels (73 x 73 x 3 mm) were prepared using a three-chambered gradient former, similar to that of Margolis and Wrigley (8), so that total gel concentration (%T) increased linearly from 3% at the top (origin) of the gel to 40% at the bottom: the proportion of bis (%C) remained at 4% for the top 25 mm of gel (5 to 15% T), and thereafter %C increased linearly to 12.5% at the end of the gel (15).

Blood-protein electrophoresis

HX gels provided zone sharpening conditions for the full range of serum proteins - from the very large protein complexes of the hemoglobin-haptoglobin series to very small proteins (about 10,000 daltons).

Resolution of the serum proteins in the HX gel could be further improved by providing a preliminary separation, largely based on their charge properties, by electrophoresis in a gel (5% polyacrylamide) with poor sieving properties. The resulting two-dimensional separation (Fig. 2) distinguished the immunoglobulins as continuous streaks and resolved several components that are otherwise obscured by the large albumin component. Likely identities for several components are: 1. IgM, 2. low-density lipoprotein, 3. alpha-2 macroglobulin, 4. IgG, 3. IgE, 6. IgA, 7. ceruloplasmin, 8. transferrins, 9. albumin.

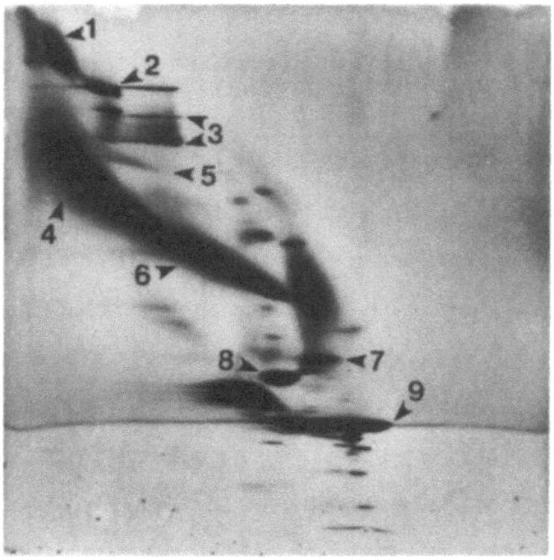

Figure 2. Two-dimensional electrophoresis of horse serum proteins, initially in 5% polyacrylamide at pH 8.3 for 45 min. at 150 volts (origin at top, to anode at right), followed by vertical anodic electro-phoresis in the second dimension at pH 8.3 in a HX gel for 2,600 volt hours.

Size Estimation for Denatured and Native Proteins

Gradient gels have proved particularly suitable for determining the sizes of polypeptides in the presence of SDS (16,17). In this connection, the use of gradient gels has avoided pitfalls associated with homogeneous gels (16). The added zone sharpening and retentive capabilities of HX gradient gels are likely to further enhance this application, particularly for small proteins (18) and for glycoproteins (19).

In addition, gradient gel electrophoresis offers the possibility of estimating molecular size in native proteins, without reduction or SDS treatment, if electrophoresis is continued sufficiently long for charge differences to be minimized. Examination of native proteins provides the advantage of permitting zone detection according to biological activity, and of avoiding the loss of information about quaternary structure. With the HX gradient gel, this approach can now be applied to uncharacterized protein mixtures without the risk of losing small protein components (8,12,16). Figure 3 shows the relationship between molecular weight (on a log scale) and migration distance (after 2,400 volt hours) for a series of globular proteins of known molecular weights (Fig. 1). The relationship is almost linear (regression coefficient of 0.99) . A line-of-best-fit has been calculated and included in Fig. 3. The points could alternatively be joined to form a shallow sigmoidal curve. This curve should probably be used in calculating the molecular sizes of unknown proteins. The linearity or otherwise of this relationship must depend on the shape of the gradient of pore size, produced by the variation in gel composition down the electrophoretic medium. The results of Figure 3 contrasts with a previous recommendation (16), using 4 to 30% gradient gels, that this approach only be used with proteins of over 67,000 daltons. In another analysis (12) of this approach to size determination, this limit was suggested as 45,000, and an equation was derived for incorporating total polymer concentration and time of electrophoresis. In such estimations using HX gels, the degree of crosslinking may also have to be taken into account.

The HX gels now provide a basis for characterizing the minor proteins of plasma smaller than albumin. It is difficult at this stage to definitely identify these proteins (Fig. 2). There is for example, the possibility that at least some of these zones may arise from proteolysis of larger proteins, and this cannot be excluded. However estimates of their molecular sizes are consistent with the following possible identities (20,21). The bands immediately below albumin would be classed as prealbumins (40,000-55,000); the next two components may be thyroxin-binding protein (36,000) and beta-2 glycoprotein III (33,000); another smaller protein (about 21,000) may be

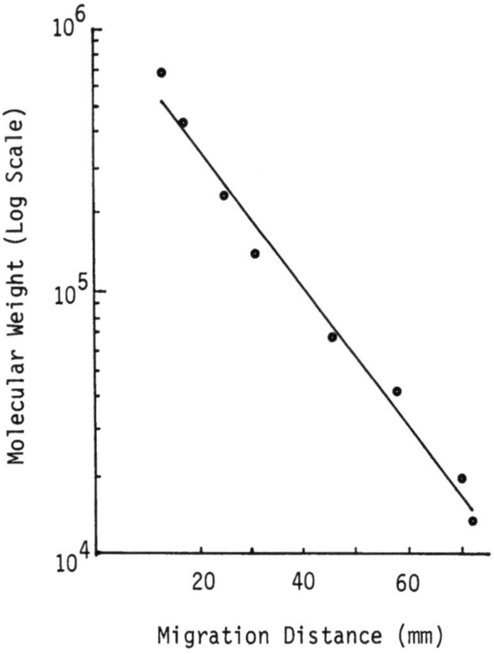

Figure 3. Relationship between size (log scale) and migration distance for proteins (Fig. 1) in a HX gradient gel after 2,400 volt hours at pH 8.3. The line-of-best-fit is shown.

retinol-binding protein. The protein of highest mobility may be either lysozyme (15,000) or beta-2 microglobulin (11,600).

References

1. Fawcett, J.S., Morris, C.J.O.R.: Separation Science 1, 9-26, (1966)
2. Rodbard, D., Levitov, C., Chrambach, A.: Separation Science 7, 705-723 (1972)
3. Hjerten, S.: Archiv. Biochem. Biophys. 99, 466-475 (1962)
4. Blattler, D.P., Garner, F., Van Slyke, K., Bradley, A.: J. Chromatogr. 64, 147-155 (1972)

5. Tanaka, T.: Sci. Amer. 244, 110-123 (1981)

6. Righetti, P.G., Brost, B.C.W., Snyder, R.S.: J. Biochem. Biophys. Methods 4, 347-363 (1981)

7. Gelfi, C., Righetti, P.G.: Electrophoresis 2, 220-228 (1981)

8. Margolis, J., Wrigley, C.W.: J. Chromatogr. 106, 204-209 (1975)

9. Margolis, J., Kendrick, K.G.: Anal. Biochem. 25, 347-362 (1968)

10. Slater, G.G.: Anal. Chem. 41, 1039-1041 (1969)

11. Rodbard, D., Kapadia, G., Chrambach, A.: Anal. Biochem. 40, 135-157 (1971)

12. Rothe, G.M., Purkhanbaba, H. : Electrophoresis 3, 33-42, 43-48 (1982)

13. O'Farrell, P.H.: J. Biol. Chem. 250, 4007-4021 (1975)

14. du Cros, D.L., Wrigley, C.W.: J. Sci. Fd Agric. 30, 785-794 (1979)

15. Margolis, J. : Application for Patent, No PF Q2000 (1981)

16. Lasky, M.: In Electrophoresis '78 (Catsimpoolas, N. ed.), Elsevier, North Holland, pp. 195-210 (1978)

17. Lambin, P.: Anal. Biochem. 85, 114-125 (1978)

18. Swank, R.T., Munkres, K.D.: Anal. Biochem. 39, 462-477 (1971)

19. Poduslo, J.F. : Anal. Biochem. 114, 131-139 (1981)

20. Schultz, H.E., Haremans, J.F. : Molecular Biology of Human Proteins, I. Elsevier Publ. Co. (1966)

21. Putman, F.W. : The Plasma Proteins, I. Academic Press, Inc., (1975)

HIGH RESOLUTION DNA SEQUENCING SYSTEM

W. Ansorge and R. Barker
European Molecular Biology Laboratory, Postfach 10.2209, D-6900 Heidelberg

Introduction

The use of the rapid DNA sequencing techniques (1,2) demonstrates the fast spreading of this technology employed as a research tool both by universities and industry. The last step in the sequencing procedure is the separation of the oligonucleotides on denaturing polyacrylamide gels. Once the sample has been prepared, it is the interest of the biologist to obtain from the gel electrophoresis and the subsequent autoradiography all the sequence information that can be obtained from the sample, i.e. the extent of the obtainable sequence information should be limited only by the quality of the sample and not by the resolving power of the electrophoresis. In that case, the way to increase the readable sequence length <u>per gel</u> is to improve the sample quality in the region of the higher base pairs so that the electrophoresis limit is reached. It is of interest to push this limit as far as possible, ideally until some "ultimate limit" in the resolving power of the polyacrylamide gel is attained.

Reducing gel thickness results in narrower separated bands as well as in less radiation scattering in the gel, and both of these effects contribute to the sharper bands obtained by autoradiography as demonstrated by Sanger and Coulson (3) on thin (0.4 mm) gels. Three modifications to the DNA sequencing gels which all result in improved oligonucleotides resolution as visualised by autoradigraphy (0.2 mm thin gels, novel

Electrophoresis '82

straight-forward drying of the gel bound chemically to its support, uniform gel temperature resulting from the use of a special thermostating plate) were reported by us (4-8). Here we discuss tests on further optimisation of the resolution on DNA sequencing gels. Samples of pBR 322 DNA of sufficient quality (they seem properly labelled even around 1000 base pairs) were prepared and used in the tests. The effect of physico-chemical properties of gels and buffers on resolution were tested e.g. ionic strength and pH of buffers, different buffer systems, acrylamide concentration, crosslinker concentration, type of crosslinker, temperature of polymerization, denaturing conditions (urea vs. no urea, formamide, SDS), gel length and thickness. Tested were as well different running conditions like electric field, gel temperature, dimension of sample slots. Gels 0.1 mm - 0.2 mm thick and up to 1.2 m long were cast and tested routinely. The sliding technique (4) modified for casting of these gels is described.

We were able to count up to 600 base pairs in the new system using 1 m long gels. At present this number seems to be near the limit of the resolving power of the polyacrylamide gel.

Materials and Methods

1. Materials

Plasmid DNA pBR 322, EcoR1, Bam H1, DNA-polymerase (Klenow) were all from Boehringer, $dAT^{32}P$ (High Specific Activity) from Amersham, acrylamide from Polyscience, BDH and BioRad, bisacrylamide from Bio Rad and Serva, urea "ultra pure" from Schwarz/Mann, ammonium persulphate, and TEMED from Bio Rad, Repelcoat from Hopkins & Williams, Rumford, England and γ-methacryloxypropyl trimethoxy silane from Wacker Chemie, Munich, FRG.

All other reagents were of analytical grade and all solutions and glassware used for the preparation of samples were sterilised before use.

Sample preparation. To resolve more than 500 base pairs in a sample of DNA the labelling of the fragment (3' end) and the Maxam and Gilbert reaction times were modified as described in (13).

Electrophoresis

The glass plates (all floated glass) for electrophoresis were as described in (6,7) and up to 1.2 m long, buffers and solutions as described in (8). Spacers and sample well formers were cut from the same piece of PTFE (0.1-0.2 mm thick) or the more rigid spacers (0.1-0.2 mm thick) from LKB, which do not inhibit gel polymerisation, were used. Thermostatting plates (up to 1.2 m long) were used as described in (6,7). The water circulating bath used for the plates was LKB 2209 Multitemp, which seems to be electrically well insulated and ground leakage current is low even at higher voltages (above 3000 Volts) across the thermostating plate. Application of samples into the slots was done as in (6,7) on 0.2 mm gels and special thin-walled capillaries (outer diameter less than 0.1 mm) were ordered from A. Müller, Glas- und Vakuumtechnik, W. Berlin, for sample application on 0.1 mm gels. Power supplies were LKB 2103, LKB 2209, Pharmacia ECPS 3000/150, Shandon for high (up to 10 kV) voltage. Gel solutions used are given in table II.

Treatment of the glass plates with silane and repelcoat was done as in (6,7). For other possible coupling silanes see

(10). Prior to gel casting, the plates were polished with dry lint free Kleenex paper. The repelcoated plate should be very carefully washed with ethanol prior to use as some spots in the autoradiographs could be traced to excess repelcoat solution.

Buffer chamber for running of two to three gels was designed. In this way four to six gels could be run in parallel from a power supply with two inputs. The gels on the thermostating plate were mounted vertically on the apparatus and two to three of them connected in series to a circulating water bath. Temperature of the water bath was raised in about 30 min from room temperature to 50°C in order to avoid thermal shock to the thermostating plate. During this time pre-electrophoresis was carried out at about 3000 V, and samples applied. Typically the gels were then electrophoresed at 3000 V for 12 hours (1 m long) or 14.5 hours (1.2 m long) overnight without sample re-application. With voltages of 5000-6000 V the times were correspondingly shorter. One can choose three gels of different concentrations, run them in parallel for the same time and obtain suitable sequence overlap from the gels up to 500-600 base pairs. A chart of DNA electrophoretic mobility on gels with different concentrations is helpful [13]. We have used gel concentrations of 4%, 6% and 12-14% for the overlap at 50°C on 1 m long gels. After electrophoresis the plates are separated. The plate with the gel is fixed (and urea washed out) in 10% acetic acid for 15 minutes, followed by a brief surface wash with water and the gel is air dried. The film is placed directly on the gel, autoradiography carried out for 1-7 days at room temperature without an intensifying screen, using KODAK X-OMAT or FUJI RX(SAFETY) films.

Gel moulding using the sliding technique

The method is used as described in (4,6). To facilitate the casting of long gels, the whole setup is inclined at an angle of about 10-15 degrees and the upper cover glass plate is sliding "up the hill". In this way the gravitational effect on the gel helps to prevent air bubbles from entering from below during gel casting. The same purpose serves also another modification of the technique. To the back of plate B (ref. 6) is added an additional assisting piece (of delrin or glass) of the same width as plate B, but about 1 mm thinner than plate B and about 5 cm long. The back spacer made of PTFE is then cut so large that it extends over the assisting piece on to the plate B (6). Because this piece is 1 mm thinner than plate B, there is a small space where gel solution can accumulate during casting which prevents the air bubbles from being drawn between the glass plates. It is advantageous to place in this 1 mm space a nylon net 0.3-0.4 mm thick (Small Parts, Inc., Miami, Fla.) which stabilises the gel solution in that space and helps to keep the back edge of plate B wet. When the sliding is finished the setup with the glass plates containing the gel is placed horizontally, the plates clamped at the bottom end, the sample comb is inserted and clamping finished. The casting of a 1.2 m long gel (with clamping) was carried out in about two minutes. The gel solution usually polymerizes in about seven to ten minutes, this time giving well polymerized sample slots which are important for band sharpness.

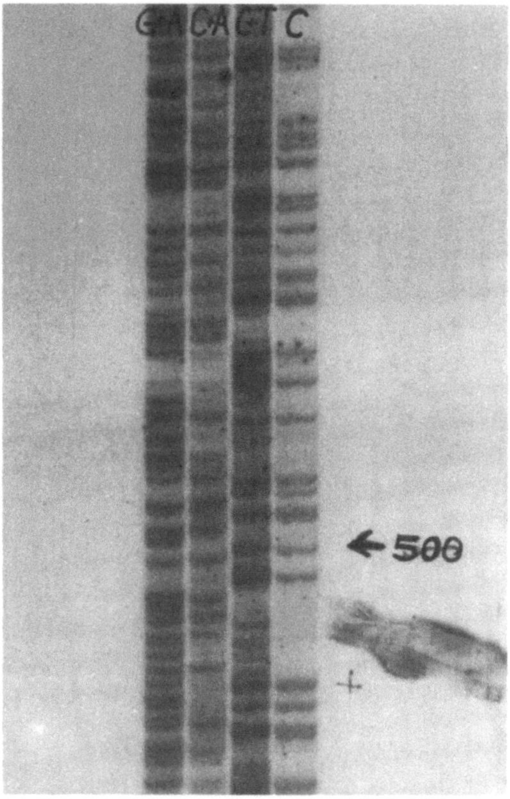

Fig. 1: Section of autoradiogram of a 0.2 mm thick, 1 m long, 4% polyacrylamide gel, showing the region of 500 base pairs (50°C, 30 V/cm, 10 mA). Sample of pBR 322 DNA.

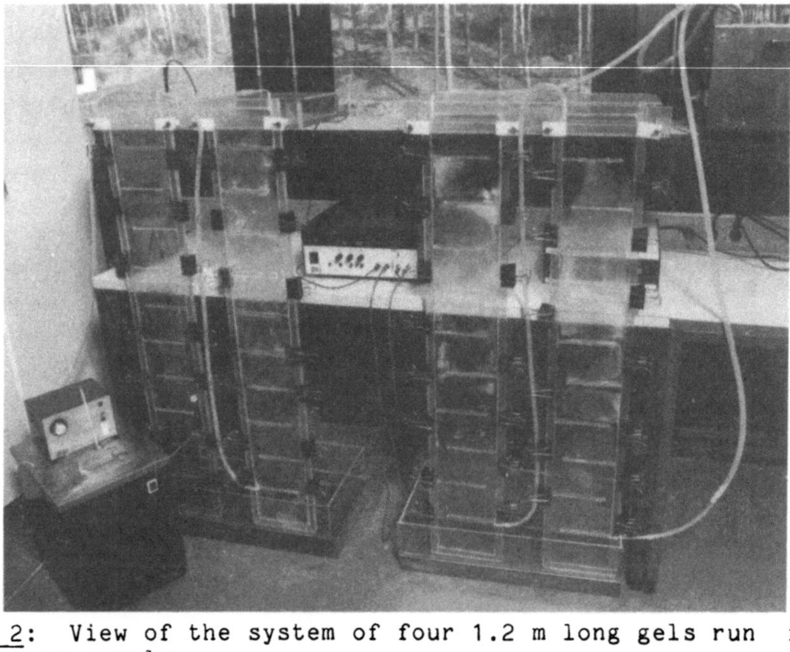

Fig. 2: View of the system of four 1.2 m long gels run from one power supply.

Results and Discussion

Ionic strength of the gel buffer. The range tested was 0.01-0.5 M TBE. On 0.2 mm thick gels, lowering the ionic strength below 0.1 M resulted in decreased resolution. On gels 0.1 mm thick, the resolution did not seem to be decreased at 0.03 M. The band resolution was improved when the ionic strength was increased to 0.2 M. With the thermally controlled 0.1-0.2 mm gels this increase in molarity is practicable, but this may not be the case for the thicker (0.4 mm) gels due to the larger currents and resulting Joule's heat.

pH of buffer. The pH was varied between 5.0 and 10.0. Better resolution, compared with the standard pH 8.3, was obtained at pH 8.8 (measured at room temperature). In more alkaline buffers (around pH 10) it seems that some loosening of the gel bound to the glass supporting plate is taking place.

The effect of other parameters on resolution will be discussed in (13).

System of 1 m long gels. As described under "electrophoresis" above and shown in fig. 4, buffer chambers for two 20 cm wide glass plates were designed. In this way four gels can be run in parallel from a power supply with two inputs. Several thermostating plates can be connected in series to a circulating water bath. It is important to change regularly the water in the bath (if possible for each run and use preferentially double distilled water), since several weeks or even days old water may increase the leakage current to ground through the thermostating plate and the water bath and cause a trip out of the power supply. All samples are applied at the same time, sample re-application is not needed. The time required for setting up and operating the sequencing system is two to three hours per day. Any gel transition effects (which may occur with increasing running time) around the sample slots do not affect the resolution of the bands in the lower part of the gel.

With good DNA samples we could read about 500 bases, with exceptionally good samples up to 650 bases. At present this number seems to be near the limit of the resolving power of the polyacrylamide gel.

Acknowledgment

We thank Kai Simons for support throughout this work and Henrik Garoff and Sune Kvist for advice and discussions of the procedure for preparation of DNA samples for sequencing.

References

1. Sanger, F. and Coulson, A.R. (1975) J. Mol. Biol. 94, 441-448.

2. Maxam, A.M. and Gilbert, W. (1977) Proc. Nat. Acad. Sci. USA 74, 560-564.

3. Sanger, F. and Coulson, A.R. (1978) FEBS Lett. 87, 107-110.

4. Ansorge, W. and De Maeyer, L. (1980) J. Chromatogr. 202, 45-53.

5. Ansorge, W. (1980) in Proceedings of the electrophoresis forum (Radola, B.J., ed.). pp. 67-74. Technical University, Munich.

6. Garoff, H. and Ansorge, W. (1981) Anal. Biochem. 115, 450-457.

7. Ansorge, W. and Garof, H., Electrophoresis '81. Editors: Allen, Arnaud 1981, Walter de Gruyter & Co., pp. 635-645.

8. Air, G.M., Sanger, F. and Coulson, A.R. (1976) J. Mol. Biol. 108, 519-533.

9. Wyckhoff, A., Rodbard, A. and Chrambach, A. (1977) Anal. Biochem. 78, 459-482.

10. Ansorge, W. Report about protein separation and visualisation in the Proceedings of the International Conference on Electrophoresis, Athens, April 1982.This volume pp 235-242.

11. Tanaka, T. Physics today, January 1981, p. 18-20.

12. Richards, E. G. and Lecanidou, R. (1971) Anal. Biochem. 40, p. 43.

13. Ansorge, W., to be submitted.

POLYMERIZATION KINETICS OF POLYACRYLAMIDE GELS: EFFECTS OF DIFFERENT CROSS-LINKERS, TEMPERATURE AND CATALYSTS.

Pier Giorgio Righetti, Cecilia Gelfi

Department of Biochemistry, University of Milano, Via Celoria 2, Milano 20133, Italy

Adriana Bianchi Bosisio

Clinical Chemistry Laboratory, S. Carlo Borromeo Hospital, Via S. Pio 5, Milano, Italy

Introduction

In view of the increasing importance of polyacrylamide gels in protein fractionation, we have started an extensive investigation on the porosity of hydrophilic gels (1), on their structure (2) and on their polymerization kinetics (3-5). Although it has been suggested that the discovery of un-supercoiled agarose "sounds the death knell for the use of cross-linked polyacrylamide in macromolecular separations" (6), we believe it might take ages before we can partake to the funeral of this matrix (will NIH provide the burial grounds?). In fact, polyacrylamide might last longer than any other matrix, especially in view of the revolutionary technique of "immobilized pH gradients" for isoelectric fractionations (7). We will summarize here the data obtained so far and indicate a new strategy for optimal polyacrylamide gel polymerization. Our aim has been to describe a set of conditions allowing at least 90-95% polymerization efficiency (i. e. conversion of free monomers into polymer chains). Lower conversions cannot be accepted, not only to ensure reproducible gel structure from run to run, but also to minimize the risk of large pools of unreacted monomers in the gel, capable of modifying, at appropriate pH, functional groups of proteins (such as $-NH_2$ termini, cysteinyl, hystidyl, tyrosyl and lysyl residues).

Electrophoresis '82

Materials and Methods

A convenient way to follow polymerization kinetics and efficiency is to monitor the disappearance of double bonds in ultraviolet (UV) light. Watkin and Miller (8) had proposed spectrophotometric monitoring at 200 nm, the absorption band of an amide unsaturated in α-β position, but a more comfortable way is to work in the neighborhood of 280 nm (9). In the case of opaque gels (highly cross-linked or polymerized below 20°C) a better estimate of reaction rate is obtained by measuring the scattered light (Tyndall effect) at 600 nm (3). In alternative to, or as a check of, spectrophotometric data, polymerization efficiency can be evaluated by chemical titration of unreacted double bonds *via* permanganate oxidation (10). All these methodologies have been described in detail in refs. (3-5) and (10).

Results and discussion

1) Effect of different cross-linkers

In addition to the original couple acrylamide-Bis (11) a host of new cross-linkers, which impart new, peculiar properties to polyacrylamide gels, have been described: DHEBA (12), DATD (13), EDA (14), PEGDA (1) and BAC (15). Their structural data have been listed in Table I. In addition to these cross-linkers, there exists a series of Bis-homologues (di-, tri-, tetra-, penta-, hexa-, hepta-, octa-, nona-, deca- and dodecano-) N,N'-methylene bisacrylamides, available from Polysciences (Warrington, Pa., USA) which, however, do not appear to have been used so far in polyacrylamide gels (at the light of present knowledge, we suspect they would impart strong hydrophobic properties to polyacrylamide matrices). Our interest in these cross-linkers was two-fold: a) to follow their different reactivity rates, if any; b) to follow their behaviour when used at progressively higher levels in a gel (highly cross-linked or high %C gels). It is known, in fact, that gels containing more than 10%C and up to 50%C become progressively more porous, thus being permeable to macromolecules in the multimillion molecular mass ($\bar{M}_r$) (16).

Fig. 1 shows the reaction kinetics of DHEBA cross-linked gels, in the 5% to 20%C range. Two phenomena can be observed from these curves: a) as %C increases, the lag period (the time before the onset of polymerization) lengt-

TABLE I

STRUCTURAL DATA AND CHEMICAL FORMULAE OF CROSS-LINKERS

Name	Abbr.	Chemical formula	M_r	Chain length
N,N'-methylene bisacrylamide	Bis (MBA)	$CH_2=CH-C(=O)-NH-CH_2-NH-C(=O)-CH=CH_2$	154	9
Ethylene diacrylate	EDA	$CH_2=CH-C(=O)-O-CH_2-CH_2-O-C(=O)-CH=CH_2$	170	10
N,N'-(1,2-Dihydroxyethylene) bisacrylamide	DHEBA	$CH_2=CH-C(=O)-NH-CH(OH)-CH(OH)-NH-C(=O)-CH=CH_2$	200	10
N,N'-Diallyltartardiamide	DATD	$CH_2=CH-CH_2-NH-C(=O)-CH(OH)-CH(OH)-C(=O)-NH-CH_2-CH=CH_2$	288	12
N,N',N''-Triallylcitric triamide	TACT	$CH_2=CH-CH_2-NH-C(=O)-CH_2-C(OH)(C(=O)NH-CH_2-CH=CH_2)-CH_2-C(=O)-NH-CH_2-CH=CH_2$	292	12—13
Poly(ethylene glycol) diacrylate 200	$PEGDA_{200}$	$CH_2=CH-C(=O)-O-CH_2CH_2OCH_2CH_2O-C(=O)-CH=CH_2$	214	13
N,N'-Bisacrylylcystamine	BAC	$CH_2=CH-C(=O)-NH-CH_2-CH_2-S-S-CH_2CH_2NHC(=O)CH=CH_2$	260	14
Poly(ethylene glycol) diacrylate 400	$PEGDA_{400}$	$CH_2=CH-C(=O)-O-CH_2CH_2OCH_2CH_2OCH_2CH_2-O-CH_2CH_2OCH_2CH_2OCH_2CH_2-O-C(=O)-CH=CH_2$	400	25

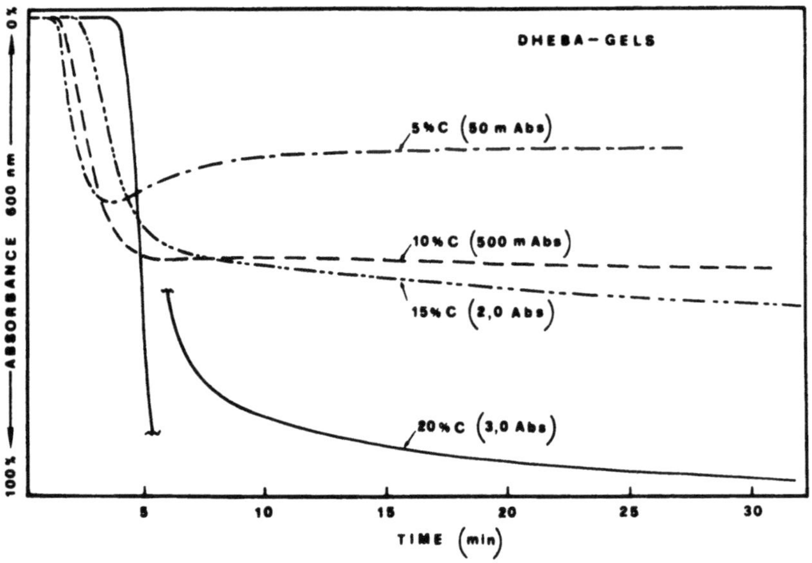

Fig. 1. Polymerization kinetics of DHEBA gels, from 5% to 20%C. The reaction was followed by measuring the Tyndall effect due to gel turbidity. Since we had to use a whole range of full-scale expansions, no absorbance scale is given in the ordinate; the direction of light absorption is expressed in percentile values. In the case of 20%C gels, the reaction was started with 2.0 A full-scale, and then, after 5 min, 1 A was added to compress the scale to 3 A, hence the broken tracing. In each curve, the full-scale expansion value is given in brackets (from Gelfi and Righetti, see ref. 3).

hens; b) as %C increases, the overall reaction rate becomes progressively slower. This appears to be a general phenomenon for all cross-linkers examined and seems to be due to increased formation of faulty topologies, or irregularities in gel structure, such as loops (chains connected with both ends to the same cross-linkers) and doublets (chains connecting the same pair of junctions). As %C is increased, loops and doublets become the most frequent topological elements, which eventually grow into beads or spheres. These highly packed regions of gel structural elements are responsible for the progressive pore size increase in high %C gels, and are probably so condensed that the free, unreacted monomers have difficulties in penetrating them, resulting in a much reduced reaction rate. Thus, when using high-cross link gels (above 15%C), polymerization should be allowed to proceed overnight for good conversion of monomers into polymer chains.

The order of reactivity of the various cross-links appears to be: Bis≈DHEBA >EDA≈BAC>>DATD. The last cross-link (DATD) was found to actually be an inhi-

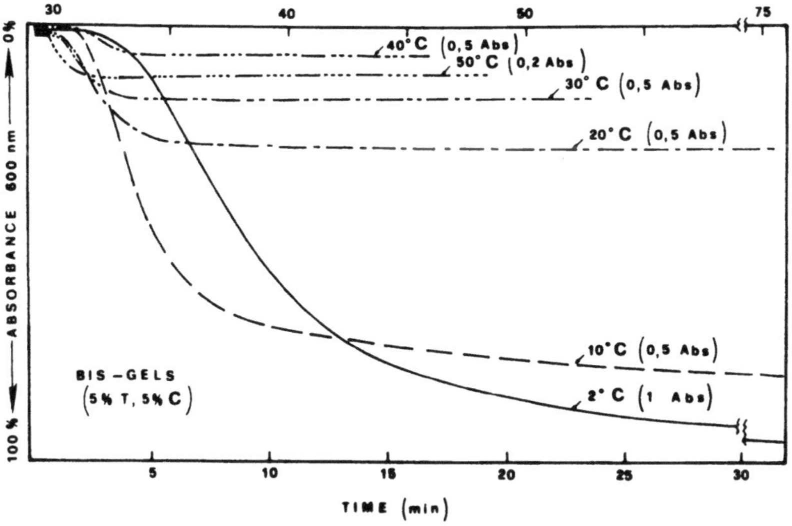

Fig. 2. Polymerization kinetics of Bis gels (5%T, 5%C) in the 2°-50°C temperature range. The upper time scale refers only to the 2°C gels; all other kinetics are measured in the lower time scale. Notice the strong opacity decrements of the same gels when the polymerization temperature is increased from 2° to 50°C (from Gelfi and Righetti, see ref. 4).

bitor of gel polymerization, leading to highly unpolymerized gels, especially at high %C values. The low reactivity of allyl compounds has been shown to be due to "degradative chain transfer" (17). The hydrogen atoms attached to the carbon in α to the double bond are responsible for this chain transfer: the resulting allyl radical is less reactive and/or has less tendency to reinitiate a new polymer chain because of an ability to stabilize itself by resonance and, therefore, the chain transfer is essentially a termination reaction. The resulting "pseudo-gel" is indeed a viscous glue full of unreacted pendant allyl groups (18).

2) Effect of temperature

At fixed %C, when the temperature is increased from 2°C up to 50°C, three phenomena are evident (see Fig. 2): a) the **process** of incorporation of monomers is markedly accellerated; b) the gel turbidity is strongly decreased and c) the lag time before the onset of polymerization is highly diminished. The opacity decrements follow the same pattern with all cross-links produ-

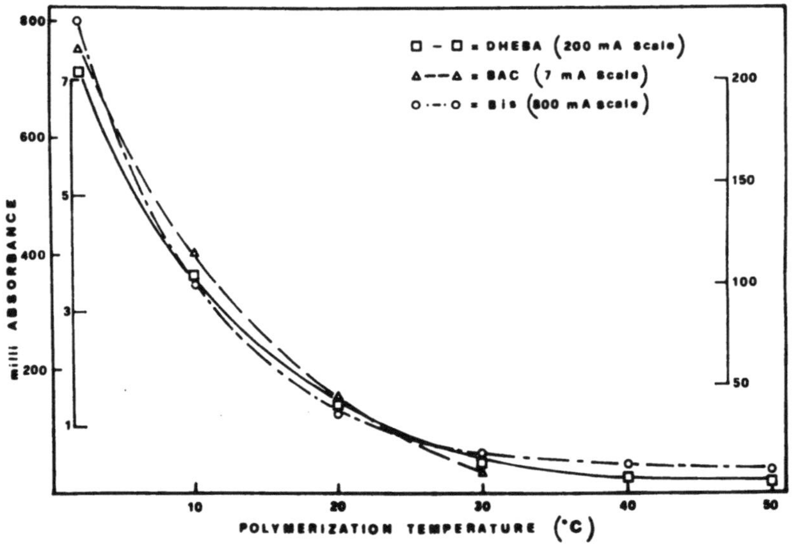

Fig. 3. Plot of turbidity decrements *vs.* polymerization temperature for Bis, DHEBA and BAC gels. The data have been taken from polymerization kinetics such as the ones shown in Fig. 2 (from Gelfi and Righetti, see ref. 4).

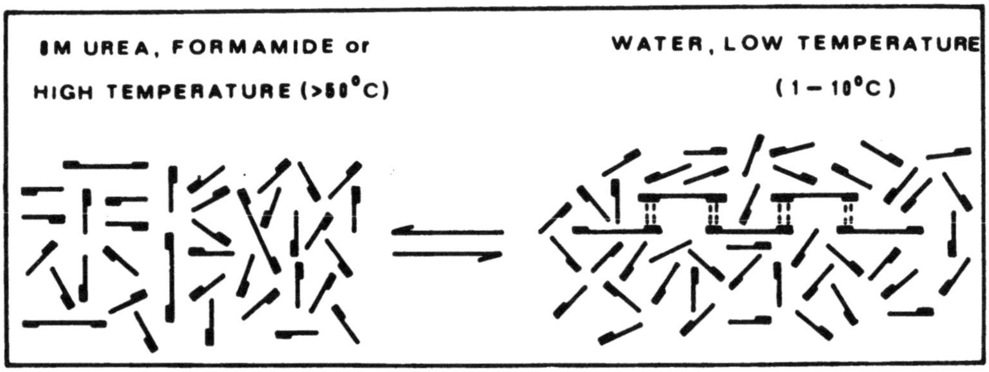

Fig. 4. Model on the distribution in solution of acrylamide (▬—) and Bis (▬—▬) molecules as a function of different variables. It is assumed that Bis molecules in water, at 1°C, are extensively H-bonded (four H-bonds /molecule) (right) while acrylamide molecules are distributed at random. A completely homogeneous (random) distribution for both comonomers is only obtained in 8M urea, formamide or at high (>50°C) temperature (left) (from Gelfi and Righetti, see ref. 4).

cing turbid gels (Bis, DHEBA and BAC) (see Fig. 3) suggesting a common mechanism of behaviour: practically in the temperature range 40-60°C these three gel types are fully transparent, indicating formation of homogeneous gels (in which singlet topologies are predominant) (3, 4). The interpretation: at low temperatures, water is not any longer an ideal solvent for Bis, DHEBA

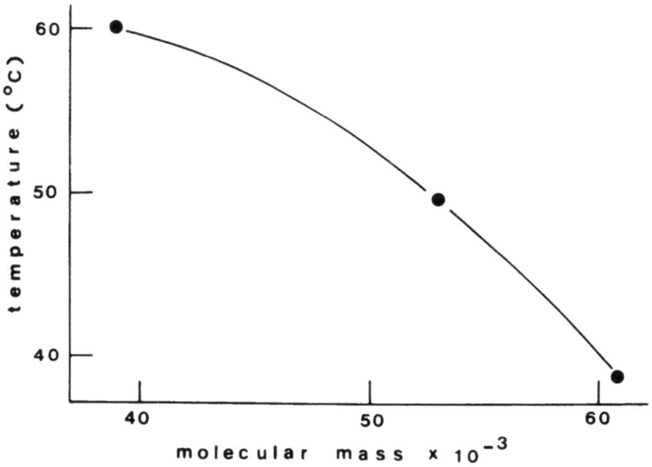

Fig. 5. Plot of molecular mass decrements of polyacrylamide chains *vs.* temperature increments. Polymerization conditions: solvent: dimethyl formamide; catalyst: perchloric acid, in the absence of cross-link. Molecular mass measurements by viscosity average data.

and BAC: these cross-links, instead of being randomly distributed in the liquid phase, form H-bonds which lead to clustering (Fig. 4, right side). At high temperatures or in Θ solvents (8M urea, formamide) the H-bonds are broken and the cross-links are randomly dispersed in solution (Fig. 4, left). We suggest 50°C as the ideal temperature for gel polymerization: this leads to gels with homogeneous distribution of cross-links, and to very high conversion rates (better than 95%). This temperature seems to be also ideal in the production of immobilized pH gradients: Immobiline chemicals, which have highly diverging polymerization rates below 20°C, exhibit highly converging kinetics at 50°C, and more uniform incorporation efficiencies (∿90%) (Ek, K. and Righetti, P.G., unpublished). This leads to production of highly predictable grafted pH gradients.

Higher temperatures could decrease the chain length of polyacrylamide by enhancing the dissociation of persulphate to radicals (19, 20), which would increase the concentration of chain initiator and thus the frequency of chain initiations (in general, at steady state, the concentrations of free radicals in homogeneous systems rarely exceed 10^{-9} to 10^{-8}M) (21). From data tabulated by Bhadani *et al.* (22) we have been able to demonstrate that this is indeed the case (see Fig. 5): under their particular polymerization conditions (dimethyl formamide as solvent and $HClO_4$ as catalyst) polyacrylamide with $\overline{M}_r$ = 61,000 was produced at 40°C, while 39,000 $\overline{M}_r$ species were generated at 60°C. This needs not be a deterrent, however: given an energy of activation for this reaction of the order of 15-20 kcal/mole (22) (quite high, in-

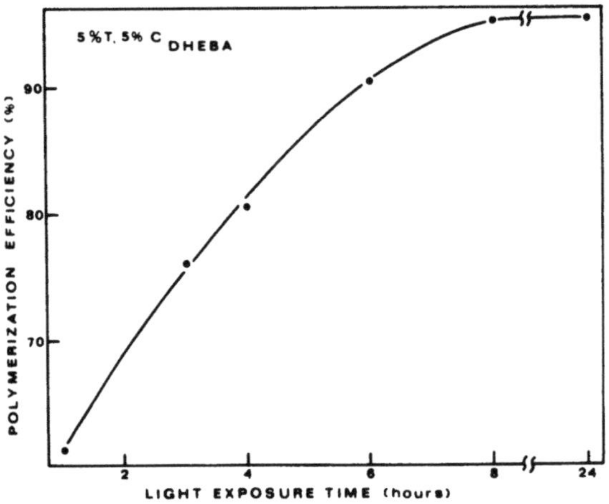

Fig. 6. Polymerization kinetics of a 5%T, 5%C DHEBA gel as a function of light exposure (two six W daylight tubes) time, followed by direct titration of unreacted double bonds in the gel with perganganate. Note that at least 8 hours exposure are required to achieve 95% polymerization efficiency (from Righetti *et al.*, see ref. 5).

deed) and a decomposition half life for persulphate of 130 hours/50°C (21) (extremely high, as compared with any other initiator) logic suggests that acrylamide should be polymerized at least at 50°C and that polymerization at 2-4°C, as suggested in the paleo-hystory of biochemistry, should be completely abandoned.

3) Effect of catalysts

While the couple ammonium (or potassium) persulphate - TEMED (N,N,N',N'-tetramethyl ethylen diamine) has been extensively characterized, other authors have suggested photopolymerization with riboflavin (or riboflavin-5'-phosphate) - TEMED (23, 24). We had noticed that photopolymerized gels had always a gluey and soft appearance, and thus investigated the efficiency of this process. As shown in Fig. 6, conditions generally accepted for photopolymerization (1 hour light exposure) barely affect 60% conversion of monomers into the polymer. 8 hours of illumination are required to obtain 95% polymerization efficiency and even so, the gels have still completely different visco-elastic properties than chemically polymerized gels (5). Even in the production of immobilized pH gradients, photopolymerization has disastrous effects on the separation process (Gianazza, E. and Righetti, P.G., unpublished). Thus photopolymerization should only be used with great caution or simply avoided in favour of highly controlled chemical polymerization.

Conclusions

Are the church bells really tolling the death of polyacrylamide gels (6)? Fear not, gents: properly polymerized gels, with the right cross-linkers, right temperature and right catalysts, have a long, long life expectancy.

References

1. Righetti, P.G., Brost, B.C.W., Snyder, R.S.: J. Biochem. Biophys. Methods 4, 347-363 (1981)
2. Righetti, P.G.: In Electrophoresis '81, Allen, R.C., Arnaud, P., eds., de Gruyter, Berlin, pp. 3-16 (1981)
3. Gelfi, C., Righetti, P.G.: Electrophoresis 2, 213-219 (1981)
4. Gelfi, C., Righetti, P.G.: Electrophoresis 2, 220-228 (1981)
5. Righetti, P.G., Gelfi, C., Bianchi Bosisio, A.: Electrophoresis 2, 291-295 (1981)
6. Buzàs, Z., Chrambach, A.: Electrophoresis 3, 130-134 (1982)
7. Bjellqvist, B., Ek, K., Righetti, P.G., Gianazza, E., Görg, A., Postel, W., Westermeier, R.: J. Biochem. Biophys. Methods 6 (1982) in press
8. Watkin, J.E., Miller, R.A.: Anal. Biochem. 34, 424-430 (1970)
9. Pegon, Y., Quincy, Cl.: J. Chromatogr. 100, 11-18 (1974)
10. Bianchi Bosisio, A., Loeherlein, C., Snyder, R.S., Righetti, P.G.: J. Chromatogr. 189, 317-330 (1980)
11. Raymond, S., Weintraub, L.: Science 130, 711-713
12? O'Connell, P.B.H., Brady, C.J.: Anal. Biochem. 76, 63-70 (1976)
13. Anker, H.S.: FEBS Letters 7, 293-296 (1970)
14. Paus, P.N.: Anal. Biochem. 42, 372-376 (1971)
15. Hansen, J.N., Pheiffer, B.H., Boehnert, J.A.:Anal. Biochem.105,192-201 (1980)
16. Chrambach, A., Jovin, T.M., Svendsen, P.J., Rodbard, D.: In Methods of Protein Separation, vol.II, Catsimpoolas, N., ed., Plenum Press, New York, pp. 27-144 (1976)
17. Bartlett, P.D., Altschul, R.: J. Amer. Chem. Soc. 67, 812-816 (1945)
18. Matsumoto, A., Oiwa, M.: J. Polym. Sci. Lett. 18, 421-425 (1980)
19. Riggs, J.P., Rodriguez, F.: J. Polym. Sci., A-1, 5, 3151-3165 (1967)
20. Gressel, J., Rosner, A.; Cohen, N.: Anal. Biochem. 69, 83-91 (1975)
21. Cooper, W., Pope, G.: in Polymerization Process, vol. 29 of the series High Polymers, Schildknrcht, C.E., Skeist, I.,eds., Wiley-Interscience,

New York, pp. 1-35 (1977)

22. Bhadani, S.N., Prasad, Y.K., Swapnakundu, S.: J. Polym. Sci. 18, 1459-1469 (1980)

23. Oster, G.K., Oster, G., Prati, G.: J. Amer. Chem. Soc. 79, 595-598 (1957)

24. Brackenridge, C.J., Bachelard, H.S.: J. Chromatogr. 41, 242-249 (1969)

ISOELECTRIC FOCUSING OF IMMUNOGLOBULINS

Ingmar Olsson and Torgny Låås
Pharmacia Fine Chemicals
Uppsala, Sweden

Introduction

Normal, healthy individuals produce large numbers, probably millions, of different antibodies. The majority of antibodies in the human belong to the immunoglobulin G class (IgG) and can be found in different body fluids, especially serum and cerebrospinal fluid (CSF). The different IgG species are characterized by differences in the amino acid composition in the variable part of the molecule giving each IgG species its particular specificity. In most cases these amino acid differences will affect the charge properties of the IgG and the different species may be considered as charge isomers and they should be separable by techniques such as ion- exchange chromatography, electrophoresis and isoelectric focusing (IEF). Of particular interest in this case is polyacrylamide gel isoelectric focusing (PAGIF) because of its extremely high resolving power.

As there is an extremely large number of IgG charge isomers in a normal serum, where each one is present in only a very low concentration, it would be expected that the pattern seen on IEF would be a continuous distribution over a certain pH range. What is obtained, however, is a number of protein bands covering most of the alkaline region. The origin of these bands has not been fully elucidated. Conductivity discontinuities, discontinuities in the pH gradient of protein-carrier ampholyte complexes have been suggested as possible

Electrophoresis '82

reasons (1, 2).

Besides beeing a challenge in itself, this situation is particularly unsatisfactory since much valuable diagnostic information should be available from the IgG pattern, provided it can be interpreted without ambiguity.

For example, some myeloma patients express IgG in the form of a series of a few charge isomers originating from one clone of cells, "monoclonal bands". These charge isomers may be caused by different amounts of sialic acid or deamidations (3).

Other myeloma patients as well as patients with inflammatory diseases of the nervous system may produce more complicated IgG pattern due to the presence of oligoclonal immunoglobulins. As an example, multiple sclerosis patients express oligoclonal IgG in their CSF (4).

Whereas the interpretation of a few distinct monoclonal bands offers no problem, the interpretation of oligoclonal IgG bands may be seriously complicated by the presence of IgG bands also in the normal sample (2).

Also for optimal resolution the pH gradient should cover exactly the whole IgG isoelectric spectra. This is not the case with conventional procedures now used for PAGIF.

Realizing the potential importance of PAGIF for the analysis of immunoglobulins, we therefore set out to study the IgG pattern obtained with standard PAGIF procedures with the hope that a deeper understanding of the physico-chemical background would enable us to develop an improved procedure for PAGIF of human IgG. Minimizing the problems discussed above.

Materials and methods

Equipment

IEF was performed on a Flat Bed Apparatus FBE 3000 using an electrophoresis Constant Power Supply ECPS 3000 equipped with a Volthour Integrator VH-1, both from Pharmacia Fine Chemicals AB.

Gels

Thin layer polyacrylamide gels of dimensions 1x100x200 mm were cast according to Pharmalyte "Instruction for use" (Pharmacia Fine Chemicals AB). The following carrier ampholytes were used: Pharmalyte^{TM}3-10, Ampholine^{TM} 3.5-9.5 (a mixture of five intervals as recommended by LKB, Bromma, Sweden), Servalyt^{TM} 3-10 (analytical grade from Serva, Heidelberg, BRD) and a mixture of Pharmalyte intervals (0.8 ml Pharmalyte 3-10, 0.6 ml Pharmalyte 6.5-9, 0.5 ml Pharmalyte 4-6.5 and 75 mg arginine per 30 ml).

Samples

The following samples were used: Rabbit anti-human albumin and rabbit anti-human C3 complement factor (Pharmacia Fine Chemicals AB). Human serum from normal individuals and myeloma patients were kind gifts from Mats Inganäs, BioCell Laboratories, Uppsala.

Running conditions

In all runs 1M sodium hydroxide and 0.04 M aspartic acid were used as catholyte and anolyte respectively. The gels were pre-run at 15 watts for 400 volthours. The samples were then applied 1.5 cm from the anode and were focused at 30 watts for a total of 3800 volthours. The gels were fixed in 10 % trichloroacetic acid and stained in Coomassie Blue R250.

Methods of measurements
The pH-gradients were measured with the broad pI calibration kit (Pharmacia Fine Chemicals AB) and also with an Ingold surface electrode connected to a Beckman pH meter.

The field strength was measured with a volt-meter using a measuring probe made from two stainless steel rods 0.25 mm thick, fixed 0.5 mm from each other.

Results and Discussion

To see if there was any connection between field strength distribution and protein pattern of focused IgG, normal human serum as well as pure IgG from man and rabbit were focused under the same conditions in Pharmalyte, Ampholine and Servalyt. Protein patterns, field strength distributions and pH-gradients were compared. (Fig. 1).

The banding pattern obtained from all carrier ampholytes was superficially similar, and in perticular there was a concentration of bands in the neutral region. The even distribution of fine bands is interrupted of a few heavy bands surrounded by regions without protein. For each type of carrier ampholyte the pH of these blank regions was always the same, irrespective of what sample was used, be it man or rabbit.

On the other hand, when comparing the different carrier ampholytes it was found that the pH of these blank regions was different: with Pharmalyte pH 6.5, Ampholine pH 6.3 and Servalyt pH 6.8 (Fig. 1). Thus, pattern seen cannot truly represent the IgG isoelectric spectra.

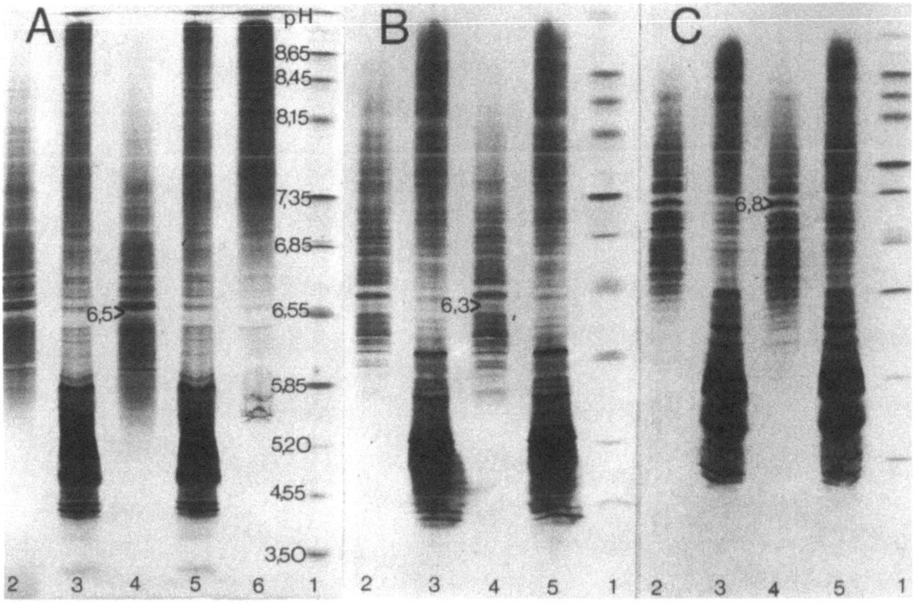

Fig. 1. Protein patterns after IEF in PAA-gel T5C3 in three different carrier ampholytes: A. Pharmalyte, B. Ampholine, C. Servalyt. Running conditions were 30 Watts, 2200 Volts and 3800 Volthours.
Samples: 1. pI kit 3-10 from Pharmacia Fine Chemicals AB, 2. Anti-albumin, IgG fraction from rabbits, 3. Human serum, 4. Anti C_3 complement factor IgG fraction from rabbits, 5. Human serum, 6. Human IgG fraction (Kabi). The pH at the most striking interruptions, in the even distributions of fine bands, were for the different carrier ampholytes: A 6.5, B. 6.3, C. 6.8 and are marked with arrows in the figure.

Nor is the number of bands obtained enough to be a reflection of all the different IgG species present in a healthy individual. Evidently, groups of IgG molecules with close pI values are concentrated into fine bands.

Figure 2.

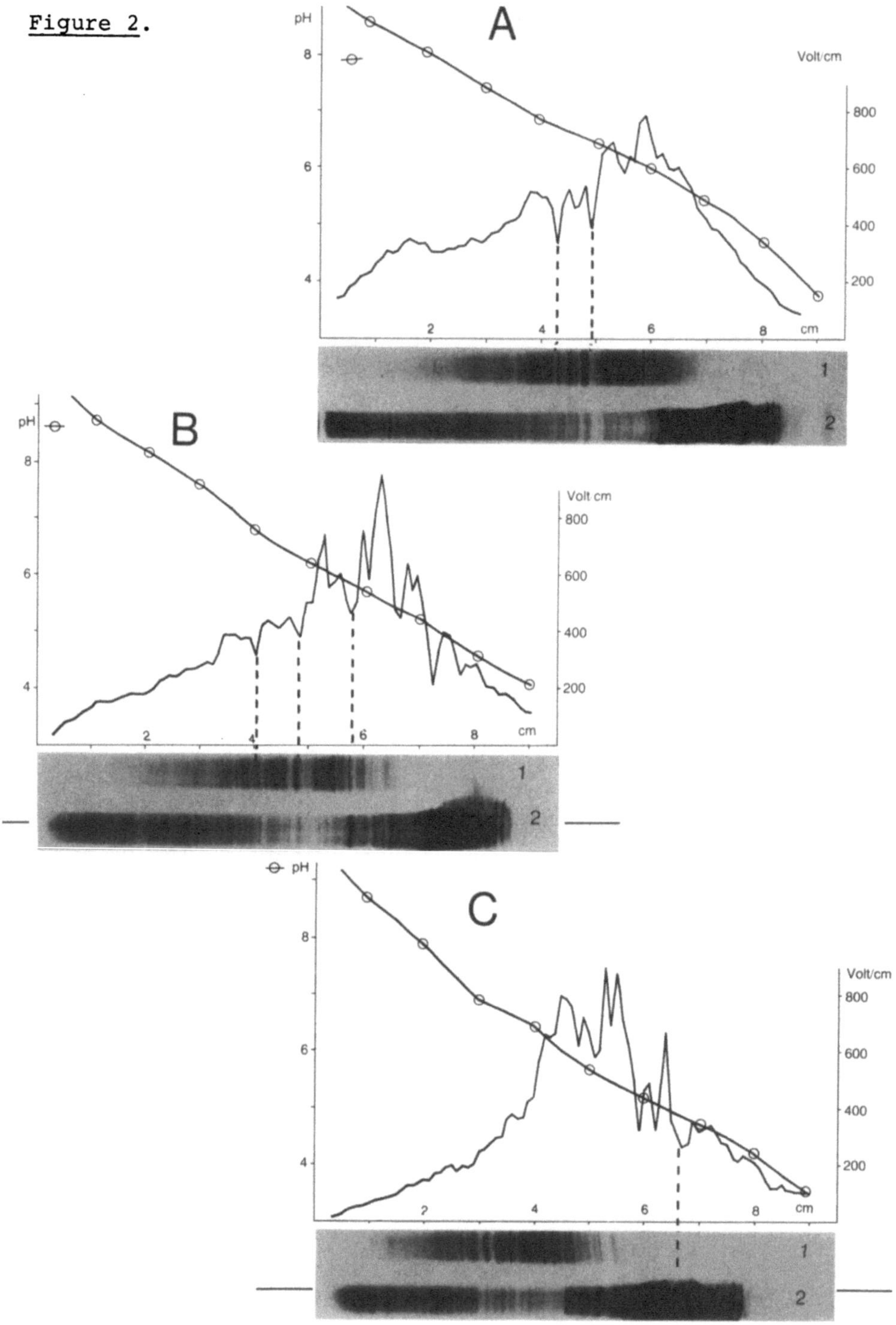

<u>Fig. 2</u>. Field strength distributions and pH-gradients compared with the IEF protein patterns of 1. Rabbit IgG and 2. Human serum, in three different carrier ampholytes: A. Pharmalyte, B. Ampholine, C. Servalyt. Positions where the "holes", in the even distribution of fine bands coincide with steep changes in the field strength are marked with dotted lines.

In case there exists a connection between discontinuities in field strength and apparent IgG banding pattern, which has been suggested by Felgenhauer and Mormann (2), we measured the voltage drop at 0.5 mm intervals across the pH gradient. It is probable that this distance is short enough to detect even very sharp discontinuities in the field strength distributions.

However, no relationship between the even distribution of fine bands and field strength distribution was found. On the other hand, the observed blank regions frequently, but not always coincided with minimas in the field strength (Fig. 2).

Since good carrier ampholyte molecules by definition focuses well and have high (compared to "bad" carrier ampholytes) conductances at their pI, a sharp field strength minima would be the expected result from one "good carrier ampholyte" present in amounts enough to dominate its area.

Supporting evidence for this hypothesis was found when we added amino acids as "bad ampholytes" in order to improve IgG separation by "bridging conductance gaps", which was suggested by Chrambach et al. (5). The amino acid mixture contained as one component histidine to 0.01 M. Histidine is actually a rather good carrier ampholyte with a pI of 7.47. After running such gels, the carrier ampholytes were precipitated with 1 % picric acid or Amido Black. On identical gels run in parallell the proteins were stained in the normal way. It appeared that histidine displaced all other carrier ampholytes as well as proteins from an area, 1 cm wide at pH 7.45 (Fig. 3).

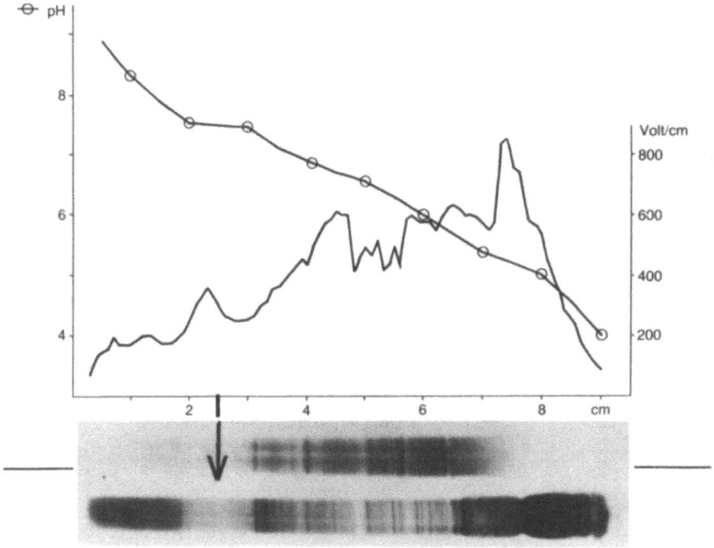

Fig. 3. Protein patterns, field strength distribution and pH-gradient after focusing in Pharmalyte 3-10 supplemented with 0.01 M concentrations of glycine, icoleucine, histidine, phenylalanine, proline, serine and valine (5).
Samples: 1. Rabbit IgG 2. Human serum. Running conditions: 30 Watts, 2200 Volts, 3800 Volthours. By the addition of amino acids new interruptions are created, in the previously "even" distributions of protein bands. The pH plateau and accompanying displacement of proteins at pH 7.45 (see arrow) is created by histidine with its relatively high buffering capacity.

Since no compulsory connections were found between the blank spaces and field strength discontinuities we conclude that both these phenomena are consequences of some other, more fundamental function that is not always expressed in the field strength distribution.

We postulate that this is the presence of a high concentration of one particular carrier ampholyte which has a high buffering capacity. This carrier ampholyte displaces other amphoteric molecules and creates a plateau in the pH gradient. This is sometimes, but not always accompanied by a measureable minima in field strength.

From these experiments it appears that it is important to have an even concentration of a large number of different carrier ampholytes so that no particular one can dominate an area.

This can be achieved by increasing the number of carrier ampholytes within a certain pH range. The relative effect of each individual will thus decrease.

Each Pharmalyte interval is synthesized separately and Pharmalyte 3-10 is a mixture of the 2.5-5, 5-8 and 8-10.5 intervals. Addition of Pharmalyte 4-6.5 and 6.5-9 to the 3-10 intervals would significantly contribute to a more balanced mixture of carrier ampholyte molecules. Furthermore, arginine was added to extend the alkaline end of the pH gradient.

This carrier ampholyte mixture was found to have the expected properties and was used to detect abnormal IgG patterns in human serum (Fig. 4).

Increasing the number of carrier ampholyte species decreased both the blank spaces previously seen in the even distribution of protein bands as well as the protein bands themselves (Fig. 4).

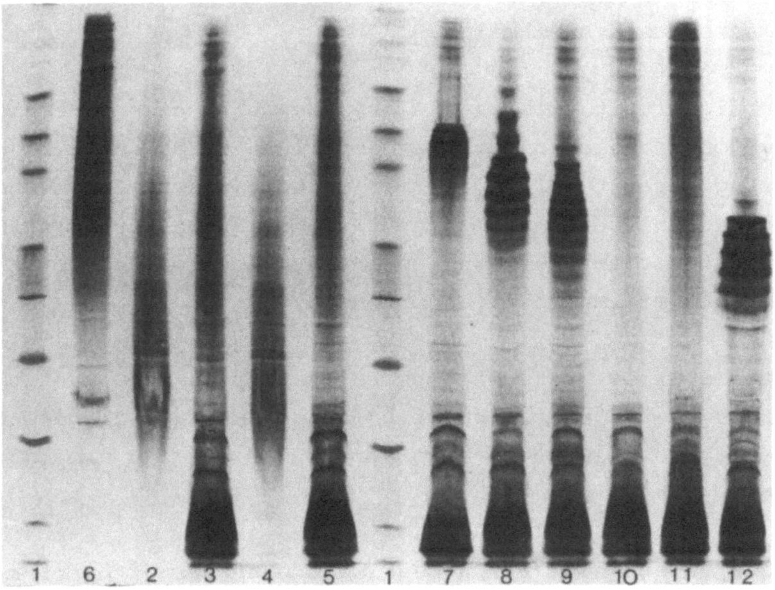

Fig. 4. IgG patterns obtained after PAGIF in the specially developed carrier ampholyte mixture of Pharmalyte 3-10, 6.5-9, 4-6.5 and arginine. Running conditions: 30 Watts, 2200 Volts and 3800 Volthours.
Samples 1-6 are the same as in Fig. 1, samples 7-12 are serum from myeloma patients.

Also the field strength distribution was smoothened out (Fig. 5). The ripples that are normally seen on the gel surface were also reduced.

We believe that the procedure for PAGIF of IgG suggested here will facilitate un unambiguous interpretation of IgG isoelectric spectra by minimizing the presence of technical artefacts in the form of bands not representating IgG individuals as well as protein empty areas. Also the pH gradient

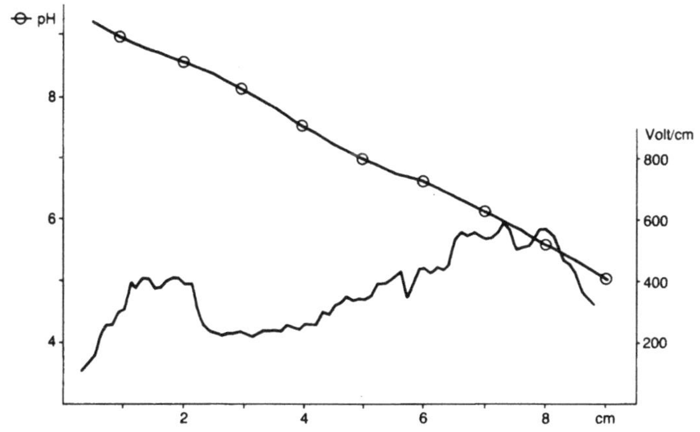

Fig. 5. Field strength and pH-gradient after focusing in a mixture of Pharmalyte 3-10, 6.5-9, 4-6.5 and arginine. Running conditions: 30 Watts, 2200 Volts and 3800 Volthours.

closer coincides with the normal IgG isoelectric distribution for increased resolution.

References

1. Felgenhauer, K., Park, S.J.: Ann. N.Y. Acad. Sci. 209, 147-153 (1973).
2. Felgenhauer, K., Mohrmann, H., in Allen, R.C. and Arnaud, P. (Editors): Electrophoresis '81, Walter de Gruyter & Co., Berlin · New York, 427-432 (1981).
3. Bouman, H., Meincke, G., Havsteen, B.: Z. Immun.-Forsch. Bd. 150, 370-377 (1975).
4. Mehta, P.D., Patrick, B.A., Wisniewski, H.M.: J. Clin. Lab. Immunol. 6, 17-22 (1981).
5. Chrambach, A., An der Lan, B., Mohrmann, H., Felgenhauer, K.: Electrophoresis 2, 5-6, 279-287 (1981).

SYNTHETIC AMIDE-TYPE OLIGOMERS AS POTENTIAL CARRIER AMPHOLYTES

C.M. Paleos, Department of Chemistry, Nuclear Research Center "Demokritos", Athens, Greece

A. Vellios and D. Stathakos, Laboratory for Enzyme Research, Department of Biology, Nuclear Research Center "Demokritos", Athens, Greece

Summary

In this work the synthesis of a novel class of amide-type oligomeric ampholytes is reported. The method of their synthesis involves the interaction of oligoamines with cyclid anhydrides and specifically in this study, with succinic anhydride. The resolving capacity of these ampholytes was evaluated in connection with their mode of synthesis using a mixture of partially oxidized human hemoglobin and methyl red.

Introduction

The development of a facile method for the synthesis of ampholytes for use in isoelectric focusing (1,2) is of great interest. In designing the synthesis of a new class of ampholytes, one should consider that these compounds should have the following properties (1a,3,4): a) Good buffering capacity and conductivity, b) low molecular weight, c) high solubility in water at the isoelectric point, d) low light absorption above 260 nm, e) different chemical properties from the proteins to be separated. Substances generally fulfilling the properties outlined above were initially prepared by Vesterberg (4). Their synthesis was accomplished by reacting acrylic acid with certain oligoamines. A similar experimental procedure was

Electrophoresis '82

adopted for the synthesis of the same type of ampholytes by Vinogradov (5) and Righetti et al. (6). In the present work the previously employed oligoamines or similar amines were allowed to interact with succinic anhydride for the formation of amide-type oligomeric ampholytes of predictable molecular weight. In addition some data on the evaluation of these ampholytes for isoelectric focusing are also reported.

Experimental

Materials: Tetraethylenepentamine (DOW) was vacuum distilled whereas N,N'-bis-(3-aminopropyl)ethylene diamine and a mixture of hexamines, consisting primarily of 4.8.11.15 tetrazaoctadecane 1.18 diamine (BASF) were used as purchased. Acrylamide (Eastman Kodak Co.) was recrystallized from chloroform and bisacrylamide (Eastman Kodak Co.) from acetone (7).N,N,N',N'tetramethylethylenediamine, also from Kodak, was used as obtained. Ampholines covering the pH area 3.5-10 were purchased from LKB Produkter AB, Sweden. Partially oxidized human hemoglobin was obtained by scaling up the analytical data of Bunn and Drysdale (8).

Ampholyte Synthesis: A typical reaction for the synthesis of amide-type ampholytes is illustrated by the interaction of succinic anhydride with tetraethylenepentamine as shown in the scheme below.

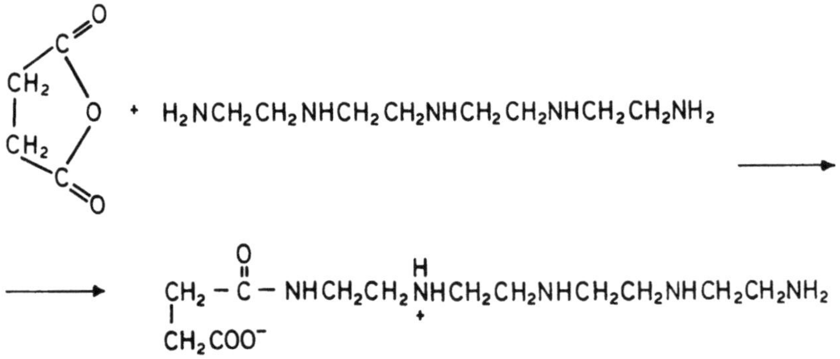

The experimental procedure in which the above reactants are involved is described below. The molarities of reactants employed may be modified as will be further discussed.

0.01 mole (1.00 g) of succinic anhydride, dissolved in dry dioxane was added slowly to 0.004 mole (1.13 g) of tetraethylenepentamine which was also dissolved in dioxane. The ampholyte precipitated instantaneously and dioxane was distilled under reduced pressure. To the remaining hydrophilic material the required quantity of water was added in order to obtain a 40% ($\frac{w}{w}$) solution. Traces of dioxane remaining in this solution were removed by flushing it with nitrogen.

Acrylamide gel isoelectric focusing: The gels which contained 5% ($\frac{w}{v}$) acrylamide, 1% ($\frac{w}{v}$) cross-linking agent, 2% ($\frac{w}{v}$) of ampholytes and 5% ($\frac{v}{v}$) glycerol were polymerized with the addition of 1% ($\frac{w}{v}$) ammonium persulfate and 0.5% ($\frac{v}{v}$) of N,N,N',N'tetramethylethylenediamine in pyrex tubes of 10 cm x 0.3 cm. These gels were pre-run for 30 min before applying the samples in order to remove excess of persulfate. Electrofocusing was performed at approx. 2° C with a LKB 2103 Power Supply at a constant voltage of 500 V for a period of about five hours. The anolyte was 0.01 M H_3PO_4 and the catholyte 0.02 M NaOH (in the upper chamber). Samples, i.e. a mixture of partially oxidized human hemoglobin and methyl red in a 10% aqueous sucrose solution were layered on top of the gels. The final pH gradients which were established upon electrofocusing were determined by cutting the gel in 0.5 cm slices. These were placed in water for 30 min and subsequently the pH was measured.

Results and Discussion

The synthesis of amide-type ampholyte oligomers is based on a known reaction of cyclic anhydrides and amines. In the present case succinic anhydride is reacting with certain oligoamines in such a molar ratio that free amino groups are left intact.

In this way each carboxylic group which is formed may protonate a free amino group. It is therefore evident that for each anhydride molecule reacting two, at least, amino groups are required for the synthesis of these ampholytes. Our experience suggests that anhydride should be slowly added to the oligoamine solution. If, on the contrary, oligoamines are added to the anhydride solution, it might be possible that almost complete amidation of the amine could occur at the first stage, at least, of the reaction. What is therefore attempted is to minimize the number of amino groups available for reaction, in order to avoid obtaining fully protonated amines at the end of the experiment. In Table I the molar quantities of succinic anhydride and oligoamines used for the synthesis of some amide-type ampholytes are shown. The performance of these ampholytes, for isoelectric focusing, will presumably depend on the stoichiometry of the reactants.

TABLE I. Ampholyte samples employed in polyacrylamide gel electrofocusing including the molar quantities of reactants used for their synthesis

Sample No.	Succinic Anhydride (moles)	Tetraethylene-pentamine (moles)	N,N,(3-aminopropyl)ethylene diamine (moles)	Hexamines (moles)
1	0.010	0.006	-	
2	0.020	0.006	0.0025	
3	0.026	0.006	0.003	
4	0.015	0.008	-	
5	0.010	0.008	-	
6	0.021	0.008	0.003	
7	0.016	0.008	0.003	
8	0.018	0.008	0.003	
9	A mixture of equal volumes of samples Nos. 1-8			
10	0.010			0.004
11	0.010			0.004
12	0.010			0.003

The main advantage of these ampholytes is their facile synthesis. They are prepared instantaneously under the described mild experimental conditions. The ampholytes are extremely hygroscopic and they easily dissolve in water forming solutions of high conductivity. Their solutions are light yellow to yellow having a relatively low UV absorption around 280 nm. The shade of their color is intensified as the amount of succinic anhydride used for their synthesis is increased. A parallel series of carrier ampholytes, not described in this paper, was prepared in DMSO in which the reaction products are soluble at about 100° C during their synthesis. The synthesized ampholytes were also evaluated as far as their resolving capacity is concerned by using a mixture of partially oxidized human hemoglobin (8) and methyl red.

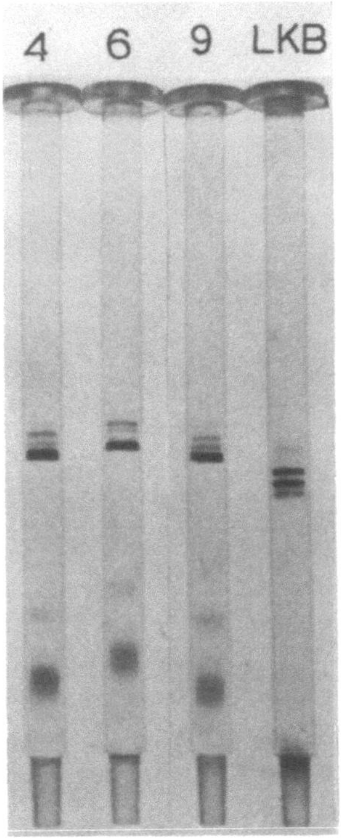

Fig. 1 Polyacrylamide gel isoelectric focusing patterns of partially oxidized human hemoglobin and methyl red (samples 4,6,9 of Table I and LKB - Ampholine 3,5-10).

The isoelectric focusing patterns of samples Nos 4,6 and 9 which showed relatively better performance are shown in Fig.1, compared to a commercial sample (LKB, pH 3.5-10). It seems that for an acceptable resolving capacity of the ampholytes the ratio of the amino groups to succinic anhydride should be about 2.5/1.

The pH gradients of ampholyte samples Nos 4,6 and 9, obtained from the polyacrylamide isoelectric focusing, are shown in Fig. 2. Those obtained with hexamines as reactants (samples Nos 10,11 and 12) are shown in Fig. 3. It is interesting to note that when the hexamines were employed in the reaction with succinic anhydride, linear pH gradients were obtained at high pHs (between 8 and 10). This may be due to the large number of isomers of these amines which produce a significant number

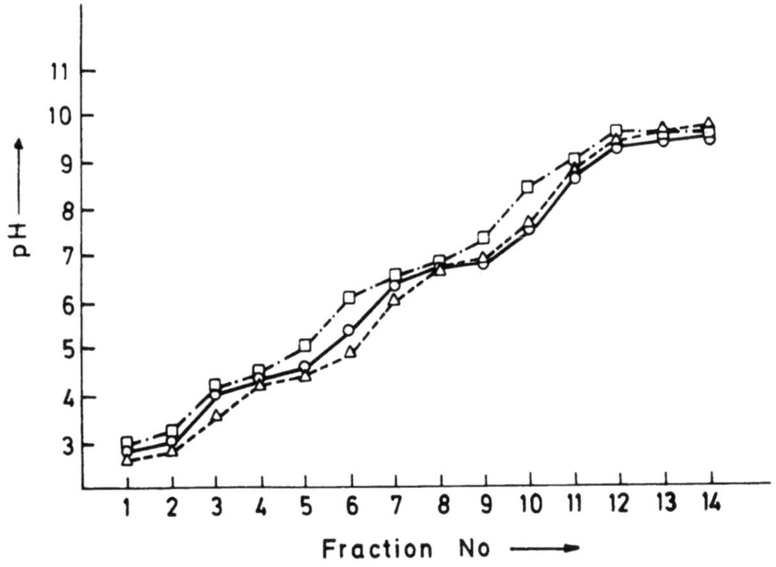

Fig. 2 The pH gradients of amide-type ampholytes; Concentration in gel 2%. o-o sample No 4, Δ-Δ sample No 6, □ — □ sample No 9 (see Table I).

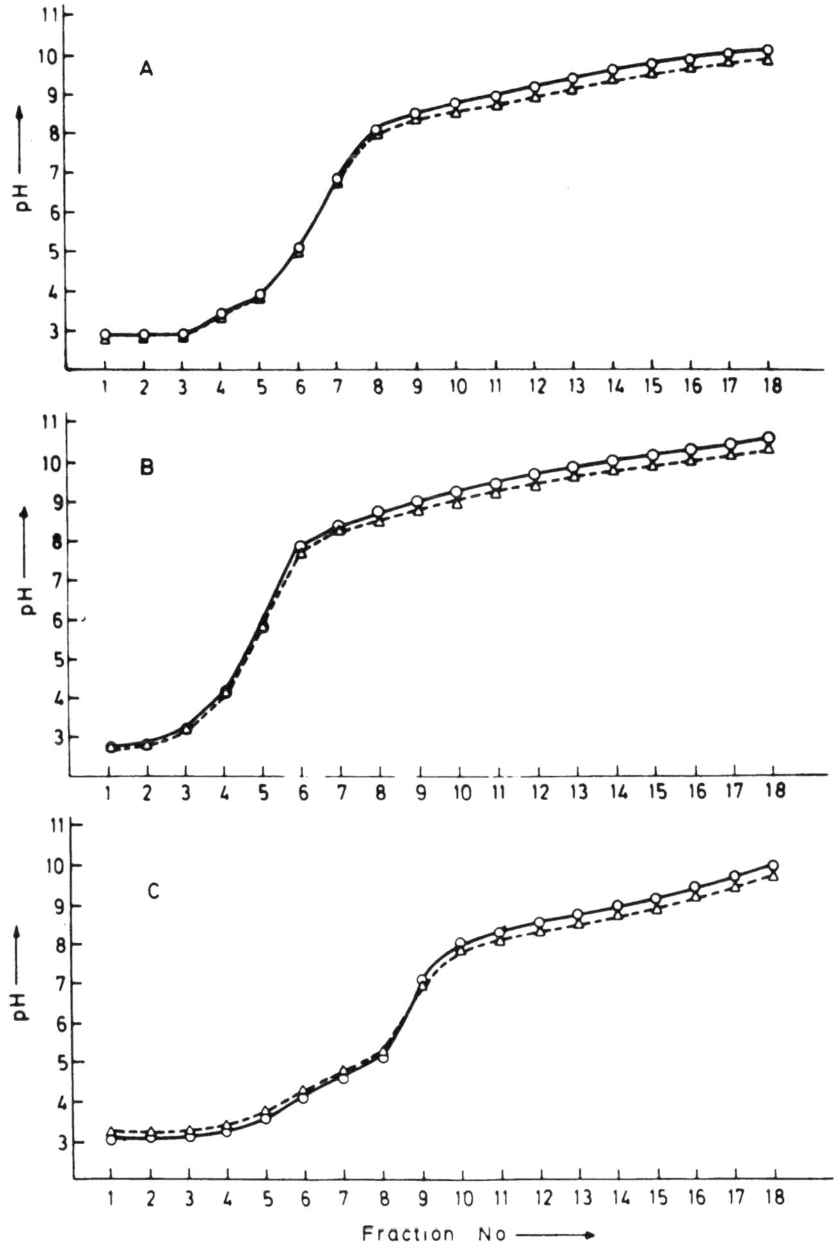

Fig. 3 pH gradients of amide-type ampholytes; concentration in gel 1%. o-o pH values after slicing; Δ-Δ pH values after 5 days at 4° C.
A: sample 10; B: sample 11; C: sample 12 (see Table I).

of amide-type ampholytes with closely spaced pIs (5) in this range. No hydrolysis was observed in the fractions up to 3 days at 4°. A certain acid shift of the pIs between pH 8 and 10 was observed after 5 days.
In conclusion, the oligomers tested, prepared by using a molar ratio of amino groups to succinic anhydride ranging from 2 to 3 behave as potentially useful carrier ampholytes. By further variation of the synthesis conditions, now in progress, and appropriate mixing of oligomers better pH gradients may be obtained.

Acknowledgement: We thank Mr. I. Lagouros for his skillful technical assistance.

References

1. a) Righetti, P.G., Drysdale, J.W.: J. Chromatogr. 98, 271-321 (1974).
   b) Allen, R.C.: J. Chromatogr. 146, 1-32 (1978).
   c) Carlionet, R., Martin, J.P., Sesboue, R., Madec, P.J., Lefebvre, F.: J. Chromatogr. 176, 89-101 (1979).
2. Righetti, P.G., Gianazza, E.: J. Chromatogr. 184, 415-456 (1980).
3. Svensson, H.: Acta Chem. Scand. 16, 456-466 (1964).
4. Vesterberg, O.: Acta Chem. Scand. 23, 2653-2666 (1969).
5. Vinogradov, S.N., Lowekron, S., Andonian, M.R., Bagshaw, J., Felgenhauer, K., Pak, S.J.: Biochem. Biophys. Res. Comm. 54, 501-506 (1973).
6. Righetti, P.G., Pagani, M., Gianazza, E.: J. Chromatogr. 109, 341-356 (1975).
7. Loening, V.E.: Biochem. J. 102, 251-257 (1967).
8. Bunn, H.F., Drysdale, J.W.: Biochem. Biophys. Acta 229, 51-57 (1971).

A NEW POLYACRYLAMIDE GEL FORMULATION ALLOWING OVEN DRYING OF HIGH PERCENTAGE SLAB GELS ON GELBOND® PAG

Samuel Nochumson and Sharon G. Gibson
FMC Marine Colloids Division, BioProducts Department, Rockland, ME 04841

Introduction

It is frequently necessary to dry down polyacrylamide slab gels to a thin film for use in autoradiography,[1] fluorography,[2] densitometry, or for a permanent notebook record. However, following high resolution electrophoresis, dry down of polyacrylamide slab gels can often result in cracking, shrinkage, and distorted bands. A commonly-used method to circumvent this problem has been to dry polyacrylamide slab gels under vacuum in the presence of a heat source.[1] Although acceptable results can be achieved using this methodology, it still has its limitations. Thus, variable results can be obtained due to factors such as the percentage of polyacrylamide and gel thickness. In an effort to make slab gel drying problem free, an agarose-based derivative, AcrylAide, has been developed as a replacement for the cross-linker in polyacrylamide gels. Slab gels prepared on GelBond® PAG using this new formulation can be oven dried directly to a thin film without cracking, shrinkage, or distorted bands. In addition, resolution comparable to conventional polyacrylamide/bisacrylamide gels can be obtained using this new formulation.

Materials and Methods

GelBond® PAG and AcrylAide are products of FMC Marine Colloids Division (Rockland, ME). All protein molecular weight markers,

Electrophoresis '82

acrylamide, N,N-methylene-bisacrylamide, and Trizma base were obtained from Sigma Chemical Company (St. Louis, MO). Sodium dodecyl sulfate was from BDH Chemicals (Poole, England) and glycine from BioRad Laboratories (Richmond, CA).

Discontinuous SDS electrophoresis was performed by the method of Laemmli[3] using the Studier chamber[4] manufactured by the Aquebogue Repair and Machine Shop (Aquebogue, NY). All polyacrylamide gels were cast at a thickness of 1.2 mm on the hydrophilic side of GelBond PAG (140 mm x 160 mm), which was included in the casting apparatus. A 2% AcrylAide solution was prepared in 2x separating gel buffer (1.5M Tris-HCl, pH 8.8, and 0.2% SDS). This solution was mixed with an equal volume of either a 30% or 40% acrylamide solution in order to form respective 15% acrylamide, 1% AcrylAide or 20% acrylamide, 1% AcrylAide gels. Polymerization was initiated with ammonium persulfate at 0.5 mg/mL and catalyzed with tetramethyl-ethylenediamine 0.25 µl/mL. After formation of the separating gel, a 5%T, 3%C stacking gel containing sample wells was cast in 0.125M Tris-HCl (pH 6.8) and 0.1% SDS. Protein molecular weight markers were prepared in sample buffer (0.0625M Tris-HCl, pH 6.8, 2% SDS, 10% glycerol, 5% 2-mercaptoethanol, and 0.001% bromophenol blue) at 1 mg/mL and placed in a boiling water bath for 1.5 minutes for complete dissociation of subunits. A 5 µL sample was applied to each sample well, and electrophoresis was performed at 25 mamp for 2.5 hours in the electrode buffer (0.025M Tris, 0.192M glycine, 0.1% SDS, pH 8.3).

Following electrophoresis, the gels, firmly attached to GelBond PAG, were removed from between the glass plates and stained overnight in 0.05% Coomassie Brilliant Blue R-250 in 25% isopropanol and 10% glacial acetic acid. The gels were destained in 45% methanol and 10% glacial acetic acid. In some cases, gels were silver stained by the method of Merril et al[5]. After the gels were completely destained, they were placed in a 7% acetic acid, 5% glycerol solution for at least two hours and then dried down to a film in a forced-air oven at 60°C.

Results and Discussion

We have developed an agarose-based derivative, AcrylAide, which, when substituted for methylene-bisacrylamide, will form a gel with polymerizing acrylamide solutions. This formulation is significantly different than composite gels of agarose and polyacrylamide, which have been prepared in order to blend certain qualities of each medium.[6] That is, the high gel strength of agarose at low concentrations is used to complement the weak gels formed by low percentage acrylamide/bisacrylamide solutions. AcrylAide, on the other hand, is used as a replacement for bisacrylamide with high percentage acrylamide solutions to form gels which can be conveniently oven dried to a thin film Thus, AcrylAide/polyacrylamide gels cast on GelBond PAG can easily be oven dried without cracking, shrinkage, or distortion of protein band patterns. This can be seen in Figs. 1 and 2 in which a 15% acrylamide/1% AcrylAide and 20% acrylamide/1% AcrylAide gels have been oven dried following discontinuous SDS electrophoresis. In contrast, a 15%T, 2.6%C conventional polyacrylamide gel exhibits severe cracking following electrophoresis and oven drying (Fig. 3).

Substitution of bisacrylamide with AcrylAide to form polyacrylamide gels does not change the molecular sieving properties appreciably from conventional gels as seen by the log MW versus R_f graph in Fig. 4. The results indicate that a 15% acrylamide/ 1% AcrylAide gel gives comparable results as a 15%T, 2.6%C acrylamide/bisacrylamide gel. In addition, AcrylAide has been used successfully with silver staining procedures[5] without any interference due to background staining. Also, unlike conventional agaroses which gel in solution, AcrylAide is a nongelling medium so that solutions will remain fluid when stored at room

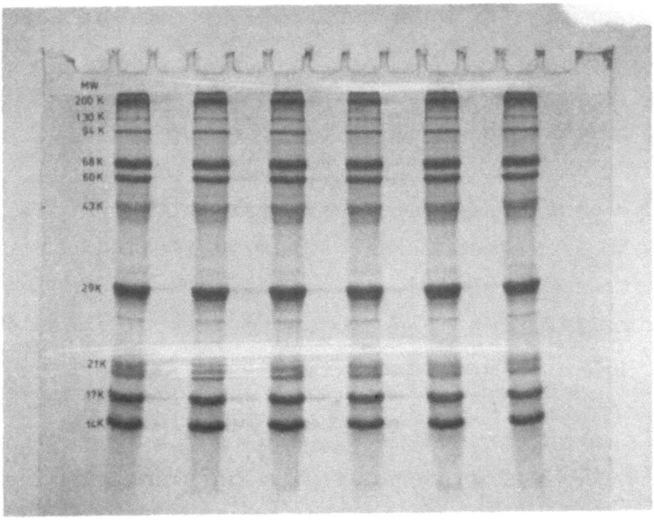

Fig. 1. Oven dried 15% polyacrylamide, 1% AcrylAide gel. Protein mixture contains myosin, β-galactosidase, phosphorylase b, bovine serum albumin, catalase, ovalbumin, carbonic anhydrase, soybean trypsin inhibitor, myoglobin, and lysozyme. (See Materials and Methods for running conditions.)

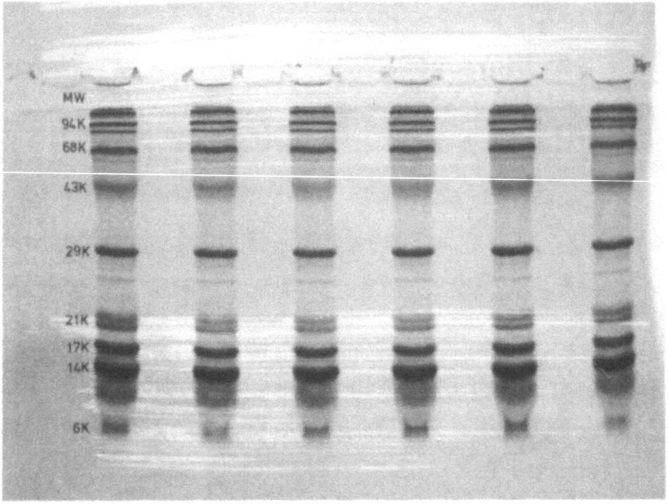

Fig. 2. Oven dried 20% polyacrylamide, 1% AcrylAide gel. Protein mixture is the same as in Fig. 1, plus insulin.

Fig. 3. Oven dried 15%T, 2.6%C polyacrylamide/bisacrylamide gel.

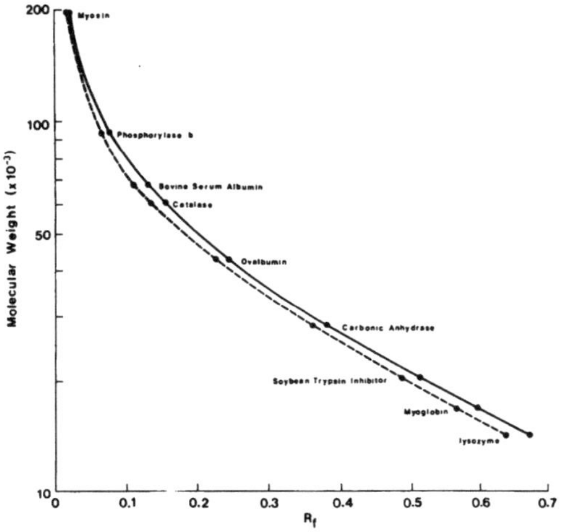

Fig. 4. Comparison between the molecular weight separations achieved with a 15%T, 1% AcrylAide gel and a conventional 15%T, 2.6%C gel.

temperature. Thus, the main advantage of using AcrylAide/polyacrylamide gels with GelBond PAG is for convenient oven drying of gels for a permanent laboratory notebook record, autoradiography, or fluorography.

References

1. Lim, R., Huang, J. J., Davis, G. A., Anal. Biochem. 29, 48 (1969).
2. Laskey, R. A. and Mills, A. D., Eur. J. Biochem. 56, 335 (1975).
3. Laemmli, U. K., Nature 227, 680 (1970).
4. Studier, F. W., J. Mol. Biol. 79 237 (1973).
5. Merril, C. R., Goldman, D., Sedman, S. A., Ebert, M. H., Science 221, 1437 (1981).
6. Dahlberg, A. E., Dingman, C. W., and Peacock, A. C., J. Mol. Biol. 41, 139 (1969).

RECOVERY OF PROTEINS FROM GEL SLABS BY A COMBINATION OF DISPLACEMENT ELECTROPHORESIS AND ISOELECTRIC FOCUSING

Stellan Hjertén

Institute of Biochemistry, University of Uppsala, Biomedical Center,
Box 576, S-751 23 Uppsala, Sweden

Zhao-qian Liu

Department of Biology, University of Peking, Beijing, China

Su-lian Zhao

Institute of Biological Products, Chengdu, Sichuan Province, China

Introduction

There is a great need for preparative high-resolving separation methods that permit isolation of proteins on the µg-scale, *i.e.*, on the scale of many analytical experiments. A very suitable method for such fractionations is polyacrylamide gel electrophoresis, provided that the proteins can be recovered in a high yield and without loss in resolution. We present here a recovery method which fulfills these requirements and in addition affords further purification and at the same time permits a determination of the isoelectric points of the proteins. This recovery technique will be only briefly described here and more in detail elsewhere (1).

An outline of the recovery technique

For the fractionation of the proteins we use gels of polyacrylamide cast in glass tubes with an inner diameter of 0.5 cm and a length of 10 cm. The proteins are localized by scanning at 280 and 310 nm as described previously (2) or by staining with Coomassie Brilliant Blue. The unstained or stained protein zones are cut out and placed above a sucrose gradient containing Pharmalyte (a carrier ampholyte from Pharmacia Fine Chemicals, Uppsala, Sweden) as outlined in Fig. 1. The gel slice G is held in place

Electrophoresis '82

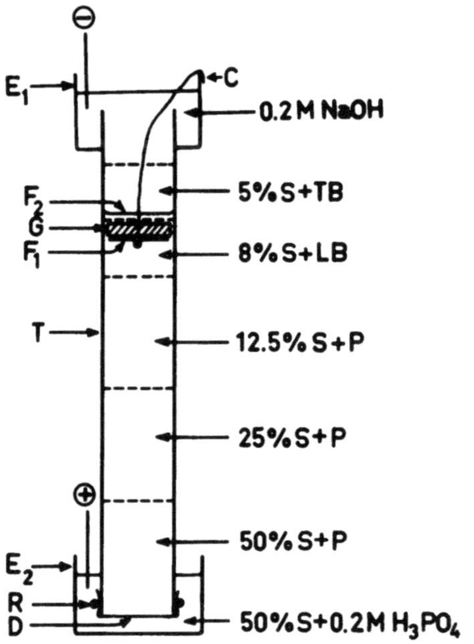

Fig. 1. A schematic drawing of the experimental conditions used for the recovery of a protein from a gel slice G by a combination of displacement electrophoresis and isoelectric focusing.

T = glass tube, E_1, E_2 = electrode vessels. F_1, F_2 = filter papers. C = cotton thread. S = sucrose solution. P = Pharmalyte (diluted 1:40). R = rubber band. D = dialysis membrane

by two circular filter papers F_1 and F_2 and is surrounded by the leading buffer (LB) in a displacement electrophoresis system. By displacement electrophoresis (isotachophoresis) the protein migrates out of the gel and into the sucrose gradient column for an isoelectric focusing step.
In this step further fractionation can occur and the protein is also prevented from migrating out of the column. When the steady state has been reached the gel slice G is removed and 2-mm fractions are withdrawn from the top of the graded glass tube T and transferred to a series of test tubes. The pH, the absorbance at 280 nm, the biological activity, etc, can then be determined in each fraction and plotted against the fraction number.

An experiment illustrating the efficiency of the recovery technique

By this technique we have recovered both unstained and Coomassie Brilliant Blue-stained proteins. An example is given in Fig. 2. The red protein phycoerythrin was submitted to electrophoresis in a polyacrylamide gel in the presence of 0.01% sodium dodecyl sulfate (SDS) in 0.05 M Tris-HAc, pH 8.0. After completion of the run the SDS was removed by washing in

HAc-methanol-H_2O (1:1:8, solution S) overnight. The gel was then washed with water several times and stained with 0.05% Coomassie Brilliant Blue R250 in solution S at 45°C for 1.5 hours. Following destaining in solution S the blue-stained phycoerythrin was cut out and recovered from the gel slice G, as outlined in Fig. 1, with the exception that all sucrose solutions in the column now contained 8 M urea to dissociate the stain from the protein. As leading electrolyte (LB) we used 0.01 M acetic acid, titrated with Tris to pH 8.0, and as terminating electrolyte (TB) 0.02 M glycine titrated to pH 9.0 with $Ba(OH)_2$ (3, 4). After an experimental time of 10 hours at 100 volts the column was photographed. As shown in Fig. 2 the phycoerythrin was eluted from the gel and split into three components (pI values around 5.8) well separated from the stain.

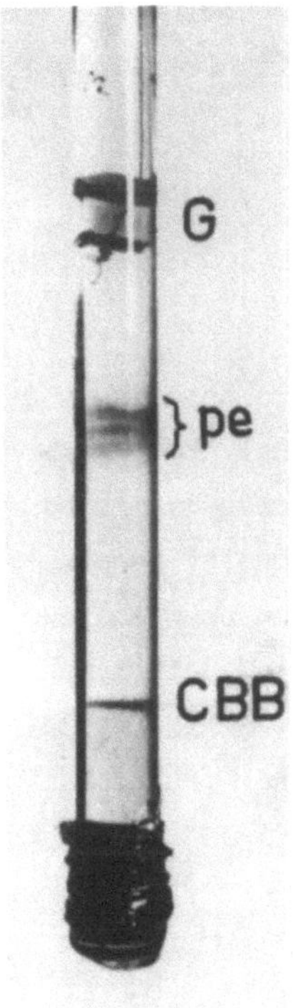

Fig. 2. Recovery of a stained protein by a combination of displacement electrophoresis and isoelectric focusing.

Following a polyacrylamide gel electrophoresis in SDS of the protein phycoerythrin the gel was stained with Coomassie Brilliant Blue. The blue-stained phycoerythrin band was sliced out. The gel slice G was placed above a sucrose gradient column of the same composition as that shown in Fig. 1, with the exception that the column also contained 8 M urea. The photography, which was taken after an experimental time of 10 hours (100 volts), shows that the phycoerythrin (pe) was eluted from the gel and fractionated into three components well separated from the stain (CBB).

Discussion

We have previously reported three different methods for the recovery of proteins following polyacrylamide gel electrophoresis (5, 6, 4; see also ref. 7). The new method outlined in this paper will be described in detail elsewhere (1) together with some other recovery techniques.

Most of our recovery experiments refer to native, unstained proteins, including enzymes. Fig. 2 illustrates, however, that even stained, denatured proteins can be recovered in a narrow zone and freed from the stain. This means that any stained protein band obtained in analytical polyacrylamide gel electrophoresis can be further analyzed, for instance by determination of the amino acid composition.

Also proteins fractionated by isoelectric focusing in polyacrylamide gels can be recovered by the method described.

Summary

After localization of the protein zones following electrophoresis in a polyacrylamide gel the protein band of interest is cut out. By displacement electrophoresis (isotachophoresis) the protein is eluted from the gel slice and concentrated in a narrow stationary zone by isoelectric focusing in a sucrose gradient. The protein band can then be withdrawn from the sucrose gradient column and its isoelectric point determined. The isoelectric focusing step has the advantages (1) to prevent the protein from migrating out of the column and (2) to afford further fractionation of the protein.

Acknowledgement

This work has been supported by grants from the Swedish Natural Science Research Council.

References

1. Hjertén, S., Liu Zhao-qian and Zhao Su-lian: J. Biochem. Biophys. Methods, to be published.
2. Fries, E. and Hjertén, S.: Anal. Biochem. 64, 466-476 (1975).
3. Routs, R.J., *Ph.D. Dissertation*, Solna skriv- och stenograftjänst AB, Solna, Sweden, 1971, pp. 76-78.
4. Öfverstedt, L.-G., Johansson, G., Fröman, G., Hjertén, S.: Electrophoresis 2, 168-173 (1981).
5. Hjertén, S. in Methodological Developments in Biochemistry (Ed. E. Reid), Vol. 2: Preparative Techniques. Longman, London, 1973, pp. 39-48.
6. Hjertén, S.: Biochim. Biophys. Acta 237, 395-403 (1971).
7. Wachlicht, H., Chrambach, A.: Anal. Biochem. 84, 533-538 (1978).

AUTOMATIC EVALUATION OF ONE-DIMENSIONAL ELECTROPHEROGRAMS BY REFINED MODELLING

Harald Kronberg, Habib Mohammed Yakin, Hans-Georg Zimmer, Volker Neuhoff

Max-Planck-Institut für experimentelle Medizin, Forschungsstelle Neurochemie, D-3400 Göttingen, Germany

Introduction

There is a need for a more accurate and precise automatic quantitation of one-dimensional electropherograms as well in routine evaluations as in research. Digital signal processing by a small laboratory computer is an adequate means to cope with noisy measuring values, an inhomogeneous background and rather confluent protein bands. A procedure for one-dimensional evaluations has been designed and implemented as a BASIC-program, which can easily be adapted to any equipment consisting of a scanner that provides the optical density of the electropherogram along the scanning line, and a laboratory computer, into which these data are transferred by analog-to-digital conversion.

The algorithms are outlined in an intuitive way, thus enabling the biochemist to assess the bearings of the procedure; details of the mathematical formalisms will be published elsewhere (1). The automatic evaluation is based on a standard method of data acquisition.

Noise reduction

Gel-scanners provide noisy measuring values A_i (i=0,...,N) as a result of the stochastic nature of the detected photons and of

Electrophoresis '82

impurities within the gel. Hence reduction of noise is an essential step of digital signal processing, especially in view of the automatic detection of protein bands being based on the determination of relative extrema of the optical density course. In order to reduce noise by averaging neighboring samples without distortion of the true protein band signal, standard sampling shall be done about ten times as frequently as the theoretical minimum. This minimum may be deduced from the sampling theorem and calls for 3 to 5 samples within a Gaussian shaped band, accordingly thirty to fifty samples per peak are required.

Fig. 1 is a sketch of the digital smoothing filter representing the triangularly distributed weights of a moving average. Every sample A_i taken within the scanning line will be replaced by the weighted mean $\bar{A}_i = \frac{1}{16} \sum_{\nu=-3}^{+3} (4-|\nu|) A_{i+\nu}$ of itself and its 4 neighbors. The advantage of this algorithm over the usual evenly weighted arithmetic mean is a stronger reduction of noise at high spatial frequencies and at the same time half the distortion of the true protein band signal. The following steps of data processing always refer to noise-reduced measuring values $\bar{A}_i$.

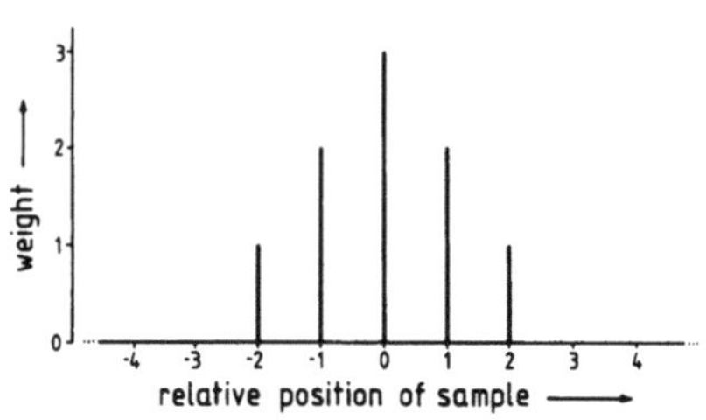

Fig. 1: Local distribution of weights of moving average.

Baseline correction

The upper diagram of Fig. 2 illustrates a realistic scanning line through a one-dimensional electropherogram. If the peak slopes do not reach the horizontal dashed line of zero optical density, the problem of accurately integrating the area arises. As long as there is no exact knowledge about the unspecific background absorbance, however, it is impossible to specify the

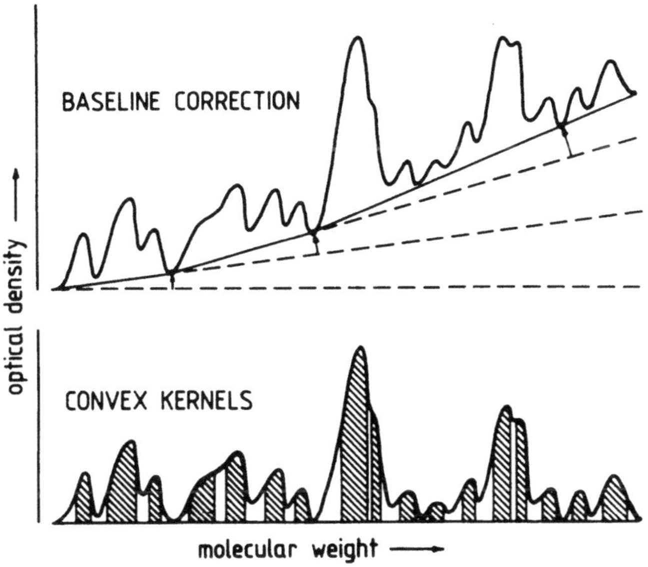

Fig. 2: Upper diagram: optical density along the scanning line (solid line), inhomogeneous background modelled by a monotonous distance-minimizing baseline (thin line). Lower diagram: result of baseline correction and segmentation by convex kernels (hatched areas).

ideal baseline correction.

Standardizing data acquisition is a means to provide information about an inhomogeneous background. Therefore scanning shall start at the clearest side of the protein trace and proceed in the more dense direction, avoiding to start or stop within a peak. Normally this is scanning from low to high molecular weight. This procedure allows for an immediate overload check of the densitometer. Then it is obvious that the absolute minimum at the start of the scanning line represents zero optical density. From there the baseline cannot but rise. The relative minima provide additional information about the course of the baseline, since they cannot fall below that line.

Thereby a procedure of linear programming (2) presents itself as a method of background correction, which results in a piecewise linear baseline that minimizes the distance to the measuring values under the condition, that it may touch but does not cut the minima. Although this procedure only allows for compensation of a monotonous background, it is optimal in the context of the available information. Fig. 2 illustrates the construction of the piecewise linear baseline with progressively steeper slope as indicated by the thin line sections. It may be compared to a thread being stretched form the starting point to the last sample around every "outstanding solid" minimum. The algorithm first detects all relative minima $(i_j, \overline{Amin}_j)$, j=0,.. ..,M, within the optical density course. The kth minimum $(i_k, \overline{Amin}_k)$ which yields the slightest slope of a straight line to the starting point $\underset{j=0,...,M}{Min} (\overline{Amin}_j/i_j) = \overline{Amin}_k/i_k$ is called a reference minimum, as it defines the first section of the baseline. Then all measuring values are referred to this extrapolated line and background correction proceeds iteratively from each previous reference minimum. The lower diagram represents the result of baseline correction.

Peak segmentation

In order to determine the peak area it is common practice to model the peaks by Gaussian profiles. However, protein bands often deviate significantly from ideal shapes. Then fitting tends to produce artificial secondary bands, since it is frequently misapplied as a method of simultaneous peak segmentation. Therefore fitting shall be preceded by an independent step of segmentation, which defines each protein band to be fitted. Every visible protein band will be detected, if the definition encounters all peaks that are separated by relative minima or at least by a slight shoulder on a slope.

The hatched kernels of protein bands in the lower diagram of fig. 2 exhibit the desired features. These kernels are sections with convex curvature of the optical density profile. Fig. 3 illustrates the operation of checking the measuring values for convex or concave curvature. Every central sample $\overline{A}_i$ within a moving neighborhood (called "window") belongs to a convex kernel, if it is greater than the arithmetic mean of the bounding samples $\overline{A}_i - \frac{1}{2}(\overline{A}_{i-1}+\overline{A}_{i+1})>\varepsilon$, $(i=0,\ldots,N,$ $\varepsilon \geqslant 0)$, indicated by an upright arrow. The size of the window (2l+1) is roughly adapted to the narrowest protein band. Convex kernels are separated by concave sections of the profile (downward arrow). The curvature check can be interpreted as the output of a digital highpass filter.

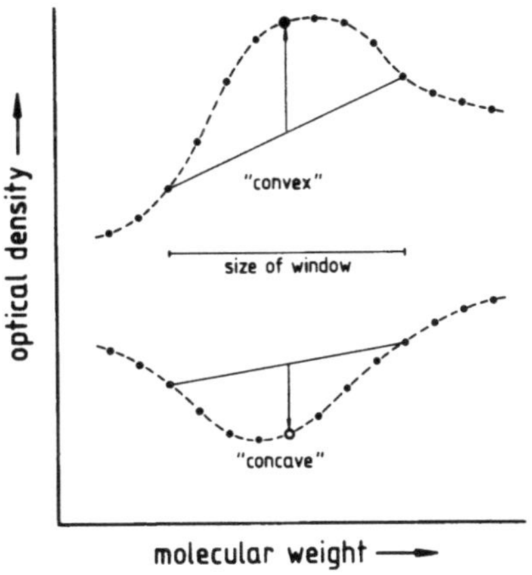

Fig. 3: Scheme of curvature check by moving digital highpass filter.

Refined fit

During inspection of the curvature each sample within a convex kernel is tested for a relative maximum, whereas the samples of concave sections are tested for minima. Position and value of the maxima within convex kernels are estimates of the mean and peak values of Gaussian profiles that shall be fitted to the protein bands. The minima determine the outermost limits of peak samples which may contribute to the least squares fit. If a minimum, however, is greater than 60% of any one adjacent maximum, then it is considered to be irrelevant and is cancelled from the list of minima. The lower sketch of a double peak in fig. 4 (solid line) illustrates such a case: the minimum of the concave region between the two peaks is even greater than the maximum of the right peak.

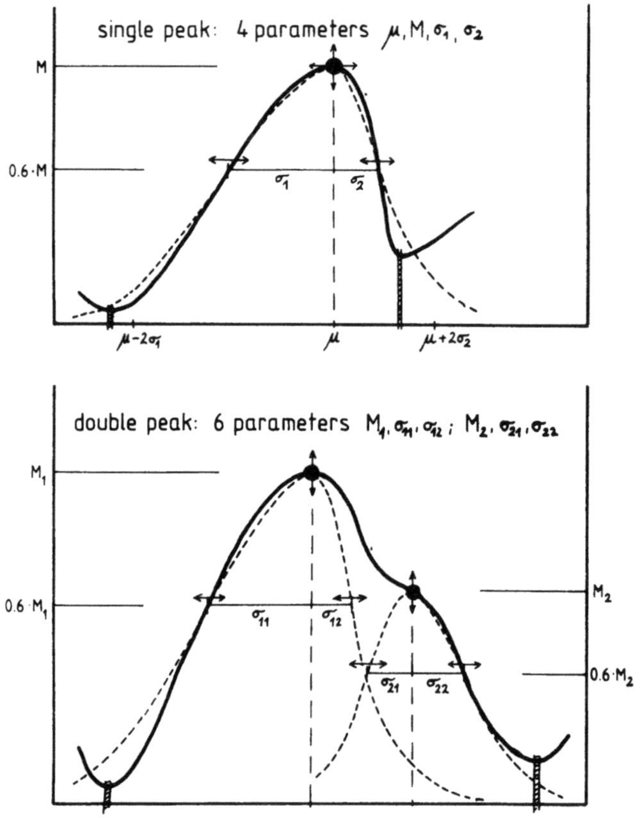

Fig. 4: Half-sided fit of Gaussian profiles. Fat curve: course of optical density, thin dashed line: sketch of fit. Arrows indicate possible corrections by the fit, hatched bars denote relevant minima. Upper diagram: four-parameter fit of single peak; lower diagram: six-parameter fit of double peak.

Hence it is possible to distinguish between the task to fit a single peak -namely in case of just one maximum being enclosed by subsequent minima (upper sketch in fig. 4)-, to fit a double peak if two maxima are enclosed or to fit a multiple peak, if there are even more maxima between adjacent relevant minima.

In the upper diagram of fig. 4 the fit of a single peak is indicated as a thin dashed line. As protein bands may be rather asymmetric, an iterative procedure is applied that fits two half Gaussian profiles with different halfwidths σ_1, σ_2 but the same mean μ and peak value M (1,3,4). The procedure starts with the maximum of the kernel as an estimate of the common mean and peak value and with an estimate of the halfwidths at 60% of the maximum. These 4 parameters ($\mu, M, \sigma_1, \sigma_2$) are optimized by iterative corrections as indicated by arrows, which result from a system of linear equations that take into account all measuring values up to twice the estimated halfwidth or up to the next minimum, according to the narrowest boundary. The boundaries are marked by hatched bars. After about 6 iterations the corrections become negligible.

The lower diagram of fig. 4 illustrates the fit of a double peak. In this case 4 half Gaussian profiles have to be determined. It turns out that the iterations of the fit have little effect upon the positions μ_1 and μ_2 of the peaks as compared to the remaining parameters. Therefore the double peak problem is defined as a 6-parameter fit of just two peak values M_1, M_2 and 4 halfwidths σ_{11}, σ_{12}, σ_{21}, σ_{22}. The algorithm is analogous to the single peak fit, starting with an estimate of two symmetric profiles with $\sigma_{12} = \sigma_{11}$, $\sigma_{21} = \sigma_{22}$.

The multiple peak problem first is reduced to the double peak problem. This is achieved by iteratively subtracting symmetric protein band estimates from right to left. The estimates are based on the respective outer right slopes. But since the right slope of the resulting double peak may be rather erroneous, only the left fit is accepted. Thus the multiple peak problem is reduced by one peak and the program proceeds as before until all protein bands are fitted.

Once all protein bands are fitted, the peak areas are calculated directly from the respective peak maxima M and two halfwidths σ_1, σ_2 by the simple formula $F = \sqrt{\frac{\pi}{2}} \cdot M \cdot (\sigma_1 + \sigma_2)$.

Discussion

In an attempt to test the entire procedure for its quantitative performance, artificial highly distorted peaks have been superimposed and taken as measuring values. The automatic evaluation proved to reproduce the asymmetries even in case of rather confluent bands, the peak areas being accurate within less than 2%. The procedure is already applied as well in clinical as in research analysis.

References

1. Yakin, H.M., Zimmer, H.-G., Neuhoff, V.: to be published in Electrophoresis
2. Künzi, H.P., Tzschach, H.G., Zehnder, C.A.: Numerische Methoden der mathematischen Optimierung, Teubner, Stuttgart 1967
3. Stockhausen, M.: Mathematische Behandlung naturwissenschaftlicher Probleme, pt. 2, p.32-33, UTB Steinkopff, Darmstadt 1980
4. Gellert, W., Küstner, H., Hellwich, M., Kästner, H. (Eds.): Kleine Enzyklopädie Mathematik p.696-699, Harri Deutsch, Thun u. Frankfurt 1977

QUANTITATIVE EVALUATION OF CAF - PATTERN BY VIDEO DENSITOMETRY

K.G. Schmidt, M. Raff, H. Blenke, G. Polony
Institut für Chemische Verfahrenstechnik, Universität Stuttgart
D-7000 Stuttgart, Germany

Introduction

Computer controlled tv - systems are proper tools for digital picture processing due to their high performance in data recording. We used this advantage for quantitative evaluation of cellulose acetate electrophoretograms. Limits and performance of our system with respect to quantitative densitometry of CAF-pattern will be discussed.

1. Description of the system

The investigations were carried out with an available picture digitizer with display (1) connected to a computer (2). Figure 1 shows a simplified block diagram of the system. The main parts are: video camera, picture display, computer with magnetic disc drives and output devices.
The lens system of the camera forms an image of the sample on the target of the camera tube. The video signal is transmitted through a video amplifier chain to the digitizer as well as to the display. The digitized data and the corresponding space coordinates are saved on the magnetic discs by the computer and are available for further image processing. The analog video image on the display can be supplemented by graphics or restored images from the computer. A joystick cursor allows to take over space coordinates of the picture. This makes interactive operating very easy.
Main technical data: The target of the used vidicon tube has a

Electrophoresis '82

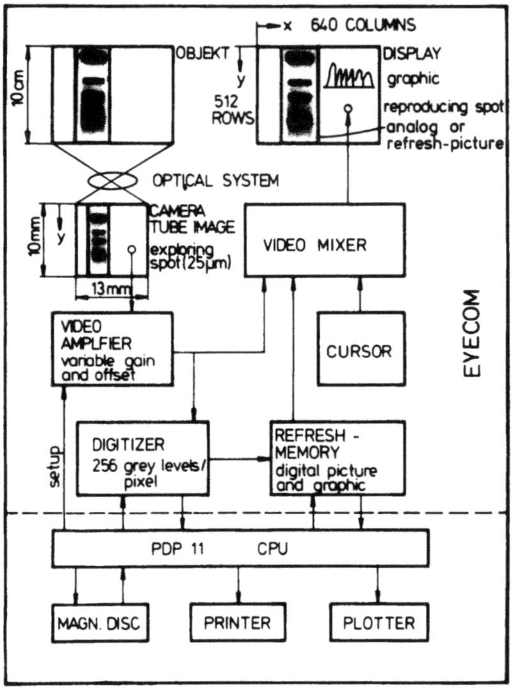

Fig. 1:
Block diagram of
the system

size of 10 mm x 13 mm. Scanning of the image is performed by video frequency with 512 lines and 640 columns. That is, one line covers about 20 microns at the target. With a scale of reproduction 10 : 1 this is a theoretical distance of 0.2 mm at the sample. The diameter of the scanning spot is approximately 25 microns. The intensity values coming from the video amplifier are digitized with 8 bits. This comes up to 256 theoretical grey values. The relation between these measured grey values z and the absorbance E, respectively the optical density o.D., is given by the intensity transfer characteristics (ITF) of the video tube, which is shown in figure 2. For ideal condi-

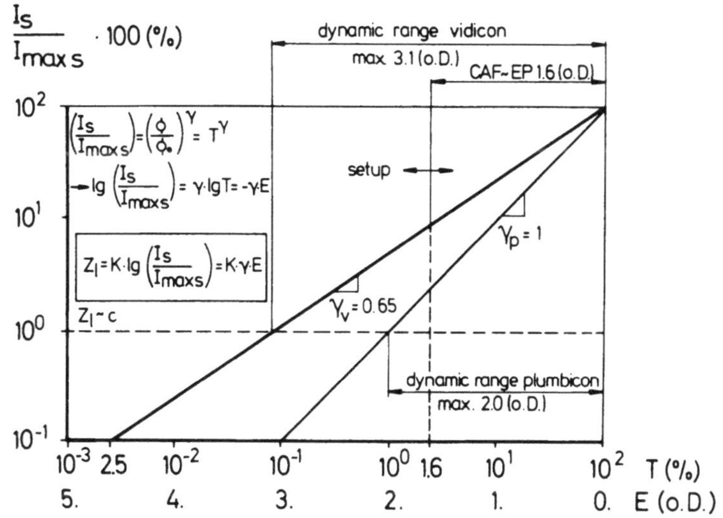

Fig. 2: Intensity transfer characteristics

tions relative intensity is an exponential function with the slope factor γ. The relation between the grey values z and the absorbance E then becomes a linear function. Factor K depends on the variable gain of the amplifier (to be set program controlled as well as the mean dark current).
The dynamic range of the camera tube is limited by the electronic noise. For the used vidicon the amount is about three percent (3). On the other hand this noise can be reduced by a factor of $1/\sqrt{n}$, when digitizing n times. In this way we reach an amount of one percent for electronic noise. The result is a dynamic range of E = 3.1 o.D. whereas a CAF electrophoretogram has a range of only 1.6 o.D.

Furthermore the resolution of the system has to be discussed. Resolution can be tested by scanning a pattern of alternate and equally spaced black-and-white stripes. Their number on a length

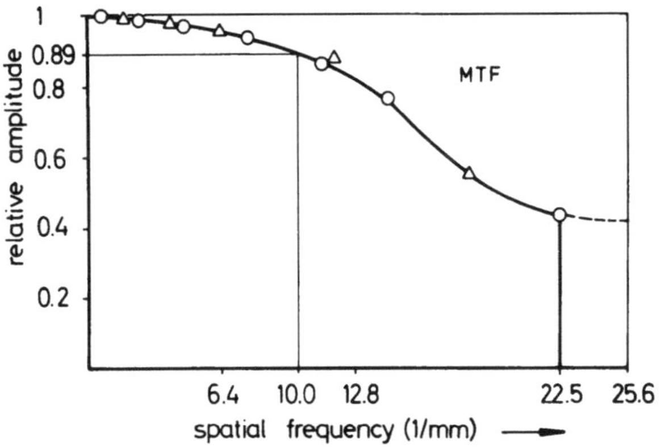

Fig. 3: Modulation transfer characteristic

of one millimeter is called spatial frequency. The detection of one black-and-white sequence needs at least two picture elements (pixels). Our system has 512 pixels for one column. The spatial frequency is 25/mm. For such cases, in which the spatial frequency is low, the video signal alternates from full black to full white level. With higher frequencies, the separation becomes smaller than the scanning spot and the video signal reduces in amplitude. The relationship between the change of the relative amplitude with spatial frequency is given by the modulation transfer function (MTF). We measured it for our system (6). Fig. 3 shows that resolution decreases at higher frequencies. For our cases we note: 1. For a frequency of 10/mm the full intensity resolution is only 11% less. This is important for evaluation of CAF - EP. 2. Restriction to low intensity resolution increases the spatial frequency up to 22.5/mm. The structure resolution for a sample of 10 cm length becomes 0.45 millimeters (4).

Statement 1 shall be discussed with respect to a CAF-phoretogram. Figure 4 shows a plot of a single CAF-EP produced by the system. Right hand up we see a histogram of the spatial frequencies of our density profile. This graph is the result of a fourier analysis. It shows, that the profile of serum EP can be reproduced by a sum of sine and cosine functions in which the highest frequency is about 2.5/mm. From this result, in connec-

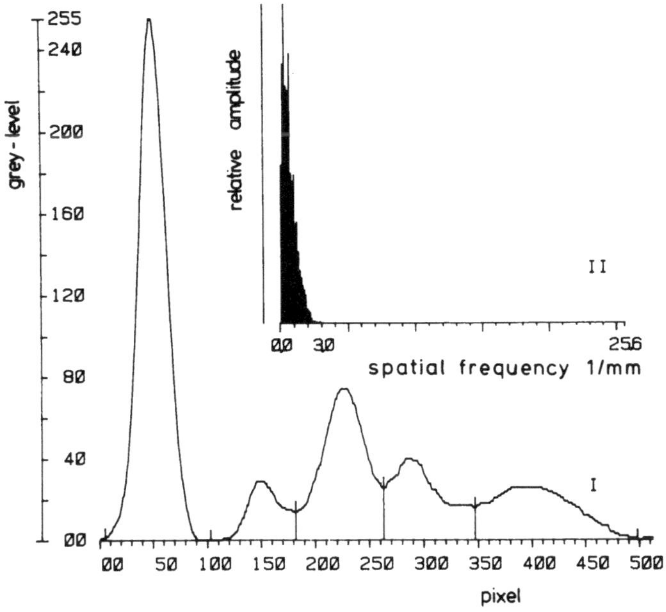

Fig. 4: CAF - EP profile with analysis of typical spatial frequencies

tion with statement 1 we derive, that up to four samples can be arranged one above the other on the object table. Thus 128 pixels of a column belong to one profile.

The result, that 128 equidistant points of a serum CAF-EP are sufficient for evaluation, could be verified by densitometric investigations.

Summary of part 1.

- Dynamic range is about E = 3.1 o.D.
- Calculated grey values and o.D. give a linear function. Hence grey values are a linear function of colour concentration.
- For one profile 128 pixels/column may be sufficient.

2. Improvement of measurements

Measurements can be improved by reduction of systematic errors (5):

- Signal to noise ratio
- Dark current
- Inhomogeneity of measuring equipment
- Unspecific inhomogeneity

As mentioned above, the signal to noise ratio is increased by multiple digitizations of a column. For each sample we used 15 columns. The other errors influence the ideal signal by additive or multiplicative components. The ideal grey value will be a function of light intensity Φ only (formula 1). In the real case the measured grey value becomes a function of the position (x,y) and the local intensity Φ (formula 2). It can be shown, that the measured value can be described by formula 3 (7). The function d (x,y) corresponds to the local dark current. The shading function s (x,y,Φ) is a perturbation function, which modulates the ideal function. The shading function takes into account errors produced by inhomogeneity of the measuring equipment, optical stimulation, vignetting of the optical system and differences in local target sensitivity.

ideal $$z_i = z_i(\Phi) \quad (1)$$

real $$z = z(x,y,\Phi) \quad (2)$$

assumption $$z(x,y,\Phi) = s(x,y,\Phi) \cdot z_i(\Phi) + d(x,y) \quad (3)$$

dark values for $\Phi = 0 \quad z_i(\Phi) = 0$

$$z(x,y,0) = d(x,y) \quad (4)$$

correction

$$z(x,y,\Phi) - z(x,y,0) = s(x,y,\Phi) \cdot z_i(\Phi) \quad (5)$$

shadingfunction for $\Phi = \Phi_c = \text{const.}$

$$s(x,y,\Phi_c) = \frac{z(x,y,\Phi_c) - z(x,y,0)}{z_i(\Phi_c)} \quad (6)$$

$$z_i(\Phi) = \left[z(x,y,\Phi) - z(x,y,0)\right] \cdot \frac{z_i(\Phi_c)}{z(x,y,\Phi_c) - z(x,y,0)} \quad (7)$$

Compensation of local dark current is possible by recording a picture for $\Phi = 0$ (closed lens) and point for point correction of the measured grey values (formula 4,5). The shading function is measured by taking a picture of constant brightness (Φ_C): formula 6. After shading correction, formula 7, we get the searched grey value. As pointed out, this value is proportional to the colour concentration of the sample.
The influence of the systematical errors over the total target is shown in figure 5. Instead of a flat table we get a hilly scenery. The efficiency of the corrections is shown in figure 6. The dotted lines give the measured profile of grey values along the 512 pixels of a column at different intensities. After correction the curves become almost straight lines.

Further improvement of measurements with respect to unspecific inhomogeneities can be done by median filtering (5) and will not be discussed here.

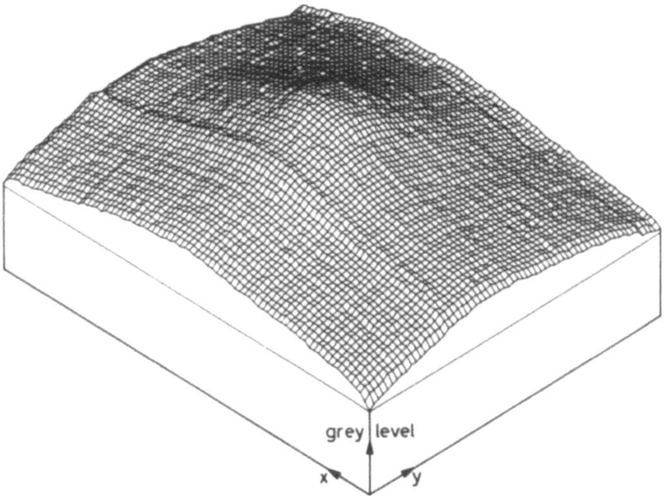

Fig. 5: Sum of systematical errors cause different readings over the target

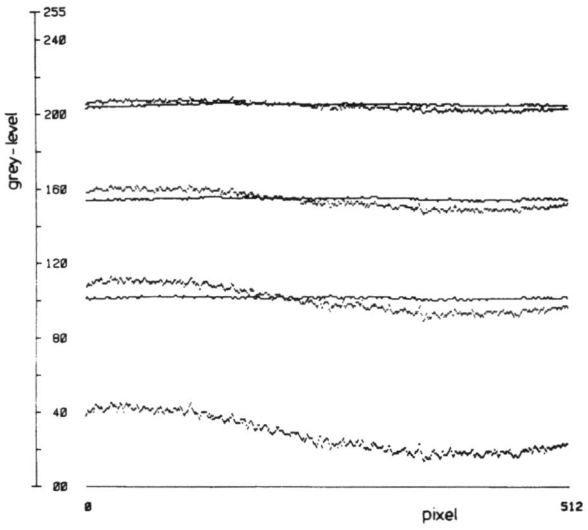

Fig. 6: Profiles from pictures of constant brightness before (dotted lines) and after correction

3. Discussion

The given explanations show, that video scanners are suitable for quantitative evaluation of CAF-EP, when preprocessing of video data takes place. For this purpose we compared photometric densitometry and video densitometry of standard brightness tablets (e.g. Kodak No. 2) as well as of CAF-EP under equal conditions (apertural width and length, monochromasia etc.). We could not find relevant differences in the standard deviations of both methods.
In competition with available automatic densitometers we generated programs, which allow interactive and automatic processing and combinations of both e.g. for baseline correction, limit detection etc.

Hence we evaluated up to 64 samples in one picture with the additional advantage of reproducing the original analog picture on the screen. Thus storage and handling of picture data without loss of information with time makes further comparison examinations possible (8).
The evaluation of CAF-EP was chosen as a well known example of clinical analysis to demonstrate limiting conditions and performance of video densitometers. Under these given limitations we can imagine, that similar densitometric problems in one or two dimensions can be solved.

References

1. Eye Com, Picture Digitizer and Display 108 Pt Spatial Data Syst., Goleta California
2. PDP 11/34, Digital Equipment, RT 11 operating system
3. Eye Com Handbook (1), 1977
4. Raff, M.: Electrophoresis 1982, this edition
5. Schmidt, K.G., Polony, G., Raff, M.: Biomedizinische Technik 25, 47 - 51 (1980)
6. Raff, M., Polony, G., Blenke, H., Schmidt, K.G.: Medizin Technik 81, 137, Schiele & Schön GmbH, Berlin 1981
7. Erhardt, R.: Analyse und Entwurf optoelektronischer Systeme zur schnellen digitalen Bilddatenerfassung und -verarbeitung, Thesis Universität Stuttgart 1980
8. Polony, G., Raff, M., Blenke, H., Schmidt, K.G.: Medizin Technik 81, 138, Schiele & Schön GmbH, Berlin 1981

IDENTIFICATION OF FUNCTIONAL PROPERTIES ASSOCIATED WITH INDIVIDUAL PROTEINS BY COMPUTER-BASED PATTERN ANALYSIS

D.L. du Cros, C.W. Wrigley
CSIRO Wheat Research Unit, North Ryde, NSW 2113, Australia

and R.A. Hare
Department of Agriculture, Tamworth, NSW 2340, Australia

Introduction

Electrophoretic techniques have proved extremely popular because of their ability to resolve many components from a crude mixture of proteins. Furthermore, information about the chemical composition and biological activities of individual components has been obtainable by applying a range of specific staining procedures directly to the gel or to the components after transferring them to an absorptive membrane. Thus for enzymes, in particular, considerable information can be obtained quickly by electrophoretic analysis of many samples without the need for tedious preparative fractionation.

However with certain other functional properties of proteins, it has not been possible to devise specific methods of detection to identify the components, relevant to the function of interest, following gel electrophoresis. One case of particular concern to us was the identification of individual wheat-flour proteins associated with either poor or superior performance in the manufacture of bread or pasta goods. The relevant properties are known to be associated with the gluten proteins of wheat-flour dough (1), but attempts to associate overall quality with protein composition have proved difficult because of the complexity of both quality and protein (2-5).

Electrophoresis '82

Without any direct method of detection in this case, methods of statistical association were sought to permit the comparison of quality attributes and protein composition in a large number of individuals with diverse genetic backgrounds. The aim was to find a consistent pattern of association between a certain protein or proteins and a specific aspect of processing quality. Such a finding would not, in itself, constitute proof that the protein(s) identified had conferred the quality attribute, but it would at least direct attention towards components worthy of further chemical characterization. Even if the protein-quality association proved to be merely due to a close genetic linkage between the genes coding for the protein and for another quality-conferring component, the information would still be of great practical value in the segregation for quality in wheat breeding.

The statistical association approach is likely to be applicable to many other cases where methods for directly detecting functional properties following electrophoresis are not available. The plans to compile a complete catalogue of human cell proteins (6) offer the opportunity of attempting to associate specific diseases or metabolic irregularities with individual protein components. More specifically, this approach might be applied to alpha-1 antitrypsin components (7), separated by gel isoelectric focusing, in determining their possible relationship to disease states such as pulmonary emphysema, neonatal cirrhosis, and a range of chronic inflammatory diseases (8,9).

Data for Analysis

In the example to be described, pattern analysis (10) was used to identify specific proteins associated with pasta-making quality (11) in durum wheat (*Triticum turgidum* L. var. *durum*).

As the main objective of the project was to devise a simple electrophoresis-based method for selecting good-quality individuals in a breeding program, the data were obtained from 103 fixed F7 lines from an Australian durum-wheat breeding program. Such a population might be expected to cover a wider and more relevant range of diversity than groups of commercial cultivars examined in previous studies with a similar objective (12,13).

The gliadin proteins of the wheat grain were extracted from semolina samples (100 mg) with 1 M urea solution (0.6 ml). The clarified extract was applied to a polyacrylamide gradient gel (2.5 to 13%, from Gradient Pty Ltd, Pyrmont 2009, Australia) for cathodic electrophoresis in sodium lactate (pH3, 4mM Na) at 25°C for 2.5 hours at 200 volts (14). The electrophoretic pattern of each line, obtained by staining with Coomassie brilliant blue G250 (15), was coded as a set of ordinal attributes by allocating a number, based on the absence of staining (1) or presence in increasing intensity (2,3 or 4), to each of the 38 band positions shown in Fig. 1. The gliadin composition of line 40316, for example, commences 1/2/2/2/3/3/3/1/2/3/... . Compositional data was obtained in this form for the full set of 103 breeding lines.

Superior pasta-making quality is related to dough strength. Processing quality was therefore determined by quantitative assessment of dough properties. Semolina, obtained from a grain sample of each line, was made into dough for testing in a mixograph by AACC Method 54-40 (15). The resulting mixograph trace was characterized according to the height (H) of the curve after 7 min. mixing, the maximum height (M), the time (T) to maximum height, and the area (A) under the curve. In addition, each line was assessed as being either strong or weak according to its overall dough properties.

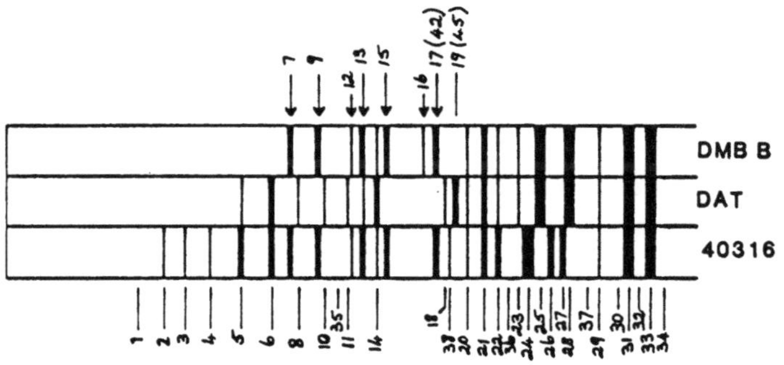

Fig. 1. Diagrammatic representation of the electrophoretic patterns of gliadin proteins from three durum wheats, a breeder's line 40316, the cultivar Durati (DAT) and the B biotype of cv Duramba, reproduced from (10).

Pattern Analysis of Data

Relationships between the 103 breeding lines with respect to their gliadin-protein composition were first determined by applying agglomerative classification (10, 16) to the protein-composition data using the program MULCLAS, available as part of the TAXON package on the Cyber 76 computer of the CSIRO Division of Computing Research, Canberra. This classification permitted the sorting of the breeding lines into groups with similar electrophoretic patterns. A comparison of these groups with a separate classification of the lines according to quality indicated that protein composition was closely related to processing quality. When the protein-based classification was truncated at the ten-group level, four of these groups contained all the weak wheats, and the strong-dough lines were in the remaining groups.

Secondly, inverse pattern analysis was used to examine relationships between the proteins themselves. This involved analysis of the protein composition of the 103 lines using the TAXON program designated INVER (17,18), plus TWONET (19) to show relationships as a network. Figure 2 shows the network which resulted from this approach. It also includes the grain quality information - four dough-property characteristics as

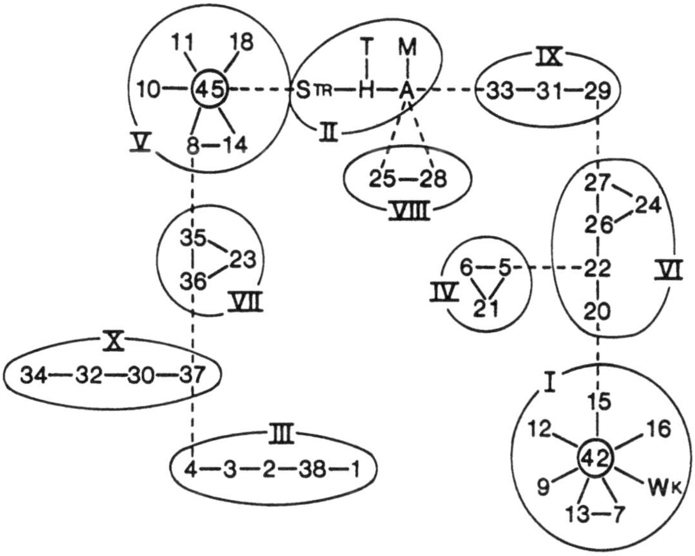

Fig. 2. Representation, as a network, of associations between gliadin bands (numbered 1 to 38 in Fig. 1), dough weakness (WK) and strength (STR), and mixograph estimates of dough properties. Dotted lines indicate weaker associations than solid linking lines.

numerical attributes from mixograph testing, and two ordinal two-state attributes (dough strength and weakness).
The network (Fig. 2) groups together those gliadin components that often occurred together in the same phenotype. The bands most closely associated with gliadin 42 (arrowed in Fig. 1) occurred within a reasonably narrow range of mobilities. This is probably because they have homologies in amino acid sequence due to their being the products of tightly linked genes, which have arisen by amplification of a single ancestral gene (18, 20). Similar explanations apply to other groups of proteins indicated by the network.

As well as pointing to relationships between the proteins themselves, the network (Fig. 2) has indicated that the four mixograph attributes are closely related to one another and to strength (group II), but distantly related to dough weakness (group I). More importantly, dough strength is related to band 45 (groups II and V), while weakness occurred together with gliadin 42 and its associated proteins (group I). A closer examination of the data revealed that all weak lines had gliadin 42, and that 90% of the strong-dough lines had gliadin 45. Similar associations of dough properties with these respective gliadins have also been reported by others (12, 13) based on studies of cultivated durum varieties. Whereas these simple relationships could be readily evaluated in their comparisons and our own, a satisfactory examination of the full compositional and mixograph data would not have been possible without the computer programs described.

Conclusion

The techniques of pattern analysis described have been borrowed from taxonomists, ecologists and agriculturalists (10,21), whose main interests have been in such attributes as morphological characteristics and yields of plant or animal material. But these workers, too, are also becoming interested in integrating their conventional approaches with biochemical information such as is provided by electrophoretic techniques. An example of one such application has been given. In addition, the methods have been used in studying protein-quality relationships in bread wheats (16,18,22) and in using seed protein and esterase isoenzyme patterns to delineate geographical relationships between some hundreds of introduced samples of the herbaceous legume Stylosanthes (10,23). The procedures are likely to also find valuable applications in other fields, such as clinical biochemistry, where electrophoretic methods are used.

References

1. Wall, J.S. : in Recent Advances in the Biochemistry of Cereals. Eds D.L. Laidmen, R.G. Wyn Jones. Academic Press, New York. Pp. 275-311 (1979).

2. Huebner, F.R., Rothfus, J.A. : Cereal Chem. 45, 242-253 (1968).

3. Huebner, F.R. : J. Agric. Food Chem. 18, 256-259 (1970).

4. Orth, R.A., Bushuk, W. : Cereal Chem. 49, 268-275 (1972).

5. Orth, R.A., Bushuk, W. : Cereal Chem. 50, 191-197 (1973).

6. Clark, B.F.C. : Nature 292, 491-492 (1981).

7. Morse, J.O. : New England J. Med. 299, 1045-1048, 1099-1105 (1978).

8. Fagerhol, M.K., Cox, D.W.: Adv. Human Genet. 11, 1-62 (1981).

9. Clark, P., Breit, S.N., Dawkins, R.L., Penny, R. : Amer. J. Med. Genet. (in press).

10. Williams, W.T. : Pattern Analysis in Agricultural Science. CSIRO,Melbourne (1976).

11. du Cros, D.L., Wrigley, C.W., Hare, R.A. : Aust. J. Agric. Res. 33, 429-442 (1982).

12. Damidaux, R., Autran, J.C., Grignac, P., Feillet, P. : C.R. Acad. Sci. Paris, Ser. D 287, 701-704 (1978).

13. Kosmolak, F.G., Dexter, J.E.,Matsuo, R.R., Leisle, D., Marchylo, B.A.: Can. J. Plant Sci. 60, 427-432 (1980).

14. du Cros, D.L. Wrigley, C.W. : J. Sci. Food Agric. 30, 785-794 (1979).

15. Cereal Laboratory Methods, 7th Edition. American Association of Cereal Chemists, St Paul, Minn., U.S.A. (1962).

16. Wrigley, C.W., Robinson, P.J., Williams, W.T. : J. Sci. Food Agric. 32, 433-442 (1981).

17. Lance, G.N., Williams, W.T. : Aust. Comput. J. 11, 27-28 (1979).

18. Wrigley, C.W., Robinson, P.J., Williams, W.T. : Aust. J. Agric. Res. 33, 409-418, 419-427 (1982).

19. Williams, W.T. : Aust. Comput. J. 12, 70 (1980).

20. Mecham, D.K., Kasarda, D.D., Qualset, C.O. : Biochem. Genet. 16, 831-853 (1978).

21. Williams, W.T., Burt, R.L., Pengelly, B.C., Robinson, P.J.: Agro-Ecosystems 6, 99-109 (1980).

22. Wrigley, C.W., Lawrence, G.J., Shepherd, K.W. : Aust. J. Plant Physiol. 9, 15-30 (1982).

23. Robinson, P.J., Burt, R.L., Williams, W.T.: Agro-Ecosystems 6, 111-118 (1980).

EVALUATION OF ISOELECTRIC FOCUSING PATTERN BY VIDEO DENSITOMETRY

M. Raff, G. Polony, H. Blenke, K.G. Schmidt
Institut für Chemische Verfahrenstechnik, Universität Stuttgart
D-7000 Stuttgart, Germany

Introduction

In the evaluation of IEF-pattern the determination of isoelectric points (IP) is as important as the calculation of protein proportions. To estimate relative concentrations of proteins, a high resolution of spot and intensity is implied, whereas for registration of gradient also an exact determination of pH-values is necessary. This pH-measuring is usually done during the separation process. For the calculation of IP the positions of detected maxima in profile curve have to be transfered to gradient curve.
It shall be shown, that with a videosystem (1,2,3,4) it is possible to do all the evaluation after the focusing procedure.

Methods

An actual application resulted from the cooperation with Perthen and Müller-Beißenhirtz in the comparison of human serum and cerebrospinal fluid (CSF) proteins by isoelectric focusing with the intension of early understanding the multiple sclerosis (5,6). For this purpose however the separation process should be examined to produce comparable results.
Therefore, with the flap-technique of Radola (7), we prepared polyacrylamide gels with the following parameters:

thickness of gel layer: 0.1, 0.5, 1 mm

Electrophoresis '82

composition of gels:		5% T, 3% C,
		3% T, 4% C,
		3% T, 20% C
urea	:	0, 3, 6, 9 mol
pH-range	:	Servalyte 4-9 T, 3-10 AG

Best results with regard to resolution and quality of focused zones were reached by a thin layer gel (0.5 mm) with 3% T, 4% C containing 6 mol urea. With these gels we had no trouble with streaking, smearing or agglomeration at application site (see fig. 1). Our experiences with the other gels mentioned above, and with a series of identical gels, containing (6 mol) and without urea, are described in (8).

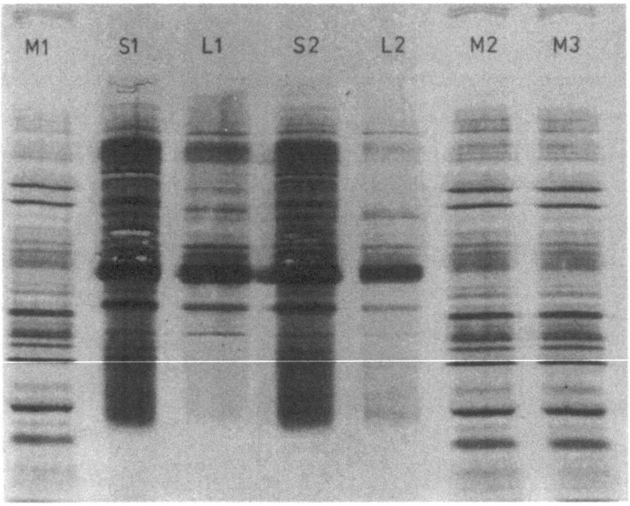

Fig. 1: Thin layer isoelectric focusing of human serum (S), cerebrospinal fluid (L) and marker proteins (M) in 500 mm polyacrylamide gel (more details see methods).

Quantitative evaluation

The evaluation was done by a videosystem, the marginal conditions of which are described by Schmidt (9) with regard to intensity and spot resolution. Since focusing pattern, compared with electrophoresis on cellulose acetate shows much finer structures, the image cannot be reduced, so that the whole 512 pixels are needed for intensity measurements of focused pherograms. Thus in one picture (640 x 512 pixels) up to 15 patterns may be included.
The condition of an exact intensity measuring is a relative amplitude of a = 1 in modulation transfer function MTF (9).
After fast fourier transformation of the profile curve of marker proteins e.g., the spectrum of spatial frequencies shows relevant frequencies even besides a relative amplitude of
$a \leq 1$ (fig. 2).

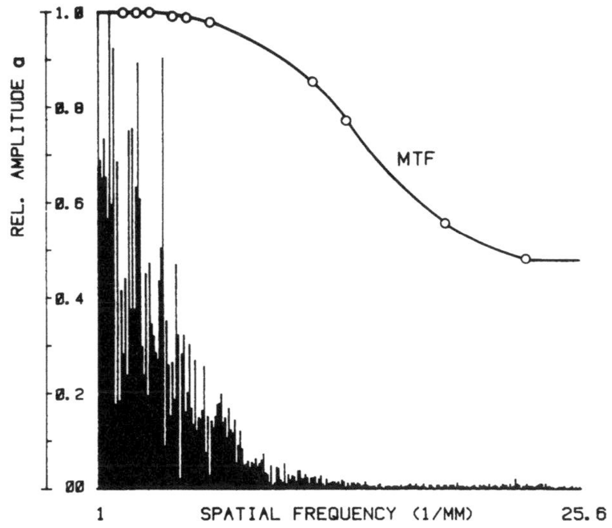

Fig. 2: Spatial frequencies of a marker protein profile curve (see fig. 4), and modulation transfer function (MTF) of the TV-system used for evaluation.

Elimination of those frequencies, and retransforming of the remaining ones shows little variations in profile curve, because only very small peaks are touched by this interference. Anyway the main part of those high frequencies in the spectrum results from noise. So this low pass filtering is another possibility to improve the signal to noise ratio, instead of continued digitizing or median filtering (15).
Exemplary fig. 3 shows the quantitative evaluation of human serum proteins after isoelectric focusing. For the calculation of relative properties the criteria of minima detection cut a great figure. Here too the fourier transformation could be helpful if so called 'fused peaks' (see arrows fig. 3) are of special interest. Because low frequencies in a spectrum like that of fig. 2 describe elementary modulation, a band pass filtering (elimination of low <u>and</u> high frequencies) effects a better recognition of those peaks.

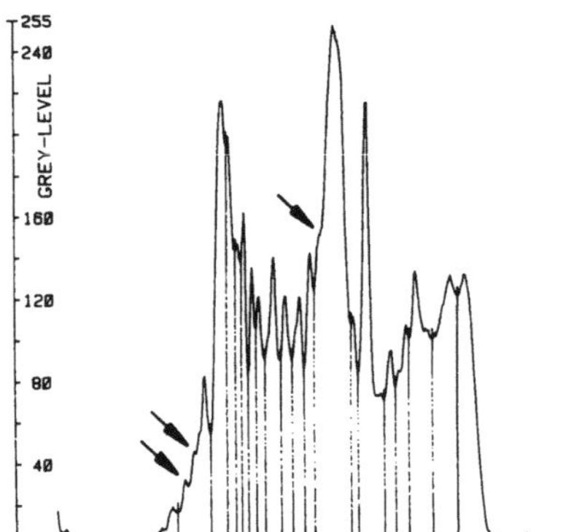

NO	LIMIT	AREA	REL. %
1	125	371	1.0
2	159	1510	4.1
3	176	2684	7.4
4	185	1543	4.2
5	191	858	2.4
6	198	924	2.5
7	206	898	2.5
8	215	954	2.6
9	231	1749	4.8
10	243	1255	3.4
11	255	1255	3.4
12	265	1205	3.3
13	302	6910	18.9
14	309	710	1.9
15	336	3014	8.3
16	347	923	2.5
17	361	1277	3.5
18	385	2656	7.3
19	410	2921	8.0
20	480	2879	7.9

Fig. 3: Quantitative evaluation of focused human serum proteins (fig. 1 : S 1).

The resolution of scanner with respect to MTF is improvable, if only peak identification is important. Schmidt demonstrated that in an object of 10 cm height the finest structure to be separated is 0.45 mm (9). If this object is a focusing phoretogram with a pH-range of 4-9, proteins with pI-values differing by only 0.023 pH could be resolved. This is the dimension of Vesterberg and Svensson's theoretically found Δ pH for myoglobins (10). Of course this resolution of the scanner is dependent on ampholyt range and image scale up.

pH-gradient

Determination of pH-gradient: pH (x), where x means distance from anode, in thin layer gels is usually done by surface electrodes. Because of diffusion of the ampholytes during the measurement, the value is dependant on measuring duration or attitude of electrode type respectively. Moreover the trouble by justifying such an electrode on wide pH-ranges cannot be neglected, because for a range of 5 pH-units it runs up to 0.23 pH-units (8).

Regarding a mixture of protein markers (SERVA), there is a relatively high variation of pI-values, if focusing conditions change (see table 1). Main influences are focusing time and interference of CO_2 (11) as well as urea in gel or applicated protein (10).

For these reasons the comparison of IP from different investigations is only convenient, if the values can be abstracted by the parameters of the experiment. This should be possible if a mixture of carefully examined markers would be focused together with other samples, serving as a standard reference.

	MARKER PROTEINS	SERVA	DELINC. /11/	ALLEN /13/	MALM. /14/	RADOL /12/	UFR. /8/		STACH.-B. /10/
	urea	no	no	no	no	no	no	6m.	6m.
1	amyl.-glycosidase	3.5	3.8	3.5	-	-	3.2	3.6	-
2	ferritin	4.4	-	-	-	4.38	4.25	4.75	5.34
3	bovine serum albumin	4.7	4.9	-	4.7	4.85	4.5	5.05	-
4	β-lacto-globulin	5.34	-	5.2	5.1	5.14	5.0	5.55	-
5	conalbumin	5.9	5.9	-	5.9	5.88	6.05	6.45	6.4
6	horse myoglobin	7.3	7.4	7.35	7.3	7.32	7.7	8.2	7.44
7	sperm whale myoglobin	8.3	8.3	-	8.1	8.19	8.65	9.2	7.87
8	ribonuclease	9.45	9.3	-	9.5	8.92	9.6	10.1	8.41
9	cytochrome c	10.65	10.1	10.25	-	9.30	10.7	11.1	9.01

Table 1: Isoelectric points of a mixture of marker proteins

Determination of gradient and isoelectric points

Supposing the phoretograms of marker, serum and CSF proteins represent such a group of samples, determination of isoelectric points by the videosystem can be realized very easily. At first profile curve of marker proteins must be determined either from the digitized values or by directly scanning the original pattern. As a result of heterogenities there are more peaks identified in profile, than markers are included (see fig. 4). That is why a competent evaluator should effect their detection. With the help of a cursor he assigns interactively the positions of the maxima of the marker proteins, which are automatically attached to the respective isoelectric points. With that the bases of gradient curve are laid up and serve now for mathematical formulation of the function pH (x) e.g. by linear interpolation between these bases or by regression with the method of least squares.

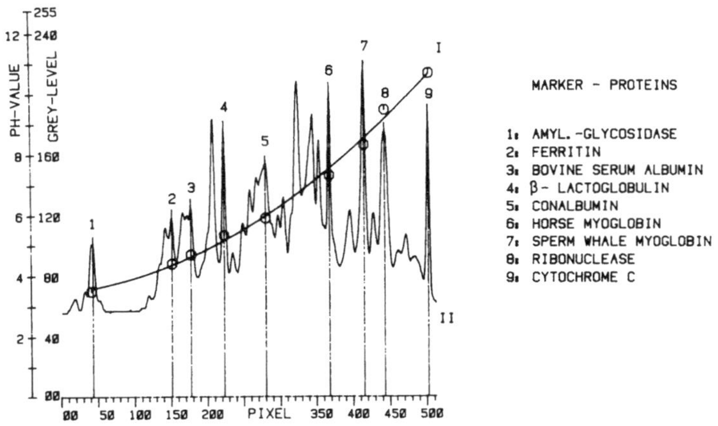

Fig. 4: Profile curve of marker proteins (fig. 1 : M 1) and calculated gradient curve.

Thus gradient curve in fig. 4 can be described by a function in the form of:

$$pH\ (x) = a + b\,x + c\,x^2 \qquad (1)$$

which is used for further calculation of isoelectric points.

Distance (x) from anode can easily be transformed in picture elements (pix) by the equation:

$$x\ (cm) = \frac{h\,(cm)}{512}\ pix, \qquad (2)$$

where h means height of the object. If whereabouts (pix) of resolved peaks are calculated in each case between two limits (see fig. 5), isoelectric points can be computed by inserting these 'pix-values' found into equations (1) and (2). An example of such a complete evaluation for the group of samples mentioned is demonstrated in fig. 5.

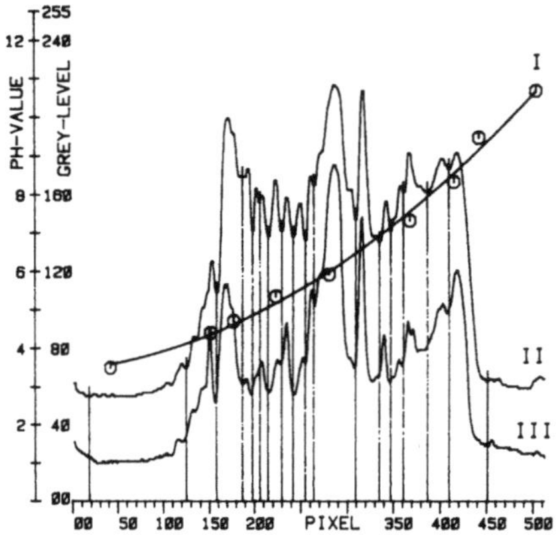

	SERUM: II		CSF: III	
NO.	PIX.	IP	PIX.	IP
1	119	4.07	115	4.04
2	152	4.36	150	4.34
3	170	4.54	167	4.51
4	192	4.79	190	4.76
5	201	4.89		
6	207	4.96	207	4.96
7	222	5.15	225	5.19
8	234	5.31	235	5.32
9	248	5.50	252	5.58
10	260	5.68	261	5.69
11	285	6.06	285	6.06
12	316	6.57	316	6.57
13	341	7.02	340	7.00
14	357	7.32	356	7.30
15	366	7.50	366	7.50
16	401	8.22	403	8.26
17	417	8.57	418	8.59
18			455	9.44

GRADIENT-FUNCTION (I)

$$PH(PIX) = 3.446 + 0.00245 \cdot PIX + 0.000024 \cdot PIX^2$$

Fig. 5: Profile curves of human serum and cerebrospinal fluid pattern (fig. 1 : S 1 and L 1) and gradient curve, with which isoelectric points have been calculated.

Discussion

Evaluation of IEF pattern by a digital video scanner is easy, fast and comfortable, because devices like digitizer, refresh memory, graphic, joystick and peripherals ideally combine the possibilities of interactive with automatic measuring methods. In protein analysis, with the challenge for individual, as well as for routine examinations this is very important.

In case of gradient determination, by aid of a mixture of marker proteins, it is even possible to overcome the disadvantages of the measurement with a surface pH-electrode.

Since furthermore also two dimensional electrophoretic separations can be evaluated (17) the TV-scanner is a proper tool in clinical chemistry.

References

1. Schmidt, K.G., Polony, G., Raff, M.: Biomedizinische Technik 25, 47-51 (1980)
2. Raff, M., Polony, G., Blenke, H., Schmidt, K.G.: Medizin Technik 81, 137, Schiele & Schön GmbH, Berlin (1981)
3. Polony, G., Raff, M., Blenke, H., Schmidt, K.G.: Medizin Technik 81, 138, Schiele & Schön GmbH, Berlin (1981)
4. Eye Com Handbook (1), 1977
5. Perthen, B., Müller-Beißenhirtz, W.: in Electrophoresis 79, 695-699, de Gruyter (1979)
6. Stibler, H.: Journal of the Neur. Sciences 36, 273-288(1978)
7. Radola, B.J.: Electrophoresis 1, 43-56 (1980)
8. Ufrecht, R.: Studienarbeit Nr. 340 (1982). Institut für Chem. Verfahrenstechnik, Universität Stuttgart, Germany
9. Schmidt, K.G., Raff, M., Blenke, H., Polony, G.: in Electrophoresis 82, de Gruyter (1982)
10. Stachelberger, H.: in Electrophorese-Forum 80, 193-198 Editor: Radola, B.J. (1980)
11. Delincée, H., Radola, B.J.: Anal. Biochem. 90, 609-623(1978)
12. Radola, B.J.: Biochem. Biophys. Act. **295**, 412-428 (1973)
13. Allen, R.C., Christopher, J., Ovincz, L., Allen, R.C. jr., Liu, P.: in Electrophorese-Forum 80, 116-125 Editor: Radola, B.J. (1980)
14. Malmud, D., Drysdale, J.: Anal. Biochem. 86, 620-647 (1978)
15. Erhardt, R.: Analyse und Entwurf optoelektronischer Systeme zur schnellen digitalen Bilddatenerfassung und -verarbeitung, Thesis Universität Stuttgart 1980
16. Rilbe, H.: Ann. N.Y. Acad.Sci. 209, 11-22 (1973)
17. Lemkin, P.F., Lipkin, L.E.: in Electrophoresis 81, 401-411 de Gruyter (1981)

EFFECTIVE FIXATION AND RAPID STAINING OF PROTEINS IN SDS-PAGE.

Rapid Staining for SDS-PAGE

Birgit C. An der Lan

Bureau of Biologics, Division of Bacterial Products, Food and Drug Administration, Bethesda, MD 20205

J. V. Sullivan

Division of Research Services, Biomedical Engineering and Instrumentation Branch, National Institutes of Health, Bethesda, MD 20205

Andreas Chrambach

Endocrinology and Reproduction Research Branch, National Institute of Child Health and Human Development, National Institutes of Health, Bethesda MD 20205, USA.

Abstract

In SDS-PAGE, using conventional staining of proteins on cylindrical or slab gels of several millimeter thickness, electrophoresis and pattern development cannot be achieved in a single workday, because the presence of SDS interferes with the customary fast "no-background" staining procedures of PAGE available in the absence of SDS. Furthermore, the conventional procedures of protein staining in SDS-PAGE rely on protein fixatives which, for many proteins (e.g. membrane or acid soluble proteins), cannot be regarded as effective. To solve these problems, 1) the fixative, 12.5% TCA, was used with 45% methanol to remove SDS from the gel cylinders prior to staining. The function of methanol in the fixative is to increase the CMC of SDS to a value which ensures the monomeric rather than micellar structure of the detergent, and therefore the rapid diffusion of SDS from the gel; 2) the perforated gel holders in the diffusion apparatus of Duhamel et al, 1980, were

Electrophoresis '82

modified by making them of TCA-resistant plastic. 3) The minimum time for SDS-diffusion from the gel compatible with protein staining by the procedure of Diezel et al., 1972, was determined, using standard proteins. This time was found to be the same as that needed for the red color of the moving boundary front marker, pyronin Y-SDS, to completely disappear from the gel. Representative minimum times for SDS removal are 2 h for a 5 %T, 15 $\%C_{DATD}$ Gel and 3 h for a 9 %T, 2 $\%C_{Bis}$ gel, using gels at a steady-state level of 0.03% SDS (Wyckoff et al., 1977). At least 1 more hour is needed to diffuse the methanol out of the gel by equilibration with 12.5% TCA, but Coomassie G-250 can be added during this equilibration step. Thus, for the development of protein patterns on polyacrylamide gels 3.5 and 4.5 h were required at the 2 gel concentrations respectively. 4) The presence of charcoal does not increase the rate of removal of SDS from the gels, but it helps to clarify the fixative by adsorption of insoluble contaminants that otherwise tend to adhere to the gels.

Introduction

Conventionally, stained gel patterns in polyacrylamide gel electrophoresis in SDS-containing buffers (SDS-PAGE) are obtained by fixing in methanolic acetic acid solutions and staining with Coomassie Brilliant Blue R 250 [e.g. (1,2)]. Such procedures have 2 major defects: 1) The methanolic acetic acid mixtures are not universally effective protein fixatives. Thus, protein zones soluble in these media can disappear during staining. An example is a preparation of mouse liver microsomal proteins shown in Fig.5 of (3). 2) The gel background stains and lengthy destaining is required, preventing one from obtaining the gel pattern on the day of electrophoresis.

To overcome the first problem, the universally effective protein fixative, trichloroacetic acid (TCA), was applied to SDS-PAGE under conditions which provide for the efficient removal of SDS from the gel. Removal was brought about by efficient stirring of the fixative in which the gel was suspended, using a modification of an apparatus described by Duhamel et al. (4). Furthermore, SDS removal from the gel required increasing the critical micelle concentration (CMC) of SDS by the addition of methanol to TCA, thus keeping the detergent in its monomeric form (5). Once SDS was removed, the protein staining procedure of Diezel et al. (6) could be applied which gives no stained background, is carried out in 1 h on a gel equilibrated with TCA and provides sufficient TCA levels in the gel to maintain effective fixation of proteins during storage. This report describes the resulting staining procedure which allows for protein staining in SDS-PAGE on the same day as electrophoresis.

Materials and Methods

1) Modification of the destainer of Duhamel *et al.* (4): The diffusion destainer of Duhamel *et al.* (4) was modified as follows: a) The perforated gel holders, Part A, Fig.1, (for cylindrical gels of 6 mm) were constructed from 10 ml polymethylpentene measuring cylinders (Markson Cat. No. R-5923). Polymethylpentene was found resistant to cracking or clouding after three weeks' continuous exposure to 12.5% TCA, 45% methanol. By contrast, other plastics, such as polycarbonate (Lexan) or polystyrene, became cloudy after a few hours. Longer times were not tested, and it is therefore probably advisable not to store the holders in this solvent permanently. The hole size was 5 mm in diameter. The gel holders were numbered by 1/12, for gel identification, using a heated stencil. b) The reservoir (Part B, Fig.1) was a 1 l beaker, cut to eliminate the pouring rim, which could fit under a tightly sealing lid (Part C). c) The lid (Part C) was turned on a lathe from a Teflon block. d) The adsorbent (charcoal) drum (Part D) was constructed from polypropylene beakers as described by Duhamel, except that the Nylon mesh was replaced with TCA resistant Propyltex mesh (Tetko Inc., 420, Saw Mill River Road, Elmsford, NY 10623, Mesh No. 5-40-420, width 55). The mesh is not very pliable. It therefore needs to be heated in boiling water so that it will mould around the polypropylene cylinders. The drum was not more than half filled with activated charcoal (14 x 40 mesh, Hoefer Scientific, Cat. No. SE 535).

2) SDS-PAGE: SDS-PAGE was carried out in 6 mm gel tubes using the procedure of Wyckoff *et al.* (7), which employs 0.03% SDS (BDH) in the upper buffer reservoir only [as compared with 1% in the more commonly used procedure of Weber and Osborn (1)]. For standard SDS-proteins the high molecular weight kit of Bio-Rad (Cat.No.161-0303) was used and the mixture boiled according to the manufacturer's specifications.

3) Staining of SDS-proteins: Proteins were fixed in methanolic solutions of either acetic acid or TCA, as described under Results. Proteins were stained by the procedure of Diezel *et al.* (6).

Results

1) The conventional fixative for SDS-proteins, acetic acid, can be replaced by 12.5% TCA, which is a more effective fixative, especially for acid soluble proteins: Large molecular weight (MW) standard proteins (Bio-Rad kit) were subjected to SDS-PAGE [9 %T, 2 $\%C_{Bis}$, discontinuous

buffer system 4217.II.IV. (7), 25°C, 6 mm diameter gels] and fixed for 3 h in the presence of a charcoal drum (contained in Part D, Fig.1) A) in 9.2% acetic acid, 45% methanol (1), and B) in 12.5% TCA. They were then equilibrated with 12.5% TCA for 60 min (until gels cleared) and stained by the procedure of Diezel et al. (6). Staining intensities in the 2 cases were indistinguishable (Fig.2).

2) Staining intensity is a function of methanol concentration in the fixative. Large MW standard proteins (Bio-Rad), separated by SDS-PAGE on gels (1.2 ml Resolving Gel/tube) as described in section 1) were fixed for 3 h in 650 ml 12.5% TCA containing 0, 20 or 45% methanol, equilibrated with TCA and stained as described in section 1). Staining is most intense, and therefore resolution is best, in the gel fixed in the presence of 45% methanol (Fig.3).

3) A fixation time of 3 h is sufficient for 5 %T, 15 $\%C_{DATD}$ gels. Standard proteins, stacked by SDS-PAGE as described in section 1), fixed under condition A), section 1), followed by equilibration with 12.5% TCA for 3.5, 5.5 or 16 h and staining as described in section 1), give stained patterns which are indistinguishable (data not shown). The red color of pyronin-Y-SDS, used to mark the moving boundary, was no longer visible after 3 h.

4) Stain intensities are independent of the presence of charcoal during fixation. Polyacrylamide gels of standard proteins (as in section 1) were fixed under conditions A), section 1), for 3 h in the presence and absence of a charcoal drum (Part D, Fig.1). The drum was half filled with charcoal. The gels were then equilibrated with TCA and stained as described in section 1). There was no difference in staining intensity or resolution between gels fixed in the presence or absence of charcoal (data not shown). However, the presence of charcoal often results in cleaner gels (there appears to be some TCA-precipitable material which adheres to the gels and is removed by filtration through the charcoal drum).

5) Recommended fixation and staining procedure for SDS-PAGE. After electrophoresis the gel is placed into the gel holder (Part A, Fig.1) and immersed in the assembled fixation apparatus (Fig.1) filled with 650 ml 12.5% TCA, 45% methanol. Fixation is carried out with magnetic stirring for 3 h with a charcoal drum in place. Fixation is continued until the red zone of pyronin Y-SDS (the moving boundary marker) has diffused from the gel. An alternative way to gauge the disappearance of SDS from the gel would be to monitor the TCA for $BaCl_2$ precipitable sulfate; but this would entail changing the TCA/methanol solution repeatedly. After fixation the gels are removed from their holders

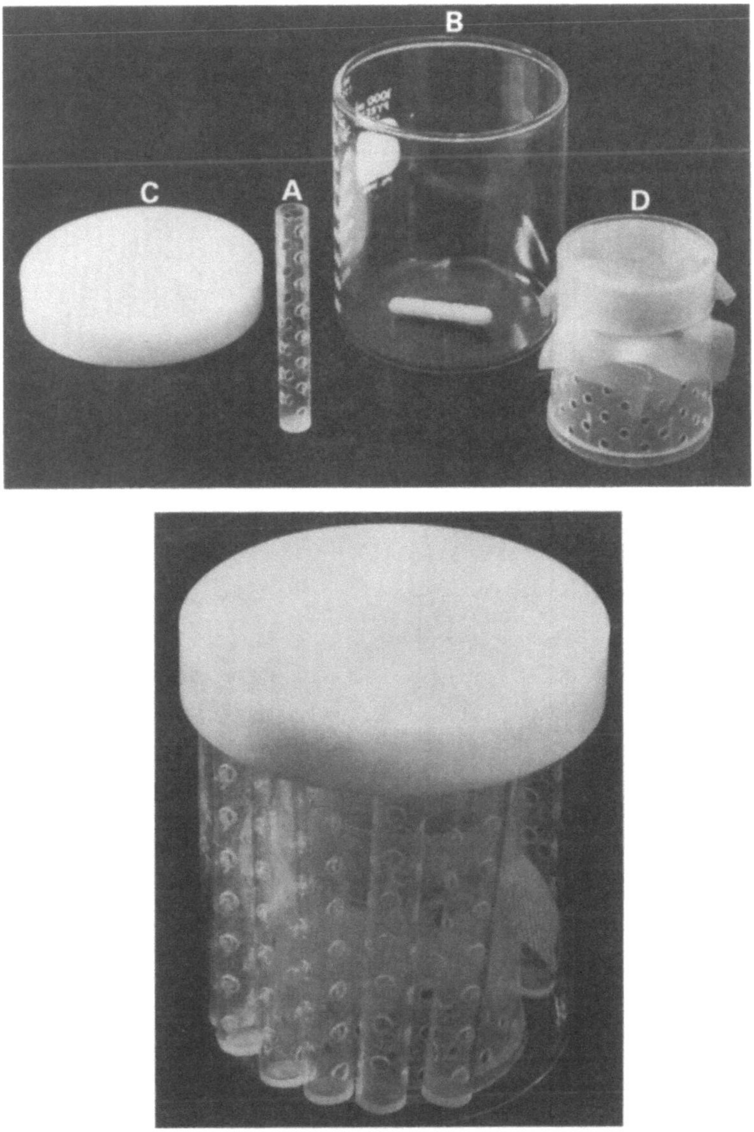

A) Perforated holder for 6 mm diameter gels. Length: 105 mm. Hole diameter: 5 mm. B) Beaker (1000 ml) with ground surface, containing a magnetic stirring bar of 50 mm length. C) Teflon top. D) Polypropylene perforated drum with a floor of Propyltex netting at 3/4 height, held in place with an inner polypropylene sleeve, and a Propyltex top held in place by a polypropylene sleeve (all polypropylene parts are cut from 250 ml Tricorner beakers, Fisher Scientific, Cat.No.02-593-50C). The space enclosed by the netting serves as the adsorbent (charcoal) reservoir. Height: 80 mm. Bottom panel: Assembled apparatus.

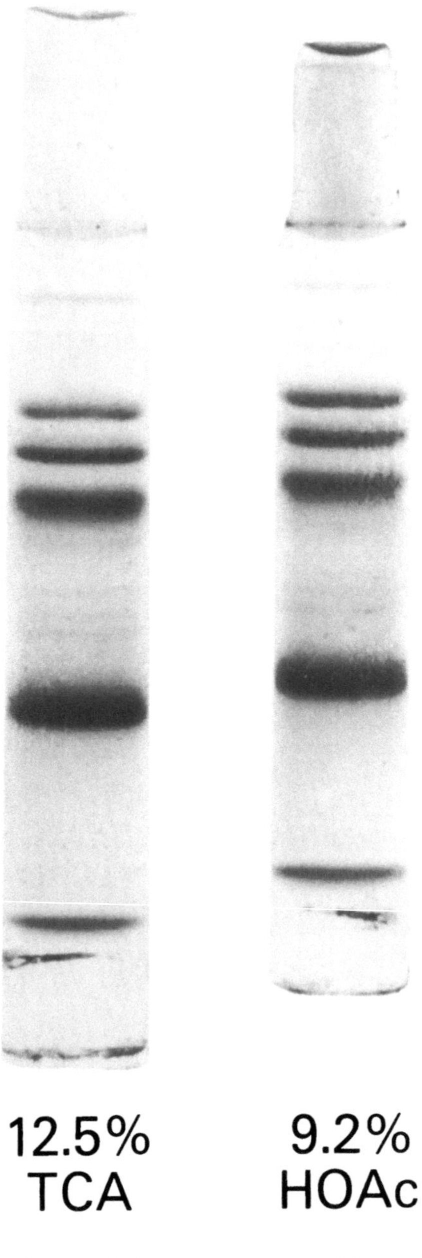

Fig.2: SDS-Protein (Bio-Rad large MW standards) patterns in SDS-PAGE [9 %T, 2 $\%C_{Bis}$, discontinuous buffer system 4217.II.IV.(7) with 0.03% SDS in the Upper Buffer only, 25°] developed by fixation for 3 h in 12.5% TCA, 45% methanol, or in 9.2% acetic acid, 45% methanol, as indicated in the figure, followed by removal of methanol for 1 h in 12.5% TCA and staining with Coomassie Blue G-250 for 1 h.

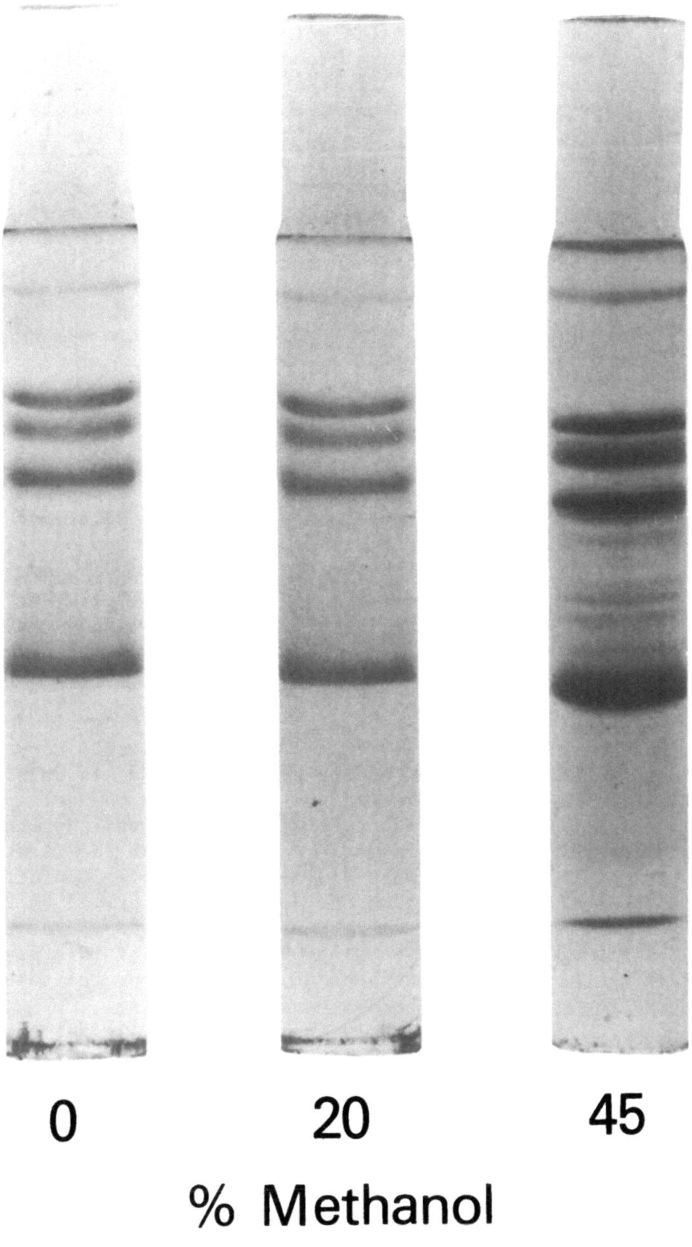

Fig.3: SDS-protein patterns obtained as in Fig.2 except that SDS was removed from the gel in 12.5% TCA containing variable proportions of methanol as shown in the figure.

and placed in 50 ml measuring cylinders containing 20 - 50 ml 12.5% TCA, depending on the length of the gel (i.e. enough TCA to cover the gel). At this point the gels turn cloudy. Equilibration with TCA is continued until the gels clear. In the representative case of a 9 %T, 2 $\%C_{Bis}$ gel clearing took 1 h. More concentrated gels should require longer times. After equilibration the gel is stained for 1 h by adding 2 ml 0.25% aqueous Coomassie Brilliant Blue G-250 to the cylinders (i.e. final concentration 0.01%), and inverting a few times to mix the solutions, as described by Diezel et al. (6). The stain can also be added at the beginning of equilibration without detriment. Gels are stored in tubes of not more than 5-6-fold the gel volume, containing 5% acetic acid. No destaining is required.

Discussion

To date, SDS-PAGE of proteins has been plagued by lengthy, overnight destaining procedures which precluded the visualization of gel patterns on the same day as that on which the electrophoresis was carried out. To overcome this problem, and to be able to apply one of the staining procedures used in PAGE of native proteins, which does not require destaining [e.g. (6), (8)], it is necessary to remove SDS from the gel prior to staining. Such removal depends on converting the micellar form of the detergent, which prevails at the SDS concentrations in the gel (0.03% to 1%), and which has a particle size very similar to that of an average sized protein (5), to the monomeric form. Conversion is achieved by increasing the CMC of SDS by adding solvents such as methanol to the fixative. The rate of diffusion of SDS from the gel appears to be mainly a function of the flow rate of the fixative. To achieve an efficient flow rate, both rapid magnetic stirring of the fixative and good contact of the fixative with the gel are needed. Contact appears better in the perforated tubes of the apparatus described by Duhamel et al (4) than in the commercial diffusion destainer (Hoefer Scientific, Cat. No. SE 530; Bio-Rad Cat. No. 165-0910). In part this may be due to the fact that in the Duhamel apparatus, the perforated tubes slowly rotate under the effect of vigorous magnetic stirring.

Charcoal does adsorb SDS rapidly from solution (4), but surprisingly it appears without a significant effect on the diffusion rate of SDS from gels. Apparently, the adsorption from solution is not rate determining, but rather the diffusion rate from the gel. Nonetheless, there is a small advantage in using charcoal in the drum of the diffusion apparatus: It appears to adsorb TCA precipitable impurities that tend to adhere to gels when they are fixed in 12.5% TCA.

The possibility of using TCA rather than aqueous acetic acid solutions as the protein fixative is the second attractive feature of the proposed fixation and staining procedure for SDS-PAGE. It is a far more general and effective medium for protein fixation than acetic acid, particularly for membrane and acid soluble proteins (such as mouse liver microsomal proteins and filamentous hemagglutinin) which tend to solubilize with time of exposure to the conventional acetic acid fixatives, or fail to fix altogether [e.g. see Fig.5 of (3)]

The time required for diffusion of SDS from gels increases with a) the concentration of SDS in the gel and b) gel concentration. a) For the purpose of making the staining time as short as possible, an SDS procedure using the minimum amount, such as that of Wyckoff et al. (7), is preferable to those that include 0.1% SDS in the gel, such as (1), (2) and (9). The minimum concentration of SDS required is that necessary to saturate the protein, which may differ for different proteins. This should be determined by varying the SDS concentration until constant mobility is achieved, as was done for several standard proteins [Fig.10 of (7)]. b) Longer diffusion times may be necessary for gels that are more concentrated than the 9 %T used in this study. These can be gauged by the time taken for Pyronin-Y-SDS to diffuse from the gel, as described above.

References

1) Weber, K. and Osborn, M.: J. Biol. Chem. 244, 4406-4412 (1969).

2) Fairbanks, G., Steck, T. L. Wallach, D. F. H.: Biochemistry 10, 2606-2617 (1971).

3) Hjelmeland, L. M., Nebert D. W., Chrambach, A.: Anal. Biochem. 95, 201-208 (1979).

4) Duhamel, R. C., Meezan, E., Brendel, K.: J. Biochem. Biophys. Methods 4, 73-80 (1981).

5) Hjelmeland, L. M., Nebert, D. W., Chrambach, A.: Electrophoresis '78 (Catsimpoolas, N., ed.) Elsevier, North Holland Publ. Co., Amsterdam-New York NY, pp. 29-56 (1978).

6) Diezel, W., Kopperschlaeger, G., Hoffman, E.: Anal. Biochem. 48, 617-620 (1972).

7) Wyckoff, M., Rodbard, D., Chrambach, A.: Anal. Biochem. 78, 459-482 (1977).

8) Chrambach, A., Reisfeld, R. A., Wyckoff, M., Zaccari, J.: Anal. Biochem. 20, 150-154 (1967).

9) Laemmli, U.K.: Nature 227, 680-685 (1970).

FAST VISUALISATION OF PROTEIN BANDS BY IMPREGNATION IN POTASSIUM PERMANGANATE AND SILVER NITRATE

W. Ansorge
EMBL, Meyerhofstr. 1, Postfach 10.2209, D-6900 Heidelberg, Fed. Rep. of Germany

Introduction

Several techniques using silver for detection of protein bands have been reported in the last few years, see for example recent reviews (1,2).

A silver visualisation technique using impregnation in a weak solution of $KMnO_4$ and a copper salt was presented by us at the conference on Electrophoresis in Charleston, 1981. Here we show this technique with some modifications which help to reduce significantly the total time needed for the process (down to about one hour for 1 mm thick gel). The results have shown the technique to be reproducible and applicable in the gel thickness range of 0.05-2 mm. (Gels thicker than 2 mm have not been tested, but it is very likely that the process can be applied to them as well.) Amounts as low as 2×10^{-10} of protein per band were detected. Protein bands of different colours are obtained.

The characteristics of this technique show that its chemical process differs from those used in the other silver detection techniques. As with all of the silver mehods, it is important to use the same process for reproducibility, as addition of other chemicals to some steps can change not only colours of the bands but also their relative intensity. Further evaluation is necessary to see if all proteins are detected by the silver method. The technique has been applied with good results to the continuous, discontinuous and two-dimensional protein and DNA separations.

Electrophoresis '82

Advantages and applications of ultrathin (< 0.2 mm) gels are described in (3-8). Wider application had been limited by problems in the preparation of larger bubble-free gels and in their handling. With the sliding technique and gel casting device (3-6), gels in the thickness range of 0.01-0.2 mm and up to 1m length (or longer) can be reliably prepared. The technique and device are also used to cast ultrathin films of agarose and agarose-polyacrylamide mixture with a heated thermostatting plate. The separation procedure developed for protein separation on ultrathin gels is very fast (about 40 min on a 20 cm long plate), gives sharp bands and is very sensitive. The sensitivity is further increased by the silver staining technique, performed in a short time (∿25 minutes) on the very thin gel films.

Materials and Methods

$CuCl_2 \times 2 H_2O$, $KMnO_4$, K_2CO_3, 37% formaldehyde solution (all analytical grade) were from MERCK, Darmstadt, FRG. Diethoxydimethylsilane from FLUKA AG, Buchs, Switzerland. Trichloromethylsilane from MERCK-Schuchardt, Munich, FRG. The methacryloxypropyl trimethoxysilane was from WACKER Chemie, Munich, FRG. Acrylamide was from BDH, N,N'-methylene bisacrylamide from SERVA. REPELCOAT was obtained from Hopkin & Williams Ltd., Rumford.

The electrophoresis apparatus was essentially as described in (5), buffers and solutions as described in (3). The 1 mm and thicker gels were cast with the standard technique by pouring.

The ultrathin gels were formed between two glass plates (200 x 200 x 4 mm), one of which had a notch (165 x 25 mm) cut out from the top to allow contact between the gel and the buffer of the upper electrophoresis chamber. All glass plates used were cut from floated glass. Spacers (500 x 15 mm) and the well formers for 4 mm wide and 8 mm deep slots were cut from the same piece of PTFE. A cheap thermostatting plate was designed and constructed meeting the high requirements on surface planarity and heat transfer capacity. The water was heated and circulated using a water thermostat from LKB. Application of samples into the slots was done with thin-walled glass capillaries (outer diameter $\leq$ 0.1 mm) for X-ray analyses (A. Müller, Glas- und Vakuumtechnik, Berlin). An LKB 2103 unit (2000 V) power supply was used.

Treatment of glass plates. To bind the ultrathin gel onto its support (glass or polyester) it was treated as shown in (3,5). On glass the whole treatment is done at room temperature. Three silanes were found to provide binding of the polyacrylamide gel onto the glass plates (diethoxydimethylsilane, trichloromethylsilane, and methacryloxypropyl- trimethoxysilane). From these the last one, the relatively non-hazardous methacryloxypropyltrimethoxysilane has been used extensively, as reported in (3). It also seems to give the best background in the silver staining technique. REPELCOAT was used to repel gels from the other surface in contact. The plate was wet at room temperature with concentrated REPELCOAT solution, dried in a fume hood for 5 min and then washed with ethanol before use.

Removal of the covalently bound gel from a glass plate. The glass plate with bound gel was placed overnight in 10% DECONEX (BORER Chemie AG, Solothurn, Switzerland) solution (or 1/2 hour in concentrated DECONEX) and then brushed under tap water. Before reuse it was further washed with deionized water and finally with ethanol.

Visualisation of protein bands. After electrophoresis, gels are treated as shown in table I. We use volumes of 200-300 ml and a shaker is used to agitate the solutions (at about 40 rpm) with gels thicker than 0.2 mm. The agitation is not needed for ultrathin gels; with agitation time is halved.

Table 1. Procedure for visualisation of protein bands. (Solution A is 10% ethanol, 5% acetic acid)

Step:	gel thickness 1 mm	0.2 mm
1: 50% methanol, 12% TCA, 2% $CuCl_2$	20 min	5 min
2: Solution A	10 min	5 min
3: 0.01% $KMnO_4$	10 min	5 min
4: Solution A	10 min	2 min
5:10% ethanol	10 min	5 min
6: H_2O	10 min	5 min
7: 0.1% $AgNO_3$	10 min	5 min
8: H_2O dip about 20 sec, followed by 10% K_2CO_3 for 1 min		Developer 9 for 20 sec
9: Developer 0.01% formaldehyde, 2% K_2CO_3	3-4 min	2-3 min
10: Solution A	1-2 min	20 sec

Then gel into H_2O for 15 min, store and seal in a plastic file cover or dry the gel. When overdeveloped, the stained background (and the bands) can be reduced using a solution of about 0.5% $Na_2S_2O_3$, followed by H_2O.

As shown in (4), after the development in step 9 and brief (about 10 minutes) water wash, one may repeat the procedure from silver nitrate (step 7) for even greater sensitivity.

We found considerably increased sensitivity especially for low molecular weight proteins, when 20-25% TCA is used in the first step, when 0.1% formaldehyde is added to step 1 and 0.01% KOH to step 3.

The visualisation process reported by us at the Electrophoresis Conference in Charleston, 1981, differed from the procedure in table I mainly in the total time of the procedure (steps 1, 2, 5 and 7 were considerably longer) and in the use of $CuSO_4$ in step 3 instead of $CuCl_2$ in step 1.

Results and Discussion

Comments on the visualisation procedure. For gels thicker than 1 mm, the same procedure (Table 1) is used, but duration of the steps may be increased (by about five minutes), and especially the times of steps 1 and 5 seem to be important. We found that when the gel is left in the solution of step 1 for a long time, the stacking gel may turn green in the development. With ultrathin gels, the first and second steps should not be applied for longer than 5 minutes, as a longer wash may result in lower band resolution. For gels thinner than 0.2 mm step 1 and step 2 should be shorter than 5 minutes.

When gels thicker than 0.2 mm cannot be stained immediately after the electrophoresis run, it is recommended to fix the gel for 15 minutes in 50% meth/20% TCA and store it in 10% ethanol/5% acetic acid. The visualisation procedure is then started again with step 1.

The procedure can be significantly shortened by cutting down each step (except steps 3 and 8) by about 5 minutes, at the cost of darker background. Good results were obtained with changing trays for each step, as well as with replacing solutions in the different steps when using only one tray.

Further results and detailed discussion of the visualising technique will be given in 10.

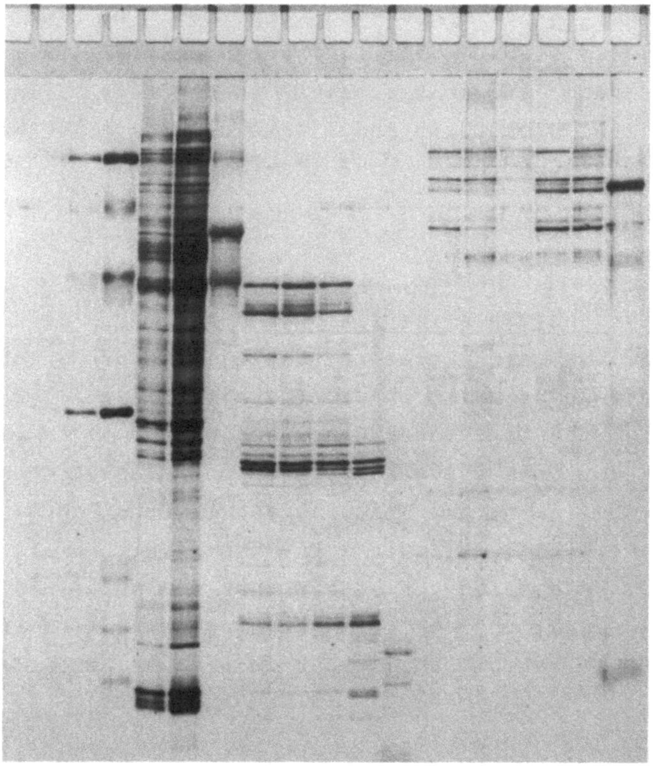

Fig. 1: Cellular extract proteins separated and visualised on 0.2 mm thick gel. Up to 20 Watts power can be applied to the 20 x 20 cm thermostated gel, running time 30-40 minutes.

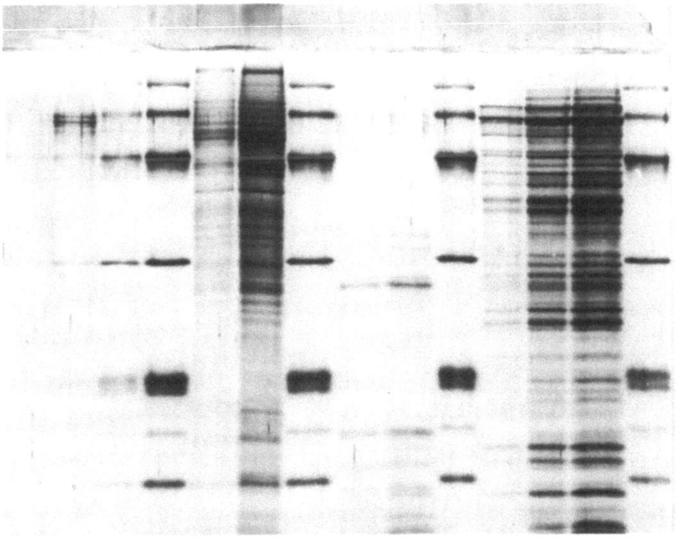

Fig. 2: Cellular extract proteins separated and visualised on 1 mm thick gel. Up to 12 Watts power can be applied to the 20 x 20 cm thermostated gel, running time about 2 hours at 300 V.

Acknowledgment

I thank Kai Simons for support, Graham Warren for discussions and John Dickson for technical assistance in the course of the work.

References

1. M. Poehling and V. Neuhoff, ELECTROPHORESIS (1981), Vol 2, 141-147

2. D.C. Ochs, E.H. McConkey and D.W. Sammons, ELECTROPHORESIS (1981), Vol. 2, 304-307

3. W. Ansorge and L. De Maeyer, Chrom. 202, 45-53 (1980)

4. W. Ansorge: Proceedings of the Electrophoresis Forum, Ed. B.J. Radola, Technical University, Munich, October 1980

5. W. Ansorge and H. Garoff: Proceedings of Eelectrophoresis '81, Ed. Allen, Arnaud. Walter de Gruyter & Co., 635-646 (1981)

6. H. Garoff, W. Ansorge: Anal. Biochem. 115, 450-457 (1981)

7. Görg, A.: Proceedings of Electrophoresis '81, Ed. Allen, Arnaud. Walter de Gruyter & Co., Berlin (1981).

8. B. Radola: Proceedings of Electrophoresis '81, Ed. Allen, Arnaud. Walter de Gruyter & Co., Berlin (1981).

9. Metal ions in biological systems. Vol. 3, p. 62, Ed. H. Siegel. Marcel Dekker, Inc., New York, 1974.

10. W. Ansorge, to be submitted

SILVER STAIN IMMUNOFIXATION FOR α_2HS-GLYCOPROTEIN: A NEW METHOD FOR DETECTION OF PROTEIN HETEROGENEITY

Diane Wilson Cox and Brenda J. Andrews

Research Institute, The Hospital for Sick Children and University of Toronto, Toronto, Canada

Introduction

The high resolution provided by isoelectric focusing (IEF) in acrylamide has allowed the detection of new genetic polymorphisms in serum proteins such as transferrin (1), vitamin-D binding protein or group specific protein (Gc) (2), and α_1-antitrypsin (3). Transferrin and α_1-antitrypsin are present in sufficiently large amounts that the gel can be stained with a protein stain such as Coomassie Blue R250 and the prominent bands of these two proteins can be detected. The use of immunofixation, either directly in the gel (4), or applied on a cellulose acetate membrane (5), followed by Coomassie Blue R250 staining of the immune precipitate, has made possible the identification of heterogeneity in a specific serum protein assayed in a mixture of proteins found in whole serum.

We have used the principal of immunofixation in polyacrylamide gels followed by silver staining of the immune precipitate. This method, which we call 'silver stain immunofixation', makes possible the examination of heterogeneity of any protein for which anti-serum is available, even when the protein is as low as 0.3 µg/ml. The technique is equally useful for proteins present in higher amounts in serum but for which low titer anti-sera are available. We have applied this technique to the study of genetic heterogeneity of α_2HS glycoprotein (α_2HSG).

α_2HSG is present in serum at a concentration of approximately 40-80 mg/dl; the molecular weight of its single polypeptide chain is about 50,000 daltons, total carbohydrate is 13.4%, sialic acid content is 4.1% (6). The

Electrophoresis '82

function of this serum protein is not entirely clear. However it appears to have opsonic properties, stimulating phagocytosis of mouse macrophages (7) and of human monocytes (8). Serum concentrations of α_2HSG are decreased in patients with malignancy (9).

In a study of two dimensional electrophoresis of plasma proteins, Anderson and Anderson (10) have described two electrophoretic variants of human α_2HSG, which they named N and L. Using silver stain immunofixation, we have observed polymorphism of α_2HSG, with three codominant alleles.

Materials and Methods

Serum was obtained from clotted venous blood from 68 unrelated normal individuals. Sera were obtained from 21 family units with two or more offspring. Sera were stored at -20°C for no longer than six months or at -70°C for longer periods.

The sera were diluted 1:88 with 0.1M β-alanine. In some tests, sera were treated for two hours at room temperature with a final concentration of 15 mM dithioerythritol and were diluted 1:44 with 0.1M β-alanine prior to electrophoresis. Samples of 15 µl were applied to the IEF gel on 3 x 9mm LKB IEF application strips, about 1cm from the cathode strip. IEF acrylamide gels were prepared as for α_1-antitrypsin (PI) typing (3), with slight modifications: C = 3%, T = 6%. Gels 1mm thick were poured on polyester sheets treated with Silane A 174 (Pharmacia). The final concentration of ampholyte was 2%, using a 2:1 mixture of pH 3.5-5 and pH 4-6 ampholine from LKB-Produkter AB. The run was carried out on an LKB 2117 Multiphor apparatus with an LKB 2301 power supply.

Electrode strips were soaked in 0.1M β-alanine for the cathode and 0.04M L-glutamic acid at the anode. The total time of the electrophoretic run was three hours: one hour prefocusing at maximum settings of 1,000V, 45MA, 30W, with a further two hours after sample application at maximum settings of 1400V, 45MA, 30W.

After the electrophoretic run, undiluted α_2HSG anti-serum (Lot #001-1, Atlantic Antibodies, Westbrook, Maine) was applied on the gel surface. The gel was incubated for two hours in a humidity chamber and was washed in 0.9% NaCl for 72 hours. The immunoprecipitate remaining in the gel was stained according to the method of Oakley et al. (11) with elimination of the first fixation solution.

Results

A series of bands, precipitated by the α_2HSG anti-serum, was observed in the gel, focusing between pH 4.7 and 4.9 (Fig. 1). Four different band patterns were observed. These were interpreted as resulting from various combinations of three alleles at the locus which we have designated, according to nomenclature guidelines (12), as *HSGA*.

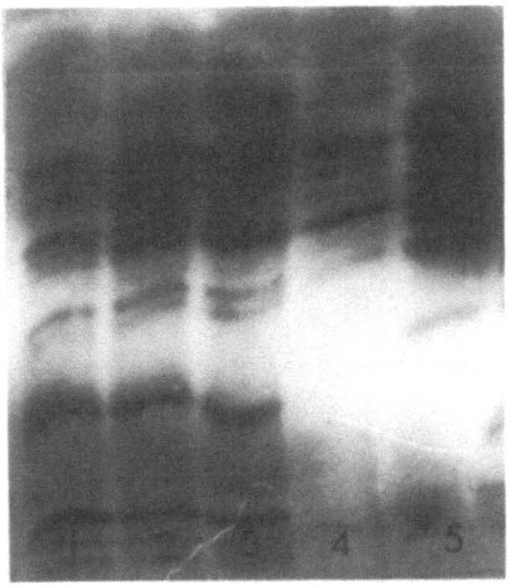

Fig. 1. Electrophoretic pattern of α_2HS glycoprotein by IEF using silver stain immunofixation. Phenotypes are as follows: 1: HSGA 2-2, 2: mixture of equal amounts of HSGA 1-1 and 2-2, 3: HSGA 1-2, 4: HSGA 1-1, 5: HSGA 1-3. Anode is at top.

A study of 21 families with a total of 67 offspring was consistent with the hypothesis of three alleles: *HSGA*1*, *HSGA*2*, and *HSGA*3*. These allele products resulted in four observable phenotypes. The band patterns of some of these phenotypes are shown in Figure 1. The other possible phenotype, HSGA 2-3 was not observed in this series.

Mixing of a serum of type HSGA 1-1 with HSGA 2-2 produced a pattern indistinguishable from that of HSGA 1-2 (Fig. 1).

Phenotypes of α_2HSG observed in 68 normal controls are shown in Table 1.

TABLE 1

PHENOTYPES OF α_2HS GLYCOPROTEIN

		Phenotype				
	Total	1-1	1-2	2-2	1-3	2-3
Observed	68	27	31	8	2	0
Expected	68	26.7	30.5	8.7	1.3	0.7

The numbers expected if the population is in Hardy Weinberg equilibrium are very close to the observed. The HSGA allele frequencies were calculated to be *HSGA*1:* 0.640, *HSGA*2:* 0.345, *HSGA*3:* 0.015.

Sera typed according to our system were compared with sera generously provided by Dr. N. L. Anderson. HSGA 1-1 was identical to L and HSGA 2-2 was identical to N, as observed by 2-dimensional electrophoresis (10).

Discussion

The technique of silver stain immunoelectrophoresis has made possible the visualization of bands in an IEF gel which could not be observed using Coomassie Blue R250 protein staining. For α_2HSG, the limitation is not in the amount of the serum protein, but in the titer of the commercial antiserum. It has not been possible to obtain an antibody of higher titer. Concentration of the anti-sera would make the typing excessively expensive.

Two of the genetic variants described appeared to be similar to those described previously in 2-dimensional electrophoresis (10). Gene frequencies are very similar to those obtained from 2D electrophoresis: L (*HSGA*1*) 0.64 and N (*HSGA*2*) 0.36 (13). However the third allele, *HSGA*3*, was not observed previously and perhaps could not be distinguished on the 2D gels. Approximately 50 samples can be run on a single IEF gel, in contrast to one sample per 2D gel. The ease of typing a large number of samples and the high degree of polymorphism will make this protein suitable for forensic studies.

The technique of silver stain immunofixation should have wide application in the study of heterogeneity, genetic or non-genetic, in proteins present in mixtures at a low concentration and for which antibody is available. Protein concentrations in the range of 0.3 µg/ml can be detected. The method is also suitable for situations in which only low titer anti-serum is available.

References

1. Thymann, M.: Hum. Genet. 43, 225-229 (1978).
2. Constans, J., Viau, M.: Science 198, 1070-1071 (1977).
3. Cox, D.W.: Am. J. Hum. Genet. 33, 354-365 (1981).
4. Johnson, A.M.: J. Lab. Clin. Med. 87, 152-163 (1976).
5. Arnaud, P., Wilson, G.B., Koistinen, J., Fudenberg, H.H.: J. Immunol. Methods 16, 221-231 (1977).
6. Putnam, F.W.: In Putnam, F.W. (Ed.), The Plasma Proteins, Vol. 1, 1-55, Academic Press, New York (1975).
7. Lewis, J.G., Andre, C.M.: Immunology 39, 317-321 (1980).
8. Lewis, J.G., Andre, C.M.: Immunology 42, 481-487 (1981).
9. Baskies, A.M., Chretien, P.B., Weiss, J.F., Makuch, R.W., Beveridge, R.A., Catalona, W.J., Spiegel, H.E.: Cancer 45, 3050-3060 (1980).
10. Anderson, N.L., Anderson, N.G.: Proc. Nat. Acad. Sci. 74, 5421-5425 (1977).
11. Oakley, B.R., Kirsch, D.R. Morris, N.R.: Analyt. Biochem. 105, 361-363 (1980).
12. Shows, T.B., Alper, C.A., Bootsma, D. et al.: Cytogenet. Cell Genet. 25, 96-116 (1979).
13. Anderson, N.D., Anderson, N.G.: Biochem. Biophys. Research Comm. 88, 258-265 (1979).

DEMONSTRATION OF OLIGOCLONAL IGG IN THE UNCONCENTRATED CSF BY SILVER STAIN

Ulrich Wurster
Labor der Neurologischen Klinik der Medizinischen Hochschule Hannover, Postfach 610180
D 3000 Hannover-61

Introduction

The low protein ($>$0,1 g/l) content of cerebrospinal fluid (CSF) requires a concentration step before the actual electrophoretic separation. Several attempts have been made to abolish the need for prior concentration with its risk for unrepresentative protein loss and other artefacts. This has been sought to achieve by either applying larger sample volumes (disc electrophoresis; Cellogel) or using more sensitive dyes such as nigrosin, c.f.(1). The great potential of a silver-based stain for this problem has been recognized already ten years ago by Kerenyi and Gallyas (2), but their argentometric method of staining agar gels proved to be too cumbersome and unreliable to gain wider acceptance. With the current upsurge of silver staining methods, it appeared logical to test their usefulness for the staining of CSF proteins after IEF.

IEF is the most powerful method today for the demonstration of oligoclonal IgG, produced by the CNS in a variety of neurological diseases. Although not entirely specific for multiple sclerosis (MS), oligoclonal bands in the CSF are of great diagnostic value in clinically uncertain cases, with over 90 % of the MS patients displaying these bands after IEF (for a recent review, see (1)).

Electrophoresis '82

Materials and Methods

Coomassie: 0,015 ml CSF concentrated to 2,0 g/l IgG (Centriflo CF 25, Amicon) and serum diluted accordingly with distilled water are pipetted on pre-cut filter papers provided with the ready prepared electrofocusing gels (PAG plate, pH 3,5-9,5, LKB, and placed in the middle of the gel. Isoelectrofocusing was carried out at 7°C on a Multiphor (LKB): 50 mA, 1200 V, 25 W for 100 min; 1700 V, 40 W for a further 20 min. Gel fixation and staining with Coomassie Blue G employed the procedure D of Vesterberg et al. (3).
Silver: 5x10 mm basins were cut from disposable semi-micro cuvets (Sarstedt) and dipped in siliconoil to prevent leaking of the 0,03 ml sample diluted to 0,02 g/l IgG. Of the 5 silver staining methods (4), (5), (6), (7), (8), tested, the first 3 had given promising results in preliminary experiments. Oakley's (4) procedure is preferred for routine work because of the almost colourless background. Staining is performed in glass-dishes under gentle agitation (30 cycles/min) exept for the fixation, overnight wash and conservation steps.

Traces of the 15 % (w/v) trichloroacetic acid (TCA)/5 % (w/v) sulfosalicylic acid (SSA) fixation (30 min, 60°C) are rinsed off with distilled water. The gel is then washed with 400 ml 50 % (v/v) methanol/10 % (v/v) acetic acid for 40 min and with 5 % methanol/7 % acetic acid for 20 min. After soaking in 200 ml 10 % glutaraldehyde (Merck Nr. 4239) for 30 min it is washed for 3 times 30 min with freshly distilled water and left standing in water overnight. After another water change the next day, the gel is impregnated with an ammoniacal silver solution for 40 min (1,0 ml 24 % (w/v) silver nitrate (Merck) are added dropwise to a vigorously mixed solution of 40 ml 1 M NaOH and 2,8 ml 25 % NH_4OH and made up to 200 ml with water). Prior to development with 400 ml 0,005 % (w/v) citric acid containing 0,2 ml formaldehyde (37 % with 10 % methanol), a 10 min water wash is employed. Staining is stopped after 15-20 min by adding directly to the developer 200 ml of 10 % ethanol/0,5 % acetic acid under continued shaking for 5 min. This solution is then discarded and the gel is washed 4 times over a 30 min period with 200 ml portions of stopper. It is left standing in stopping solution for 1 h or longer before it is soaked for at least 1 h in the preserving solution (stopper plus 10 % glycerol) and dried overnight. After the gel has been conserved by a rolled-on plastic sheet, it can be filed and stored without loss of intensity.

Results and Discussion

Current silver staining methods have been developed for 0,8-1,5 mm thick, notably 2 D, slab gels. Therefore one would expect no problems in the application of these procedures to 1,0 mm thick PAG plates. However, since PAG plates are backed by a thin plastic foil, the various solutions can penetrate the gel in one direction only, thus longer times are needed. Furthermore, expansion or contraction of the gel in certain solutions will cause parts of the gel to come off its support, which will not only alter their staining behaviour, but also increase the risk of tearing during agitation. Excessive dehydration with 50 % methanol as suggested by Wray et al. (8) leads to considerable shrinkage of the gel. These authors have also reported on the silver staining of PAG plates, but a photographic destainer (which is hard to control) had to be used to remove small dark artifacts from the gel surface. A common feature of the silver-based staining methods is their use of 50 % alcohol, either alone (8) or in combination with 10-12 % acetic acid for the fixation of proteins. Only the latest version of Merril's (9) stain

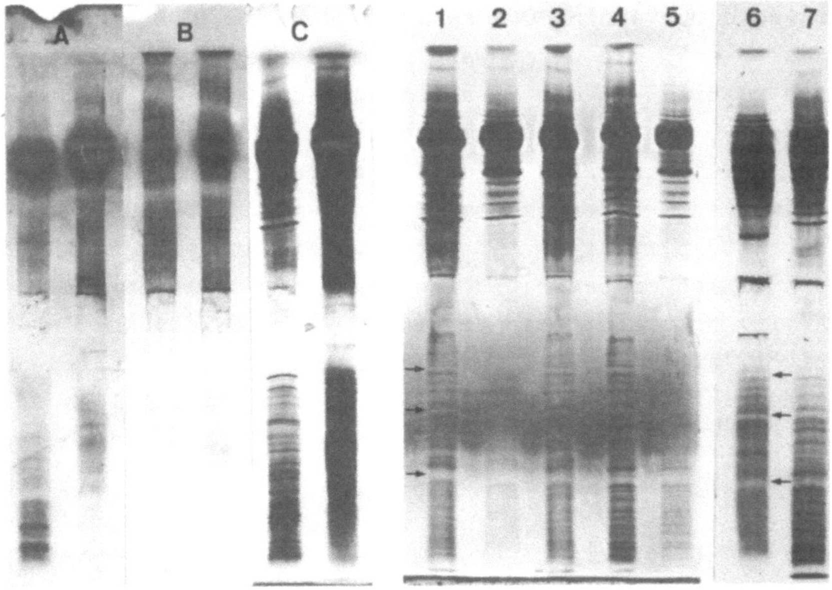

Fig. 1 Effect of various fixatives. A: 50 % ethanol/10 % acetic acid, 2 h, 22°C. B: 50 % methanol/12 % acetic acid, 1 h, 22°C. C: 15 % TCA/5 % SSA, 30 min, 60°C. Left sample: MS CSF 0,06 g/l IgG, right sample: serum 0,2 g/l IgG. 0,015ml were applied on filter paper strips. Silver stain after (5).

Fig. 2 Effect of sample application. 1-5: Silver stain after (4) at 0,32% silver. 1-3: normal CSF, 0,01 g/l IgG. 1: 0,03 ml in basin, 2: 0,02 ml on paper, 3: 0,02 ml in basin. 4-5: MS CSF, 0,02 g/l IgG. 4: 0,02 ml in basin, 5: 0,02 ml on paper. 6-7: Coomassie Blue G, 2,0 g/l IgG, 0,015 ml on paper. 6: MS serum, 7: MS CSF, same as 5. The gaps in the normal IgG pattern have been marked by arrows.

offers 20 % TCA as an alternative. Attempts to fix PAG plates with mixtures of alcohol/acetic acid yielded disastrous results. As Fig.1 vividly illustrates most of the proteins have been leached out. Even those proteins that are fixed to some extent, are surrounded by large haloes underscoring the wash-out effect of these fixatives. On the other hand splendid preservation of the focused proteins was obtained when the TCA/SSA fixation method employed for Coomassie Blue staining was tried. These observations are in accord with the results of Frey and Radola (10) who found 20 % TCA to be the most effective fixative for ultra-thin layer IEF.

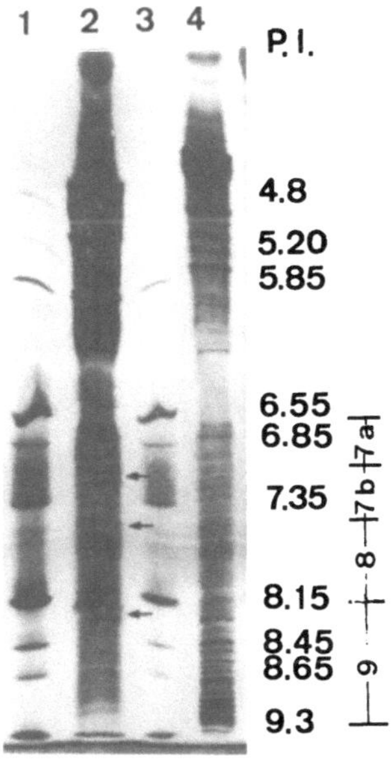

Fig.3 pH gradient profile. Silver stain after (6). 1: high pI marker proteins (Pharmacia, 1+49) with 0,0002 g/l human serumalbumin (Behring), 2: native CSF, 0,019 g/l IgG, 3: same as 1 but in 1+149 dilution, 4: MS CSF diluted 1+4,8 to 0,02 g/l IgG. The pI of the marker proteins are given on the right scale, together with the 4 zones demarcated by the gaps (arrows) in the normal IgG pattern.

The demonstration of a few major oligoclonal IgG bands of undiluted native CSF from MS patients with elevated IgG was quickly achieved by either of the 3 selected silver stains. It proved difficult however, to stain normal CSF and the most basic of the oligoclonal bands, as well as the CSF specific γ-trace band at pH 9,3, so prominent after Coomassie-Blue (Fig. 6). Smyth and Wadström (11) had reported on the adsorptive properties of filter paper strips and it was suspected that cation exchange effects became significant at the minimal amount of protein (>3 ug) applied. Indeed, on pipetting of the samples into plastic basins, directly on the gel's surface, considerable more bands, including the most basic ones, can be recognized in Fig. 2 at enhanced overall intensity, so that a sample with only 0,01 g/l IgG can be readily evaluated.

Artefactual staining occurred in all 3 modifications of the silver stain in the form of continuous wavy bands especially around pH 8 (Figs. 2,3,4). The bands probably represent stained species of ampholytes which have not been removed from the gel despite the extensive washings. Due to their brown-red colour they can be easily discerned from the IgG bands which in our hands are either brown-black (4) or green-black (5) or blue-grey (6). In the latter 2 modifications other colours may be noted, ranging from a bright red for

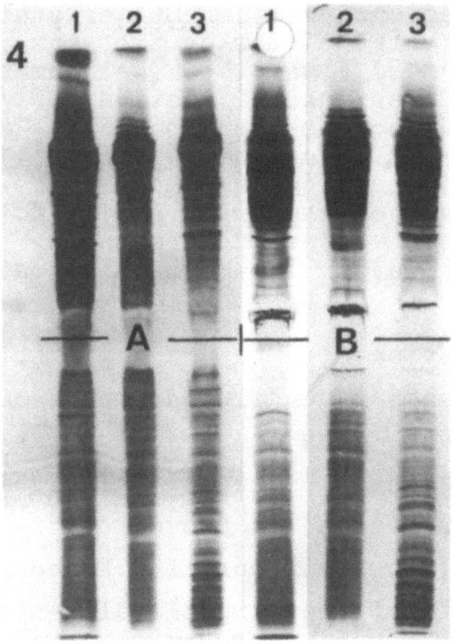

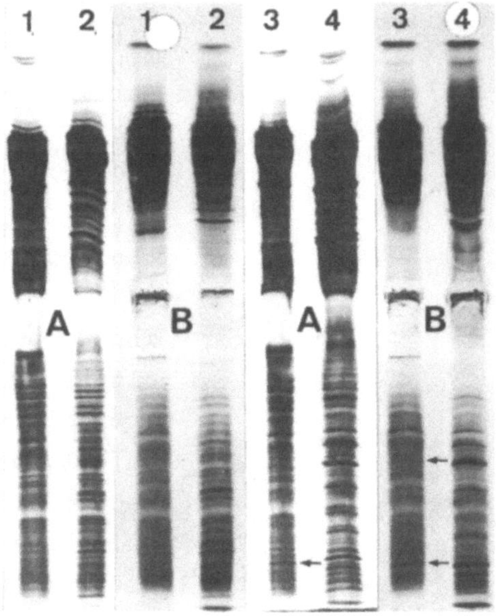

Fig. 4 Iso-pH-lines. A: Silver staining after (5). 1: native CSF, 0,02 g/l IgG, 2: MS serum, 0,02 g/l IgG, 3: MS CSF diluted 1+3,9 to 0,02 g/l IgG, B: Coomassie Blue G, 2,0 g/l IgG.

Fig. 5 Comparison between silver and Coomassie stain. A: silver staining after (4), 0,12 % silver. 1-4: 0,02 g/l IgG, 1,3: MS serum; note the presence of strong CSF bands also in the serum in lane 3 (arrows), 2: MS CSF, dilution 1+3,6, 4: MS CSF, dilution 1+0,1. B: Coomassie blue G, 1-4: 2,0 g/l IgG.

albumin to blue for transferrins. The silver concentration of 0,8 % recommended by Oakley et al. (4), has been lowered to 0,12 %, which not only reduced cost, but also resulted in an almost clear general background. Timing of development is now less critical (15-20 min) and staining sensitivity against all CSF proteins seems to be unimpaired with the exception of albumin, which is actually a welcome effect, because the adjacent proteins may now be visualized as well. Although the two other methods provide a similar level of sensitivity (2 ng human serumalbumin/lane) their yellow background gives a less intense contrast. On the other hand they are free from the annoying but harmless phenomenon of "gel rolling-up during development" which sometimes occurs in the routine version.

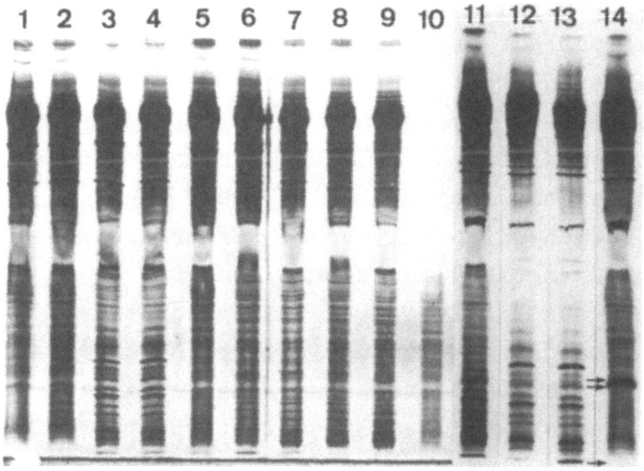

Fig. 6 Changes in IgG patterns of sequential CSF samples. 1-10: silver staining after (4), 0,12 % silver. All samples were diluted to 0,02 g/l IgG. 1-4: MS. 1: 9/77, 2: 11/79, 3: 9/80, 4: 8/81, 5-9: Meningoencephalitis. 5: 31.8.81, 1+4 6: 7.9.81, 1+4,4, 7: 14.9.81, 1+4,7, 8: 21.9.81, 1+9,4, 9: 23.10.81, 1+14,5, 10: human IgG (Miles). 11-14: Coomassie Blue G, 11: same as 2 but conc CSF with 2,1 g/l IgG was applied (basin), 12/13: same as 3/4, 1,7 g/l IgG (paper), 14: conc. MS CSF stored for 2 years at $-20^{\circ}C$, 2,5 g/l IgG (basin).
Note the absence of γ-trace (arrow) in samples 1, 2 and 14 due to storage at $-20^{\circ}C$ and its transformation to a more acidic band (double arrow). The gradual disappearance of γ-trace in samples 5-9 is caused by the increasing dilution of the CSF.

Under the assumption that the artefactual lines really identify carrier ampholytes, they represent iso-pH-lines and can be exploited to follow the course of the pH gradient. Since the lines show a tendency to be displaced to the acidic side in channels with higher protein or salt load, accurate pI values of proteins in a given sample cannot be determined via comparison with neighbouring marker proteins (Fig. 3). A similar decrease of the pH in the alkaline region was described for tube gels, but at much higher protein load (12). Thus it seems more probable that the increased amount of salt contained in undiluted native CSF is responsible for the observed effect. Plotting of the pI of the marker proteins in Fig. 3 against their distance from the cathode gives a linear pH gradient, which may however, be displaced in parallel from gel to gel. Both the imperfect reproducibility and the occasional crowding of bands in front of the cathode may be attributed to uncontrollable minor variations in the cathodic drift. Comparison between different gels and evaluation of a given pattern is considerably facilitated by the 3 gaps interrupting the distribution of normal serum, CSF or isolated human IgG (Figs. 2,3,6). In the 4 zones thus demarcated and designated as 7a, 7b, 8 and 9 (Fig. 3), a fairly constant number of rather broad bands on a diffuse background can regulary be observed. The sharp and often numerous (up to 40) oligoclonal bands contrast quite clearly against this normal pattern. Of course serum is always run side by side and a minimum of 3 CSF unique bands is required for a positive rating.

The marked heterogeneity of normal IgG seen after spreaded electrofocusing (13), confirmed by the present study, appears quite disturbing in view of the enormous variability of polyclonal IgG. Thomas and Hoden (13) have built up very high field strengths over 20 h so that insufficient focusing times are unlikely. Discontinuities of the pH gradient can be ruled out by the observation that oligoclonal bands show up also in the normally blank spaces. Considering the recent calculations of Rodwell (14) that 480 unique charged amino acid compositions of the V_H region will gather in only 41 pH bands, the assumption of a uniform continuum for polyclonal IgG may not be valid. Instead, the IgG molecules tend to cluster at certain positions, giving the impression of distinct banding. On densitometric analysis of Coomassie stained gels and on visual inspection of silver gels the presence of considerable material between the bands becomes evident. Chrambach et al. (15) have expressed the view that as long as distinct bands appear in the pattern of normal IgG, the interpretation of patterns from patients with certain neurological diseases remains ambiguous. Such a statement cannot be accepted in the light of the mostly very clear-cut differences between the constant pattern of normal IgG on the one side and the conspicuous oligoclonal banding on the other. Nevertheless the not absolute specificity of the assay necessitates that a diagnosis of MS is made in the appropriate clinical context anyway.

Comparison of the patterns of conc. CSF stained with Coomassie and native CSF of the same patient stained with silver, reveals almost identical oligoclonal IgG bands, thereby ruling out suspected concentration artefacts (Figs. 4,5). On closer inspection the higher resolution of the silver stain and its ability to depict even very faint bands, which cannot be seen in concentrated CSF because of overlap, can be noticed. The formerly overloaded albumin area has also become more transparent and several proteins can be detected in the CSF with no counterpart in the serum.

An open question in MS research is the constancy of an individual oligoclonal pattern (16). As stated above, limited reproducibility among different gels demands the simultaneous separation of serial specimens of the same patient on the same gel. Silver staining now offers the opportunity to analyze frozen aliquots of CSF of earlier spinal taps or CSF from laboratory animals with EAE too small for concentration. While less apparent after Coomassie, it is quite obvious from the silver stain in Fig. 6 that a certain oligoclonal pattern in MS can be stable over years and then change drastically. In contrast, comparatively rapid alterations were taking place in a case of meningoencephalitis after tick bite.

The protein nature of the silver stained oligoclonal bands has been proved by their congruency with Coomassie Blue, but how can one be certain that they are indeed IgG? CSF-proteins focusing in the pH region over 7,0 are generally believed to represent IgG. This has been repeatedly demonstrated by immunofixation, a method which in its peroxidase-conjugated form, can identify oligoclonal IgG also in native CSF (17). However, because of its limited sensitivity and complicated handling it cannot be regarded as an alternative to the silver stain for routine applications. With the possibility to use unconcentrated CSF for the demonstration of oligoclonal IgG on ready made IEF gels over a distance of 4 cm, the requirements for a standard assay seem to have been met, which should enable the method to gradually replace the less sensitive traditional agar gel electrophoresis.

Acknowledgment:
With support of the HERTIE Foundation, Frankfurt.

References

1. Delmotte, P., Carton, H.: Clin. Neurol. Neurosurg. 83, 183-207 (1981).
2. Kerenyi, L., Gallyas, F.: Clin. Chim. Acta 38, 465-467 (1972).
3. Vesterberg, O., Hansen, L., Sjosten, A.: Biochim. Biophys. Acta 491, 160-166 (1977).
4. Oakley, B.R., Kirsch, D.R., Morris, N.R.: Anal. Biochem 105, 361-363 (1980).
5. Merril, C.R., Goldman, D., Sedman, D.A., Ebert, M.H.: Science 211, 1437-1438 (1981).
6. Morrisey, J.H.: Anal. Biochem. 117, 307-310 (1981).
7. Sammons, D.W., Adams, L.D., Nishizawa, E.E.: Electrophoresis 2, 135-141 (1981).
8. Wray, W., Boulikas, T., Wray, V.P., Hancock, R.: Anal. Biochem. 118, 197-203 (1981).
9. Merril, C.R., Goldman, D., Van Keuren, M.L.: Electrophoresis 3, 17-23 (1982).
10. Frey, M.D., Radola, B.J.: Electrophoresis 3, 27-32 (1982).
11. Smyth, C.J., Wadström, T.: Anal. Biochem. 65, 137-152 (1975).
12. Cantrell, S.J., Babitch J.A., Torres, S.: Anal. Biochem. 116, 168-173 (1981).
13. Thomas, J.M., Hodes, M.E.: Anal. Biochem. 90, 596-608 (1978).
14. Rodwell, J.D.: Anal Biochem. 119, 440-449 (1982).
15. Chrambach, A., An der Lan, B., Mohrmann, H., Felgenhauer, K.: Electrophoresis 2, 279-287 (1981).
16. Mattson, D.H., Roos, R.P., Arnason, B.G.W.: Ann. Neurol. 9, 34-41 (1981).
17. Mattson, D.H., Roos, R.P., Arnason, B.G.W.: J. Neurosci. Meth. 3, 67-75 (1980).

AFFINITY ELECTROPHORETIC SYSTEMS COMPARED WITH THE CORRESPONDING AFFINITY COLUMN SYSTEMS - GENERAL PRINCIPLES

J. Hau,

Laboratory Animal Unit, University of Odense
5ooo Odense C, Denmark

P. Larsen, T. C. Bøg-Hansen

The Protein Laboratory, University of Copenhagen
22oo Copenhagen N, Denmark

B. Teisner

Institute of Medical Microbiology, University of Odense
5ooo Odense C, Denmark

Marianne Nilsson

DAKO Immunoglobulins
2ooo Copenhagen F, Denmark

Introduction

During the last decade we have used analytical electrophoresis systems extensively and it has become evident that analytical electrophoretic systems in several instances are able to replace time consuming and material consuming systems such as affinity column chromatography and affinity batch procedures. Using lectin in the gels Bøg-Hansen (1) introduced the term affinity immunoelectrophoresis for the analytical electrophoretic method used as prediction for the corresponding preparative experiments. This presentation deals with a generalization of the method and demonstrates the use of other affinity electrophoretic systems compared to the corresponding column experiments.

Electrophoresis '82

Lectins

The use of free lectins as ligands incorporated in the electrophoresis gels has been extensively reviewed (2,3).In general, proteins not shifted in migration velocity by the presence of lectin in the gel are the proteins which appear in the void of the corresponding column experiment. Proteins shifted by the lectin in the electrophoresis are similarly affected in their flow through the corresponding lectin column. Proteins completely precipitated by the lectin in the electrophoresis gel are bound to the lectin on the column and must be eluted by a specific sugar (Fig.1). Table I shows some examples of the correspondance between reactions of proteins in crossed affino immunoelectrophoresis and lectin affinity chromatography.

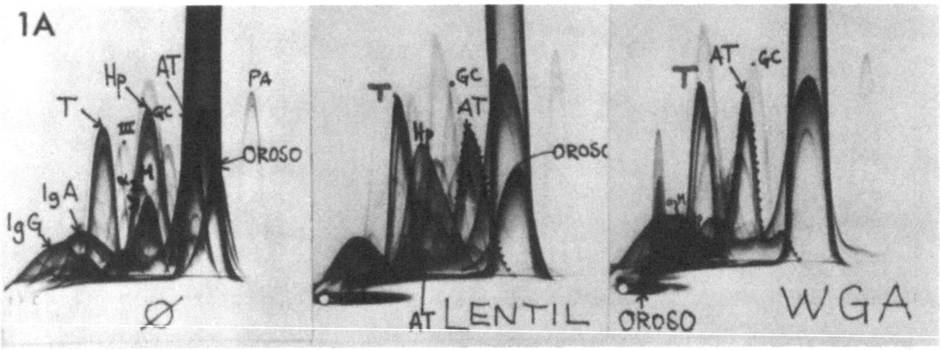

Fig. 1 A. Crossed lectin affinity electrophoreses. Wells contained 5 µl human serum electrophorized first lo V/cm, 4o min (anode to the right) in an empty gel (ø) a gel containing Lentil (loo µg/cm^2) and a gel containing WGA (loo µg/cm^2). The second dimension gel contained anti total human serum.

1 B

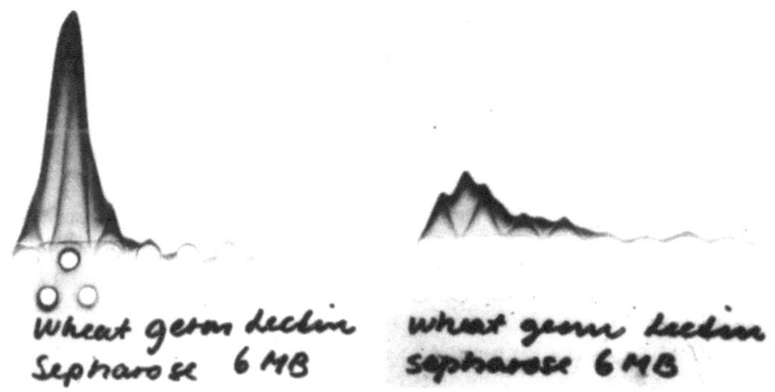

Fig. 1 B. Fused rocket immunoelectrophoresis analysis of column chromatography experiment (WGA-Sepharose 6MB) using a sample of human serum and antibody against α_2-Macroglobulin (1 $\mu l/cm^2$). The peak to the left represents the α_2-Macroglobulin not bound to the gel and the peak to the right represents the α_2-Macroglobulin bound to the gel and eluted with a specific sugar.

Protein A.

The interaction of serum proteins with proteins A has been extensively used for preparative purposes in column experiments. Immunoglobulins from various species have different affinities for protein A and a very good correlation between column experiments and electrophoretic experiments have been found in the present study (Fig. 2, Table II). The crossed affinity electrophoresis was found to be a more sensitive method to predict results of the corresponding column experiment than the modification of the Grabar immunoelectrophoretic analysis.

2A

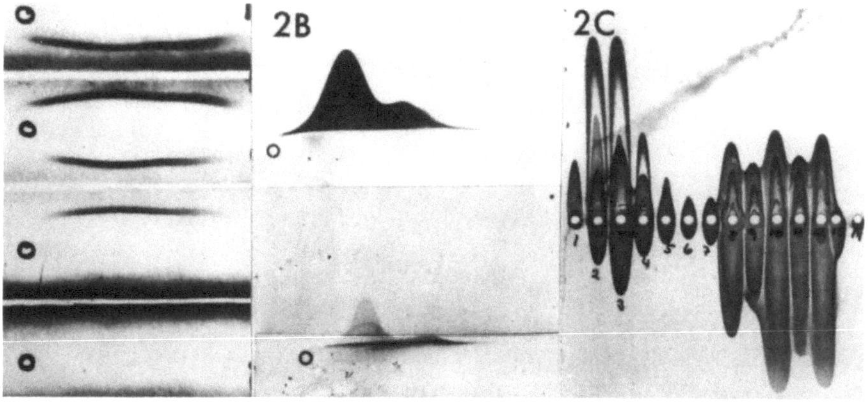

Fig. 2.

A. One dimensional affinity electrophoretic analysis analogous to immunoelectrophoretic analysis a.m. Grabar. Wells contained 5 µl Protein A (lo g/l) (Pharmacia Fine Chemicals) which was electrophorized for 4o min at lo V/cm (anode to the right). Troughs contained loo µl serum, from top: Human, guinea pig, cow, sheep; diffusion for 24 h. Only section of plates is shown.

B. Crossed two dimensional affinity electrophoresis. Wells contained 5 µl Protein A, electrophorized first lo V/cm for 4o min (anode to the right). The second dimension gel contained 1 µl serum (top: Human; bottom: Sheep) per 15 ml gel. The second dimension electrophoresis was 2 V/cm for 24 h (anode at the top).

C. Analysis of column chromatography experiment by immunoelcctro-assay (rocket immunoelectrophoresis). The column contained o.5 µl Protein A Sepharose (Pharmacia). Sample applied: Human serum corresponding to 15 mg immunoglobulins. Elution with a pH gradient, phosphate pH 7 to acetate pH 3.5; flow 2 ml/sec (details to bc published elsewhere, P. Larsen).

Heparin.

Using free heparin in the first dimension gel in crossed immunoelectrophoresis it is possible to obtain information useful for the corresponding heparin sepharose column experiment. Because of the high negative charge of heparin, proteins interacting with heparin were shifted anodically in the electrophoretic migration. This increase in migration velocity was due to the increase in negative charge caused by the binding of heparin. In column experiments the proteins which interacted with heparin in immunoelectrophoresis were retarded or bound to the heparin-sepharose gel (Fig.3). Table III is a list of proteins tested in heparin crossed immunoelectrophoresis and their behaviour in heparin column experiments. It seems that the use of free heparin in the electrophoretic system can predict whether the protein will interact with heparin on the beads but not the mode of interaction and the degree of affinity.

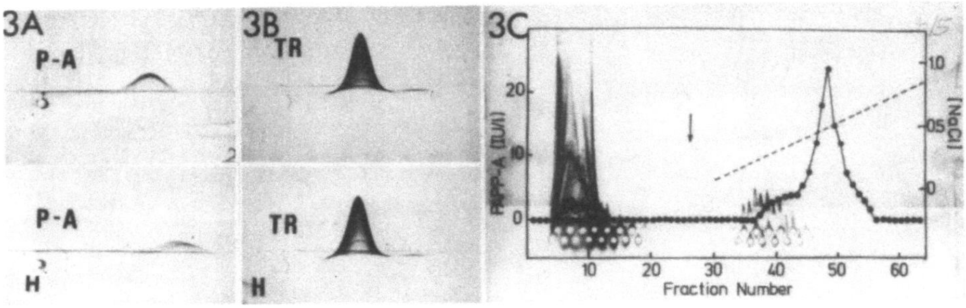

Fig. 3.

A. Crossed heparin affinity electrophoresis with empty first dimension gel (P-A) and free heparin (2o i.u./ml (P-A, H). The antigen was human pregnancy serum and the antibody anti-pregnancy-associated plasma protein A (PAPP-A).

B. As in A) but with anti-α_1-antitrypsin instead of anti-PAPP-A.

C. Fused rocket immunoelectrophoresis analysis of fractions from a heparin column experiment employing anti-total human pregnancy serum as the antibody. The elution gradient covered the interval o.15 M to 1.o M NaCl. PAPP-A was quantified by electro-immunoassay.

Conclusion

Three examples of electrophoretic experiments with ligands, commonly used in group specific affinity column experiments have been brought forward in this paper. The comparison between lectin affino immunoelectrophoreses and the corresponding column experiments has been made by other groups as well and Bjerrum (8) has performed experiments with hydrophobic gels. At our laboratories we are presently calibrating predictive analytical electrophoretic systems with the use of steroid bound gels and hydrophobic ligands. Together these results indicate that virtually all analytical column experiments employing gels coupled with various ligands may be substituted by electrophoretic analyses. In the electrophoretic systems the ligand is either used in its free form or coupled to a gel matrix suitable for the electrophoretic conditions required. With the application of analytical affinity electrophoresis it is possible to obtain information, which is not easily obtainable otherways. The method is not preparative in its basic form. However, other analytical methods in combination with the electrophoresis may yield sufficient information, so that purification of a protein may not be necessary in order to characterize it. Combinations of other analytical methods with electrophoresis has been described e. g. molecular weight estimation (6), isoelectric point determination (5), binding characterization (9), enzyme characterization (1o) and evaluation of amphiphilic properties (11).

Table I. Lectin-protein interaction. Precipitate retardation in crossed affinity immunoelectrophoresis compared to the retention in column experiments (Fig. 1).

Lectin	Protein	Electrophoresis	Chromatography	Ref.
Con A	Albumin	0	0	4
-	Prealbumin	0	0	4
-	gc-globulin	0	0	4
-	Transferrin	R	B	4
-	Ovosomucoid	0 & R (h)	0, B (h)	4
-	Haptoglobin	P	B	4
-	α_2-Macroglobulin	P	B	4
-	Cholinesterase	P	B	4
-	Hemopexin	P	B	4
-	Pregnancy-associated murine protein 1 (PAMP-1)	P	B	5
-	PAMP-2	0, R (h)	0, B (h)	5
-	PAMP-3	R	B	6
-	AFP	0, R, R (h)	0, B, B (h)	3
-	Pregnancy-specific β_1-glycoprotein	P	B	7
-	Pregnancy-associated plasma protein A	P	B	7
WGA	Albumin	0	0	4
-	Prealbumin	0	0	4
-	gc-globulin	0	0	4
-	Transferrin	0	0	4
-	Haptoglobin	0, R (h)	0, B (h)	4
-	α_2-Macroglobulin	0, R (h)	0, B (h)	4
-	Cholinesterase	0, R (h)	0, B (h)	4
Lentil	Albumin	0	0	4
-	Prealbumin	0	0	4
-	gc-globulin	0	0	4
-	Transferrin	0	0	4
-	Ovosomucoid	0	0	4
-	Haptoglobin	0, R (h)	0, B (h)	4
-	α_2-Macroglobulin	0, R (h)	0, B (h)	4
-	Cholinesterase	0, R (h)	0, B (h)	4

0 indicates no retardation or no binding, R indicates retardation, B indicates binding, P indicates precipitation in the first dimensional gel and (h) indicates heterogencity.

Table II. Protein-A-Immunoglobulin Interaction

Precipitate formation in one dimensional affinity electrophoretic analysis (analogous to immunoelectrophoretic analysis a.m. Grabar) (1-D) and in two dimensional crossed affinity electrophoreses (2-D) compared to the retention in column experiments (Fig. 2).

Serum	Electrophoresis		Chromatography
	1-D	2-D	
Human	strong	strong	strong
Monkey	strong	strong	ND
Guinea pig	strong	strong	strong
Dog	strong	strong	ND
Swine	strong	strong	ND
Cat	strong	strong	ND
Rabbit	ND	weak	weak
Cow	very weak	weak	ND
Horse	0	weak	ND
Rat	weak	weak (diffuse)	
Fetal calf	0	0	ND
Chicken	0	0	ND

0 = no precipitate formation
ND = not determined
Experimental details see legend to Fig. 2.

Table III. Comparison of reaction in crossed heparin affinity immuno-electrophoresis and in chromatography with heparin.

Protein	Electrophoresis		Chromatography	
	Interaction	Migration %	Interaction	Elution
Factor B	strong	+ 191	retention	PW
Antithrombin III	strong	+ 1o4	binding	salt o.9 M
C4	strong	+ 88	retention	PW
PAPP-A	strong	+ 21	binding	salt o.6 M
α_1-Antitrypsin	0	+ 2	0	void
α_2-Macroglobulin	0	+ 1	0	void
PZP	0	- 4	0	void
SP-1	0	- 2	0	void
Albumin	0	- 1	0	void
Transferrin	0	- 5	0	void
IgG +)	0	0/-9	0	void

0 = no interaction; PW = elution by prolonged washing; salt = eluted with salt (molarity given); void = eluted in void volume.
+) IgG represented by anodic and cathodic peak.

We have described and discussed the molecular background for microheterogeneity and for the detection and quantification of microheterogeneity forms in lectin affinity electrophoresis as well as the technical problems in a discussion paper (12).

Other aspects of analytical affinity electrophoresis such as detection and quantification of microheterogeneity forms of serum proteins (13) and of AFP (14) are published in these proceedings.

Acknowledgements

This work was supported by The Danish Medical Research Council (Project Numbers 92o7, 2o724 and 12-ooo84) and The Harboe Foundation. The authors thank Pia Jensen, Jette Brandt, Bente Kristiansen, Karen Rambusch and Lindy Benson for expert technical performance.

References

1. Bøg-Hansen, T. C.: Anal. Biochem. 56, 48o-488 (1973)
2. Bøg-Hansen, T. C.: J. Chrom. Libr. 18 B, 173-192 (1981)
3. Kerckaert, J.-P., Bayerd, B., Biserte, G.: Biochim. Biophys. Acta 576, 99-1o8, (1979).
4. Nilsson, Marianne, Bøg-Hansen, T. C.: Protides Biol. Fluids Proc. Collog. 27, 599-6o2 (1979).
5. Hau, J., Svendsen, P., Teisner, B., Thomsen Pedersen, G.: J. Reprod. Fert., 58, 389-393 (198o).
6. Hau, J., Svendsen, P., Teisner, P., Brandt, Jette: Biol. Reprod. 24, 163-169 (1981).
7. This work
8. Bjerrum, O. J.: Anal. Biochem., 9o, 331-348 (1978).
9. Bjerrum, O. J., Ramlau, J., Bock, E., Bøg-Hansen, T. C.: In: Membrane Receptors. Series B, vol. 11, 116-156.
1o. Bøg-Hansen, T. C., Brogren, C.-H.: J. Instit. Brew. 8o, 443-446 (1974)
11. Bhakdi, S., Bhakdi-Lehnen, Birgit, Bjerrum, O. J.: Biochim. Biophys. Acta, 47o, 35-44 (1977).
12. Andersen, M. M., Hau, J., Bøg-Hansen, T. C.: In: Lectins Biol. Biochem., Clin. Biochem., II, 777-786 (1982).
13. Hinnerfeldt, F., Albrechtsen, J., Bøg-Hansen, T. C.: These Proceedings
14. Bøg-Hansen, T. C., Breborowicz, J.: These Proceedings.

DEMONSTRATION AND QUANTIFICATION OF MICROHETEROGENEITY FORMS. LECTIN AFFINITY ELECTROPHORESIS OF HUMAN α-FETOPROTEIN.

T.C.Bøg-Hansen and J.Breborowicz

The Protein Laboratory, University of Copenhagen, Denmark, and Department of Pathological Anatomy, Academy of Medicine, Poznan, Poland

The occurrence of microheterogeneity in glycoproteins seems to be biologically determined. Examples of this is α-fetoprotein (AFP) and ferritin (see 11 for references). Also the amount of each individual microheterogeneity form of a glycoprotein seems to be under biological regulation. Therefore, an attractive goal would be a reliable and simple methodology to permit qualitative as well as quantitative determination of microheterogeneity forms. With such a methodology, the microheterogeneity forms could be closer monitored during biological processes, like tumor growth and fetal malformation and development. For instance, enough information is accumulated about AFP to say that separation of AFP into two or three components with LCA will indicate whether it is produced by the liver or by the yolk sac (7,8,15).

During our studies we have used analytical affinity electrophoresis systems extensively, and it occurs that analytical elec-

Electrophoresis '82

trophoretic systems are easy to use and may be able to substitute for more time consuming and material consuming preparative systems as affinity column chromatography or affinity batch procedures, see for instance the short review by Hau et al. (9) or (4,6). A short paper on analytical affinity electrophoresis has appeared in this series of proceedings (3).

Another chapter in these proceedings deals with AFP microheterogeneity forms (14). In our earlier studies we have dealed with detection and occurrence of AFP microheterogeneity forms (7,8,9,12). The purpose of this study was to investigate further possibilities for their detection and quantification.

ONE-DIMENSIONAL ROCKET IMMUNOELECTROPHORESIS WITH LECTIN

We performed rocket immunoelectrophoresis with an intermediate gel with lectins. When we used LCA in the intermediate gel we obtained separate precipitates corresponding to the three individual microheterogeneity forms. Fig. 1 shows three experiments with an AFP sample containing three microheterogeneity forms (left wells) and an AFP sample containing two microheterogeneity forms (right wells). The number of microheterogeneity forms were identical to the number observed in crossed immunoelectrophoresis, see (14). That individual microheterogeneity forms could be observed in a one-dimensional immunoelectrophoresis was somewhat surprising, since the microheterogeneity forms of AFP appear to be totally immunochemically identical and fuse to form a coherent precipitate in crossed immunoelectrophoresis with lectin in the first dimension gel.

In all immunological precipitation techniques, a single precipitate is expected from the reaction of a monospecific antiserum with its respective antigen. However, sometimes splitting of precipitation lines can be observed as an artefact, for instance in double immunodiffusion. Two lines may be formed if the antigen is applied twice to the same well, provided that there is a significant delay between the applications. Similar experiments with repeated application of the same antigen show double precipitates in rocket immunoelectrophoresis (1). In our experiments, the lectin may have a similar effect, as AFP variants migrate with a different speed through the lectin-containing intermediate gel and divide into a number of zones corresponding to the microheterogeneity forms detected by the lectin. First the unretarded AFP variant meets the antibody front and the first precipitate is formed. Due to the

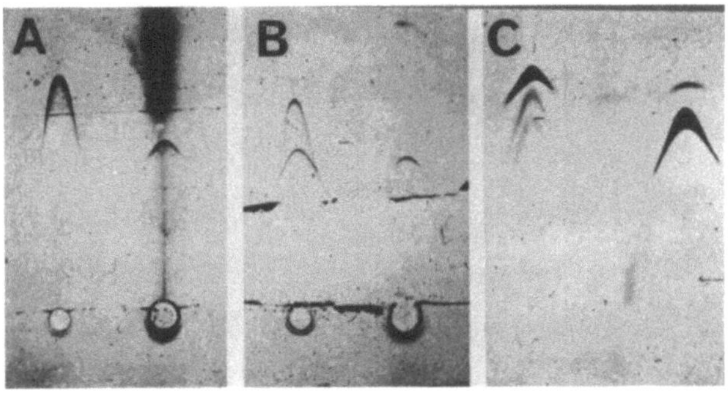

Figure 1. One-dimensional immunoelectrophoresis of human α-fetoprotein (AFP) with two microheterogeneity forms (right wells) and with three microheterogeneity forms (left wells).
A. Control experiment with a blank intermediate gel.
B. Immobilized LCA in the intermediate gel.
C. Free LCA in the intermediate gel.

electroendosmosis, the antibodies may pass the precipitate after it is saturated; then the antibodies will meet the next AFP variant to precipitate either below, with or above the first precipitate, depending upon the ratio of antigen to antibody. In Fig. 1C the strongly binding form gives an immunoprecipitate either above (left well) or below (right well) the non-binding form.

The spliting into distinct immunoprecipitates was observed when the intermediate gel (containing 1.8 mg conA/ml or 0.25 mg LCA/ml) was at least 3 cm wide. When the intermediate gel was 6-7 cm wide, the immunoprecipitates were separated enough to permit at least a semiquantitative determination of each microheterogeneity form. Both with immobilized lectin (Fig. 1B) and with free lectin (Fig. C) in the intermediate gel, this phenomenon could be observed.

DISCUSSION

With the introduction and development of the analytical affinity electrophoretic analyses (see 4,6 for reviews) we believe that a useful tool became available for investigation of glycoprotein microheterogeneity. Interestingly, the fluctuations in glycorprotein microheterogeneity have been studied during health and disease for ferritin, orosomucoid = α-1-acid glycoprotein, inter-α-trypsin inhibitor and other glycoproteins (see 11). Most studied is AFP as a parameter of fetal neural tube defects (13,16,17) as a tumor marker for AFP-producing tumors (8,18). Especially, crossed immunoelectrophoresis with lectin in the first dimension gel is very sensitive for detection of microheterogeneity forms (6). Thus with crossed immunoelectrophoresis AFP shows extensive microheterogeneity as shown also by Toftager-Larsen who found up to five individual and characteristic fractions of AFP with Ricinus communis agglutinin (15). The striking conclusion of these studies is that the glycoprotein microheterogeneity is an expression of biological processes.

The reason for the present study is that crossed immunoelectrophoresis with lectin in the first dimension gel is a rather work-intensive method. In our attempts to develop an easy method for routine screening of AFP microheterogeneity forms for diagnosis, we found that the one-dimensional immunoelectrophoresis with high amounts of lectin in an intermediate gel may be worthwhile to explore further. It is essential that conclusions drawn from such experiments are highly reliable. In ongoing studies we aim to test the methodology for highly reliable routine detection and quantification of glycoprotein microheterogeneity forms.

Acknowledgments. The studies were supported by grants to TCBH from the Danish Medical Research Council. The presentation was made possible by a grant from the Harboe Foundation to The Protein Laboratory.

REFERENCES

1. Axelsen, N.H. (ed.) Immunoprecipitation Techniques in Gel. Scand.J.Immunol.Suppl.10, 1982 in press.

2. Bøg-Hansen, T.C. Anal. Biochem. 56, 480-488.

3. Bøg-Hansen, T.C. in B.J.Radola (ed.) Electrophoresis '79. Advanced Methods, Biochemical and Clinical Applications, pp. 193-202. De Gruyter, Berlin, 1980.

4. Bøg-Hansen, T.C. in W.H.Scouten (ed.) Solid Phase Biochemistry. Analytical and Synthetic Aspects. J.Wiley and Sons, New York, 1982 in press.

5. Bøg-Hansen, T.C., Bjerrum, O.J. and Ramlau, J. Scand J.Immunol. 4, Suppl. 2, 141-147 (1975).

6. Bøg-Hansen, T.C. and Hau, J. in Z.Deyl (ed.) Journal of Chromatography Library, Vol. 18B, Elsevier, Amsterdam, 1982 in press.

7. Breborowicz, J. and Mackiewicz, A. in T.C.Bøg-Hansen (ed.) Lectins, Biology, Biochemistry, Clinical Biochemistry Vol. 1, De Gruyter, Berlin, 1981, pp. 303-314.

8. Breborowicz, J., Mackiewicz, and Breborowicz, D. Scand.J.Immunol. 14, 15 (1981).

9. Hau, J., Larsen, P., Nilsson, M., Teisner, B., and Bøg-Hansen, T.C. These proceedings, pp. 261-269.

10. Hau, J., Westergaard, J.G., Ipsen, L., Teisner, B., Bøg-Hansen, T.C., and Søndersaard, K. in T.C.Bøg-Hansen (ed.) Lectins, Biology, Biochemistry, Clinical Biochemistry Vol. 2, De Gruyter, Berlin, 1982 pp. 457-466.

11. Hinnerfeldt, F., Albrechtsen, J. and Bøg-Hansen, T.C. These Proceedings.

12. Mackiewicz, A. Breborowicz, J. Oncodev. Biol. Med., 1, 251-261 (1980).

13. Nørgaard-Pedersen, B., Toftager-Larsen, K., Philip, J., Hindersson, P. Clin. Genet 17, 355-361 (1980).

14. Taketa, K., Toguchi, E., Izumi, M., and Takeo, K. These proceedings.

15. Toftager-Larsen, K. in T.C.Bøg-Hansen (ed.) Lectins, Biology, Biochemistry, Clinical Biochemistry Vol. 2, De Gruyter, Berlin, 1982, pp. 433-444.

16. Toftager-Larsen, K., Kjaersgaard, E., Jacobsen, J.C., Nørgaard-Pedersen, B. Clin. Chem. 26, 1656-1659 (1980).

17. Toftager-Larsen, K., Nørgaard-Pedersen, B. in T.C.Bøg-Hansen (ed.) Lectins, Biology, Biochemistry, Clinical Biochemistry Vol. 1, De Gruyter, Berlin, 1981, pp. 293-302.

18. Toftager-Larsen, K., Petersen, P.L., and Nørgaard-Pedersen, B. in T.C.Bøg-Hansen (ed.) Lectins, Biology, Biochemistry, Clinical Biochemistry Vol. 1, De Gruyter, Berlin, 1981, pp.283-292.

STUDIES ON HETEROGENEITY OF ANTI-DNP ANTIBODY BY MEANS OF TWO DIMENSIONAL AFFINITY ELECTROPHORESIS

Kazusuke Takeo, Ryosuke Suzuno, Masanori Fujimoto, Tatehiko Tanaka, and Akira Kuwahara
Department of Biochemistry, Yamaguchi University School of Medicine
755-Ube. Japan

Introduction

Resolution of heterogeneous antibodies into each individual immunogobulin is essential for elucidation of molecular interaction between antigen and antibody. For this purpose, isoelectric focusing has generally been used (1-5). In spite of its high resolving power, however, satisfactory resolution of immunoglobulins still has not been attained.

We previously discovered an interaction between phosphorylases and glycogen by means of affinity electrophoresis (6). While liver and muscle phosphorylases in rabbits could not be distinguished by disc gel electrophoresis, both enzymes have been sharply separated by affinity electrophoresis according to the difference in their affinity to glycogen. A similar result has been obtained in the case of dextran specific myeloma protein (7). By this method, IgA type myeloma proteins have been resolved into two fractions. For separation of highly heterogeneous protein mixtures, each having the same molecular weight, a two-dimensional technique, using, in the first dimension, the differences in their charged states and, in the second dimension, the difference in their biological activity, should be effective. Thus, anti-Dnp antibodies were separated in the first

Electrophoresis '82

direction by isoelectric focusing and in the second direction by affinity electrophoresis. We called this method two-dimensional affinity electrophoresis.

Preparation of anti-Dnp antibody and water soluble affinity ligand.

Rabbits were immunized with Dnp-human gamma-globulin injected into the foodpads followed by intracutaneous injections of the antigen on the back of the animals at monthly intervals. The animals were bled 10 days after the 6th boost. The anti-Dnp-antibodies were purified from the plasma by ammonium sulfate fractionation followed by affinity chromatography with a Dnp-lysyl-Sepharose column (8,9).

The water soluble affinity ligand , Dnp-polyacrylamide conjugate, was prepared by coupling dinitrofluorobenzene to a water soluble acrylamide-allyamine (10 : 1 w/w) copolymer. Concentrations of Dnp residue in the affinity ligand were calculated from the absorption at 360 nm (8), using Dnp-glycine as the standard.

The two antibodies (I and II) prepared in the different rabbits migrated to the gamma-globulin fraction with standard disc electrophoresis. SDS-disc electrophoresis and the Ouctarlony procedure showed that both antibodies I and II were pure and were almost completely composed of IgG species.

Isoelectric focusing patterns of anti-Dnp antibodies.

Isoelectric focusing was carried out in a 5% polyacrylamide gel (3 cm x 15 cm x 1 mm) containing 5.3 M urea, 10% sucrose and 2% of Ampholine, with a pH range of 3.5 to 10. Antibody

I separted into approximately 30 fractions being predominated by three fractions at pI values 5, 5.5,6, while antibody II separated into approximately 35 protein fractions being predominated by three fractions at pI values 5.5, 5.9 and 6.3.

Fig. 1 Two dimensional affinity electrophoresis of anti-Dnp antibody.

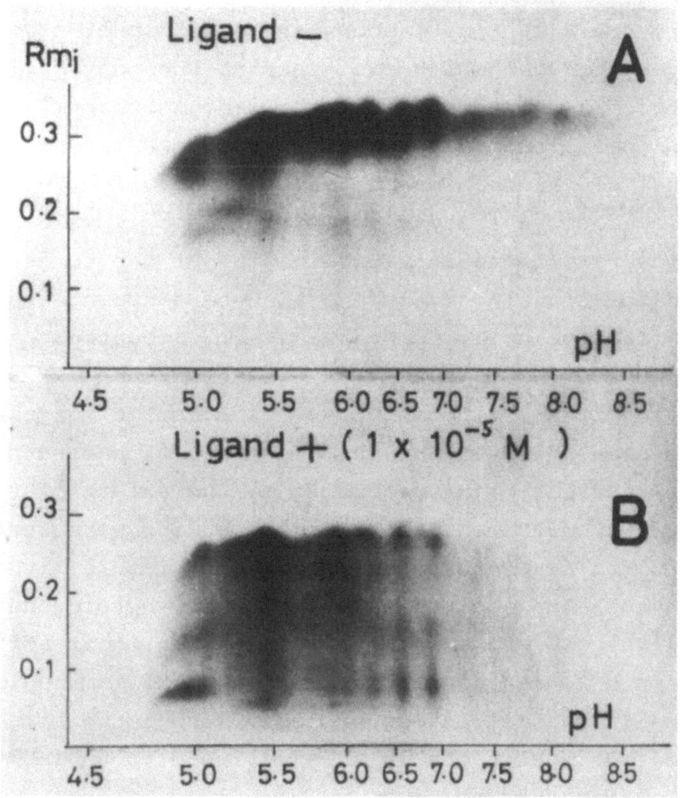

A and B: Electrophoresis patterns obtained with the gels in the absence and presence of Dnp-polyacrylamide conjugate at a concentration of 0.01 mM.

Two-dimensional affinity electrophoresis.

A 3 mm strip from the isoelectric focusing gel was applied in the affinity electrophoresis. The run was carried out using the acidic buffer system described by Reisfeld et al (10), in 5% polyacrylamide gel (13.5 cm x 10 cm x 2 mm) containing varying concentrations of the Dnp-polyacrylamide conjugate until the tracking methylene blue band migrated 10 cm through the gel. Protein was visualized by silver staining. Picture A in Fig. 1 shows the electrophoresis pattern obtained by the gel which contained no Dnp-polyacrylamide conjugate. In this condition, the immoglobulin migrated as a broad line. The immunoglobulin fractions having alkaline or neutral pI values migrated at the same velocity, while the immunoglobulin fractions having acidic pI values migrated slower. The relative mobilities of the former fractions to the tracking methylene blue band amounted to 0.34, and the fraction having pI value at 5.0 amounted to 0.29.

As seen in Picture B, when Dnp-polyacrylamide conjugate (0.01 mM) was added to the gel, mobility of anti-Dnp immunoglobulins decreased and according to the extent of their affinity to the hapten, the immunoglobulins were further resolved into several spots. From this picture, it is clear that the anti-Dnp antibodies are not only heterogeneous in their pI values, but also heterogeneous in their affinity to the hapten. We can count over 100 spots of immunoglobulin on the gel. The resolution of the immunoglobulin is not yet complete, but most spots in this gel seem to be individual immunoglobulins.

Determination of the dissociation constants between anti-Dnp immunoglobulins and Dnp-polyacrylamide conjugate and the affinity distribution pattern.

Acccording to the affinity equation, Eq. 1 (6), dissociation constant, K_d, can be calculated, where Rm_o and Rm_i are the relative mobility of immunoglobulin in the absence and in

$$\frac{1}{Rm_i} = \frac{1}{Rm_o} \left(1 + \frac{c}{K_d} \right) \text{..........} \quad (1)$$

the presence of affinity ligand. But in order to simplify the determination of dissociation constants for every spot from over 100 spots of anti-Dnp immunoglobulins, the dissociation constants calculated by Eq. 1 were converted to a negative logarithmic scale, or pK_d, and directly plotted on the affinity pattern as seen in Fig. 2. From this pK_d

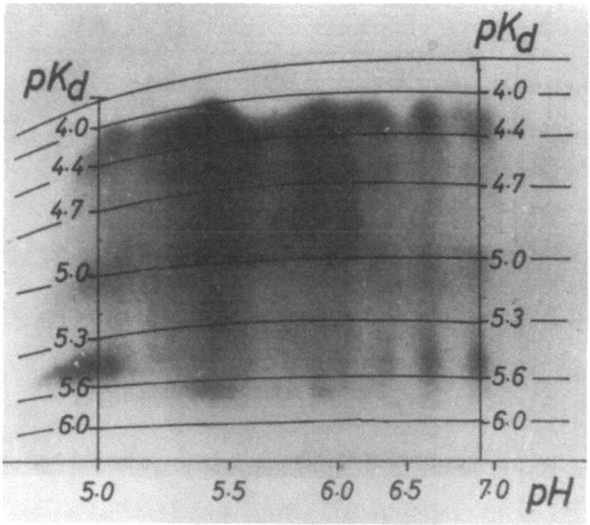

Fig. 2 pK_d - distribution pattern of two dimensional affinity electrophoresis.

distribution pattern, we can easily estimate the K_d value of each individual spot. K_d values obtained ranged from 10^{-4} M to 10^{-6} M.

Fig. 3 shows densitometric distribution of antibody I on the pK_d scale. The curve has four maximums at pK_d values 4.5, 4.7, 4.9, and 5.4. At pK_d value 5.7 one shoulder was observed. This pK_d distribution curve represents one of the quantitative evaluations of the heterogeneity of an antibody.

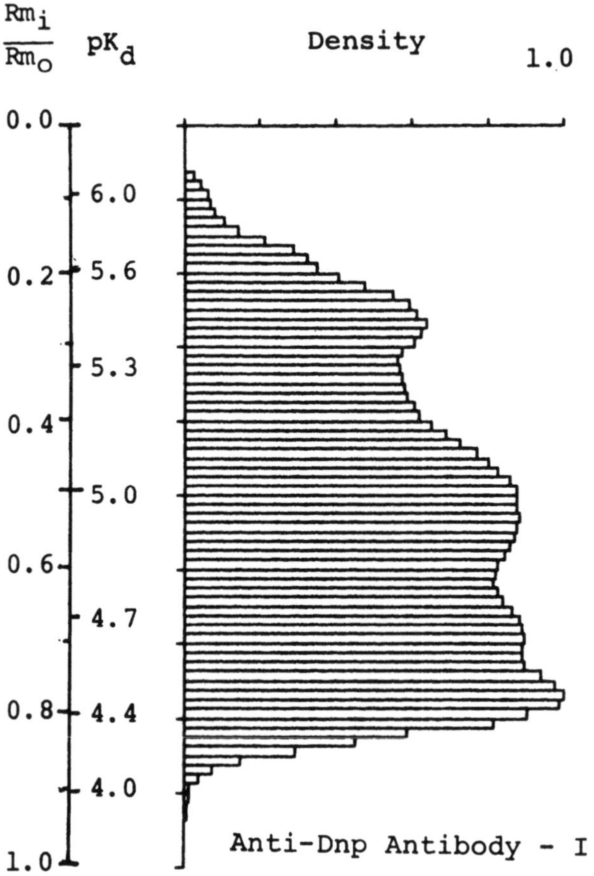

Fig. 3 Densitometric pattern of two dimensional affinity electrophoresis in pK_d scale.

Conclusion.

A new resolution technique for highly heterogeneous antibodies has been developed using isoelectric focusing and affinity electrophoresis in two-dimensions. Using this method, rabbit anti-Dnp antibodies were separated into over 100 spots. Most of the spots seem to be individual immunoglobulins. From the distribution patterns of the anti-Dnp antibodies obtained using two-dimensional affinity electrophoresis, heterogeneity of antibodies can be quantitatively evaluated.

References

1. Awden, Z. L., Willamson, A. R., Askonas, B. A.: Nature 219, 66 - 67 (1967).
2. Williamson, A. R.: Eur. J. Eur. Immunol. 1, 390 - 394 (1971).
3. Freedman, M. H., Painter, R. H.: J. Biol. Chem. 246, 4340 - 4349 (1971).
4. Painter, R. H., Freedman, M. H.: J. Biol. Chem. 246, 6692 - 6699 (1971).
5. Cisar, J., Kabat, E. A., Dornaer, M. M., Liao, J.: J. Exp. Med. 142, 435 - 459 (1975).
6. Takeo, K., Nakamura, S.: Arch. Biochem. Biophys. 153, 1 - 7 (1972).
7. Takeo, K., Kabat, E. A.: J. Immunol. 121, 2305 - 2310 (1978).
8. Eisen, H. N.: Methods Med. Res. 10, 94 - 102 (1964).
9. Eisen, H. N., Gray, W., Little, J. R., Simms, E. S.: Methods Immunol. Immunochem. 1, 351 - 360 (1967).
10. Reisfeld, R. A., Lewis, U. J., Williams, D. E.: Nature 201, 281 (1962).

DETECTION OF ALLERGENS FOLLOWING GEL ELECTROPHORESIS

R. Sutton*, C.W. Wrigley and B.A. Baldo

CSIRO Wheat Research Unit, North Ryde, N.S.W. Australia.
*Present address: CSIRO Division of Animal Production, Prospect, N.S.W. Australia.

Introduction

While the best analytical fractionation of proteins is often achieved with various forms of polyacrylamide gel electrophoresis, it has not been extensively used with specific enzyme- or antibody-staining techniques. This is because the small pores in polyacrylamide restrict the diffusion of reagents into the gel and because agents such as SDS or carrier ampholytes, included in the gels, may inhibit binding.

Methods which enable the detection of biological activity following gel electrophoresis have been reviewed recently (1). We have chosen to use the electrophoretic transfer method or 'western blotting' whereby protein zones are transferred from the polyacrylamide separation gel on to a nitrocellulose membrane (2,3,4,5,6,7). The protein zones, fixed to the membrane are then immediately accessible to reagents, with no restrictions on molecule size or reaction time due to spreading of the protein zones by diffusion. Moreover, the nitrocellulose membrane is more versatile than a gel in terms of the detection procedures which may be used and it can be dried more easily for autoradiography.

The transfer method is particularly useful in the characterization of allergens because IgE does not precipitate

Electrophoresis '82

antigen. Previously, IgE-binding zones in gels have been detected either by elution of the proteins from gel sections and testing by RAST (8), or by crossed radioimmunoelectrophoresis (CRIE) (9). The latter procedure involves electrophoresis in a second dimension in agarose containing antiserum in order to precipitate the proteins in the gel prior to their reaction with IgE from human serum. Thus, allergens which are not precipitated by the antisera (either because they have a low molecular weight and are not precipitated by antibody, or because specific antibodies are not present) are not detected by CRIE.

Types of Electrophoretic Separation Used

The electrophoretic transfer technique can be applied following many different types of electrophoretic procedures to transfer the protein zones from the gel on to nitrocellulose membrane. We have applied 'western blotting' to:

(a) Gradient gels with a polyacrylamide concentration range of 3% to 27% (70 x 70 x 3 mm, Gradient Pty Ltd, Pyrmont, NSW, Australia), run at pH 3.1 in a sodium lactate buffer (5 mM Na) (10) and transferred to a membrane on the cathodic side with 0.7% acetic acid as the electrolyte. These gels, which are 3 mm thick, require about 90 minutes for electrophoresis to ensure transfer of all the protein zones.

(b) Gradient gels with a polyacrylamide concentration range of 3% to 40%T and 12.5%C (from Gradient Pty Ltd, Pyrmont, NSW, Australia; see full description in this volume by Wrigley, Campbell and Margolis). These were run at either pH 3.1 or pH 8.3 and transferred cathodically in 0.7% acetic acid for 90 minutes.

(c) Isoelectric focusing gels with 7.5% polyacrylamide, 1 M urea, 2% carrier ampholytes with a pH range 3.5 to 10 (Pharmacia, Uppsala, Sweden), (90 x 120 x 1.5 mm) (10) transferred cathodically for 60 minutes in 0.7% acetic acid.

(d) Two-dimensional separations where proteins were fractionated in the first dimension by isoelectric focusing (pH 5-9) in tube gels and in the second dimension using a gradient gel containing 3 to 27% polyacrylamide. In this case, the transfer conditions were the same as in (a) above.

(e) SDS gel electrophoresis with a discontinuous buffer system (11,12). For SDS electrophoresis, transfer was anodic using a dilute Tris-glycine-SDS buffer which was found to be more effective for transferring the high-molecular-weight wheat proteins than the Tris-glycine-methanol buffer recommended previously (2).

Electrophoretic Transfer

Proteins were transferred from the separating gel onto a nitrocellulose membrane (Type GWSP pore size 0.22 um, Millipore, Massachusetts, USA), by procedures based on those of Towbin et al. (2). The unstained gel was placed in direct contact with a sheet of wetted membrane, and together they were successively placed between 2 pieces of Whatman 3M paper, 2 Scotchbrite scouring pads and then rigid perforated plates (Pharmacia gel holders for electrophoretic destaining). In order to avoid trapping any bubbles between the layers, the 'sandwich' was put together under water. It was then secured with rubber bands and placed between the electrode plates of an electrophoretic destainer (Pharmacia AB Sweden). Power (36 V/2.5cm) was applied for 60 min in the case of thin (1.5mm) gels or 90 min for thick (3mm) gels.

For transferring proteins from gradient or isoelectric gels, the electrode solution was 0.7% acetic acid, and the membrane was placed on the cathodic side of the gel. For SDS gels, the membrane was placed anodically and the buffer was 10 mM Tris, 77 mM glycine, 0.1% SDS buffer, pH 8.3. After SDS transfer, the membrane was washed in PBS buffer with 10% methanol to remove the SDS.

A second membrane can be placed behind the first to detect protein bands, which after saturating the first membrane, migrate through to the second. When the same sample was applied across the whole gel and the transfer membrane was cut into several strips, one strip was stained to evaluate the effectiveness of transfer.

Detection of Protein Zones

Membranes were stained in amido black (0.15% in a water, methanol, acetic acid mix of 65:25:10) for 10 minutes and destained for 20 minutes in the same solvent. The gel used for transfer was stained afterwards in an appropriate Coomassie Blue stain (10) to check that all the protein zones had transferred from the gel.

Alternatively, the positions of proteins electrophoretically transferred to the membrane was revealed negatively as white areas against a pink background by using a 1% solution of haemoglobin in PBS instead of bovine serum albumin to saturate absorbtive sites on the membrane (see below). Although this procedure was less sensitive than conventional staining, it did not involve denaturation of the protein on the membrane, thus permitting subsequent allergen detection on the same membrane.

Detection of IgE-Binding Zones

The unstained membrane was washed for 10 minutes in phosphate-buffered saline (50 mM sodium phosphate buffer, 0.9% NaCl, pH 6.8) (PBS). The remaining absorbtive sites on the membrane were saturated by immersion of the membrane in 1% bovine serum albumin in PBS. The membrane was washed once with PBS and in

RAST buffer (Pharmacia AB, Sweden). RAST buffer can be replaced by PBS at pH 7.4 with 0.05% human serum albumin and 0.5% Tween 20. The membrane was then incubated for 16 to 24 hours at room temperature with serum, usually diluted 1 in 7 with PBS, with occasional rocking to ensure mixing. (A membrane strip from one gel lane requires 1 to 1.5 ml of diluted serum.) Unbound serum proteins were removed by washing with PBS and with RAST buffer.

The membrane was then incubated for 6 to 16 hours at room temperature with occasional rocking with ^{125}I-antihuman IgE (Pharmacia AB Sweden) made up as directed and diluted 1 in 7 with PBS (approximately 0.2 ug, 5 x 10^5 cpm per lane). Alternatively, membranes were incubated with ^{125}I-Protein A (Amersham International, U.K.) (approx. 30 ng, 3 x 10^5 to 2 x 10^6 cpm) to detect zones that bound IgG. The membranes were washed three to five times with RAST buffer and air dried.

The ^{125}I-labelled zones were detected by the sensitive indirect auto-radiography technique of Laskey and Mills (13), using preflashed Xomatic S film (Kodak, USA), a calcium tungstate intensifying screen (Cronex High Plus, Dupont, USA) and storage in dry ice (-70°C).

Applications of Western Blotting to the Study of Allergens

Because several samples can be run on a single slab gel and transferred together, the western blotting technique is ideal for comparative studies. Firstly, if the same sample is applied uniformly across a gel, the nitrocellulose transfer membrane can be cut into strips, each of which can be reacted with a different serum sample. Using this approach, it was shown that IgE from sera from different individuals with bakers' asthma bind to different sets of protein bands, following both isoelectric focusing and gradient gel electrophoresis. When sera from asthmatic patients having house dust mite allergy were reacted with transfer membranes carrying house dust mite protein zones, a more consistent pattern of IgE-binding proteins was observed, although some differences were discernible between individuals. Many allergenic zones were identified for both mite and flour extracts (14,15).

Replicate transfer membrane strips can also be used to compare the specificities of the different immunoglobulin classes in a serum sample by using different antisera. This approach was used to identify the wheat-flour proteins that bind serum IgE and IgG from the same individual. IgE bound mainly to albumin proteins, whereas the IgG bound mainly to gliadin components.

Thirdly, this technique allows the direct comparison of the IgE-binding proteins present in different allergen sources. By running gradient and isoelectric focusing gels with wheat, rye and barley flour extracts in adjacent lanes, we were able to show that serum IgE from a baker allergic to wheat flour bound to a number of protein zones in rye and barley extracts, some of which had the same mobility as the wheat allergens. These common IgE-binding bands may be homologous proteins responsible for clinically observed crossreactions between cereal flours.

Finally, by allowing the investigator to pinpoint the isoelectric point and relative size and shape of proteins which bind IgE from many allergic individuals, western blotting provides a useful and analytical examination of an allergen source prior to preparative fractionation (15). The information obtained from the western blot can be used to select the specific allergen(s) for purification and to devise a strategy for fractionation. The transfer procedure then serves to monitor fractionation and eventually to check on purity.

References

1. Anderton, B.H., Thorpe, R.C.: Immunol. Today 1, 122-127 (1980).
2. Towbin, H., Staehelin, T., Gordon, J.: Proc. Nat. Acad. Sci. U.S.A. 76, 4350-4354 (1979).
3. Wrigley, C.W.: Proceedings Sydney Allergen Group 1, 98-105 (1981).
4. Legocki, R.P., Verma, D.P.S.: Anal. Biochem. 111, 385-392 (1981).
5. Burnette, W.N.: Anal. Biochem. 112, 195-203 (1981).
6. Glass, W.F., Briggs, R.C., Hnilica, L.S.: Anal. Biochem. 115, 219-224 (1981).
7. Sutton, R., Wrigley, C.W., Baldo, B.A.: 1982, J. Immunol. Methods (in press).
8. Ceska, M.: in Biological and Biomedical Applications of Isoelectric Focusing (eds. N. Catsimpoolas and J. Drysdale), Plenum Press, New York, p. 57-106 (1977).
9. Weeke, B., Lowenstein, H.: in A Manual of Quantitative Immunoelectrophoresis (eds. N.H. Axelson, J. Kroll and B. Weeke), Universitetforlaget, Oslo, p. 149-153 (1973).
10. du Cros, D.L., Wrigley, C.W.: J. Sci. Food Agric. 30, 785-794 (1979).
11. King, J., Laemmli, U.K.: J. Mol. Biol. 62, 465-477 (1971).
12. Lawrence, G.J., Shepherd, K.W.: Aust. J. Biol. Sci. 33, 221-233 (1980).
13. Laskey, R.A., Mills, A.D.: FEBS Lett. 82, 314-316 (1977).
14. Krilis, S., Baldo, B.A., Sutton, R.: Submitted for publication.
15. Sutton, R., Hill, D.J., Baldo, B.A., Wrigley, C.W.: Clin. Allergy 12, 63-74 (1982).
16. Baldo, B.A., Sutton, R., Wrigley, C.W.: Prog. Allergy 30, 1-66 (1982).

QUALITATIVE CHARACTERISTICS OF CAPILLARY ISOTACHOPHORESIS

Identification of isotachophoretic zones

Frans Everaerts, Jetse Reijenga and Theo Verheggen,
Laboratory of Instrumental Analysis, Eindhoven University of Technology, P.O. Box 513, 5600 MB Eindhoven, The Netherlands

Introduction

The basic qualitative characteristics of ionic species, which can be deduced from the signals of universal detectors, is a rather complex numerical value of the absolute mobility of the ionic species, separated in consecutive zones generally according to their effective mobilities. Differences in pK_a values however can cause socalled enforced isotachophoretic zones in which e.g. anions with a low pK_a value and hence a low pH of the steady-state zone, migrate in front of the anions with a high pK_a value. In literature this is called a pH-barrier (1).
It is clear that such enforced isotachophoretic zones, although found by exception in e.g. biological fluids, obscures the isotachopherograms of complicated mixtures (e.g. serum, urine).
With calculated effective mobilities, signals from universal detectors in general can be evaluated. In practice, first a series of relative qualitative characteristics, in a well defined operational system, need to be collected.
Nowadays the potential gradient (or conductivity) of the steady-state zones is measured, as these pass the sensing electrodes (Pt/Ir 10%) of the probe. Because of the step-wise character these values are called step-heights. Thermometric step-heights, derived from thermocouples or thermistors are

Electrophoresis '82

seldom used yet because of the low resolution power of this detector.
Suggestions are made in literature (1,2,3) to use, instead of the absolute step height, relative step heights making use of the conductivity, respectively potential gradient , of the leading electrolyte zone and/or a zone of an internal standard, e.g. perchlorate, chlorate.
The availability of a photometric detector, which registers the passage of e.g. UV absorbing zones, makes further identification simpler. Especially if this photometric detector not only has a defined wavelength (which is common in commercial available instruments), but has a combination of wavelengths at a fixed point in the separation compartment or even with which a complete UV spectrum can be obtained (4).

Disturbing factors in qualitative interpretation

A sample, to be analysed according the isotachophoretic principle, can finally be detected in consecutive zones, which not all originate from the sample. If model mixtures are used, e.g. to verify theory, to check applicability of an equipment or to test operational systems, such false zones are easily recognised and the origin can be explained. For more complicated mixtures such zones may obscure the qualitative and even can have an influence on the quantitative interpretation.
Therefore it is important to check the operational electrolytes on its impurities, both in the leading electrolyte and the terminating electrolyte.
Small amounts of impurities in the terminating electrolyte, with higher effective mobilities in the operational system chosen than that of the terminating ions, will be pushed forwards by the large potential gradient in the terminating compartment. These migrate through preceding zones originating from the sample injected and will dilute these zones, while creating zones of impurities at the proper separation bounda-

ries according to their effective mobilities. These impurity zones cause in fact a "moving boundary" effect (1): become elongated with time depending on both the effective mobilities and the concentrations of these impurities and separands present. The final length of the disturbing zones depends on the time required for analysis, as is normal for experiments on the moving boundary principle. The sample zones, which are "diluted" by impurities (impurity), migrate thus with a velocity lower than the isotachophoretic principle. The choice of the leading electrolyte and/or impure terminating electrolyte can even cause that only one isotachophoretic zone boundary is formed: the "leading electrolyte-impurity from the terminating electrolyte" boundary or no isotachophoretic zone boundary, if an impurity has an effective mobility higher than the leading ion. However in a well chosen operational system these effects can be neglected. In a similar way, impurities of the leading electrolyte will remain behind, if their effective mobilities are lower than that of the leading ion and form on a similar way zones according the moving boundary principle. If a program for data reduction is available (5), in two analyses, injecting the sample with different volumes, zones which elongation (reduction) is not according the linear calibration principle, can be omitted. The elongation of the real sample zones by the impurities can be neglected, supposed the impurity to be present in a low concentration. This simplifies qualitative and quantitative interpretation.

Another disturbing factor in qualitative analysis is the so-called mixed zone. If this mixed zone is a steady-state one (6,7,8), the ions migrate with equal effective mobility at equal pH of the zones, it only can be recognised if the sample is analysed at a pH at which the ions migrate with different effective mobilities. In fact the problem is somewhat similar to the enforced zones, which can even migrate in reversed order of effective mobilities due to a pH barrier. The stationary mixed zone, only present during the isotachophoretic separation process, always has a step-height, which is in

between the step-heights of the ions which form this zone. Its concentration however is always adapted to the concentration of the leading electrolyte. As in moving boundary electrophoresis the ultimate step-height of the stationary mixed zone is determined by both the effective mobilities and the concentrations of all participating ions. The stationary mixed zones easily can be understood if the sample is injected in different volumes, because the elongation (reduction) is not according the linear calibration principle. The moment at which a stationary mixed zone disappears depends on resolution and load capacity (7,8).
Uncontrolled complex formation can cause false zone(s), which make(s) qualitative and quantitative interpretation problematic.
In Fig. 1 a separation of a model mixture is shown (9). Due to the presence of a sample ion with a higher effective mobility than that of the leading ion and two ionic species present in the terminating reservoir the electropherogram shows that three different electrophoretic principles can be created in one single run, all, of course, following Kohlrausch(10) Regulating Function. Fig. 1 stipulates moreover the importance of having an universal and specific detector available. By this "false zones" in the isotachoferogram more easy can be detected.
Another false zone is shown in Fig. 2. Designedly a dead volume has been created in the isotachophoretic equipment. After refilling the equipment with fresh leading electrolyte, a part of the terminating electrolyte is left half-way the separation compartment. This causes in the next-following experiment a zone of "terminator-ions", migrating in the leading electrolyte ahead of the isophoretic sample zones. The velocity is much less than the isotachophoretic one, because this zone migrate according the zone electrophoretic principal.
Some of the sample zones did overrun this terminating zone. For this the length of separation column was chosen such that the zone of A´5MP was visibly diluted and not a real sample zone

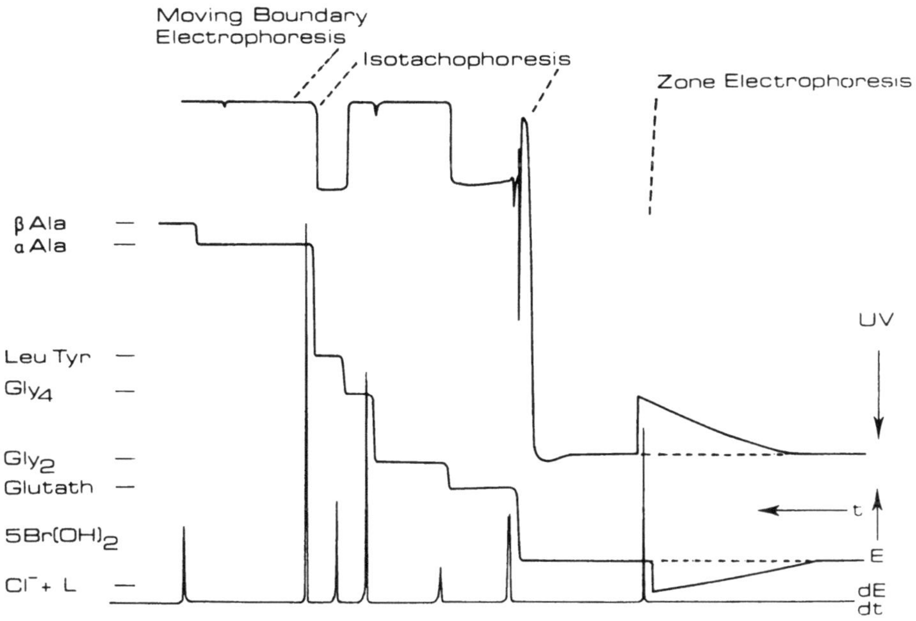

Fig. 1. Different electrophoretic configurations in one experiment. UV = UV absorption at 254 nm, E = Electric field strength, t = time. β-Ala = β-alanine, α-Ala = α-alanine, Leu-Tyr = leucine-tyrosine, Gly (4) = glycylglycylglycylglycine, Gly (2) = glycylglycine, Glutath = glytathion, $5Br(OH)_2$ = 5-bromo-2,4-dihydroxybenzoate, Cl = chloride.
The separation chamber originally has been filled with a solution of 0.004 M 5-Br-2,4-diOH-$C_6H_2COO^-$ and $HOC_2H_4N^+H$, pH = 9. 0.05% Polivinylalcohol (MOWIOL), purified with a mixed-bed ionexchanger (Merck, Darmstadt, FRG), has been added for stabilising purpose. $Ba(OH)_2$ has been added in the reservoir containing α-alanine and β-alanine, pH = ca 10.

whilst passing the UV-detector. This clearly is shown in the UV-trace. If not well understood it can puzzle the interpretation of complex mixture to be analysed. A similar disturbance (11) is observed during an isotachophoretic analysis, registered with measuring electrodes corrected via the separation column via a salt bridge (e.g. KCl). The shape of the UV-trace suggest however that a "dilution" of the A´5MP zone is caused by a less mobile ion (12).
Fig. 2 stipulates that a good construction of isotachophoretic

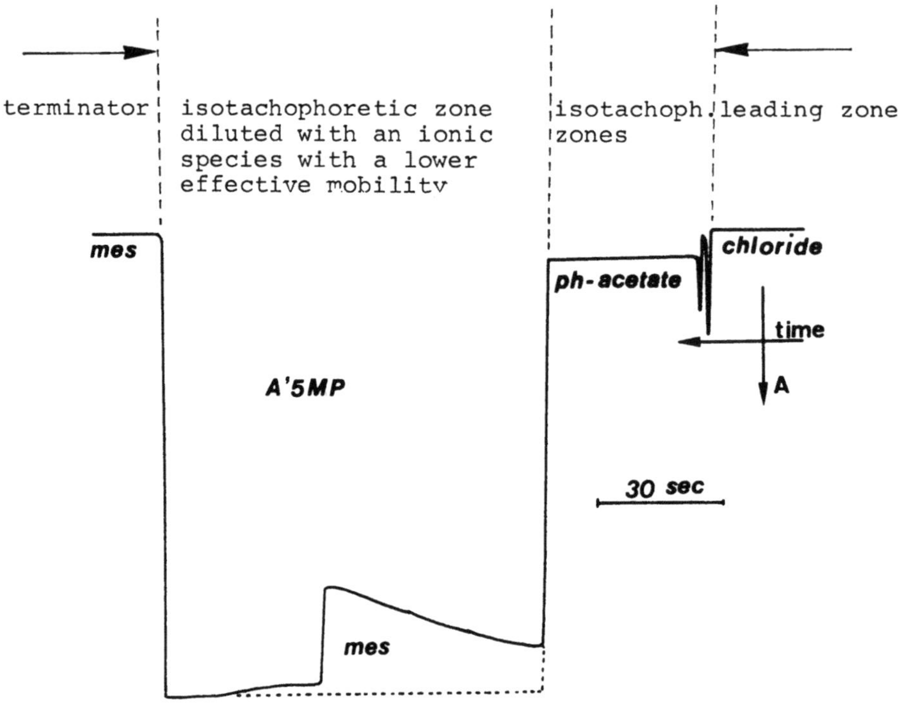

Fig. 2. A disturbance in the qualitative registration due to insufficient rinsing of the various compartments. The terminator morpholinoethanesulfonic acid, remained after rinsing in the separation compartment, migrates zoneelectrophoretically in the isotachophoretic zone of A´5MP, whilst detected.

equipment is necessary and such zones therefore seldom will be found.

Qualitative characteristics, conclusions

For universal detectors the qualitative characteristics are closely linked to the effective mobilities of the separands. These effective mobilities can be varied by selecting numerous properties of the leading electrolyte.
Generally the pH and concentration of the leading electrolyte

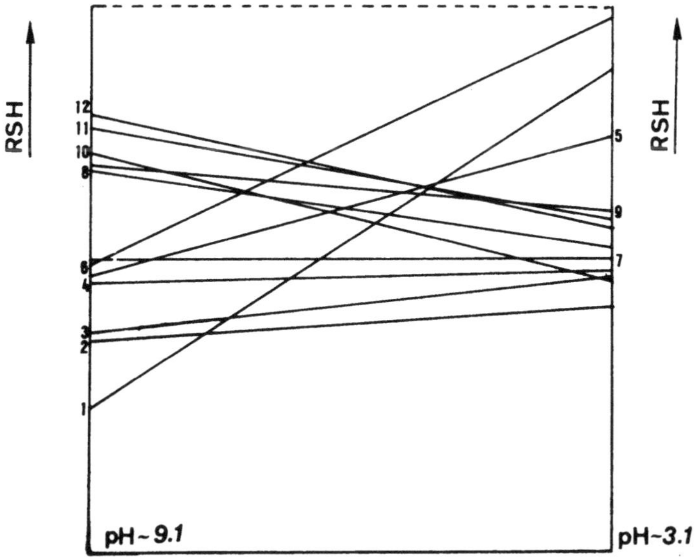

Fig. 3. Relative stepheights (RSH) of some aminoacids as a function of pH. For operational conditions is referred to Table I. 1 = ^{35}J-Tyrosine, 2 = Serine, 3 = Threonine, 4 = Methionine, 5 = Tyrosine, 6 = Histidine, 7 = Phenylalanine, 8 = Valine, 9 = Tryptophan, 10 = Alanine, 11 = Isoleucine, 12 = Leucine.

are chosen to create an optimum in separation parameters for the separands. Complexation is seldom used yet. The information on effective mobilities, step-heights, reduced step-heights, in well defined operational systems are reported in scientific literature, a list of data is in preparation. In Fig. 3 reduced step-heights are plotted of some amino-acids. Often a qualitative and quantitative analysis is asked for the endproducts of a native protein being exhaustively hydrolysed to a mixture of its component amino. While left hand side reduced step-heights refer to the analyses of amino acids separated at pH = 8.6, the right-hand side information is obtained at pH 3.8. The operational conditions are listed in

Table 1. Operational systems for isotachophoretic separations of anions at pH = 8.6 and pH = 3.75.

	pH = 8.6	pH = 3.75
Leading anion	BHBA*	Chloride
Concentration (M)	0.01	0.01
Counter ion	Ammediol**	BALA✶
Additive	HEC (0.3%)	HEC• (0.03%)
Current density (A/cm^2)	0.032	0.032
Terminating anion	BALA✶	Caproate
pH	ca 10.5	ca 4
Additive	$Ba(OH)_2$	none

*BHBA = 5-Br-2,4-diOH-$C_6H_2COO^-$, **Ammediol = 2-amino-2-methyl-1,3-propanediol, ✶BALA = bèta-alanine , •HEC = hydroxyethyl-cellulose.

For BHBA in common systems also Chloride can be chosen, which will not visualise the presence of chloride or carbonate at this pH. Hydroxyethylcellulose was purchased from Polyscience Inc. (Warrington,PA,USA) and was purified by ion-exchanger No.V (Merck,Darmstadt,GFR). All other chemicals were of analytical grade and purchased from Merck. Bèta-alanine and ammediol were further purified, for the system at pH = 8.6, by recrystallization with methanol.

Table I.

New transparant materials be proved against alcohols and ketons, of which conductivity or potential gradient detectors with inner diameters of 0.2 mm can be constructed, are available. This will open new fields of applications and makes the proper selection of operation conditions more complicated, but principly inexhaustible (6).

The high resolution of these detectors with small cellvolume will need an automatic data handling with respect to both qualitative and quantitative information, necessary (12).

Data reduction is often necessary and increases the analytical value complicated isotachopherograms ("fingerprints").

Especially for routine analysis data reduction is recommended. The simultaneous detection of UV absorbing zones at various wavelengths makes the identification of isotachophoretic zones simpler, especially if UV-ratios becomes available as scientific information (12).

References

1. Everaerts F.M., Beckers J.L., Verheggen Th.P.E.M., Isotachophoresis, J. Chromatog Libr., Volume 6, Elsevier Scientific Publishing Company, Amsterdam, Oxford and New York, (1976).
2. Haruki J. and Mizuno J., J. Chromatog., 119, 605 (1976).
3. Everaerts F.M., Mikkers F.E.P. and Verheggen Th.P.E.M., in Separation and Purification Methods, Perry E.S., Oss van C.J. and Gruska E. (Editors), Marcel Dekker, Inc., New York, 6(2), (1977).
4. Vacík J. and Everaerts F.M., in Electrophoresis, Deyl Z. (Ed.), J. Chromatog. Libr., Volume 18, Elsevier Scientific Publishing Company, Amsterdam, Oxford and New York, (1979).
5. Reijenga J., Internal Report, Eindhoven University of Technology, Eindhoven, The Netherlands, (1981).
6. Wielders J.P.M., Thesis, Eindhoven University of Technology, Eindhoven, The Netherlands, (1978).
7. Mikkers F., Everaerts F. and Peek J., J. Chromatog., 168, 293, (1979).
8. Mikkers F., Everaerts F. and Peek J., J. Chromatog., 168, 317, (1979).
9. Mikkers F.E.P., Thesis, Eindhoven University of Technology, Eindhoven, The Netherlands, (1981).
10. Kohlrausch F., Ann. Phys. (Leipzig), 62, 209, (1897).
11. Wiel van de H., Dutch States Mines, Geleen, The Netherlands, personal communication, (1974).
12. Mikkers F. and Everaerts F., in Analytical Isotachophoresis, Everaerts F.M. (Editor), Proc. 2nd. Int. Symp. Isotachophoresis, Anal. Chem. Symp. Ser., Vol. 6, Elsevier Scientific Publishing Companay, Amsterdam, Oxford and New York, (1980).

FREE FLOW FIELD STEP FOCUSING - A NEW METHOD FOR PREPARATIVE PROTEIN ISOLATION

Horst Wagner, Reinhard Kessler
Fachrichtung Anorganische Analytik und Radiochemie der Universität des Saarlandes
D-6600 Saarbrücken, Germany

The electrophoretic separation methods, such as zone electrophoresis, isoelectric focusing or isotachophoresis, are primarily analytical methods, with respect to their development and main applications, with versatility and very high resolution as the decisive advantages. It is obvious to use these methods for preparative separations. In many cases the corresponding analytical apparatus has only been scaled up. The disadvantages, such as worse cooling and thereby increased diffusion, have to be dealt with. Another disadvantage is mainly the discontinuous nature of such methods.

In principle, the continuous methods are particulary suited, for preparative range, as they guarantee the throughput of appropriate amounts. Here especially the system of the free flow electrophoresis has to be mentioned, the main application of which is the preparative separation of cells (1, 2, 3). The collection of the fractions is done by means of a multichannel peristaltic pump, so that mixing of the zones already separated does not take place and a loss in selectivity is avoided. Moreover there is no dilution, as it occurs in the case of the usual elution of the fractions. In addition there are no longer disturbing interactions of a supporting medium with the substances to be separated, as there is no carrier.

Electrophoresis '82

An important point for free flow electrophoresis is the residence time. The aim should always be to use a method with residence times as short as possible. For that reason, and based on our experience with the continuous free flow ion focusing for the separation and concentration of metal cations (4, 5, 6), we have developed the free flow field step focusing for the preparative isolation of proteins.

The principle of the separation, which is very simple as such, can be best explained by the flow chart of the electrolyte system (fig. 1).

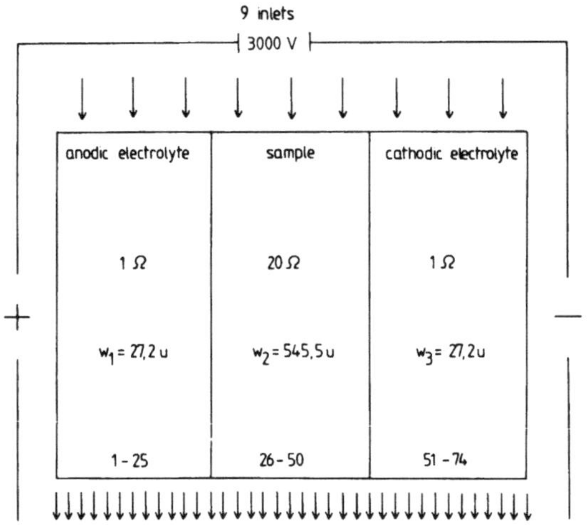

Fig. 1: Flow chart of the discontinuous electrolyte system

Two electrolyte solutions are used, one of them on the side of the anode, the other on the side of the cathode, with the sample solution between them. The values shown in this graph should only be looked at as an example for the relationship between the conditions in the electrolyte system. In this arrangement the three solutions are pumped through the separation chamber (fig. 2). The electrical field is applied perpendicular to the flow direction. It is important with this method, that the two electrolyte solutions must have a conductivity about 20 times higher than that of the sample solution. On account of this difference in conductivity the applied voltage drops down by 90% in the range of the sample solution. Charged particles in the range of the sample solution, move very fast as cations or anions in the direction of the corresponding electrodes. When the ions reach the boundary between the sample and electrolyte solution, i.e. the range of low electrical field, their mobility decreases drastically. The result is a focusing of the substances at the boundary of the different solutions. With longer duration time of the experiment the focused zones slowly migrate into the range of the electrolyte solutions, as focusing at the boundary is not a steady state.

For separations an additional regulation of the migration direction of the substances to be separated is necessary. With the separation of metal ions, the change in charge is achieved by means of suitable complexing agents forming anionic complexes. With amphoteric substances such as proteins, the direction of migration may be controlled by pH. If the pH of the sample solution exceeds the pI value of the protein, the protein will migrate to the anode; if the pH lies below the value of pI, migration towards the cathode will result. In order to stabilize the pH of the sample solution during electrophoresis, H^+ or OH^- ions have to be supplied permanently. For this reason the pH of the electrolyte solutions is also very important. If the sample solution is to have a pH value below 7, an acid must be added to the electrolyte solution of the side of the anode. If the sample solution has a pH value above 7, a base must be added to the electrolyte solution of the side of the cathode. The different mobilities in the different solutions have to be controled in order that not too many H^+ or OH^- ions are supplied. An additional stabilization is achieved by the addition of buffer substances. In certain cases it is possible that, due to the differences in pH of the sample and electrolyte solutions, the isoelectric point of a protein is reached within the range of the boundary between the solutions. Then we have not only a focusing at the boundary between solutions of different conductivity, but also isoelectric focusing at the same time. Interferences with the separation will happen only if precipitation takes place.

The experimental arrangement (fig. 2) consists of the storage containers for the electrolyte and sample solutions, the separation chamber and the fraction collector.

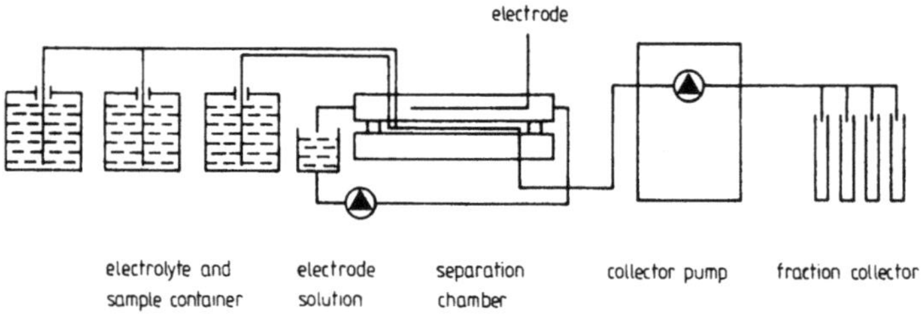

Fig. 2: Experimental arrangement

The electrolyte and sample solutions are pumped through the separation slot by means of a multichannel pump and are collected by a fraction collector. By means of another pump the electrode solution is continuously renewed so that no depletion of the electrolyte within the electrode compartment will take place. The capillary separation slot consists of two cooling plates which are arranged one above the other. The upper plate is made of plexiglas, the lower one is made of brass. Electrical isolation is achieved by means of a teflon foil. The distance of the separation slot is fixed by inserting a distance spacer. The apparatus is commercially available by Bender & Hobein, Munich.

In general good focusing will be achieved by an optimum combination of conductivities, pH of the electrolyte and sample solutions as well as the voltage applied and the residence time. Therefore, these parameters are first discussed by looking at the focusing of single proteins.

The process of focusing can be clearly illustrated by the example of the enrichment of the cationic dye methylene blue (fig. 3).

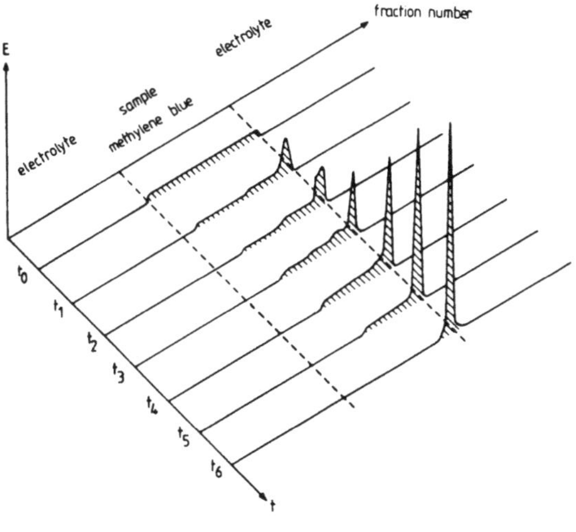

Fig. 3: 3-dimensional-picture of the focusing of methylene blue

The distribution of methylene blue across the cross section of the chamber during the run is ascertained by experiments with different residence times.

The solutions used show a stepwise conductivity profile depending on the concentrations chosen. In order to reduce the mobility at each step of conductivity effectively, a certain minimum difference in conductivities is necessary (fig. 4).

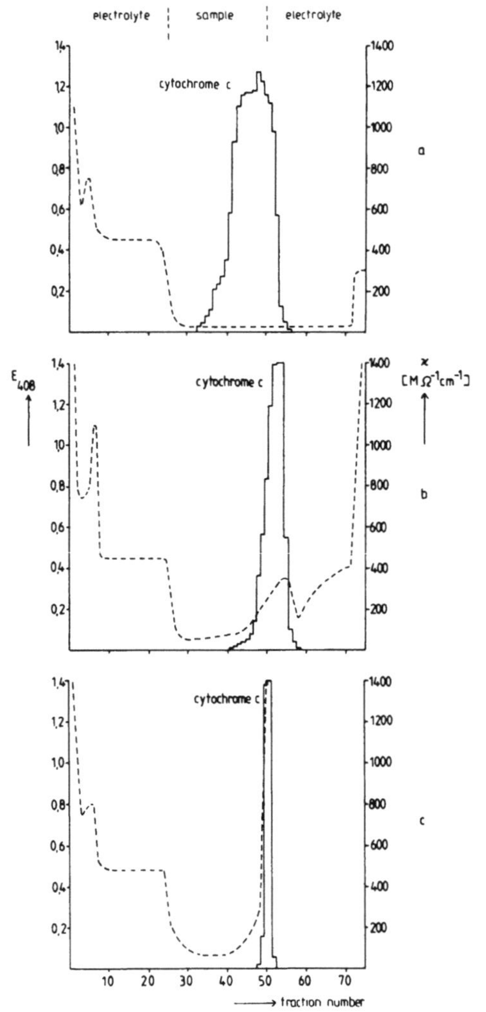

Fig. 4: Dependence of focusing of cytochrome c on the concentration of KCl in the cathodic electrolyte solution
Anodic electrolyte: HCl, pH=3.0.
Sample: 200mg/l cytochrome c, pH=7.5
Cathodic electrolyte: NaOH, pH=10.5
Voltage 1500 V; residence time 30 sec.
Concentration of KCl:
a) o mol/l
b) 10^{-3} mol/l
c) $2 \cdot 10^{-2}$ mol/l

NaCl or KCl solutions are used as electrolyte solutions in appropriate concentrations; a physiological NaCl solution can be used too. The addition of salt above the minimum amount needed does not lead to an improvement of focusing. The side on which a protein is focused is determined by the pH of the sample solution (fig. 5).

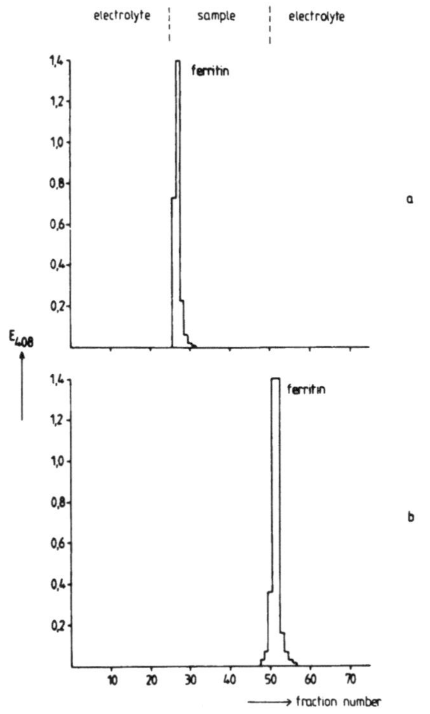

Fig. 5: Dependence of focusing of ferritin on pH of sample solution.
Anodic electrolyte: KCl 10^{-2} mol/l, pH=6.5 .
Sample: 100 mg/l ferritin
a) pH=7.0
b) pH=3.0
Cathodic electrolyte: KCl 10^{-2} mol/l pH=6.5 .
Voltage 1000 V;
residence time 120 sec.

The residence time depends first of all on the voltage applied: it is shortened more and more, the higher the voltage is. Because of the increasing heat production the rise in voltage is limited by the effectiveness of the cooling of the apparatus. Residence times of 20 seconds up to 3 minutes are sufficient, while times of about 1 hour are necessary with isoelectric focusing and isotachophoresis. Due to these small residence times the throughput is 10 to 100 times higher than that with the other methods.

By maintaining the optimum experimental conditions focusings are possible, where 99% of the protein applied can be found within two fractions. The concentration factor depends on the width of the sample-inlet section. The average concentration factors hitherto achieved are about 60.

With this method separations of mixtures of two components (fig. 6), the separation of one protein out of a mixture (fig. 7), as well as the separation of several proteins at the same time can be carried out. Complex mixtures can be separated by repeated runs with varied separation conditions (fig. 8). As in focusing single proteins, the separations are controled by the pH of the solution.

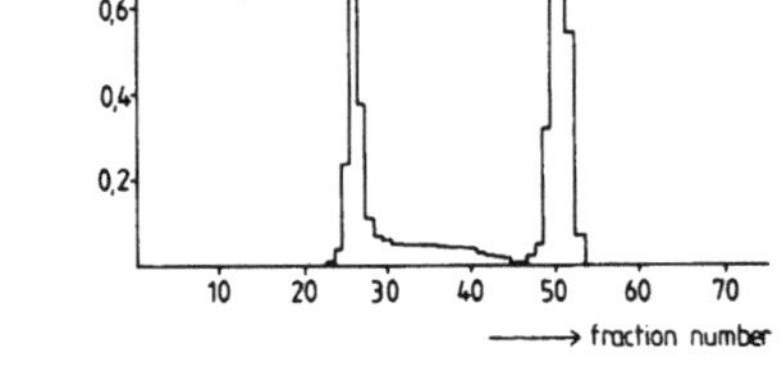

Fig. 6: Separation of a protein mixture
Anodic electrolyte: KCl 10^{-2} mol/l, pH=6.5
Sample: 50 mg/l ferritin and 50 mg/l cytochrome c, pH=7.0
Cathodic electrolyte: KCl 10^{-2} mol/l, pH=6.5
Voltage 1000 V;
residence time 180 sec.

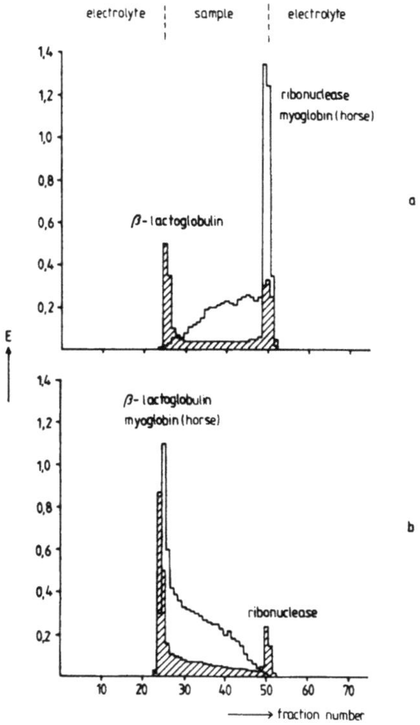

Fig. 7: Separation of a protein from a mixture
Anodic electrolyte: KCl 10^{-2} mol/l, pH=8.0
Sample: 40 mg/l ß-lactoglobulin, myoglobin and ribonuclease
Cathodic electrolyte: KCl 10^{-2} mol/l
a) pH=6.0 b) pH=8.0 (sample)
Voltage 1000 V;
residence time 180 sec.

The protein to be separated must have the highest or the lowest value of pI.

After having shown on certain examples a method for separation of proteins by focusing on conductivity boundaries the advantages should be summarized:

1) No need of carrier ampholytes, therefore low costs.
2) High throughput due to short residence times of the solutions in the separation chamber
3) Easy control of the separation by pH
4) No restrictions by low solubilities at the isoelectric point, as it is not necessary to reach that point.

5) After the separation the protein is concentrated.

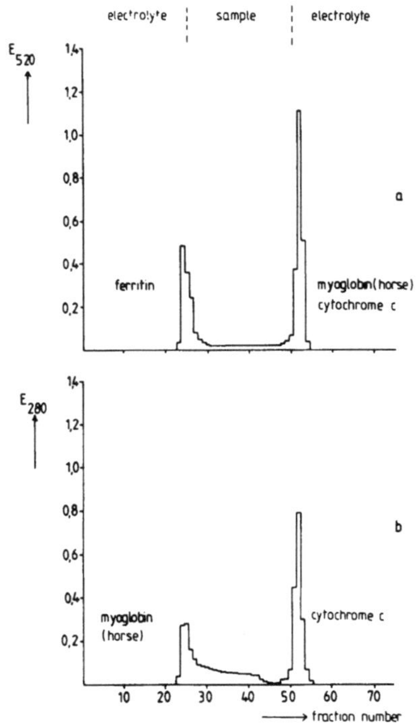

Fig. 8: Separation of a protein mixture

a) Anodic electrolyte: KCl 10^{-2} mol/l,pH=6.5
Sample:ferritin, myoglobin, cytochrome c, pH=6.5
Cathodic electrolyte:KCl 10^{-2} mol/l,pH=6.5
Voltage 1800 V

b) Anodic electrolyte:KCl 10^{-2} mol/l pH=6.5
Sample:fractions Nr 51-53 pH=8.2
Cathodic electrolyte:KCl 10^{-2} mol/l,pH=6.5
Voltage 1000 V

References

1. Hannig, K.: Proteins and biological fluids 15, 523 (1967)
2. Hannig, K.: Methods in microbiology 513. Ed.: Norris, J.R. and Ribbon, D. W. Academic Press, London 1971
3. Hannig, K., Heidrich, H. G.: Cell separation methods. Ed.: Bloemendal, H. Elsevier/North-Holland Biomedical Press, Amsterdam 1977
4. Wagner, H., Neupert, D.: J. Chromatog. 147, 281 (1978)
5. Wagner, H., Neupert, D.: J. Chromatog. 156, 219 (1978)
6. Wagner, H., Blatt, H.: Fresenius Z. Anal. Chem. 306, 208 (1981)

ASECS: ANTIGEN-SPECIFIC ELECTROPHORETIC CELL SEPARATION

Ernil Hansen and Kurt Hannig
Max-Planck-Institut für Biochemie
D-8033 Martinsried, FRG

Introduction

We describe a new cell separation method that combines the high specificity of antibody reactions with the high separation capacity of continuous free-flow electrophoresis. The method is based on the fact that immunoglobulins bear a much lower net negative charge than cell surfaces. Thus, the electrophoretic mobility of cell populations can be decreased by reaction with appropriate antibodies. While this effect of antibodies on cells has been studied in a number of analytical investigations (for review see Ref.1), we have used it for preparative separation of human lymphocyte subpopulations.

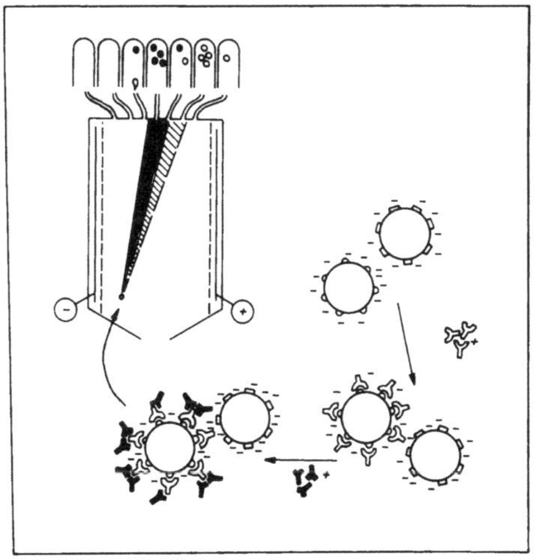

Fig.1 ASECS: The principle

Electrophoresis '82

Results

Human lymphocytes prepared from defibrinated blood by Ficoll gradient centrifugation were subjected to free-flow electrophoresis according to Hannig (2). While cell electrophoresis has been applied very successfully to the separation of lymphocyte subpopulations in mice and rats, it has failed in humans (3). Accordingly, a rather homogeneous distribution of human lymphocytes in cell electrophoresis was observed.
After reaction with relevant antibodies, however, antigen-positive lymphocyte populations appeared deflected to lower electrophoretic mobility in electrophoresis (Fig.2). They then could be isolated with high recovery, purity and vitality. Sandwich techniques, optimal concentrations of antibodies and non-capping conditions were used. Fluorescence labelled second antibody allowed rapid detection of antigen-positive cells in the fractions of electrophoresis.

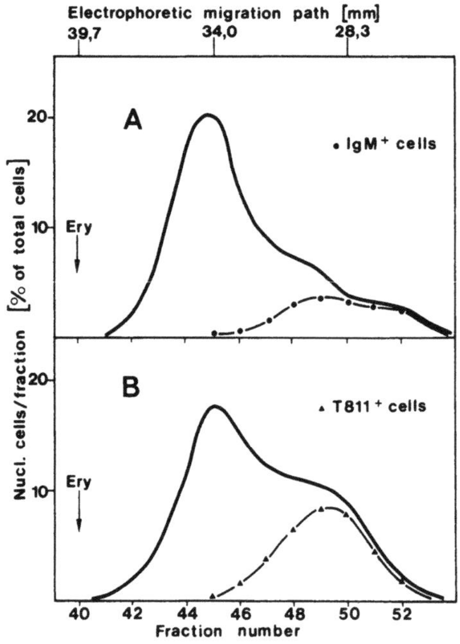

Fig.2 ASECS: Separation of human lymphocyte subpopulations

This way B cells were separated from human lymphocyte preparations after incubation with anti-IgM antibody and FITC-labelled anti-rabbit Ig antiserum (Fig.2 A). Similarily, a T cell subpopulation reactive with the monoclonal antibody t811 (4) was isolated using TRITC-labelled goat-anti-mouse Ig and rabbit-anti-goat Ig antisera (Fig.2 B).
Flow cytometric analysis showed that the cells with the strongest fluorescence intensity had the lowest electrophoretic mobility. Thus, electrophoresis separates antibody labelled cells according to their antigen density. With a flow rate of 100 000 cells per second this method has a very high separation capacity compared to other cell separation techniques involving antibodies.

Conclusion

The described new method greatly expands the separation potential of free-flow electrophoresis. It should be applicable to a wide range of cells and other bioparticles for which specific antibodies are available.

References

1. Sherbet, G. V.: The Biophysical Characterization of the Cell Surface, Acad. Press, N.Y., chap. 4, p. 36 (1978).
2. Hannig, K.: in Techniques of Biochem. and Biophys. Morphology, eds. O. Glick and R. M. Rosenbaum, Willey and Sons Inc., N.Y., London, Sidney, Toronto, Vol. 1, p. 191 (1972).
3. Stein, G., Flad, H. D., Pabst, R., Trepel, F.: Biomedicine 19, 388 (1973).
4. Rieber, E. P., Lohmeyer, J., Schendel, D. J., Riethmüller, G.: Hybridoma Vol. 1, p. 59 (1981).

PREPARATIVE DENSITY GRADIENT ELECTROPHORESIS OF CELLS AND CELL ORGANELLES. A NEW SEPARATION CHAMBER.

A. Tulp, A. Timmerman and M.G. Barnhoorn
The Netherlands Cancer Institute, Department of Biochemistry, Plesmanlaan 121, 1066 CX Amsterdam, The Netherlands

Introduction

Electrophoretic mobility of mammalian cells is caused by the negative surface charge mainly stemming from sialic acid coupled to glycoproteins. Boltz et al. (1) and Catsimpoolas and coworkers (2) have been pioneers in the field of preparative density gradient electrophoresis of cells.
Based upon the results of these authors, we have changed a low gravity separation design (3) that possesses excellent hydrodynamic characteristics, into an electrophoresis chamber that combines relatively low cost with good resolution (4). Cells in aqueous suspension sediment under the influence of ambient gravity, centrifugal force and electrical force according to:

$$\frac{dp}{dt} = \frac{2}{9} \frac{(\rho_p - \rho_m) r^2}{\eta} \left(g + \omega^2 x\right) + \frac{\zeta D}{4\pi\eta} E \ldots \qquad [1]$$

For symbols see (4,5). It follows from [1] that acceptable separation times at weak physical forces are obtained if a zonal sample is thinly layered over a liquid column, in which differential sedimentation takes place. Very thin sample layers (0.3 mm) can be applied in the low g separation chamber developed recently (3,5). In general, an n-fold reduction in sample layer thickness results in an n-fold reduction in separation time.
In this paper an electrophoresis device is described for the separation of cells and cell organelles at low electric field strength and at unit gravity.

Electrophoresis '82

Materials and Methods

Preparation of cells has been described *in extenso* elsewhere (4).
Plasma membrane vesicles derived from Novikoff tumor cells were prepared as described by Emmelot *et al.* (6).
Mycobacterium bovis was a gift from Dr. P. Das (RIV, Bilthoven, The Netherlands). Bacteria were electrophoresed in the presence of 0.1% Triton X-100. Electrophoresis was carried out at a constant current of 90 mA and 27 Volt in a buffer containing 0.26 M glycine, 21.2 mM triethanolamine pH 7.2 at 4°C. Flat gauge platinum - iridium electrodes (Ø 8 cm) were obtained from Drijfhout en Zoons, Amsterdam, The Netherlands; Cellophane sheets from LKB, Bromma, Sweden; neuraminidase from Behringwerke AG, Marburg, W-Germany, and Blue Dextran and Ficoll 70,000 from Pharmacia, Uppsala, Sweden.

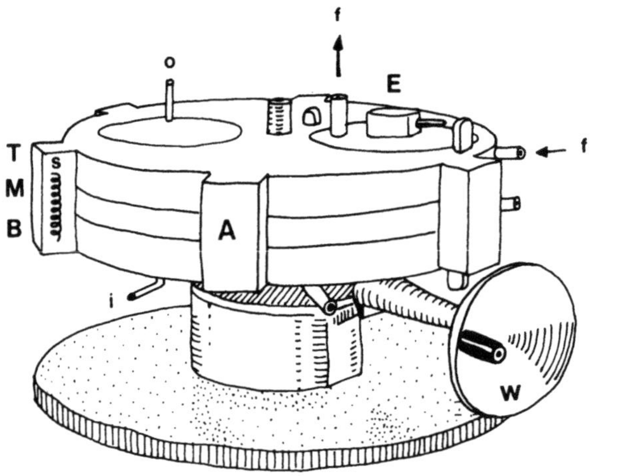

Fig. 1 The electrophoresis separation apparatus. For a description see text.
T, M and B top, middle and bottom plate respectively;
E = electrode; s = spring; A = connector; W = worm gear;
f = flushing channel; i = inlet; o = outlet.

Instrument

Fig. 1 shows a drawing of the electrophoresis separation apparatus. A more detailed description of the device and its exact dimensions are given in (4). Briefly, the device is made of three Perspex circular plates of 24 cm diameter; top (T) and bottom (B) plates are clamped together by 6 connectors (A) in which strings (S) provide the necessary tension. Bottom and top plate can move simultaneously while the middle plate remains stationary thanks to a wormgear mechanism (W). In the middle plate a hollow separation cylinder is present (height 1.5 cm). Bottom and top plate each have a cone with flow deflector, that when arranged in line with the hollow separation cylinder, form a unit gravity sedimentation chamber, Fig. 2A. After the chamber has been filled with density gradient, particle zonal layer and overlay, in the order as described for the 1 g chamber (3), the top and bottom cones are translated in 2 min sideways, their place being taken by a

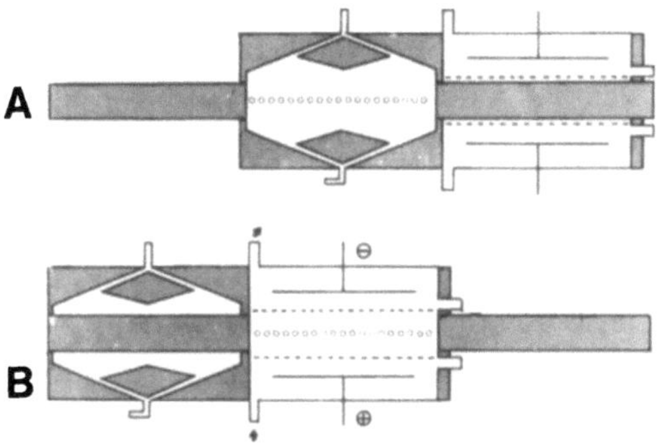

Fig. 2 Movement of top and bottom plates relative to the stationary middle plate.
A. Top and bottom cone are in line with the cylindrical separation chamber. Open circles represent sample layer.
B. The electrode compartments are in line with the separation chamber. Interrupted lines represent Cellophane membranes.

top and bottom electrode compartment, Fig. 2B. The density gradient is isolated hydrodynamically but not electrically from the electrode compartments by Cellophane membranes. Since the electrode compartments were interconnected by a tube of 160 cm length, hydrostatic pressure in bottom and top electrode compartments was almost equal, garanteeing that the Cellophane membrane remained flat. Upon initiation of electrophoresis at a constant current of 90-100 mA, the electrode compartments were flushed at 100 ml/min from a constant level reservoir with electrophoresis buffer in order to remove Joule heat, pH drifts and products of electrophoresis. After electrophoresis, the electrode compartments moved simultaneously sideways, their place to be taken by the top and bottom cones. By introducing dense cushion liquid via the bottom cone of the chamber, the liquid contents of the separation chamber were fractionated via the top as described previously (3,5).

Experiments

The device was first tested with various biological particles and test substances by visual inspection of the migration of zonal sample layers. First it was established that convective disturbances in the Ficoll gradient caused by Joule heating, were absent since no visible schlieren patterns developed at 100 mA. Fig. 3 shows that a Blue Dextran layer of 1 mm thickness, sandwiched between a Ficoll density gradient, remained sharp and stationary for at least one h, indicating that electroosmosis was virtually absent under these conditions. Fig. 3 also shows that a suspension of mouse erythrocytes migrated at uniform velocity for 15 min. Then the polarity of the electrodes was switched so that the erythrocytes migrated back to their initial layering position. A control experiment in the absence of electrophoresis showed that due to the gravity force alone the erythrocytes sedimented less than 0.2 mm into the gradient (curve designated by 1 g).

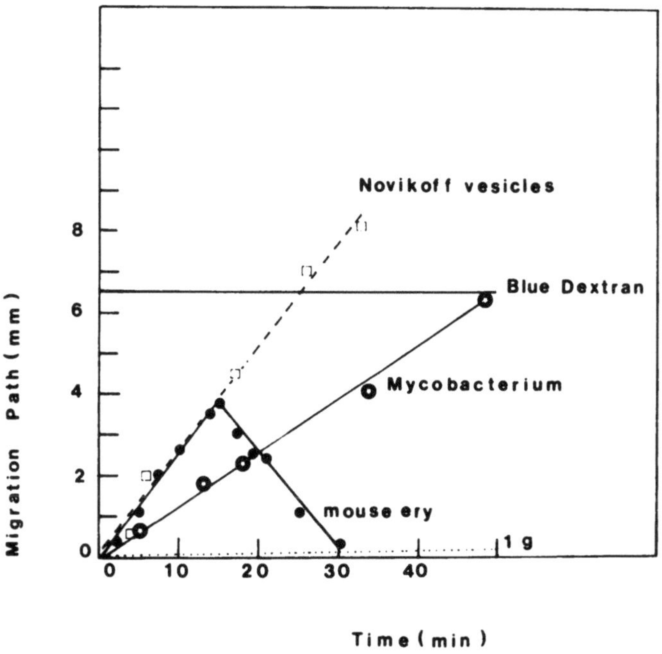

Fig. 3 Electrophoretic migration of Blue Dextran and biological particles. A 1 mm thick sample was layered over a linear 1 to 7% Ficoll gradient. The migration was followed by visual inspection. The temperature was 4°C, the constant current 100 mA.

A suspension of Mycobacteria - in the presence of Triton to prevent aggregation - migrated much more slowly and in 50 min; the migration path was 6,5 mm.
Plasma membrane vesicles derived from Novikoff tumor cells migrated linearly about 8 mm in half an hour.
As prototypic mammalian cells, human erythrocytes were chosen for their large mobility difference (7). In order to lower the migration velocity due to gravity, the Ficoll concentration was varied from 5 to 11%. Within 5 min after initiation of electrophoresis, two bands of erythrocytes became visible and in fact separation could already be terminated at that stage. The distance between these bands increased with time and after

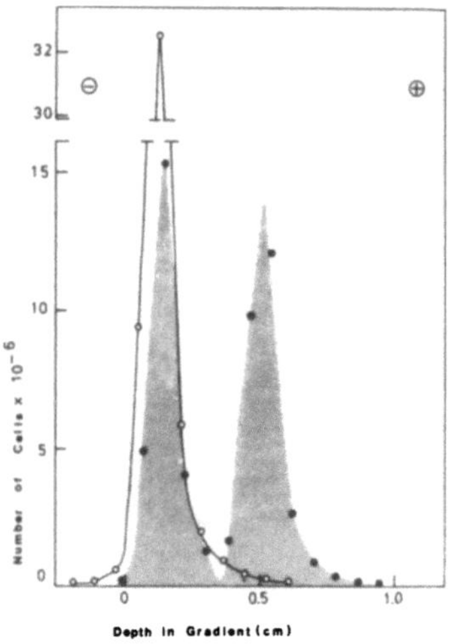

Fig. 4 Electrophoretic separation of treated and untreated human erythrocytes. A mixture of 5 x 10^7 untreated and 5 x 10^7 neuraminidase-treated erythrocytes was layered 1 mm thick on a 5 to 11% Ficoll gradient and electrophoresed at 4°C at a current of 90 mA.
(● ●) electrophoresis profile
(○——○) sedimentation at 1 g.

25 min the liquid interval between the erythrocyte bands was completely transparent. Fig. 4 shows the separation of a mixture of 10^8 treated and untreated erythrocytes. Separation was effectuated within 0.7 cm vertical distance, underlining the excellent hydrodynamic properties of the device. To the eye, the two erythrocyte bands were relatively widely spaced so it was thought justified to intrapolate the electrophoretic profile (shaded part) to its full depth in spite of the fact that fractions of 4 ml each were collected. The modal untreated erythrocytes migrated 3.77 times more rapidly than the treated red cells. Fig. 4 also shows that in the absence of electrophoresis the erythrocyte mixture sedimented as one sharp band 0.12 cm into the gradient due to the gravity force alone.

Discussion

The electrophoretic part of equation [1] can be rewritten

$$T = \int_{0}^{H} \frac{4\pi\eta(x)}{\zeta D(x)} \cdot \frac{q\ \sigma(x)}{i}\, dx \ldots \qquad [2]$$

where T = time to reach a position H in the density gradient; σ = conductivity; q = surface of electrophoresis chamber and i = current.

By inserting known values of the physical constants in equation [2], a prediction on resolving power can be made using a small programmable calculator. Several factors in equation [2] contribute to a reduction of separation time:

i A density gradient solute of low intrinsic viscosity, such as Ficoll 70,000 is preferred above Ficoll 400,000 (1,2).

ii Due to the dipolar nature of glycine, it is likely that the electrophoresis buffer used contributes in a positive sense to the dielectric constant D.

iii At low conductivity σ , the ζ potential of cells is increased since ζ is proportional to the square root of conductivity.

iv By far the greatest effect on separation time comes from the possibility to layer very thin zonal samples. Whereas Boltz et al. (1) and Catsimpoolas and coworkers (2) applied relatively thick sample layers (0.4 cm), the mere fact that the present apparatus permits layering of zonal suspensions of 0.03 cm thickness, reduces separation time almost 10-fold.

The present electrophoresis chamber will be useful in particular with classes of cells that were first sorted according to size in the low g separation chamber described previously (3,5). After separation of a cell mixture into defined size classes, a uniform size class can then be electrophoresed at 2.7 Volt/cm within one h, while the electric force may be directed parallel or antiparallel to the gravity field. The potential of this two-step procedure is at present under active investigation.

Acknowledgments

Ms Anne Veerman is gratefully thanked for the accurate preparation of the manuscript.

References

1. Boltz, R.C., Todd, P., Streibel, M.J. and Louie, M.K.: Prep. Biochem. 3, 383-401 (1973).
2. Griffith, A.L., Catsimpoolas, N. and Wortis, H.W.: Life Sciences, 16, 1693-1702 (1975).
3. Tulp, A., Collard, J.G., Hart, A.A.M., and Aten, J.A.: Anal. Biochem. 105, 246-256, (1980).
4. Tulp, A., Barnhoorn, M.G., and Timmerman, A.: submitted for publication.
5. Tulp, A., Kooi, M.W., Kipp, J.B.A., Barnhoorn, M.G., and Polak, F.: Anal. Biochem. 117, 354-365 (1981).
6. Emmelot, P., Bos, C.J., van Hoeven, R.P., and Blitterswijk, W.J.: Methods Enzymol. 31, 75-90 (1974).
7. Seaman, G.V.F., and Uhlenbruck, G.: Archiv. Biochem. Biophys. 100, 493-502 (1963).

SECTION III TWO-DIMENSIONAL SEPARATIONS

MOLECULAR PROBES FOR HUMAN GENETIC DISEASES BY TWO-DIMENSIONAL PROTEIN ELECTROPHORESIS AND SILVER STAINING

Carl R. Merril, David Goldman*, Margaret L. Van Keuren**, and Michael H. Ebert*

Laboratory of General and Comparative Biochemistry
National Institute of Mental Health
Bethesda, Maryland 20205

*Laboratory of Clinical Science
National Institute of Mental Health
Bethesda, Maryland 20205

**National Institute on Alcohol Abuse and Alcoholism
Bethesda, Maryland 20205

Introduction

The development of molecular probes to study genetic diseases can be achieved primarily by studying either the cellular proteins, DNA, or metabolic pathways. Application of technologies to facilitate these studies can be performed at any of three levels: studies of the primary genetic error, observation of secondary metabolic effects, or use of genetic linkage (Table 1).

A brief review of the approach to developing molecular probes is given in this report. Major stress is placed on two-dimensional electrophoresis of proteins (2DE) and our experiences with this approach.

Electrophoresis '82

Table 1. Molecular probes for human genetic disease.

	Primary genetic errors	Secondary metabolic effects	Genetic linkages
DNA probe with DNA restriction endonuclease analysis	+		+
Protein electrophoresis	+	+	+
Enzymatic analysis	+	+	

Detection of Primary Metabolic Errors

One of the first successes in demonstrating a primary molecular alteration in a genetic disease was the demonstration, by one-dimensional electrophoresis, of an abnormal form of hemoglobin in victims of sickle cell anemia by Pauling and co-workers (1). By quantitative analysis of the electrophoretic separations, they found that individuals with sickle cell trait contain the two hemoglobins in the ratio 40 percent sickle to 60 percent normal. Normal individuals and individuals with sickle cell anemia contain 100 percent normal or sickle cell hemoglobin respectively. These results were in agreement with genetic investigation (2).

Methods for identification of proteins affected by primary mutational events have evolved and proliferated since the work of Pauling and his co-workers. One of the newest methods, two-dimensional electrophoresis (2DE), can separate more than a thousand proteins from a single tissue on the basis of their isoelectric points and masses. An example of a

2DE analysis of mutational events is illustrated by a study of isogenic strains of E. coli. Each strain contains an amber mutation which causes premature chain termination in an enzyme involved in galactose metabolism (Fig. 1). The use of three strains with three different enzyme mutations is analogous to the use of three simultaneous equations with three unknowns. Absence of a spot appearing just below E. coli EF-tu protein in the strain with an amber mutation in the gene for galactokinase (K)

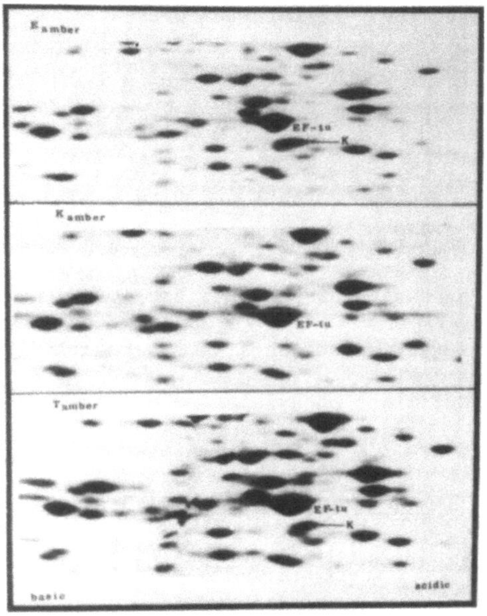

Fig. 1. Isogenic strains of E. coli. Each strain contains an amber mutation in an enzyme of the Leloir pathway for galactose metabolism. Note the absence of galactokinase, indicated as polypeptide (K) under elongation factor-tu (EF-tu) in the strain which is galactokinase minus. This identification of galactokinase (K) was corroborated by antibody precipitation and other genetic and biochemical experiments.

allowed the tentative identification of this spot as the kinase (K) protein. This identification was corroborated by antibody precipitation and additional biochemical genetic experiments (3). We have also utilized 2DE in a similar manner to demonstrate a primary mutation in the β-actin gene in a human fibroblast strain (4, 5).

Development of DNA probes combined with the use of DNA restriction endonuclease digests, now permit detection of deletions of the hemoglobin (Hb) α-gene in α-thalassemia and deletions of Hb-β and Hb-δ-genes in Hb δβ-thalassemia (6). As more DNA probes become available, this approach will help in the diagnosis and understanding of many genetic diseases.

Classic metabolic and enzymatic analyses have lead to the understanding of the metabolic lesions in many diseases, such as the Lesch-Nyhan syndrome (deficient hypoxanthine phosphoribosyltransferase activity), galactosemia (deficient galactose-1-phosphate uridyltransferase) and phenylketonuria (deficient phenylalanine hydroxylase activity). These examples are a few of many such diseases in which the metabolic error was discovered by the utilization of enzymatic assays. Such classic biochemical investigations will continue to provide molecular probes as diseases are investigated with increased awareness of biochemical mechanisms and pathways.

Secondary Metabolic Effects

Secondary effects of a primary mutation. These effects are often due to feedback control mechanisms, post-translational modifications, or altered protein stabilities. In the Lesch-Nyhan syndrome, in which the primary defect is a deficiency of hypoxanthine phosphoribosyltransferase activity, five other enzymes have been found to be altered (7). Use of 2DE electrophoresis revealed eleven quantitative protein differences in lymphocytes (in a survey of 400 proteins in each lymphocyte 2DE) (Fig. 2). The differences were present in the patterns of all the Lesch-Nyhan patients examined and were not found in any normal individuals (7).

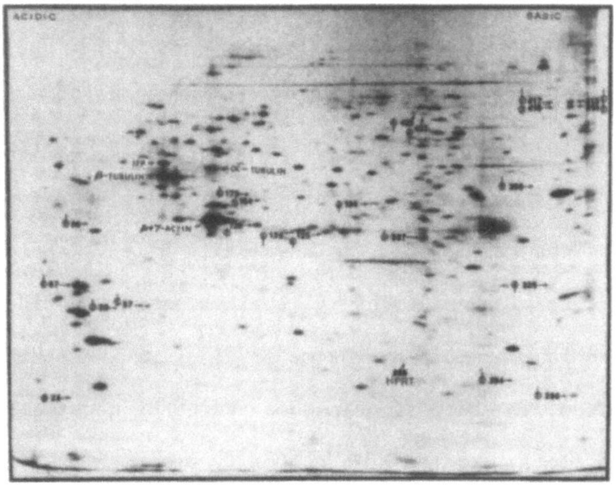

Fig. 2. Demonstration of proteins quantitatively altered in lymphocytes from patients with the Lesch-Nyhan syndrome. The direction of the quantitative alteration is indicated by the arrows by the numbered spots on the figure.

Secondary protein modulations have also been observed in fibroblasts with altered numbers of Chromosome 21 (8, 9, 10), cells exposed to hormones (11, 12) or heat shock (13) and cells which have undergone neoplastic transformation (4, 14). Specific secondary metabolic effects may serve as an aid in the diagnosis and understanding of the pathophysiology of metabolic diseases. A correlative catalog of these effects may provide significant new insights when various inborn errors with similar phenotypic expression are compared.

Genetic Linkage

Human genetic diseases may be studied by linkage analysis. Full utilization of this approach will require the development of a high resolution genetic map of the human genome. This map may then be related to the physical map of the chromosome. As an example of this approach, a child with mental retardation and bilateral retinoblastoma was found to have a deletion in the long arm of Chromosome 13 (15). This linkage of a deletion in Chromosome 13 and retinoblastoma and other abnormalities was later found in family studies of a number of other cases.

An early attempt to develop a biochemically based genetic map involved a screen utilizing numerous quantitative enzyme assays for enzymes which were mapped to chromosomal locations. This approach was largely abandoned when it was found that an extra copy of chromosome 21 increased the enzyme activity of glucose-6-phosphate dehydrogenase (G6PD), which is an enzyme known to be coded for by the X chromosome (16).

Recently, the use of polymorphic markers, including blood groups, histocompatability antigens, and electrophoretically polymorphic proteins has been found to be useful for establishing genetic linkage. Polymorphisms are defined as genetic variants which occur with an allelic frequency of 1% or more. Highly polymorphic regions may, in general, be regions which have less constraint for a precise DNA sequence. In this regard, DNA polymorphism studies have demonstrated greater variation in intervening sequences than in coding sequences; ie., variations in the hemoglobins are the same in carp as man despite their evolutionary distance (17), and the excised of proinsulin varies more than the functional portions (17).

A number of laboratories are developing, by recombinant techniques, random single copy DNA probes for the detection of DNA restriction fragment length polymorphisms (RFLPS). Discovery of a sufficient number of RFLPS will facilitate the construction of a high resolution genetic linkage map (18).

The construction of this map may be complimented by use of protein 2DE which has the capacity to screen large numbers of protein polymorphisms. Polymorphic variants which have been described using 2DE include serum haptoglobin, Gc-globulin, α-1 antitrypsin, α-2HS glycoprotein, arginine-rich lipoprotein, and the unidentified serum protein G-4 (19), a polymorphic protein of brain, PC-1 Duarte (20) and two lymphocyte proteins which show genetic transmission (21). Furthermore, 2DE resolved 13 of 17 isoenzymes produced from 5 loci (22).

Discovery of a sufficient number of polymorphisms will ensure successful linkage analysis for diseases which are primarily monogenic. Barriers to the discovery of a sufficient number are technical, not theoretical. Polymorphisms of proteins and of DNA sequence are present in enormous numbers. Harris (23) found that 24 of 104 human enzyme loci were polymorphic in a European caucasian population, for an average heterozygosity of 6.3%. Based on the observed frequency of protein polymorphism and the approximation that one third of base pair alterations produce a change in protein charge, a lower limit estimate for DNA sequence polymorphism is 0.001/base pair (18).

Detection of protein polymorphisms with quantitative two-dimensional electrophoresis was evaluated in a study of lymphocyte proteins from 28 individuals. One hundred and eighty-six of the most dense polypeptide spots from each autoradiogram of ^{14}C-labeled, phytohemaglutinin-stimulated lymphocytes were measured for variation in position or density. Seven probable polymorphisms were identified among six of these proteins for a a calculated minimum average heterozygosity of 1.2% (Fig. 3). Five proteins had heterozygosities of greater than 0.36, indicating that many rarer polymorphic variants will be detected, even in the small subset of visualized lymphocyte proteins we studied (24). These protein variations were diagnosed as probable polymorphisms because they met the following criteria: 1) all variants displayed charge alterations, 2) a phenotypic heterozygous state was observed for each variant in which two proteins could be observed, each with 50 percent of quantity of the protein found in the homozygous state, 3) two sets of monozygotic twins shared the same phenotype for each polymorphism, 4) at least three phenotypic (two homozygous and a heterozygous) states were observed for each polymorphism. Confirmation of these polymorphisms will depend on

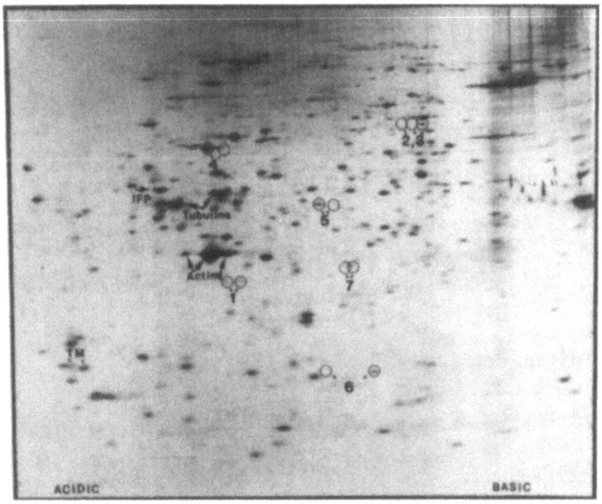

Fig. 3. Position of seven possible protein polymorphisms in human lymphocytes in a two-dimensional electrophoretogram. The positions of the polymorphic variants are indicated by circles in the photograph.

observation of genetic transmission in family studies which are in progress.

The ability to observe multiple polymorphisms in a subset of the proteins visible on the lymphocyte electrophoretogram raises the possibility that two-dimensional electrophoresis may provide a powerful tool for human genetic linkage studies. Because more than 1000 polypeptides may be visualized, we estimated from this study that 40 common and 200 more rare polymorphisms could be rapidly scored on each electrophoretogram. Sixty-three might be observed to be in the heterozygous state in any individual.

Studies with 2DE have yielded conflicting rates of polymorphism. In human fibroblasts, average rates of heterozygosity of less than 1% were found (25,26), suggesting that large classes of proteins were being visualized that show little charge variation. Glycolytic enzymes (27), ribosomal core proteins (28), and structural proteins such as actin (29) show low levels of polymorphism.

Our rate of minimum heterozygosity (1.2%) is in closer agreement with other studies. A 4% rate of average heterozygosity was observed in Drosophila melanogaster (30) and a 2% rate was observed in house mice (31). As rarer polymorphisms are detected, our calculated rate of average heterozygosity will increase.

Polymorphic proteins may be mapped to chromosomal locations by analysis of proteins in hybrid cells with a single human chromosome (as compared to those in the non-human parent cell) (8, 9), by looking for gene dosage effects in cell lines with varying ploidy of a chromosome (8, 9), or by using an antibody prepared against a single protein to isolate a protein mRNA complex. This mRNA can be transcribed into cDNA with reverse transcriptase. The cDNA could be used as a DNA polymorphism probe and could also determine chromosomal location by in-situ hybridization.

The length of the human genome is 3300 centimorgans (cM) (32). Our analysis has shown that 200 randomly located polymorphisms would cover almost 70% of the genome (24). Analysis of 1000 proteins per tissue 2DE electrophoretogram may allow more than 40 commonly polymorphic proteins and 200 rarer ones to be rapidly evaluated on a single 2DE gel. Other tissues, such as fibroblasts and erythrocytes and the use of subcellular fractions will provide many additional protein markers for linkage analysis. When a polymorphic locus is close to a genetic disease locus, the two loci will rarely be separated by recombination and the polymorphism may be useful as a marker and to locate the disease locus.

Silver Staining

The use of silver as an ultra-sensitive protein stain came about after an extensive search for new protein stains was undertaken in our laboratory in August of 1978. We tried heavy metal stains (such as uranium and lead), fluorescent stains (including fluorescamine and 2-methoxy-2,4-diphenyl-3-(2H)-furanone (MDPF), Hoffman La Roche, NJ) and chemiluminescent stains. Sensitivities of these stains was less than, or at best, equivalent to Coomassie blue. It wasn't until application of a modified histological silver stain was attempted in December, 1978, that we obtained a stain which was 100 fold more sensitive than Coomassie blue protein stains (33, 34).

At least one other histological silver stain has been adapted for use in staining proteins in polyacrylamide gels (35-37). Unfortunately, histologically-derived silver stains gave, in some cases, variable results (36), used large amounts of silver or other expensive reagents (33-37) or took hours to perform (33, 34, 38).

We recognized that the ability of silver to form images had been used in both histology, to visualize tissue structures, and in photography, to record levels of light. In both cases, ionic silver is reduced to metallic silver. In contrast to histological stains, which have in many cases become complex, silver-based photographic chemistry has become relatively simple. Photographic image development consists of three steps: activation of silver crystals by light, selective reduction of the activated silver crystal (by a "developer"), followed by removal of non-reduced silver (in photography by a "fixer"), resulting in a negative image. We found in May, 1980, that a negative image could also be produced in a polyacrylamide gel containing proteins. If the gel is soaked 10 min in silver nitrate (2 gm/100 ml water), rinsed with water, then reduced with a photographic developer for 10 min (Eastman Kodak D76 diluted 1:10 works well). Unreduced silver ions should then be removed, just as in photography by 5 min in a photographic fixer. All of the above steps must be made in the dark. Permanence is achieved by three 20 min water washes. A gel image obtained in this manner contains clear regions in portions of the electrophoretogram containing protein while the rest of the gel is grayish brown (Fig. 4). To achieve a positive image in both photography (as when using slide film) and in electrophoretic gels, "photoreversal" of the image must be accomplished. Photoreversal may be achieved by either exposure to light during image development or by chemical means. Our first positive image photoreversal silver stain used potassium ferryicyanide and light to facilitate the photoreversal (39). Later, we found greater sensitivity could be achieved with potassium dichromate (40), without additional light (41).

The mechanism of silver staining is unknown. However, the fact that without photoreversal procedures a negative image is obtained (Fig. 4) suggests that staining may be based on a "Donnan like" equilibrium effect.

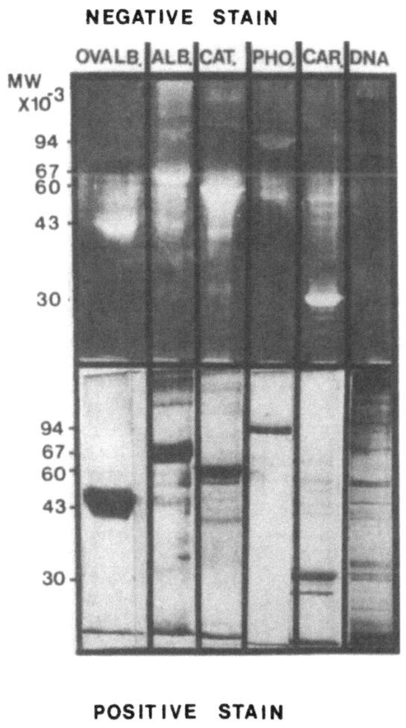

Fig. 4. Illustration of negative and positive silver staining of proteins in a polyacrylamide gel. A negative image results if ionic silver is reduced to metallic silver in the dark. However, if photoreversal chemicals or light are utilized, a positive image results.

In the Donnan equilibrium, charged colloidal substances, such as proteins, alter the distribution of mobile ions, such as silver. If silver stains are based on a Donnan equilibrium, the specific effect for silver could be overwhelmed by increasing the total ionic concentration. We have found that by adding ammonium nitrate to the silver nitrate solution (to a final concentration of 1 M), the silver staining effect is eliminated in both photochemical and histological silver staining of proteins in gels. We have also found that ammonium nitrate will block histological silver staining of brain tissue, suggesting that histological tissue staining may also be based in part on a physical "Donnan like" effect.

A Donnan equilibrium effect is also consistant with the ability of silver to stain other charged polymers such as DNA (42, 43) and lipopolysaccharides (44) in polyacrylamide gels. Knowledge of the mechanism of silver stains may facilitate further increases in sensitivity and applications.

Acknowledgments

This research was supported by the National Institute of Mental Health and the National Institute on Alcohol Abuse and Alcoholism. We would like to acknowledge Dinah Darragh, Sandra Rucker and Maxine Steyer for their secretarial assistance.

References

1. Pauling, L., Itano, H. A., Singer, S. J., Wells, I. C.: Science 110, 543-548 (1949).

2. Neel, J. V.: Science 110, 64 (1949).

3. Merril, C. R., Gottesman, M. E., Adhya, S. L.: J. Bact. 147, 875-887 (1981).

4. Leavitt, J., Kakunaga, T. J.: J. Biol. Chem. 255, 1650-1661 (1980).

5. Leavitt, J., Bushar, G., Kakunaga, T., Hamada, H., Hirakawa, T., Goldman, D., Merril, C.: Cell 28, 259-268 (1982).

6. Orkin, S. H., Alter, B. P., Itay, C., Mahoney, M. J., Lazarus, H., Hobbins, J. C., Nathan, D. G.: N. Engl. J. Med. 299, 166-172 (1978).

7. Merril, C. R., Goldman, D., Ebert, M.: Proc. Natl. Acad. Sci. U. S. A. U. S. A. 78, 6471-6475 (1981).

8. Van Keuren, M. L., Goldman, D., Merril, C. R.: Annals N. Y. Academy, in press.

9. Van Keuren, M. L., Goldman, D., Merril, C. R.: In Electrophoresis '81, Allen, R. C., Arnaud, P. (eds), de Gruyter, Berlin, pp. 355-369 (1981).

10. Weil, J., Epstein, C. J.: Am. J. Hum. Genet. 7, 478-488 (1979).

11. Garrels J. I.: J. Biol. Chem. 254, 7978-7985 (1979).

12. Nikodem, V. M., Trus, B. L., Rall, J. E.: Proc. Natl. Acad. Sci. U. S. A. 78, 4411-4415 (1981).

13. Miller, M. J., Xuong, N. H., Geiduschek, E. P.: Proc. Natl. Acad. Sci. U. S. A. 76, 5222-5225 (1979).

14. Leavitt, J., Goldman, D., Merril, C., Kakunaga, T.: Carcinogenesis 3, 61-70 (1982).

15. Lele, K. P., Penrose, L. S., Stallard, H. B.: Ann. Hum. Genet. 27, 171-172 (1963).

16. Vogel, F. and Motulsky, A. G.: Human Genetics, 273, Springer Verlag, New York (1979).

17. Crow, J. F.: In Population and Biological Aspects of Human Mutation, Hook, E. B., Porter, I. H. (eds), Academic Press, New York, pp. 3-15 (1981).

18. Botstein, D., White, R. L., Skolnick, M, Davis, R. W.: Am. J. Hum. Genet. 32, 314-331 (1980).

19. Anderson, L., Anderson, N. G.: Proc. Natl. Acad. Sci. U. S. A. 74, 5421-5425 (1977).

20. Comings, D.: Nature (London) 277, 28-32 (1977).

21. Hamaguchi, H., Ohta, A., Mukai, R., Yabe, T., Yamada, M.: Human Genet. 59, 215-220 (1977).

22. Wanner, L. A., Neel, J. V., Meisler, M. H.: Am. J. Hum. Genet. 34, 209-215 (1982).

23. Harris, H.: In The Principles of Biochemical Genetics, 3rd Ed., Elsevier/North Holland Biomedical Press, New York, pp. 316-405 (1980).

24. Goldman, D., Merril, C. R.: Proc. Natl. Acad. Sci. U. S. A., in press.

25. Walton, K. E., Styer, D., Gruenstein, E. I.: J. Bio. Chem. 254, 7951-7960 (1979).

26. McConkey, E. H., Taylor, B. J., Phan, D.: Proc. Natl. Acad. Sci. U. S. A. 76, 6500-6504 (1979).

27. Cohen, P. T. W., Omenn, G. S., Motulsky, A. G., Chen, S. H., Giblett, E. R.: Nature New Biol. 241, 229-231 (1973).

28. Bucknall, W. E., Kazazian, H. H., Clement, L. T., Childs, B.: Johns Hopkins Med. J. 137, 123-126 (1975).

29. Vandekerckhove, J., Weber, K.: J. Mol. Biol. 126, 783-802 (1978).

30. Brown, A. J. L., Langley, C. H.: Proc. Natl. Acad. Sci. U. S. A. 76, 2381-2384 (1979).

31. Racine, R. R., Langley, C. H.: Nature (London) 283, 855-857 (1979).

32. Renwick, J. H.: Ann. Rev. Genet. 5, 81-120 (1971).

33. Switzer, R. C., Merril, C. R., Shifrin, S.: Anal. Biochem. 98, 231-237 (1979).

34. Merril, C. R., Switzer, R. C., Van Keuren, M. L.: Proc. Natl. Acad. Sci. U. S. A. 76, 4335-4339 (1979).

35. Kerenyi, L. Gallyas, F.: Clin. Chim. Acta 38, 465-467 (1972).

36. Verheecki, P.: J. Neurol. 209, 59-63 (1975).

37. Karcher, D., Lowenthal, A., Van Soon, G.: Acta Neurol. Belg. 79, 335-337 (1979).

38. Oakley, B. R., Kirsch, D. R., Morris, R.: Anal. Biochem. 105, 361-363 (1980).

39. Merril, C. R. Dunau, M. L., Goldman, D.: Anal. Biochem. 110, 201-207 (1981).

40. Merril, C. R., Goldman, D., Sedman, S. A., Ebert, M. H.: Science 211, 1437-1438 (1981).

41. Merril, C. R., Goldman, D., Van Keuren, M. L.: Electrophoresis 3, 17-23 (1982).

42. Somerville, L. L., Wang, K.: Biochem. Biophys. Res. Commun. 102, 53-58 (1981).

43. Goldman, D., Merril, C. R.: Electrophoresis 3, 24-26 (1982).

44. Tsai, C. M., Frasch, C. E.: Anal. Biochem. 119, 115-119 (1982).

TWO-DIMENSIONAL ELECTROPHORESIS OF PROTEINS: USE OF CELLULOSE ACETATE MEMBRANE AND ISOELECTRIC FOCUSING GEL

Helena Geada and Carlos Manso
Instituto de Química Fisiológica, Faculdade de Medicina de Lisboa

J.A. Filipe da Silva and Luís Wandschneider
Laboratório de Bioquímica e Clínica VI, Instituto Português de Oncologia de Francisco Gentil, Lisboa, Portugal

Summary

A simplified technique of high resolution two-dimensional electrophoresis has been developed for the study of protein microheterogeneity,using a common cellulose acetate electrophoresis followed by a high resolution flat-bed polyacrylamide isoelectric focusing. This method has been used for serum protein studies, in normal and pathological cases, and also for protein studies of neoplastic and non neoplastic adjacent human mammary tissues. This technique detects the microheterogeneity of all proteins and gives an easy and high resolution interpretation of protein patterns for clinical diagnosis.

Introduction

Two-dimensional gel electrophoresis described by O'Farrell(1) combines isoelectric focusing and sodium dodecyl sulfate polyacrylamide gel electrophoresis for the high resolution separation of proteins. Anderson and Anderson (2) adapted this technique to the analysis of human plasma proteins and developed the ISO-DALT system (3,4). Several groups have reported

Electrophoresis '82

two-dimensional techniques based on O'Farrell method (5,6,7) and others have developed a two-dimensional agarose/gradient polyacrylamide electrophoresis (8,9) . Recently, Emerson et al (10) used an horizontal two-dimensional electrophoresis per - formed in agarose isoelectric focusing and sodium dodecyl sulfate gel electrophoresis in an exponential gradient for characterization of plasma proteins.

We have developed an horizontal high resolution two-dimensional technique for the study of protein microheterogeneity, using a common cellulose acetate electrophoresis in the first dimension, followed by a polyacrylamide gel isoelectric focusing, in the second dimension. We have used cellulose acetate in the first dimension, because this is the most common electrophoretic technique used in all routine medical laboratories and based on this electrophoresis we detected the microheterogeneity of all proteins by a high resolution isoelectric focusing gel.

Materials and Methods

Serum was obtained from blood donors and patients with neoplastic mammary gland disease. Mammary tissue was obtained from patients who were subjected to surgery,and immediatly placed at -20°C. Tissues were defrozen at 4°C, cut in fragments, washed 3 times in saline. 1gr tissue was homogenized with 3 volumes of saline in a POTTER with a Teflon embolous for 20 minutes, sonicated for 5 minutes in a 40 K C MSE sonicator and centrifuged at 5 000G for 30 minutes.

Cellulose acetate electrophoresis - we have used cellogel membranes in 0.04M sodium barbital buffer. Electrophoresis was performed at 300V during 35 minutes. Sample amount depended on protein concentration. For serum samples,1.5µl and 10µl were used, respectively, for protein staining and for isoelectric focusing. For samples with less amount of protein, as homogenized mammary tissues, 10µl were applied for protein staining and up to 50µl for isoelectric focusing. Proteins were stained in a 0.5% Ponceau S solution in 5% trichloracetic acid for 5 minutes.

Isoelectric focusing - the gel solution consists of 2ml acrylamide (29.1%), 2ml bisacrylamide (0.9%), 7ml sucrose (20.5%), 600µl LKB Ampholytes and 400µl ammonium persulphate (1%). LKB Ampholyte mixtures will depend on the sample to study. We have used pH 4 - 6, pH 4 -6/5 -8 in equal volumes or 1/1/1.5 ampholytes pH 4 - 6 / 5 - 8 / 3.5 - 10, using 20µl N,N,N',N',-tetramethyl--ethylene diamine (TEMED) in this last case. 1M Na OH and 1M H_3PO_4 were used as cathodic and anodic solutions. The gel solution was poured between two glass plates (150 x 125mm) separated by a 0.6mm U-frame gasket. After cellulose acetate electrophoresis, two strips were cutted just to obtain protein separation spot and were laid on the polyacrylamide gel surface, near the cathodic end, with proteins in contact with the gel. Electrophoresis was performed initially at 15mA up to 1 000V, during 3 hours. Upon completion, the gel was stained with Coomassie Brillant Blue (11) at 80°C, during 1 hour for serum samples and 3 hours for mammary tissue proteins. Gels were destained overnight.

Results

Serum proteins were studied with different pH ampholyte mixtures to detect the best gradient for the study of protein microheterogeneity. Ampholytes pH 4 - 6 (Fig 1), pH 4 - 6 / 5 - 8 (Fig 2) and pH 4 - 6 / 5 - 8/ 3.5 - 10 (Fig 3) have been used. The positions of the major plasma proteins were identified by compa -

rison with proteins on cellulose acetate strip, as is indicated in Fig 3, where the albumin, α_1-, α_2-, β- and γ zones are defined in the acetate membrane.

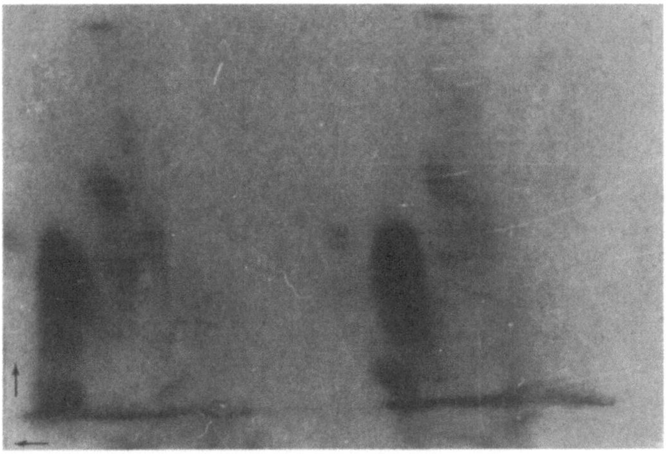

Fig. 1 - Serum proteins separated by two-dimensional electrophoresis with ampholytes pH 4-6. Albumin and α_1-globulins are well defined.

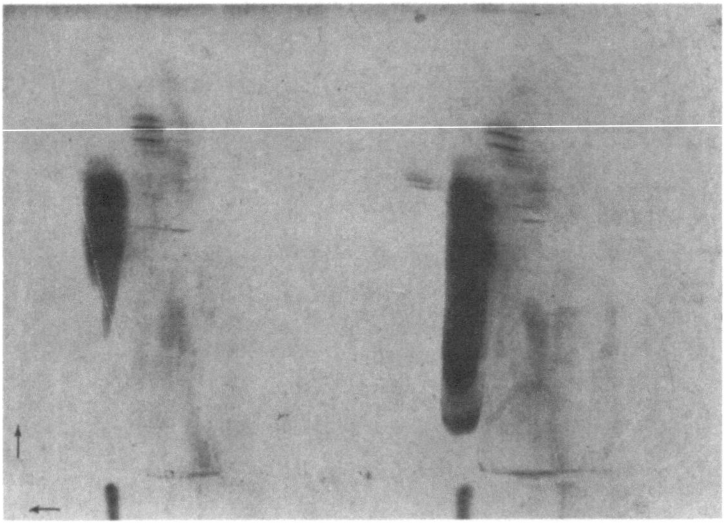

Fig. 2 - Serum proteins separated by two-dimensional electrophoresis with ampholytes pH 4-6/5-8. Albumin, α and β-globulins are well defined.

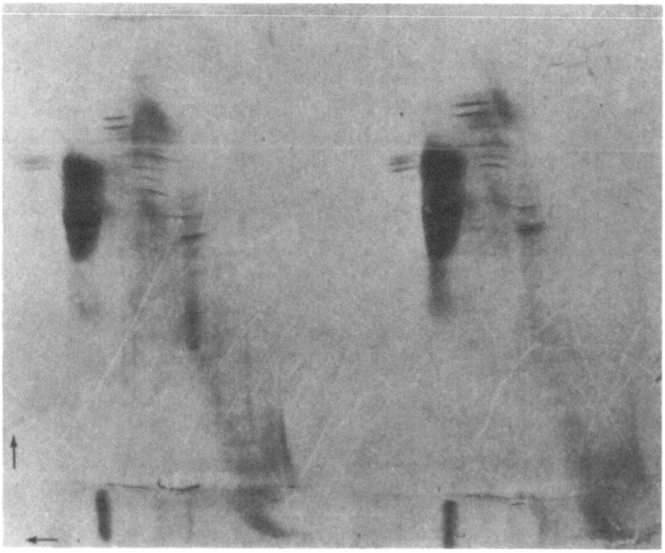

Fig. 3 - Serum proteins separated by two-dimensional electrophoresis with ampholytes pH 4-6/5-8/3.5-10. Cellulose acetate strips can be detected at the cathodic end of the gel. Microheterogeneity of all proteins is clearly defined.

According to the ampholyte mixture we have the microheterogeneity of different proteins. Using ampholytes pH 4-6, pre-albumin, albumin and specially α_1-globulin zones are well defined. We can detect also α_2-globulin proteins.

With ampholytes pH 4-6/5-8, we can see clearly all the zones mentioned above and also β-globulin zone proteins. Using ampholytes pH 4-6/5-8/3.5-10, we detect the microheterogeneity of all proteins, including the γ-globulin zone. Cellulose acetate strips are not removed during isoelectric focusing because this will disturb the pH gradient. We can see in all pictures an horizontal line, near the cathodic end of the gel, which indicates the place where the cellulose acetate strips are laid on the gel.

In Fig 4, the use of pH gradient 4-6/5-8 for protein study of 2 high β/γ globulin sera is shown, in which we can detect abnormal patterns, as indicated by arrows.

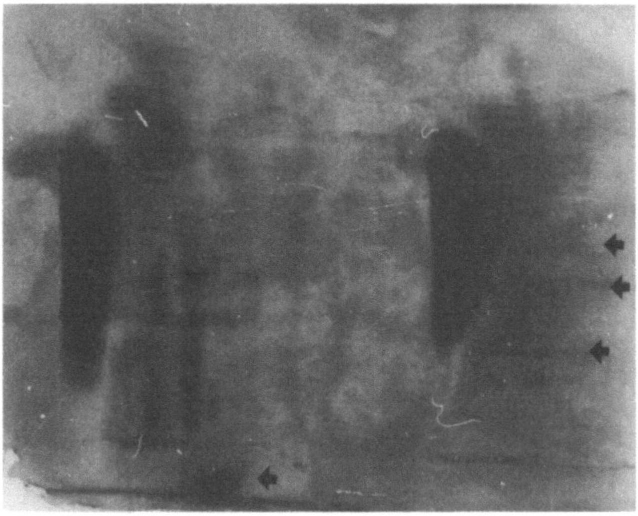

Fig. 4 - Serum proteins of two different high β/γ globulin sera separated by two-dimensional electrophoresis with ampholytes pH 4-6/5-8. Abnormal patterns are indicated by arrows.

We have also applied this technique to the protein study of homogenized neoplastic and non neoplastic adjacent human mammary tissue, to detect any possible protein alteration between these two samples (Fig 5).
In this case we have used pH 4-6/5-8/3.5-10 ampholyte mixture, and a deposit of 50µl of non neoplastic tissue and 10µl of neoplastic tissue in cellulose acetate strips. Sample application will depend on protein concentration.

To search for neoplastic protein markers in plasma, this technique was also used to compare blood donor and patient blood samples using pH 4-6/5-8/3.5-10 ampholyte mixture (Fig 6).

In all these cases, we tried to look for protein zones, in which we could observe alteration of protein microheterogeneity, significant alteration in protein amount or detection of a specific protein.

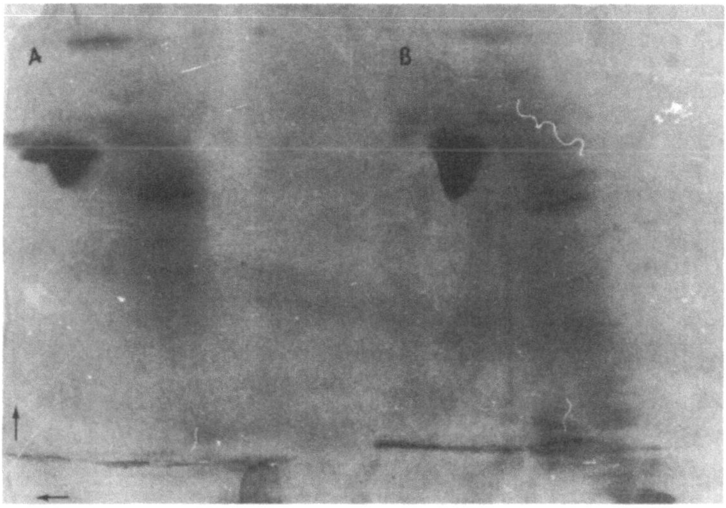

Fig. 5 - Protein studies of homogenized non neoplastic adjacent (A) and neoplastic (B) human mammary tissue separated with pH 4-6/5-8/ /3.5-10 ampholyte mixture.

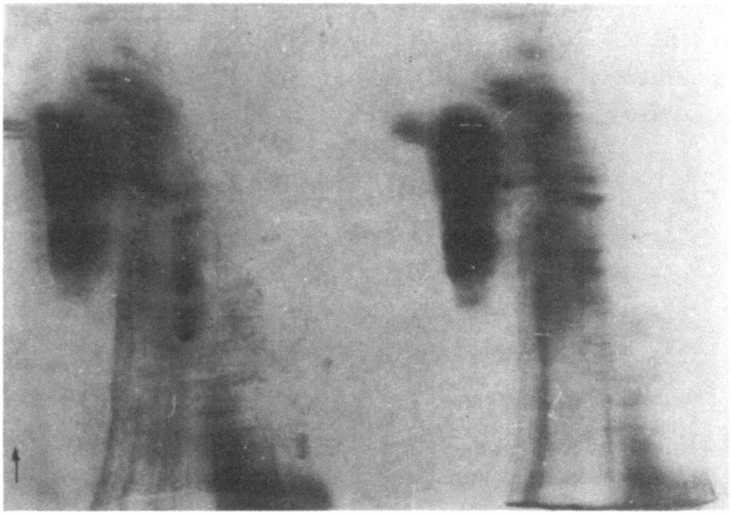

Fig. 6 - Blood donor and patient serum samples studied by two-dimensional electrophoresis with pH 4-6/5-8/3.5-10 ampholyte mixture.

Discussion

Two-dimensional techniques have been shown to give the greatest available resolution for characterization of protein mixtures. However, with some of these techniques, samples are denatured or it is necessary to use albumin-depleted plasma to clarify the entire pattern. With the simple technique we have presented here we used non-denatured samples. Besides, we can eliminate albumin from other proteins simply by doing a cellulose acetate electrophoresis in the first dimension, which is a very simple method and can perform the separation of proteins in well known zones. So, after isoelectric focusing, we will not have to deal with the albumin spot overlaying most protein spots. Likewise we can identify protein microheterogeneity without thinking in albumin.

It is also very easy to lay the cellulose acetate membrane on the surface of the polyacrylamide gel, taking care to avoid any preliminary contact of proteins from the strip, which will be imediately impressed on the gel. Also both electrophoreses were performed in horizontal position, which in case of isoelectric focusing permits direct measurement of the pH gradient in the gel. The position of proteins from cellulose acetate strip and its isoelectric point will permit identification of proteins. All the electrophoretic technique is performed in 3 hours and 35 minutes, and with the width of the gel used different samples can be compared directly side by side, under identical conditions.

With the ampholyte mixtures used we can study microheterogeneity of all proteins. Specially with pH 4-6/5-8/3.5-10, all proteins are clearly visible. With ampholyte mixtures pH 4-6/ /5-8 almost all proteins of albumin, α_1-, α_2- and β-globulins are clearly identified. For the study of α_1-globulin zone we can perform the isoelectric focusing gel with ampholytes pH 4-6.

This technique has been applied to the study of abnormal sera and homogenized human mammary tissues giving stimulating results to continue its development.

Acknowledgements

This investigation was supported in part by Centro de Metabolismo e Endocrinologia (Mb L2) INIC.

References

1. O'Farrell, P. H.: J. Biol. Chem. 250, 4007-4021 (1975).
2. Anderson, L., Anderson, N. G.: Proc. Natl. Acad. Sci. 74, 5421-5425 (1977).
3. Anderson, N. G., Anderson, N. L.: Anal. Biochem. 85, 331-340 (1978).
4. Anderson, N. L., Anderson, N. G.: Anal. Biochem. 85, 341-354 (1978).
5. Jäckle, H.: Anal. Biochem. 98, 81-84 (1979).
6. Latner, A. L., Marshall, T., Gambie, M.: Clin. Chim. Acta 103, 51-59 (1980).
7. Clark, P. M. S., Kricka, L. J., Gomo, A. R. Z., Whitehead, T. P., Bullock, D. G., Saidi, H.: Clin. Chim. Acta 103, 219-228 (1980).
8. Felgenhauer, K.: J. Chromat. 173, 299-311 (1979).
9. Felgenhauer, K., Hagedorn, D.: Clin. Chim. Acta 100, 121-132 (1980).
10. Emerson, D. L., Chapuis-Cellier, C., Arnaud, P.: Electrophoresis 1, 159-163 (1980).
11. Vesterberg, O.: Biochim. Biophys. Acta 257, 11-19 (1972).

ISOELECTRIC FOCUSING IN IMMOBILIZED pH GRADIENTS. III: APPLICATION TO TWO-DIMENSIONAL SEPARATIONS.

Angelika Görg, Wilhelm Postel, Rainer Westermeier
Lehrstuhl für Allgemeine Lebensmitteltechnologie der Technischen Universität, München, D-8050 Freising-Weihenstephan, G.F.R.

Bengt Bjellqvist, Kristina Ek
LKB Produkter AB, Box 305, 16126 Bromma, Sweden

Elisabetta Gianazza and Pier Giorgio Righetti
Department of Biochemistry, University of Milano, Via Celoria 2, Milano 20133, Italy

Introduction

Two dimensional (2-D) separations of complex protein mixtures have become a quite popular and extremely powerful tool in biochemical separations. The common strategy today is to discriminate molecules on the basis of net charge (or free mobility) in the first dimension, followed by fractionation according to size in the second dimension. The most generally adopted method, at present, is to perform isoelectric focusing (IEF) in the first dimension, followed by sodium dodecyl sulphate (SDS) electrophoresis perpendicular to the first run. The technique received a new impetus in 1975, when O'Farrell (1) described a simple method for performing IEF-SDS runs, and coupled this technique to very sensitive detection methods, such as autoradiography. Alternatively, 2-D maps can be revealed by the silver stain (2). We have recently described a modification of this technique, by which both runs are performed in gel slabs, cast as ultrathin matrices, and run horizontally (3, 4). This allows a better transfer of protein zones from the 1st to the 2nd dimension and improved separations due to reduction of vertical thermal gradients in the gel thickness. However, one of the major problems of 2-D techniques is the reproducibility of 2-D maps from run to run. Costancy of spot position and intensity in different experiments is a prerequisite for

Electrophoresis '82

any genetic or clinical study, where the appearance of "new" bands can be of paramount importance in detecting a mutation or the insurgence of a pathological condition in a patient. Unfortunately, present day IEF systems are beliguered by the "cathodic drift" (5), a form of pH gradient instability which results in a net movement (and eventual loss) of protein zones towards the cathode. Thus, if all the experimental conditions are not rigorously controlled from run to run, it is quite difficult to obtain reproducible 2-D maps. By exploiting the recently described technique of "immobilized pH gradients" (6) we report here a new method for developing high-resolution, highly reproducible 2-D maps. Applications to the analysis of seed proteins and zymogram techniques will be presented.

Materials and Methods

Ampholine carrier ampholytes pH 3.5-10 and Immobilines^{Tm} pK 4.2 and 6.2 were from LKB Produkter AB, Bromma, Sweden. Acrylamide, N,N'-methylene bis-acrylamide (Bis), N,N,N',N'-tetramethylethylendiamine (TEMED) and ammonium persulphate (AP) were from Bio Rad Labs., Richmond, Calif., USA.

Immobilized pH gradients. All gels were made to contain 5%T, 3%C_{Bis} (nomenclature of Hjertén) (7) and 8M urea. Two types of immobilized pH gradients were run: one spanning a pH range 4.7-6.0 and a narrower pH 5.2-5.8 range (as measured after IEF with a surface electrode). For the wider range, the dense acidic solution was made to contain 20% glycerol, 15 mM Immobiline pK 4.4 and 3.5 mM Immobiline^{Tm} pK 6.2 (resulting in a pH of 4.8), while the light basic solution contained no glycerol, 3.6 mM Immobiline^{Tm} pK 4.4 and 15 mM Immobiline^{Tm} pK 6.2 (final pH 6.9). For the narrow pH range, the dense acidic solution contained 20% glycerol, 20 mM Immobiline^{Tm} pK 4.4 and 12 mM Immobiline^{Tm} pK 6.2 (final pH 5.4), while the light basic solution had a mixture of 12 mM Immobiline^{Tm} pK 4.4 and 20 mM Immobiline^{Tm} pK 6.2 (actual pH 6.4). All pH measurements were made in the presence of 8M urea, and no corrections were applied (8). The discrepancy between the nominal pH ranges read in solution and the actual pH ranges found in the gel after IEF might be due in part to the different types of electrodes used (bulb vs. surface electrode) and to the different Immobiline environments (liquid vs. gel phase). The samples (acetone powder of ground, whole seeds) were dissolved in 0.2M Tris-Gly buffer, pH 8.6 at a concentration of 200 mg/ml; 20 µl of this solution were applied per slot, at the cathode in the present experiments, but at the anode in all subsequent runs, since this markedly reduces streaking. The gel was pre-run for 60 min or until the visible salt front had passed the application pockets. The running conditions (for a gel 0.5 mm thick, 4 cm wide): 10W (max) power; 20 mA (max) current and 2500 volts, for 16 hours at 10°C. The anolyte was 10 mM H_3PO_4 and the catholyte 100 mM Gly.

2-D electrophoresis. The second dimension run was in a slab 360 µm thin, containing an acrylamide gradient from 4% to 22.5%T, with constant 2.5%C_{Bis}, 0.1% SDS and 375 mM Tris-HCl, pH 8.8. After fixing and staining, the IEF strip was washed in water for 10 min and then equilibrated for 30 to 60 min in 1% SDS, 2% β-mercaptoethanol and 375 mM Tris-HCl, pH 8.8. The run lasted 3 hours at 10°C, 50 mA, 30W (max) and 600V (max). For handling of gel strips, transfer from the 1-D to the 2-D run, see refs. (3) and (4).

Zymograms. Staining for trypsin-inhibitor after IEF in Immobiline^{Tm} gradients was as follows (9): the gel was incubated for 15 min at 37°C in 200 ml of 100 mM phosphate buffer, pH 7.6, containing 60 mg trypsin. It was then rinsed 6 times in distilled water and left to incubate in a dry basin for an additional 15 min at 37°C. The zymogram was developed by immersion in 180 ml of freshly made 50 mM phosphate buffer, pH 7.6, containing 100 mg of fast blue B. To this solution, just prior to use, are added 20 ml of dimethyl formamide containing 50 mg of N-acetyl-DL-phenyl-alanine-β-naphthyl ester. Color development takes several hours. The reaction is stopped in methanol/acetic acid/H_2O (25:10:65 v/v).

Results

Fig. 1 (left) shows the IEF profiles of different legume species in a conventional Ampholine pH gradient. Since most proteins focus in the acidic zone, especially in the pH 5-6 interval, as also generally demonstrated for all protein species (10), we have expanded this range by using Immobilines covering first a narrow pH 4.7-6.0 interval (Fig. 1, right) and then a narrower pH 5.2-5.8 region (Fig. 1, lower panel). The ultra-narrow pH range se-

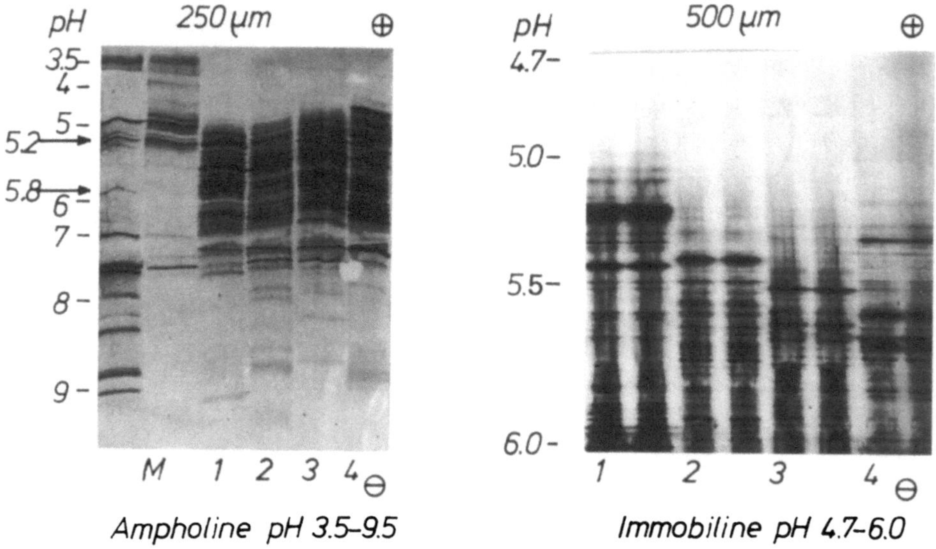

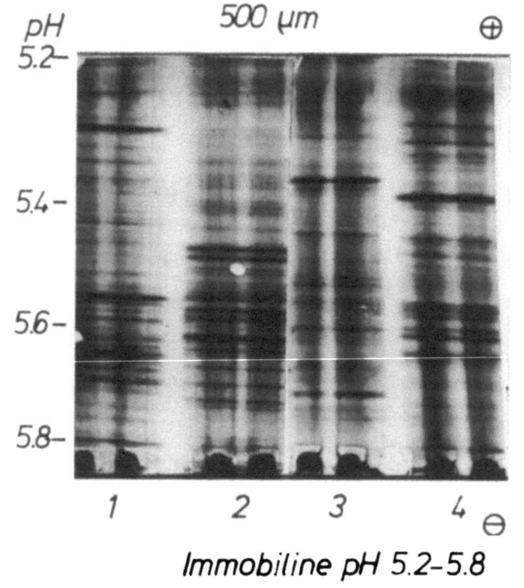

Fig. 1. Upper left: IEF in 5%T, 3%C gels, containing 8M urea and 2% Ampholine pH 3.5-10. Running conditions:1.5 hours, 10°C, 1500 V (max). Upper right: IEF in Immobiline pH 4.7-6.0:prefocusing for 1 hour, IEF overnight, 10°C, 2500 V. Lower box: IEF in Immobiline pH 5.2-5.8: prefocusing 1 hour, IEF overnight, 10°C, 2500 V. pH gradients measured with a surface electrode. Gels stained with Coomassie Brilliant Blue R-250. Samples: 1: pea (*Pisum sativum*); 2: broad bean (*Vicia faba*); 3: lentil (*lens culinaris*); 4: chick pea (*Cicer arietinum*); M: pI marker proteins from Pharmacia (wide and narrow ranges). The ground seeds were extracted with 25 mM Tris-Gly buffer, pH 8.6.

parates the most interesting part of the legume proteins with the highest charge heterogeneity and clearly shows species-specific patterns which are not readily apparent in the wider pH intervals. This progressive increment of resolution can also be appreciated in the scans of Fig. 2.

Within the same family of legumes, IEF in ultra-narrow Immobiline^{Tm} gra-

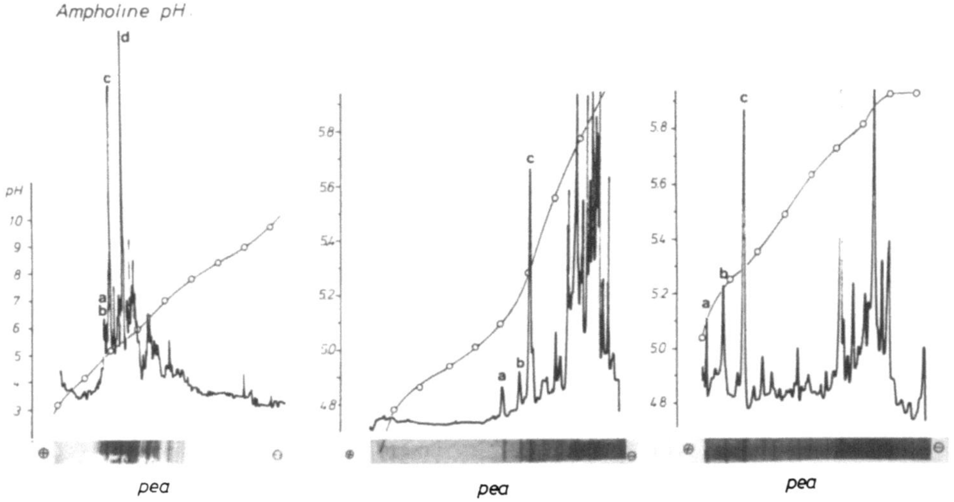

Fig. 2. Densitometer scans of the IEF tracks of sample No. 1 (see Fig. 1). Laser densitometer (LKB 2202 Ultroscan). pH gradients measured with a surface electrode. Letters a-e refer to the identical protein peaks to demonstrate the pH enlargement in ultra-narrow pH gradients.

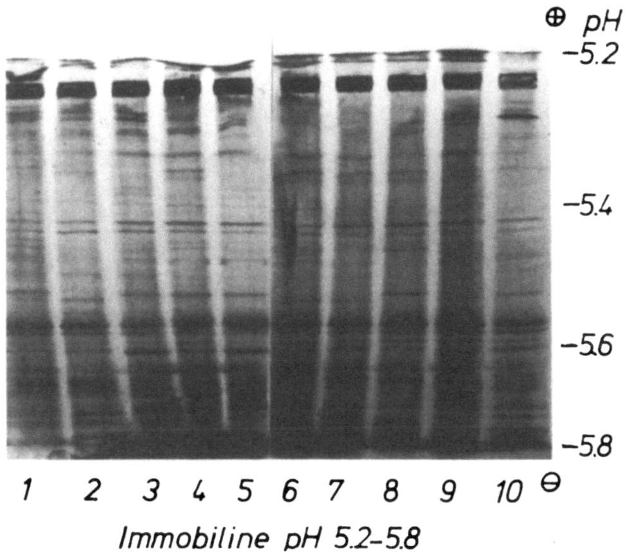

Fig. 3.
IEF in Immobiline[Tm] pH 5.2-5.8. The gel (5%T, 3%C, 0.5 mm thick) contains 8M urea. Running conditions: prefocusing for 1 hour, IEF overnight 10°C, 2500V. Gel stained with Coomassie Blue R-250. Samples: 10 different varieties of pea: 1: Kearly; 2: Mini; 3: Nugget; 4: Obtusy; 5: Platinum; 6: Progress No. 9; 7: Salout; 8: Sprite; 9: Teconor and 10: Turon. The ground seeds were extracted with 25 mM Tris-Gly buffer, pH 8.6. Clear genetic differences can be observed in the IEF pattern.

dients can also readily display genetic differences among different varieties. This point is illustrated in Fig. 3 in the case of some common pea varieties: the IEF profiles clearly show differences in band position and relative intensities which would have escaped detection in conventional, 2 pH

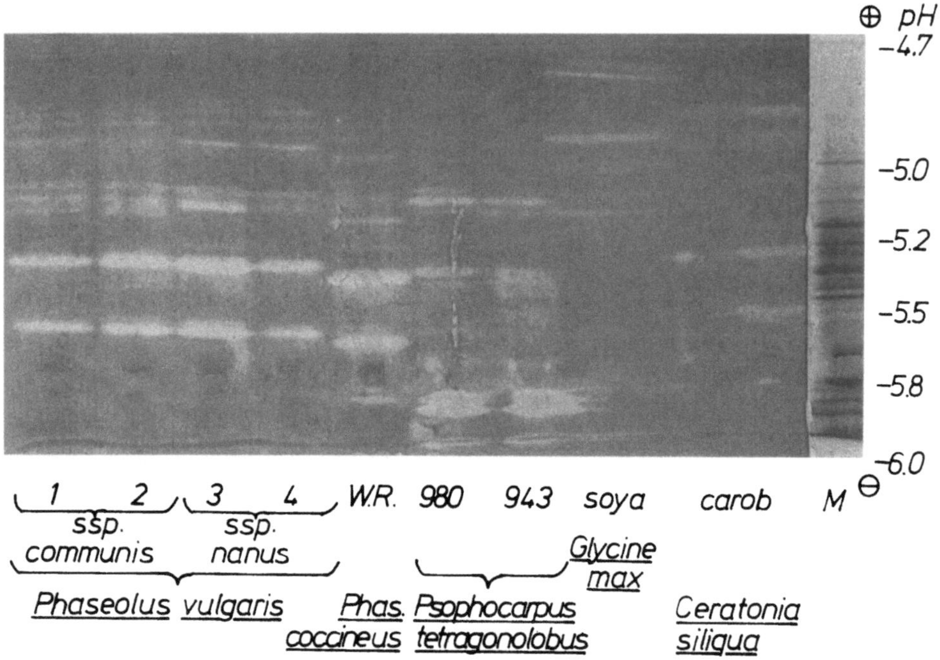

Fig. 4. Trypsin inhibitors of different legume species and varieties. IEF in Immobiline pH 4.7-6.0 in 5%T, 3%C gels, 0.5 mm thick, containing 8M urea. pH gradient measured with a surface electrode. Running conditions: prefocusing 1 hour, IEF overnight, 10°C, 2500 V. Staining procedure according to ref. (9). Sample varieties: 1: Mombacher Speck; 2: Zucker Perl Princess; 3: Sotexa; 4: Felix; W.R.: Weisse Riesen; 980 and 943: different varieties of winged beans. For sample prepration, see Fig. 3.

unit intervals of carrier ampholytes. Notwithstanding the rather flat pH gradient along the separation path, the bands are extremely sharp due to the possibility of applying high voltage gradients (a minimum of 250V/cm, but as much as 1000V/cm can be tolerated in this system) (6, 11).

Immobiline^{Tm} pH gradients also seem to be fully compatible with zymogram techniques, as shown in Fig. 4. The zones of activity of trypsin inhibitor from different legumes are completely developed and seem to behave as after conventional electrophoresis or IEF in Ampholine-generated pH gradients. Again, due to the high resolving power of ultra-narrow pH gradients, species specific isozyme patterns are readily apparent, as well as genetic variants within different varieties of the same legume.

An important application of Immobiline^{Tm} pH gradients would be their use

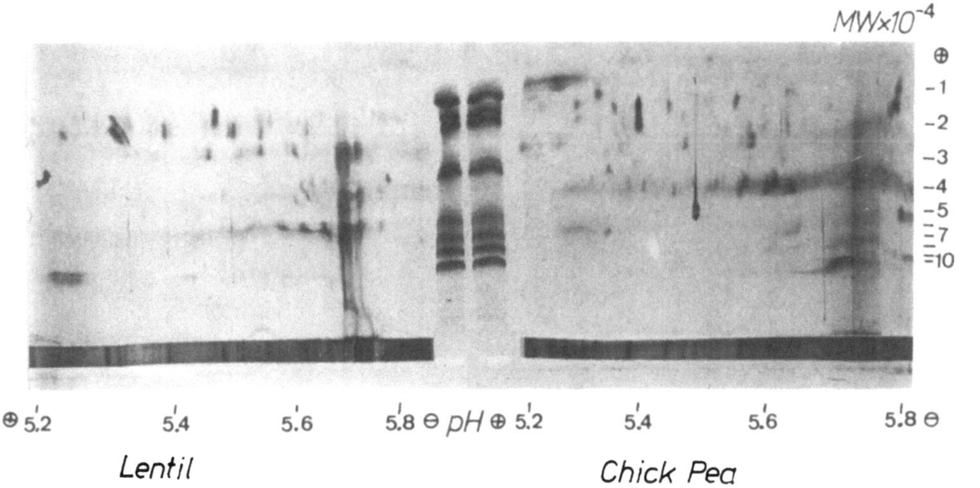

Fig. 5. Ultrathin-layer horizontal high-resolution 2-D electrophoresis. 1st dimension: IEF in Immobiline pH 5.2-5.8, in 5%T, 3%C gels, 0.5 mm thick, containing 8M urea. Running conditions: prefocusing for 1 hour, IEF overnight, 10°C, 2500 V. After fixing and staining, the IEF strips were rinsed 10 min in water, equilibrated for 60 min in SDS buffer, blotted and loaded on the 2-D gel. 2nd dimension: SDS-PAGE in a 360 µm thin pore gradient gel (4-22.5%T, 2.5%C). Running conditions: 3 hours, 10°C, 50 mA, 30W max, 600V.

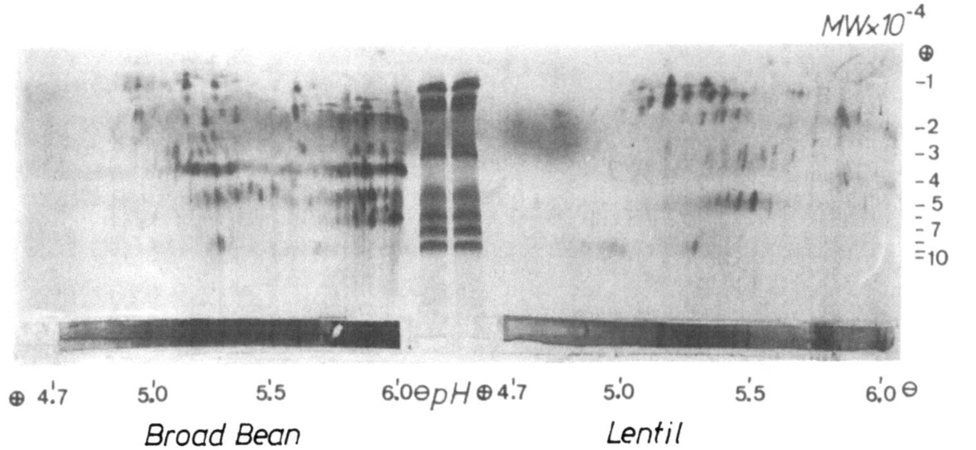

Fig. 6. Ultrathin-layer horizontal high-resolution 2-D electrophoresis. IEF in Immobiline pH 4.7-6.0. For gel properties and running conditions see Fig. 5.

in the first dimension of 2-D techniques, due to their unique properties of being completely drift-free. Examples of these separations are shown in Figs. 5 and 6. Fig. 5 shows the 2-D patterns of proteins extracted from lentil and chick pea, originating from a ultra-narrow pH gradient in the first dimension. The high resolution achieved along the pI axis helps to spread the spots in the 2-D plane. In Fig. 6, the 2-D maps, originating from a broader pH gradient in the first dimension, of broad beans and lentils are compared. Crowding of the spots in the 2-D plane, resulting from a compression of the protein bands in the IEF run, is clearly visible. In both cases it appears as though, in contrast to the broad charge heterogeneity, both legumes exhibit a restricted size dispersion, since horizontal strings of spots appear, suggesting charge isomers of the same mass subunits (12, 13).

Discussion

Several advantages are inherent to the present technique for generation of 2-D maps:

1) highly reproducible 2-D maps are produced since there is no drift in the IEf dimension. This has been a bad nuisance in the past, since too many variables had to be controlled to avoid shifts of spot positions from run to run. With Immobiline^{Tm} pH gradients, full control of the first dimension run is now available: the gradient is insensitive to salts in the sample and is totally independent of running conditions, such as voltage and time (provided the product of the two is enough for reaching the pI position for all proteins in the sample).

2) due to the ease of generating ultra-narrow pH gradients(tipically lower than ½ pH unit) it is possible to perform "blow-up" experiments, i.e. to enlarge considerably overcrowded regions for maximizing the separation. This cannot be done with conventional carrier ampholytes since, below a minimum pH interval (in general greater than ½ pH unit) too few amphoteric species remain to ensure an even buffering power and conductivity along the separation path. In this system, it has been demonstrated that two proteins, differing in pI by as little as 0.001 pH unit (6), can be resolved: such a resolving power has not been achieved previously.

It appears also that Immobiline^{Tm} pH gradients are compatible with detection techniques commonly employed in the biochemical field. For instance the fact that zymogram stains can be developed in conventional ways (see Fig.4) suggests that Immobiline^{Tm} gels should also allow detection by immunotechniques or by specific stains (e.g. for glyco-, lipo- or metallo-proteins). In the case of 2-D techniques, it appears that a full transfer of protein zones from the 1st to the 2nd dimension is possible, since there is no streaking in the 2-D map (see Figs. 5 and 6). However, prior to the SDS run, the IEF strip should be incubated in the SDS denaturing solution for a much longer time (30-60 min) than after conventional IEF runs (only 2-4 min). This suggests that a protein focused in a grafted pH gradient might require considerably longer times to interact with SDS micelles or that diffusion of SDS in the gel layer is hampered by the fixed charges. Further work is in progress to elucidate these aspects.

References

1. O'Farrell, P.: J. Biol. Chem. 250, 4007-4021 (1975).
2. Merril, C.R.,Switzer, R.C., Van Keuren, M.L.: Proc. Natl. Acad. Sci. US 76, 4335-4339 (1979)
3. Görg, A., Postel, W., Westermeier, R., Gianazza, E., Righetti, P.G.: J. Biochem. Biophys. Methods 3, 273-284 (1980)
4. Görg, A., Postel, W., Westermeier, R., Gianazza, E., Righetti, P.G.: In Electrophoresis '81, Allen, R.C., Arnaud, P.,eds., de Gruyter, Berlin, pp. 257-270 (1981)
5. Righetti, P.G., Drysdale, J.W.: Ann. N.Y. Acad. Sci. 209, 163-186 (1973)
6. Bjellqvist, B., Ek, K., Righetti, P.G., Gianazza, E., Görg, A., Postel, W.: In Electrophoresis '82, Stathakos, D., ed., de Gruyter, Berlin, pp. 61-74 (1983)
7. Hjertén, S.: Arch. Biochem. Biophys. Suppl; 1, 147-151 (1962)
8. Gianazza E., Righetti, P.G., Bordi, S., Papeschi, G.: In Electrofocusing and Isotachophoresis, Radola, B.J., Graesslin, D.,eds., de Gruyter, Berlin, pp. 173-179 (1977)
9. Uriel, J., Berges, J.: Nature 218, 578-579 (1968)
10. Gianazza, E., Righetti, P.G.: J. Chromatogr. 193, 1-8 (1980)
11. Righetti, P.G., Gianazza, E., Bjellqvist, B., Ek, K., Görg, A., Westermeier, R.: In Electrophoresis '82, Stathakos, D.,ed., de Guyter, Berlin pp. 75-82 (1983)
12. Görg, A., Postel, W., Westermeier, R.: Z. Lebensm. Unters. Forsch. (1982) in press
13. Westermeier, P., Postel, W., Görg, A., Telek, L.: In Electrophoresis '81, Allen,R.C.,Arnaud,P.,eds.,de Gruyter, Berlin, pp. 281-287 (1981)

TWO-DIMENSIONAL ELECTROPHORESIS IN AGAROSE GELS

Susan E. Coulson and Richard B. Cook

FMC Corporation, Marine Colloids Division, BioProducts Department, Rockland, Maine 04841

Introduction

Two-dimensional gels are typically run in polyacrylamide in both dimensions, or a combination of an agarose first dimension followed by a polyacrylamide separating gel. These gels provide an excellent means of resolving protein subunits, denatured proteins, or native proteins between 10,000 and 200,000 daltons. Unfortunately, polyacrylamide gels are difficult to cast and reproduce at concentrations <5%. As a result, many high molecular weight proteins cannot be examined in polyacrylamide gels.

In contrast, agarose gels have traditionally been regarded as too macroporous for use in the MW range from 20,000-200,000 daltons. They have, nevertheless, been shown to be ideal for the electrophoretic separation of high MW nucleic acids ($\leq 100 \times 10^6$ daltons) as a function of molecular size. SeaPrep® agarose, however, has been reported to have the sieving properties of polyacrylamide gels when used at 5-7½ concentrations under non-denaturing conditions.[1] More recently, it has been shown that 2-10% SeaPrep® agarose can resolve standard proteins varying in MW from 17,800 to 670,000.[2] This makes possible the substitution of SeaPrep® agarose for low concentration polyacrylamide gels in 2-D applications.

The use of SeaPrep gradient gels has not previously been reported. This report focuses on the potential of SeaPrep®

Electrophoresis '82

gradient gels for the study of a continuous range of both high and low molecular weight substances which could not previously be examined in a single gel medium. The application of Sea-Prep® gradient gels to 2-D gels should expand the range of proteins which can be studied and, therefore, catalogued in the Human Protein Index now being compiled.[3]

Materials and Methods

IsoGel™ agarose, SeaPrep™ agarose, IsoGel Ampholytes (pH 3.5-9.5) and GelBond™ film are products of FMC, Marine Colloids Division (Rockland, Maine). A mixture of high and low MW proteins was prepared from: hemocyanin (Limulus; Sigma H-1757), α_2 macroglobulin (bovine; Boehringer-Mannheim), thyroglobulin (porcine; Sigma T-1126),catalase (bovine; Sigma C-100), IgG (human myeloma; courtesy of Dr. Wayne Hoffmann, Arthritis Research Center, Dallas, Texas), myoglobin (whale; Sigma M-0380) and myoglobin (horse; Sigma M-0630). The mixture was formulated to contain equal amounts of each protein and 6µL containing 70µg of this mixture was applied to each 1st dimension gel. A plastic sample application mask (available from Marine Colloids) was used for both the first dimension and analytical IEF gels. The long dimension of the rectangular sample application slot (2 x 10mm) was oriented perpendicular to the electrodes for 1st dimension gels but parallel to the electrodes for analytical gels. This orientation facilitated introduction of the sample into the 2nd dimension gel as a narrow band.

The 1st dimension gel slice was sectioned from the IsoGel IEF plate by using a paper cutter. Holding the edges of the Gel-Bond film. The slice was then inverted onto the 2nd dimension gel leaving just enough space for the cathodal wick (5mm).

The agarose gradient gels were cast on GelBond film using a

Hoeffer XPO 77 exponential gradient maker. The two jacketed reservoirs were maintained at 50°C while mixing and pouring the two agarose solutions: a 1% SeaPlaque agarose in 0.05M Tris·HCl buffer (pH 9.0) and 1% SeaPlaque with 4% SeaPrep agarose in 0.375M Tris·HCl buffer (pH9.0). A GelBond NF release sheet was used in the casting frame to facilitate disassembly. The polyacrylamide IEF and 2nd dimensions polyacrylamide gels were cast on GelBond PAG film (Marine Colloids) using standard ammonium persulfate polymerization procedures.

Results and Discussion

The protein mixture was formulated to span a wide range of pI (5 pH units) and molecular weight (> $3x10^6$ daltons). The analytical isoelectric focusing patterns displayed considerable microheterogeneity for all mixture components in both agarose and polyacrylamide gels. Some differences in the IEF resolutions of certain proteins (α_2 M and C; see Figures 1&1) were observed between the two gel media. These differences could result from the sieving effect that a 5% polyacrylamide IEF gel could exert on proteins having high molecular weights (>200,000 daltons). In contrast, a 1% IsoGel agarose IEF gel has been shown capable of achieving literature pI values for a wide range of both low and high molecular weight proteins.[4] In any case, it was noted that differences in resolution and band position due to sieving influences in the 1st dimension would tend to compromise the mapping concept implicit in 2-D electrophoresis.

Profound differences in the morphology of the 2nd dimension spots were observed between the two gel media. The SeaPrep/SeaPlaque agarose gradient gel produced more spherical 2nd dimension spots for certain low (M_w, M_h) and high molecular weights (T) proteins. The more compressed spot morphology in the agarose gradient was rationalized on the basis of the greater ease of sample migration from the first to the 2nd dimension gels when both are made with 1% agarose as opposed to 5% polyacrylamide. Other proteins, having even higher molecular weights (α_2M, H), were observed to migrate significant distances into the agarose gels whereas they remained either at the origin or were substantially retarded in the 5% polyacrylamide gel (compare region X in Figures 1&2). Similarly, the migration of the human myeloma proteins (IgG) has been restricted more in the 5% polyacrylamide gels (Figure 2 versus Figure 1).

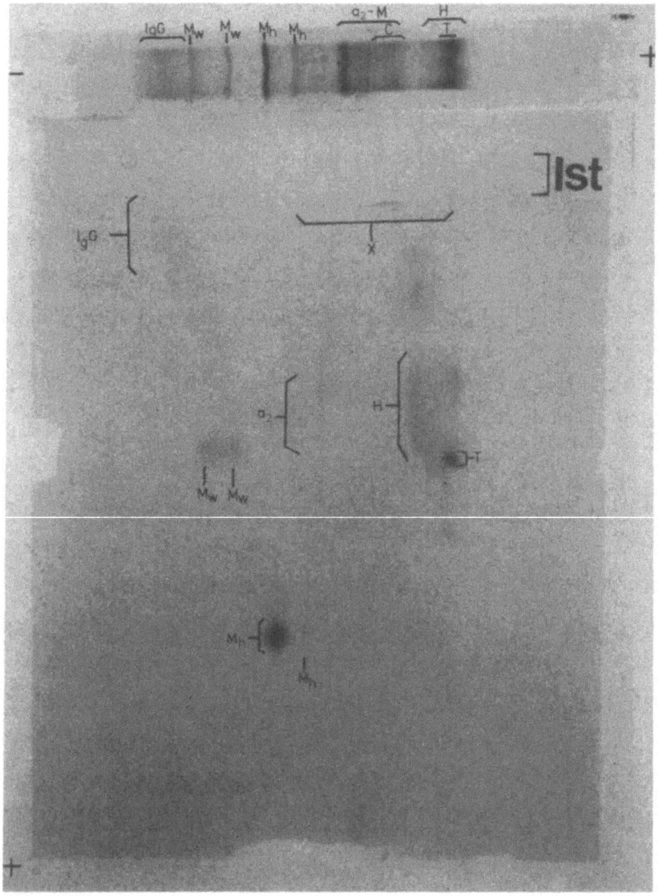

Figure I. A non-SDS horizontal slab gel was cast from 1% SeaPlaque in 0.5M Tris·HCl, pH 9.0 and 1% SeaPlaque + 4% SeaPrep 15/45® in 0.375 Tris·HCl, pH 9.0. The electrode buffer was 0.05M Tris·HCl, 0.192M glycine, pH 8.3. The gel was run for 3 hrs. at 25mamps.

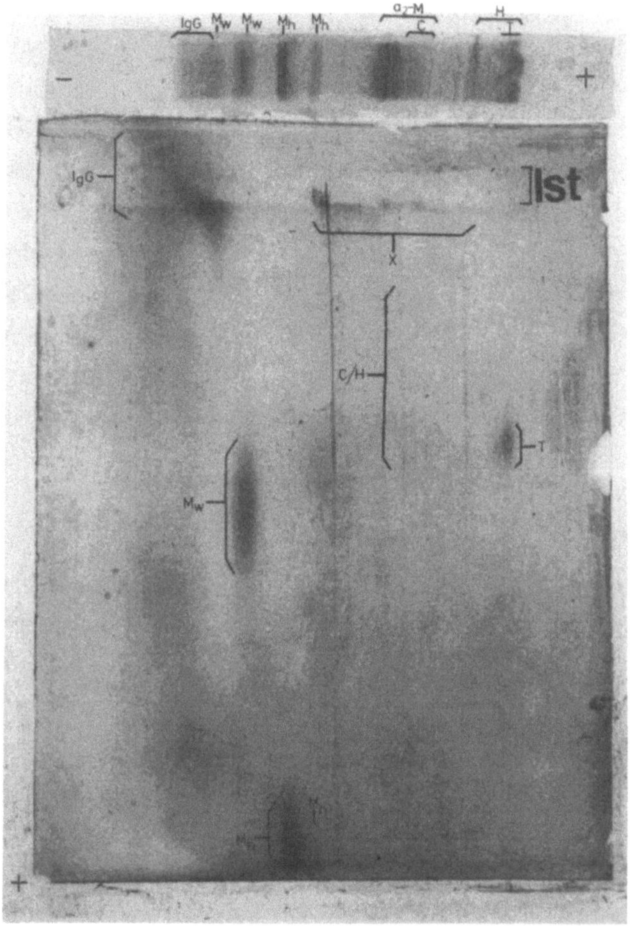

Figure II. A non-SDS horizontal slab gel was cast from 5%T 3%C polyacrylamide in 0.5M Tris·HCl pH 9.0. The electrode buffer was 0.05 M Tris·HCl, 0.192 M glycine, pH 8.3. The chamber platen was cooled to 15°C and the gel was run at 25 mamps for 2 hours, 25 minutes. Myoglobin protein markers and 0.5% Bromophenol blue were used to monitor the progress of the run.

TABLE 1

Protein	Source	MW (daltons)	Subunits No.	Subunits MW (daltons)	pI (empirical)
Hemocyanin*	Limulus	3.3×10^6	48	68,000	3.78-4.21
Thyroglobulin	Porcine	669,000	2	335,000	3.89-4.0
α_2-Macroglobulin**	Bovine Plasma	725,000	4	(181,000)	4.46-5.30
Catalase	Bovine Liver	232,000	4	57,500	4.67-4.73
Myoglobin	Whole Skeletal Muscle	17,000	0	- - -	8.20 major 7.70 minor
Myoglobin	Horse Skeletal Muscle	17,500	0	- - -	7.40 major 7.00 minor
IgG	Human Myeloma	160,000	0	- - -	8.06-8.62

*Eur. J. Biochem: 95, 399 (1979)

**Barret, H.A.: Methods in Enzymology, 80, 737 (1981)

All remaining molecular weights were drawn from Biological Chemistry, H. Mahler and E. Cordes, P. 176, 2nd edition, Harper & Row Publishers, NY (1971)

It must be recognized that since the gels shown were not run in the presence of SDS, the protein migration distance in the 2nd dimension cannot be directly related to molecular weight. Instead, the migration distance of each protein in the 2nd dimension gel is a function of the difference between its pI and the buffer pH. A 2nd dimension Tris™ buffer of pH 9.0 was chosen because its pH exceeded the highest pI in the sample mixture (IgG myeloma). Even under these conditions, however, slight differences in pI produced dramatic differences in 2nd dimension mobility. For example, the horse myoglobin (M_h) moves much farther than the whale myoglobin (M_w) due to their difference in pI not molecular weight. Similarly, thyroglobulin (T) and its subunits move much further in the 2nd dimension than IgG even though it has a molecular weight 3-6 times that of IgG. When Laemmli SDS techniques are used in the agarose gradient gel, however, all proteins in the mixture were found to move anodally as a function of their SDS micelle size (results not shown). Under the SDS conditions, no exclusion of α_2M, catalase or hemocyanin was observed at the origin as is seen in Figure 2 and to a lesser extent in Figure 1.

The primary advantage of the use of SeaPrep gradient gels is their ability to examine high molecular weight proteins in their native state. The fact that migration distance in the 2nd dimension does not reflect molecular weight differences for proteins having significantly different isoelectric points does not obviate the utility of this technique. Convenient and authoritative tables of pI and molecular weight have been published for a wide spectrum of proteins.[5] Judicious selection of protein markers having the same pI but different molecular weights than the test proteins of interest will permit the accurate measurement of molecular weight as a function of migration distance in the 2nd dimension. A buffer pH can be chosen which will maximize the anodal (or cathodal) mobility of the proteins under investigation. Using this technique even ultra-high molecular weight proteins should be amenable to 2-D electrophoresis in their native state. Such proteins can, therefore, be described and included in the Human Protein Index based on their non-denatured properties. This capability could have important application to the detection of numerous disease processes which are thought to result in or develop from high molecular weight molecules, aggregates or complexes.

References

1. Nochumson, S., Cook, R. B., and Williams, K.: A New Extremely Low Gelling Temperature Agarose and its Applications, Electrophoresis '80; Walter de Gruyter, Munich, Germany.

2. Buzas Z. and Chrambach, A.: Un-Supercoiled Agarose With a Degree of Molecular Sieving Similar to That of Cross-Linked Polyacrylamide, Electrophoresis (in press).

3. Anderson, N. L., Edwards, J. J., Giometti, C. S., et al.: A High Resolution Two-Dimensional Electrophoretic Mapping of Human Proteins, Walter de Gruyter, Munich, Germany, Electrophoresis '79, 313.

4. Harper, D.: Isoelectric Focusing in Agarose Gels, Walter de Gruyter, Munich, Germany, Electrophoresis '81, 205.

5. Righetti, P., and Caravaggio, T.: Isoelectric Points and Molecular Weights of Proteins: A Table, J. of Chromatography, 127, 1 (1976).

ASPECTS OF HIGH RESOLUTION TWO DIMENSIONAL GEL ELECTROPHORESIS : ATTEMPTS TO INCREASE RESOLUTION

A.H.M. Burghes, M.J. Dunn, J.A. Witkowski and V. Dubowitz

Jerry Lewis Muscle Research Centre, Department of Paediatrics and Neonatal Medicine, Hammersmith Hospital, London W12 OHS, U.K.

1. Introduction

In any cell type up to 8000 proteins are synthesized. Two dimensional gel electrophoresis provides the capability of resolving such complex mixtures (1). Polymorphic variation appears low using this technique (2), thus facilitating comparison between cell lines. In studies of Duchenne muscular dystrophy (DMD) we have used one dimensional methods to compare normal and DMD tissue (3 - 6), and have begun analysis by 2D-PAGE (6). However, for optimal utilization of 2D-PAGE it is important to maximise resolution and to have facilities for both quantitative and qualitative evaluation.

2. Features of 2D-PAGE for optimal resolution

(1) Operation of the two dimensions on different physicochemical principles, i.e. charge and molecular weight. (2) Complete solubility and disaggregation of all sample proteins with their complete entry into the gel. (3) pH gradient stability over long times to allow simultaneous resolution of basic and acidic proteins at equilibrium. (3) The separate use of non-equilibrium methods to determine mobility variants, i.e. substitutions involving amino acids with pK_a = pI (7). (5) pH gradient spreading and optimization of ampholytes. (6) High field strengths to obtain sharp bands in the IEF dimension. (7) Gradient shape engineering and the use of long SDS gels. (8) Minimization of diffusion and protein loss during equilibration and complete elution of proteins from the IEF gel. (9) Minimization of lateral diffusion in SDS-PAGE. (10) Sensitive detection systems

Electrophoresis '82

that minimize spreading. (11) Quantitation and gel evaluation.

3. Solubility and Disaggregation

For maximum electrophoretic resolution it is necessary to dissociate proteins into their component polypeptides, maintain them in solution and prevent interactions between them. As the forces responsible for the folding of proteins are the same as those for protein-protein interaction, agents able to disrupt these interactions are generally denaturing. Various solubilization procedures have been compared for IEF using fibroblasts (4). Solubilization in different buffers containing urea and NP-40 yielded identical patterns. When SDS solubilization was used no difference in the profiles was observed with or without competition buffer. When SDS procedures were compared to those using urea and NP-40, the only difference found was that less material remained at the point of sample application with a corresponding increase in two bands in the middle of the gel. NP-40 was found to be removed from the gel when SDS was present in the sample, resulting in proteins being exposed to a detergent-free environment which may lead to their precipitation. The zwitterionic detergent, sulphobetaine, in combination with NP-40 was not as effective as 8M urea/NP-40 and sulphobetaine was not compatible with high levels of urea. The 8M urea/NP-40 system would therefore appear to be the best procedure available at the current time, but approximately 10% of the original sample was found to remain at the origin of these gels. It may be possible to increase sample dissaggretation and entry into the gel using alternative denaturing agents either alone or in combination.

Most procedures have utilized detergents with the polyoxyethylene head group. The table lists some of these detergents and detergents with alternative head groups. It is not known whether any of these head groups interact with ampholytes or whether such interactions interfere with solubilization or focusing of proteins. Alkylolamides (Rewocid DU185) precipitated on the addition of ampholytes, probably due to ester formation with the hydroxyl group. This raises the possibility that the terminal hydroxyl group of Triton-type detergents might react with ampholytes.

TABLE OF DETERGENTS

Chemical or trade name	Head group structure	Head group name
Non-ionic		
Lubrol	$-(O-CH_2-CH_2)_n-OH$	Polyoxyethylene
Triton	 $-(O-CH_2-CH_2)_n-OH$	Polyoxyethylene
Mercaptans	$-S-(O-CH_2-CH_2)_n-OH$	Polyoxyethylene
Ammonyx LO	$CH_3-N(\rightarrow O)-CH_3$	Amine oxide
Rewocid DU 185	$C(=O)-N(H)-CH_2-CH_2-OH$	Alkylolamides
α Octylglyceryl-ether	$-CH(OH)-CH(OH)-CH_2(OH)$	Glycerol
Decyl-β-D-gluco-pyranoside	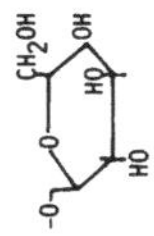	Sugar
Dodecyl-β-D-maltoside	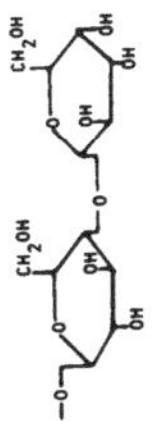	Sugar
Dodecyl urea	$-N(H)-C(=O)-NH_2$	Amide
Zwitterionic		
Zwittergent	$CH_3-N^+(CH_3)-CH_2-CH_2-CH_2-S(=O)(=O)-O^-$	Sulphobetaine
Phosphoryl-choline	$-O-P(O^-)(=O)-O-CH_2-CH_2-N^+(CH_3)_2-CH_3$	Phosphoryl-choline

Sulphobetaine, although not precipitating in TCA alone, did precipitate when ampholytes were also present (as does SDS - S.E. Coulson, personnal communication). Addition of sodium sulphate to the precipitated sulpho-betaine resulted in clearing, implying sulphate interaction with the ampholytes. The glucoside-type detergents did not precipitate in mixtures of TCA and amphclytes, implying a lack of interaction. It is interesting to note that sulphobetaine has its headgroup charge inverted compared to phospholipids. Dodecyl urea would be particularly interesting in regard to its solubilizing ability.

4. pH Gradient Stability and the Resolution of Basic Proteins

Conventionally 2D-PAGE is carried out utilizing rod IEF gels (1). There is considerable pH gradient instability associated with the cathodic region and the pH gradients do not extend much above pH 7.0, which has led to the development of transient state non-equilibrium systems (8). In an attempt to improve the focusing of cathodic proteins in the equilibrium system, the tubes were treated to reduce EEO by a method which results in a firmly bound layer of methyl-cellulose (6). The sample and overlay buffer contained arginine (0.1M) and the anolyte was 0.04M aspartic acid, 5 mM H_3PO_4 and 8M urea (6). The catholyte was 2.5% ethylendiamine. Fig. 1 shows a separation of fibroblasts labelled with [^{35}S]-methionine. The pH gradient, measured in the presence of 8M urea, extended to 10.5. However, the cathodic bands still showed poor resolution and were streaky, and increasing the focusing time did not result in any improvement. Streaking may be due to difficulty of the proteins in traversing the cathodic region due to poor distribution of ampholytes in this region. Flat bed gels run on silanized glass plates with Pharmalyte 8-10.5 showed severe thinning implying poor conductivity in this region (4). However, it was noted that in pH 3-10 flat bed gels very sharp cathodic bands were obtained. As the samples in this case were applied about 1 cm from the cathode, the proteins did not have to cross regions of poor conductivity at either end of the gel. Also the use of thin gels permits the application of high field strengths (4).

To apply the advantages of flat bed IEF to 2D-PAGE a special plastic was developed, that would bind polyacrylamide in the presence of 8M urea and 2% NP-40, in which polyester sheets were treated with Dow Corning Prime Coat 1200 followed by reaction with silane A174 (4). C4T4 gels (0.5 mm thick and 15 cm long) containing 8M urea, 2% (w/v) NP-40 and 4% (w/v) Pharmalyte 3-10 were cast on these supports. The anolyte was 0.04M aspartic acid, 5 mM H_3PO_4 and the catholyte was 1M NaOH, 0.25M $Ca(OH)_2$, 0.025M arginine and 0.025M lysine (spun to remove precipitate). The gels were run for a total of 15500 V.hr. The gel was then covered in Saran-Wrap and the appropriate tracks cut out, frozen and stored at -70°C. The wrapping was removed before equilibration for 5 min. in 8 ml of SDS equilibration buffer (1). The gel strip was then carefully applied through a buffer layer to prevent air bubbles being trapped and sealed in place with agarose on top of a 7-20% gradient acrylamide SDS gel (1.4 mm thick). Both sharp cathodic and anodic spots were obtained (Fig. 2) and the pH gradient extended to pH 10. In addition the gel being bound to a support prevented the occurrence of artefacts due to gel stretching. Short equilibration times should minimize diffusion, but some streaking of the high molecular weight proteins was observed. It should be noted

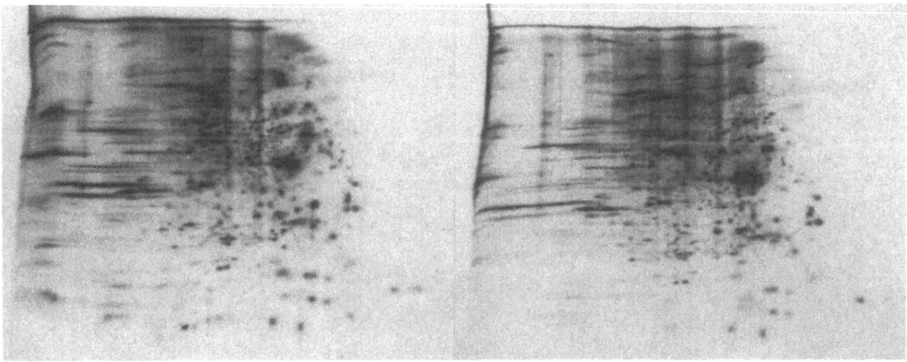

a b

FIG. 1 Separation of normal (A) and DMD (B) [^{35}S]-methionine labelled fibroblast proteins. Rod IEF gels (16.5 x 0.3 mm), containing 8M urea, 2% NP-40, 4% Pharmalyte 3-10 were used in the first dimension. The gels were prerun at 200V for 1 hr, samples (1.5 x 10^6 cpm) applied and run for a total of 10,000 V.hr. The gels were equilibrated for 40 min. (3) and applied to 7 to 20% gradient SDS gels (1.4 mm thick). 10-day exposure using LKB Ultrofilm.

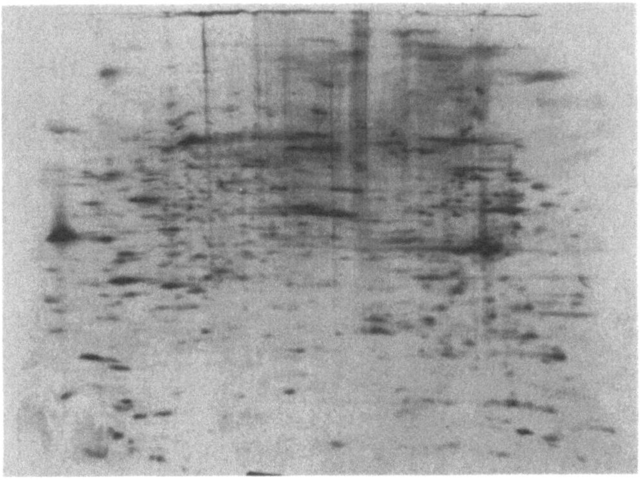

FIG. 2 Separation of [^{35}S]-labelled proteins using the flat-bed technique. The IEF gel was run at 15°C in a CO_2-free N_2 atmosphere using platinum ribbon electrodes. The gel was prefocused at 600V for 45 min. The samples (750,000 cpm) were applied and run overnight at 800V limiting, then the voltage was increased (1000V for 33 min, 1500V for 40 min, 2000V for 15 min) until a total of 15,500 V.hr was reached. Details of the second SDS dimension are given in the text. 20-day exposure using Kodak X-Omat H film.

that the system resolved the alkaline proteins that are resolved by NEPHGE (9), but as the two systems can resolve different types of charge mutation the two techniques are complementary (7). No qualitative differences between normal and DMD patterns have been observed using either the flat bed or the rod gel systems.

5. Optimization of Ampholytes and pH Gradient Spreading

Fig. 3 shows [^{35}S]-methionine-labelled fibroblast proteins focused on C4T4 IEF gels (0.5 mm thick, 10 cm long) containing 8M urea/2% NP-40 and using several commercial ampholytes and cast on plastic supports. Their respective pH gradients are shown in Fig. 4. Servalyt gave better resolution of the acidic proteins whereas Pharmalyte and Ampholine were better in the alkaline and middle range. However, Ampholine gave rise to inappropriate spreading of the alkaline region. As commercial ampholytes are not synthesized by the same procedure, it is predicted that mixtures can give better separation due to the presence of more ampholyte species and that different mixtures will be optimal for each pH region (10). Fig. 5 shows a C4T4 gel (0.5 mm thick and 10 cm long), cast on a silanized glass plate, containing 8M urea, 2% NP-40 and an ampholyte mixture (1.7% Servalyt 2-11, 1.7% Pharmalyte 3-10 and 0.6% Pharmalyte 5-8). The narrow range ampholyte was added to create pH gradient spreading in the pH 5-8 region (Fig. 4). These procedures resulted in improved resolution.

A region of a two-dimensional gel obtained using a C4T4 IEF gel (0.5 mm thick and 15 cm long) containing 8M urea, 2% NP-40 and the mixture of ampholytes is shown in Fig. 6. In this case the focusing gel was equilibrated in 8 ml of equilibration buffer (0.1875M Tris, pH 8.8, 3% SDS, 5% β-mercaptoethanol) for 7 min. The focusing gel was then covered with a polyester strip and applied to the top of a 7-20% gradient SDS gel (0.7 mm thick) without a stacking gel. Very sharp, well-resolved spots were obtained.

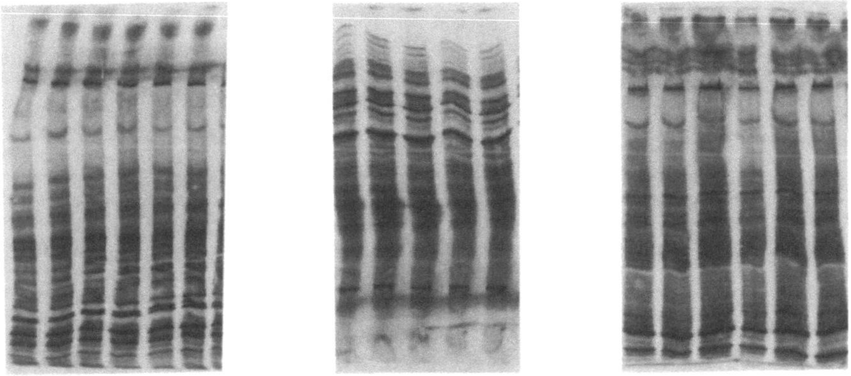

Fig. 3 : IEF gels of fibroblast proteins using various commercial ampholytes. (A) 4% LKB Ampholine pH 3.5-9.5 for agarose and polyacrylamide gel IEF. (B) 4% Servalyt 2-11. (C) 4% Pharmalyte 3-10. The gels were prefocused at 400 V for 45 min. The samples (200,000 cpm) were applied and run at 15W, 800 V limiting for 33 min. and then 15W, 1500 V limiting for 8500 V.hr. The pH gradient was measured with a surface electrode and the gel refocused at 3000 V limiting to give a total of 8700 V.hr for the run. The autoradiograph was exposed for 1 day using Kodak X-Omat H film with an intensifying screen.

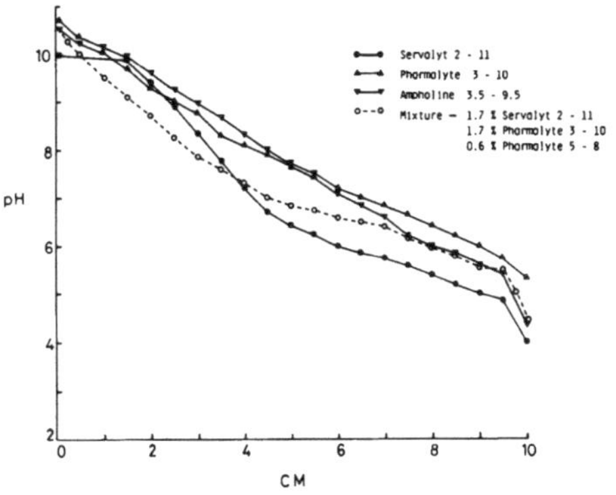

Fig. 4 : pH gradient of IEF gels shown in Fig. 3 and 5.

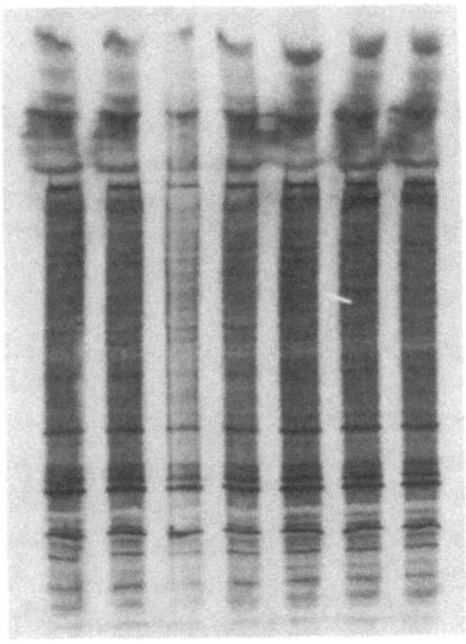

Fig. 5 : IEF gel of [^{35}S]-methionine labelled fibroblast proteins using the ampholyte mixture described in the text. The gel was run as detailed in the legend to Fig.

7. Quantitative Analysis

Analysis of complex 2D maps is a difficult but essential task. Fig. 7 shows an analysis of a region of a gel which has been scanned with a flat bed Joyce-Loebl Chromoscan 3. It was found that a 50 µm step was optimal for spot separation. Methods for analysis using this system are under development. Dual labelling methods (2) should allow a thorough analysis of maps of normal and dystrophic proteins both qualitatively and quantitatively and overcome the problems of gel normalization. Recently a missing spot in DMD fibroblasts has been reported (11), but so far we have not observed any changes associated with the disease. It might be expected that more than one change would occur in the patterns due to secondary metabolic effects (12). Although clones from carriers could be used in order to decrease polymorphic variability, it is not clear whether all genes on the short arm of the X chromosome undergo X-inactivation (13). Further investigation involving cellular sub-fractionation and/or selective solubilization should assist localization and characterization of protein changes associated with disease.

Acknowledgements

This work was supported by the British Medical Research Council and the Muscular Dystrophy Group of Great Britain. We would like to thank Mrs. L.J. White for her photographic expertise, Ms. J. Carter for her excellent technical assistance and Mrs. C. Trand for typing the manuscript. We are grateful to Mr. W. Burnip of Vickers Instruments for the loan of the Joyce-Loebl Chromoscan 3, to Mr. D. Fairley of LKB for the gift of special IEF Ampholine, and to Rewo Chemicals Ltd for the gift of Rewocid DU 185.

References

1. O'Farrell, P.H.: J. Biol. Chem. 250, 4007-4021 (1975).
2. McConkey, E.H., Taylor, B.J. and Phan, D: Proc. Natl. Acad. Sci. USA 76, 6500-6504 (1979).

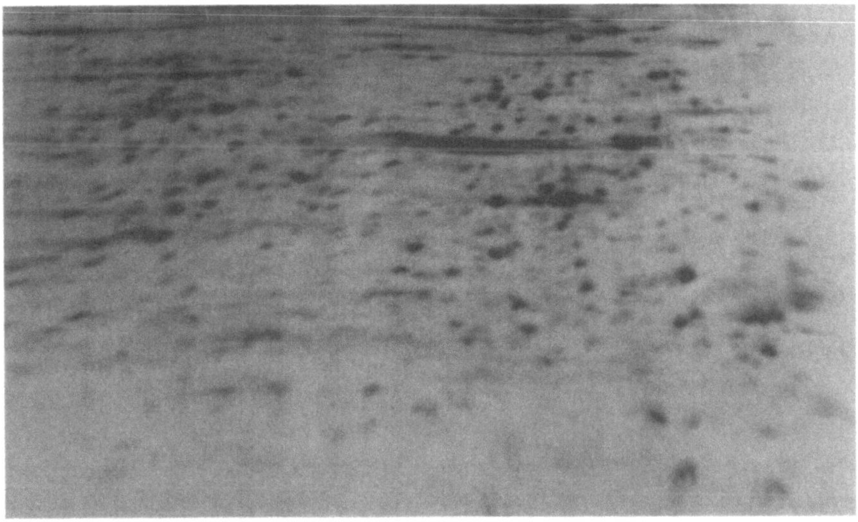

FIG. 6 Region of a 2-dimensional gel of [^{35}S]-methionine labelled fibroblast proteins. The first dimension IEF gel was cast on a plastic support and contained the ampholyte mixture described in the text. The gel was run as detailed in Fig. 2 for a total of 15,500 V.hr. The second dimension gel was run as described in the text. 15-day exposure using Kodak X-Omat H film in a Kodak cassette.

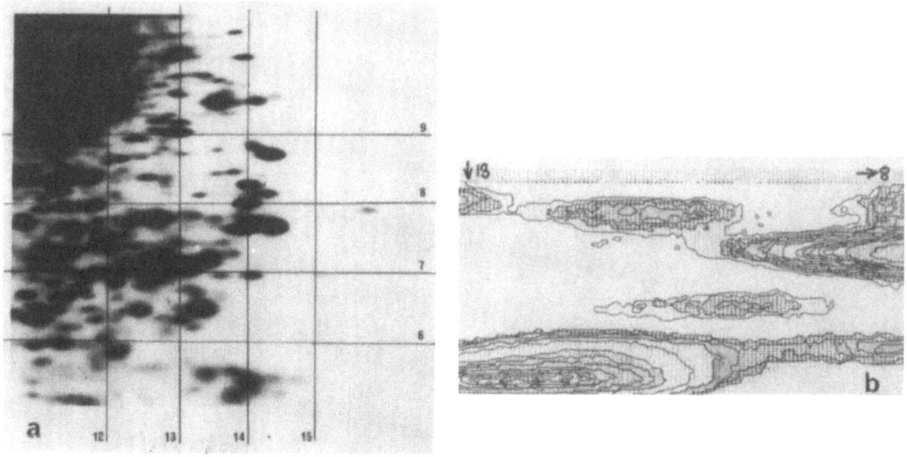

FIG. 7 Preliminary analysis of 2-dimensional electrophoretic maps using a Joyce-Loebl Chromoscan 3. (A) Enlargement of the region of the gel analysed. (B) Scan with 50 μm steps.

3. Burghes, A.H.M., Dunn, M.J., Statham, H.E. and Dubowitz, V: Electrophoresis 3 (I: in press) (1982).

4. Burghes, A.H.M., Dunn, M.J., Statham, H.E. and Dubowitz, V: Electrophoresis 3 (II: in press) (1982).

5. Dunn, M.J., Burghes, A.H.M., Thompson, B.J., Statham, H.E., Witkowski, J.A. and Dubowitz, V: (this volume).

6. Burghes, A.H.M., Dunn, M.J., Statham, H.E. and Dubowitz, V: in Electrophoresis '81 (Eds. R.C. Allen and P. Arnaud), pp. 295-308, de Gruyter, Berlin (1981).

7. Righetti, P.G: J. Chromatogr. 173, 1-5 (1979).

8. O'Farrell, P.Z., Goodman, H.M. and O'Farrell, P.H: Cell 12, 1133-1142 (1977).

9. Bravo, R., Bellatin, J. and Celis, J.E: Cell Biol. Int. Reports 5, 93-96 (1981).

10. Righetti, P.G. and Gianazza, E: J. Chromatogr. 184, 415-416 (1980).

11. Wrogemann, K., Rosenmann, E., Kreis, C., Dobbs, M. and Thompson, R.G: presented at the Sixth National Scientific Workshop of the Muscular Dystrophy Association of Canada (1982). To be published Canad. J. Neurol. Sci.

12. Merril, C.R., Goldman, D. and Ebert, M: in Electrophoresis '81 (Eds. R.C. Allen and P. Arnaud), pp. 343-353, de Gruyter, Berlin (1981).

13. Mohandas, T., Shapiro, L.J., Sparkes, R.S. and Sparkes, M.C: Proc. Natl. Acad. Sci. U.S.A. 76, 5779-5783 (1979).

2D-ELECTROPHORESIS OF PROTEINS IN UNICELLULAR ALGA E. GRACILIS AND YEAST S. UVARUM

Joachim Das
Institute of Biochemistry, School of Medicine, University of Kiel, Olshausenstr. 40-60, D-2300 Kiel, FRG

Introduction

2dimensional-IEF-SDS-polyacrylamide gel electrophoresis has been developed into a potent method for protein mapping and visualisation of protein differences in complex biological systems (1). A great variety of human and animal organs have already been investigated. Allthough first results were obtained with 2d-electrophoresis of plant material, only few botanical samples have since been analysed. Among these are unpigmented parts of plants, e. g. potatoes (2) and wheat grain (3), and depigmented leafs of tobacco (4). Recently, a method has been published (5) for the separation of yeast proteins by ultrathin layer 2d-electrophoresis.

The intention of this work is to test three types of 2d-IEF-SDS-PAGE - two of which are well established for the separation of animal proteins (6,7). We have analysed proteins from whole cell lysates of the yeast <u>Saccharomyces uvarum</u> and the unicellular green alga <u>Euglena gracilis</u>. The three variants of the method differ mainly in the size of the gels and the appropriate electric current conditions. Two of the geltypes are in two dimensions of the same size; but one is rather thick, whereas the other is thin. The third gel is small in all three dimensions.

We wanted to find the gel type which shows the larger number

Electrophoresis '82

of protein spots in a high resolution which enables the eye or an automat to resolve and identify spots in the 2d maps.

Materials and Methods

Reagents

Acrylamide (4x crystallized) and N,N'-methylen-bis-acrylamide (2x crystallized) were bought from Carl Roth, Karlsruhe, FRG. Servalyt (pH 5-7), N,N,N',N'-tetramethyl-ethylene-diamine, ammonium persulfate, Triton X-100 and Coomassie Brillant Blue R 250 were obtained from Serva, Heidelberg, FRG. The Ampholines (pH 3.5-10 and 4-6) were purchased from LKB, Bromma, Sweden, and Sephadex G 200 superfine from Pharmacia, Uppsala, Sweden. All other chemicals were products of the highest quality from Merck, Darmstadt, FRG.

Preparation of samples

The procedure follows with some modifications the method of O'FARRELL (8). Aerobically grown cells (10^7) of Saccharomyces uvarum and autotrophically grown cells (10^8) of Euglena gracilis were suspended in 0.1 ml of sonication buffer containing 10 mM Tris x HCl (pH 7.4) and 5 mM $MgCl_2$. The cells were disrupted on ice by 10 successive sonications of 30 sec. each at a setting of 3 on a Branson Sonifier B-12. Each sonication was followed by a cooling interval of 30 sec.. To the disrupted cells were added urea to a final concentration of 9.5 M as well as 0.1 ml of a lysis buffer containing 9.5 M urea, 2% Triton X-100 and 2% ampholytes (3 parts Servalyt pH 5-7, 1 part Ampholine pH 4-6, 2 parts Ampholine pH 3.5-10). The mixture was lysed overnight at 4°C with shacking. Subsequently it was stored at -20°C. The protein concentration was determined by the micro-biuret method (9). Before the protein determination, the alga lysate was washed several times with cold acetone to remove chlorophyll. Prior to IEF, the samples were in most experiments supplemented with a Sephadex mixture (6) without mercaptoethanole and ampholytes.

2d-electrophoretic methods

1. method (gel size: 72 x 65 x 3 mm). The method of gel preparation and electrophoresis was that of KLOSE & FELLER (6), but the ampholyte composition was modified. We used 3 parts of pH 5-7, 1 part pH 4-6 and 2 parts pH 3.5-10. Eighty µl of a mixture of sample and Sephadex were placed on the gel and 70 µl of an overlie solution. During IEF, the pH-gradient in the gel ranged from pH 4.5 to 8. The SDS-electrophoresis was performed at constant power produced by an ISCO-power supply (Type 493) as follows: 5 min. 5W, 5 min. 10W, 5 min. 15W, 5 min. 20W, 5 min. 25W and 30 min. 30W per 2 gel cells. The electrode buffer contained 0.39% Tris, 1.84% glycine and 0.16% SDS.

2. method (gel size: 35 x 25 x 0.5 mm). The micro-method follows essentially that of POEHLING & NEUHOFF (7). All the solutions used were the same as before. For IEF, 10 µl capillaries (intra-end, Brand, Wertheim, FRG) with I. D. of 0.6 mm and a length of 32 mm were filled with gel solution by capillary forces to a height of 25 mm and pressed vertically into a cushion of plasticine covered with parafilm (American Can Co.). On the next day, water was removed from the gel and the remainder of the tube was filled with about 3 µl of sample devoid of Sephadex using a micro-syringe. The upper chamber of the disc-electrophoresis apparatus (Desaga, Heidelberg, FRG) was prepared in a special way: Each hole for holding the original electrophoresis tubes was closed with adhesive tape, which was perforated with a needle. The perforation had to be so small, that the micro-tube could be clamped vertically (sample above) in it. After attachment of the tubes (reaching from the upper chamber into the lower electrode buffer), it was sealed to the tape by drops of hot vacuum fat. The voltage conditions during

Table 1. Resolution and number of spots obtained on 2d-gels after separation of proteins of whole cell lysates of E. gracilis and S. uvarum. (suffic. = sufficient)

Sample	E. gracilis			S. uvarum		
Gel size	72x65x 3 mm	72x65x 0.8 mm	35x25x 0.5 mm	72x65x 3 mm	72x65x 0.8 mm	35x25x 0.5 mm
Coom. Bl. stain						
Number of spots	~150	~75	~15	~250	~150	~20
Resolution	suffic.	good	good	suffic.	good	good
Silver stain						
Number of spots	> 2000	~ 500	~ 100	> 2000	~ 500	~ 100
Resolution	bad	suffic.	bad	bad	suffic.	bad

Fig. 1: 2d-IEF-SDS-PAGE protein maps of whole cell lysates of E. gracilis obtained by the three methods described. 1. dimension: IEF; 2. dimension: SDS-electrophoresis. The horizontal rows of pictures show protein clusters of the different sized gels: A, B: 72 x 65 x 3 mm; C, D: 72 x 65 x 0.8 mm; E, F: 35 x 25 x 0.5 mm. The left hand gels are stained with Coomassie Blue, the right hand gels with silver. SDS-electrophoretic dimension is calibrated with proteins of known molecular weight: The protein markers on gels A, B (3 thick bars at both sides of the gels) are bovine serum albumin (M.W.: 67 000), Chymotrypsinogen (M.W.: 25 000) and Myoglobin (M.W.: 17 000). The protein markers on gels C, D (2 faint double lines across the gels) are bovine serum albumin and Chymotrypsinogen.

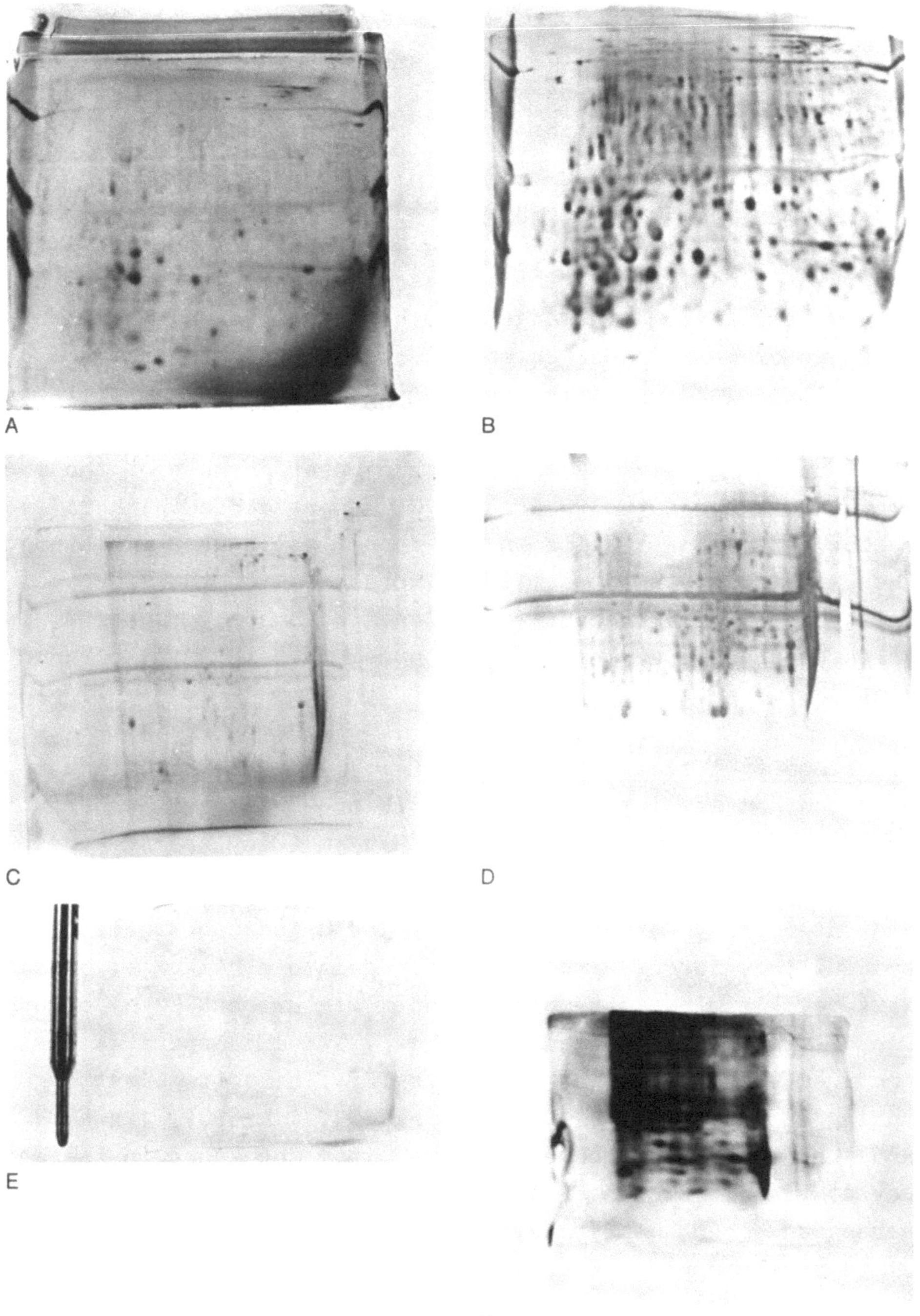
A
B
C
D
E
F

IEF were: 0.5 h 100V, 0.5 h 200V and 0.25 h 300V. After IEF the gels were removed from the tubes using plasticine and a small steel wire (7). The slab gel cells for the 2. dimension were miniaturized versions of the one previously published (6). Each cell consisted of 2 cleansed (7) glass plates (35 x 30 mm) from dia frames and 2 distance holders of the dimensions 0.5 (distance of the plates) x 2 x 30 mm prepared from a foil of 5 layers of parafilm (5). The plates and spacers were mounted with adhesive tape. The edges of the tape were covered with stearine or nail enamel to ensure absolute tightness. Each cell was filled with separation gel to a height of 25 mm, followed by 2 mm of spacer gel. The remainder of the cell was filled with embedding gel. The slots in the upper chamber of the electrophoresis apparatus (Pharmacia GE 4)were closed with silicon rubber paste. Small slits in the rubber rolls were cut along the existing slots to permit the attachment of the micro gel cells. Finally, all seams were closed with hot vacuum fat. The voltages applied during electrophoresis were: 5 min. 60V and 10 min. 120V.

3. method (gel size: 72 x 65 x 0.8 mm). All the solutions used were the same as in the 1. method. Micro-hematocrit tubes with an I. D. of 1.1 mm and a length of 75 mm (Clay-Adams, Parsippany, USA) were deheparinized for 1 h with concentrated H_2SO_4 and intensively washed with water. The dried tubes were filled by capillary forces with separation gel to a height of 50 mm. Next day, 15 µl of a sample/Sephadex mixture and 10 µl of overlie solution were applied to the gel. The IEF-apparatus was prepared the same way as for micro-IEF. The voltage settings during IEF were: 0.5 h 100V, 0.5 h 200V, 1 h 400V, 1 h 600V and 1 h 800V. The gels were removed from the tubes in the same way as the micro gels. For SDS-electrophoresis double gel-cells were used because they had the appropriate thickness for the slots of the upper chamber of the electrophoresis apparatus. The cells were in principle mounted in the same way as before.

Three of the glass plates from the 1. method were separated by distance holder prepared from a foil of 8 layers of parafilm. The components were assembled using adhesive tape. Each cell was filled with separation gel to a height of 65 mm, followed by 3 mm of spacer gel. The remainder of the cell was filled with embedding gel. Power conditions were: 5 min. 20W, 5 min. 40W and 10 min. 60W per pair of double gel cells.

Staining of the gels

The gels were stained with Coomassie Blue according to (6). The periods for fixing and staining of the thin gel could be reduced to 0.5 h. The methods of (10) and (11) were applied for silver stain.

Results and Discussion

The adequacy of any separation method rests on the fulfilment of at least two requirements: 1. The results must be reproducible, and 2. the compounds under test must be well separated. The last demand we may call the resolution of the method. If the method of separation is intended to serve as a tool for comparing different samples a third assumption is necessary: The method must deliver enough information to detect the differences in question. This means, applied to 2d-electrophoresis, that the protein maps must be reproducible, that the spots must be well separated and that the number of spots must be large. The requirement that reproducibility can be achieved by careful work is absolute, since otherwise nobody would use the method. We have tested three types of 2d-IEF-SDS-polyacrylamide gel-electrophoresis on their resolution and the number of spots they produce with samples of E. gracilis and S. uvarum. The three variations differ in the size of the gels and the applied electric current conditions. Two gels are of the same size (72 x 65 mm), but one is 3 mm, whereas the other is

0.8 mm thick. The third is miniaturised in all three dimensions (35 x 25 x 0.5 mm).

The protein maps obtained by the three methods after Coomassie Blue and silver staining are shown for E. gracilis in fig. 1 and for S. uvarum in fig. 2. The resolution and number of spots are summarized in the table.

The number of spots is due to the larger amount of sample applied to the IEF-gel highest in the protein maps of the thick gels. Especially with silver stain the number of spots increases at least 10 fold which renders a resolution that is insufficient. If the number of spots obtained with Coomassie Blue stain is too small to allow the detection of differences in the protein composition then the silver stain must be used. The loss of resolution may in some cases be compensated by an increase in the gel size, which leads to the distribution of the spots over a larger area.

The number of spots on the thinner gels is smaller because of the smaller volume of sample applied. Particularly the micro gels tend even with silver stain to supply an insufficient number of spots for an reasonably certain detection of protein differences. Allthough the spots are smaller than with the thick gels, the resolution remains the same since the size of the gels also decreases. Therefore, the micro method is only applicable to the separation of samples containing few but concentrated proteins.

Fig. 2: 2d-IEF-SDS-PAGE protein maps of whole cell lysates of S. uvarum obtained by the three methods described. Conditions as in fig. 1.

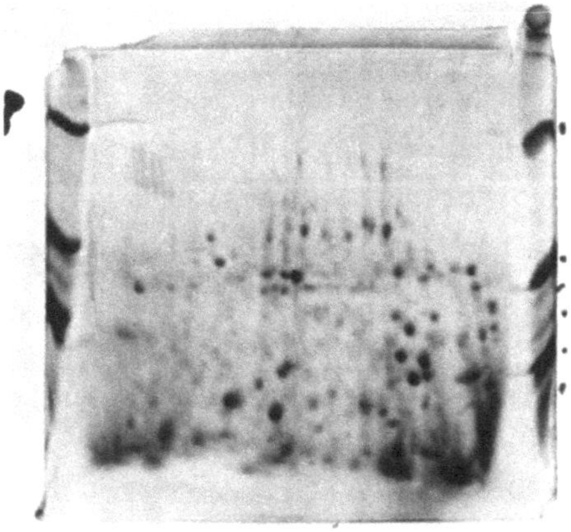

A

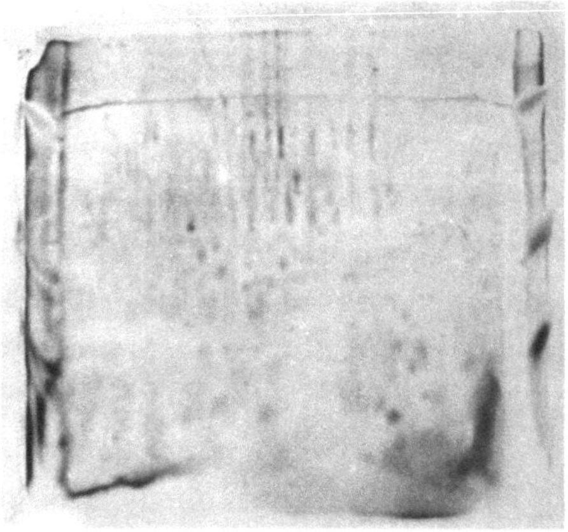

B

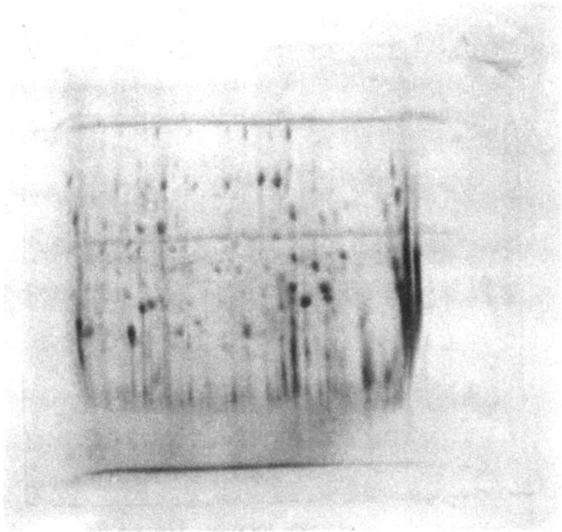

C

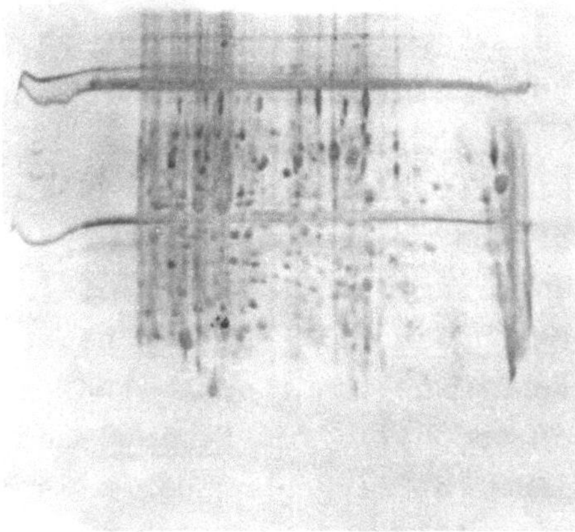

D

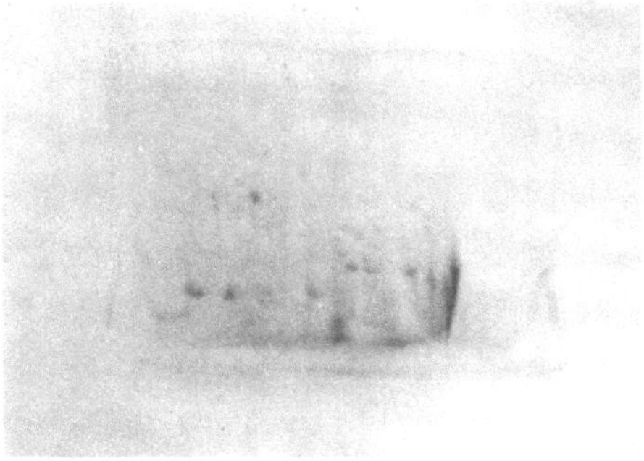

E

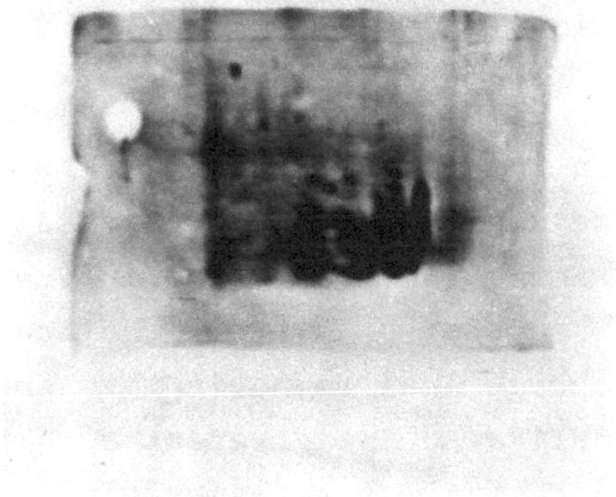

F

In contrast, the large thin gels offer a real gain in resolution. The number of spots after Coomassie Blue stain is not sufficient, but becomes it after silver staining. A higher voltage can be chosen for the IEF because the cooling of the thin gels is more effective than of the thick ones. This results in narrower IEF-bands and smaller spots after SDS-electrophoresis. Otherwise, the area of the thin gel is the same as on the thick one. Therefore, the resolution is better if the spots are small. Hence, the thin gels show best resolution of all the three gels tested. One disadvantage of the thin gels must be mentioned: The number of spots obtained after silver stain is not as great as that presented by the thick gels. However, if the number of spots is sufficient, the thin gels are commendable for comparing protein samples by 2d-electrophoresis.

Finally, I want to make a few comments on the electrophoresis of the two samples, the lysates of E. gracilis and S. uvarum. Both samples deliver protein maps with a large number of spots. By Euglena, it was not necessary to extract the chlorophyll as it was required by leaf samples of tobacco (4), since all chlorophyll from Euglena remains in the sample space during IEF. The larger number of spots on the yeast gel is probably due to the higher protein content in the yeast sample (350 mg/ml) compared to that of the alga sample (250 mg/ml).

Summary

In this investigation, three types of 2d-electrophoresis differing mainly in the size of the gels have been tested. Test samples were whole cell lysates of the alga E. gracilis and the yeast S. uvarum. The best resolution of protein spots on the gel was found in thin and relative large gels. Thick gels delivered the greatest number of spots, especially after silver stain, but the resolution was poor. The micro-method is only suitable for concentrated protein samples containing a relatively small number of proteins because the resolution is rather bad.

Acknowledgement

This work was supported by grant Bu 234/8 of the Deutsche Forschungsgemeinschaft. I express my gratitude to J. Hebbeln and U. Warthold for introducing me in 2d-electrophoresis and Drs. K. Westerbrink and Mrs. E. Postel for many helpful suggestions. Technical assistance of Mrs. I. Hahn and Mrs. M. Ionescu is greatly acknowledged.

References

1. Anderson, N. G., Anderson, N. L.: Behring Inst. Mitt. 63, 169-210 (1979)
2. Macko, V., Stegemann, H.: Hoppe Seylers Z. physiol. Chem. 350, 917-919 (1969)
3. Wrigley, C. W.: Biochem. Genet. 4, 509-516 (1970)
4. Hari, V.: Anal. Biochem. 113, 332-335 (1981)
5. Görk, A., Postel, W., Westermeier, R, Gianazza, E., Righetti, P. G.: In: "Electrophoresis '81", Ed. R. C. Allen, Walter de Gruyter & Co., Berlin, 259-270 (1981)
6. Klose, J., Feller, M.: Electrophoresis 2, 12-24 (1981)
7. Poehling, H. M., Neuhoff, V.: Electrophoresis 1, 90-102 (1980)
8. O'Farrell, P. H.: J. Biol. Chem. 250, 4007-4021 (1975)
9. Alt, J., Krisch, K., Hirsch, P.: J. Gen. Microbiol. 87, 260-272 (1975)
10. Oakley, B. R., Kirsch, D. R., Morris, N. R.: Anal. Biochem. 105, 361-363 (1980)
11. Switzer, R. C., Merril, C. R., Shifrin, S.: Anal. Biochem. 98, 231-237 (1979)

POLYPEPTIDE COMPOSITION OF DIFFERENT CHLOROPLAST FRACTIONS FROM BARLEY ANALYZED BY 1- AND 2-DIMENSIONAL GEL ELECTROPHORESIS.

R. Valcke, J.P. Noben & M. Van Poucke.

Limburgs Universitair Centrum, Dept. S.B.M., Universitaire Campus
B-3610 Diepenbeek, Belgium.

Introduction

In 1975, O'Farrell (17) introduced the sensitive 2-D-technique for the separation of proteins based on their pI on an IEF gel in the first dimension and their molecular weight on a SDS-polyacrylamide slab gel in the second dimension. Several authors have used the 2-D technique for analysing the polypeptide pattern of thylakoid membranes (3) and other subchloroplast fractions (4,16). Their results clearly indicate the complexity of the polypeptide pattern.

Recently, Gilbert and Beutow (9) analyzed the chloroplast polypeptides from Euglena gracilis. They compared the polypeptide pattern obtained by one- and two dimensional gel analysis and improved the resolution using linear polyacrylamide gradients in the second dimension. They described the polypeptide composition of a rough chloroplast pellet without any reference to specific subfractions. However, one of the most important problems in this kind of analysis is to know how the pattern obtained corresponds with the polypeptide composition of the subfractions. Therefore, using a combination of isopycnic density centrifugation on Percoll gradients and discontinuous sucrose gradient centrifugation, we first purified different fractions, e.g. the stroma, the envelope and the thylakoids from barley chloroplasts. The polypeptide composition of these fractions were then analysed using IEF on polyacrylamide gels in the first dimension followed by SDS-PAGE on polyacrylamide gradient gels in the second dimension.

Electrophoresis '82

Materials and Methods

Hordeum vulgare var. Menuet was grown in a phytotron as outlined before (20). Pure chloroplasts were prepared as described by J.J.Morgenthaler (14). Isolation of the different chloroplast fractions was performed according to R. Douce (5). Iso-electric focusing of the polypeptides was carried out on tube gels 3mm in diameter and 9cm long as described (9). After focusing the gels were removed from the tubes, equilibrated for 90 min in 10% glycerol, 0.0625M Tris-HCl buffer pH 6.8, 2.3% SDS and 5% 2-mercaptoethanol and mounted with agarose on slab gels (7.5 to 20% acrylamide gradient stabilized with a 5 to 17.5% sucrose gradient). The discontinuous buffer system of Neville (15) was used. Electrophoresis was performed at 10°C during 16h at 20mA. After electrophoresis the gels were stained with 0.25% Coomassie Brillant Blue R-250 and destained as described (7). Negative staining of the membranes for electron microscopy was performed with 1% uranyl acetate during 10 sec.. After drying, the samples were examined in a Philips EM 400 electron microscope. Absorption spectra were recorded on a Beckman 5260 spectrophotometer. RuBPC as a reference was purified (2,8,11,19).

Results

Isopycnic density gradient centrifugation on Percoll gradients shows two distinct green bands (fig.1). Based on phase contrast and electron microscopy (20) the intact chloroplasts are located in band II. This band was collected, washed several times, subjected to osmotic shock and centrifuged on discontinuous sucrose gradients. The separation obtained is shown in fig.2. On top of the gradient remained the stromal proteins. They were concentrated by adding an equal volume of 20% trichloroacetic acid. After centrifugation at 5000g for 10 min,the pellet was resuspended in a mixture of cold ethanol-ether (1:1,v/v) followed by washing in pure ether. The final pellet was resuspended in 50mM Tricine-KOH buffer pH 7.5, 250mM sucrose to a protein concentration of 3 mg/ml.
On the interface between 0.6 and 0.93M sucrose was an opaque to yellow region representing the envelope membranes (E). Down the gradient were two

distinct green bands, one band on the interface between 0.93 and 1.30M sucrose, the other on the interface between 1.30 and 1.90M sucrose. This two bands were fractions of the thylakoid membranes and were called T_1 and T_2 resp..

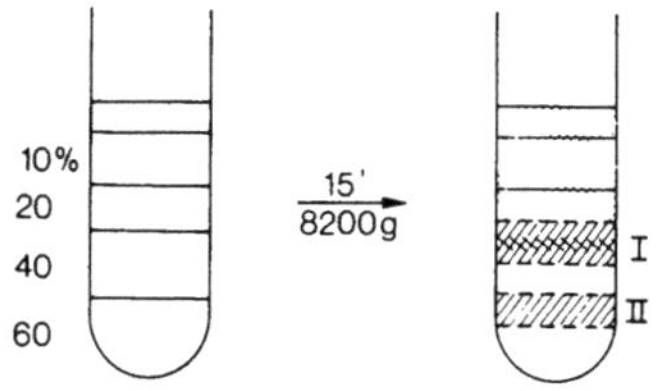

Fig.1: Isopycnic density centrifugation on discontinuous Percoll gradients (%).

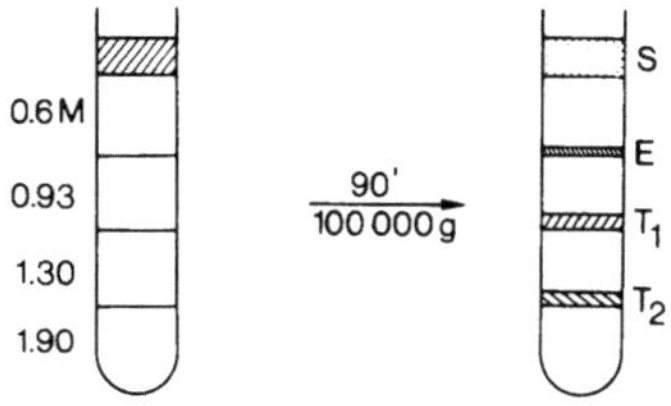

Fig.2: Density gradient centrifugation on discontinuous sucrose gradients (M).

The first characterization of the different membrane fractions was based on their appearance after negative staining (fig.3) and on their absorption spectra (fig.4).

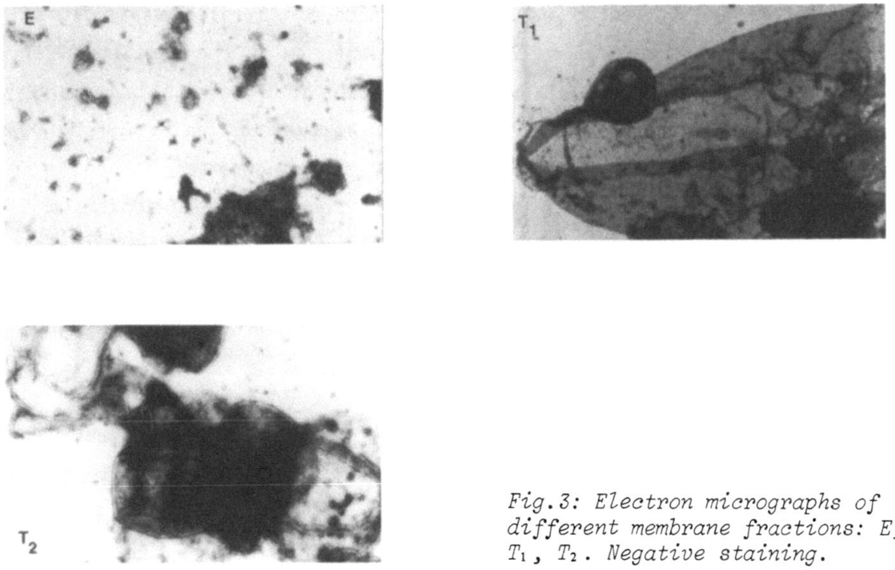

Fig.3: Electron micrographs of different membrane fractions: E, T_1, T_2. Negative staining.

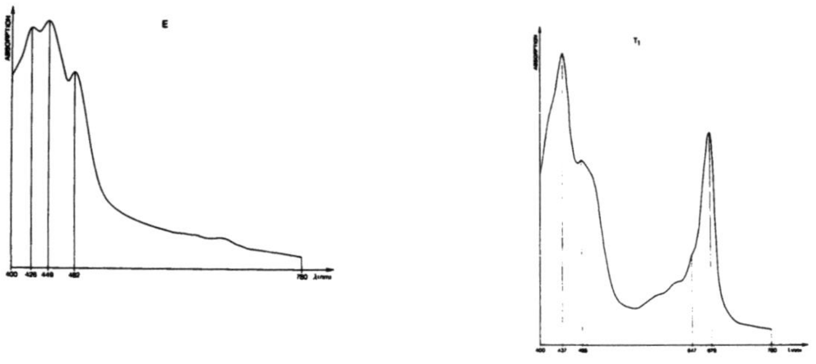

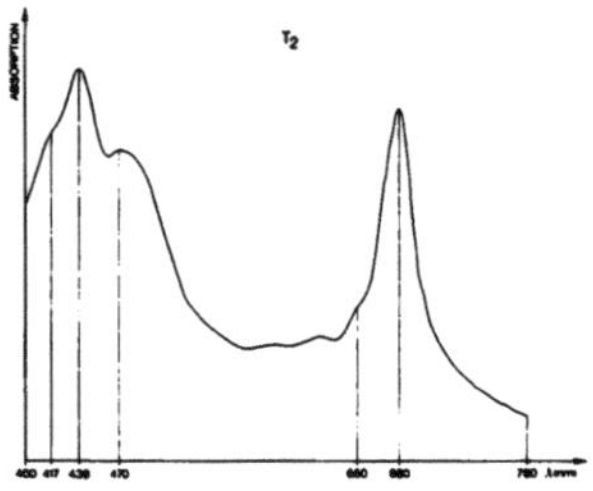

Fig.4: Absorption spectra of different membrane fractions: E, T_1, T_2. Absorption maxima given in the abscissa.

Envelopes and thylakoids were removed from the sucrose gradient by aspiration through a syringe and collected by centrifugation (60 min at 145.000g and 10 min at 48000g resp.). Each membrane sediment was resuspended in the Tricine buffer, extracted with aceton (9) and centrifuged at 5000g for 10 min. The pellet was extracted twice in anhydrous ether followed by centrifugation as before. The final pellet was resuspended in the Tricine buffer to a protein (E) or chlorophyll (T) concentration of 3 mg/ml. Stroma and membrane proteins were solubilized adding two volumes of lysis buffer (9.5M urea, 2% Triton-X-100, 2% ampholine pH 3.5-10 and 5% 2-mercaptoethanol). After vigorous mixing on a vortex mixer, the suspensions were sonicated 5 times during 10 sec. Of all fractions, the polypeptide pattern was analyzed on 1-D-SDS linear polyacrylamide gradient (7.5-20%) gel on the one hand and on a 2-D gel (IEF in the first dimension, SDS-7.5-20% polyacrylamide slab gels for the second dimension) on the other hand.

Separation of the stroma polypeptides on a 1-D-SDS-PAGE resolved about 35 bands. Iso-electric focusing reveals about 44 bands. Most of the bands focused in the neutral to acidic region (pH 7.5 to 4.5). No bands could be detected in the basic region (pH higher than 7.5). Subjected to a second run on gradient gels, the focused polypeptides resolved into a large number of spots (169) (fig.5). The most prominent spots on the 2-D gel correspond with the intense colored bands on the 1-D gel.

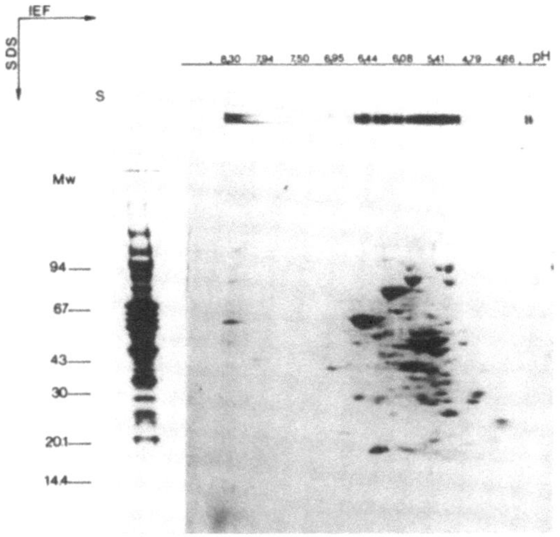

Fig.5: Polypeptide analysis of the stroma fraction. Upper gel: IEF, kathode (-) at the left, anode (+) at the right. The corresponding pH gradient is shown in the upper part of the fig.. Left gel: 1-D-SDS-PAGE. Mw in Kdaltons are given with reference to standard proteins. Middle gel: 2-D analysis (IEF in the first dimension, SDS-PAGE in the second dimension as indicated by the arrows).

Analysis of purified Ribulose Biphosphate Carboxylase (RuBPC), one of the major proteins of the stroma, gives 5 major subunits with pI 7.4,7.2,7.05, 6.90 and 6.74 and two minor subunits with pI 6.29 and 6.01. On a 1-D-SDS-PAGE the enzyme consists of two subunits with Mw of 55 kD and 15.5 kD resp.. On a 2-D gel we obtained about 15 spots. However, it is very difficult to correlate the results of the purified RuBPC with the analysis of the total stromal protein fraction. The results fit very well on the 1-D-SDS-PAGE gels: the two subunits of the purified enzyme are clearly resolved at the same po-

sitions in the total stroma protein fraction. The problem arises in the focusing experiment and consequently in the 2-D fingerprint. The 5 focused subunits mentioned for the putified protein cannot be shown at the same positions in the focusing pattern of the total protein fraction.
Concerning the thylakoid membranes we obtain two different fractions (T_1 and T_2) during the sucrose gradient centrifugation. In fig.6 the polypeptide composition of T_1 fraction is shown. A 1-D-SDS-PAGE reveals about 25 bands. In comparison with other studies (1,4,13 and 14) some of the polypeptides could be characterized as apoproteins of different chlorophyll-protein complexes (the 70,25,20,18 and 16 kD for photosystem I; the 50,48,38,31 and 27 kD for photosystem II including the light harvesting chlorophyll-protein complex). Other polypeptide bands were assigned to the chloroplast coupling factor CF_1 (the 57,56,37,21 and 14 kD bands). Iso-electric focusing of the T_1 fraction resolves about 26 bands. All the polypeptide bands focused in the neutral to acidic region (pH 7 to 4.5). Some proteins focused quite distinctly at about pH 4.76,6.20 and 7.15. Other proteins appeared as broad bands in the 5.10 to 5.90 region and in the 6.40 to 6.90 region. Interesting-

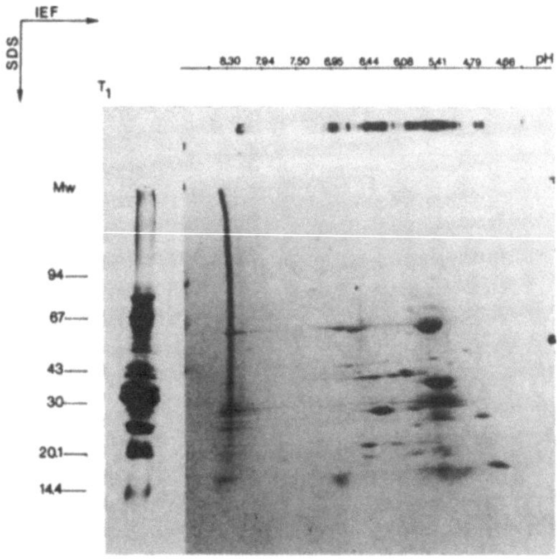

Fig.6: Polypeptide analysis of the T_1 fraction. Gel positions as in fig.5.

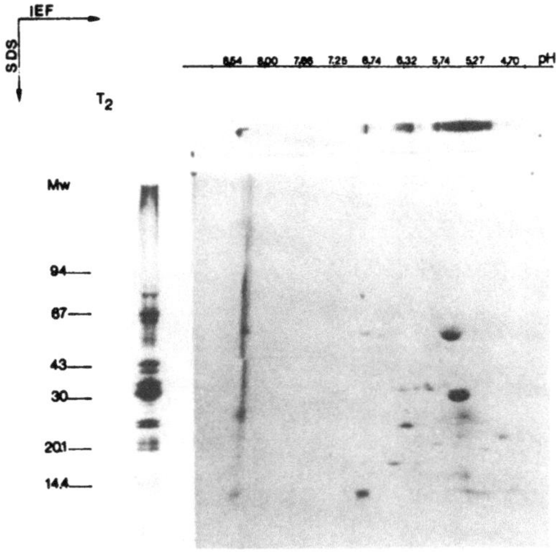

Fig.7: Polypeptide composition of T_2 *fraction. Gel positions as in fig.5.*

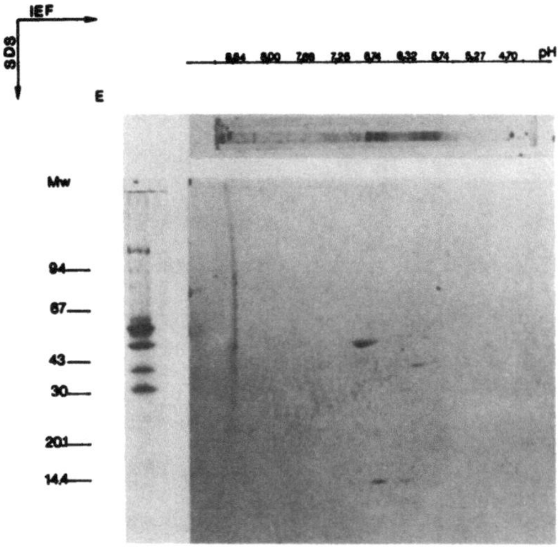

Fig.8: Polypeptide composition of E fraction. Gel positions as in fig.5.

ly, there was no great difference in the number of polypeptide bands between the IEF and the 1-D-SDS-PAGE method. Nevertheless, a 2-D fingerprint resolves about 74 different spots.
T_2 fraction (fig.7) show some important differences compared with the polypeptide composition of T_1. In the 1-D-SDS-PAGE the 70 kD band is strongly reduced. The 25 and 16 kD bands and the 40 to 55 kD region are less intense. A more dramatic difference was observed on focusing gels. T_2 fraction gave about 10 focusing bands and a 2-D fingerprint resolved 39 spots.
With our method, the polypeptide pattern of the envelope is poorly resolved (fig.8). 1-D-SDS-PAGE reveals 11 bands. IEF revealed about 6 bands and a 2-D fingerprint resolves 12 spots.

Discussion

The purpose of this investigation is to obtain a two dimensional map of the polypeptides of the different chloroplast fractions. In the analysis of the chloroplast proteins it is absolutely necessary to determine unequivocally which polypeptides belong to a certain subfraction. Therefore, the first requirement in this kind of analysis is to purify the different fractions of the organelle. The use of Percoll gradients reduces the contamination of the plastids with nuclear material and mitochondria. After recovery of the intact plastid fraction and several washings we obtain pure chloroplasts as examined by electron microscopy (20). This is taken to mean that the different fractions (S,E,T_1,T_2) obtained after osmotic shock and centrifugation on discontinuous sucrose gradients really correspond to the different parts of the intact chloroplasts.
The combination of the IEF with SDS-PAGE on gradient gels enables us to characterize each polypeptide within certain limits by its iso-electric point (pI) and its molecular weight (Mw). One has to remember that in the case of the membrane fractions it was necessary to extract the membranes by aceton-ether to obtain good resolution. Preliminary experiments have indicated that prior to IEF, extraction of the membranes with non-ionic detergents give very poor resolution due to the smearing of the pigment-protein complexes. Examination of the polypeptide pattern obtained by unidimensional SDS-PAGE

and by two dimensional fingerprinting shows that it is very difficult to correlate the banding pattern and the fingerprints with certainty. First polypeptide bands separated on a IEF gel can be resolved in multiple polypeptides on a 2-D gel and secondly, the apparent molecular weight of a specific polypeptide differs in the two procedures used. Moreover, the case of RuBPC shows that it is impossible to identify particular spots in a 2-D fingerprint of a complex stroma sample with those of a single purified component purified from that chloroplast fraction.

The composition of the envelope membrane is much more complex than shown here by our preliminary results. Recent work by R.Douce (6) indicates a broad spectrum of polypeptides, the composition of which is as complex as that of the thylakoid membranes. The differences between the results of R.Douce and our own may be due to the solubilization procedure.

Abbreviations

pI: iso-electric point; Mw: molecular weight; IEF: iso-electric focusing; PAGE: poly-acrylamide gel electrophoresis; SDS: sodium dodecyl sulfate; 1-D: one-dimensional; 2-D: two-dimensional.

Acknowledgments

The authors wish to thank Prof.Dr.D.Dekegel, Pasteur Institute of Brabant and Free University of Brussels, for critical discussion and Miss B. Vanacken for technical assistance.

References

1. Bengis, C., Nelson, N.: J. Biol. Chem. 250, 2783-2788 (1975).
2. Blair, E., Ellis, R.J.: Biochim. Biophys. Acta 319, 223-234 (1973).
3. Boschetti, A. et al.: in Chloroplast Development, Ed. Akoyunoglou G. Elsevier North-Holland, 195-200 (1978).

4. Chua, N.H., Gillham, N.W.: J. Cell Biol. 74, 441-452 (1977).

5. Douce, R.: Adv. Bot. Res. 7, 1-116 (1979).

6. Douce, R.: Meeting Society of Experimental Biology, Leiden (Nl) (1982).

7. Fairbanks, G. et al.: Biochem. 10, 2606-2617 (1971).

8. Feierabend, J., Wildner, G.: Arch. Biochem. Biophys. 186, 283-291 (1978).

9. Gilbert, C.W., Buetow, D.E.: Plant Physiol. 67, 623-628 (1981).

10. Høyer-Hansen, G. et al.: Carlsberg Res. Comm. 44, 337-351 (1979).

11. Kleinkopf, G.E. et al.: Plant Physiol. 46, 204-207 (1970).

12. Machold, O. et al.: Carlsberg Res. Comm. 44, 235-254 (1979).

13. Machold, O.: Biochem. Physiol. Pflanzen 176, 805-827 (1981).

14. Morgenthaler, J.J. et al.: Arch. Biochem. Biophys. 168, 289-301 (1975).

15. Neville, D.M.: J. Biol. Chem. 246, 6328-6334 (1971).

16. Novak-Hofer, I., Siegenthaler, P.A.: Biochim. Biophys. Acta 468, 461-471 (1977).

17. O'Farrell, P.H.: J. Biol. Chem. 250, 4007-4021 (1975).

18. Siegenthaler, P.A., Novak-Hofer, I.: in Bioenergetics of Membranes, Ed. Packer, L. et al. 269-286 (1977).

19. Strøbaek, S., Gibbons, G.C.: Carlsberg Res. Comm. 41, 57-72 (1976).

20. Valcke, R., Van Poucke, M.: Proc. V Int. Congr. Photosynthesis, Ed. Akoyunoglou, G. vol. III, 285-293 (1981).

TWO-DIMENSIONAL ELECTROPHORESIS OF SIMIAN VIRUS 40 (SV40) TUMOR ANTIGEN

K. Palme[1] and R. Henning

Dept. Biochemistry, University of Ulm, P.O.Box 4066, D-7900 Ulm/Donau,FRG.
[1] Present address: The Salk Institute, P.O.Box 85800, San Diego, California 92138.

Introduction

Simian Virus 40 (SV40) is a DNA tumor virus coding besides structural polypeptides regulatory proteins, the tumor antigens, which are responsible for several important biological functions. Large SV40 tumor antigen (T-Ag), a 94 kd phosphoprotein, is present in virus infected cells as well as in SV40 transformed cells. Besides initiation of viral DNA replication it regulates early and late viral transcription and plays an important role in the mechanism of SV40 induced neoplastic transformation (for review see 1). Although it is unclear at this time how the multiple functions of this protein are regulated at the molecular level it might be that posttranslational modifications like phosphorylation influence some of its activities. Changes in the phosphorylation state have been found (2,3). Thus, it was reasonable to analyze the charge pattern of T-Ag by high resolving techniques like isoelectric focusing and two-dimensional electrophoresis.

Several attempts to separate immunoprecipitated T-Ag by two-dimensional electrophoresis have been described (4,5,6). Crawford and O'Farrell, for example showed that T-Ag precipitated nearly completely at the application point of the isoelectric focusing gels (4). Only after alkylation and SDS treatment it could be shown as a broad smear in two-dimensional gels. These results indicate that as long unknown effects can influence the two-dimensional electrophoresis of immunoprecipitated T-Ag. Artifacts induced by the isoelectric focusing system are well known(7), but additional changes of the

Electrophoresis '82

isoelectric focusing pattern by interaction of proteins itself or with other ligands can not be excluded. Since we were able to separate charge isomers of purified T-Ag (8), interaction of T-Ag with components copurified during immunoprecipitation seemed to modify its charge pattern.

In order to understand what might influence the separation of immunoprecipitated T-Ag during isoelectric focusing we analyzed carefully the conditions of sample preparation of T-Ag for two-dimensional electrophoresis. Our results show that host cell components like RNA, Calcium and cellular proteins, the non viral tumor antigens (NVT) coprecipitated with T-Ag in the immunoprecipitates can either completely prevent the separation of T-Ag by two-dimensional electrophoresis or cause smearing effects. After blocking of these interactions with RNase A, EGTA and NEM T-Ag could indeed be focused in reproducible charge patterns between pH 6.0-6.5.

Materials and methods

Cells and sera

SV80 cells, SV40 transformed human epithelial fibroblasts were grown in Dulbecco's modification of Eagles minimal essential medium (DMEM) containing 5% fetal calf serum at 37° C in a 5% CO_2 containing atmosphere. Hamster SV40 tumor sera were obtained from hamsters bearing tumors induced by SV40 transformed hamster cells (H65/90 B). 3-4 weeks after injection of 10^6 H65/90 B cells the hamsters were bled and the sera titrated against nuclear T-Ag by immunofluorescence on SV3T3 cells.

Labeling of cells and immunoprecipitation

Approximately 5 x 10^6 SV80 cells were washed twice with either methionine- or phosphate-free DMEM and then labeled for 4 h with either 50 uCi ^{35}S-methionine (Amersham/Buchler, specific activity 800 uCi/mmol) in one ml

DMEM without methionine or with 100 uCi ^{32}P-phosphate (NEN, specific activity 1000 mCi/mmol) in one ml phosphate-free DMEM. The labeled monolayer cells were washed with phosphate-buffered saline (PBS), scraped off the plate and approximately 5 x 10^6 cells were lysed with 1 ml of extraction buffer containing 0.1 M Tris-HCl, pH 9.0, 0.1 M NaCl, 0.5% NP40 (Fluka,Buchs, Switzerland) and 1% Trasylol for 30 min on ice. The lysate was clarified at 800 x g for 2 min and at 105000 x g for 30 min at 4° C. After precipitation of the lysates with 10 ul normal hamster serum and 200 ul of a 10 % suspension of inactivated and formaldehyd fixed S.aureus (1 h, 4° C) T-Ag was immunoprecipitated by adding 10 ul SV40 tumor serum and 200 ul of S.aureus overnight as described by Kessler (9). After washing the immunoprecipitates with 0.05 M Tris-HCl, pH 7.4, 0.15 M NaCl, 0.005 M EDTA, 1% sucrose and 1% NP40 the pellet was washed once with 0.1 M Tris-HCl, pH 7.0, 0.1 M EDTA and once with 0.001 M Tris-HCl, pH 7.0.

Digestion of RNA adsorbed to immunoprecipitates

RNA was digested by incubation of the immunoprecipitates with 100 ul of 0.5 M Tris-HCl, pH 7.0, 10% Trasylol, containing 0.5% or 0.8% (w/v) RNase A (Serva, Heidelberg, FRG). After incubation overnight at 20° C the immunoprecipitates were pelleted, washed once with 0.1 M Tris-HCl, pH 7.0, 0.1 M EDTA and with water. The pellets were eluted with 100 ul 0.01 M Tris-HCl pH 7.0, containing 0.3% SDS (BDH Ltd., Poole, England), 1% 2-mercaptoethanol, 0.05% RNase A by incubation for 1 h at 20° C. After lyophilisation the samples were reduced for 30 min at 20° C by 100 ul of 0.01 M Tris-HCl, pH 7.0, 0.01 M dithiothreitol (DTT), 2% NP40, containing 9.5 M urea. After addition of 20 ul buffer containing 6.3% NP40, 9.3 M urea, 13% ampholine (LKB, Munich, FRG) pH 3-10 and 5% 2-mercaptoethanol the samples were subjected directly to isoelectric focusing or stored frozen at - 20° C.

Two-dimensional gel electrophoresis

Immunoprecipitates from 5 x 10^6 SV80 cells containing approximately 1 ug of T-Ag were subjected to equilibrium isoelectric focusing in the first dimension. Isoelectric focusing was performed as described previously (10). The stock solution contained freshly dissolved urea, deionized with Biorad mixed-bed-ion-exchange resin (Biorad AG X 8 D) to remove ionic impurities (conductivity 10-15 umho). The polyacrylamide focusing gels (140 mm x 2 mm, 3.3 % acrylamide) contained 9.3 M urea, 2% NP40, 2% ampholine (LKB), pH 3-10. The lower electrode reservoir was filled completely with degased phosphoric acid (0.1 M) and the upper chamber with degased NaOH (0.2 M). Samples were loaded on the cathodic end of the gels. Isoelectric focusing was started at 4° C at constant voltage of 50 V for 1 h, continued at 250 V for 12 h, for 7 h at 800 V and finally the gels were hyperfocused for 1 h at 1000 V. For pH measurements the frozen gels were sliced in 2 mm segments and eluted with degased water containing 0.01 M KCl.
The second dimension was run according to O'Farrell (10) employing the Laemmli system (11) exept that the frozen gels were loaded directly without equilibration to the second dimension slab gels (0.1 x 10 x 13 cm). The

focusing gels were anchored into place with 1% agarose in 2% SDS, 0.065 M Tris-HCl, pH 6.8, containing 0.005% bromophenol blue. After constant current electrophoresis at 7 mA through the stacking gel (4% acrylamide) and 12 mA through the separation gel (usually 11.5% acrylamide) the gels were either stained with Coomassie blue R 250 or processed for fluorography and exposed to Kodak X-Omat XR-5 film.

Results

T-Ag was isolated by indirect immunoprecipitation from extracts of ^{35}S-methionine or ^{32}P-phospate labeled SV80 cells. After elution of T-Ag with 9.3 M urea, 2% NP40 and 10% 2-mercaptoethanol from the immunoprecipitates the eluates were analyzed by two-dimensional electrophoresis. Fig.1 shows the separation of the immunoprecipitated proteins. Cellular proteins coprecipitated with T-Ag are well resolved in our gel system. In contrast most of the T-Ag precipitated at the application point and only part of it entered the gel forming a smear between pH 7.8-5.0.

As it was possible to separate purified T-Ag in stable charge isomers by isoelectric focusing (8), we assumed that interactions of T-Ag with components of the cell extract cause these precipitation effects. Therefore we analyzed the possible association of cell extract components to T-Ag containing immunoprecipitates. As a DNA binding protein T-Ag might be immunoprecipitated as a complex with nucleic acids. Further interactions between T-Ag and the host cell coded non viral T-Ag have been shown (for refs. see 12), but formation of disulfide bridges between T-Ag polypeptide chains or interaction of T-Ag with divalent cations could also induce aggregation of T-Ag molecules.

After labeling of SV80 cells with ^{3}H-thymidine or ^{3}H-uridine we and others (E. Khandijan and R. Weil, personal communication, 6) found only ^{3}H-uridine labeled RNA associated with the T-Ag immunoprecipitates. There was apperently no specific RNA adsorbed to the T-Ag immunoprecipitates, but bound RNA

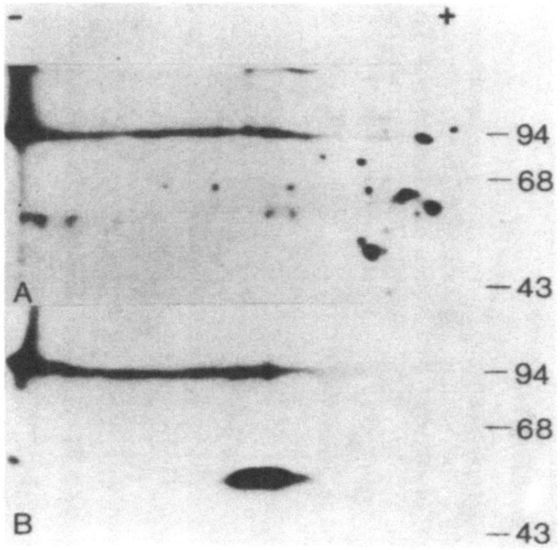

Fig. 1 Two-dimensional electrophoresis of T-Ag immunoprecipitates

5×10^6 SV80 cells were labeled with either 50 uCi ^{35}S-methionine or 100 uCi ^{32}P-phosphate for 4 h and T-Ag was immunoprecipitated from the detergent lysates as described in Materials and Methods. After elution with 9.3 M urea, 2% NP40, 2% ampholine (LKB) pH 3-10 the samples were analyzed by two-dimensional electrophoresis. Molecular weight standards: phosphorylase a (94 kd), bovine serum albumin (68 kd), ovalbumin (43 kd).

A ^{35}S-methionine labeled

B ^{32}P-phosphate labeled

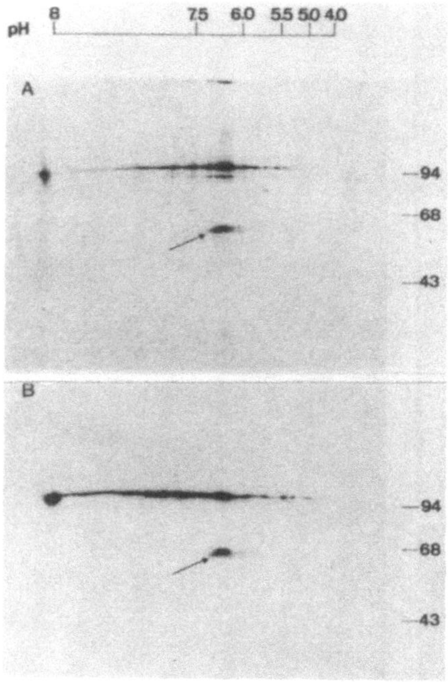

Fig. 2 Two-dimensional electrophoresis of T-Ag immunoprecipitates after incubation with RNase A

The washed immunoprecipitates were incubated with 0.05% RNase A (see Materials and Methods) and then analyzed by two-dimensional electrophoresis.

A ^{35}S-methionine labeled

B ^{32}P-phosphate labeled

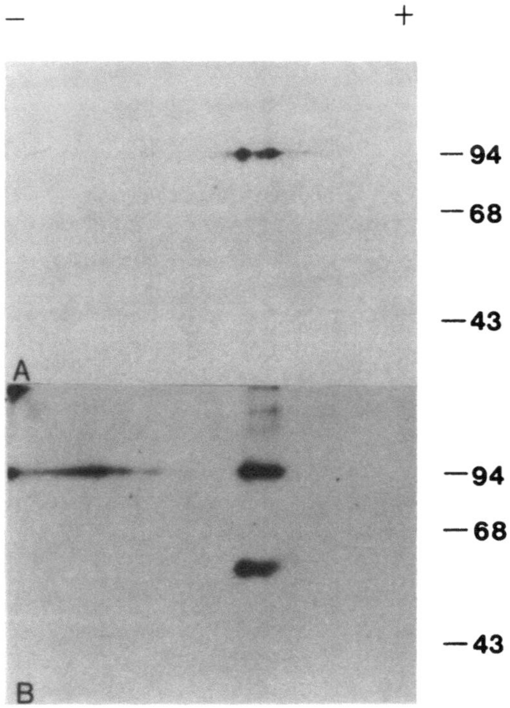

Fig. 3 Two-dimensional electrophoresis of T-Ag after complete RNase A digestion

T-Ag was isolated from SV80 cells by indirect immunoprecipitation (see Fig. 1). The washed immunoprecipitates were incubated with RNase A (0.8% w/v) and after elution of the immunoprecipitates the eluates were separated by two-dimensional electrophoresis.

A ^{35}S-methionine labeled
B ^{32}P-phosphate labeled

prevented completely the two-dimensional electrophoresis of T-Ag. To remove RNA associated with the T-Ag immunoprecipitates we incubated the immunoprecipitates with various amounts of RNase A. Fig.2 shows that after incubation of the immunoprecipitate with 0.05% RNase A most of the T-Ag was electrophoresed into the focusing gels. Several phosphorylated T-Ag forms were resolved between pH 7.8-5. NVT focused as a doublet at pH 6.4-6.2. The precipitate at the application point indicated that the RNase A digestion was not complete. After raising the RNase A concentration to 0.8% T-Ag was electrophoresed nearly completely into the focusing gels and was separated

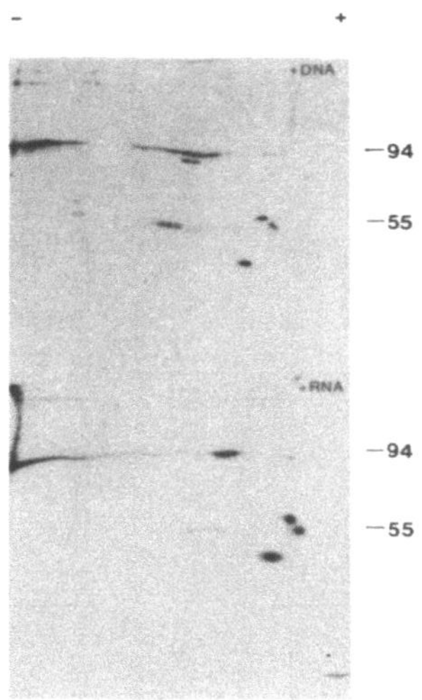

Fig. 4 Influence of nucleic acids on the two-dimensional electrophoresis of T-Ag

T-Ag was isolated from ^{35}S-methionine labeled SV80 cells by indirect immunoprecipitation and incubated with 0.8% RNase A. After elution 20 ug of either calf thymus DNA (Fig.4 A) or calf liver RNA (Fig.4 B) was added and after incubation for 30 min at 20° C the eluates were separated by two-dimensional electrophoresis.

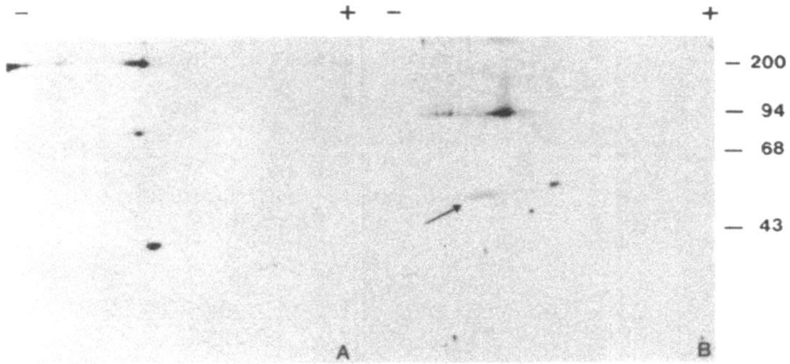

Fig. 5 Influence of alkylation on the two-dimensional electrophoresis of T Ag

T-Ag was isolated from ^{35}S-methionine labeled SV80 cells by indirect immunoprecipitation (see Fig.1). After RNase A incubation the immunoprecipitates were eluted with 9.3 M urea, 2% NP40, 0.01 M Tris-HCl, pH 7.0, and 0.01 M DTT for 1h at 4° C.

A Two-dimensional electrophpresis pattern of the eluate after incubation for 4 h at 4° C.
B Two-dimensional electrophoretic pattern of the eluate after alkylation with 0.08 M N-ethylmaleimide for 4 h at 4° C.

now in 2-3 spots at pH 6.4-6.2 (Fig.3). NVT focused under these conditions with pI's identical to T-Ag. To show more directly that T-Ag-nucleic acid complexes are responsible for precipitation and induction of charge heterogeneity we added in reconstitution experiments calf liver RNA or calf thymus DNA to T-Ag containing RNA free urea eluates. After binding of these nucleic acids we observed again smearing effects and more heterogenous forms of T-Ag (Fig.4).

The T-Ag polypeptide chain contains several cysteines (13,14). Therefore interactions occuring in vivo or during sample preparation can also influence its mobility during two-dimensional electrophoresis. Even under strong reducing conditions we observed a protein with a molecular weight of 230 kd with varying concentration having similar pI's like T-Ag (Fig.5 A). After reduction and alkylation of the sample with NEM this 230 kd protein was completely dissociated and monomeric T-Ag was resolved at pH 6.4-6.2 (Fig.5 B). Interestingly NVT was now also detectable but focused at slightly more alkaline pH. This indicates that T-Ag and NVT focused together as a stable complex which may be formed by artificial oxidation of sulfhydryl groups, however after blocking of their interactions both proteins are resolved according to their original pI's.

It is known that purified T-Ag can form under non optimal conditions instable charge patterns due to association equilibria (8,15). Therefore we had to prove that under optimal conditions immunoprecipitated T-Ag was focused as stably charged molecule. In two-dimensional refocusing experiments all stably charged proteins run in the second dimension on a diagonal whereas instably charged proteins show new and different patterns. To show whether there are differences in the stability of the charge pattern of cellular proteins and immunoprecipitated T-Ag we focused ^{35}S-labeled SV80 proteins and T-Ag in urea containing isoelectric focusing slab gels. After equilibrium focusing the gel strips containing the focused bands were cut out, laid on a second slab gel and refocused under identical conditions. As expected for stably charged proteins all cellular proteins had in the second dimension the same isoelectric points and run on a diagonal (Fig.6 B). Fig.6 D shows the pattern of immunoprecipitated T-Ag, which was refocused in the second mension with identical pI's thus having a stable charge pattern.

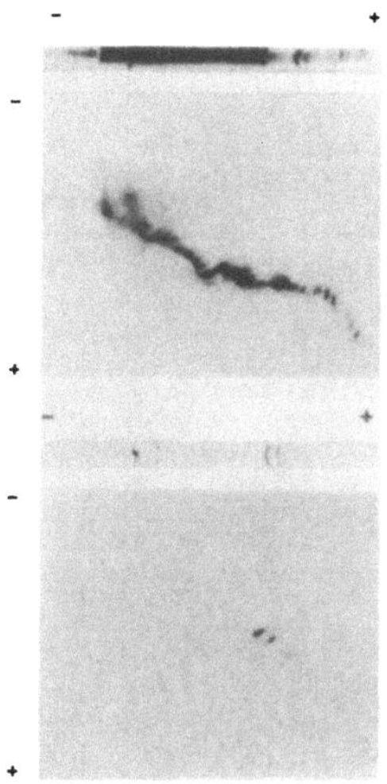

Fig. 6 Two-dimensional isoelectric refocusing of SV80 proteins and immunoprecipitated T-Ag

SV80 cells were labeled with ^{35}S-methionine and lysed as described in Fig.1. An aliquot of the 105000 x g supernatant was lyophilised and solubilised in 9.5 M urea, 10% 2-mercaptoethanol, 2% NP40 and 2% ampholine (LKB) pH 3-10 (SV80 proteins). A second aliquot of the 105000 x g supernatant was immunoprecipitated, incubated with RNase A and after elution the samples were focused in slab gels containing 8.8 M urea, 2% NP40 and 2% ampholine (LKB)pH 3-10 over 6 cm at 20W and 10° C. For refocusing the gel strip containing the focused bands was cut out, laid on a second slab gel and refocused under identical conditions.

A first and B second focusing dimension of SV80 proteins
C first and D second focusing dimension of immunoprecipitated T-Ag

Discussion

In this study we analyzed sytematically aggregation phenomena observed during two-dimensional electrophoresis of T-Ag containing immunoprecipitates. However, in contrast to several reports (4,5,6,16) T-Ag can be separated in distinct charge isomers by two-dimensional electrophoresis. The precipitation effects induced by RNA may be a more general problem in two-dimensional analysis of nucleic acid binding proteins (6) and indeed, similar problems have been observed during two-dimensional analysis of adenovirus 72 kd DNA binding protein (17). We observed during two-dimensional electrophoresis of T-Ag additional formation of oligomers by binding of divalent cations to T-Ag. Depending on the calcium concentration T-Ag was either shifted to more alkaline pI's or totally precipitated at the application point, although the separation of cellular control proteins was not influenced. Chelating reagents like EGTA dissolved these aggregates completely (unpublished observation). Ampholytes have chelating properties, but they are not able to dissociate polypeptides crosslinked by calcium. Therefore, the subunits of the adenylate cyclase crosslinked by calcium focus at pI 6.3, however after complexing of calcium with EGTA both subunits dissociate having now pI's of 7.3 and 6.1 (18).

Altogether these observations demonstrate that aggregations can severely influence the two-dimensional separation of T-Ag and other proteins and document the need of carefully controlling the conditions of sample preparation.

References

1. Tooze, J.1980. DNA Tumor Viruses. Vol.2.2nd ed. Cold Spring Harbor, New York.
2. Edwards, C.A. et al.1979. J. Virol. 29, 753-762.
3. Montenarh, M. & R. Henning.1980. FEBS Letters 114, 107-110.
4. Crawford, L.V., P.Z.O'Farrell.1979. J.Virol. 29, 587-596.
5. Fanning, E. et al.1981. J. gen. Virol. 55, 367-378.
6. Khandjian, E. et al.1982. Proc.Natl.Acad. Sci. U.S.A. 79, 1139-1143.
7. Gianazza, E. & P.G. Righetti. 1980.In Electrophoresis '79. W.de Gruyter,129-14
8. Palme, K. & R. Henning.1980. FEBS Letters 118, 229-232.
9. Kessler, S.W.1975. J.Immunol. 115, 1617-1624.
10. O'Farrell, P.H.1975. J.Biol.Chem. 250, 4007-4021.
11. Laemmli, U.K.1970. Nature 227, 680-685.
12. McCormic,F. & E. Harlow.1980 J.Virol. 34, 213-224.
13. Fiers, W.R. et al.1978. Nature 273, 113-120.
14. Reddy, V.B. et al.1978 Science 200, 494-502.
15. Palme, K. & R. Henning.1980. In Electrophoresis '79. W.de Gruyter, 141-150.
16. Crawford, L.V. et al.1980.Cold Spring Harbor Symp. Quant. Biol.,XLV, 129-140.
17. Klein, H. et al.1979. J.Biol.Chem. 254, 11051-11060.
18. Wahrmann, J.P. et al.1980. Biochim. Biophys Acta 612, 421-452.

TWO-DIMENSIONAL ELECTROPHORESIS OF FIBRINOGEN AND FIBRIN SUB-UNIT CHAINS

Brita Teige, Frank Brosstad, Bjørnar Olaisen and Randi Helland
Institute of Forensic Medicine, and Research Institute for Internal Medicine, University of Oslo, Rikshospitalet, Oslo 1, Norway

Introduction

Fibrinogen is a plasma protein consisting of two identical sets of three chains Aα, Bβ and γ (1). In the clotting process fibrinogen converts to fibrin. Most steps in this process are known, but they are difficult to visualize with conventional one-dimensional electrophoretic methods. In order to study in more detail the fibrinogen molecule and its conversion to fibrin we have therefore applied two-dimensional electrophoresis.

Materials and Methods

Fibrinogen is prepared from human plasma by beta-alanine precipitation (2). Fibrin monomers are made by treating fibrinogen with reptilase or thrombin in the presence of EDTA (3). Crosslinked fibrin chains are obtained by treating fibrinogen with thrombin in the presence of factor XIII and Ca^{++} (4). Two-dimensional separations of reduced samples are performed according to the procedure described by Anderson and Anderson (5,6) on ISO-DALT equipment purchased from Electro Nucleonics. In the first dimension a pH gradient from 3.5 to 10 is used. The gels are stained with Coomassie Brilliant Blue and/or silver nitrate according to R.P.Tracy (7).

Electrophoresis '82

Results and Discussion

Fibrinogen

Reduced fibrinogen showing the three different chains Aα, Bβ and γ is seen in fig. 1. All three chains are well separated in both dimensions, and different subpopulations within each chain can easily be identified.

The Aα-chain shows marked heterogeneity. It appears as several spots which differ both in isoelectric point and in SDS-molecular weight.

The Bβ-chain is seen as four distinct spots differing in quantity and isoelectric point, but with the same SDS-molecular weight around 62 000. The two small spots seen right under the Bβ-chains are related to this chain, since observed genetic variation in the Bβ-chain appears in these spots as well (8). The two spots probably represent a fragment of the Bβ-chain with a molecular weight around 5 000 less than the original chain. Such a fragment is formed when the proteolytic plasma enzyme, plasmin, splits the Bβ-chain at amino acid number 42 (9).

The γ-chain is mainly seen as two spots with different isoelectric point but with the same SDS-molecular weight about 53 000. A replica of the γ-chain spots is seen a little above and to the left of them. This is the γ'-chain, a molecular form of the γ-chain which has an additional carboxy-terminal fragment of about 2 000 (10,11,12). Traces of γ-dimers are also present in the sample.

Neuraminidase treated fibrinogen

In order to evaluate to what extent the degree of heterogeneity observed in the chains is caused by sialic acid, fibrinogen was treated with neuraminidase which eliminates sialic acid from proteins. The result of this treatment is seen in fig. 2. No major changes can be seen in the Aα-chain pattern.

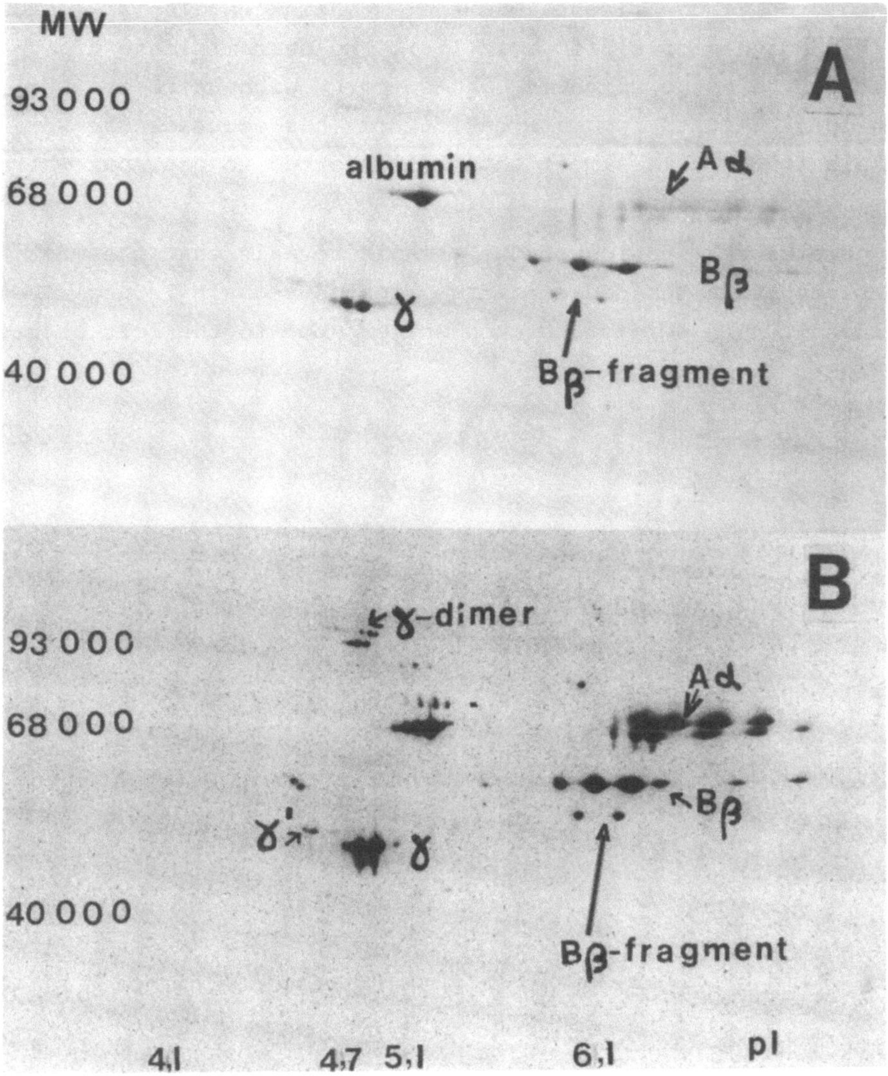

Fig. 1. Two-dimensional electrophoresis of reduced fibrinogen. A: Stained with Coomassie Brilliant Blue. B: Stained with silver nitrate.

According to the litterature the Aα-chain contains only negligible amounts of sialic acid (13). The neuraminidase treatment affects the Bβ-chain, and compared with untreated Bβ-chain a shift to two more basic forms has occured. The Bβ-chain fragment is almost totally converted to one spot which is more basic than the two original ones.

A careful examination of the γ-chain reveals that the most basic spot in the neuraminidase treated γ-chain is new. The slightly more acidic and much weaker spot to the left corre-

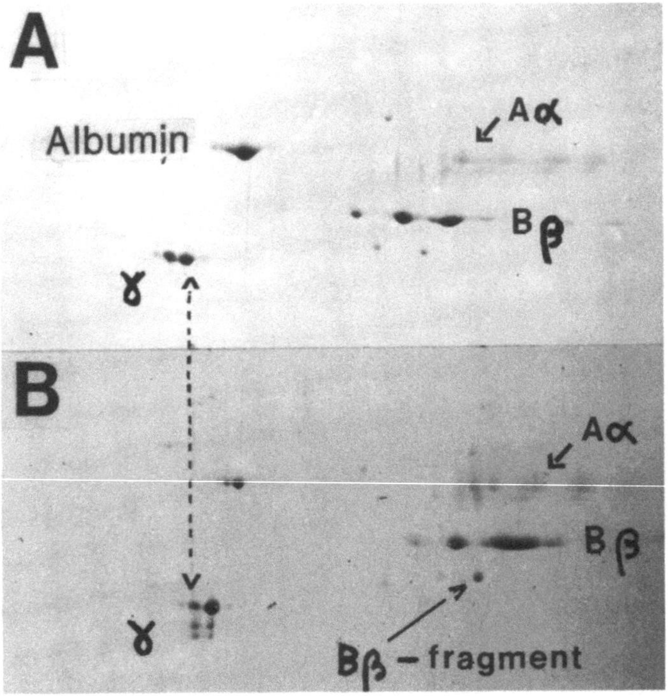

Fig. 2. A comparison between the two-dimensional electrophoretic pattern of reduced preparations of untreated and neuraminidase treated fibrinogen. The gels are stained with Coomassie Brilliant Blue. A: Untreated fibrinogen. B: Neuraminidase treated fibrinogen.

sponds to the most basic spot in the untreated γ-chain. Both in neuraminidase treated Bβ-chains and γ-chains the two main spots probably represent chains with zero and one sialic acid. The corresponding two main spots seen in untreated fibrinogen represent the chains which are known from the litterature to contain one or two sialic acids (13).

Fibrin monomers

Before crosslinking and polymerization to fibrin, release of fibrinopeptides takes place (1). Fibrinopeptide A is a small peptide containing the first 17 amino acids from the N-terminal end of the Aα-chain. The snake venom reptilase specifically removes the fibrinopeptide A, leaving a fibrinogen molecule called des AA fibrin monomer (3). This molecule contains a more basic and slightly lighter chain, the α-chain, but unchanged Bβ- and γ-chains. This is illustrated in fig. 3.

The proteolytic enzyme thrombin sequentially removes both fibrinopeptide A and fibrinopeptide B (3). Fibrinopeptide B is another small peptide which consists of the first 15 amino acids from the N-terminal end of the Bβ-chain. The transformation from Bβ-chain to β-chain, indicating the loss of fibrinopeptide B, is very easily observed in samples containing both chains, as seen in fig. 4. After removal of fibrinopeptide A and fibrinopeptide B the remaining fibrinogen molecule is called des AABB fibrin monomer. In this molecule both the Aα-chain and the Bβ-chain have been changed to more basic and slightly lighter forms, illustrated in fig. 5.

As expected, the two spots believed to be Bβ-chain fragments without the first 42 amino acids, are unaffected by the reptilase and thrombin treatment.

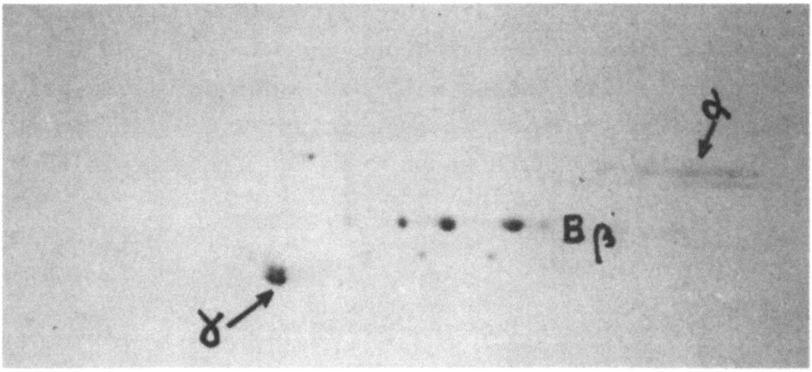

Fig. 3. Two-dimensional electrophoresis of reduced preparations of reptilase treated fibrinogen (des AA fibrin monomer). The gel is stained with Coomassie Brilliant Blue.

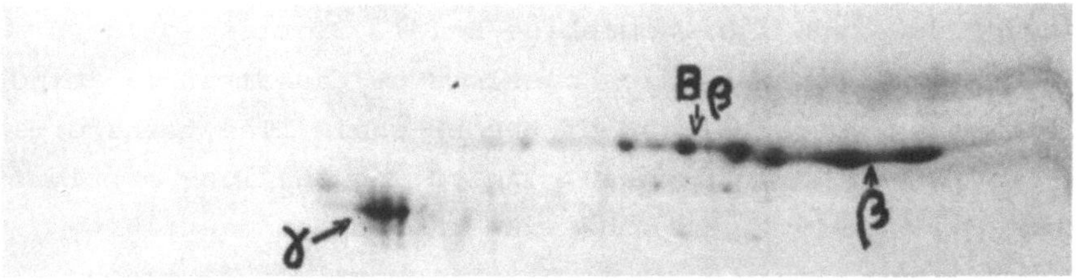

Fig. 4. Two-dimensional electrophoresis of a reduced fibrin monomer sample showing the conversion from des AA to des AABB. The gel is stained with Coomassie Brilliant Blue.

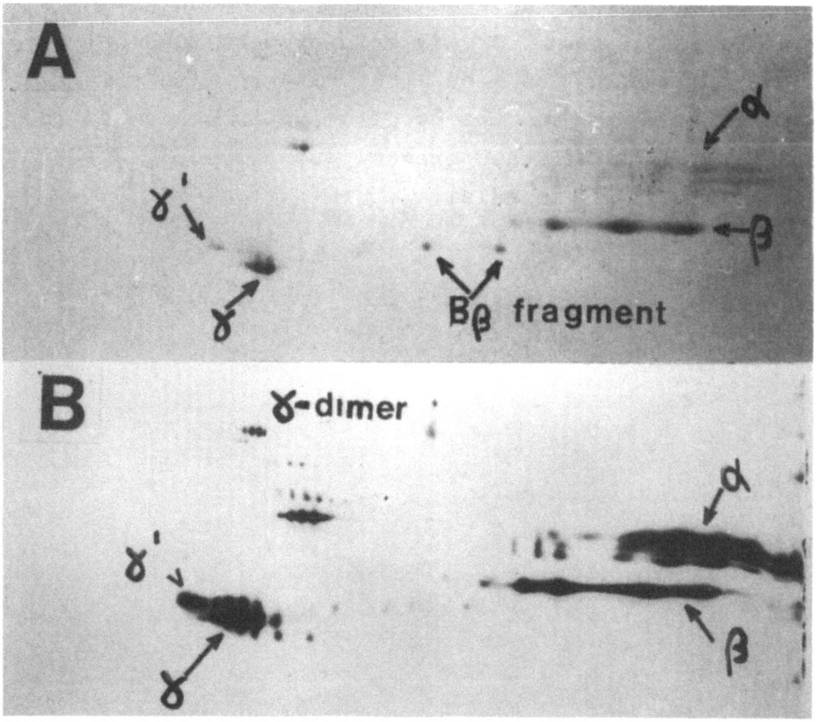

Fig. 5. Two-dimensional electrophoresis of reduced preparations of thrombin treated fibrinogen (des AABB fibrin monomers). A: Stained with Coomassie Brilliant Blue. B: Stained with silver nitrate.

Crosslinking

Fibrin monomers crosslink in the presence of factor XIII and Ca^{++} (4). In the crosslinking process a dimerization of γ-chains occurs. This results in an intermediate polymer which further develops into a fibrin network involving the α-chain and the β-chain (1). The crosslinking process may be monitored by the appearance of γ-dimers and by the loss of fibrinopeptide A and B. In partially crosslinked preparations both γ-dimers and γ-monomers as well as Bβ, β and α-chains are seen. In totally crosslinked fibrin preparations only γ-dimers and β-chains are seen, fig. 6. Even with silver staining (not shown) no traces of α-chains are seen. The α-chains are at this stage probably so polymerized that they will not move into the gels.

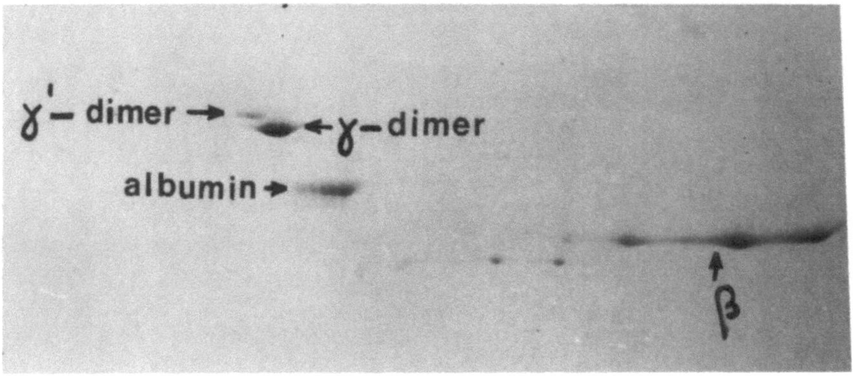

Fig. 6. Two-dimensional electrophoresis of a reduced sample of crosslinked fibrin. The gel is stained with Coomassie Brilliant Blue.

Conclusions

Two-dimensional electrophoresis of reduced fibrinogen from single individuals visualizes the three fibrinogen chains Aα, Bβ and γ. Heterogeneity within each chain is revealed in detail.

The method clearly demonstrates changes in the fibrinogen molecule caused by treating it with neuraminidase, reptilase or thrombin. The γ-chain dimerization in the presence of factor XIII, can easily be followed. This makes the technique useful in examinations of purified fibrinogen samples, and in samples where it is important to know the degree of fibrinopeptide release and crosslinking.

References

1. Doolittle, R.S.: Sci.Amer. *245*, 92-101 (1981).
2. Jacobsen E., Kierulf, P.: Thromb.Res. *3*, 145 (1973).
3. Brosstad, F., Kierulf, P., Gravem, K., Godal, H.C.: Thromb. Res. *13*, 715-723 (1978).
4. Brosstad, F.: Thromb.Res. *13*, 799-803 (1978).
5. Anderson, N.G., Anderson, N.L.: Anal.Biochem. *85*, 331-340 (1978).
6. Anderson, N.L., Anderson, N.G.: Anal.Biochem. *85*, 341-354 (1978).
7. Tracy, R.P.: Personal communication.
8. Brosstad, F., Teige, B., Olaisen, B., Gravem, K., Stormorken, H. This Conference.
9. Marder, V.J., Schulman, R., Carrol, W.R.: J.Biol.Chem. *244*, 2111-2119 (1969).
10. Mosesson, H.W., Finlayson, J.S., Umfleet, R.A.: J.Biol. Chem. *247*, 5223-5227 (1972).
11. Wolfenstein-Todel, C., Mosesson, H.W.: Proc.Natl.Acad. Sci. U.S.A. *77*, 5069-5073 (1980).
12. Kuyas, C., Haeberli, A., Straub, P.W.: J.Biol.Chem. *257*, 1107-1109 (1982).
13. Gati, W.P., Straub, P.W.:J.Biol.Chem. *253*, 1315-1321(1978).

NUMERICAL TAXONOMY OF TWO-DIMENSIONAL PROTEIN MAPS: A RATIONAL BIOCHEMICAL APPROACH TO TUMOR CHARACTERIZATION.

K. Westerbrink, B. Havsteen

Biochemisches Institut im Fachbereich Medizin, Christian-Albrechts-Universität Kiel, D-2300 Kiel (Federal Republic of Germany)

and K. Groenier

Andragogisch Instituut, and Instituut voor Huisartsgeneeskunde, Rijksuniversiteit Groningen, Groningen (The Netherlands).

INTRODUCTION.

The exploitation of the full potentialities of the high-resolution 2D protein mapping[a] is mainly impeded by problems in the interpretation and evaluation of the complicated protein patterns, which are obtained. Only few laboratories yet have the facilities for (video)densitometric, digitalization or image analysis procedures. Though time-consuming, the simplest practicable method, still, is to explore the protein maps by visual inspection. Spots, which can be detected by the naked eye, are marked and diagrammed. The comparison of protein maps can also be performed manually, e.g. by layering of sheets of transparent copies of the individual diagrams. Thus, three numerical values for any pair of maps are obtai-

[a]Abbreviations used: 1D, one-dimensional, resp. first dimension; 2D, two-dimensional, resp. second dimension; IEF, isoelectric focusing; SDS-PAGE, polyacralamide gel electrophoresis in the presence of sodiumdodecylsulphate; L, liver; S, spleen; LN, lymph node; N, morphologically normal, i.e. non-tumor sample; CBCC, centroblastic-centrocytic lymphoma; PC, plasmacytoma; IC, immunocytoma; CLL, chronic lymphatic leukemia; MDS, multidimensional scaling.

Electrophoresis '82

(legend to Table 1, opposite page)

TABLE 1. Dissimilarity matrix of a set of 13 protein maps. The empirical values of the dissimilarity index δ , defined as described under 'Experimental', are tabulated for every possible pair of maps in the set. The randomly chosen samples are arbitrarily numbered. All samples are from different individuals, with the exception of samples nr 5 and 7, which represent protein preparations from one lymph node of the same patient, electrophoretically analyzed in different runs and separately evaluated.

TISSUE	LN											S	L
DIAGNOSIS	N	N	N	CBCC	CLL	PC	CLL	IC	N	N	N	N	N
SPOT NR (m)	169	114	84	128	116	91	81	107	118	112	98	88	89
SAMPLE NR	1	2	3	4	5	6	7	8	9	10	11	12	13
2	2.68												
3	3.22	2.42											
4	3.30	2.46	3.61										
5	3.67	4.11	4.71	3.28									
6	3.13	2.80	2.89	3.21	2.91								
7	3.17	3.53	3.46	2.80	1.94	2.74							
8	3.12	3.70	3.78	3.70	2.91	2.60	1.85						
9	2.34	2.27	2.88	2.57	3.59	3.02	2.69	3.09					
10	2.60	1.94	2.84	3.53	4.70	2.98	3.71	3.98	2.59				
11	2.61	2.37	3.04	2.42	4.63	2.57	2.65	2.87	2.09	2.04			
12	6.41	6.77	5.88	4.84	6.85	5.88	4.83	5.96	5.44	5.45	4.81		
13	7.90	5.34	5.92	6.23	10.39	8.00	7.95	7.52	6.39	5.93	5.68	9.41	

TABLE 1.
(legend on opposite page)

ned: the total number of spots detectable in each of the two protein maps, and the number of spots common to both. These numbers can be used to calculate an 'overall (dis)similarity index' for any pair of maps. In this paper it is shown, that such data suffice to apply the principles and techniques of numerical taxonomy, as formulated in the classical work of Sokal and Sneath (1), with the objective of classifying protein maps into groups, the grouping being exclusively based on the information contained in the protein maps themselves. Described here are the first results on the applicability of these procedures to the problem of lymphoma characterization, based on a set of protein maps of crude membrane protein enriched fractions of normal human lymphatic tissue samples and such of lymphoma patients. As all procedures described are essentially independent on the nature and source of the samples, they may find application in a variety of other taxonomical problems, both in the field of medicine, as in biology.

EXPERIMENTAL.

Tissue samples. The tissue material consisted of normal human tissues (lymph node, spleen, liver), as well as lymph node samples from patients with lymphomas of various types. The designation 'normal' refers to non-cancerous tissue; inflammatory material can be included. The samples were stored frozen at -20°C.

Preparation of crude membrane protein fractions. The procedure applied to obtain an operationally well defined crude membrane protein enriched protein fraction from the tissue samples has been described previously (2).

2D-electrophoresis. Protein mapping of the protein fractions was performed essentially according to O'Farrell (3) and Klose (4), i.e. 1D-IEF/2D-SDS-PAGE, with several modifications (technical details will be published elsewhere).

Dissimilarity matrix. The 'overall dissimilarity' between two protein maps was defined and determined as follows. Maps were compared by visual inspection of transparent copies of drawn diagrams of the original patterns. Let the total number of spots detectable by the naked eye in individual maps be m(A), m(B), m(C), etc. and the number of spots common to a pair of maps be $m(A \cap B)$, $m(A \cap C)$, etc. Subsequently the dissimilarity index δ was calculated, defined as the numerical ratio of union and intersection of a pair of spot sets, e.g. maps A,B:

$$\delta(A,B) = \frac{m(A) + m(B) - m(A \cap B)}{m(A \cap B)} = \frac{m(A \cup B)}{m(A \cap B)} \qquad (i)$$

The major improvement of this measure over the commonly used 'matching percentage' is, that it takes into account that the total number of spots in different maps are not necessarily equal. The value of δ ranges from 1, for identical maps, to infinity, for maps which have no spots in common.

For 11 randomly chosen protein maps of the crude membrane protein fractions from both normal lymph nodes and those of lymphoma patients, the value of δ for every possible pair of maps was determined. These values were compiled into a 'dissimilarity matrix', on which several clustering procedures were applied (see below). One representative map of a protein fraction obtained with the same fractionation procedure as was used for the lymph nodes, from a set of normal spleen maps, and one such a map of a normal liver were likewise evaluated and included in the dissimilarity matrix.

Cluster analysis. A hierarchical method was used, as described by Johnson (5) and available as the computer program HICLUS (6). Three criterion functions were used: the 'maximum' (diameter or maximum linkage) method, the 'minimum' (connectedness or single linkage) method, and the 'average linkage' method. Application of the method in the field of biochemistry has been described (7).

Multidimensional scaling. The objective of MDS is to map a set of objects into a set of points in a metric space, such that objects, which are similar (in some empirical sense; for our definition see above) are close together in the space, and objects which are dissimilar are distant from each other in the space. The results shown in this paper were obtained by applying the computer program MINISSA (10). The 'fit' of the mathematical solution with the empirical data is expressed as a 'badness of fit' coefficient, usually called 'stress' as defined by Kruskal (8,9).

RESULTS AND CONCLUSIONS.

Table 1 shows the dissimilarity matrix, constructed as described under 'Experimental'. The data matrix can essentially be extended to any number of maps. The required number of maps to be pairwise compared, however, becomes rapidly rather large. For a matrix of X samples X(X-1)/2 values of δ have to be determined. Superficial visual inspection of the matrix is not very revealing. More sophisticated mathematical techniques are required to obtain a graphical representation of the structure of the data.

The results of hierarchical cluster analysis of the matrix data are shown in Figure 1A-C. Dendrograms were constructed with three different criterion functions: the maximum variant solution (Figure 1A), the minimum variant solution (Figure 1B) and the average linkage solution (Figure 1C). The number of clusters found at different (dis)similarity levels varies.

(legend to Figure 1, opposite page)

FIGURE 1. Hierarchical clustering solutions of the matrix data of Table 1. A: The maximum variant solution of the complete data matrix. B: The minimum variant solution of the lymph node samples. C: The average linkage solution of the lymph node samples. Open circles at the end of each branch represent morphological normal tissue samples, closed circles pathological samples. The numbering of the samples corresponds to that of Table 1. For the diagnoses of the pathological samples reference is also made to Table 1. Crosses (x) indicate clustering configurations at some dissimilarity levels ( ), as described in the text.

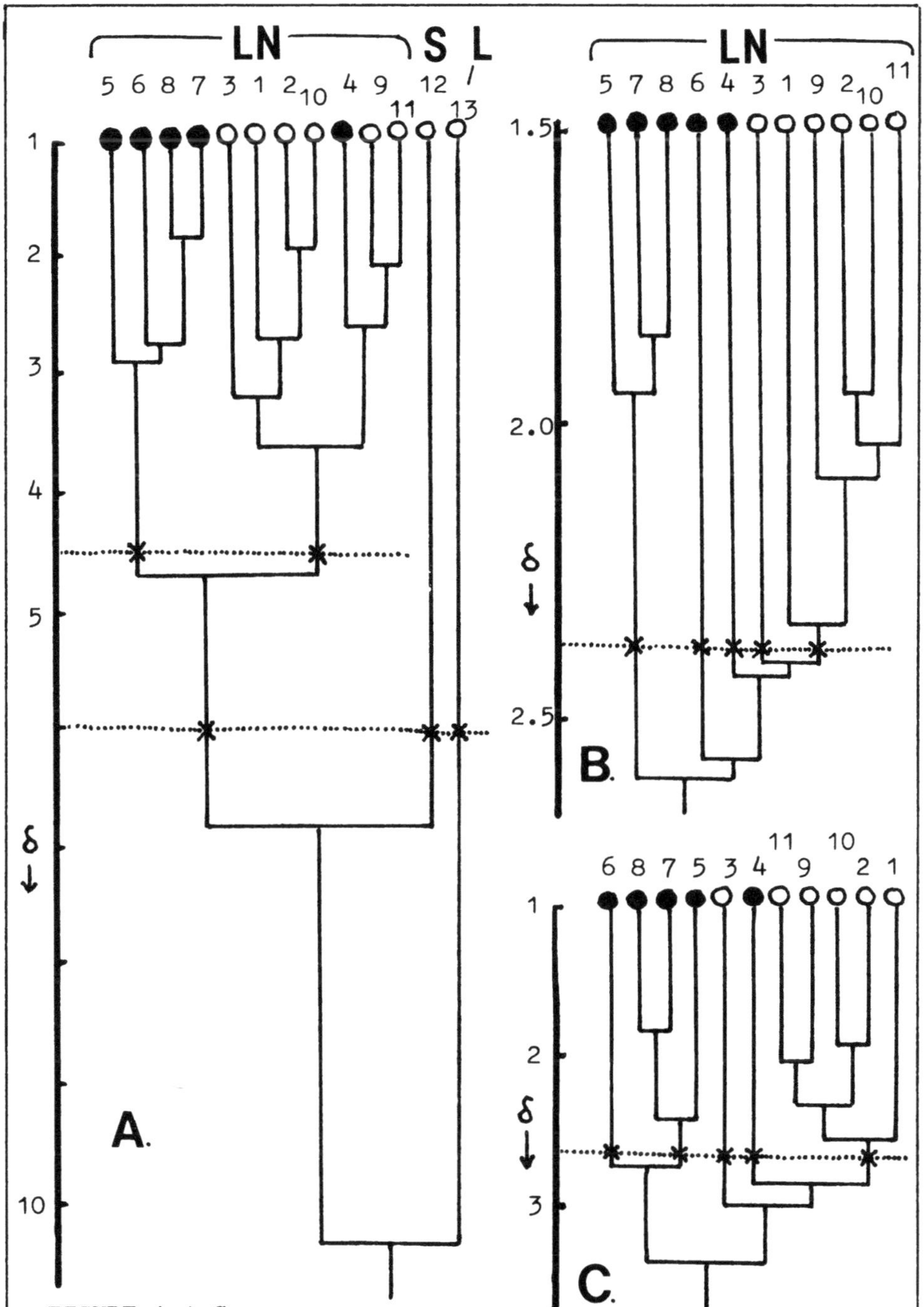
LN
S
L
5 6 8 7 3 1 2 10 4 9 11 12 13
1
2
3
4
5
δ
10
A.
LN
5 7 8 6 4 3 1 9 2 10 11
1.5
2.0
δ
2.5
B.
6 8 7 5 3 4 11 9 10 2 1
1
2
δ
3
C.

FIGURE 1.A-C.

If, e.g., in Figure 1A the level of $\delta = 6.0$ is chosen, three 'branches' are found. One is leading to the liver sample (nr 13), a second one to the spleen sample (nr 12), and the third one comprises all of the lymph node samples analyzed. This represents one graphical demonstration of the tissue specificity of the protein maps. Within the group of lymph nodes, at the level of $\delta = 4.5$ (Figure 1A), two clusters are formed. One comprises the pathological samples nr 8, 7, 6 and 5, the other one all normal samples, with the exception of the pathological sample nr 4. Thus, a striking clustering of the normal and pathological samples is seen. This represents one graphical demonstration of the sensitivity of the protein map information in respect to pathological change within one tissue. A further reduction of the dissimilarity level, in this graph, in order to increase the number of tissue 'types', shows the limits of the method. Further splitting of the group of four pathological samples leads to the meaningless result, that samples nr 5 and 7, two protein maps of the same patient (see legend to Table 1), appear in different clusters. Hence, for every algorithm applied, and dependent on the accuracy of the determination of δ, a minimal dissimilarity level has to be defined above which value the clustering branches reflect actual and meaningful differences between the samples analyzed.

The minimum variant solution (Figure 1B) also displays a

(legend to Figure 2, opposite page)

<u>FIGURE 2. Multidimensional scaling solutions of the data matrix of Table 1.</u> A: One-dimensional solution of the complete matrix. Normal and pathological lymph nodes cluster in the area LN1-11. B: Two-dimensional representation of the complete matrix. Normal and pathological lymph nodes are not differentiated, all of them cluster in the area LN 1-11. C: One-dimensional solution of the lymph node samples. Closed circles represent pathological samples with indicated diagnoses, open circles represent morphologically normal samples. The numbering of the samples corresponds to that of Table 1 and Figure1. D: Two-dimensional representation of the lymph node samples. Indicated are the sample numbering and the diagnoses. The areas enclosed by the continuous lines (——) represent the optimal solution of the hierarchical cluster analysis of Figure 1C. In addition, two contour levels, according to the 'contour mapping procedure' (1) are drawn. Crosses (x) indicate the centroids of the normal and tumor samples. The sum of the average distances of the actual samples from the centroid of both sets is smaller than the distance between the two centroids. The scaling of the coordinate system can be arbitrarily chosen; only the spatial distribution of the samples is of relevance for the interpretation of the diagram.

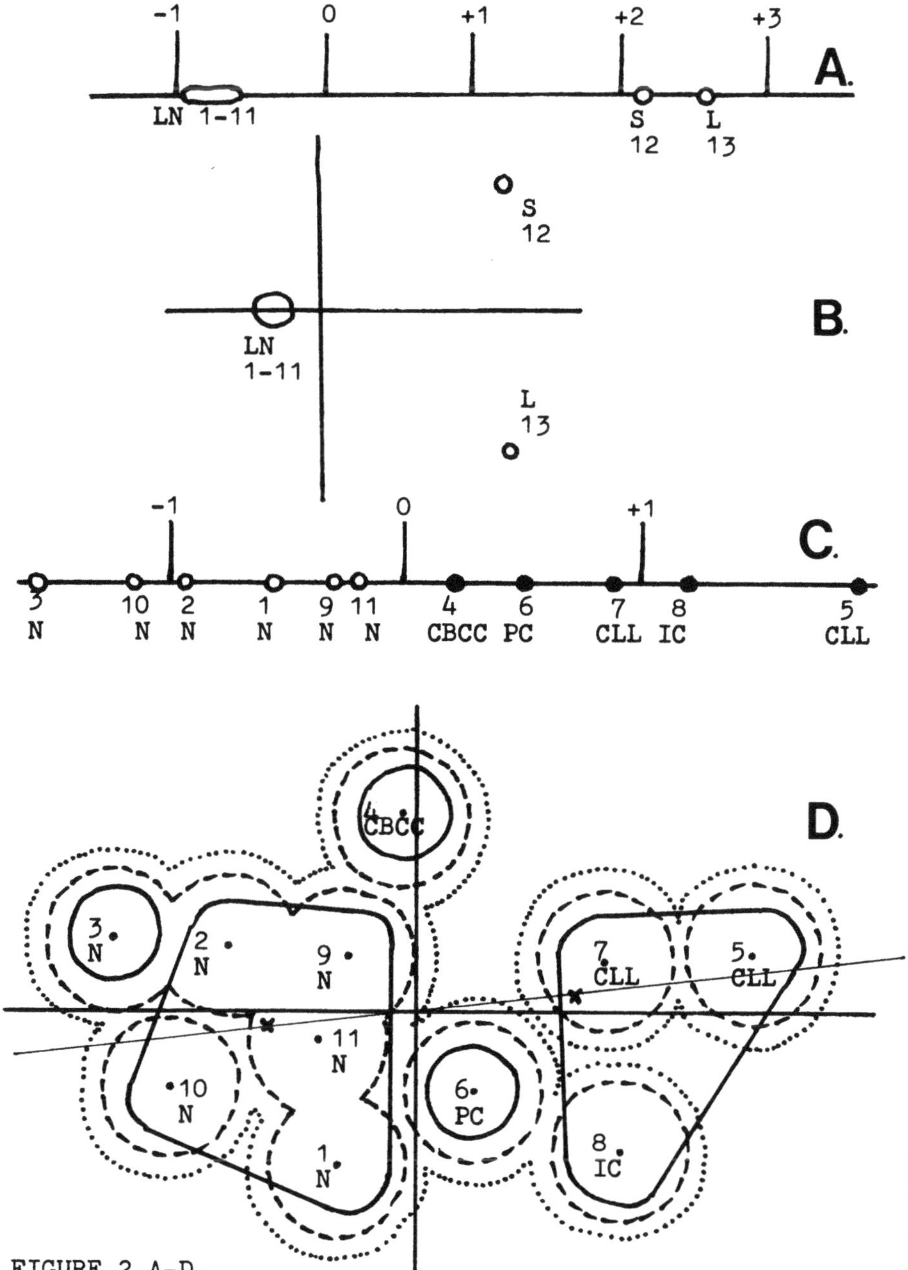
-1
0
+1
+2
+3
A.
LN 1-11
S
12
L
13
S
12
B.
LN
1-11
L
13
-1
0
+1
C.
3
N
10
N
2
N
1
N
9
N
11
N
4
CBCC
6
PC
7
CLL
8
IC
5
CLL
D.
4
CBCC
3
N
2
N
9
N
7
CLL
5
CLL
11
N
10
N
6
PC
1
N
8
IC

FIGURE 2.A-D.

clear clustering, slightly different from the maximum variant solution. At a level of $\delta = 2.4$ five clusters are found within the group of lymph nodes. One comprises the normal lymph nodes 11, 10, 2, 9 and 1, one leads to the normal lymph node nr 3, one to the pathological lymph node nr 4 (CBCC), one to the plasmacytoma sample nr 6, and one comprises the pathological samples 8, 7 and 5. The average linkage solution (Figure 1C), which can be regarded as an optimal solution between the extremes of Figures 1A and 1B, shows the same five 'types'.
Application of multidimensional scaling (MDS) to our data matrix leads to the graphical representations shown in Figure 2. The one-dimensional solution of the complete data matrix is shown in Figure 2A. This picture represents another graphical demonstration of the tissue specificity of the protein maps. Figure 2C shows the one-dimensional solution for the set of lymph nodes separately. A remarkable similarity between this pattern and the ones of Figure 1 is seen. The normal tissue samples cluster at one end of the straight line, all pathological tissue samples at the other end. The amount of 'stress' to achieve these solutions are 18.9 and 24.7%, respectively. The two-dimensional solutions shown in Figure 2B (including liver and spleen) and Figure 2D (lymph nodes only) needed a stress of 3.5 and 13.7%, respectively, and are therefore to be considered as better representations of the data. Figure 2B shows the evident clustering of the different tissue samples, Figure 2D the grouping within the set of lymph node samples. Useful in the delimitation of clusters in such a spatial configuration is the embedding of the solution derived from cluster analysis in the configuration space. Figure 2D, therefore, shows the 2-dimensional MDS-solution with the hierarchical solution of Figure 1C at the level of five clusters embedded in it. In addition, two contours according to the 'contour mapping procedure' (1) are indicated.
From the striking correlation between overall protein map information and morphological data, it is concluded, that application of the principles and techniques of numerical taxonomy can be very useful in the interpretation of complex 2D protein maps, not only in tumor characterization, but also in a variety of other research areas in medicine and biology.

ACKNOWLEDGEMENTS.

This work was supported in part by grants of the Deutsche Forschungsgemeinschaft awarded to Sonderforschungsbereich 111, University of Kiel. Thanks are due to Mrs. Edith Postel for excellent technical assistance.

REFERENCES.

1. Sneath, P.H.A. and Sokal, R.R.: Numerical taxonomy, Freeman, San Francisco-London, 1973.
2. Westerbrink, K., Warthold, U., Hebbeln, J., Postel, E. and Havsteen, B.: J. Cancer Res. Clin. Oncol. 101, 43-51 (1981).
3. O'Farrell, P.: J. Biol. Chem. 250, 4007-4021 (1975).
4. Klose, J.: Humangenetik 26, 231-243 (1975).
5. Johnson, C.J.: Psychometrika 32, 241-254 (1967).
6. Johnson, C.J. HICLUS, a computer program for hierarchical cluster analysis, Edinburgh version, University of Edinburgh, 1975.
7. Fitch, W.M. and Margoliash, E.: Science 155, 279-284 (1967).
8. Kruskal, J.B.: Psychometrika 29, 1-27 (1964).
9. Kruskal, J.B.: Psychometrika 29, 115-129 (1964).
10. Roskam, E.E. and Lingoes, J.C. MINISSA, a computer program for multidimensional scaling, Edinburgh version. University of Edinburgh, 1975.

ISO-DALT ANALYSIS OF URINARY PROTEINS IN RENAL FAILURE

P M S Clark, G M Power and T P Whitehead
Department of Clinical Chemistry, Wolfson Research Laboratories
Queen Elizabeth Medical Centre, Birmingham B15 2TH, UK.

Abstract

The Iso-Dalt technique was used in the analysis of concentrated urines from patients with acute and chronic renal failure, patients following renal transplantation and from healthy controls.

Generally, although there were some proteins excreted by all groups of patients, eg, albumin, transferrin, IgG gamma and light chains, α_1-acid glycoprotein, the excretion of other proteins was variable even within groups of patients with the same renal pathology. A decrease in the number of proteins with time after transplantation was demonstrated and conversely an increase preceding an episode of graft rejection was found.

In conclusion the pattern of urinary protein excretion in renal disease is both variable and complex. Further clinical studies and developments in data handling will be necessary before these changes can be of diagnostic value.

Introduction

Normal human urine contains many proteins in trace amounts which originate from : serum, the kidney itself or the uro-genital tract or from a disease process (Anderson et al, 1979). In renal disease urine total protein levels may be used to

Electrophoresis '82

assess the level of kidney damage and in addition the relative distribution of high and low molecular weight proteins may indicate the type of renal pathology. With glomerular damage high molecular weight proteins (> 40,000 Daltons) are excreted whereas lower molecular weight proteins are excreted in tubular disease.

The objective of this study was to investigate the proteinuria associated with various renal pathologies and following renal transplantation using the Iso-Dalt technique in order to assess the molecular weight distributions of the excreted proteins and to identify any novel proteins that might be of diagnostic or prognostic use.

Methods

Subjects

24 h urine collections were obtained from ten subjects with renal failure of various pathologies and from several patients following renal transplantation. The original renal pathology in the latter group varied. For some patients a series of 24 h urine collections was obtained.

Urine Preparation

Aliquots (500 ml) were taken from the 24 h urine collections from all patients. Samples were stored frozen (-20 ^{o}C) until analysed.

Aliquots (500 ml) were dialysed against 4 l distilled water, using 18/32" dialysis tubing (The Scientific Instrument Centre Ltd, London) overnight at 4 ^{o}C. Samples were then filtered through Whatman No 1 filter paper using a vacuum water pump. The samples were then placed in dialysis tubing and concentrated against PEG 6000 (400 g/100 ml). Concentrated samples (50 ml volumes) were stored in amber, rubber stoppered vials at -20 ^{o}C. They were then freeze-dried and reconstituted with distilled water (5 ml) prior to analysis.

Analysis

Iso-Dalt analysis was carried out according to the method of Anderson et al (1978a and b), using ampholytes of pH 3.5 - 10 for the first dimension isoelectric focusing and an acrylamide gradient of 10 - 20% in the second dimension electrophoresis.

Results and Discussion

The Iso-Dalt analysis showed that a range of urinary protein patterns is found in hypertensive renal failure, both serum and other unidentified proteins being excreted. The degree of proteinuria seemed to reflect the clinical severity of the hypertension and associated renal failure (Fig 1).

The proteinuria associated with other types of renal failure was more severe but of a variable pattern. In one case of acute renal failure (Fig 2) there was a noticeable absence of albumin and a predominance of low molecular weight proteins. Some of the unidentified proteins (Fig 3, numbered 1-8) were also found in the urines of the transplant patients.

In the renal transplant patients a moderate to severe proteinuria was found. In some patients a reduction in the number of proteins found paralleled clinical improvement (Fig 4) though this was not always the case. Particularly noticeable was the reduction in low molecular weight proteins and Ig light chains. The latter may be due to immunosupressive therapy.

The patient who suffered acute and complete rejection of his graft showed a massive proteinuria with a predominance of high

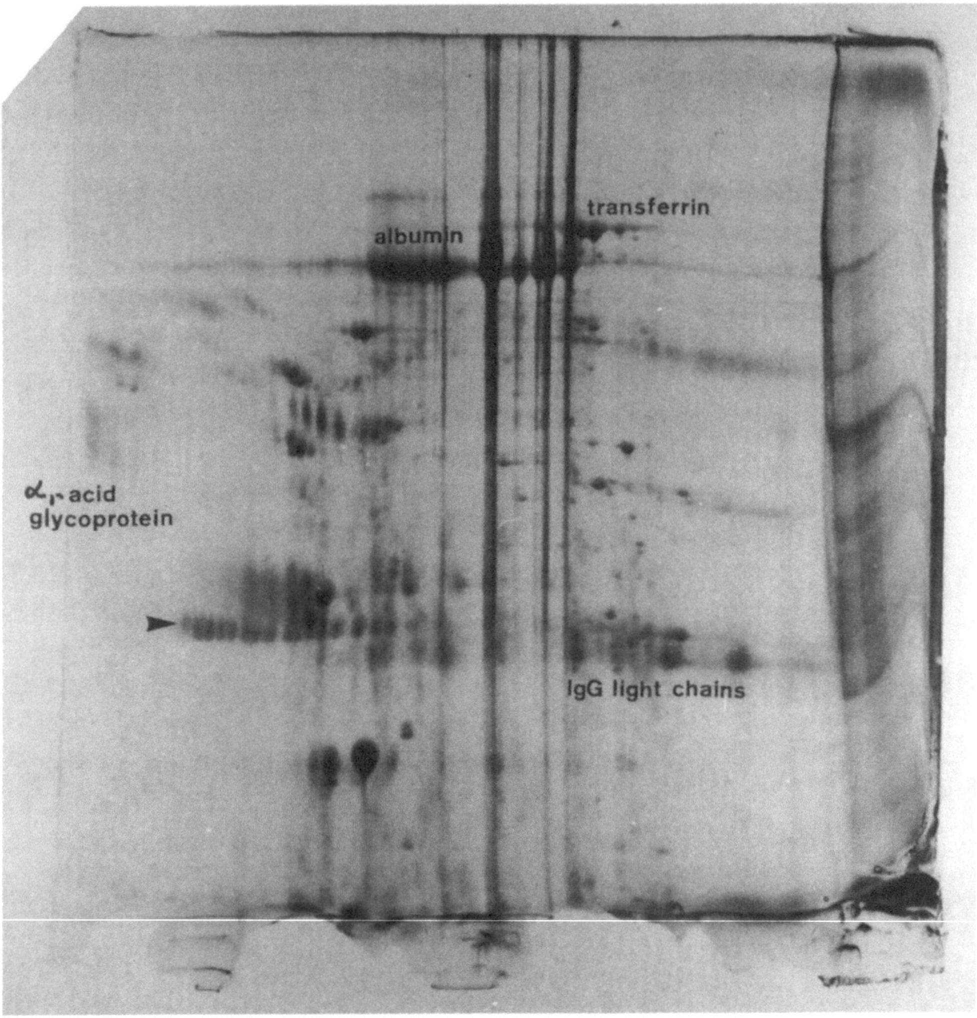

Figure 1 Iso-Dalt gel of urine proteins from patient with severe hypertensive renal failure. ▶ = carbonic anhydrase standard.

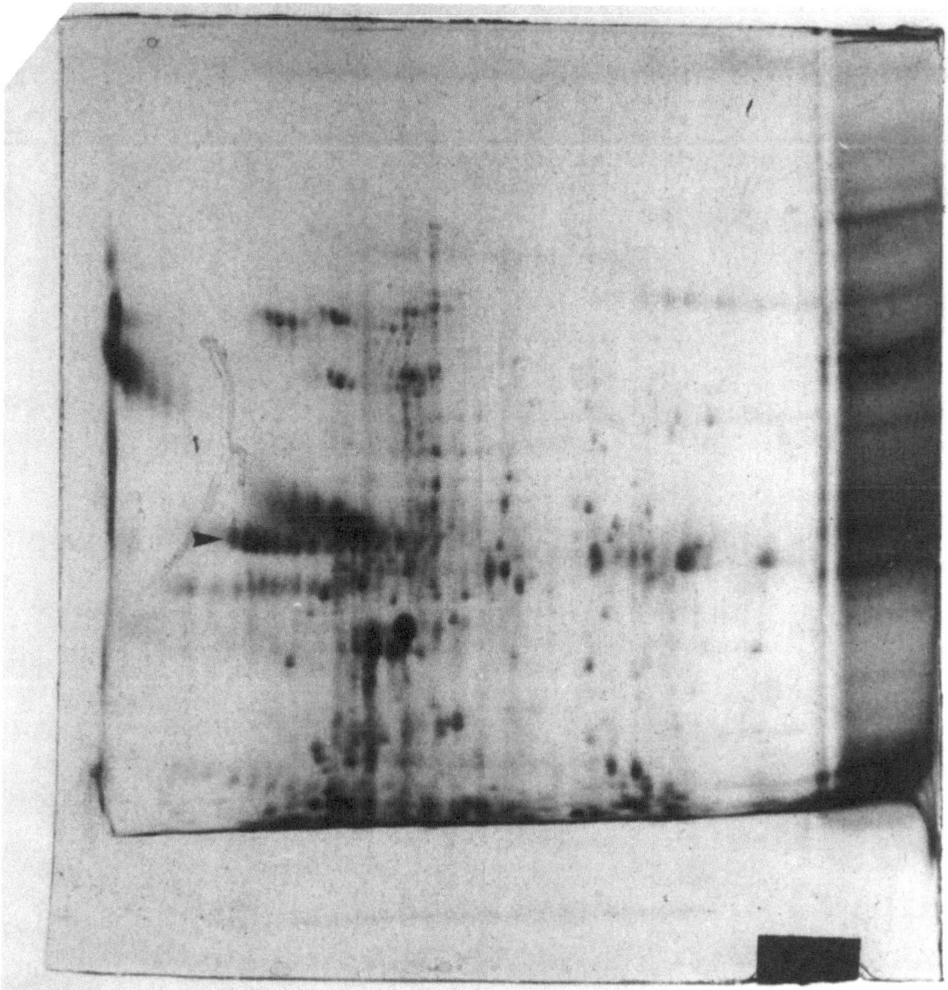

Figure 2 Iso-Dalt gel of urine proteins from patient with acute renal failure.

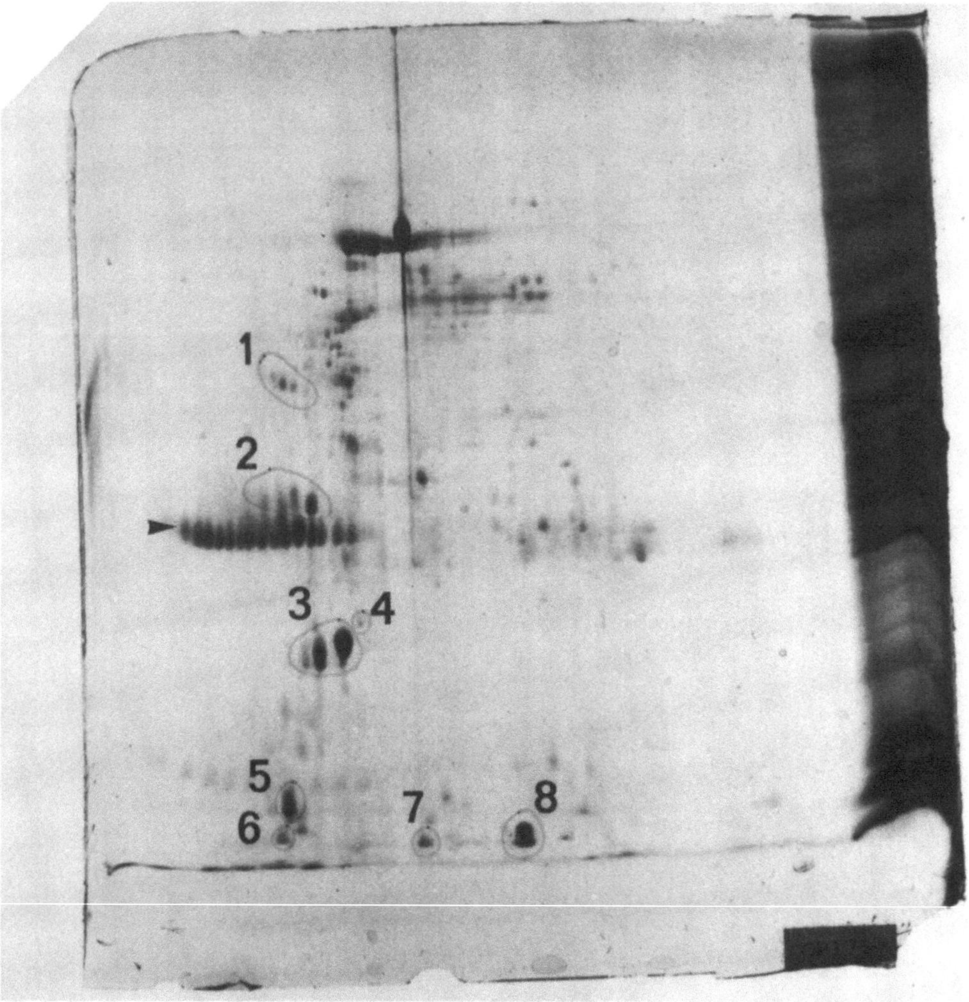

Figure 3 Iso-Dalt gel of urine proteins from patient with recurring pyelonephritis and staghorn calculi leading to renal failure.

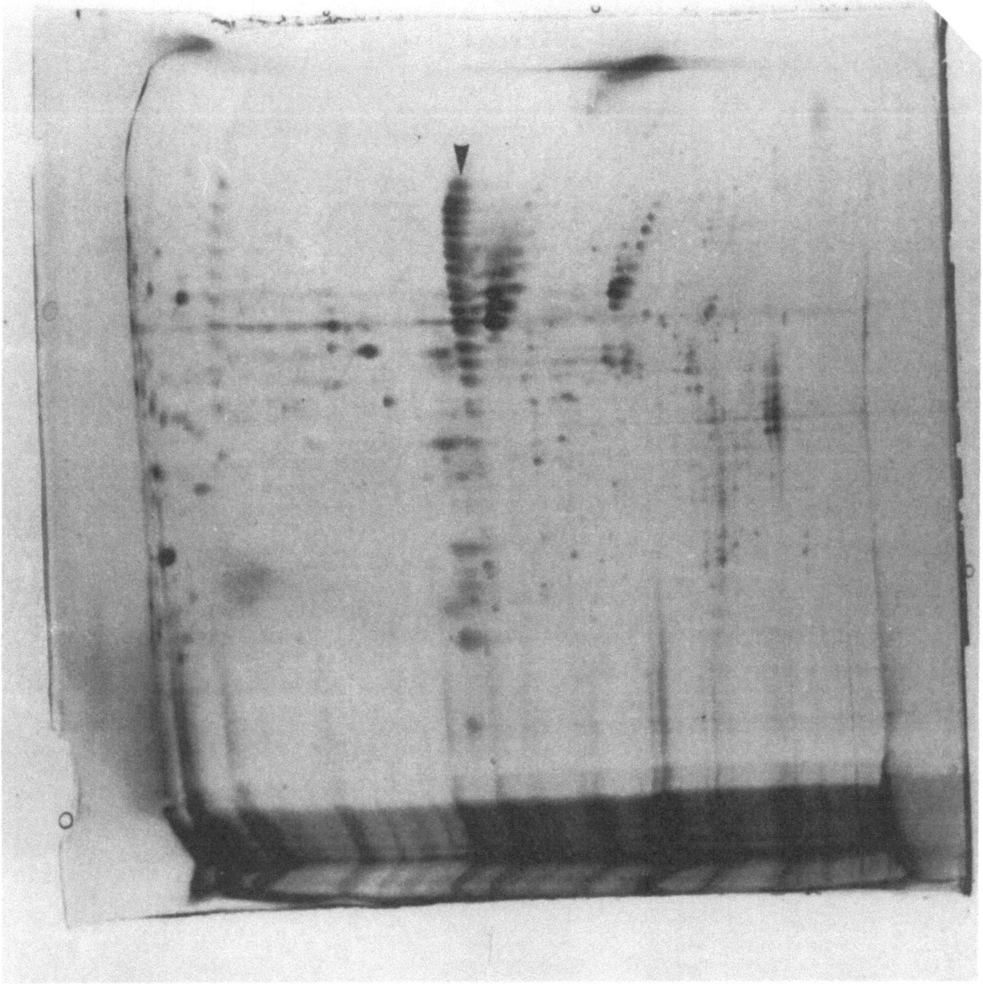

Figure 4 Iso-Dalt gel of urine proteins from patient with renal transplant (a) 10 weeks post-transplant (b) 11 weeks post-transplant and improved clinical condition.

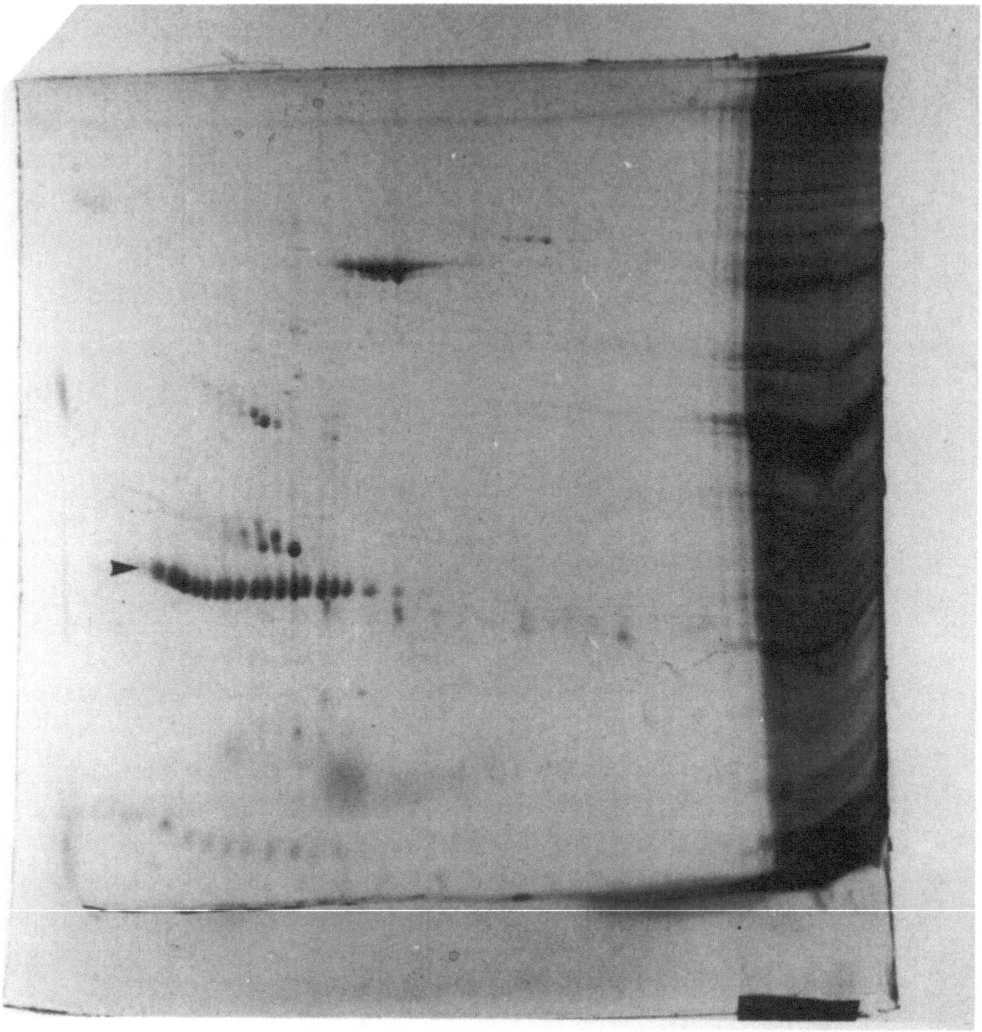

(b)

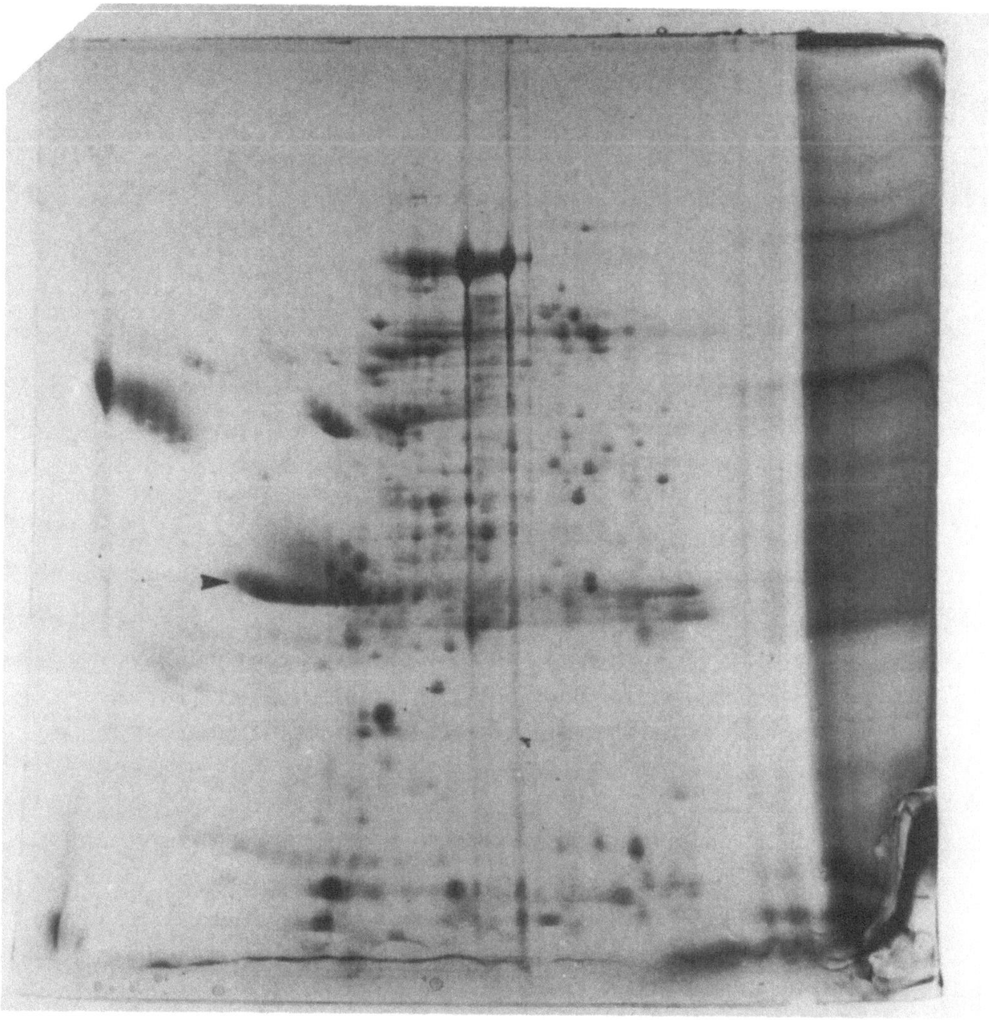

Figure 5 Iso-Dalt gel of urine proteins from patient with renal transplant. Biopsy of transplanted kidney indicated tubular damage due to cyclosporin therapy.

molecular weight proteins and the presence of a protein complex only found in one other patient who had previously suffered a mild episode of graft rejection.

A further patient showed a severe proteinuria (Fig 5) reflecting a more complicated clinical picture. There were more proteins found in the middle molecular weight/basic region of the gel and also more low molecular weight proteins in the second specimen studied. This might be due to the progression of the tubular damage due to cyclosporin therapy which had been demonstrated on renal graft biopsy.

In summary although there were some proteins excreted by all groups of patients eg, albumin, transferrin, IgG gamma and light chains, α_1-acid glycoprotein, the excretion of other proteins was variable even within groups of patients with similar renal pathologies. A decrease in the number of proteins excreted with time after transplantation was demonstrated and in some, though not all, patients an increase preceeding an episode of graft rejection was found. Further clinical studies are necessary to determine the diagnostic value of these findings.

Acknowledgements

The financial support of the Department of Health and Social Security is gratefully acknowledged.

References

Anderson, N.G. and Anderson, N.L., Anal. Biochem., 85, 331-341 (1978a)

Anderson, N.L. and Anderson, N G, Anal. Biochem., 85, 341-354 (1978b).

Anderson, N.G, Anderson, N.L and Tollakson, S.L, Clin. Chem., 25, 1199-1210 (1979).

BINDING OF ANTICANCER DIBROMO-DULCITOL AND DIANHYDRO-DULCITOL (GALACTITOL) TO NUCLEAR PROTEINS IN ASCITES TUMOUR CELLS

Anna Fónagy, Egon J. Hidvégi
'Frédéric Joliot-Curie" National Research Institute for Radiobiology and Radiohygiene
H-1775 Budapest, Hungary

András Jeney
I. Institute of Pathology and Experimental Cancer Research, Semmelweis University Medical School, Budapest, Hungary

Jenő Szamos, László Institóris
Cancer Research Laboratory, Chinoin Pharmaceutical and Chemical Works Ltd., Budapest, Hungary

Introduction

Dibromo-dulcitol (DBD) and its diepoxy derivative dianhydro-dulcitol (DAD, or elsewhere dianhydro-galactitol, DAG) are widely used anticancer agents (1-6). Studies proved inhibition of cell proliferation and the precursor incorporation into DNA due to DBD or DAD treatment, though to various extent in different types of cells studied (7-13). The inhibitory action has currently been regarded as a result of covalent binding of DBD or DAD to chromatin components (14-16). Experiments also showed reduction in the amount and synthesis of chromatin proteins and impaired interaction between DNA and histone H1 (9, 16). This paper reports on binding of labelled DBD or DAD to certain nuclear proteins of ascites tumour cells. Specific binding to these proteins, involved in regulation of transcription, might explain the altered chromatin function of tumour cells on treatment.

Electrophoresis '82

Results

Novikoff hepatoma ascites tumour strains (Chester Beatty Research Institute, London) were used 4 days after transplantation of 10^7 cells on 100-200 g CFY female rats. ^{3}H-DAD and ^{3}H-DBD were labelled at the carbon of position one of the hexitol skeleton as reported previously (17). Specific radioactivity of ^{3}H-DAD and ^{3}H-DBD were 0.5 mCi (18.5 MBq)/mg. ^{3}H-DBD were administered i.p. in 2%(v/v) Tween-80 suspension. ^{3}H-DAD was dissolved in isotonic saline and given i.p. Nuclei were isolated from washed cells by citric acid technique (18, 19). Nuclei were extracted immediately after preparation with 0.2 M H_2SO_4 at 0 °C (19), protein was precipitated in ethanol and dried in vacuum. Proteins (500 µg) were separated in the first dimension by the method of Panyim and Chalkley (20). Gels were stained with Coomassie Brilliant Blue, scanned with Chromoscan at 620 nm. Stained gels were cut into 1 mm slices, digested in 0.5 ml Soluene-100 at 60 °C for 3 h and radioactivity measured in toluene based scintillation fluid by Nuclear Chicago Mark III. equipment. Slab gel electrophoresis in the second dimension was performed in the presence of SDS by the method of Orrick et al (21). Labelled proteins were detected by fluorography (22, 23). DNA was isolated from nuclei by a modified Marmour method (24) using proteinase-K and RNase digestion. RNA was isolated from nuclei by the method of Steele and Busch (25, 26).

Table I, shows the specific radioactivity of nucleic acids and proteins of nucleus after in vivo treatment of ascites cells with ^{3}H-DBD for 3 h and 20 h. In each experiment significantly more counts were found in acid soluble nuclear proteins than in nucleic acids. Slight amount of radioactivity was detected only in isolated DNA. As time elapsed the labelling increased in nucleic acids while decreased in proteins. The majority of the radioactivity derived from ^{3}H-DBD showed association to some high molecular weight protein fractions and histones. After ^{3}H-DBD labelling for 20 h, the radioactivity was less in each protein fraction than after 3 h labelling. All the histones ex-

Table I
Binding of ^{3}H-DBD to nuclear RNA, DNA and acid soluble protein

^{3}H-DBD		SPECIFIC RADIOACTIVITY		
dose	labelling time	nuclear RNA	DNA	nuclear protein
10 mg/kg	3 h	263 (0.77)	ND	3680 (10.8)
10 mg/kg	20 h	612 (1.79)	ND	1940 (5.6)
50 mg/kg	3 h	662 (1.94)	190 (0.56)	5927 (17.3)
50 mg/kg	20 h	1942 (5.68)	441 (1.29)	2646 (7.7)

Animals bearing Novikoff hepatoma ascites tumour cells were injected with H-3-DBD in doses of 10 mg/kg or 50 mg/kg body weight, i.p. Three or 20 h later the specific radioactivity of nucleic acids or proteins extracted from nuclei was measured. The specific radioactivity is expressed in dpm/mg nucleic acids or proteins and data in pmole/mg are also given in parenthesis. ND: means that the radioactivity of the sample was less than 50 dpm/mg.

cepting histone H1 were labelled higher with ^{3}H-DAD than with ^{3}H-DBD. Comparing the binding of DBD and DAD to histones one may speculate on possibilities, namely, that DBD is more likely to be bound to histone H1, being in the internucleosomal linker region, whereas DAD to histones H3, H2A, H2B, H4 in the core particle. Both drugs resulted high labelling of two high molecular weight protein groups between fractions 18-21 and 5-9, respectively (Fig.1).

Further on, the ^{3}H-DAD labelled acid extractable nuclear proteins were fractionated by two-dimensional polyacrylamide gel electrophoresis using SDS in the second dimension (Fig.2). Radioactivity was detected by fluorography in the high molecular weight protein groups, in histones H3, H2A, H2B, H4 and in protein A10. The molecular weights of labelled drug, binding high molecular weight protein groups (marked on figure with arrow heads) were calculated and found to be $55\text{-}75 \times 10^3$ and $90\text{-}100 \times 10^3$ (Fig.2B).

On ^{3}H-DBD tratment for 20 h and especially on ^{3}H-DAD for 3 h, the amount of the protein became prominent between 42-44 fractions in one dimensional gel (Fig.1) i.e. between histones H1

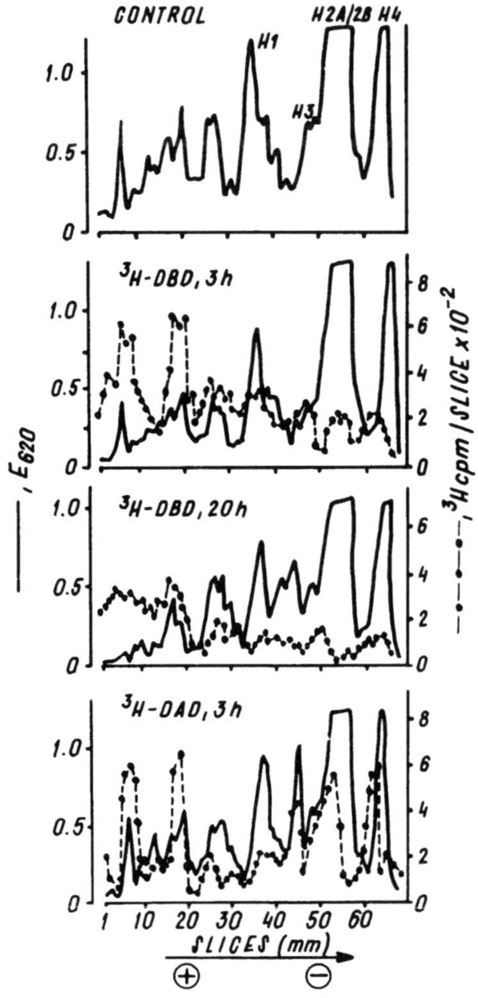

Figure 1. Binding of ^{3}H-DBD and ^{3}H-DAD to acid soluble nuclear proteins. One dimensional gel electrophoresis. Rats bearing Novikoff hepatoma ascites tumour cells were labelled with 10 mg/kg H-3-DBD for 1 or 20 h, or with 10 mg/kg H-3-DAD for 3 h, i.p. Proteins were fractionated on 10%(w/v) acid-urea polyacrylamide gel. Solid line shows the Chromoscan tracing at 620 nm, dotted line is the radioactivity in 1 mm slices. Direction of migration is indicated. The pattern shows histones at the faster moving end and high molecular weight nuclear proteins at the left part of the pattern.

and H3, as compared to untreated control. This protein fraction showed also a significant drug binding. In the two-dimension separation this fraction corresponds mainly to protein A10 (21, 27) which became prominent on drug treatment (cf.Fig.2A and B). The existence of this protein in Novikoff hepatoma ascites cells has already been reported (21, 27), but we found it only a small or undetectable amount if cells were harvested before 5 days after transplantation (Fig.2A). However, it became prominent at some later stage of cell growth (data unpublished) and on this cytostatic drug treatment. This suggests that the protein A10 could be a degradative product of cell growth or it may relate to cell proliferation control.
The low value of binding to DNA (Table I) does not exclude that DNA binds these drugs. Mass spectrometry proved covalent binding of DAD derivative to guanine of DNA. One of the identified derivatives 1,6-dideoxy-1,6-di(guanine-7-yl) galactitol, originated from the crosslink of two guanines of antiparallel DNA chains (28). Sequences of GGCC in DNA especially bind DAD,

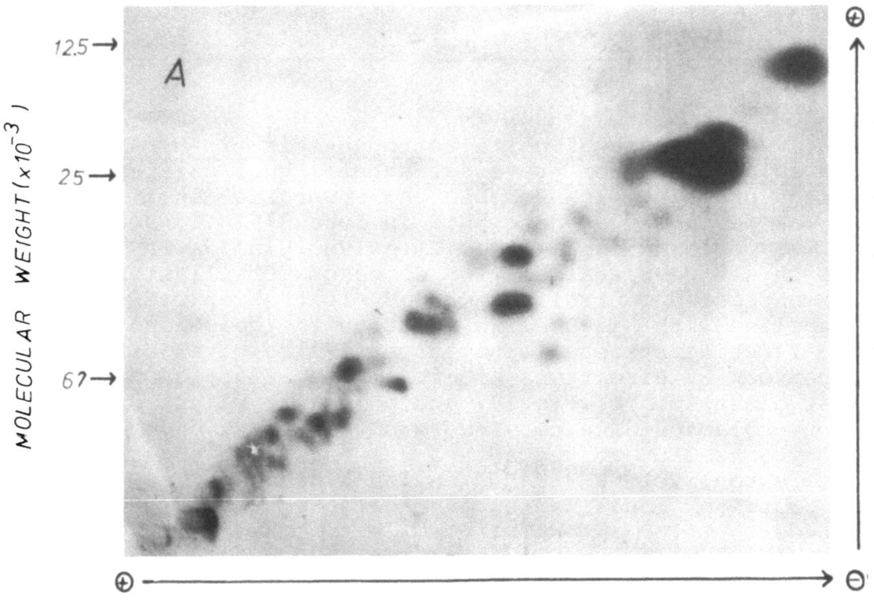

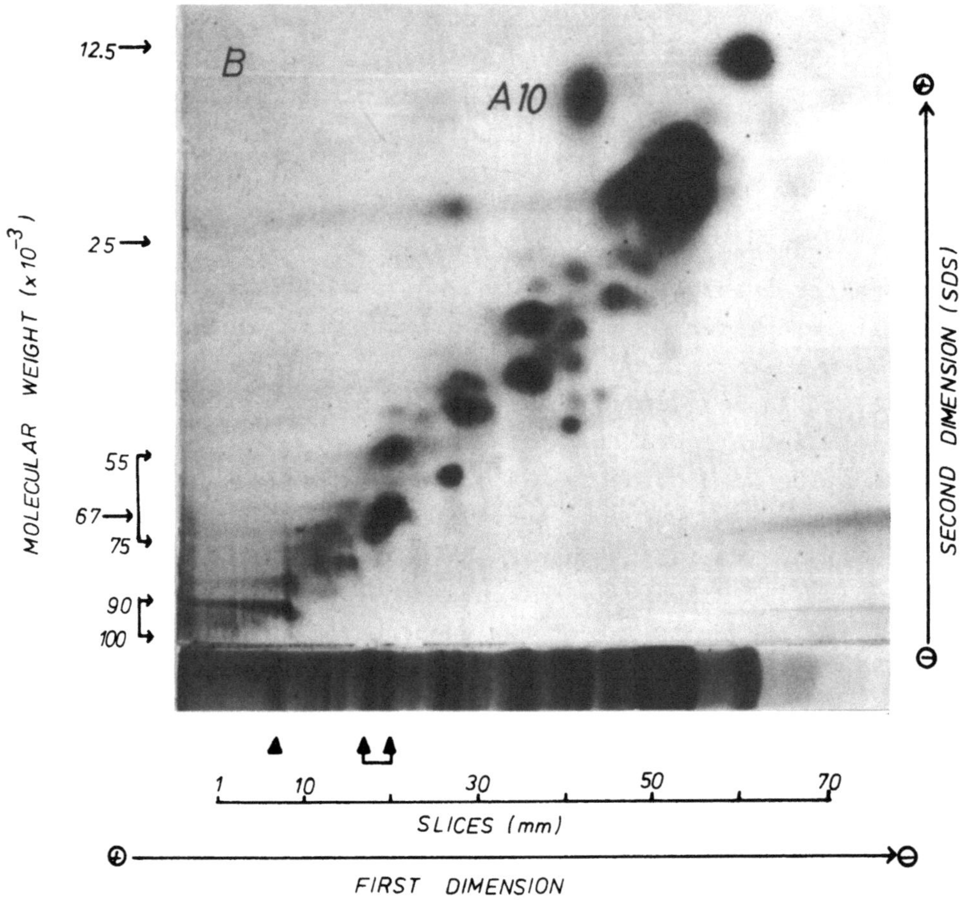

Figure 2. Two-dimensional gel electrophoresis of ^{3}H-DAD labelled acid soluble nuclear proteins. Labelling of acid extractable nuclear proteins with H-3-DAD and separation in first dimension was performed as described in legend Fig.1. Second dimension was performed on a 12%(w/v) polyacrylamide SDS slab gel. Proteins in gel were stained with Coomassie Brilliant Blue. A, is an untreated control pattern. B, is the separation of proteins labelled with H-3-DAD. In addition to the second dimension the first dimensional separation is also given at the bottom of pattern B. Directions of migrations, fraction numbers in the first dimension (cf.Fig.1), and molecular weight of protein markers (β-galactosidase, bovine serum albumine, chymotrypsinogen and cytochrom C) at the second dimension are shown by arrows. The extensively H-3-DAD labelled high molecular weight nuclear proteins, localized by fluorography are indicated with arrow heads.

since this adduct inhibited the cleavage of the restriction sites for enzyme BspI, though it does not inhibit restriction enzymes of other soecificity (29).

The significantly higher binding ability of DBD and DAD to certain nuclear proteins than to nucleic acids, however, suggests that these proteins play an important role in the action of these drugs. Namely, when histones and particularly the proteins of two high molecular weight non-histone groups bind the drug, they became modified, by this means, they are altered in their function, in the regulation of gene transcription.

References

1. Sellei, C., Eckhardt, S., Horváth, I.P., Kralovánszky, J., Institóris, L.: Cancer Chemother. Rep. 53, 377-383 (1969).
2. Németh, L., Institóris, L., Somfai, S., Gál, F., Pályi, I., Sugár, J., Csuka, O., Szentirmay, Z., Kellner, B.: Cancer Chemother. Rep. 56, 593-602 (1972).
3. Andrews, N.C., Weiss, A.J. Wilson, W., Nealon, T.: Cancer Chemother. Rep. 58, 653-660 (1974).
4. Eagan, R.T., Moertel, C.G., Hahn, R.G., Schutt, A.J.: J. Natl. Cancer Inst. 56, 179-181 (1976).
5. DeJager, R., Brugarolas, A., Hansen, H., Cavalli, F., Wennerholm, A., Kenis, Y., Alberto, P.: Proc. Am. Soc. Clin. Onc. 18, 295-301 (1977).
6. Mischler, N.E., Earhart, R.H., Carr, B., Tormey, D.C.: Cancer Treatment Review 6, 191-204 (1980).
7. Hidvégi, E.J., Lónai, P., Holland, J., Antoni, F., Institóris, L., Horváth, I.P.: Biochem. Pharmac. 16, 2143-2153 (1967).
8. Vályi-Nagy, T., Jeney, A., Szabó, J., Szabó, I., Institóris, L.: Eur. J. Cancer 5, 403-414 (1969).
9. Jeney, A., Szabó, I., Vályi-Nagy, T., Institóris, L., Szabó, J.: Eur. J. Cancer 6, 297-302 (1970).
10. Börzsönyi, M., Lapis, K., Varga, L., Várterész, V.: Arzneimittel-Forsch. 19, 669-671 (1969).
11. Hidvégi, E.J., Sebestyén, J., Szabó, L.D., Köteles, G.J., Institóris, L.: Biochem. Pharmac. 25, 1705-1710 (1976).

12. Baló-Banga, J.M., Institóris, L., Hidvégi, E.J.: Studia Biophys. 50, 73-81 (1975).

13. Pályi, I., Cancer Chemother. Rep. 59, 493-499 (1975).

14. Institóris, E., Holczinger, L., Bánfi, D.: Z. Krebsforsch. 82, 101-107 (1974).

15. Jeney, A., Dzurillay, É., Lapis, K., Institóris, L.: Chemotherapy, Eds. Hellmann, K., Connors, T.A. Vol. 8, pp. 145-151, Plenum Publ. Corp., New York (1976).

16. Jeney, A., Dzurillay, É., Lapis, K., Valkó, É.: Chem.-Biol. Interactions 26, 349-361 (1979).

17. Institóris, L., Dzurillay, É., Pethes, G.: Z. Krebsforsch. 79, 49-57 (1973).

18. Higashi, K., Narayan, K.S., Adams, H.R., Busch, H.: Cancer Res. 26, 1582-1590 (1966).

19. Taylor, C.W., Yeoman, L.C., Daskal, I., Busch, H.: Exptl. Cell. Res. 82, 215-226 (1973).

20. Panyim, S., Chalkley, R.: Biochemistry 8, 3972 (1969).

21. Orrick, L.R., Olson, M.O.J., Busch, H.: Proc. Natl. Acad. Sci. USA 70, 1316-1328 (1973).

22. Bonner, W.M., Laskey, R.A.: Eur. J. Biochem. 46, 83 (1974).

23. Laskey, R.A., Mills, A.D.: Eur. J. Biochem. 56, 335 (1975).

24. Marmur, J.: J. Mol. Biol. 3, 208-219 (1961).

25. Steele, W.J., Okamura, N., Busch, H.: J. Biol. Chem. 240, 1742-1749 (1965).

26. Steele, W.J., Busch, H.: Methods in Cancer Research, Ed. Busch, H. Vol. III, pp. 61-92, Academic Press, New York (1967).

27. Yeoman, L.C., Taylor, W.C., Busch, H.: Biochem. Biophys. Res. Commun. 51, 956-966 (1973).

28. Institóris, E., Tamás, J.: Biochem. J. 185, 659-666 (1980).

29. Financsek, I., Fónagy, A., Hidvégi, E.J.: Proc. 6th Internatl. Rad. Res. p. 234, Tokyo (1979).

A SIMPLE ONE DIMENSIONAL GEL PROCEDURE FOR DETERMINATION OF THE EXTENT OF PHOSPHORYLATION OF RIBOSOMAL PROTEIN S6

Peter Nielsen, Keith Manchester, Harry Towbin, Julian Gordon and George Thomas

Friedrich Miescher-Institut, P.O. Box 273, CH-4002 Basel, Switzerland

Introduction

The extent of phosphorylation of the ribosomal protein S6 is usually measured using two dimensional gel systems which apparently separate the mixture of phosphorylated derivatives according to the number of phosphates each derivative contains (1,2). Most two dimensional systems require several hundred µg total ribosomal protein per sample, making analysis difficult for cells/tissues where the ribosomal content is low. Moreover, the ribosomal proteins need to be extracted free from RNA.

We describe here the use of a modified one dimensional Laemmli SDS gel which separates the different phosphorylated forms of S6 in one dimension. The protein S6 is specifically stained by use of a monoclonal antibody after electrophoretic transfer to nitrocellulose.

Procedure

Two dimensional basic gels were modified Kaltschmidt and Wittman gels as described (3) except that the electrophoresis time in the first dimension was 14.5 h at 225 V and in the second dimension 19 h at 150 V. The one dimensional urea-SDS gel was that of Laemmli (4) with urea added to the stacking and separating gel solution as well as to the sample buffer to give a

Electrophoresis '82

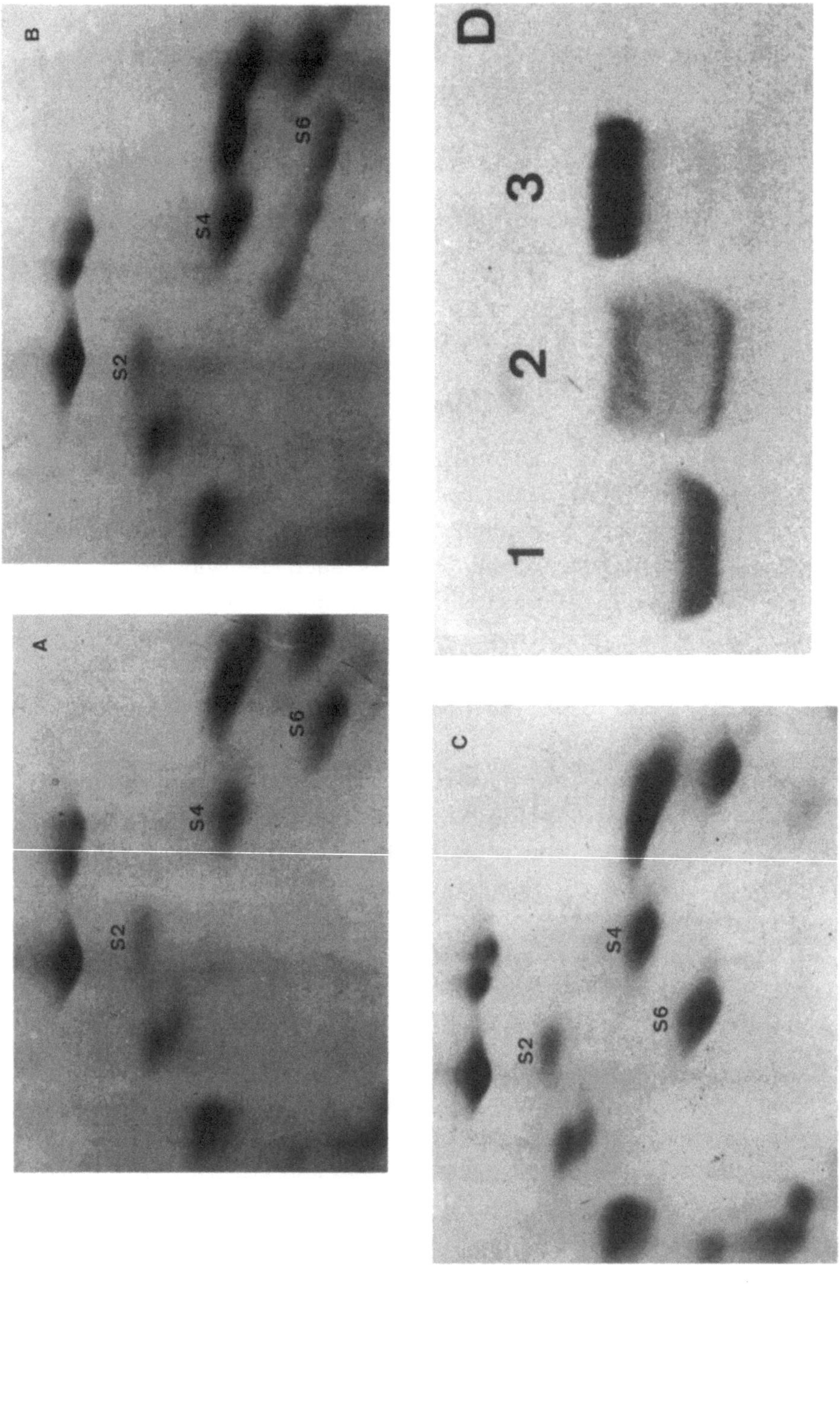

Fig. 1

Fig. 1. Separation of phosphorylated S6 derivatives on one and two dimensional gels. Confluent 3T3 monolayers were stimulated by the addition of 2.5% or 10% foetal calf serum. Two h later cells were harvested, ribosomal proteins prepared and separated using either two dimensional gels (A-C) or a one dimensional urea-SDS gel (D). The protein in the two dimensional gels (200 µg per gel) was stained with Coomassie blue and the protein in the one dimensional gel (20 µg per sample) was transferred to nitrocellulose and stained immunologically. For the two dimensional gels electrophoresis relative to the figure was from left to right in the first dimension and top to bottom in the second dimension. A, untreated culture; B, 2.5% foetal calf serum; C, 10% foetal calf serum; D, lane 1, same sample as in A; lane 2, same sample as in B; lane 3, same sample as in C.

final concentration of 6 M. The separating gel was 12% acrylamide and 0.34% bisacrylamide, pH 8.8, and gels were run for 18 h at 120 V and 4°.

Electrophoretic transfer of protein from gels to nitrocellulose was as described by Towbin et al. (5) except that the Tris-glycine buffer was modified by inclusion of 0.01% SDS and exclusion of methanol. Electrophoresis was for 15 min at 60 V. The nitrocellulose blot was saturated by incubation with shaking with 10% horse serum in TBS (150 mM NaCl, 20 mM Tris-Cl, pH 7.5) for 1 h at room temperature. It was subsequently incubated with shaking at ambient temperature for 12-16 h in TBS which contained 2% horse serum and a monoclonal antibody to S6 (a 1:1000 dilution of ascitic fluid), then rinsed several times with TBS. After further incubation for 2 h with a 1:1000 dilution of rabbit anti-mouse antibodies coupled to horseradish peroxidase (DAKO, Copenhagen) in 2% horse serum in TBS, the blot was rinsed several times with TBS and stained in a mixture of 0.6 ml of 0.3% 4-chloro-1-naphthol in methanol and 5 µl of 30% H_2O_2 per 10 ml of TBS. The staining reaction was stopped by rinsing the blot with distilled water.

Ribosomal proteins for two dimensional gels were extracted from polysomes by magnesium precipitation and acetic acid solubilisation according to Hardy et al. (6). Approx. 200 μg of protein (corresponding to the protein extracted from 5 A_{260} of polysomes) was applied per gel. For the one dimensional gels, 0.5 A_{260} of polysomes per sample were added directly to the urea-containing sample buffer.

Results

Fig. 1 compares the profiles on one and two dimensional gel systems of ribosomal proteins from 3T3 cells where the level of S6 phosphorylation was known to be low (serum arrested),

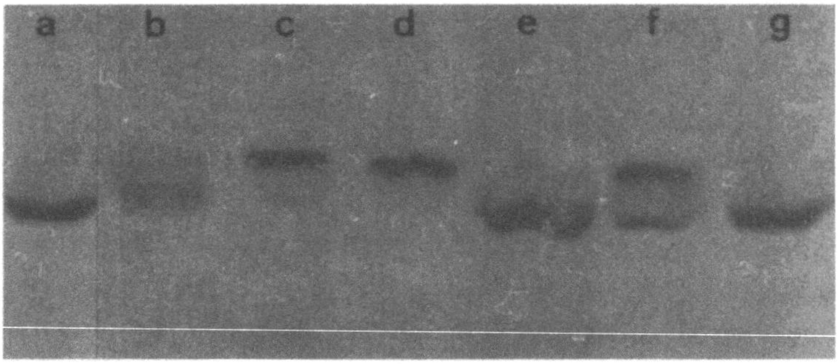

Fig. 2 (above). Effects of cycloheximide on S6 phosphorylation in various rat tissues. The anti S6 antibody stained protein blot of a one dimensional urea-SDS gel is shown for liver following saline injection (a), and for liver (b), heart (c), diaphragm (d), kidney (e), skeletal muscle (f) and brain (g), following cycloheximide injection. In all tissues examined from saline treated rats, S6 migrated to the non-phosphorylated position (only the result from the liver of saline-treated rats is shown in lane a).

intermediate (2.5% serum stimulated) or high (10% serum stimulation) (G. Thomas, unpublished). Minimally phosphorylated S6 migrated as a single Coomassie-blue stained spot in the two dimensional system (Fig. 1A) and as a single anti-body stained band in the one dimensional system (Fig. 1D). Maximally phosphorylated S6 migrated as a single spot displaced towards the origin (anode) in both dimensions of the two dimensional system (Fig. 1C), and in the one dimensional system as a single band migrating more slowly than the minimally phosphorylated S6 (Fig. 1D). A distribution of stained S6 between the two extreme positions is seen in both gel systems with the preparation which contained S6 phosphorylated to an intermediate level (Figs 1B and 1D).

Use of the one dimensional procedure allows the relatively easy screening of S6 phosphorylation in a variety of tissues. For example, it has been previously reported that injection of cycloheximide causes stimulation of S6 phosphorylation in rat liver (7). Fig. 2 verifies this and extends the findings to other tissues and organs. Varying degrees of stimulation of S6 phosphorylation were observed in liver (Fig. 2B), heart (Fig. 2C), diaphragm (Fig. 2D), kidney (Fig. 2E), skeletal muscle (Fig. 2F), but not in brain (Fig. 2G). Interestingly in heart and diaphragm the phosphorylation of S6 in response to cycloheximide was particularly strong. Injection of buffer with no cycloheximide resulted in no observable phosphorylation of S6 in liver (Fig. 2A), nor in any of the other tissues examined in Fig. 2 (not shown).

Diabetes has been reported to stimulate S6 phosphorylation (8). However, we have been unable to detect any sign of S6 phosphorylation in liver, kidney and skeletal muscle of diabetic rats (Fig. 3), whereas some evidence of S6 phosphorylation was seen in liver, though not in other tissues, of diabetic and normal rats treated with insulin. This stimulation of S6 phosphorylation in liver following insulin injection was subsequently

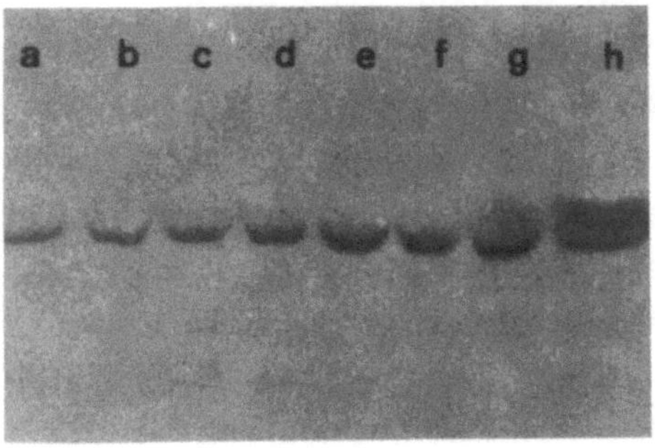

Fig. 3. Analysis of S6 phosphorylation in diabetic and insulin treated rats. A protein blot is shown which was immunologically stained for S6 derived from (a) kidney, (c) skeletal muscle and (f) liver of diabetic rats, and (b) kidney, (d) muscle and (g) liver from insulin-treated diabetic rats. Control rat liver is shown in (e), and as an internal control S6 from liver of cycloheximide-treated rats is included in lane h.

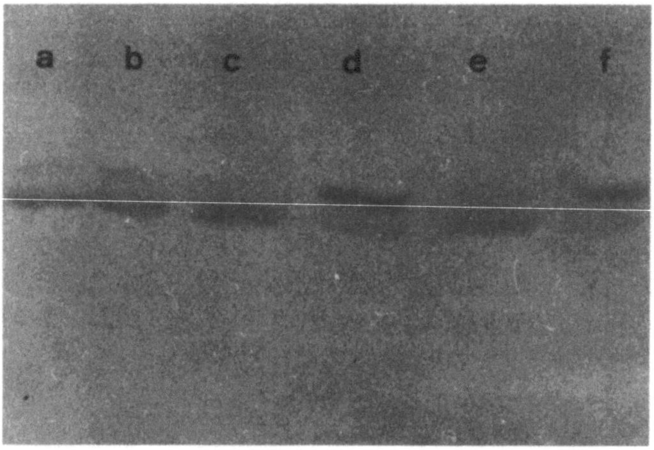

Fig. 4. Phosphorylation of S6 following denervation of rat diaphragm. An immunologically stained protein blot of S6 derived from the denervated and the contralateral non-denervated hemidiaphragms of rats is shown. Lanes b, d and f are from denervated diaphragm 3 h, 6 h and 5 days after denervation respectively. Lanes a, c and e are the corresponding non-denervated portions of the diaphragm 3 h, 6 h and 5 days after the operation.

confirmed by two dimensional gels (not shown).

That the method is of particular use for assay of S6 phosphorylation in tissues of relatively small mass and low ribosome concentration is shown by the results in Fig. 4 which demonstrate the intense phosphorylation of S6 observed in denervated diaphragm muscle following unilateral denervation. Following nerve section the denervated diaphragm undergoes a transient hypertrophy (9) coincident with the phosphorylation of S6.

Discussion

The advantages of the one dimensional method over the standard two dimensional procedure are 1) simultaneous analysis of many samples on one gel, 2) lower amount of material (20 µg total ribosomal protein) required for each sample probably because of less loss during electrophoresis and because of sensitive staining procedure, 3) minimal sample purification before loading, and 4) greater speed due to less sample preparation and only one gel dimension to run.

An important step in the procedure is the transfer to nitrocellulose. Experiment shows that only approx. 25% of S6 run on the gel ends up on the nitrocellulose. Any improvement to recovery would further increase the sensitivity of the method.

References

1. Traugh, J. A., Porter, G. G.: Biochemistry 15, 610-616 (1976).
2. Gressner, A. M., Wool, I. G.: J. Biol. Chem. 249, 6917-6925 (1974).
3. Ramjoue, H. -P., Gordon, J.: J. Biol. Chem. 252, 9065-9070 (1978).
4. Laemmli, U. K.: Nature (London) 227, 680-685 (1970).

5. Towbin, H., Staehelin, T., Gordon, J.: Proc. Natl. Acad. Sci. USA 76, 4350-4354 (1979).

6. Hardy, S. J. S., Kurland, C. G., Voynow, P., Mora, G.: Biochemistry 8, 2897-2905 (1969).

7. Gressner, A. M., Wool, I. G.: Biochem. Biophys. Res. Commun. 60, 1482-1490 (1974).

8. Gressner, A. M., Wool, I. G.: Nature (London) 259, 148-150 (1976).

9. Manchester, K. L., Harris, E. J.: Biochem. J. 108, 177-183 (1968).

SECTION IV BIOLOGICAL AND BIOMEDICAL APPLICATIONS

COMPARISON OF ACTIVITY AND PROTEIN PROFILES OF ENDOTOXINS EXTRACTED AFTER VARIOUS INTERVALS OF BACTERIAL GROWTH ON SOLID MEDIUM

Bernhard Ditter, Renate Urbaschek, Werner Hörhammer*, Bernhard Urbaschek

Department of Immunology and Serology, Hygiene Institute, University of Heidelberg, Klinikum Mannheim; *Microbiological Quality Control, KNOLL AG, Ludwigshafen, West Germany

Robert C. Allen

Departments of Pathology and Laboratory Animal Medicine, Medical University of South Carolina, Charleston, SC 29464, USA

Introduction

Endotoxins prepared between and within laboratories, often show markedly different biological activities when tested. Careful control of all procedures, from bacterial growth conditions through each step of the preparative procedure does not appear to alleviate much less obviate this problem. Schlecht (1,2) studied differences in content, chemical composition and serological characteristics of phenol/water LPS (3) extracted from different phases of bacterial growth in submersed cultures. Besides the interest to study biological activities of Boivin endotoxin preparations (4,5) extracted at varying periods of bacterial growth, it was the object of this preliminary study to assess if marker proteins could be separated and detected in endotoxin extracts, to determine if a correlation could be made with any such markers, and to determine if qualitative or quantitative changes or a combination of both could be used to assess endotoxin activity.

Since the amounts of protein in endotoxin are very slight, a sensitive assay method was required. The work of Switzer (6), Merril (7) and Allen (8) utilizing diamine silver staining to visualize proteins at the nanogram or lower levels makes such an approach possible.

Electrophoresis '82

Methods and Materials

Preparation of endotoxins

Two hours, 4h, 6h, 10h, 12h, 24h, and 36h after culturing E. coli 0111 in Kolle dishes (in order to suppress the growth of smooth forms) on standard I nutrient agar (Merck, Darmstadt), endotoxins were extracted according to the trichloroacetic acid method of Boivin (4,5). All extraction procedures were standardized according to the wet weight of bacteria harvested. The growth curve illustrated in Fig. 1 was determined by optical density and measured after suspending the bacteria in each Kolle dish in 15 ml saline followed by washing and diluting 1:1000. The colony-forming units were determined from further dilution of these suspensions. As compared to a bacterial growth curve from batch cultures no characteristic logarithmic growth was obtained under these conditions. The relationship between the dry weight of the bacteria to the yield of endotoxin was 2.55% ± 0.25 without any obvious correlation to the time of harvesting the bacteria.

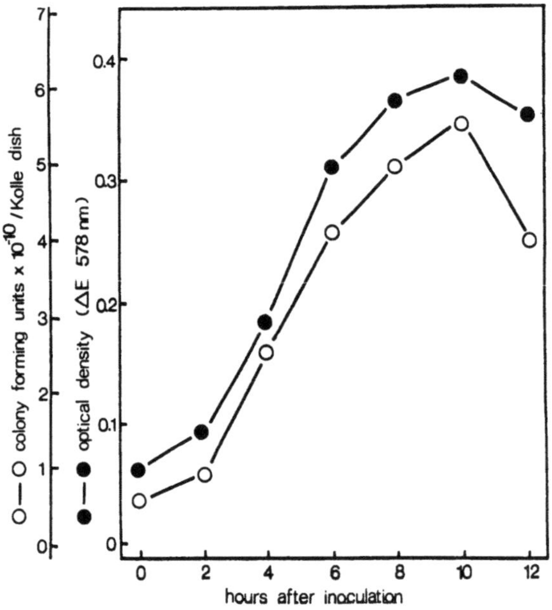

Figure 1. Optical density and colony forming units as a function of culture incubation time.

The protein content of the endotoxins determined by the method of Bradford (9) ranged from 0.8% to 1.3% with an average of 0.92% ± 0.16% of the dry weight.

Toxicity

To determine the toxicity of the different endotoxin preparations - female, 8 to 9 weeks old, NMRI mice were used. Death occurred within 48 hours after injection of 500 µg, 250 µg, 200 µg, 150 µg, 125 µg, 100 µg, and 62.5 µg in groups of ten mice each so that for each endotoxin preparation 70 mice were used. The LD 50 was calculated according to the method of Reed and Muench (10).

Limulus-Amebocyte-Lysate (LAL) Test

A sensitive in vitro assay the LAL-test was used to study the question whether the endotoxin preparations differ in their LAL-activity and if this correlates with their biological activities in more complex experimental models in vivo.

Semiquantitative Tube Test: All endotoxin preparations 10 pg, 20 pg, 40 pg, and 60 pg were suspended in 1 ml pyrogenfree and Limulus-negative distilled water. From each sample 0.1 ml was pipetted into 12 x 75 mm tubes made pyrogenfree by 2 hours exposure to 240° C. One tube without endotoxin served as negative control. After addition of 0.1 ml LAL (endotoxin sensitivity 60 pg/ml) to each sample the tubes were kept vibration-free for one hour at 37° C. Total gelation, indicated by a firm clot after slowly inverting the tubes 180°, was registered as positive reaction. The lack of gelation, or the collapse of the clot after inversion of the tube was considered as negative.

Quantitative Microtiter Test: In order to achieve a subtle quantitative determination of the samples, a quantitative microtiter test was developed. The method has been described in detail (11,12). All endotoxin preparations and two-fold dilutions of the standard endotoxins ranging from 200 pg/ml to 0.39 pg/ml in a volume of 50 µl were pipetted with a multi-channel pipette into Linbro U-bottom microtiter plates (Flow Laboratories). Fifty µl of LAL was added to each well to initiate the reaction. An 8-channel ^{R}titertek multiskan (kindly provided by Flow Laboratories) was used to measure the increase in optical density at 405 nm. The [R]titertek multiskan was kept in a 37° C humified incubator. Measurements were taken at five-minute intervals up to 70 minutes. The maximal increase in optical density within 5 minutes, termed △OD max 5', was determined. A standard curve was obtained with the △OD max 5' of EC2, a reference endotoxin from the FDA, Bethesda, and [R]Novo-Pyrexal, Hermal Chemie (Fig. 2). These were compared with the LAL-activity of given concentrations of the endotoxins under investigation.

Determination of colony-stimulating factor (CSF) and white blood cell count (WBC):

Female 9 to 10 weeks old NMRI mice were injected intravenously with 5 μg/mouse of the different endotoxin preparations. Two hours after injection 10 mice per group were bled from the orbital plexus, the blood pooled per group, and the serum frozen at -20° C until assayed for CSF content. In order to obtain the serum CSF value 50 μl of each sample was plated three times in 35 mm cultures dishes (Falcon) with 1 ml of a femoral bone marrow suspension of 1×10^5 nucleated cells/ml in a semisolid agar according to the method of Bradley (13) and Pluznik (14). The plates were kept for 7 days in a humidified incubator at 37° C and 7.5% CO_2. Accumulations of more than 50 cells were counted as colony forming units, CFUc; the CSF content was expressed as colonies per 1×10^5 bone marrow cells. White blood cell count was determined using a [R]Coulter Counter, ZBI, (Coulter Electronics).

Determination of pyrogenicity in rabbits

The pyrogenicity of the endotoxin preparations extracted from the different growth periods was determined in rabbits according to the European Pharmacopeia. Thermal elements were connected with the central measuring devices of the Microbiological Quality Control (Knoll AG, Ludwigshafen, precision ± 0.05° C). During the test period, the female rabbits were kept in a separate room at a constant temperature without food or water. Rectal temperatures were measured at 5-minute intervals from 90 minutes before until 4 hours after i.v. injection of 1 ml/kg of the endotoxin samples into 3 rabbits each. First, the 10-hour-endotoxin preparation and [R]Novo-Pyrexal were compared in concentrations of 0.1 ng, 1.0 ng, and 10 ng per kg of body weight. Then 1.0 ng/kg of the different endotoxin preparations were tested to compare the pyrogenicity of each. The total increase in temperature as compared to the pre-injection temperature was expressed by addition of the maximal increase in temperature as compared to the pre-injection temperature of the three rabbits in each group.

Isoelectric focusing

Samples were adjusted to 1 µg protein per µl volume and then extracted with carbon tetrachloride using 10 µl of CCl_4 to 40 µl of endotoxin in distilled water. The mixture was agitated on a Vortex mixer for one minute and allowed to settle for 5 to 10 minutes. The aqueous layer was then used for sample application of 5 µl samples on 3 x 5 mm filter paper tabs.

Isoelectric focusing was carried out on 200 µm thick ultrathin-layer 6% polyacrylamide gels covalently bound to glass plates with ^{R}Polyfix 1000. Separations were carried out on pH 3-10 gradients using 4% ^{R}Pharmalyte (Pharmacia) and pH 3.5 to 10 Ampholine (LKB Produkter) using a "Cold Focus" apparatus (MRA Corporation) with a 6 cm electrode distance.

The former seperation was carried out at 1.0 watt for 15 minutes, 2.0 watts for 15 minutes, and 3.0 watts-15 minutes, 5.0 watts-10 minutes, 10 watts-10 minutes, and 20 watts for 2-3 minutes. Voltage gradients ranged from 2.5 V/cm at the start to 550 V/cm at the end of the focusing to final sharpen the bands. Separations with ampholine were carried out similarly, but with power not exceeding 10 watts and the voltage gradient ranging from 30 V/cm to 500 V/cm.

Following isoelectric focusing, the gels and proteins therein were fixed for 10 minutes in 20% TCA, washed 5 minutes in water, dried at 60° C and stained with Coomassie Brilliant Blue R250 at 55° C (8) for 3 minutes. Following destaining, they were dried and then later rehydrated in 3/10-minute changes of 10% alcohol at 55° C. They were then counterstained with silver diamine as previously described (8).

Results and Discussion

Toxicity

All endotoxin preparations were highly toxic, two-to threefold more toxic than compared to data found in the literature. Considering the calculated 50 median lethal doses (LD 50) - 109.5 µg $\pm$ 3.78 per mouse - it can be concluded that the different endotoxin preparations showed minor differences in toxic activity (Table 1). In comparison the LD 50 of ^{R}Novo-Pyrexal used as standard endotoxin is 260 µg per mouse (Hanke, K., Hermal-Chemie, personal communication).

Table 1: LD 50 following injection of different endotoxin preparations in mice within 48 hours

Endotoxins prepared at different times of bacterial growth	LD 50 endotoxin concentration per mouse i.v.
2 h	114 µg
4 h	107 µg
6 h	112 µg
8 h	108 µg
10 h	112 µg
12 h	104 µg

Limulus-Amebocyte-Lysate (LAL) activity

The results of semiquantitative tube test used for comparison of the activity of definite amounts of different endotoxin preparations are summarized in Table 2. At the concentration of 60 pg/ml and 40 pg/ml, all endotoxin preparations were positive. The standard endotoxins resulted in negative reactions with 20 pg/ml whereas with the exception of the 2 hours-endotoxin all preparations were positive. At lower concentrations, 10 pg/ml all reactions were negative. These results indicate that the standard endotoxins both are less active than the endotoxins extracted by us except for the 2 hours-endotoxin preparation. It is also obvious that - which we experienced repeatedly - the LAL with the endotoxin sensitivity of 60 pg/ml is more sensitive than declared by the producer.

Table 2: Semiquantitative tube test with Limulus Amebocyte Lysate (LAL) of different endotoxin preparations in comparison of standard endotoxins

Endotoxin concentration	LAL - reaction							
	Standard endotoxins		Endotoxins prepared at different times of bacterial growth					
pg/ml	EC-2	RNovo-Pyrexal	2h	4h	6h	8h	10h	12h
60	+	+	+	+	+	+	+	+
40	+	+	+	+	+	+	+	+
20	-	-	-	+	+	+	+	+
10	-	-	-	-	-	-	-	-

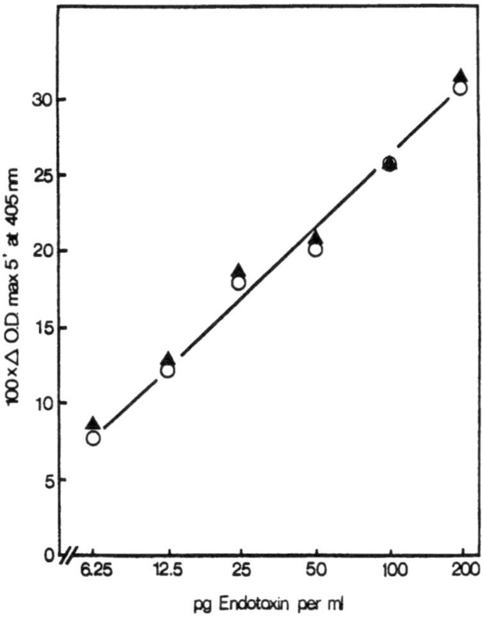

Figure 2. The ΔOD max 5' of the standard endotoxins, EC-2 (○), and RNovo-Pyrexal (▲) at given endotoxin concentrations.

The quantitative assay also was used to determine differences in activity of the endotoxin preparations, extracted after different times of bacterial growth. In Figure 2, the standard curves of EC-2 and ®Novo-Pyrexal are illustrated showing a linearity within a dose range from 6.25 pg/ml to 200 pg/ml. Both endotoxin standards show almost identical activities.

The results from the dilution steps of the endotoxin samples revealed, as compared with the standard endotoxin curves, that all samples except for the 2 hours-endotoxin preparation show higher LAL-activities at a given concentration, 100 pg/ml (Fig. 3). The highest activities were in the 8- and 10-hour-endotoxin preparations, which parallels the intensity of the protein subfraction.

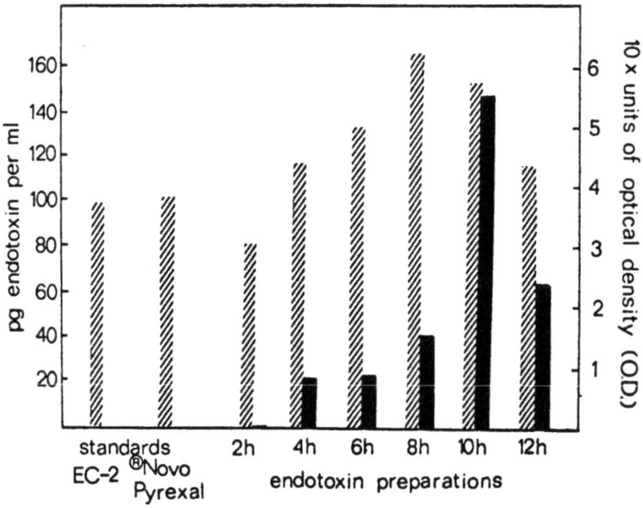

Figure 3. Limulus-Amebocyte-Lysate (LAL) activity of 100 pg/ml endotoxins prepared at different times of bacterial growth in comparison to standard endotoxins and the optical density of the variable peak of 2.1 µg of protein from the endotoxin samples determined by densitometric scanning.

CSF and WBC

The CSF content of serum from mice obtained 2 hours after injection with 5 µg of the different endotoxin preparations, summarized in Table 3, expresses the similarity of results between the different groups. The high activity of all endotoxin preparations was seen from experiments, in which 50 ng/mouse of these preparations induced an evaluating of serum CSF with an average of 99.6 ± 8.7 colonies/1 x 10^5 bone marrow cells. The serum CSF of control mice was 11.0 ± 2.31. Also, the peripheral white blood cell count which shows the characteristic endotoxin-induced decrease did not reveal any differences comparing the effect of the endotoxins prepared at different times of bacterial growth. Although no major differences exist the highest CSF level and the lowest WCB is caused by injection of the 8-hour endotoxin preparation.

Table 3. Colony-stimulating factor and white blood cell count of mice, 2 hours after intravenous injection with 5 µg/mouse of different endotoxins preparations ± se

Endotoxins prepared at different times of bacterial growth	Colonies/10^5 bone marrow cells	White blood cell count
2 h	97.7 ± 3.52	2530 ± 119
4 h	101.1 ± 3.49	3070 ± 425
6 h	97.6 ± 3.04	2980 ± 441
8 h	107.4 ± 3.40	2420 ± 139
10 h	104.5 ± 4.23	2780 ± 388
12 h	101.0 ± 2.47	3170 ± 535
controls	11.0 ± 2.31	5160 ± 640

Pyrogenicity

The pyrogenicity study using different concentrations of the 10-hour endotoxin (Table 4) showed that 1.0 ng/kg is a pyrogenic threshold dose according to the European Pharmacopeia (ranging from 1.15 to 2.65° C) which then was used for comparison of the different endotoxin samples. [R]Novo-

Pyrexal has similar pyrogenic activities as the 10-hour endotoxin. No significant differences were found comparing the pyrogenicity of the endotoxins extracted at different times of bacterial growth, the values are all within the range of pyrogenic threshold doses (Table 5).

Table 4: Determination of pyrogenicity in rabbits at 3 endotoxin concentrations

i.v. doses ng endotoxin/ml per kg body weight	°C increase in temperature of three rabbits 10 hours endotoxin	Novo Pyrexal
0.1	0.6	0.8
1.0	1.4	2.1
10.0	3.8	3.2

Table 5: Determination of pyrogenicity in rabbits with 1 ng/kg of different endotoxin preparations

	endotoxins extracted at different times of bacterial growth					
	2 h	4 h	6 h	8 h	10 h	12 h
total increase in temperature (°C) of three rabbits	1.3	2.1	1.5	1.0	1.3	1.5

Isoelectric focusing

The protein patterns following delipidation indicated a marked qualitative and quantitative change with time of culture incubation. The maximum endotoxin activity was demonstrated at 8 hours while maximal protein changes were shown in 10-hour cultures (Fig. 4a,b). The more acidic proteins showed the greatest change with time with the three major proteins present at 2 hours of culture growth with isoelectric points of 4.68, 4.98, 5.36 increasing up to 10 hours and then decreasing up to 36 hours of culture growth. Two new, more acidic proteins appear with isoelectric points of approximately 4.53 and 4.40 (Fig. 4a,b). The protein band at 5.40 (marked with the arrow), and additionally the protein band with an isoelectric

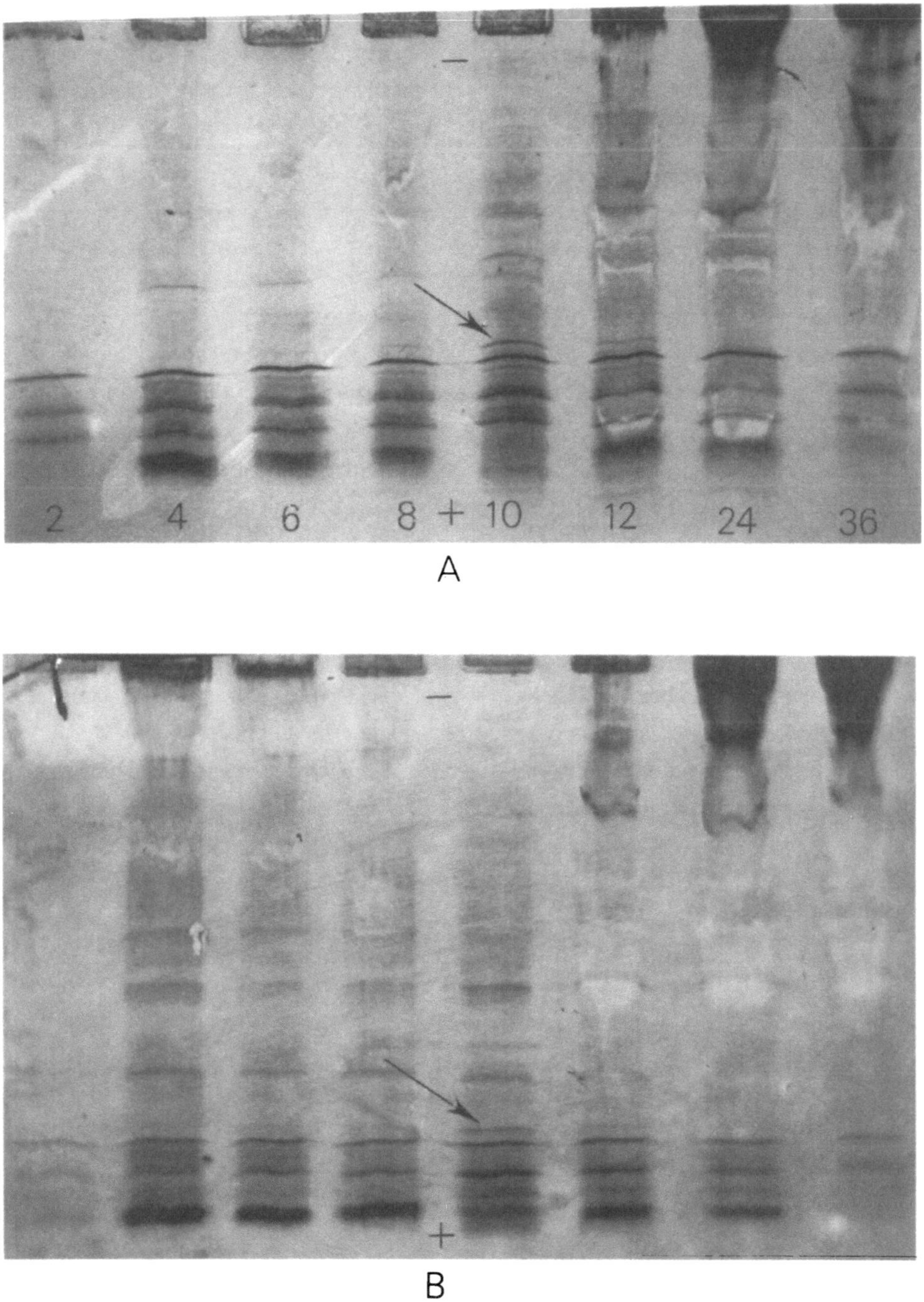

Figure 4a and 4b: Endotoxin protein profiles of 2-36 hour cultures. A, separated on pH 3-10 Pharmalyte and B, separated on 3.5-10 Ampholine; using 6 per cent T gels. Staining was performed with diamine silver.

point of 6.37 also increase over the first 10 hours of incubation and then decrease. Stains with silver diamine indicated a yellow color in these latter bands caused by either a higher concentration, or more likely, from the appearance of the bands less complete delipidation under the standard conditions employed.

Conclusion

Endotoxins extracted at different times after bacterial growth on solid medium did not differ significantly - in the concentrations used - in their biological activities, such as acute toxicity in mice, serum colony stimulating factor, decrease in white blood cell counts and pyrogenicity tested in rabbits. Also the 24 h and 36 h endotoxin - not described - had similar biological activities. However, differences were found in the most sensitive parameters studied, in protein pattern of the endotoxins determined by isoelectric focusing and in their LAL activity. With isoelectric focusing, distinct quantitative changes with culture incubation times of greater than 2 hours are demonstrated. Maximal protein changes were apparent in 10-hour cultures. The endotoxin activity in the quantitative LAL microtiter test increased with endotoxins extracted over the first 8 hours of bacterial growth with a subsequent decrease. A similar course was seen with protein changes of the endotoxins which peak in the 10-hour endotoxin. Whether the changes observed in the band with an isoelectric point of 5.40 correlate with activity or whether the point in time at which the major acid band at 4.68 is reduced with the appearance of the two new, more acidic proteins is not clear from this study. However, the approach, of using endotoxin protein markers to assess endotoxic properties would appear to have a valid application for preliminary assessment of lots of endotoxin since the method is both rapid and inexpensive.

Acknowledgements

The authors wish to acknowledge the excellent technical assistance of Ms. Margaret Ann Simmons, Mrs. Joyce Christopher and Ms. Ruth Breunig.

The limulus amebocyte lysate used was kindly provided by Dr. St. Watson, Associates of Cape Cod, Woodshole, Massachusetts.

This investigation was supported by the Deutsche Forschungsgemeinschaft within the SFB 90.

References

1. Schlecht, S.: Zbl. Bakt. Hyg., I. Abt. Orig. A 232: 61-72 (1975).
2. Schlecht, S. und Fromme, I.: Zbl. Bakt. Hyg., I. Abt. Orig. A 233: 199-222 (1975).
3. Westphal, O., Lüderitz, O. und Bister, F.: Z. Naturforsch. 7b: 148-155 (1952).
4. Boivin, A., Mesrobeanu, J., and Mesrobeanu, L.: C. R. Soc. Biol. 113: 490-492 (1933).
5. Boivin, A., Mesrobeanu, J., and Mesrobeanu, L.: C. R. Soc. Biol. 114: 307-310 (1933).
6. Switzer, R.C., Merril, C.R., and Shifrin, S.: Ann. Biochem. 98: 231-237 (1979).
7. Merril, C.R., Switzer, R.C., and VanKeuren, M.L.: Proc. nat. Acad. Sci., Wash. 76: 4335-4339 (1979).
8. Allen, R.C.: Electrophoresis 1: 32-37 (1980).
9. Bradford, M.M.: Analyt. Biochem. 72: 248-254 (1976).
10. Reed, L.J. and Muench, H.: Amer. J. Hyg. 27: 493-497 (1938).
11. Ditter, B.: Inaugural-Dissertation, Heidelberg (1983).
12. Ditter, B., Urbaschek, R., and Urbaschek, B.: Prog. Clin. Biol. Res. 93: 385-392 (1982).
13. Bradley, R.T. and Metcalf, D.: Austr. J. exp. Biol. med. Sci. 44: 287-300 (1966).
14. Pluznik, D.H. and Sachs, L.: Cell Physiol. 66: 319-324 (1965).

POLYACRYLAMIDE GEL ELECTROFOCUSING OF BETA-LACTAMASES FROM *PROTEUS MIRABILIS*

F. Figueroa, Z. Gonzãlez-Lama, R.H. Lõpez-Orge, A. Sierra
Department of Microbiology, Colegio Universitario de Las Palmas (Divisiõn de Medicina), Las Palmas de Gran Canaria, Spain

Introduction

Beta-lactamases hydrolyse the cyclic amide bond in the beta-lactam ring of penicillins, cephalosporins and related compounds. Analytical isoelectric focusing (IEF) is a high resolution technique for the identification and characterization of beta-lactamases (1), even when more than one beta-lactamase is present in a single strain. The technique has been applied to the study of beta-lactamases in Gram-negative (2,3) and anaerobic bacteria (4,5). It may also be applied to the study of plasmid-mediated beta-lactamases which appear to be genus, species and subspecies-specific in taxonomic studies (8).

We wish to report here the separation of beta-lactamases from a strain of *Proteus mirabilis* ONPG+, isolated in our laboratory from clinical specimens by isoelectric focusing in polyacrylamide gel slabs.

Material and methods

Proteus mirabilis ONPG+ was isolated from a clinical specimen and has been identified by using the API 20E system (API system S.A. France).

It was grown in Brain-Heart Infusion (Difco lab., Detroit, Michigan, USA) at 37°C, 24h, under aeration. Cells of *Proteus mirabilis* were harvested by centrifugation at 800xg for 10 min

Electrophoresis '82

and washed several times with cold distilled water. Cell-free extracts were obtained as described elsewhere (9).

Isoelectric focusing in polyacrylamide gel was performed using the multiphor apparatus (LKB Produkter, Bromma, Sweden), pH range 3.5-9.5 . Beta-lactamase activity in the gel was located using a chromogenic cephalosporin substrate (10) (nitrocefin, Glaxo Research Ltd., Greenford, England). The pH was determined by use of a flat surface electrode at 0.5 cm intervals.

Results and discussion

Figure 1, is a photograph, with counterpart drawing for clarity, demonstrating beta-lactamase activity in cell-free crude extracts of Proteus mirabilis ONPG+ by isoelectric focusing in polyacrylamide gel. We have found ten distinct beta-lactamases, with pI's: 4.9, 5.0, 5.1, 5.2, 5.3, 5.35, 5.4, 5.5, 5.6 and 5.7 . The major component is the isozyme of pI 5.3 . The pI's of some beta-lactamases are: TEM-1 in E. coli (pI: 5.4) (11, 12), TEM-2 in Klebsiella aerogenes (pI: 5.7) (11) and RPL-11 in Pseudomonas aeruginosa (pI: 5.7) (13). Other lactamases have the same pI's as chromosomally-mediated beta-lactamases of Gram-negative bacteria: Pseudomonas aeruginosa (class II chromosomal beta-lactamase, pI: 5.3) (14), Levinea malonatica (constitutive class I chromosomal beta-lactamase, pI: 5.2) (15) and Citrobacter koseri (constitutive class I chromosomal beta-lactamase, pI: 5.0) (15). For the others (pI's: 4.9, 5.1, 5.35 and 5.5) we have not found references.

Our strain of Proteus mirabilis shows resistance to the following beta-lactam antibiotics: penicillin, ampicillin, carbenicillin, cloxacillin, oxacillin, amoxicillin, azlocillin and cephaloridine (seven penicillins and only one cephalosporin). In contrast to the beta-lactamases produced by Gram-positive organisms only a relatively small number of Gram-negative bacteria have been reported to produce chromosomally-mediated class II beta-lactamases, which one distinguishes by their predominant activity against penicillins, and lack of activity against cephalosporins. Chromosomally-mediated class II beta-lactamase has also been reported in Proteus mirabilis GN310 (16). Since strains of Proteus mirabilis usually make negligibly low levels of chromosomally-mediated beta-lactamase (17)one may suspect that high levels of beta-lactamase in this species could be due to the presence of R-factors. Our strain of Proteus mirabilis ONPG+ probably has a chromosomally-mediated class II beta-lactamase because there is a predominant activity against penicillins, and a low activity against cephalosporin. So this strain may have some plasmid-mediated beta-lactamase. These beta-lactamases are numerous, with widespread isoelectric points, and need further characterization.

Figure 1: Isoelectric focusing of beta-lactamases from Proteus mirabilis ONPG+ cell-free extract.

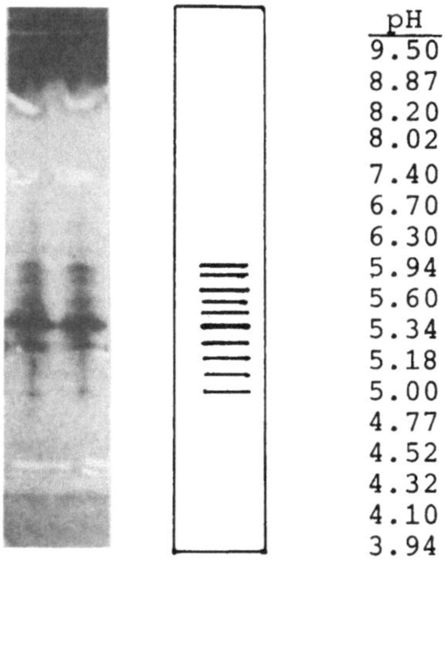

References

1. Bidwell,J.L.,Reeves,D.S.:J.Antimicrob.Chemother.6,793(1980)

2. Labia,R.,Guionie,M.,Barthélémy,M.:J.Antimicrob.Chemother. 7,49-56 (1981)

3. Tajima,M.,Masuyoshi,S.,Inoue,M.,Takenouchi,Y.,Sugawara,S., Mitsuhashi,S.:J.Gen.Microbiol.126,179-184 (1981)

4. Nord,C.E.,Olsson-Liljequist,B.:J.Antimicrob.Chemother.8 33-42 (1981)

5. Timewell,R.M.,Phillips,I.,Soderholm,J.,Nord,C.E.:Antimicrob. Agents Chemother.19,700-704 (1981)

6. Hedges,R.W.,Matthew,M.:Plasmid 2,269-278 (1979)

7. Matthew,M.,Hedges,R.W.,Smith,J.T.: J.Bacteriol.138,657-62 (1979)

8. Matthew,M.,Harris,A.M.:J.Gen.Microbiol.94,57-67 (1976)

9. González-Lama,Z.,Santana,O.E.,Betancor,P.:in Electrophoresis'81,Allen,A.ed.Walter de Gruytier and Co.Berlin.New York, pp.699-702 (1981)

10. O'Callaghan,C.H.,Morris,A.,Kirby,S.M.,Shingler,A.H.:Antimicrob.Agents Chemother.1,283-288 (1978)

11. Richmond,M.H.,Sykes,R.B.:in "Advances in Microbial Physiology"(A.H.Rose and D.W.Tempest eds.)9,31-88 Academic Press London and New York (1973)

12. Matthew,M.,Hedges,R.W.: J.Bact.125, 713-718 (1976)

13. Matthew,M.,Sykes,R.B.: J.Bact.132, 341-345 (1977)

14. Furth,A.,:Biochem.biophys.Acta.377,431-433 (1975)

15. Sykes,R.B.,Matthew,M.:J.Antimicrob.Chemother.2,115-157(1976)

16. Sawai,T.Mitsuhasi,S.,Yamagishi,S.:Jap.J.Microbiol.12,423-434 (1968)

17. Smith,J.T.:J.gen.Microbiol.55,109-120 (1969)

POLYACRYLAMIDE-STARCH GEL ELECTROFOCUSING OF CATALASE FROM PSEUDOMONAS AERUGINOSA

Z.González-Lama,M.J.Cutillas,A.M.Lamas
Departamento Microbiología.Colegio Universitario (División de Medicina)Las Palmas de Gran Canaria. Spain.

E.Meléndez-Hevia
Departamento de Bioquímica.Fac.Biológicas.La Laguna.Spain

Introduction

Adequate defense against oxygen toxicity requires efficient scavenging of both O_2^- and of H_2O_2.Superoxide dismutases,which scavenge O_2 and catalases and peroxidases,which scavenge H_2O_2, provide the necessary defenses.Catalases have been studied in yeast (1,2,3),*Aspergillus niger* (4),*E.coli* (5),*Micrococcus luteus* (5),halophilic vibrios (7),*Pseudomonas fluorescens* (8)and other microorganisms (9,10),but electrofocusing of catalases from microorganisms has not been studied. This paper describes the application of isoelectric focusing to the study of catalase from *Pseudomonas aeruginosa* using a special polyacrylamide-starch gel.

Material and methods

Pseudomonas aeruginosa was isolated from a clinical specimen and grown in Brain-Heart Infusion (Difco lab. Detroit,Michigan USA) at 37°C,24h,under aeration. Cells of *Pseudomonas aeruginosa* were harvested by centrifugation at 800xg for 10 min and washed several times with cold distilled water. Cell-free extracts were prepared as described elsewhere (11).
Isoelectric focusing was carried out in a thin layer polyacryl-

Electrophoresis '82

amide gel mixed with 1% hydrolysed starch, with an ampholine gradient pH: 3.5 - 9.5 .
Isoelectric focusing was performed using the multiphor apparatus (LKB Produkter,Bromma,Sweden).
Catalase activity in the gel was located by:
1) Soaking the slabs of gel after isoelectric focusing in a solution of 2.5 mM H_2O_2 in 10 mM phosphate buffer,pH:7.0 for 15 min.
2) Washing the gel with cold distilled water.
3) Soaking again the gel in a solution:2.5 ml of 10% KI, 10 ml of 2N H_2SO_4 and 0.05 ml of 1% ammonium molybdate.
The isozymes are seen as white zones, due to catalase activity, appearing on a dark blue background.

Results and discussion

Figure 1, is a photograph,with counterpart drawings for clarity, demonstrating catalase activity in the cell-free crude extracts of Pseudomonas aeruginosa by isoelectric focusing in polyacrylamide-starch gel. We have found four distinct catalases, with pI's: 4.2, 5.3, 5.8 and 5.9 . One problem encountered is that focused bands of catalase diffuse rapidly in the gel after development.This precludes satisfactory recording of results, since serial photographs are usually necessary to show the weaker bands which develop more slowly.
Several catalases have been described in yeast(12,13) and other microorganisms (8,14,15). Holmes and Masters (16) demonstrate the presence of four catalases in mammal liver extracts and assume the presence of four isozymes. Seah et al.(17)separated two catalases from Saccharomyces cerevisiae, designated them A and T and considered them as diferent molecular forms of the same enzyme, and other authors (14,15) used the electrophoretic mobility of catalases as a taxonomic criterion of Mycobacteria. Two different catalases were identified in E.coli (5).

The relations between catalase and the activities of other components of the hydrogen peroxidase metabolism, such as superoxide dismutase are presently under investigation (18)

Figure 1: Isoelectric focusing of catalases from _Pseudomonas aeruginosa_ cell-free extract.

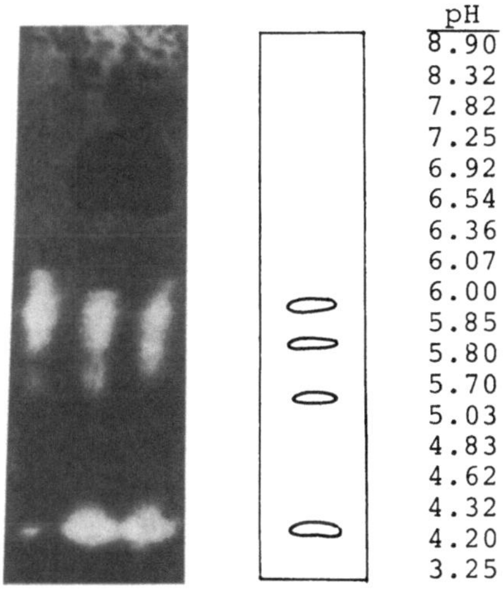

References

1. Seah,T.,Kaplan,J.G.: J.Biol. Chem. 248, 2889-2893 (1973)
2. Zimniak,P.,Hartter,E., Woloszczuk,W.,Ruis,H.:Eur.J.Biochem. 71, 393-398 (1976)
3. Berte,C.,Sels,A.:Molec.gen. Genet. 172, 45-52 (1979)
4. Gruft,H.,Ruck,R.,Traynor,J.: Can. J. Biochem. 56, 916-919 (1978)
5. Hassan,H.M.,Fridovich,I.: J. Biol. Chem. 253,6445-6450 (1978)
6. Marie,A.L.,Priess,H.,Parak,F.:Hoppe-Seyler's Physiol.Chem. 359,857-862 (1978)
7. Daily,O.P.,Debell,R.M.,Joseph,S.W.: J. Bacteriol. 134 375-380 (1978)

8. Rodriguez-Bravo,S.,Pionetti,J.M.:Biochimie, 63,535-540 (1981)

9. Wheeler,P.R.,Gregory,D.: J.Gen.Microbiol. 121,457-464(1980)

10. Austin,F.,Barbieri,J.T.,Corin,R.E.,Grigas,K.E.,Cox,C.D.:Inf. Immun. 33, 372-379 (1981)

11. González-Lama,Z.,Santana,O.E.,Betancor,P.: in"Electrophoresis'81",Allen,A. ed. Walter de Gruyter and Co. Berlin New York,pp.703-707 (1981)

12. Seah,T.,Bhatti,A.R.,Kaplan,J.G.:Fed.Proc.Ann.Soc.Exp.Biol. 31 1614 (1972)

13. Rytka,J.,Sledziewski,A.,Lukaszkiewicz,J.,Bilinski,T.: Mol. Gen.Genet. 160, 51-57 (1978)

14. Gruft,H.,Gaafar,A.H.:J.Chromatogr.56,168-171 (1971)

15. Stavri,H.,Stavri,D.: Arch.Roum.Path.Exp.Microbiol.34,85-89 (1975)

16. Holmes,R.S.,Masters,C.J.:Arch.Biochem.Biophys.109,196-199 (1965)

17. Seah,T.C.M.,Bhatti,A.R.,Kaplan,J.G.:Can.J.Bichem.51,1551-1555 (1973)

18. González-Lama,Z.,Cutillas,M.J.:Cienc.Biol. 6,213-215(1981)

FRACTIONATION OF 7,8-DIHYDROPTEROATE-SYNTHETASE FROM E.COLI BY AFFINITY CHROMATOGRAPHY AND ISOELECTRIC FOCUSING TECHNIQUE

Lothar Bock and Rainer Bartels

Department of Medicinal and Pharmaceutical Chemistry,
Borstel Research Institute, D-2061 Borstel, Germany

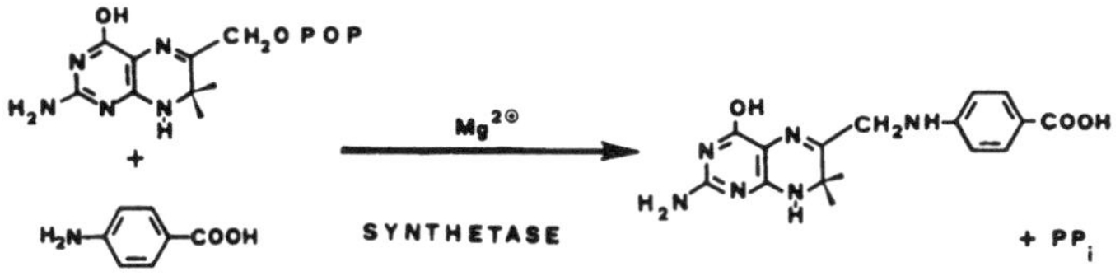

The title enzyme catalyzes the indicated reaction in the folate metabolism of microorganisms (1).

In a first fractionation step common for the two enzymes 7,8-dihydropteridinealcohol pyrophosphokinase and 7,8-dihydropteroate synthetase, the corresponding SYNTHETASE fraction from the Sephadex G-100 gel filtration step (Fraction Ia) was used as a starting material and bound to an affinity column of folate-Sepharose 4B which was equilibrated with 0.2 M tris-HCl buffer pH 8.1.

After a washing step by 0.2 M NaCl in equilibration buffer the enzyme activity was detached from the matrix with the aid of a 250 µM folate-pulse in the washing solution (Fraction II), (Fig. 1). Folate and salt were separated by gel filtration on a Sephadex G-25 column (Fraction III) yielding a degree of protein purification of about 100-fold when compared with Fraction Ia.

Enzyme activity was determined by thc tlc-method (2) and expressed as arbitrary area units of the fluorescent reaction product, dihydropteroate.

Electrophoresis '82

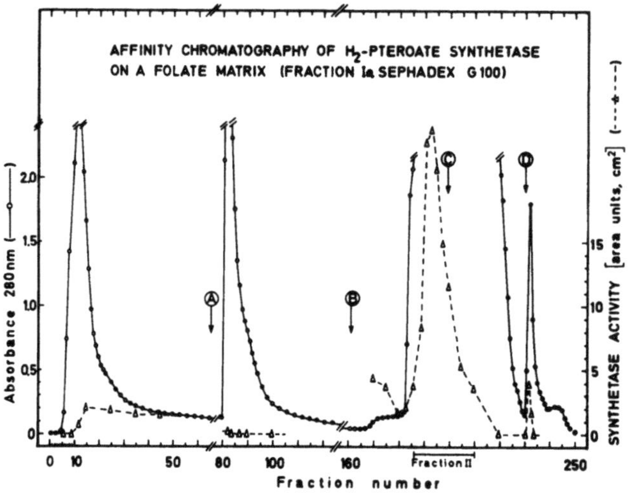

Figure 1. Affinity chromatography of dihydropteroate-synthetase on a folate matrix.
Folate-Sepharose 4B column 2 x 15 cm (3), equilibrated with 0.2 M tris-HCl, pH 8.1; sample load: 13.8 mg protein/ml in a total volume of 49 ml; flow rate: 16.5 ml x cm^{-2} x h^{-1}; fractions: 8.5 ml.
A : wash 0.2 M NaCl; B : pulse 250 µM folate + 0.2 M NaCl;
C : like B but without folate; D : wash 2 M NaCl only.
A,B,C in equilibration buffer.
Fractions with SYNTHETASE activity (Fraction II) were concentrated: a) by ultrafiltration (UF); PM 10 membrane (Amicon)
b) by lyophilization (lyo), in the presence of excess salt.
Separation of folate and excess salts by gel filtration (Sephadex G-25 column) (Fraction III).

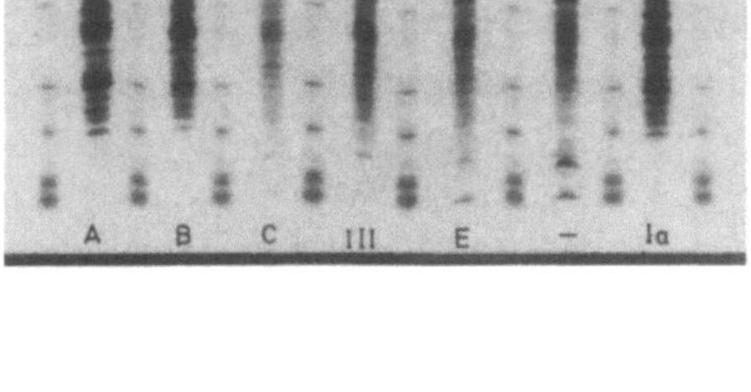

Figure 2. Ultrathin-layer IEF analysis in a 100 µm polyacrylamide gel (50 mm x 95 mm, 5% T, 3% C, 2% carrier ampholytes existing of 20% Servalyt 4-6 and 80% Servalyt 2-11), 900 Vh, 4°C
Ia: Fraction Ia (starting material)
III: Fraction III (fraction 187-210)
A : fraction 5- 79
B : fraction 80-167
C : fraction 168-186
E : fraction 211-230

Fraction III was further purified by flat bed isoelectric focusing in shallow pH gradients stabilized by agarose-Sephadex matrices according to Manrique and Lasky (4). Contact prints were stained for protein and the pH gradients were determined at room temperature. Constant zones of the gel (7 mm) were scraped off and analyzed for SYNTHETASE activity (2). SYNTHETASE activity covers a relatively wide range of pI values from about 4.8 to 5.4 consisting of several well resolved protein bands as shown by the contact prints. The main peak(s) of activity seem(s) to be dependent on the pretreatment of the sample (i.e. ± folate).

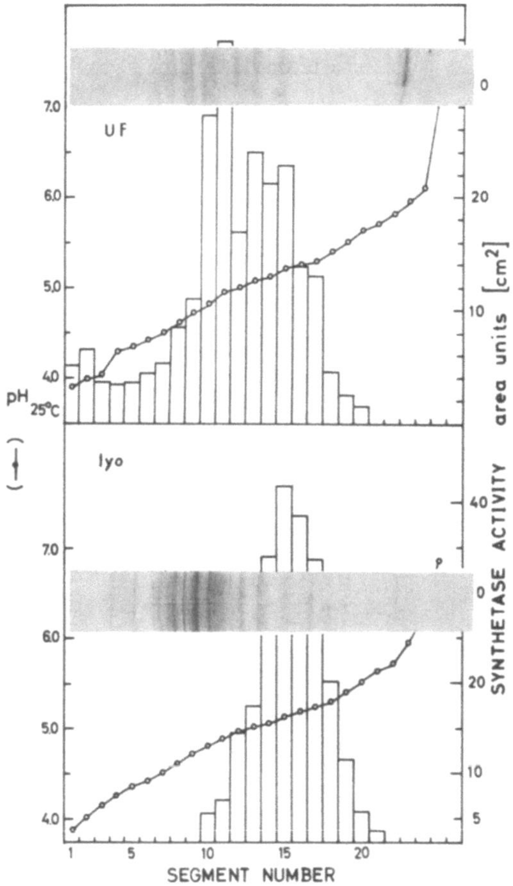

Figure 3. Flat bed IEF of Fraction III; matrix: agarose-Sephadex gel mixture 100 x 180 x 1.8 mm (see ref. 4), except that for a better sample application a slot of 65 x 2 mm was cut out; sample preparation: aliquots of Fraction III ("UF" and "lyo", respectively , adjusted to 2.5 mM tris-HCl, pH 8.1 before lyophilized for storage) were dissolved in the ampholyte fraction received from the matrix slot after prefocusing step (pH 5-6); load: 5 mg protein; ampholyte: 4 % Servalyt AG mixture consisting of 20 % pH 2-11, 40 % pH 4-6, and 40 % pH 5-7; anode: 25 mM aspartic acid, 25 mM glutamic acid; cathode: 2M ethanolamine, 25 mM arginine, 25 mM lysine; temperature: + 10°C; prefocusing: 1000 Vh (mean)
focusing: 14 500 Vh (mean)
contact print: see ref.6

Several enzymatically active zones were saved and the proteins were separated from the contaminating ampholytes by gel filtration on Sephadex G-25 M mini columns. In the following ultrathin layer IEF analysis these distinct protein bands show a rather similar pattern which again cover a broad pI range.

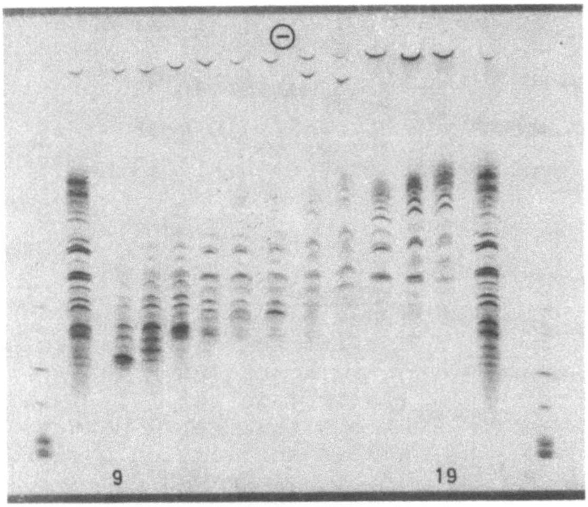

Figure 4. Ultrathin layer IEF analysis in a 100 µm polyacrylamide gel (50 mm x 90 mm, 5 % T, 3 % C, 3 % ampholytes consisting of 20 % Servalyt 4-6 and 80 % Servalyt 2-11); 970 Vh; 4°C;
segment 9 - 19 and Fraction III at both sides

(UF)

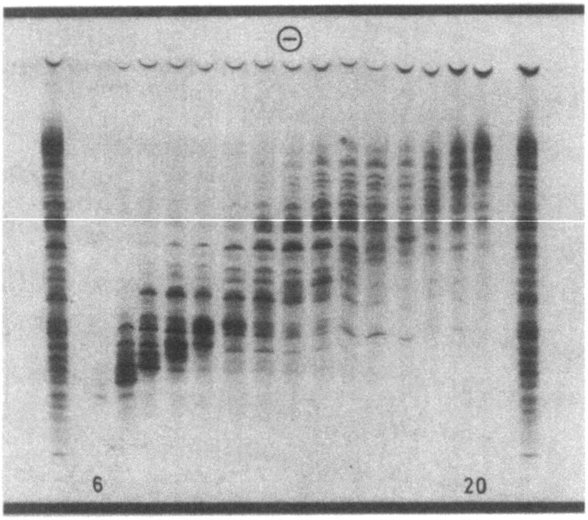

Figure 5. Ultrathin layer IEF analysis in a 100 µm polyacrylamide gel (50 mm x 90 mm, 5 % T, 3 % C, 3 % ampholytes consisting of 50 % Servalyt 4-6 and 50 % Servalyt 2-11); 1130 Vh; 4°C;
segment 6 - 20 and Fraction III at both sides

(lyo)

Therefore changes in protein conformation with or without participation of folate and/or the existence of isoenzymes is proposed for 7,8-dihydropteroate synthetase from E.coli (working hypothesis).

References

1. Richey, D.P., Brown, G.H.: J. Biol. Chem. 244, 1582-1592 (1969)
2. Bock, L., Butte, W., Richter, M., Seydel, J.K.: Anal. Biochem. 86, 238-251 (1978)
3. Then, R.L.: Anal. Biochem. 100, 122-128 (1979)
4. Manrique, A., Lasky, M.: Electrophoresis 2, 315-320 (1981)
5. Radola, B.J.: Electrophoresis 1, 43-56 (1980)
6. Radola, B.J.: Biochim. Biophys. Acta, 295, 412-428 (1973)

PURIFICATION OF 7,8-DIHYDROPTERIDINEALCOHOL PYROPHOSPHOKINASE FROM E.COLI AND CHARACTERIZATION BY ULTRATHIN LAYER IEF

Rainer Bartels and Lothar Bock

Department of Medicinal and Pharmaceutical Chemistry,
Borstel Research Institute, D-2061 Borstel, Germany

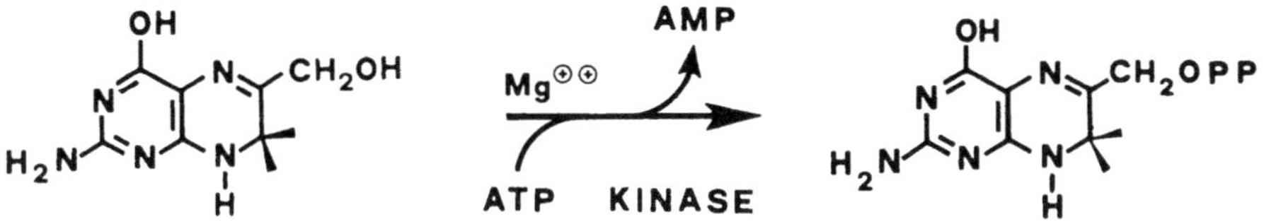

The title enzyme catalyzes the indicated reaction in the folate metabolism of microorganisms (1).

Starting with the enzymatically active fraction from a gel filtration step (Sephadex G-100, Fraction I) the 7,8-dihydropteridinealcohol pyrophosphokinase-activity (KINASE) was adsorbed onto an affinity column of folate-Sepharose 4B equilibrated with 0.01 M tris-HCl, pH 7.9.

After washing with 0.2 M NaCl in equilibration buffer the KINASE-activity subsequently was eluted within a salt gradient 0.2 to 0.5 M NaCl in equilibration buffer (Fraction II,Fig.1,2). The enzyme activity was determined by the tlc-method (2).

Fraction II was further purified by IEF on a granulated gel (4). Constant zones of the granulated gel were scraped off and analyzed for pH-gradient and KINASE-activity (Fig. 3.). Two active zones were saved and the protein was separated from the contaminating ampholytes by gel chromatography on a Sephadex G-25 M mini column (Fraction III and IV). The same procedure was used for the adjacent inactive zones.

Electrophoresis '82

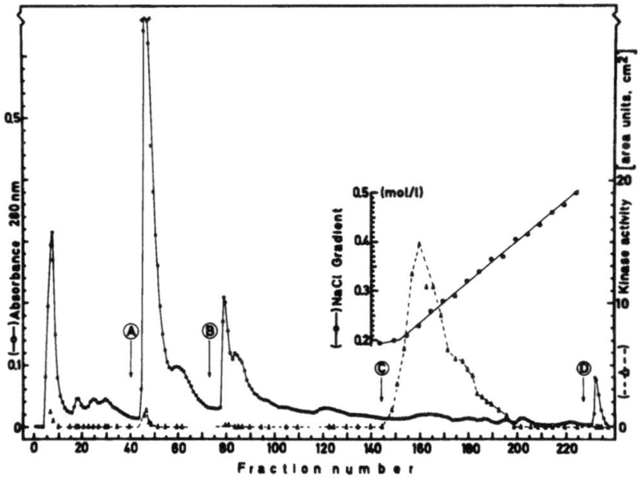

Figure 1. Affinity chromatography of Sephadex G-100 fraction of dihydropteridinealcohol pyrophosphokinase (1,2).
Folate-Sepharose 4B column 2 x 13.5 cm (3), equilibrated with 0.01 M tris-HCl, pH 7.9; sample load: 5.8 mg protein/ml in a total volume of 50 ml; flow rate: 16.5 ml x cm^{-2} x h^{-1}; fractions: 8.7 ml. Enzyme activity was determined by the tlc-method (2) and expressed as arbitrary area units of the fluorescent reaction product, dihydropteridinealcohol pyrophosphate.
A : 0.1 M NaCl wash, B : 0.2 M NaCl wash, C : start NaCl gradient, D : 2.0 M NaCl wash.

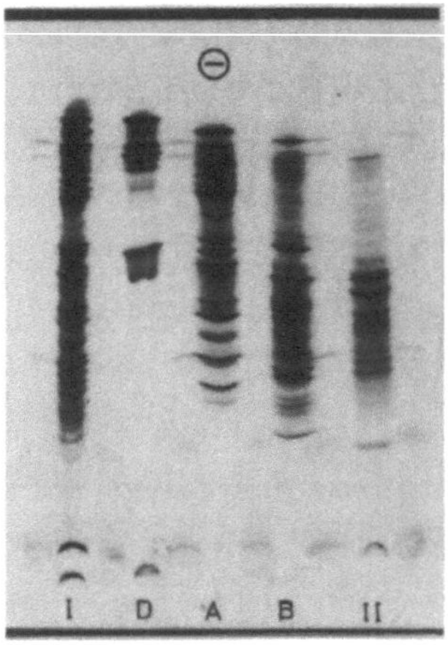

Figure 2. Ultrathin layer IEF analysis in a 100 µm polyacrylamide gel (50 mm x 95 mm, 5 % T, 3 % C, 2 % carrier ampholytes consisting of 20 % Servalyt 2-11 and 80 % Servalyt 3-5); 440 Vh; $4^{o}C$; stained with Serva Blue G.

I : Starting material, Fraction I
D : fraction 5 - 40
A : fraction 46 - 73
B : fraction 79 - 144
II: fraction 150- 172,Fraction II

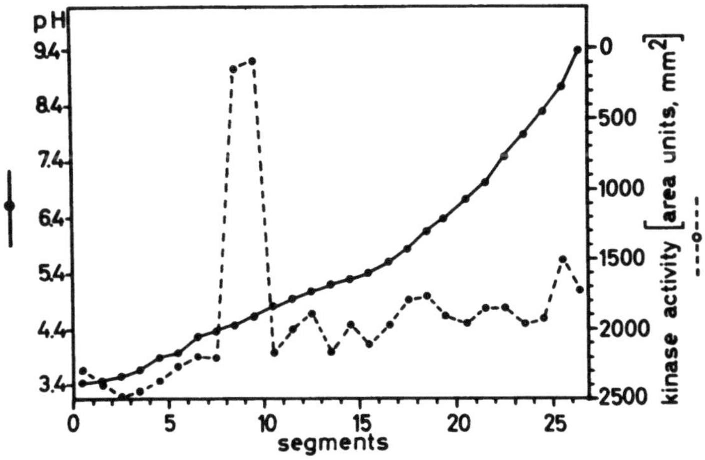

Figure 3. Flat bed IEF for Fraction II, KINASE-activity profile by analysis of the enzyme substrate (tlc-method,2); sample: Fraction II of the folate affinity column; matrix: Sephadex IEF 132 mm x 188 mm x 1 mm, 2 % carrier ampholytes (50 % Servalyt 3-5 and 50 % Servalyt 2-11); anode: 25 mM/l aspartic acid and 25 mM/l glutamic acid; cathode: 2 M/l ethanolamine, 25 mM/l arginine, and 25 mM/l lysine; focusing: 11 000 Vh at 4°C; analyzed zone: width 7 mm.
zone 9: Fraction III, zone 10: Fraction IV

In the following ultrathin layer IEF analysis the zones with KINASE-activity show a protein pattern of three main bands, one of which corresponds to the enzyme activity. In both preparative and analytical isoelectric focusing procedures, respectively, the range of isoelectric point (pI) of KINASE protein is 4.6 - 4.8 (Fig. 3 and Fig. 4).

In SDS disc-electrophoresis, KINASE Fraction IV shows two main bands, of which the faster migrating band seems to correspond to contaminating ampholytes. The molecular weight of the slower migrating band is in the range of 19 500 to 21 500 Daltons, which is similar to that value of Fraction I determined by gel filtration techniquc (2) but differs from that found by others (1).

In 8 M urea gels Fraction IV shows a pattern of 3 main and 2 side bands. (Data not shown)

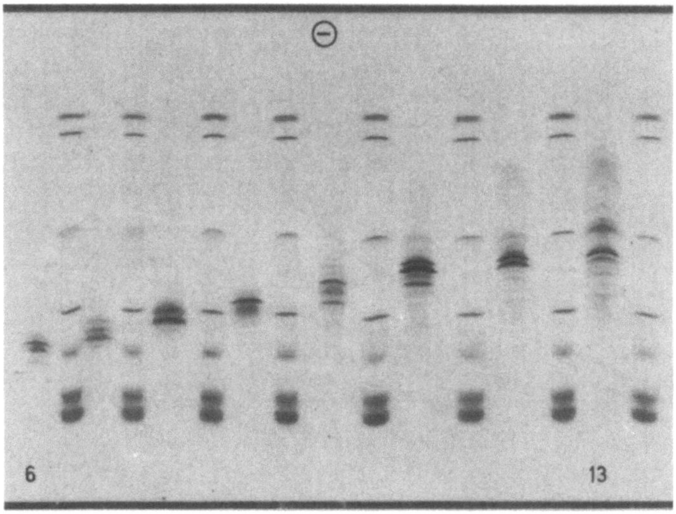

<u>Figure 4.</u> Ultrathin layer IEF analysis in a 100 µm polyacrylamide gel (50 mm x 95 mm, 5 % T, 3 % C, 2 % carrier ampholytes consisting of 30 % Servalyt 2-11 and 70 % Servalyt 3-5); 440 Vh; 4°C; stained with Serva Blue G.
Segment/zone 6 - 13 and marker proteins (Pharmacia low-pI calibration kit)

As described in the literature, the pH optimum of Fraction IV is in the range of 8.2 (Fig. 5) and the enzyme reaction was linear with time up to 90 min. The optimum of Mg^{2+} conc. was in agreement with ref. 1. The optimal concentrations for the substrates ATP (Fig.7) and dihydropteridinealcohol (Fig.8), however, show an increase of at least one order of magnitude probably due to the higher degree of purification. Thus the Mg^{2+} ion seems to be significant only for the binding to ATP and not directly for the KINASE enzyme.

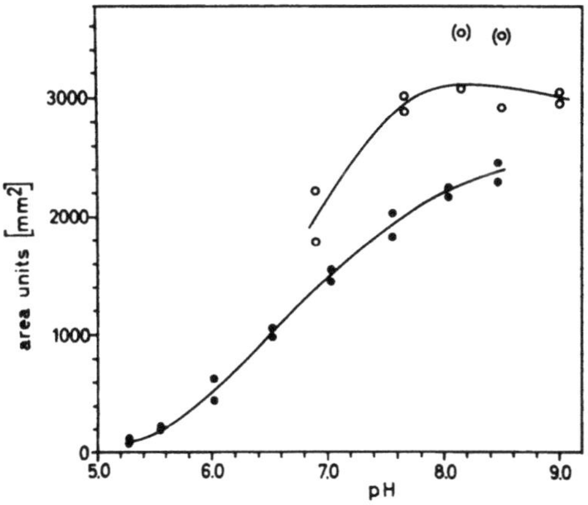

Figure 5. KINASE-activity as a function of pH; tlc-method (product) under standard conditions (2). -o- tris-HCl, -●- tris-maleate-NaOH.

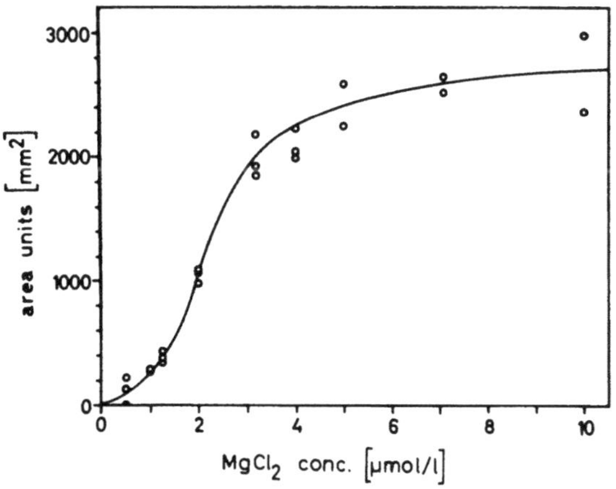

Figure 6. KINASE-activity as a function of Mg^{2+} concentration. For conditions see Fig. 5.

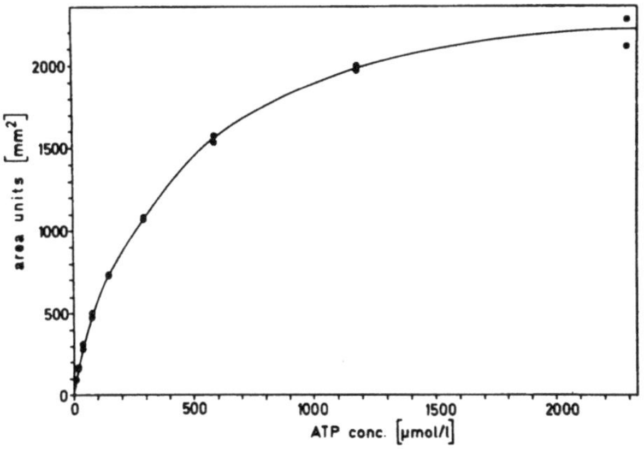

Figure 7. KINASE-activity as a function of ATP concentration. For conditions see Fig. 5.

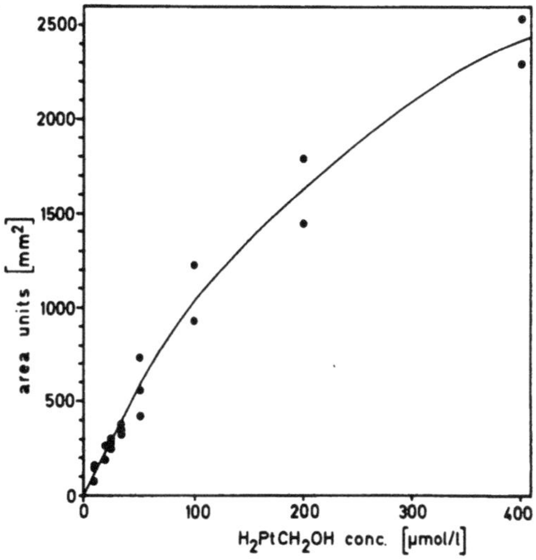

Figure 8. KINASE-activity as a function of dihydropteridine alcohol concentration. For conditions see Fig. 5.

References

1. Richey, D.P., Brown, G.H.: J.Biol.Chem. 244, 1582-1592 (1969)

2. Bock, L., Butte, W., Richter, M., Seydel, J.K.: Anal. Biochem. 86, 238-251 (1978)

3. Then, R.L.: Anal.Biochem. 100, 122-128 (1979)

4. Radola, B.J.: Biochim.Biophys.Acta 295, 412-428 (1973)

5. Radola, B.J.: Electrophoresis 1, 43-56 (1980)

ANALYSIS OF PROTEIN AND POLY-A$^+$ RNA IN NORMAL AND SULFUR-STARVED SYNECHOCOCCUS 6301

Hans-Peter Köst, Hedwig Thoma, Erich Warm und Ahlert Schmidt
Botanisches Institut der Universität München
Menzingerstr. 67, D-8000 München 19

Introduction

Synechococcus 6301 is a unicellular cyanobacterium containing huge amounts of the blue-colored biliprotein, C-phycocyanin (C-PC) besides allo-phycocyanin (APC) and colorless proteins. C-PC is composed of two subunits, an α-subunit and a ß-subunit, in a 1:1 ratio (1). It is the main constituent of the light-harvesting antenna (phycobilisomes (2)) of the cyanobacterium(3). -In all of the biliproteins investigated so far, protein and bile pigment chromophore are thioether-linked to a cysteine residue of the apoprotein (e.g. 4,5). Although some of the key steps of the synthesis of phycocyanobilin (=PCB), the prosthetic group of phycocyanins, have recently been published (6,7), the biochemistry of thioether formation is still unknown.-

Biliprotein biosynthesis can best be studied in "inducible" systems, such as a dark grown culture of Cyanidium caldarium, a unicellular thermophilic red alga which upon illumination will produce large quantities of biliproteins and chlorophyll a (8, 9, 10). - Wood and Haselkorn (11) describe a system (Anabaena) which degrades phycocyanin under nitrate starvation conditions; the biliprotein is resynthesized during regeneration with nitrate containing medium.

A very efficient system is represented by Synechococcus 63o1. Under sulfur starvation conditions (2.5 x 10^{-6} moles SO_4^{--}/l medium) phycocyanin is nearly completely degraded. It is quantitatively regenerated by sulfate (3.1 x 10^{-4} moles SO_4^{--}/l).

Electrophoresis '82

For the present work, we have chosen the latter system. Parallel to an analysis of protein content of normal and sulfur-starved cells, cyanobacterial RNA was analyzed using isotope labeling with ^{32}P.

Results

1) Induction of sulfur deficiency and reversion.
$MgSO_4$ of the original medium (Stanier et al., 12) was replaced by $MgCl_2$; sodium sulfate was added to the desired concentration (eg. $2.5x10^{-6}$ M/l). After a short logarithmic growth, (3 days), the cultures changed their color from deep blue-green to yellow-green. The absorption spectrum revealed a far-reaching degradation of phycocyanin (λ_{max}, culture = 630 nm) whereas the chlorophyll content remained unchanged (fig. 1, 0 hours). After addition of sulfate ($3.4x10^{-4}$ M/l) a rapid phycocyanin-production was initiated (Fig. 1, 2-22 hours). It has to be noted, that the phycocyanin recovery takes place prior to cell division (paper in preparation).

2) Analysis of soluble proteins by SDS-PAGE
For conditions of separation, see legend of fig. 2. Cells grown in "normal medium" ($3.1x10^{-4}$ Moles/l SO_4^{--}, A) exhibit bands in a range of 15 000 ->500 000 daltons. The most prominant are the α and ß-subunits of C-phycocyanin at ca. 15 000 and 20 500 daltons.(B)From 32 500 to about 200 000 dalton, only minute amounts of protein are visible. Proteins remaining under starvation conditions are obviously indispensible for cell vitality, what phycocyanin, to a large extent, is not.(C)Upon 12 hours regeneration, a strong band appears at 43 300 daltons, after 72 hours of regeneration however,(D)the band has nearly deteriorated. At this time, the full protein outfit has already been restored (D) (compare also fig. 1). The band at 43 300 daltons might

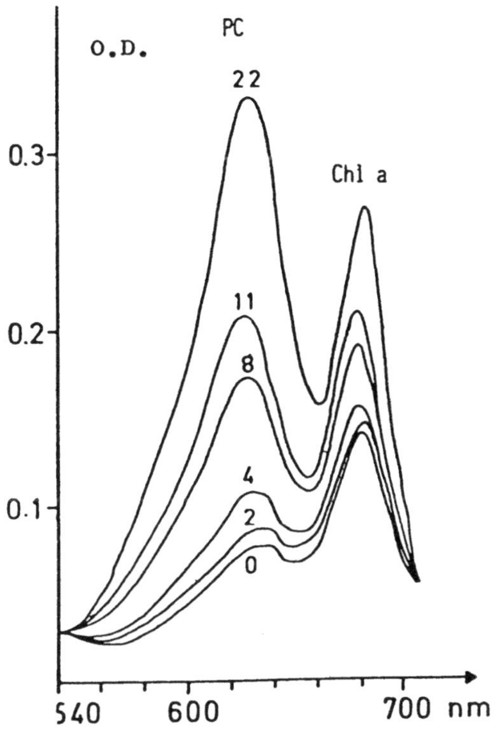

Fig. 1

Absorption spectra of Synechococcus 6301 regenerating from sulfur starvation.

The optical density O.D. is given for regeneration times of 0-22 hours (see numbers above curves). Note that O.D. 630 (phycocyanin) increases at a considerably higher rate than O.D. 675 (chlorophyll a). After 22 hours, regeneration is complete; the absorption spectrum corresponds to that of a culture grown in "normal" medium.

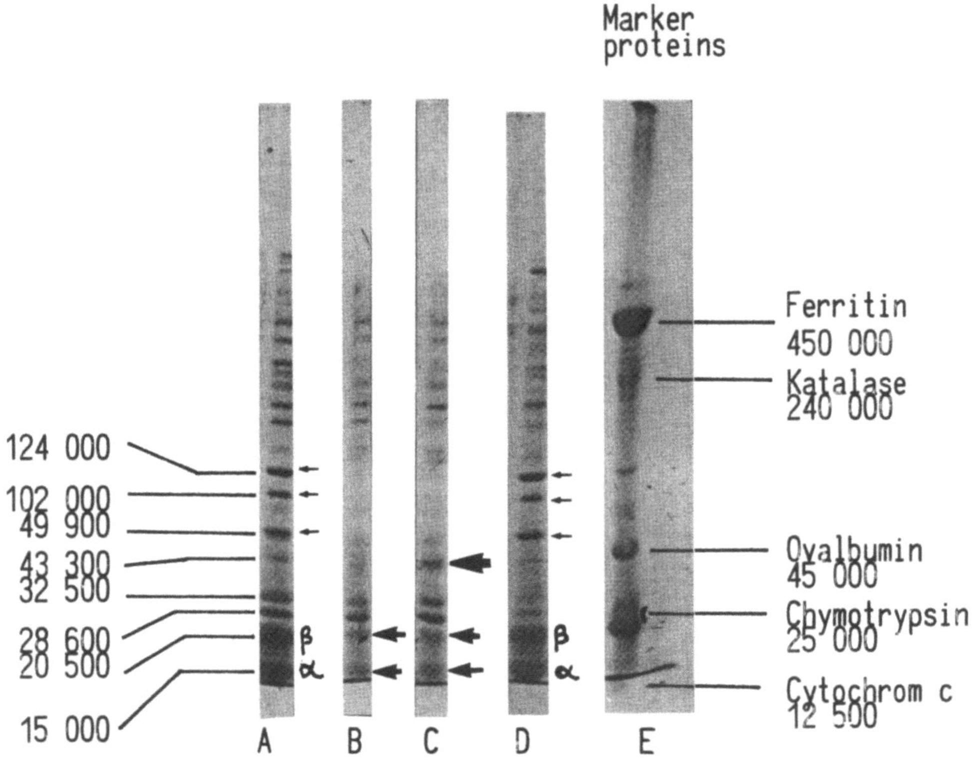

Fig. 2

SDS-PAGE of soluble proteins from Synechococcus 6301.
Gel composition: 10% Acrylamide (10.3% T, 2.6% C, 0.1% SDS, pH 8.8). Electrophoresis conditions: 15 min 120 V, 20 mA; 3.8 hours 200 V, 30 mA. Gel stain: Serva Blue G. Each lane corresponds to the separation of proteins from 2.5 mg wet weight.

Lane A: Cyanobacteria grown in normal medium ($3.1x10^{-4}M/l\ SO_4^{--}$)

B: Sulfur starved cells ($2.5x10^{-6}$ $M/l\ SO_4^{--}$)

C: Cells regenerated in normal medium, reg. time: 12 hrs

D: like C, regeneration time 72 hours

E: Molecular weight standards

Note that some proteins (eg. mol. weight 32 500, 28 600 daltons) are not degraded, wheras proteins marked with arrows (← ⟵) disappear under starvation conditions. Note band at 43 300 daltons induced upon 12 hours regeneration time. (⟵)

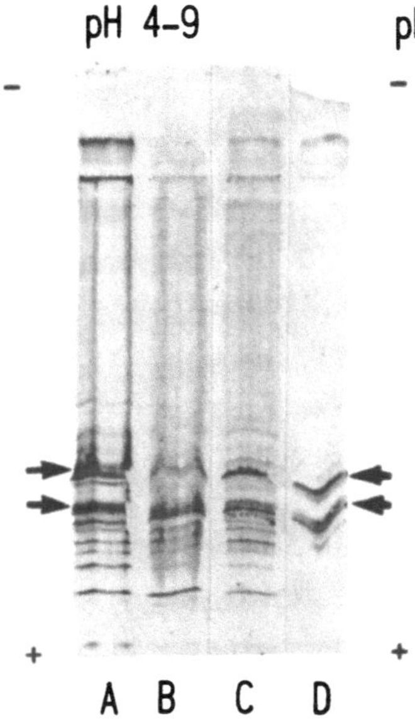

Fig. 3
7.6 M/l urea

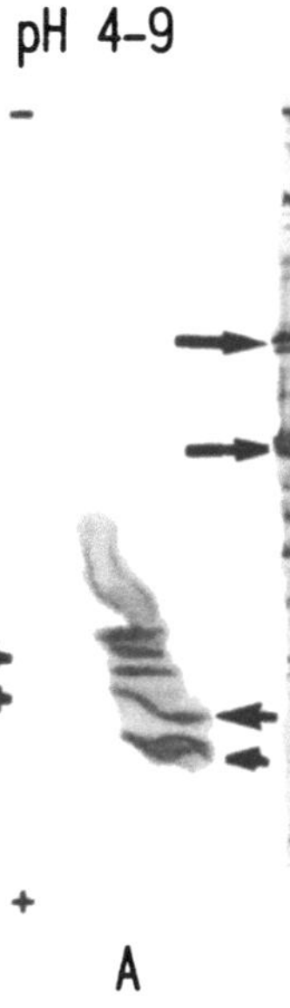

Fig. 4
2 M/l tetra-
methyl urea

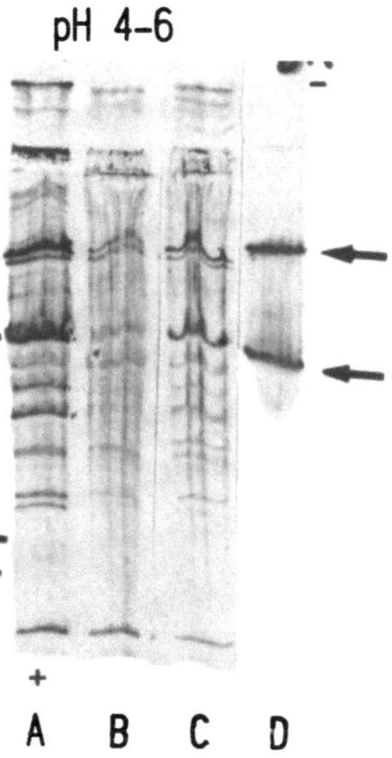

Fig. 5
7.6 M/l urea

IEF-PAGE of soluble proteins from Synechococcus 6301.
Gel composition: 4%T, 3%C, 3% Servalyte^{R}, 7.6 M/l urea (2 M/l TMU). Electrophoresis conditions: prefocussing: 400 V, 30 min; focussing: 800 V, 30 min; 1000 V, 30 min (Tetramethyl urea gels 60 min). Gel stain: Serva Blue G.

All figures: lane A: Cyanobacteria grown in normal medium (3.1×10^{-4} M/l SO_4^{--})

B: Sulfur-starved cells (2.5×10^{-6} M/l SO_4^{--})

C: Cells regenerated in normal medium (1 day)

D: purified C-phycocyanin (marker protein); a separation into a and B-subunits can be seen (arrows; also note arrows marking phycocyanin subunits in lanes A-C)

Upon regeneration conditions, C-phycocyanin subunits degraded under sulfur starvation reappear.

perhaps be assigned to a biosynthetic function, it can also be found, but very weakly, in cells growing logarithmically in "normal medium".

3) Analysis of soluble proteins by IEF-PAGE.
A separation of cellular proteins on the level of subunits was also possible by IEF on ultrathin (0.175 mm) gels containing 7.6 M/l urea (Fig. 3, 5). As a marker, electrophoretically pure C-phycocyanin from Synechococcus ($E_{620/280} \approx 5.4$) had been added. Most proteins focuse in a pH-range near the isoelectric points of the subunits (P_I α-sub = 5.26, P_I ß-sub = 5.17) (1). The separation can be improved employing narrow-range carrier ampholytes (Fig. 5) The degradation of protein bands (especially the α - and ß-subunit of PC) during sulfur starvation and their reappearance upon regeneration could clearly be monitored. A separation on IEF gels containing 2 M tetramethyl urea (TMU) (Fig. 4) is also possible although bands are not that clearly separated as in urea. One advantage of tetramethyl urea, however, is the prevention of crystallization (often observed in gels containing high concentrations of urea).

4) Labeling of cyanobacteria with ^{32}P and extraction of total crude RNA. For ^{32}P-labeling, 1.3-3 g cells ("normal" culture, sulfur-starved culture) were resuspended in 1oo ml of PO_4^{3-} free medium with 1 mCi $^{32}PO_4^{3-}$, 3.1 x 10^{-4} M/l SO_4^{--} (for "normal" cells or regenerating cells) or 2.5 x 10^{-6} M/l SO_4^{--}. After an incubation period of 8 hours, the cells were harvested by centrifugation and extracted. The procedure followed the method of Hall et al (13) with slight modifications. Cells were split at 84o-14oo bar by passing a French pressure cell (2x); the resulting suspension was incubated with proteinase K to digest ribonucleases. After a centrifugation, RNA was precipitated with 12 M LiCl. 3 resuspension steps with 2 M LiCl followed; the RNA was

now redissolved with 2o% potassium acetate and precipitated with ethanol. RNA obtained in this way usually was not pure (as checked by UV-spectroscopy) and had to be further purified by passing a cellulose column. The product thus obtained exhibited a clear miximum at 26o nm and was used for further experiments.

5) Isolation of poly-A^+ RNA and poly-A^- RNA. Oligo-dT-cellulose (Sigma) was washed with 0.05% diethyl pyrocarbonate (DEP) in water, 0.1 N NaOH, elution buffer (EB) (14) and binding buffer (14). Prior to use, the sample to be separated was subjected to a heat treatment ($6o^{o}$, 3 min.) to produce single stranded RNA, and applied. The column flow-through contained poly-A^- RNA (Fig. 6). An electrophoresis of this fraction on non-denaturing agarose-urea gels yielded two bands corresponding to 16 S and 23 S RNA indicating that degradation by ribonucleases had not taken place (Fig. 7). Fig.6 gives the ^{32}P radioactivity per fraction number of a "normal" algal culture; no principal differences could be observed with "sulfur-starved" or "regenerated" cultures.

6) Hybridization with polyuridylic acid. For hybridization, ^{32}P labeled RNA was bound to nitrocellulose and subsequently treated with poly-uridylic acid. Non hybridized RNA was removed by digestion with ribonuclease A and T_1 (Boehringer). The determination of radioactivity yielded: hybridized RNA from total: poly-A^-: 2.2%, poly-A^+: 6.5%. We consider these values as a confirmation of the presence of poly-A^+ sequences in the procaryotic Synechococcus RNA, a result which had not to be expected from the beginning.

Acknowledgement: This work was supported in part by a grant from the Deutsche Forschungsgemeinschaft (DFG). For a generous gift of tetramethyl urea (TMU), H.-P. K. is indebted to Serva GmbH, Heidelberg.

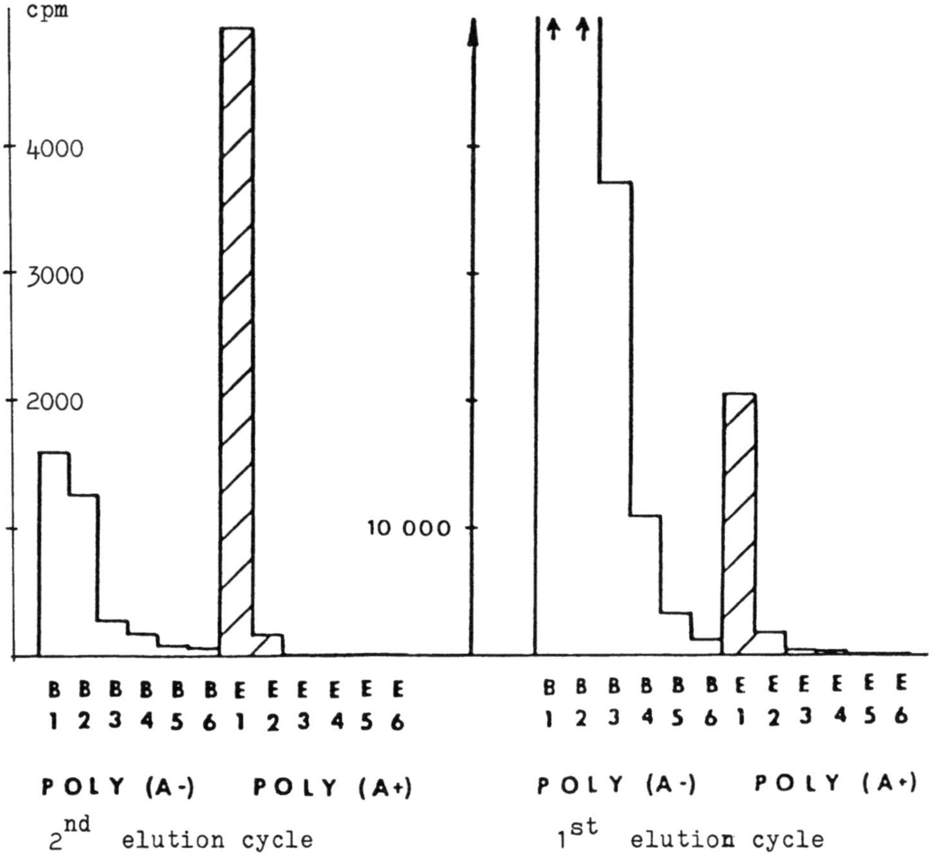

Fig. 6

Separation of ^{32}P labeled poly-A^+ and poly-A^- RNA by affinity chromatography on oligo-dT cellulose. The radioactivity per fraction is given in cpm as a function of the fraction number.

B: fractions obtained by elution with "binding buffer" (BB)
E: fractions obtained by elution with "elution buffer" (EB)

For B- and E-fractions, elution was performed with 5x1 ml, 1x5 ml.
Poly-A^+ fractions from first elution cycle (right) were re-applied on regenerated columm (second elution cycle, left).

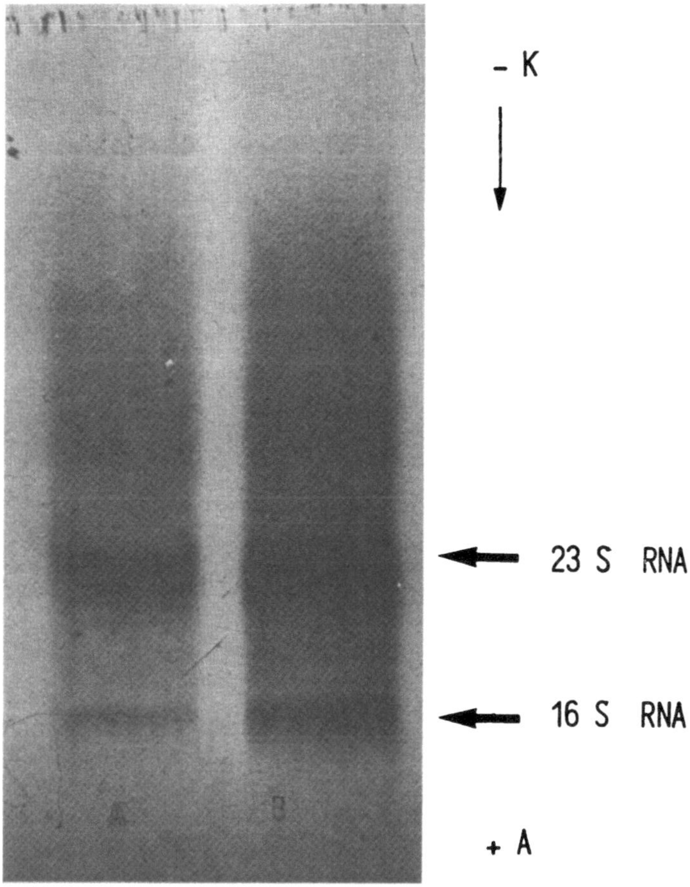

Fig.7 (2.5 x normal size)

Non-denaturing electrophoresis of poly A^--fraction on agarose-urea gels (1.5% agarose, 0.48 mm). As can be seen, the preparation was not degraded by ribonuclease activity. Two main bands at 16 S and 23 S are clearly visible (arrows).

Separation conditions:

Electrophoresis on LKB Multiphor 2117

Sample volume 5 ul; lane A: 16 000 Cpm, lane B: 20 000 Cpm

Starting voltage: 60 V (10 mA); final voltage 250 V

Autoradiography: 21 days exposure on LKB Ultrofilm ^3H

References

1. Glazer, A.N., Fang, S.: J. Biol. Chem. 248, 663 (1973).
2. Gantt, E.: BioScience 25, 781 (1975).
3. Lundell, D.J., Williams, C.R., Glazer, A.N.: J. Biol. Chem. 256, 3580 (1981).
4. Köst-Reyes, E., Köst, H.-P.: Eur. J. Biochem. 102, 83 (1979)
5. Glazer, A.N., Hixson, C.S., De Lange, R.J.: Anal. Biochem. 92, 489 (1979).
6. Brown, S.B., Holroyd, J.A., Troxler, R.F.: Biochem. J. 190, 445 (1980).
7. Brown, S.B., Holroyd, J.A., Troxler, R.F., Offner, G.D.: Biochem. J. 194, 137 (1981).
8. Troxler, R.F., Biochemistry 11, 4235 (1972).
9. Troxler, R.F., Brown, A.S., Brown, S.B.: J. Biol. Chem. 254, 3411 (1979).
10. Schuster, A., Köst, H.-P., Rüdiger, W., Eder, J.: Elektrophorese Forum '80, 2. Diskussionstagung TU München, p. 285 (1980).
11. Wood, N., Haselkorn, R.: J. Bacteriol. 141, 1375 (1980).
12. Stanier, R.Y., Kunisawa, E., Mandel, M., Cohen-Bazire, G., Bact. Rev. 35, 171 (1971).
13. Hall, T.C., Ma, Y., Buchbinder, B.U., Pyne, J.W., Sun, S.M., Bliss, F.A.: Proc. Natl. Acad. Sci. USA 75, 3196 (1978).
14. Aviv, H., Leder, P.: Proc. Natl. Acad. Sci. USA 69, 1408 (1972)

ELECTROPHORETIC MEASUREMENTS ON PURPLE MEMBRANE PARTICLES

Nezabravka Popdimitrova, Stoyl Stoylov,
Scientific Medico-biological Institute of Sofia
1431 Sofia Bulgaria

Maria Kantcheva
University of Sofia, Department of Biophysics
1000 Sofia, Bulgaria

Introduction

The purple membrane of H.Halobium is the subject of intense interest due to its photoactivated proton pumping function and to its relatively simple chemical composition. This membrane appears in the form of purple patches in the plasma membrane when the bacterium is grown in light at low oxygen tension. These patches cover about half of the total surface area of the membrane and can be separated by osmotic shock(1). Electrophoretic experiments have shown that the purple membrane carries a net negative charge at pH - values above 2,3 and its pI may be as low as 2,2 (2).
An examination of light effects on the electrophoretic migration of purple membrane fragments in a vertical tube along a sucrose gradient demonstrates that at pH 4,5 the dark adapted form of bacteriorhodopsin is less negatively charged than the light - adapted form (3). The presented experimental results did not provide a direct information about the electrical surface charge of purple membrane particles. It is the

Electrophoresis '82

purpose of this contribution to obtain data on the electrophoretic mobility of purple membrane particles at different ionic and pH environments.

Results

Microelectrophoresis allows the electrophoretic mobility of suspension particles to be experimentally measured and bring a direct information about the charges carried by the particle surface.

A preliminary microscopic observation of the purple membrane fragments (with Photomicroskop III Opton at 2000x) has demonstrated the existance of different sizes of particles. The size of the larger particles (observed in the cell electrophoresis chamber) is about 1-2 μm and they are formed probably by aggregation of several smaller particles. The ratio of large particles to small ones is pH - dependent.

The electrophoretic mobility of the purple membrane particles from H.Halobium strain R_1M_1 was determined with the aid of a Zeiss - Opton cytopherometer (rectangular chamber) purchased from Zeiss - Opton Oberkochen, Germany, at a current of 1 to 5 mA at $20^{o}C$.

Table 1 presents the electrophoretic mobility of purple membrane particles in dependence of the electrolyte concentration of the suspending medium.

The pH - course change of the electrophoretic mobility is indicated in Fig. 1. The purple membrane carries a net negative charge at pH values above 1,7 and its pI is at pH 1,5. Some of this experimental results correspond to the electron microscopic characterization of the surface charge properties of the purple membrane fragments using anionic ferritin as an indicator (2). However they can not give further information above pH 5 as no binding of ferritin occurs in that range.

Table 1.
Electrophoretic behaviour of purple membrane particles with changing ionic concentration in NaCl solution.

U ($m^2.s^{-1}.V^{-1}$)x10^{-8}	I (M)
0,8787 ± 0,0082	2.10^{-1}
1,4235 ± 0,0107	10^{-1}
1,6748 ± 0,0100	5.10^{-2}
1,7250 ± 0,0126	$2,5.10^{-2}$
1,9281 ± 0,0207	10^{-2}
2,1303 ± 0,0252	5.10^{-3}

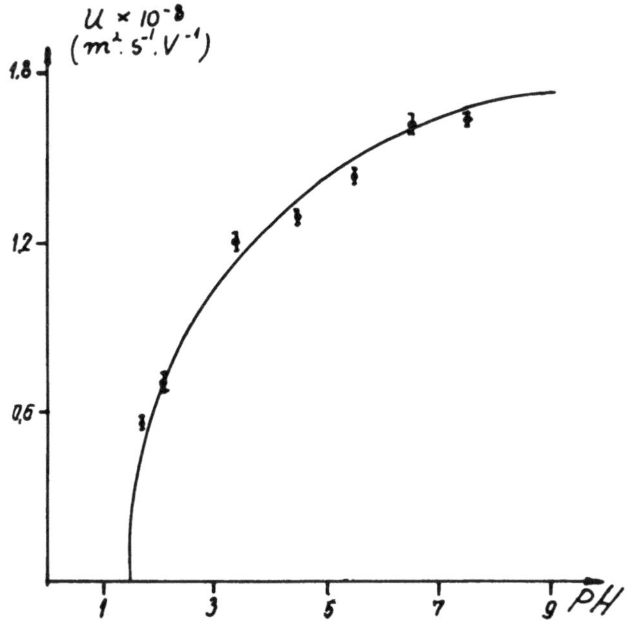

Fig. 1

References

1. Oesterhelt, D., Stoeckenius, W. : Methods in Ehzymology (Part A. Biomembranes), 31,667-678 (1974).

2. Neugebauer, D. Ch., Oesterhelt, D., Zingsheim, H. : J. Mol. Biol. 125, 123 (1978).

3. Eisenbach, L., Eisenbach, M. : Analyt.Biochem. 92, 228 (1979).

THE ACTIVITIES OF ISOLATED ALKALINE- AND ACID-DNases ON DENATURED AND NATIVE DNA

Younis Hussein and Mokhtar El-Zawahri
Department of Genetics, Assiut University, Assiut, Egypt.

Introduction

It was reported that chloroplast replication in Euglena gracilis is independent of the nuclear and cellular division and that the mechanisms effecting the control of nuclear DNA also effecting the control of plastid DNA metabolism (1). Since very little is known about DNA metabolism in this unicellular alga, the study of the deoxyribonucleases (DNases) may function in the control of Euglena DNA metabolism. DNases have long thought to perform some adjunct function in DNA duplication (2). It is suggested that extract of Euglena contain at least two DNases; one referred to as " Euglena acid DNase " and the second referred to as " Euglena alkaline DNase " (3). In the present study, further trials were done to identify the DNases present in extracts of Euglena and their activities on denaturated and native DNA. Wild type strain of Euglena gracilis was used as the source of DNases in this work. Cells were grown on a rotary shaker using artificial light at 25°C in erlenmyer flasks containing liquid medium according to the method described by Freyssient et al., (4). At the stationary phase, the cells were harvested by centrifugation, washed with 10^{-3}M Na Cl and suspended in MTN (0.5M mannitol; 0.2M tris buffer, pH 8; 0.01M Na Cl). The cell suspension was sonically disrupted for 1.5 min. using 70 watt. The sonicates were centrifuged for one hour at 16000 rpm. The supernatant fluid was collected and used as a source for DNases. Commercial DNA of Salmon sperms

Electrophoresis '82

was used as a standard DNA to calibrate the dosage of DNases activities. Stock solutions were prepared by dissolving commercial DNA in bidistilled water at concentrations of 10, 30,40, 50 , 60 , 70 , 80 , 90 and 100 µg/ml. One ml. of each concentration was added to 2 ml. reagent (1.8 gm. diphenylamine, 100 ml. glacial acetic acid, 2.75 ml. sulphric acid conc. and 0.5 ml. acetaldehyde 1.6%) which was incubated at 30°C for 18 hours. Optical density at 600 and 650 mµ were determined (5). Euglena DNA was isolated and used as substrate. The concentration of substrate DNA was determined as described by Burton(5). This purfied DNA was referred to as native DNA which denatured by 0.1 N NaOH at 60°C for 10 min. followed by rapid cooling and nutralized by HCl. DNase assay performed according to the Method described by lehman (6). Reaction mixtures for assay of Euglena acid DNase generally contained 1 ml. of native DNA (510 µg/ml.); 1 ml. citrate - phosphate buffer at pH 4, 4.6, 5, 6 and 7; 1 ml. of the enzyme source and Ca Cl_2 at a final concentration of 2 mM. Reaction mixture for assay of Euglena alkaline DNase generally contained 1 ml. DNA (510 µg/ml.); 1 ml. tris buffer to a final concentration of 0.2 M at pH 8, 9 , 9.5 and 10; 1 ml. of enzyme source and Ca Cl_2 at a final concentration of 30 mM. Reaction mixtures were incubated for 0 , 60 , 90 and 120 minutes at 37°C. The reactions were stopped by adding perchloric acid at a final concentration of 0.7 N and followed by rapid cooling in the ice and centrifuged to remove the precipitate. Aliquots of the supernatant were analyzed by the method of Burton (5) and the optical densities obtained were converted to deoxypentose equivalent by comparison with those obtained from a standard solution of DNA. All data are averages of three replicate assays.

Results

1) DNA standard curve. The estimation of standard DNA by the

relation between DNA and diphenylamine is shown in Figure 1. From the figure it can be seen that the relationship between the concentration of the commercial DNA and the optical density at 600-650 mμ is lenear. This indicates that the optical density is very nearly proportional to the DNA concentration. Thus the DNA concentration can be detected directly from the optical density.

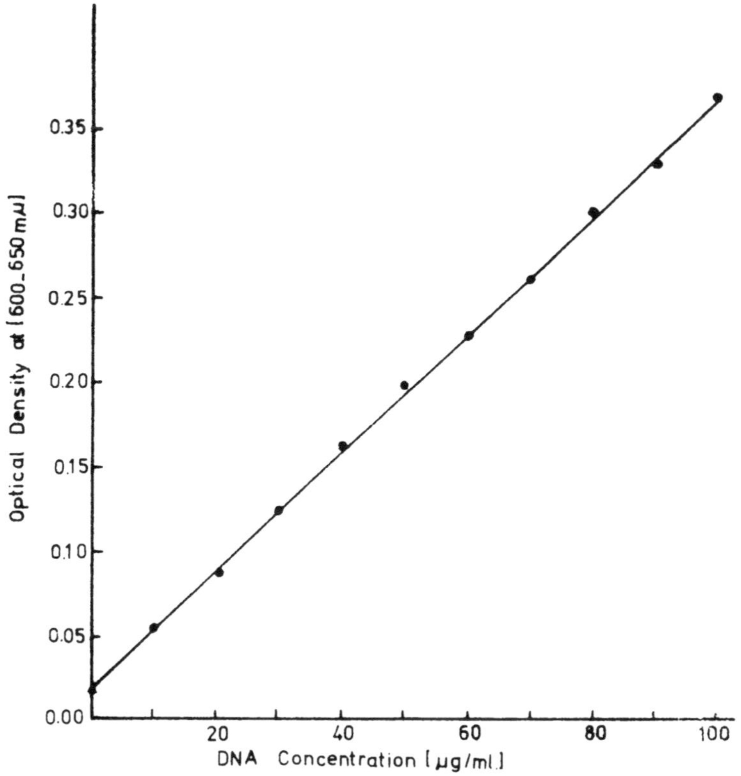

Fig. 1 : Standard curve which represents the relationship between the concentration of the commercial DNA and the optical density at 600 - 650 mμ.

2) DNase activities during time of incubations. Figure 2 indicates that Euglena alkaline DNase activity increased by increasing the time of incubation. Five hours may be of value if greater DNA concentration is required , but otherwise a two hours of incubation time is preferable , since it gives the optimal linear relationship after which it changes to plateau. Using different alkaline pH, it is seen that the activity increases by increasing the pH from 7.0 to 7.5 and 8.0.

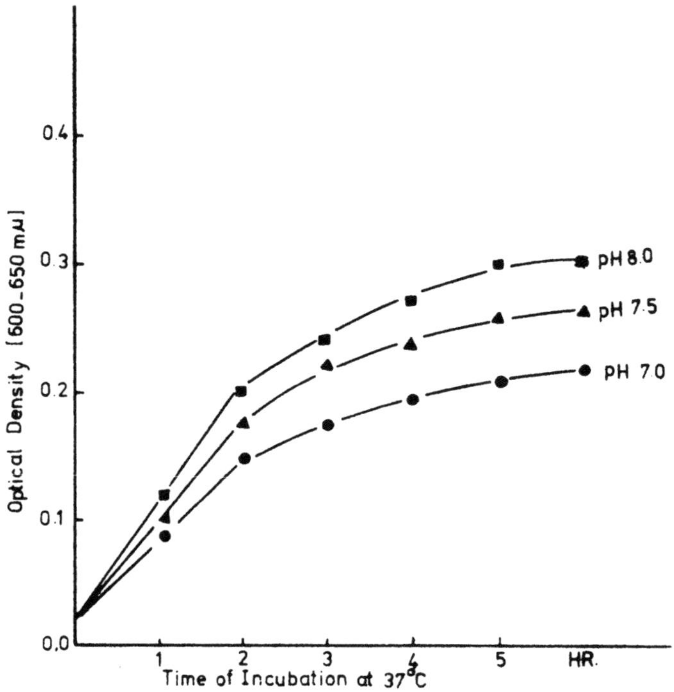

Fig. 2 : Effect of time of incubation on the activity of alkaline-DNase.

3) Effect of Temperature on DNase Activities. The effect of temperature on alkaline DNase activities is plotted in Figure 3. Using the preferable time of 2 hours at 20 ,25, 30 , 37 , 40 and 45°C, the optimum temperature for incubation was found to be 37°C, since it gives a linear relationship between the optical density and the time of incubation.

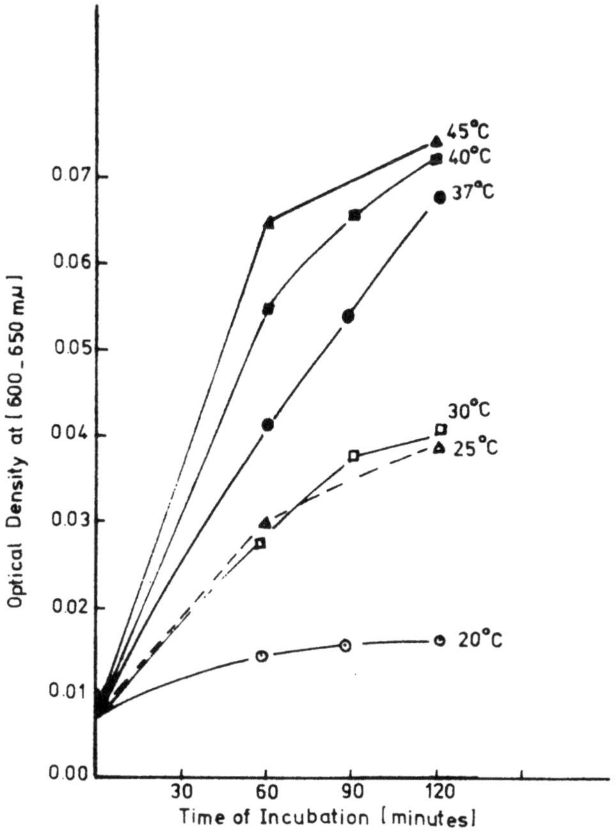

Fig. 3 : Effect of temperature on alkaline DNase activity on denatured DNA.

4) pH activity of Euglena DNases. Figure 4 shows that extract of Euglena contains at least two DNases. One with an acidic pH optimum of about 5. Euglena acid DNase exhibits a preference for native DNA. The second enzyme, Euglena alkaline DNase, has an alkaline pH optimum of about 9.5 . It exhibits very high activity with denaturated DNA and , under our conditions, alkaline DNase shows very little activity when native DNA is substituted for heat denatured DNA.

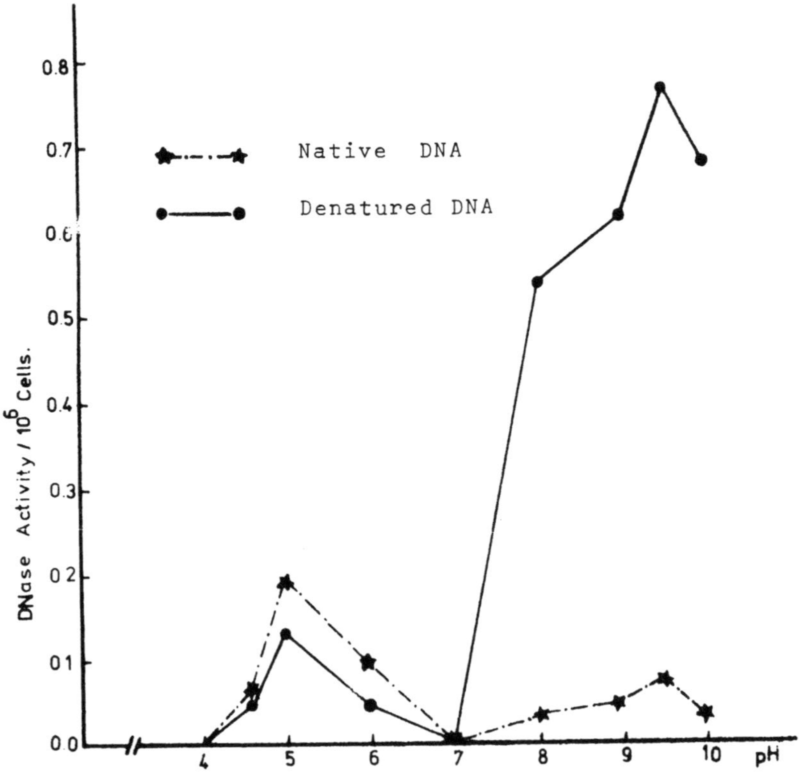

Fig. 4 : Euglena DNases pH activity curves obtained with native DNA and denatured DNA.

These results indicate that there are at least two types of DNases. Euglena acid DNase which has a pH optimum of 5 and acts prefrentially on native DNA. The second is Euglena alkaline DNase which has a pH optimum of 9.5 and acts preferentially on heat-denatured DNA. Thus, as it is seen from Figure 4, to avoid the activity of both enzymes during isolation of Euglena DNA, the optimum pH is 7. The properties and purification patterns of Euglena acid DNase are similar to DNase II-like enzyme which is common to perhaps all eukaryotic organisms (7,8). It is of interest that Euglena alkaline DNase apparently has unusual enzymic properties. It is reported that few DNases have a pH optima as high as 9.5 (9).

References

1. Carell, E.F., J. Cell Biol. 41,431 - 440 (1969).
2. Lehman, I.R., Ann. Rev. Biochem. 2,645 (1967).
3. Carell, E.F., Egan, J.M., and Pratt E.A., Arch. Biochem. Biophys. 138, 26 - 31 (1970).
4. Freyssinet G., Heizmann P. Verdier G., Trabuchet G. et Nigon V., Physiol. Veg. 10, 421 - 442 (1972).
5. Burton, K., The Biochemical Journal. 62, 315 - 323 (1956).
6. Lehman, I.R., J. Biol. Chem. 235, 1479 - 1489 (1960).
7. Cordonnier, Bernardi, G., Biochem. J. 94, 12 P (1965).
8. Russell, A.P., Comp. Biochem. Physiol. 18, 961 (1966).
9. Laskowski, M., Sr., in " Advances in Enzymology " (F.F. Nord, ed.), Vol. 29, P. 165. Wiley (Interscience), New York, 1967.

A FUNGAL PROTEINASE RESISTANT PROTEIN IN BOVINE PLASMA

John S. Chen, Maria Giulia Menesini Chen, Elio Sbardellati, Mario Bari, Carlo Ricci

P.F. del CNR, c/o Institute of Biological Chemistry, University of Siena, Siena 53100, Italy

Giuseppe Pompucci

Istituto di Fisiologia Generale, Università di Sassari, Italy

Introduction

A proteinase preparation EL25-79 isolated from Aspergillus niger (Rohm, Darmstadt) was found to have fibrinolytic activity as well as properties which prevent clot formation in both human and bovine blood and do not provoke hemolysis even at enzyme concentrations as high as 0.5% (1,2). In addition, we found that human fibrinogen is more quickly digested than human albumin by EL25-79 (3). The same study, with SDS polyacrylamide gel electrophoresis, revealed that EL25-79, at concentrations as high as 0.5%, hydrolyzed human plasma proteins to molecular weights of 9.000, 7.500 and possibly less. In this study we assess the comparative effects of EL25-79 on bovine and human plasma.

Materials and method

Human blood was obtained from healthy persons. Bovine blood was collected from the local slaughterhouse. Proteinase preparation EL25-79, isolated from Aspergillus niger, was obtained from Rohm GmbH, Darmstadt. All other chemicals were of analytical grade.

Electrophoresis '82

Both bovine and human blood were collected in tubes containing various concentrations of EL25-79 ranging from 0.004 to 0.5% in a final volume of 1 ml. Another series of experiments was carried out using Na-citrate (0.2%) treated human and bovine plasma. The reaction continued at 37°C for 30 min. After incubation, the red cells were removed by centrifugation at 500 xg for 15 min. Bovine serum was also treated with 0.1 and 0.5% EL25-79. Samples were then analyzed by SDS polyacrylamide gel electrophoresis according to the method described by Weber and Osborn (4).

Results and discussion

Fig. 1 shows the protein patterns of human plasma treated with 0.1% EL25-79. Plasma proteins (Fig. 1 a) had been cleaved into two major bands of 7.5 K and 9 K Daltons, respectively, and one minor band of about 12.000 Daltons (Fig. 1 b). Fig. 2 shows the SDS polyacrylamide gel electrophoresis analysis of bovine plasma protein treated with 0.05, 0.1 and 0.5% EL25-79. Arrows indicate a protein band which seems immune to proteolytic enzyme digestion. The molecular weight of this protein band is about 45.000 Daltons. It is also shown that this 45 K band persists throughout all enzyme concentrations tested.

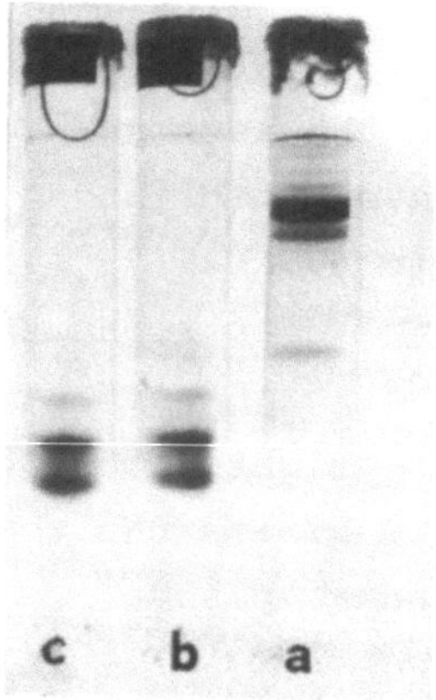

Fig. 1. SDS polyacrylamide gel electrophoresis analysis of human plasma proteins treated with fungal proteinase EL25-79. a control b, c, 0.1% EL25-79.

In order to know whether the EL25-79 resistant protein belongs to clotting proteins, bovine blood was allowed to clot for an

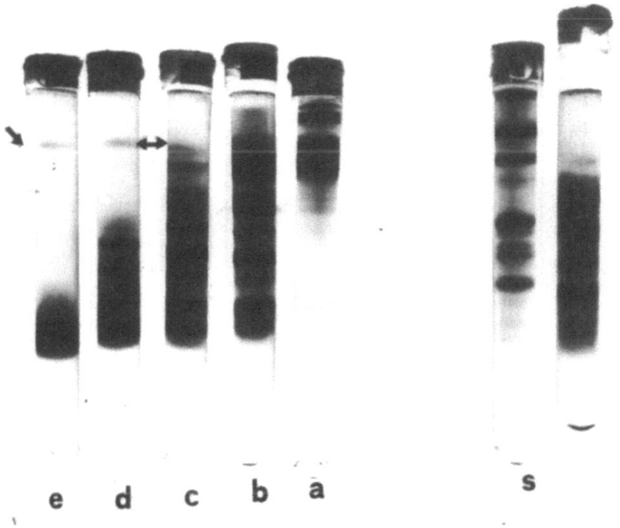

Fig. 2. SDS polyacrylamide gel electrophoresis analysis of bovine plasma proteins treated with EL25-79. a. control; b. 0.004%; c. 0.02%; d. 0.1% and e. 0.5%; S. standard (Dalton marker). Arrows indicate the 45 K protein bands.

hour at room temperature. Serum was then obtained by centrifugation at 700 xg for 30 min. Samples were treated with EL25-79 and analyzed by SDS polyacrylamide gel electrophoresis as described above. Fig. 3 shows that the protein band similar to that mentioned above appears also in enzyme-treated serum. This would imply that the proteinase resistant protein in bovine blood has no part in blood coagulation. We can also note that at the concentration of 0.5% proteinase EL25-79 cleaves both fibrinogen and albumin molecules

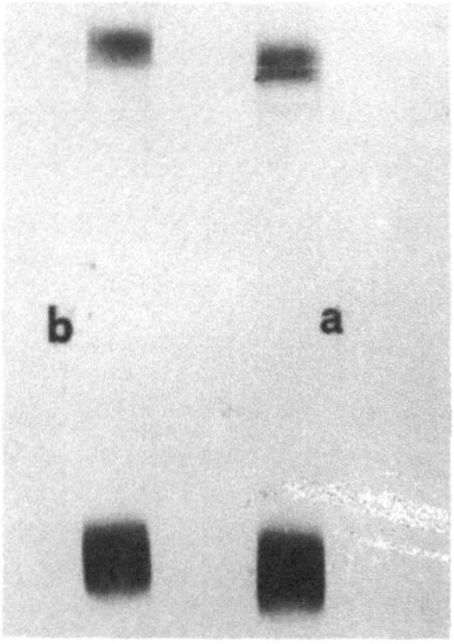

Fig. 3. SDS polyacrylamide gel electrophoresis analysis of bovine serum proteins (a), plasma (b) treated with 0.5% EL25-79.

into protein fragments with a molecular weight of less than 12 K (3). Therefore it is almost certain that this 45 K protein band, according to the data available, is a fragment of globulin molecules or one of the many existing protein factors in plasma.
Studies are in progress to determine the biological significance of these findings.

References

1. Chen, J.S., Menesini Chen, M.G., Pompucci, G., Quaglia, G.B. and Massacci, A.: Bollettino di Zoologia 48, Suppl. p. 35 (1981).

2. Menesini Chen, M.G., Micheli, V., Chen, J.S., Pompucci, G. and Sbardellati, E.: Riassunti del 27º Congresso Nazionale della Società Italiana di Biochimica, Parma, p. 235-236 (1981).

3. Chen, J.S., Menesini Chen, M.G. and Pompucci, G.: Abstract in III, International Congress of Clinical Enzymology, Salzburg, pp. 115 (1981).

4. Weber, K. and Osborn, M.: J. Biol. Chem. 244 (16), 4406-4412 (1969).

IDENTIFICATION OF FILARIAL PROTEINS, INDUCING AN IMMUNE RESPONSE IN NATURAL INFECTIONS OF RODENTS WITH DIPETALONEMA VITEAE (FILAROIDEA) BY THE "WESTERN BLOTTING" TECHNIQUE

Richard Lucius*; Ernst-Wilhard Rauterberg**; Hans-Jochen Diesfeld*

* Universität Heidelberg, Institut für Tropenhygiene und öffentliches Gesundheitswesen, INF 324, D-6900 Heidelberg FRG

** Universität Heidelberg, Institut für Immunologie und Serologie, INF 324, D-6900 Heidelberg, FRG

Abstract: Proteins obtained by SDS solubilisation of various developmental stages (male and female adults, uterus content, microfilariae, infectious larvae) of the filaria *Dipetalonema viteae* were separated by SDS-polyacrylamide gel electrophoresis (SDS-PAGE) and electrophoretically transferred onto nitrocellulose filter sheets ("Western Blotting"). The filter sheets were incubated with serum pools from either infected or uninfected hosts of different species (*Meriones unguiculatus*, *Mastomys natalensis*, golden hamster). Reacting antibodies in the sera were detected by radiolabeled Protein A and visualized by subsequent autoradiography only in the sera of infected rodents. Depending on the species of hosts and the filarial stage differing patterns of autoradiographic bands were obtained. In the sera of infected *M. unguiculatus* antibodies only against few parasite components were observed. Antibodies from infected *Mastomys natalensis* and golden hamsters reacted with a higher number of parasite proteins. It is suggested that the species of rodents tested in the present experiments reveal a species specific immune response towards solubilized proteins of *D. viteae*.

Electrophoresis '82

Introduction

One of the main problems in parasitology is due to the lack of specificity of serodiagnostic tests for worm infections. This especially helds true for filariasis. Filariasis is a group of tropical diseases, which are transmitted by arthropods as vectors. Serodiagnostic tests are urgently needed, because the diagnostic tests up to now are time consuming or relatively unreliable. Such tests could be of value not only for clinical diagnosis, but also for epidemiological and biological studies (1). This situation is reflected by the fact that the WHO demands the development of species specific serodiagnostic tests for a long time (2, 3). The tests used up to now are mainly based on whole homogenates, extracts or sections of worms as antigens. Such reagents contain a great number of components which enhances the probability of unspecific reactions. Isolation of clearly defined antigens could greatly improve the specificity of serodiagnostic tests.
In our laboratory, we have studied the localisation of antigens in several species of helminths (4, 5, 6, 7) and we have performed work about the protein composition of filariae (8). As a laboratory model, we used Dipetalonema viteae, a filaria which can be maintained in ticks and laboratory rodents. However, conventional techniques could only identify the protein composition and failed to detect proteins of importance in the host parasite interaction. In order to identify such antigens, we investigated the immune response of Meriones unguiculatus against developmental stages of D. viteae by the "Western Blotting" technique according to Burnette (9) and subsequent immunerecognition (10).

"Western Blotting" is a technique for the electrophoretical transfer of proteins from SDS-polyacrylamide gels onto nitrocellulose filter sheets. In the present study, we tried to approach the question, to what extent results about the immune response of rodents towards filariae can be extended from one laboratory model to another one. For this purpose we investigated the immune response of three species of rodents towards D. viteae by the "Western Blotting" technique.

Material and Methods

Parasites: The life cycle of D. viteae was maintained in Ornithodorus moubata and in Meriones unguiculatus and Mastomys natalensis. Adult worms were obtained from male rodents 10 - 25 weeks after the infection. Uterus content (UC) was isolated from adult female filariae as described by Diesfeld et al. (5). Microfilariae (MF) were isolated from the blood of hosts by the technique of Wegerhoff (11). Infectious larvae from ticks were isolated as described by Prüsse et al. (8).

Sera: Pooled sera were prepared from groups of infected M. unguiculatus, Mastomys natalensis and golden hamsters, respectively 18 - 23 weeks after the infection. Sera of uninfected animals served as controls. Each pool was obtained from ten different animals. At least two different pools were tested in each species.

SDS-PAGE: The worm material was solubilized in a sample buffer (63 mM Tris-HCl, pH 6.8; 10 % (v/v) glycerol; 3 % (w/v) SDS; 5 % (v/v) 2-mercaptoethanol; 0.1 % (w/v) bromphenol blue) as described by Prüsse et al. (8). The samples were subjected to electrophoresis in SDS-polyacrylamide slab gels (dimensions: 140 mm x 110 mm x 2 mm with T = 7,5 % or 10 % and C = 2,5 %) in a discontinuous buffer system (12) with the modifications of Ruppel and Cioli (13). The electrophoresis was performed in an apparatus according to Studier (14) at room temperature with 140 V and 100 mA for about 5 hours. In such experiment, proteins of known molecular weight (Pharmacia,

Electrophoresis Calibration Kit) were used for molecular weight determination.

Blotting and Immunorecognition: The electrophoretical transfer of proteins onto nitrocellulose filter sheets was performed as described by Towbin et al. (18) at 6V/cm and 200 mA for 16 hours at 4 °C in a buffer containing 25 mM Tris, 186 mM glycine and 20 % (v/v) methanol. The immunerecognition was performed according to Burnette (9) with modifications reported in a previous study (10).
Briefly, the filter sheets with the adharent proteins were incubated for 90 min in a solution of 3 % (w/v) bovine serum albumin (BSA) (99 % purity, Serva) in a solution containing 0.9 % (w/v) NaCl and 10 mM Tris-HCl, pH 7.4 (Tris-saline) in order to saturate unspecific protein binding on the nitrocellulose for 90 min. This and the following steps were performed at room temperature. After a short rinsing in Tris-saline the nitrocellulose strips were incubated for 90 min with a solution of 3 % (w/v) BSA in Tris-saline containing the sera to be tested in a final solution of 1 : 30. The strips were washed according to the following scheme: Three changes of Tris-saline, two changes of Tris-saline containing 0.05 % (v/v) Triton-X 100 (Serva), three changes of Tris-saline. Each step required five minutes and was performed under constant shaking. Subsequently, the filter strips were incubated for 90 min with a solution of 3 % (w/v) BSA in Tris-saline containing 1.5 - 2.5 µCi ^{125}J Protein A (specific activity 30 µCi/mg; Amersham-Buchler, Germany). To remove unspecific binding of radioactivity, the sheets were washed again according to the scheme mentioned above and the last step was repeated until the washing liquid did not contain any detectable radioactivity. The nitrocellulose strips were dried between filter papers, fixed to a support with scotch tape and exposed at - 70 °C to an X-ray film (Kodak) in combination with an intensifying screen (Dr. Goos, Germany).

Identification of labeled Bands: After autoradiography, the nitrocellulose sheets were stained with Coomassie Brilliantblau R 250 (Merck, Germany) in a solution of 50 % (v/v) methanol and 7 % (v/v) acetic acid for three minutes and destained in the solution without dye. Autoradiographic bands of low molecular weight were identified by direct comparison with the stained nitrocellulose replica. Labeled bands of high molecular weight were identified by comparison of the Rf values of the autoradiographies with those of the gels, which had been stained after the transfer.

Results

Separation of proteins by SDS-PAGE: The separation by SDS-PAGE of proteins of male and female filariae and of UC, MF and L_3, solubilized in a buffer containing SDS and 2-mercaptoethanol, resulted in reproducible patterns, each consisting of various bands (Fig. 1). 40-50 bands of very different intensity could be distinguished in each stage. Most of the bands occurred in all stages, however, some others were stage specific. Furthermore the density of the bands revealed some stage specific differences. When compared to other stages, the pattern of UC showed the most obvious differences.

Transfer of proteins: The Coomassie Blue staining of the nitrocellulose filters revealed a pattern of bands similar to the stained gels. However, this held true only for proteins of a molecular weight of about 100.000 and smaller. As shown by the staining of both, the filter and the polyacrylamide gel after blotting, it became obvious that the proteins with a higher molecular weight were transferred only to a lesser degree. This effect was dependent upon the "pore size" of the gel since gels of higher polyacrylamide concentration showed the impediment of the transfer to an even higher degree.

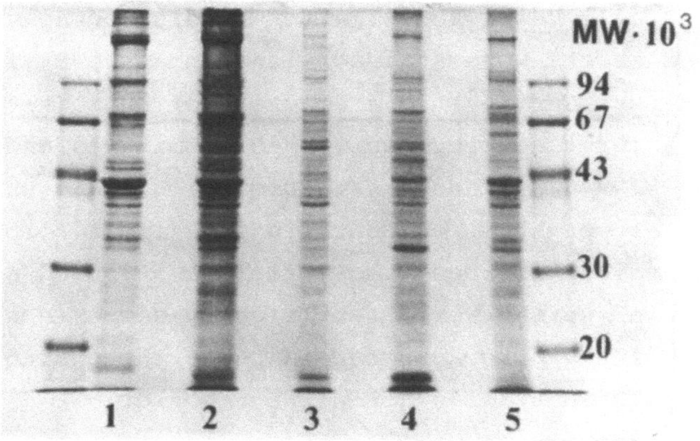

Figure 1: Proteins of various stages of D. viteae, separated by SDS-PAGE in a gel with T = 10 %. (1) males; (2) females; (3) UC; (4) MF; (5) L_3.

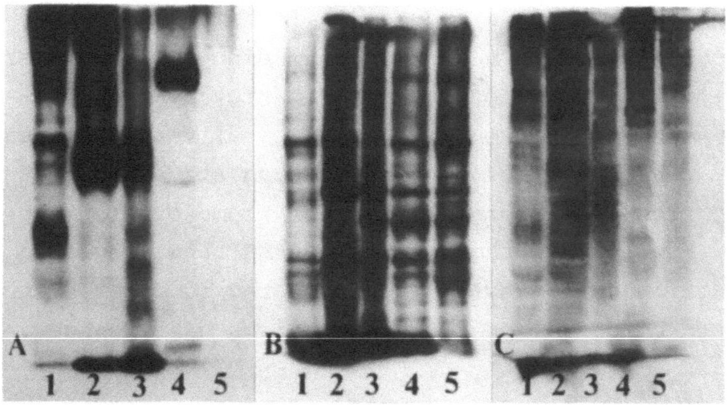

Figure 2: Autoradiographic visualization of IgG antibodies from pooled sera of three species of rodents, bound to proteins of various developmental stages of D. viteae. Separation by SDS-PAGE in a gel with T = 7.5 %. (1) males; (2) females; (3) UC; (4) MF; (5) L_3. (A) M. unguiculatus; (B) M. natalensis; (C) golden hamster.

Immunerecognition of Proteins: The interaction of immunoglobulins from infected host sera with filarial proteins was detected by radioactively labeled Protein A and subsequent autoradiography. By this technique it could be demonstrated that the infected sera contained antibodies reacting only with some of the protein bands transferred onto the nitrocellulose filters. These antibodies could only be detected in sera of infected hosts since control sera from healthy uninfected rodents failed to react. Using the same antigen preparations and the same serum pools, the autoradiographic patterns were highly reproducible. However, there were variations in the intensity of labeling, when different pools of sera were used. The autoradiographs revealed variations in the patterns of labeled bands depending on the host species and the parasite stage. When using pooled sera of infected M. unguiculatus, relatively few bands of the parasite stages were labeled. Fig. 3a shows that the pattern of autoradiographic bands differed obviously between the different stages. Proteins of adult filariae revealed a higher number of labeled bands than those of UC and the larval stages.

Pooled sera of M. natalensis reacted to more bands of parasite proteins than those of M. unguiculatus (Fig. 3b). In this host the antibodies reacted with the proteins of different stages in a highly similar pattern.

An even higher number of bands was visualized using sera of infected golden hamsters (Fig. 3c). In this case, the pattern of labeled bands of different stages showed slight variations.

The majority of autoradiographic bands could be correlated to stained protein bands. However, some of the labeled bands could not clearly be correlated. The intensity of the majority of the stained protein bands, which are reacting with antibodies, did not correspond to the intensity of the bands in the autoradiograph. Several thin protein bands of the high molecular weight region appeared as very pronounced bands in the autoradiograph, whereas some heavily stained bands were only faintly labeled.

Discussion

The present study demonstrates the successful application of a combination of "Western Blotting" technique and immunorecognition for the identification of filarial proteins reacting with antibodies from infected hosts. Pooled sera of various species of infected hosts were tested. They reacted with several protein bands, whereas control experiments with sera of uninfected hosts did not reveal a binding of antibodies to filarial proteins. Thus, it is suggestive that the antibodies are induced by the filarial infection and not by other antigens.

The experiments suggest that different species of hosts show a variation in the pattern of antibodies as a result of a different immune reaction to the same filarial parasite. This further strengthens the fact, that laboratory findings about the immune response of laboratory animals cannot be extended from one species to another one. The most striking finding of our experiments was the inverse correlation between the number of labeled bands reflecting the diversity of antibodies, with the viability of worms in different host species. As such, the lowest number of immunolabeled bands was detected with sera of infected M. unguiculatus. In this rodent, which is closely related to the natural host of D. viteae, the parasites show a good growth and production of microfilariae when compared to the situation in golden hamsters (15,16), where the the highest number of labeled bands was observed. In M. natalensis, the situation is intermediate. Thus, it is tempting to speculate that this correlation is due to the immune response of the host. However, it could also be a secondary phenomenon, based on biochemical differences in the parasite host interaction. I. e. hosts, offering a suitable microenvironment to the parasite, are possibly not confronted with as much immunogenic material of degenerating parasites as inappropriate hosts.

The applied techniques allow some suppositions about the properties of the labeled proteins. In some cases, protein bands which were very thin in the gel or could not be detected, were heavily labeled in the autoradiography. This suggests that proteins which have a very low concentration in the parasite, elicit a strong immune response in the host. Such immunogens occur in the secretions, which are shed into the host (5, 6). Their concentration in the worm is supposedly very low. On the other hand, faibly stained or undetectable bands in the gel could consist of glycoproteins with a low protein content, their staining intensity being very low. The finding, that sometimes very pronounced bands in the gel are only faintly marked could be explained by the migration of several proteins as one band, the reacting ones being represented in very low concentrations. Other separation methods could give in the future more information about the composition of such bands.

It has to be considered that the number of labeled bands possibly does not correspond to the number of reacting proteins, because the solubilisation of the proteins occurred under reducing conditions. Under such conditions, proteins can be split into their subunits, each subunit being separated in one band. Thus, antibodies against one protein could have bound to several bands of subunits. A further complication could be due to the limited transfer of some proteins, especially in the high molecular weight region and using gels of high concentration. The quantity of many proteins of high molecular weight was by far sufficient for an intensive marking of many bands, although their concentration was too low to be detected by Coomassie Blue staining. This is a proof for the high sensibility of the method, nevertheless it is possible that some more labeled bands in this region could have been detected, if their concentration on the nitrocellulose strips had been higher.

Another limiting factor could be the property of Protein A, not to bind to certain populations of IgG. Kronvall and Williams (17) described that Protein A does not bind to IgG_3 of human sera. Similar restrictions of the binding to IgG could apply for sera of the experimental animals used in this study. Furthermore, it is possible that antibodies did not react to antigens, because the antibody binding sites of antigens were changed by the experimental procedure, especially by SDS, in a way that antibodies could not recognize them any more. Within these limitations, the combination of "Western Blotting" and immunerecognition proved to be a suitable tool for investigating immunological properties of parasite proteins and the immune reaction of infected hosts by a relatively simple and rapid method.

Literature

1. WHO: Document TDR/FIL/SWG (6)/81.3 (1981)
2. WHO: Second annual report; Special Progr. for Research and Training in Tropical Diseases (1978)
3. WHO: Third Annual Report; Special Progr. for Research and Training in Tropical Diseases (1979)
4. Diesfeld, H. J.; Kirsten, C.: Tropenmed. Parasit. 26, 499 (1975)
5. Diesfeld, H. J.; Kirsten, C.; Tropenmed. Parasit. 29, 27 (1978)
6. Diesfeld, H. J.; Kirsten, C.: Trans. Roy. Soc. Trop. Med. Hyg. 73, 533 (1979)
7. Diesfeld, H. J.; Kirsten, C.; Stappert, U.: Tropenmed. Parasit. 32, 253 (1981)
8. Prüsse, A.; Diesfeld, H. J.; Vollmer, S.: in press, Helminthology (1982)
9. Burnette, W. N.: Anal. Biochem. 112, 195 (1981)
10. Lucius, R.; Rauterberg, E. W.; Diesfeld, H. J.: submitted to publication (1982)
11. Wegerhoff, P. H.: Diss., Biologie, Univ. Tübingen (1977)
12. Laemmli, U. K.: Nature 227, 680 (1970)
13. Ruppel, A.; Cioli, D.: Parasitology 75, 339 (1977)
14. Studier, F. W.: J. Mol. Biol. 79, 237 (1973)
15. Diesfeld, H. J.; Kirsten, C.; Heid, M.; Alle, W.: Zbl. Bakt. Hyg., I. Abt. Ref. 277, 115 (1982)
16. Gass, R. F.; Tanner, M.; Weiss, N.: Z. Parasitenkd. 61, 73 (1979)
17. Kronvall, G.; Williams, R. C.: J. Immunol. 103, 828 (1969)
18. Towbin, H.; Staehlin, T.; Gordon, J.: Proc. Natl. Acad. Sci. USA 76, 4350 (1979).

AN ECONOMICAL MICROSCALE COUNTER-IMMUNOELECTROPHORESIS METHOD FOR SCREENING OF ALEUTIAN DISEASE IN MINK

Elie Gillis, °Ann Gillis-Van Maele, Paul Coucke, °Annie Ponteur-Van Stichel and Yvette Galle-De Ville

Laboratorium voor Fysiologische Scheikunde, Rijksuniversiteit, Ledeganckstraat 35, B-9000 Gent, Belgium and °Industriële Hogeschool van het Rijk C.T.L., Voskenslaan 270, B-9000 Gent, Belgium

Introduction

Aleutian disease (A.D.) is a viral infection of mink(1). Most wild mink strains develop a certain degree of resistance to the infection. Farm-reared animals are often selected for colour shades of their fur and the resulting inbred lines may show persistent sensitivity to viral invasion, with concommittant loss of viability or at least a decline of fur quality. Highly evaluated "blue" mink, e.g., is prone to Aleutian disease affliction. Because both horizontal and vertical spreading of the infection is rapid through a sensitive farm population, mink farmers try to eliminate a potential virus source before visual damage to the affected animals has occured.

All infected animals produce specific antibodies against A.D.-virus. In sensitive breeds, this defence mechanism appears to be ineffective and viral disease further develops. Yet, the enhanced production of antibodies can be used as a detection method for infected animals (2). For economical reasons, the test was often performed in a non-specific way by the Lugol method, that screens out only the heavily affected part of the farm population. Sanitation of a breed, however, requires early detection of a serological reaction in virus carriers.

Electrophoresis '82

Due to its specificity and sensitivity, counter-immunoelectrophoresis is the method of choice to achieve efficient carrier elimination (3). Clinical laboratories are not inclined to fulfill this demand for immunoelectrophoretic analyses, firstly because the burden of incoming requests is very irregular and secondly because the mink farmer cannot afford to pay the regular price for this type of expertise.

We developed a method, that allows us to cope with many thousands of samples, arriving only twice a year (in weaning and pelting seasons) and expected to be analysed within a fortnight. Economizing on space, time and material investments, we were able to provide reliable analytical results at a modest price.

Materials and Methods

<u>Electrophoresis medium</u>.

Solution A (per l) pH 8,8	:	4,82 g Tris
		2,05 g barbital
		8,13 g sodiumbarbital
		0,30 g sodiumazide
Solution B (per l)	:	20 g agarose, high electroendosmosis (Serva)

<u>Gel support and well forming system</u> (fig. 1). Gels are cast on glass microscope slides (26 x 76 mm), positioned two at a time on a 4 mm thick aluminum support (fig. 2). An acrylate well former, gliding along nylon guide rods, is lowered onto the slides (fig. 3). It positions 20 siliconized glass rods, 1.7 mm dia., according to an alternating pattern of pairs (fig. 4) a few tenths of a mm above the slide surface. In a similar design, the aluminum support can take 4 slides and two gel formers.

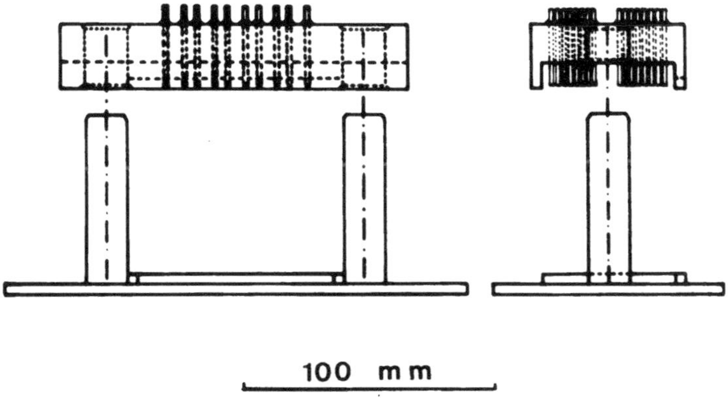

Fig. 1. Well former : front and side views.
A : aluminum base
B : nylon guide rods
C : acrylate block
D : siliconized glass rods.

Fig. 2. Positioning of microscope slides on aluminum support.

Fig. 3. Well former, with acrylate block in uppermost position, showing array of glass rods.

Gel casting procedure. The slides are numbered and dated on the matte part with a lead pencil. A thin layer of warmed solution B is spread on the slides with a soft brush. The slides are dried in an oven at 50°C.

The aluminum slide support is warmed on a platen to about 35°C. Slides are quickly positioned between guide rulers on the support and the whole is transferred to a cooling plate with active cooling switched off. Gel material, made by mixing 1 vol. A and 1 vol. B and kept at about 90°C, is pipetted on the coated glass to a level of 2 ml per slide. Immediately afterwards, the well former is lowered over the slides and active cooling of the support is switched on. After gelification, i.e. within about 5 min, the well former can be lifted off and the slides removed from the support.

Normally gel slides are stored in a humid atmosphere for at least 24 h before use. Good results can be obtained even with fresh gels, but it is more economical to make a stock of slides in a routine fashion. Pencil marks on the slides facilitate regular turn-over.

Sample application. Mink blood serum to be tested is supplied in hematocrit tubes. The capillaries are inserted with their open end in a Drummond delivery bulb and cut about 2 mm below the liquid level.

Closing the bulb air vent with the index, the operator inserts the liquid-filled end of the tube in a well at the anodic side of a pair and by slight pressure on the bulb makes liquid contact with the bottom of the well. Withdrawal of the capillary, with the bulb air vent open, assures delivery of about 1 µl of serum in the well without overflow.

The cathodic wells are filled in series with a dispensing pipette depositing 1 µl of viral antigen (United, Middleton, Wis., USA).

The first and last well pairs on each slide serve as a control, using 1 µl of standardized positive serum (United).

Electrophoresis. Slides are placed gel-side down between two paper wicks, dipping in buffer tanks. Electrophoresis is run for about 1 h at 120 V (5-6 mA per slide) in an hermetic enclosure to prevent drying of the gels. The actual running time is monitored by checking for the appearance of a visible immunoprecipitate between the control pairs. Visibility is enhanced against a dark background by oblique lighting.

Washing and staining. Non-precipitated material is removed in 0.9 % NaCl, washing the gels overnight at 4°C with very gentle agitation. The agar layer is dried at 50-60°C. Staining is performed in a solution containing 0,05 % Coomassie Brilliant Blue, 0,1 % Copper sulfate, 10 % acetic acid and 25 % ethanol. After 1 h, gels are destained with 7,5 % acetic acid, 30 % ethanol. Results can be read after about 2 h. Stained gels can be dried again and stored indefinitely in a dry atmosphere.

Results and Discussion

A positive result is read as a single sharp protein band between a well pair. Normally, there is very little stainable background in this area (fig. 4).

Strongly hemolyzed sera give rise to remanent protein staining, spreading towards the anode. The intertwined disposition of the pairs is such, that this background staining lies in between the adjacent pairs and never reaches the next pair on the same flow line.

Incomplete destaining does not seriously hamper interpretation and can be corrected, using the same slides. It sometimes happened, that one of the controls was not sufficiently visible or blurred (fig. 5, slide 55). These tests had to be repeated as a whole. All tentatively positive precipitation smears turned out to be neatly positive.

Incomplete salt rinsing before staining may result in less readability of the precipitin lines (fig. 5, slide 17). The complete test run has to be repeated.

Since only a small amount of serum is applied per test, dubious results or accidental damage to the slides do not lead to irreparable loss of information. In case the content of a

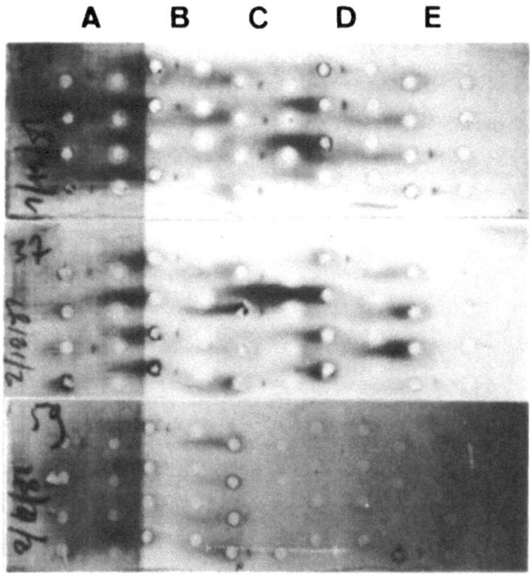

Fig. 4. Normal results of micro-scale counter-immunoelectrophoresis :
Controls (positive) : upper left and lower right well pairs.
Positive test results :

Slide	Row	Well pair (from top)
31	A	2, 4
	B	1, 2, 4
	C	1, 2, 3, 4
	D	1, 2
	E	3
37	A	3
	B	1, 3, 4
	C	1, 2, 4
	D	1
59	B	1
	C	4

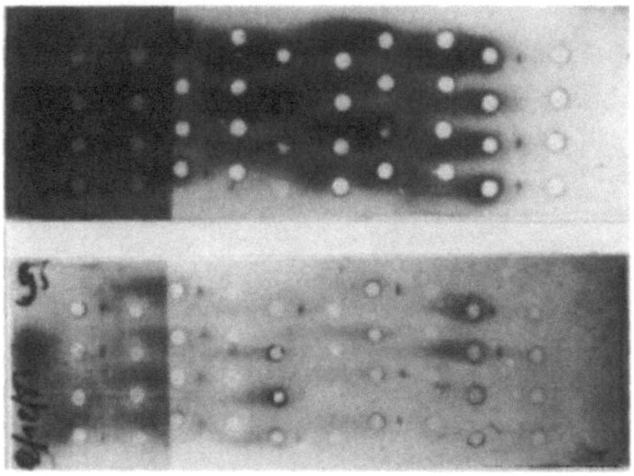

Fig. 5. Non-valid test results :
Slide 17 : incomplete salt rinsing ; test to be repeated.
Slide 55 : one control missing ; test to be repeated.

hematocrit tube has to be tested over again, a sample of 1 µl is withdrawn with a piston pipette by insertion of a fine polyethylene capillary into the lumen of the tube.

The lower limit of visibility of a precipitin line, that is to be interpreted as positive, is a matter of discussion. In the light of the farm sanitation objective, it is better to overestimate the number of positive cases, than to miss some.

When testing a batch of newly bought animals, certified to be AD-free, we found a few positive reactions (fig. 4, slide 59, besides the controls : row B, first pair and row C, fourth pair). It is not clear whether these animals were indeed frauds or whether they were freshly infected upon arrival. To determine the degree of reliability of our virosis detection method, experimental infections of previously negative animals should be performed. At the moment, however, a veterinary research program on mink is not supported in our country.

Apart from AD-viral antigen, other negatively charged antigenic molecules can be detected with this method, provided their electrophoretic mobility is not too high. In fact, the experimental set-up was tested with serum albumins (bovine or human) and the corresponding rabbit antisera. Later on, detection of urea-denatured keratins was performed. SDS-denatured keratins, however, migrate so fast, that precipitation occurs in the antiserum well.

Conclusion

The microscale counter-immunoelectrophoresis method, specifically developed for screening of Aleutian Disease virus in mink farms, is economical, fast and reliable. Precise determination of its lower detection limit awaits further research with deliberately infected animals. The technique is suitable for other negatively charged antigens.

Acknowledgement. We are grateful to Dr. J. Haagsma, Drs. B.A. Bokhout and W. Stok of C.D.I., Rotterdam, NL for valuable advice.

References

1) D.D. Porter, A.E. Larsen, H.G. Porter ; Adv. Immunol. (1980) 29, 261-286.

2) H.J. Cho, D.G. Ingram ; Can. J. Comparat. Med. (1973) 37, 217-223.

3) H.J. Cho, J. Greenfeld ; J. Clin. Microbiol. (1978) 7, 18-22.

CROSSED IMMUNOELECTROPHORESIS FROM SDS-POLYACRYLAMIDE GELS

Julio Coloma and Jose V. Castell
Laboratorio de Enzimologia Experimental. Centro de Investigacion. Hospital La Fe. Ministerio de Sanidad. Valencia 9. Spain.

Introduction

Crossed immunoelectrophoresis has produced several powerful techniques with increased resolution, sensitivity and specificity (1). Proteins are resolved according to electrophoretic mobilities (2), electrofocusing (3) or by means of gel filtration (4) in the first dimension and then run at right angles into the antibody containing agarose in the second dimension. Polyacrylamide gel electrophoresis has the advantages over agarose that samples are concentrated and improved separation is obtained when discontinuous buffer systems are used. Inclusion of SDS in the first acrylamide run (SDS-PAGE), a well established procedure, resolves proteins according to molecular weights and renders a valuable system for protein studies. It was found desirable to combine SDS-PAGE with a subsequent immunochemical analysis in the 2nd dimension. However, small amounts of free SDS severely disturb the formation of immunoprecipitates in the second dimension. This is evidenced by the non-formation of antibody-antigen complexes and the occurence of additional precipitation lines and flakes. Several attempts to overcome such difficulties lead to the use of different techniques to minimize the effect of SDS: extensive dialysis of SDS-treated samples (5), extraction with solvents (6), inclusion of a non-ionic detergent in an intermediate agarose barrier (7), and recently, crossed immunoelectrophoresis after fixation and staining of the gel (8). In our

Electrophoresis '82

hands most of these drawbacks could be avoided after effective removal of free SDS from the gel with an ion exchange resin.

Material and Methods

Electrophoresis in the first dimension was carried out according to the general procedure of Laemli (9) on polyacrylamide slab gels. Stacking gel was 5% acrylamide, 0.15% Bis, 0.1% SDS in Tris/HCl 0.1 M pH 6.8. Separating gel was 12.5% acrylamide, 0.1% Bis, 0.1% SDS in Tris/HCl 0.1 M pH 8.4; electrophoresis buffer as recommended by Laemli (9).

Removal of SDS. Gel strips were cut off and placed in a beaker with water and slowly stirred for one to two hours in the presence of a cloth bag containing MB-1 (BDH), a mixed ion exchange resin. Removal of SDS was satisfactory after one hour.

Crossed Immunoelectrophoresis (CIE). 1.25% agarose M (LKB) in Tris/barbital/glycine (20mM/0.35mM/40mM, pH 8.8, ionic strength 10^{-3}) containing 5% antiserum was used in the second dimension. As schematized in Fig. 1, the antibody containing agarose was cast in the upper part of a precoated agarose glass plate (band A), to give a volume/surface ratio of 0.14 ml/cm^2. Then the SDS-free polyacrylamide strip (band B) previously equilibrated in buffer for 10 min. was placed close to the band A and sealed with 1.25% agarose (band C). Electrophoresis was carried out at 15 V/cm for 20 h at $4^{o}C$. The same buffer was used as electrophoresis buffer. At the end of the run the acrylamide strip was removed and stained for proteins. The agarose plate was washed, dried and stained as in conventional immunoelectrophoresis.

Fig. 1. Assembly of gel layers for CIE

Results and Discussion

The presence of free SDS is the major source of artifacts in crossed immunoelectrophoresis as previously mentioned. Several procedures were tested to evaluate their ability to reduce free SDS from acrylamide gels down to a concentration which did not interfere with the 2nd dimension run (5-8). In our hands the use of reported methods to minimize SDS interferences failed, when applied to CIE. Precipitation lines were not defined and gave rise to artifacts. Washing of the polyacrylamide strip in the presence of a mixed ion-exchange resin was a convenient and effective method to remove SDS from gels. As shown in Fig. 2, an excellent alignment of bands in the acrylamide strip and peaks in the antibody containing agarose was observed. Immunoprecipitation lines were clearly defined and narrow and no artifacts were noticed. The resolution and sensitivity was high. Some bands faintly stained in the polyacrylamide gel were clearly visible after CIE. Only high molecular weight proteins (>250K) yielded diffuse peaks, though this varied according to preparations.

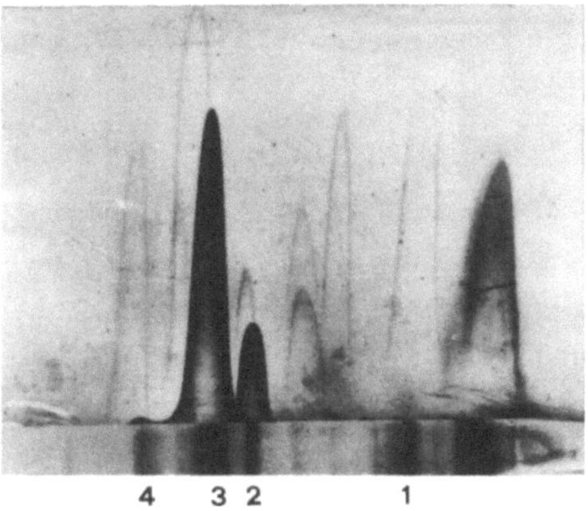

Fig. 2. Crossed immunoelectrophoresis from SDS-polyacrylamide gel. Sample was rat serum fraction 33-45% sat. $SO_4(NH_4)_2$. Antiserum to whole rat serum was raised in rabbits. 1) Haptoglobin, 2) Transferrin, 3) Albumin, 4) Prealbumin.

The staining of acrylamide gels after the 2nd dimension immunoelectrophoresis showed that after removal of free SDS, proteins still were able to move towards the agarose gel. SDS-denatured proteins were recognized by antibodies in the second dimension and defined precipitation lines were formed.

Losses and diffusion of protein bands during the removal of SDS were not important. Fig. 3 summarizes the behaviour of four different proteins along the duration of washes. Band diffusion was estimated by gel scanning as the ratio of half-peak width/peak height. After two hours washing only low molecular weight proteins showed increased diffusion. This, however, had no influence on the resolution of CIE.
As Fig. 4 shows, the protein content in the acrylamide gel decreased only lightly. For albumin (band 3) the protein loss was estimated less than 6% after two hours washing.

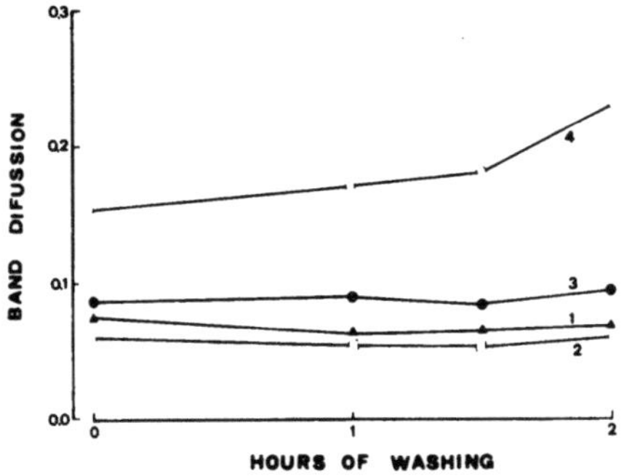

Fig. 3. Protein diffusion during SDS removal.

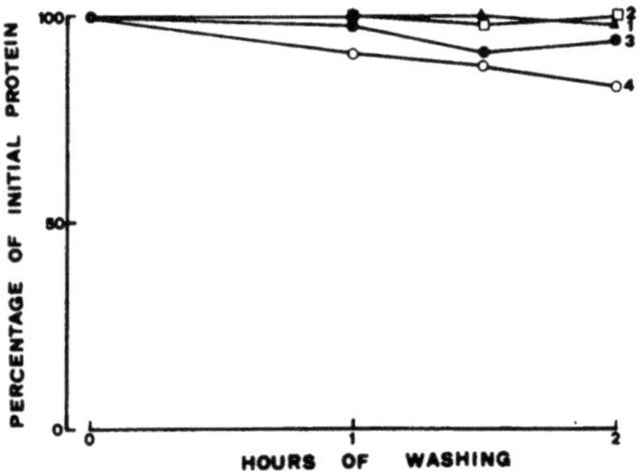

Fig. 4. Protein loss during SDS removal.

(Protein legends as in Fig. 2)

The present work shows that the use of an ion-exchange resin is a good choice to remove free SDS from gels. It was possible to combine SDS-PAGE with a subsequent immunochemical analysis with no interference caused by SDS. This procedure proved effective with soluble proteins. The advantages of this modification were: simplicity in handling and excellent alignment of proteins resolved according to molecular weights with immunoprecipitates.

References

1. Laurell, C.B.: Analyt. Biochem. 10, 358 (1965).
2. Clarke, H.G.M., Freeman, T.A.: Prot. Biol. Fluids 14, 503 (1967).
3. Söderholm, J., Smyth, C.J.: Scand. J. Immunol. 4, 107 (1975).
4. Nielsen, C.S.: Scand. J. Immunol. 4, 101 (1975).
5. Nielsen, C.S., Bjerrum, O.J.: Scand. J. Immunol. 4, 73 (1975).
6. Phillips, H.M.: Analyt. Biochem. 117, 398 (1981).
7. Converse, C.A., Papermaster, D.S.: Science 189, 469 (1975).
8. Kessler, R.E.: Analyt. Biochem. 116, 129 (1981).
9. Laemli, H.M.: Nature 227, 680 (1975).

COMPARISON OF WAGTAIL LIVER ISOZYME PATTERNS IN AGAROSE THIN LAYER ISOELECTRIC FOCUSING WITH THOSE OBTAINED BY ISOELECTRIC FOCUSING ON POLYACRYLAMIDE GELS (PAGIF)

Manfred Gemeiner, Ingrid Miller and Harald Czikeli
Institut für Medizinische Chemie
Veterinärmedizinische Universität Wien, A-1030 Vienna, Austria

Introduction

To encompass genetic polymorphism of three yellow wagtail subspecies (Motacilla flava flava, M.f.cinereocapilla, M.f.feldegg; Passeriformes, Aves) we used isoelectric focusing on ultrathin layers of polyacrylamide (1) or agarose gels (2). We are presenting here a comparison of the two methods in respect to advantages and limitations in the evaluation of the isozyme patterns of lactate dehydrogenase (LDH), alcohol dehydrogenase (ADH), malate dehydrogenase (MDH), glucose phosphate isomerase (GPI), phosphoglucomutase (PGM), acid phosphatase (ACP) and esterases.

Materials and Methods

Liver samples were obtained and processed as previously described (1). Quail liver sample homogenates were used as references.
Ultrathin (100 µm) polyacrylamide gels (T = 5.4, C = 2.8) for isoelectric focusing and gels (0.8 % agarose) for isoelectric focusing on ultrathin layers of agarose (200 µm) were prepared with the "flap technique" (3) as described previously (1,2). Focusing conditions and ampholyte mixtures for the different enzymes as described in the figures.

Results and Discussion

LDH: Focusing on ultrathin layers of polyacrylamide gels and staining with

Electrophoresis '82

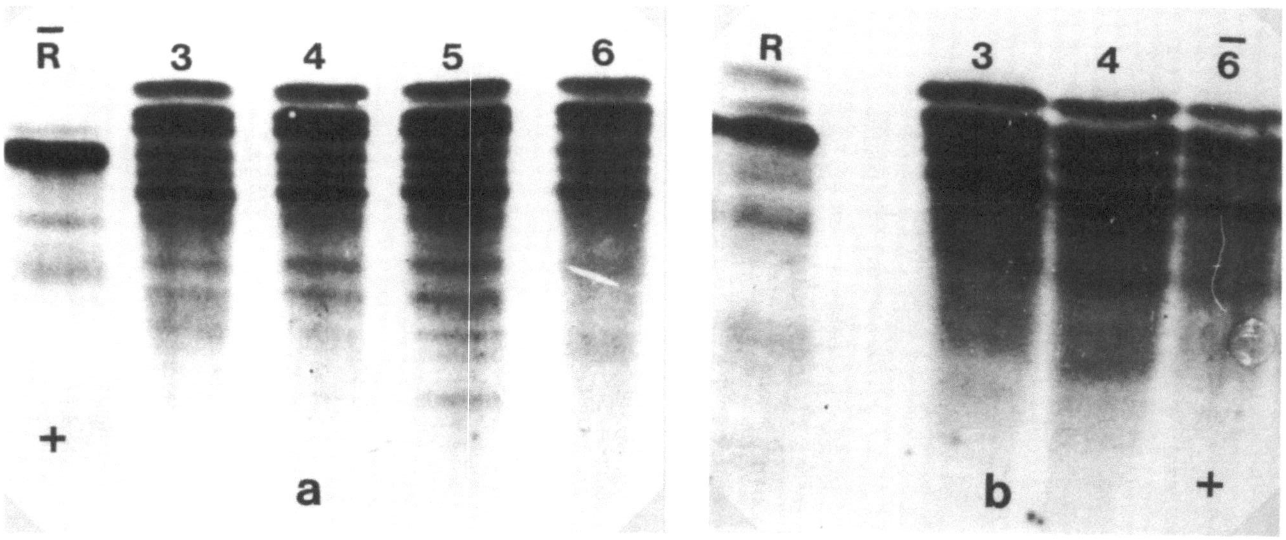

Figure 1. Focusing and staining of LDH-isozymes

a) polyacrylamide gel, Ampholine pH 7-9; prefocusing for 30 min at settings of 1500 V, 45 mA, 10 W, 10°C, corresponding to an initial voltage of 400-500 V and a final voltage of 1500 V. Focusing to steady state at 2250 V.h. Anolyte: DL-aspartic acid/L-glutamic acid, 0.025 M each. Catholyte: 1.8 M ethylenediamine.

b) agarose gel containing 10% (w/v) sorbitol, Ampholine pH 7-9 and Servalyt T pH 4-9 in the ratio of 1:1. Focusing to equilibrium at 2500 V.h, electrolytes as mentioned above.

Samples: (R) reference sample, (3)-(6) wagtail samples. 5 µl samples (liver homogenates diluted 1:5 corresponding to 10 µg protein/mm^2) were applied anodically. Staining: with calcium lactate, NAD, MTT and PMS as mentioned in (1,2); method used: agarose overlay (2 mm).

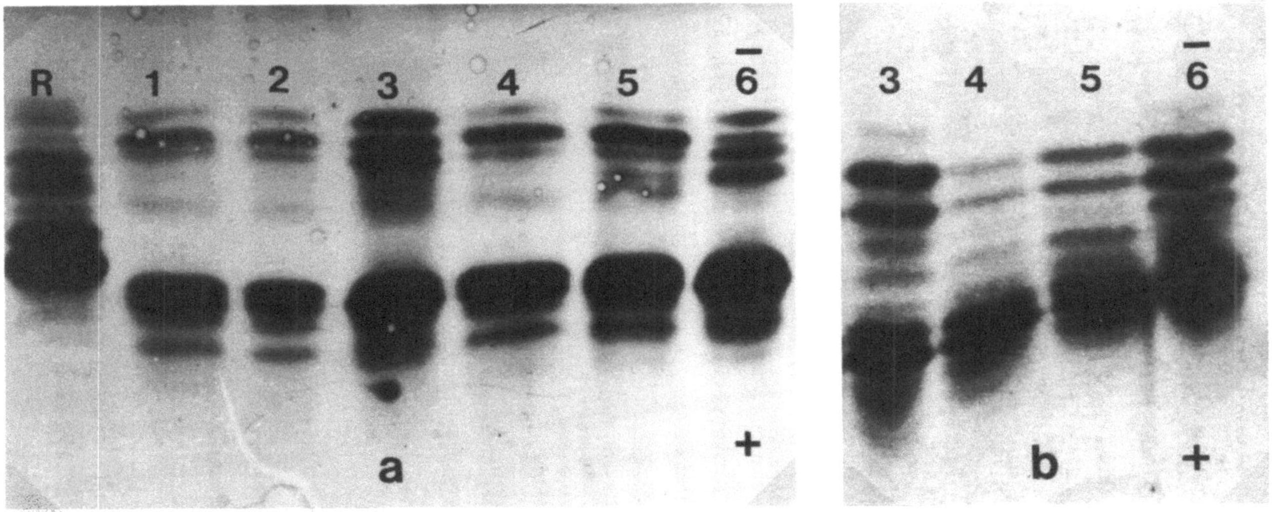

Figure 2. Focusing and staining of PGM-isozymes
a) polyacrylamide gel, b) agarose gel containing 10% (w/v) sorbitol; both gels containing Servalyt T pH 4-9. Focusing conditions as mentioned in Fig.1. Samples: (1)-(6) wagtail samples, (R) reference sample. 5 µl samples (liver homogenates diluted 1:8 corresponding to 6 µg protein/mm^2 for polyacrylamide or diluted 1:10 corresponding to 5 µg protein/mm^2 for agarose gels) were applied anodically. Staining: with α-D-glucose-1-phosphate, $MgCl_2$, glucose-6-phosphate dehydrogenase, NADP, PMS and MTT as mentioned in (1,2); method used: agarose overlay (2 mm).

an agarose overlay of 2 mm thickness gave sharp LDH-isozyme bands (Fig.1a). The different samples show strong similar band patterns in the more basic region and band differences at the anodic side. Using the same ampholyte mixture for isoelectric focusing on ultrathin layers of agarose, we obtained broader bands. The use of a slightly different ampholyte mixture brought band sharpening, nevertheless resolution of the bands was not as good as in polyacrylamide gels (Fig.1b). Staining of LDH-isozymes in solution led to weaker bands, even when higher sample concentrations were applied. With the ultrathin agarose overlay (2) we obtained good staining but more diffuse patterns.

PGM: Staining of PGM-isozymes with an agarose overlay after isoelectric focusing on polyacrylamide or agarose gels revealed the pattern shown on Fig.2a and Fig.2b. Looking at the patterns from the agarose focusing, we can notice that one of the cathodic bands seemed to comigrate with an other; this giving a different pattern to the one obtained by PAGIF. Using the ultrathin agarose overlay we obtained strong bands similar to those obtained by our standard overlay technique. Trying to better resolve the cathodic bands we also used gels containing 1 % (w/v) acrylamide and 0.8 % (w/v) agarose. The patterns obtained looked very similar to those given by agarose isoelectric focusing.

ADH: ADH-patterns after isoelectric focusing on ultrathin layers of polyacrylamide or agarose gels are shown in Fig.3a and 3b. Trying to enhance resolution in ultrathin layers of agarose we also used sucrose, sorbitol or glycerol as gel additives. Using these additives we could demonstrate the presence of an unexpected sorbitol dehydrogenase in our samples.

MDH: MDH-patterns on polyacrylamide and on agarose gels are shown in Fig.4a and 4b. Patterns in both gels differ by the appearance of a diffuse stained zone on the agarose gels and by the fact that bands are generally broader on agarose gels.

ACP: Focusing of ACP-isozymes on polyacrylamide gels and staining by the overlay technique showed good resolution and strong band staining (Fig.5a). Focusing on ultrathin layers of agarose and staining by the standard overlay technique gave patterns very similar to those obtained by PAGIF (Fig.5b).

Esterases: Focusing on polyacrylamide gels and staining in solution gave the esterase patterns presented in Fig.6a. Optimization of the ampholyte mixtures for the agarose focusing was a critical step in order to obtain

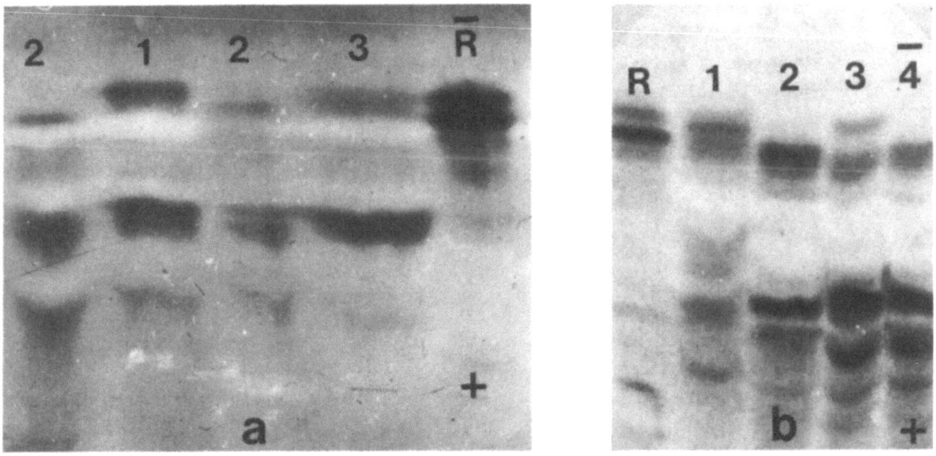

Figure 3. Focusing and staining of ADH-isozymes
a) polyacrylamide gel, Servalyt T pH 4-9 and Pharmalyte pH 8-10.5 in the ratio of 1:2; b) agarose gel containing 10% (w/v) sucrose, Servalyt T pH 4-9 and Pharmalyte pH 8-10.5 in the ratio of 1:1. Focusing conditions as mentioned in Fig.1. Samples: (R) reference sample, (1)-(4) wagtail samples. 5 µl samples (liver homogenates diluted 1:1 corresponding to 25 µg protein/mm^2) were applied anodically. Staining: agarose overlay with amyl alcohol, NAD, MTT and PMS as mentioned in (1).

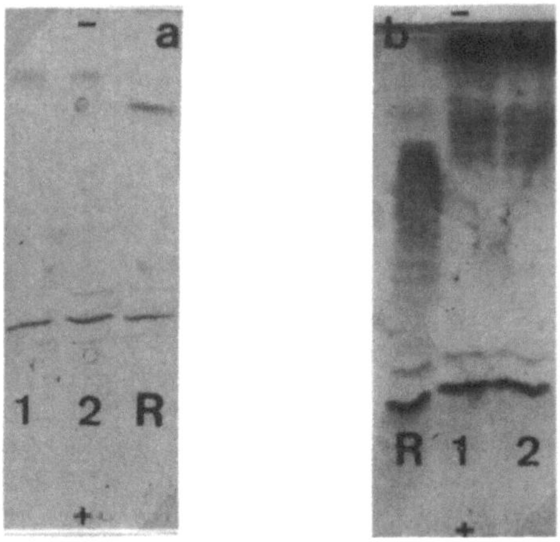

Figure 4. Focusing and staining of MDH-isozymes
a) polyacrylamide gel; b) agarose gel containing 10% (w/v) sucrose, both gels containing Servalyt T pH 4-9. Focusing conditions as mentioned in Fig.1. Samples: (R) reference sample, (1,2) wagtail samples. 5 µl samples (liver homogenates diluted 1:15 corresponding to 3 µg protein/mm^2 for polyacrylamide or diluted 1:10 corresponding to 5 µg protein/mm^2 for agarose gels) were applied anodically. Staining: agarose overlay with malic acid, NAD, pyruvate, MTT and PMS as mentioned in (1,2).

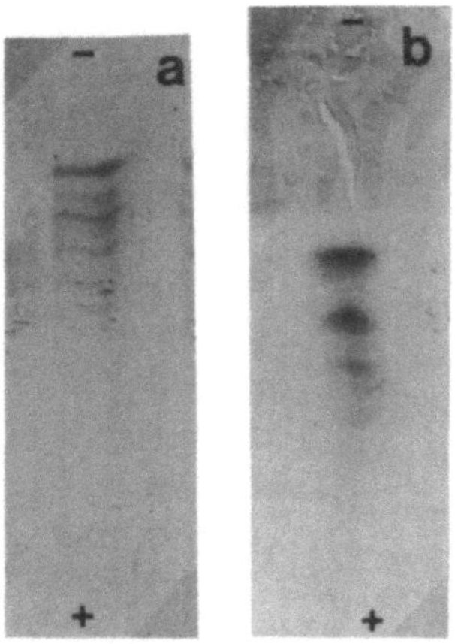

Figure 5. Focusing and staining of ACP-isozymes
a) polyacrylamide gel; b) agarose gel containing 10% (w/v) sorbitol, both gels containing Servalyt T pH 4-9 and Ampholine pH 5-7 in the ratio of 3:1. Focusing conditions as mentioned in Fig.1. Sample: wagtail sample. 5 µl samples (liver homogenates diluted 1:1 corresponding to 25 µg protein/mm^2) were applied anodically. Staining: agarose overlay with α-naphthylphosphate and Fast Blue B in citrate buffer pH 4.5, as mentioned in (2).

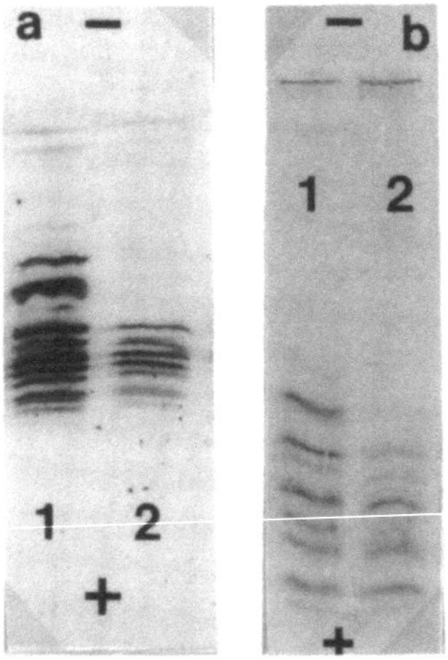

Figure 6. Focusing and staining of esterase-isozymes
a) polyacrylamide gel, Ampholine pH 3.5-5 and Servalyt T pH 4-9 in the ratio of 1:4; b) agarose gel containing 10% (w/v) sorbitol, Ampholine pH 3.5-5 and Servalyt T pH 4-9 in the ratio of 1:3. Focusing conditions as mentioned in Fig.1. Samples: (1,2) wagtail samples. 5 µl samples (liver homogenates diluted 1:4 corresponding to 12.5 µg protein/mm^2 for polyacrylamide or diluted 1:10 corresponding to 5 µg protein/mm^2 for agarose gels) were applied cathodically. Staining: staining in solution with α-naphthylacetate and Fast Blue RR in Tris-HCl buffer pH 7.0 as mentioned in (1).

Figure 7. Focusing and staining of GPI-isozymes
a) polyacrylamide gel; b) agarose gel containing 10% (w/v) sorbitol, both gels containing Servalyt T pH 4-9. Focusing conditions as mentioned in Fig.1. Samples: (R) reference sample, (1,2) wagtail samples (liver homogenates diluted 1:3 corresponding to 16 μg protein/mm^2) were applied anodically. Staining: agarose overlay with fructose-6-phosphate, NADP, glucose-6-phosphate dehydrogenase, MTT and PMS as mentioned in (1).

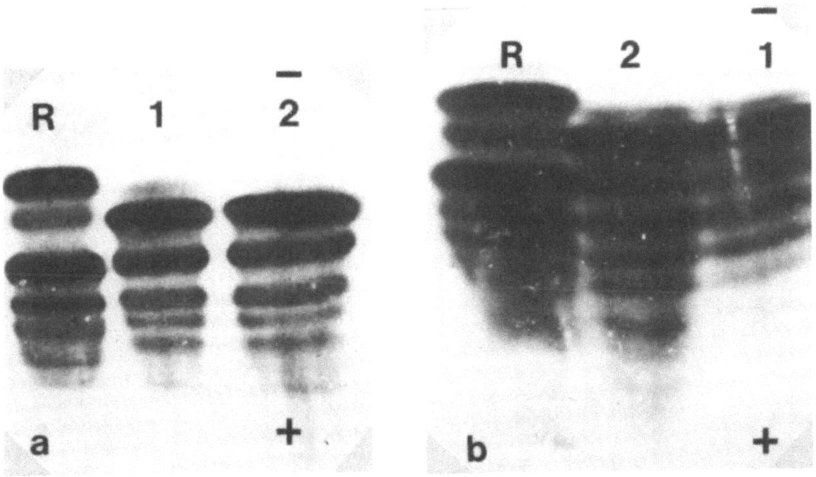

good band separation (Fig.6b). Focusing on the agarose gels gave somewhat broader bands, but due to the flatter pH-gradient in the region of interest, the resolution of the bands was good enough to overcome the broadening effect. GPI: GPI-patterns obtained by the two methods showed a high degree of similarity (Fig.7a and Fig.7b).

Conclusion

- Different pH-gradients were obtained when using same ampholyte mixtures for PAGIF and agarose IEF
- The different additives used for agarose IEF showed to have some effects on the separation (band sharpening and resolution) and the staining ("by-staining" of a sorbitol dehydrogenase).
- Isozyme patterns could be visualized by three staining procedures:
  - a thick (2 mm) agarose overlay used for LDH, ADH, MDH, PGM, GPI and ACP

  - an ultrathin agarose overlay (0.2 mm) prepared with the "flap technique", of good use for LDH and PGM but with some limitations in applicability (i.e. ADH and ACP)
  - staining in solution giving often somewhat broader bands and used for LDH, ACP and esterases
- Similar sample concentrations could be used for both methods
- PAGIF gave usually somewhat better resolution and sharper bands than agarose IEF, nevertheless isozyme patterns obtained by the two methods can be compared.

This work was supported in part by grant No.4059 from the "Fonds zur Förderung der wissenschaftlichen Forschung". We also acknowledge gratefully the apparative support of the Fond "200 Jahre Vet.Med.Universität Wien".

References

1. Gemeiner, M., Miller, I., Czikeli, H.: Electrophoresis 3, 146-151 (1982).
2. Gemeiner, M., Miller, I., Czikeli, H.: submitted.
3. Radola, B.J.: Electrophoresis 1, 43-56 (1980).

MALATE DEHYDROGENASE ISOENZYMES FROM RABBIT LENS, IRIS AND RETINA §

Johan Bours, Manfred Garbers and Otto Hockwin

Division of Biochemistry of the Eye, Institute of Experimental Ophthalmology, University of Bonn, Sigmund-Freud-Strasse 25, D-5300 Bonn-Venusberg, F.R.G.

Summary

Malate dehydrogenase (MDH) is visualized by substrate staining of the isofocused components on thin-layer polyacrylamide gels. MDH shows microheterogeneity of its components within 3 classified groups of isoenzymes. In lens, the major portion forms isoenzyme No. 1 which is focused in the region of α-crystallin at pI = 4.30 - 4.85. Some other components of lens isoenzymes No. 2 and 3 are focused at intermediate and high pI and are present in lower concentrations. In the lens cortex, the MDH-1 and -3 isoenzymes are predominant, while in the nucleus the MDH-1 remains stable and the concentrations of MDH-3 and -2 are reduced. Generally, activities of lens MDH isoenzymes decrease with increasing age.

Iris and retina show similar patterns as lens, but the major isoenzyme No. 1 is located in the region of albumin at pI = 4.60 - 5.50 and has a different appearance. The patterns of iris and retina are incomplete at an age of 1.2 months, in the 1-year-old rabbit the 3 isoenzyme groups show their full heterogeneity as a result of ageing.

§Research supported by the "Deutsche Forschungs-Gemeinschaft (Ho 249/8,12,13).

Electrophoresis '82

Introduction

The enzyme MDH (L-malate: NAD^+ oxidoreductase, E.C. 1.1.1.37) exists in threeisoenzymes with different electrophoretic mobility and isoelectric point (IEP) (1), which are: MDH-1, located in the soluble fraction from cell cytoplasma (s-MDH), MDH-3, in the mitochondrial fraction (m-MDH) and in addition a third MDH with intermediate mobility, MDH-2, has also been detected.

Markert and Møller (2) defined in 1959 the term "Isoenzyme" for enzymes with identical substrate specificity, but of different molecular forms of different electrophoretic mobility. MDH-1, -2 and -3 are according to this definition isoenzymes.

A previous study (3) dealt with thin-layer isoelectric focusing of lactate dehydrogenase activities in rabbit lens parts and intraocular tissues. Here are considered the enzyme activities of malate dehydrogenase (MDH) of the same tissues after thin-layer isoelectric focusing, followed by visualization with a specific staining method according to Dietz and Lubrano (4).

Materials and Methods

Rabbit lenses, iris and retina were dissected from enucleated eyes immediately after killing the animals. Age of the investigated rabbits was 1.2 months and 1-year. The lens wet weights were determined and the lenses were separated into nucleus and cortex by a trephane and further dissection. The tissues obtained were homogenized (Ultra-Turrax) in distilled water at 4°C and centrifuged at 9,000 x g for 1 h. The supernatants were used for further tests.

The protein content of the supernatants was determined according to Lowry et al. (5). Isoelectric focusing was made according to Bours (6). A 5% polyacrylamide gel plate was made with a crosslinking percentage of 2.6. The gel contained 1% Ampholine carrier ampholytes pH (3.5-5):(5-7):(7-9) = 1:1:1. Isofocusing was carried out in a Multiphor electrophoresis apparatus (LKB 2117) during 16 h at 100-110 V and 8-0.8 mA at 4°C (3).

The staining for MDH activities was made according to Dietz and Lubrano (4), while lactate was replaced for DL-malic acid, disodium salt. The stock solutions for staining and volumes are:

Stock Solution	Amount used (ml)
DL-Malic acid, disodium salt, M.	12.0
NAD (nicotinamide-adenine dinucleotide , (10 mg/ml	12.0
NaCl, 0.1 M	12.0
$MgCl_2$, 5 mM	12.0
Phosphate buffer 0.5 M, pH 7.4	30.0
Nitroblue tetrazolium, 1 mg/ml	30.0
Phenazine methosulfate, 1 mg/ml	3.0
	111.0 ml

The NAD solution is prepared weekly and the other solutions are stable for long periods when stored at 5°C in the dark. After isofocusing, the gels were incubated for 0.5 - 1.5 h at 37°C in the dark in 111 ml of the medium. After staining, the gels were washed with 10% acetic acid. The photos were scanned in a Chromoscan densitometer. Isoelectric points were determined at 4°C according to Bours et al. (3).

Results

MDH isoenzymes (L-malate: NAD^{+}oxidoreductase, E.C. 1.1.1.37) were detected by tetrazolium staining after isoelectric focusing. The enzyme activities in the extracts of rabbit lens, iris and retina are shown in Figure 1 and 2. For iris extract 2.6 mg protein shows maximum enzyme activity (Fig. 1c). For retina extract the maximum enzyme activity lies at 3.5 mg protein (Fig. 1f). In iris and retina the predominant MDH isoenzyme is No. 1 (Fig. 1c-g), which is the soluble anodic

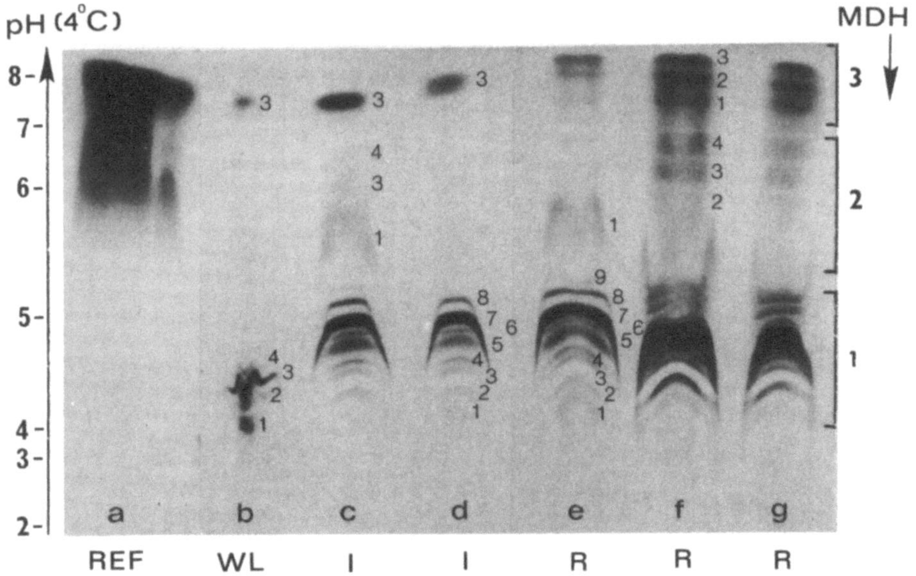

Fig. 1. Thin-layer isoelectric focusing of MDH isoenzymes from various ocular tissues of a 1-year-old rabbit. Display of microheterogeneity. a) Ref = reference MDH isoenzyme from pig heart (mitochondrial) (Boehringer Mannheim), 25 μg, b) whole lens (WL) 2.7 mg protein, c) Iris (I) 2.6 mg protein, d) Iris (I) 1.6 mg protein, e) Retina (R) 6.5 mg protein, f) Retina (R) 3.5 mg protein, g) Retina (R) 0.7 mg protein.

form from cell cytoplasm (s-MDH). The cathodic form of MDH is the isoenzyme No. 3, the mitochondrial form (m-MDH). MDH-1 appears to be firmly bound to albumin, and consists of 8-9 components of isoelectric points from 4.60 - 5.50 (Table 1). The MDH-2 has an intermediate position with IEP's of 6.00 - 7.00, and is only present in minor concentrations (Fig. 1, 2; Table 1). MDH-3, the mitochondrial form (m-MDH) shows 3 components of IEP's from 7.50 - 8.30 (Fig. 1; Table 1). m-MDH from pig heart mitochondria serves as a reference (Fig. 1a, 2h).

The MDH isoenzyme patterns from the rabbit lens show also three isoenzymes: the MDH-1 (s-MDH), the MDH-3 (m-MDH) and the MDH-2 at intermediate position (Fig. 1a, 2a-c). The MDH-1 is firmly bound to α-crystallin, and shows 4 isofocused components of IEP's from 4.30 - 4.85 (Table 1). In the rabbit lens cortex,

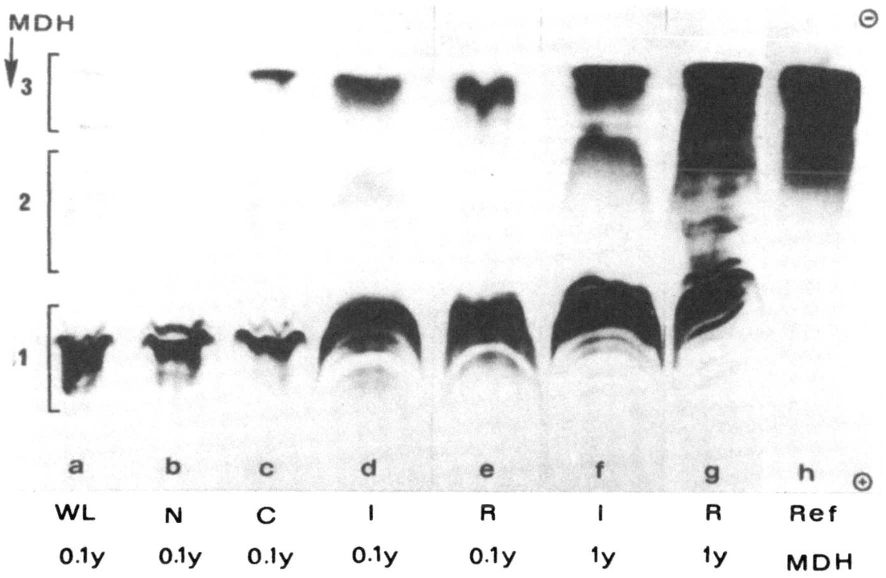

Fig.2. Thin-layer isoelectric focusing of MDH isoenzymes from various ocular tissues of 0.1 and 1.0-year-old rabbits, also related to ageing. a) Whole Lens (WL) 4.2 mg protein, 0.1-year -old, b) Nucleus (N) 4.0 mg protein, 0.1-year-old, c) Cortex (C) 3.9 mg protein, 0.1-year-old, d) Iris (I) 0.2 mg protein, 0.1-year-old, e) Retina (R) 0.2 mg protein, 0.1-year-old, f) Iris (I) 0.7 mg protein, 1-year-old, g) Retina (R) 0.7 mg protein, 1-year-old, h) Ref = reference MDH mitochondrial isoenzyme from pig heart, 25 μg.

which forms the youngest tissue in the lens that still contains mitochondria, the MDH-3 (m-MDH) is still present at an age of 0.1 year (Fig. 2c), but tends to diminish with ageing (fig. 1a).

The intensity of the stained MDH- components increases in iris and retina, when the lens extracts from 0.1- and 1.0-year -old rabbits are compared. In Fig. 2(d,e) the isoenzyme patterns of 0.1-year-old iris and retina are comparable, but in the 1-year-old iris and retina (Fig. 2f,g) the concentrations of all isoenzymes have increased considerably.

Table 1. Isoelectric points (IEP) of isofocused components of MDH isoenzymes from lens, iris and retina from 1-year-old rabbit, measured at 4°C.

MDH iso-enzyme No.	Whole Lens (Fig.2a) Component No.	IEP	%	Iris (Fig.1c) Component No.	IEP	%	Retina (Fig.1f) Component No.	IEP	%
1	1	4.30	75.9	1	4.60	61.7	1	4.60	43.1
	2	4.50		2	4.80		2	4.80	
	3	4.75		3	4.90		3	4.90	
	4	4.85		4	5.05		4	5.05	
				5	5.08		5	5.08	
				6	5.10		6	5.10	
				7	5.20		7	5.20	
				8	5.30		8	5.30	
							9	5.50	
2	1	6.00	14.3	1	6.00	14.2	1	6 00	14.7
	2	6.25		2	6.30		2	6.30	
	3	6.50		3	6.50		3	6 50	
	4	7.00		4	7.00		4	7.00	
3							1	7.50	42.2
							2	8.00	
	3	8.30	9.8	3	8.30	24.1	3	8.30	
			100.0%			100.0%			100.0%

Discussion

It is well established now that MDH consists of 3 isoenzymes (7, 8, 9). In tissues like rat liver (10), chicken brain and retina (11), rat retina (12) and human brain (13) the cytoplasmic MDH activity was greater than the mitochondrial activity. The cytoplasmic MDH has anodic mobility and IEP's from

5.0 - 5.5 (13) while the mitochondrial MDH has highest cathodic mobility (9).

The distribution of MDH isoenzymes in rabbit iris and retina is in agreement with published data of chicken retina (11) and rat retina (12).

In the rabbit lens, the anodic MDH activity is also greater than the cathodic MDH activity (14, 15, 16). Further are the activities of lens MDH isoenzymes also subjected to ageing: in the whole rabbit lens this activity is lower at 1-year- than at 0.1-year-old. This is in agreement with the decrease of MDH activity found in the bovine lens with increasing age (17,18). We found a higher MDH enzyme activity in the rabbit lens cortex (fig. 2c), which agrees with the results of Cotlier (19) who found notably high MDH activities in lens equatorial parts; Hockwin et al. (20) found in calf and bovine lens equator higher MDH activities than in the lens nucleus.

Isoelectric focusing shows always a higher resolving power than observed with any other technique, which is shown in Figures 1 and 2 as the microheterogeneity of the MDH-1 with 8-9 isofocused bands. These data show that MDH-1 is firmly bound to albumin (Fig. 1c-g, 2d-g) (14). The MDH-1 may also be bound to α-crystallin (Fig. 1b, 2a-c) since it co-precipitates with α-crystallin (14, 15, 16) during isoelectric focusing. The IEP's of MDH-1 (Fig. 1c-g, 2d-g; Table 1) (16) are well in agreement with the IEP of albumin (21, 22). It is also clearly shown in Figures 1b and 2a-c that the IEP of MDH-1 in lens is lower than that of MDH-1 in iris and retina, which agrees with the facts that the IEP of α-crystallin (4.85) (23) is lower than the IEP of albumin (4.95 - 5.50) (21).

References

1. Latner, A.L., Skillen, A.W.: Isoenzymes in biology and medicine, Acad. Press, London· New York 1968.
2. Markert, C.L., Møller, F.: Proc. Nat. Acad. Sci. U.S. 45,

753-763 (1959).

3. Bours, J., Neuhaus, H., Hockwin, O.: Albrecht v. Graefes Arch. klin. exp. Ophthal. 203, 9-19 (1977).

4. Dietz, A.A., Lubrano, T.: Analyt. Biochem. 20, 246-257 (1967).

5. Lowry, O.H., Rosebrough, N.J., Farr, A.L., Randell, R.J.: J. Biol. Chem. 193, 265-275 (1951).

6. Bours, J. : Comp. Biochem. Physiol. 65B, 215-222 (1980).

7. Yakulis, V,J., Gibson, C.W., Heller, P.: Am. J. Clin. Pathol. 38, 378-382 (1962).

8. Narang, S., Narang, N.: Comp. Biochem. Physiol. 49B, 477-490 (1974).

9. Kuan, K.N., Jones, G.L., Vestling, C. S. : Biochemistry 18, 4367-4373 (1979).

10. Burzazyan, I.B., Adunts, G.T.: Bol. Zh. Arm. 34, 336-340 (1981).

11. Pasantes-Morales, H., Klethi, J., Urban, P. F., Mandel, P: J. Neurochem. 19, 1183-1188 (1972).

12. Bonavita, V., Guarneri, V., Ponte, F. : Vision Res. 5, 113-121 (1965).

13. Taggert, R.T., Karn, R.C.: Fed. Proc. 35, 1657 (1976).

14. Garbers, M., Bours, J.: Exp. Eye Res. 25, 537 (1977).

15. Bours, J., Garbers, M., Hockwin, O.: Ophthal. Res. 12, 146-147 (1980).

16. Bours, J., Garbers, J., Hockwin, O.: Electrophoresis '79 (Ed. Radola, B.J.), 539-544 (1980).

17. Hockwin, O., Licht, W., Noll, E., Glasmacher, M.: Ber. Dtsche Ophthal. Ges. 70, 368-375 (1969).

18. Fink,H., Hockwin, O.: Albrecht v. Graefes Arch. klin. exp. Ophthal. 191, 177-195 (1974).

19. Cotlier, E.: J. Histochem. Cytochem. 12, 419-423 (1964).

20. Hockwin, O., Weimar, L., Noll, E., Licht, W.: Albrecht v. Graefes Arch. klin. exp. Ophthal. 170, 99-116 (1966).

21. Bours,J. : Exp. Eye Res. 15, 299-319 (1973a).

22. Bours, J.: Science Tools 20, 29-34 (1973b).

23. Bours, J., Brahma, S.K.: Exp. Eye Res. 16, 131-142 (1973).

THE MICROHETEROGENEITY FORMS OF HUMAN SERUM PROTEINS ARE REMARKABLY CONSTANT. A SYSTEMATIC STUDY OF REACTION WITH CON A DURING ELECTROPHORESIS

F. R. Hinnerfeldt, J. Albrechtsen and T. C. Bøg-Hansen.
The Proteinlaboratory University of Copenhagen, Sigurdsgade 34, 2200 N, Denmark.

The microheterogeneity of serum proteins is well established both in affinity chromatography and in affinity electrophoresis with lectins. Examples are shown in other chapters in these proceedings: AFP (20); pregnancy associated proteins and normal serum glycoproteins (11); orosomucoid (3) and elsewhere (1, 2, 4, 7, 8, 9, 15, 16, 17, 18, 21, 22, 23, 24).

Generally analytical affinity electrophoresis may give the same information as column affinity chromatography (6). In addition to the minuterization of the experiment, electrophoresis offers several advantages in resolution. In fact, crossed immunoelectrophoresis with lectin in the first dimension gel, appears to be the most sensitive method for determination of glycoprotein microheterogeneity (5). From our previous studies we came to the conclusion that the microheterogeneity forms of the serum proteins vary in a more or less systematic way during some pathological disorders. Therefore we decided to perform a systematic study of the variation of the microheterogeneity forms in normal serum as a basis for an evaluation of the pathological variation.

Identification of microheterogeneity forms.

When we performed crossed immunoelectrophoresis with conA in the agarose gel using identical conditions in all experiments we observed a remarkably constancy in the change of pattern

Electrophoresis '82

for the normal human serum proteins. The experiments were performed as follows, cf. Fig.1. The first dimension gel contained a standard concentration of conA (0,1mM/ml gel), enough to give a significant affinity precipitate with the precipitating glycoproteins and enough to give a significant retardation with the conA-binding glycoproteins. After the first dimension separation, the proteins were electrophoresed through an intermediate gel containing a specific displacer, α-methyl-D-glycopyranoside, which was taken into the first dimension gel by electroendosmosis (10,19) where it dissolved the affinity precipitate. Part of the affinity precipitate is still seen in Fig. 1 (AP). Proteins were identified by their characteristic immunoprecipitates with specific antibodies.

During the first dimension electrophoresis the conA-binding proteins formed complexes with conA which changed their mobility. These proteins are hemopexin, α-2-macroglobulin, haptoglobin, orosomucoid, α-1-antitrypsin, α-1-antichymotrypsin, C'3, transferrin, α-2-HS-glycoprotein and IgA. For some glycoproteins their microheterogeneity is revealed as one or more additional peaks or shoulders, e.g. orosomucoid, haptoglobin, α-1-antitrypsin, α-2-macroglobulin, transferrin and Gc globulin. The major conA precipitating glycoproteins were α-2-macroglobulin, haptoglobin, hemopexin and C'3. Table 1 shows the number of microheterogeneity forms (peaks) and their relative mobility. Significant differences in the number of peaks and their relative mobility was not found for 30 normal sera, except for α-1-antichymotrypsin and haptoglobin which only sometimes showed the minor component (peak 3 and peak 1, respectively).

The relative mobility of serum proteins without conA in the first dimension was determined in order to see if incorporation of conA in the first dimension gel had any significant influence on the standard deviation (s_x) of the relative mobility (Table 2). Comparing s_x in Table 1 with s_x in Table 2 shows that they do not differ significantly. In other words, incorporation of conA in the first dimension gel did not influence the reproduceability of the experiment.

Generally we used anti total serum protein antibodies in order to see the behavior of as many proteins as possible. Most other investigators have used specific ab in their experiments

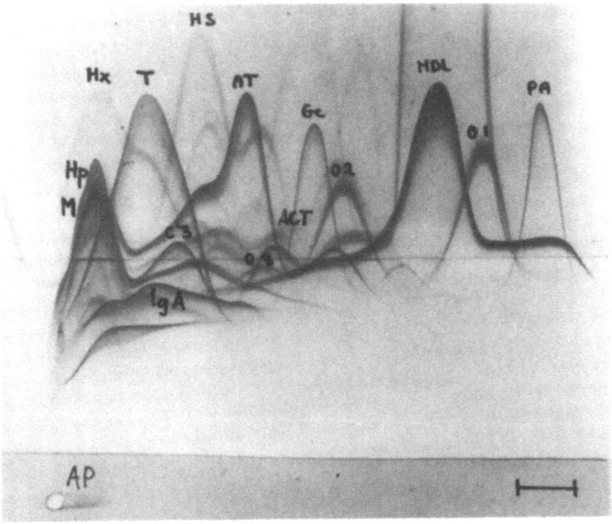

Fig. 1. Crossed immunelectrophoresis with conA in the first dimension gel and a specific displacer in the intermediate gel.
Sample: 1 µl human serum (individual doner).
Antibody: 400 µl anti total human serum proteins per 10x10 cm^2.
ConA concentration: 19.3x 10^{-6} mol/cm^2 corresponding to 3.5µl/cm^2 5% (w/v) conA in H_2O.
Specific displacer: 1.5% α-methyl-D-glycopyranoside (w/v).
Gel: 1% (v/w) agarose type HSA from Litex, Glostrup.
The plates were washed, pressed, dried and stained with Coomassie brilliant blue according to standard procedures.
First dimension: 1.5 hour at 10 V/cm. Second dimension: 2 V/cm for 18 h.
Electrophoresis buffer: Tris, barbitone, calciumlactate, pH8.6.
The bar represents 1 cm.
Abbreviations:

PA= prealbumin	M = a-2-madroglobulin
O = orosomucoid	C'3 = complement factor 3
AT = a-1-antitrypsin	Hx = hemopexin
ACT= a-1-antichymotrypsin	T = transferrin
Gc = Gc-globulin	HDL = high density lipoprotein
HS = a-2-HS-glycoprotein	
Hp = haptoglobin	

Table 1

Relative electrophoretic mobility of the microheterogeneity forms of 12 human serum proteins with free conA in the first dimension gel.

Serum protein		$\bar{x}$	s_x	$s_{\bar{x}}$	n
Prealbumin (PA):		100			
Orosomucoid (O):	Peak 1 m	86,75	2,70	0,48	32
	Peak 2	59,59	1,72	0,30	32
	Peak 3	42,71	1,95	0,35	31
α-1-antitrypsin (AT):					
	Peak 1	78,50	2,80	0,55	26
	Peak 2	57,00	2,16	0,38	32
	Peak 3 m	38,50	1,57	0,28	32
	Shoulder	28,18	1,59	0,34	22
	Peak 4	6,34	1,81	0,32	32
α-1-antichymotrypsin (ACT):					
	Peak 1	66,11	1,88	0,44	18
	Peak 2 m	47,17	0,83	0,24	12
	Peak 3 a	ND			
Gc globulin (Gc):	Peak 1	72,94	2,37	0,42	33
	Peak 2 m	57,31	3,03	0,54	33
α-2-HS-glycoprotein (HS):					
	Peak 1	41,96	1,53	0,29	33
	Peak 2 m	28,52	1,66	0,29	33
Haptoglobin (Hp):	Peak 1 a	(63,25)			4
	Peak 2	23,47	4,32	0,76	32
	Peak 3 m	6,72	2,08	0,37	33
α-2-macroglobulin (M):					
	Peak 1	17,32	2,67	0,50	28
	Peak 2 m	6,53	1,83	0,32	32
Complement factor 3 (C'3):					
	Peak 1	22,65	1,72	0,34	26
	Peak 2 m	6,42	1,86	0,38	24
Hemopexin (Hx):	Shoulder	16,44	1,99	0,37	27
	Peak	6,81	1,87	0,33	32

cont.

Table 1 continued.

Transferrin (T):	Peak 1	37,39	1,50	0,27	32
	Shoulder	25,22	1,48	0,35	18
	Peak 2	16,44	1,92	0,34	32
IgA:	Peak 1	19,63	1,36	0,34	16
	Peak 2	8,63	2,10	0,40	27
HDL:		79,50	2,06	0,40	26

The relative electrophoretic mobility, x was the mobility of the serum protein in percent of the mobility of prealbumin (PA). The migration distance was determined according to (3). For details see Fig. 1.
Abbreviations:: a = absent in most sera, m = major component, ND = not determined, $\bar{x}$ = mean, s_x = standard deviation, $s_{\bar{x}}$ = standard error of the mean.

Table 2.

Relative electrophoretic mobility of 12 human serum proteins. Without conA in the first dimension gel.

Serum protein		$\bar{x}$	s_x	$s_{\bar{x}}$	n
Prealbumin (PA):		100			
Orosomucoid (O):		87,86	1,93	0,42	21
α-1-antitrupsin (AT):		76,86	2,35	0,51	21
α-1-antichymotrypsin (ACT):		68,29	1,65	0,40	17
Gc globulin (Gc):	Peak 1 ND				
	Peak 2 m	61,33	2,09	0,45	21
α-2-HS-glycoprotein (HS):		58,85	2,39	0,53	20
Haptoglobin (Hp):		53,10	3,22	0,70	21
α-2-macroglobulin (M):		59,43	2,99	0,65	21
Complement factor 3 (C'3):		36,81	1,21	0,26	21
Hemopexin (Hx):		37,67	2,31	0,50	21
Transferrin (T):		36,00	1,00	0,22	21
IgA:		33,35	3,44	0,84	17
HDL:		79,53	2,39	0,62	15

For details se Fig. 1. Abbreviations as in Table 1.

Table 3.
Quantification of microheterogeneity forms. Planimetri of microheterogeneity forms of 12 human serum proteins with free conA in the first dimension gel.

Serum protein	conA aff.	$\bar{x}$	s_x	$s_{\bar{x}}$	n
Prealbumin (PA):					
One peak	no				
Orosomucoid (O):					
Peak 1	no	44,53	5,40	1,39	15
Peak 2	yes	40,40	4,36	1,12	15
Peak 3	yes	16,13	3,09	0,80	15
α-1-antitrypsin (AT):					
Peak 1	no	0,93	0,37	0,10	14
Peak 2	yes	13,27	1,98	0,51	15
Peak 3	yes	59,53	3,64	0,94	15
Peak 4	yes	26,27	4,84	1,26	15
α-1-antichymotrypsin (ACT):					
Peak 1	no	35	estimated		
Peak 2	yes	45	estimated		
Peak 3	yes	20	estimated		
Gc globulin (Gc):					
Peak 1	no	11,00	3,12	0,80	15
Peak 2	no	89,00	3,12	0,80	15
α-2-HS-glycoprotein (HS):					
Peak 1	yes	34,40	1,84	0,48	15
Peak 2	yes	65,60	1,84	0,48	15
Haptoglobin (Hp):					
Peak 1 a	no				
Peak 2	yes	31,60	22,47	5,80	15
Peak 3	yes	68,40	22,47	5,80	15
α-2-macroglobulin (M):					
Peak 1	yes	20	estimated		
Peak 2	yes	80	estimated		
Complement factor 3 (C'3):					
Peak 1	yes	26,87	12,81	3,31	15
Peak 2	yes	73,13	12,81	3,31	15

cont.

Table 3 continued

Hemopexin (Hx):			
Shoulder	yes	ND	
Peak	yes	ND	
Transferrin (T):			
Peak 1	no	5	estimated
Shoulder a			
Peak 2	yes	95	estimated
IgA:			
Peak 1	yes	50	estimated
Peak 2	yes	50	estimated
HDL:	no	ND	

The areas under the precipitates were measured by planimetri The area under the precipitated microheterogeneity form was expressed in percent of the total area under the precipitate of the protein. For abbreviations see Table 1.
x = relative area of the microheterogeneity form.

to look at only one precipitate at a time (1-24,except 3,4). However,for a few proteins (orosomucoid, antitrypsin, complement factor 3, hemopexin and Gc-globulin) a multispecific antibody may be used as the immunoprecipitates are very distinct and clear so that information is not lost. Other precipitates were only discernable with difficulty in the complex pattern, e.g. α-2-macroglobulin, α-1-antitrypsin, haptoglobin and complement factor 3. Finally, information was lost for α-1-antichymotrypsin. The precipitate of this protein could not be followed in its full lenght. Generally, precipitates of most of the serum proteins could not be followed in their full length. Presumably these proteins could be estimated with specific antibodies.Serum samples stored at 4°c for more than a week should not be used. The precipitate of HDL change, peak 1 of Gc globulin may increase and some precipitates dissappear.

Quantification of microheterogeneity forms.

Table 3 shows how we quantified each individual microheterogeneity form. We measured the relative area below each by planimetry. The quantative distribution found was remarkably constant. Only the distribution of orosomucoid, C'3 and haptoglobin has a standard deviation above 5. The great variation in the distribution of haptoglobin and C'3 may be explained by the difficulty in limiting each precipitate because the profiles were ill defined and did not reach the baseline. In contrast all precipitates of orosomucoid are well defined. However, this glycoprotein is an acute phase reactant, even a common cold may give several concentration rise. This could be accompanied by a shift in the microheterogeneity forms as shown earlier (3). Therefore we find that the relative distribution of microheterogeneity forms seen in the Table 3 indicates that the relative distribution of these forms are constant under non pathological conditions.

References

1. Andersen, M.M., Lihme, A., Krøll, J., Bøg-Hansen, T.C. in T.C.Bøg-Hansen (ed.) Lectins, Biology, Biochemistry, Clinical Biochemistry Vol. 1, De Gruyter, Berlin, 1981, pp. 475-486.

2. Andersen, M.M., Lihme, A., Krøll, J., Bøg-Hansen, T.C. ibid, pp. 487-496.

3. Bøg-Hansen, T.C. in B.J.Radola (ed.) Electrophoresis '79. Advanced Methods, Biochemical and Clinical Applications, pp. 193-202. De Gruyter, Berlin, 1980.

4. Bøg-Hansen, T.C., Jensen, P., Hinnerfeldt, F., Takeo, K. in T.C.Bøg-Hansen (ed.) Lectins, Biology, Biochemistry, Clinical Biochemistry Vol. 1, De Gruyter, Berlin, 1981, pp. 241-258.

5. Bøg-Hansen, T.C. and Hau, J. in Z.Deyl (ed.) Journal of Chromatography Library, Vol. 18B, Elsevier, Amsterdam, 1982 in press.

6. Bøg-Hansen, T.C. and Breborowicz, J. These proceedings.

7. Bowen, M., Raynes, J.G. and Cooper, E.H. in T.C.Bøg-Hansen (ed.) Lectins, Biology, Biochemistry, Clinical Biochemistry Vol. 2, De Gruyter, Berlin, 1981, pp. 403-412.

8. Breborowicz, J., Mackiewicz, A. ibid, pp. 303-314.

9. Breborowicz, J., Bøg-Hansen, T.C. ibid, Vol. 2, 445-456.

10. Faye, L., Salier, J.P. ibid, Vol. 2, 605-618.

11. Hau, J., Larsen, P., Nilsson, M., Teisner, B., and Bøg-Hansen, T.C. These proceedings.

12. Hau, J., Svendsen, P., Teisner, B., Pedersen, G.T. in T.C.Bøg-Hansen (ed.) Lectins, Biology, Biochemistry, Clinical Biochemistry Vol. 1, De Gruyter, Berlin, 1981 pp. 327-338.

13. Hau, J., Westergaard, J.G., Ipsen, L., Teisner, B., Bøg-Hansen, T.C., and Søndersaard, K. ibid, Vol. 2, pp. 457

14. Kerckaert, J.P., Bayard, B. ibid, Vol. 1, pp. 271-282.

15. Nicollet, I., Lebreton, J.P., Fontaine, M., Hiron, M. Biochim.Biophys.Acta 668, 235-245 (1981).

16. Nicollet, I., Lebreton, J.P., Fontaine, M., Hiron, M. in T.C.Bøg-Hansen (ed.) Lectins, Biology, Biochemistry, Clinical Biochemistry Vol. 2, De Gruyter, Berlin, 1981 pp. 404-412.

17. Nørgaard-Pedersen, B., Toftager-Larsen, K., Philip, J., Hindersson, P. Clin.Chem. 26, 1656-1659 (1980).

18. Raynes, J.G. (Leeds) personal communication.

19. Salier, J.P., Faye, L., Vergaine, D., Martin, J.P. Electrophoresis 1, 193-197 (1980).

20. Taketa, K., Toguchi, E., Izumi, M., and Takeo, K. These proceedings.

21. Toftager-Larsen, K. in T.C.Bøg-Hansen (ed.) Lectins, Biology, Biochemistry, Clinical Biochemistry Vol. 2, De Gruyter, Berlin, 1982, pp. 433-444.

22. Toftager-Larsen, K., Pedersen, P.L., Nørgaard-Pedersen, B. ibid, Vol. 1, pp.283-292.

23. Wells, C., Cooper, E.H., Glass, R.M., Bøg-Hansen, T.C. Clin.Chem.Acta 109, 59-67 (1981).

24. Wells, C.,Cooper, E.H., Glass, R.M., Bøg-Hansen, T.C. in T.C.Bøg-Hansen (ed.) Lectins, Biology, Biochemistry, Clinical Biochemistry Vol. 1, De Gruyter, Berlin, 1981, pp. 328-334.

AMPLIFICATION OF IMMUNOPRECIPITATES WITH HORSERADISH PEROXIDASE-LABELED PROTEIN A: APPLICATION TO STUDIES OF INTERACTION BETWEEN α-FETOPROTEIN AND LECTINS BY AFFINITY ELECTROPHORESIS

Kazuhisa Taketa, Eriko Toguchi and Masaki Izumi
Health Research Center, Kagawa University, Takamatsu 760, Japan

Kazusuke Takeo
Department of Biochemistry, Yamaguchi University School of Medicine, Ube 755, Japan

Introduction

By means of a newly developed technique of crossed-immuno-affino-electrophoresis (CIAE) as presented by Bøg-Hansen and his group, human α-fetoprotein (AFP) was demonstrated by Kerckaert *et al.*(1) to react differently to concanavalin A (con-A) and *Lens culinaris* hemagglutinin (LcH). Miyazaki *et al.*(2) further revealed that the reactivity of AFP produced by hepatocellular carcinoma to LcH varied over a wide range in contrast with a relatively constant reactivity of hepatoma AFP to con-A. It seems thus important to identify the molecular species of AFP responsible for the variation.
In the present study, kinetic analysis of AFP-lectin interaction was made with con-A and two closely related hemagglutinins from *Lens culinaris*, LcH-A and LcH-B, differing in their electrophoretic mobilities (3). Since immunoprecipitates in CIAE were amplified by reacting with horseradish peroxidase-conjugated protein A for color development, relatively small amounts of AFP in crude body fluids could be analyzed after sufficient dilution, thus eliminating the possible effect of other interfering glycoproteins. The results suggested the presence of three major molecular species of AFP due to differential affinities to lectins.

Electrophoresis '82

Materials and Methods

A frozen sample of ascites, obtained from a patient with primary hepatoma and having a protein concentration of 21 mg/ml and an AFP level of 50,000 ng/ml, was used throughout. AFP in ascites was separated by affinity chromatography with Con A-Sepharose (Pharmacia Fine Chemicals)into unbound, loosely bound and bound fractions. The bound fraction came through with the initial protein peak followed by the loosely bound fraction with major AFP peak during a continued washing with an equilibration buffer (0.05 M Tris-HCl, pH 7.5, containing 0.5 M NaCl, 1 mM $MgCl_2$, 1 mM $MnCl_2$, 1 mM $CaCl_2$ and 4 mM NaN_3). The bound fraction was obtained by elution with 0.2 M α-methyl-D-mannoside in the equilibration buffer.
CIAE was performed in a 1.5-mm thick, agarose gel plate made of 1% Litex, LSB (Litex Lab.) in phenobarbital buffer (pH 8.6, μ= 0.025), containing various concentrations of lectins for the first dimension of electrophoresis (10 V/cm for 1-2 h at a room temperature of 20°C) and 2.5-5 μl of anti-human AFP rabbit immunoglobulin (Dako-immunoblobulins Ltd.) /ml for the second dimension of electrophoresis (3 V/cm for 10-15 h). Five μl samples, containing 30-50 ng of AFP diluted in 0.1% bromphenol blue (BPB) in the phenobarbital buffer, were applied to narrow 5 mm-length slots. Positions of free and albumin-bound BPB migrating on the agarose plate were marked after running the first dimension electrophoresis. When samples having trace amounts of albumin were used, anti-human albumin rabbit immunoglobulin (Dako-immunoglobulins Ltd.) was included on top of the second dimension gel to locate albumin migration as a reference. A phenobarbital buffer, pH 8.6, with a μ of 0.05 was employed as electrode buffer. The migration of AFP was expressed as a mobility relative to that of albumin (R).
Immunoprecipitates in agarose gel were reacted with horseradish peroxidase-conjugated protein A (Zymed Lab.), diluted 100 folds in 0.01 M phosphate buffered saline, pH 7.2, (PBS), for 30 min at 25°C after two cycles of pressing with a filter paper

pad and washing with PBS for 30 min. Unreacted protein A was removed by further two cycles of pressing and washing, and the color was developed by reacting the plate with 0.025% 3,3'-diaminobenzidine tetra-hydrochloride(Sigma Chem. Co.) and 0.005% H_2O_2 in 0.05 M Tris-HCl, pH 7.6, for 20-30 min at 25°C. Immunoprecipitates in agarose was also stained with 0.015% Coomassie Brilliant Blue R-250 dissolved in 3% trichloroacetic acid and 5% acetic acid and destained with 0.3% acetic acid.

Results

Enzymatic amplification of immunoprecipitate in agarose gel with horseradish peroxidase-labeled protein A for color development allowed to detect as low as 1 ng of AFP in electro-immunodiffusion. Although this technique was only several folds more sensitive than the protein staining with Coomassie Brilliant Blue, it was specific to immunoprecipitates and gave a sharply contrasted stain as compared with Coomassie Brilliant Blue. The margin of immunoprecipitates became blurred, however, when the amount of antibody used was reduced below the concentration given under Materials and Methods.

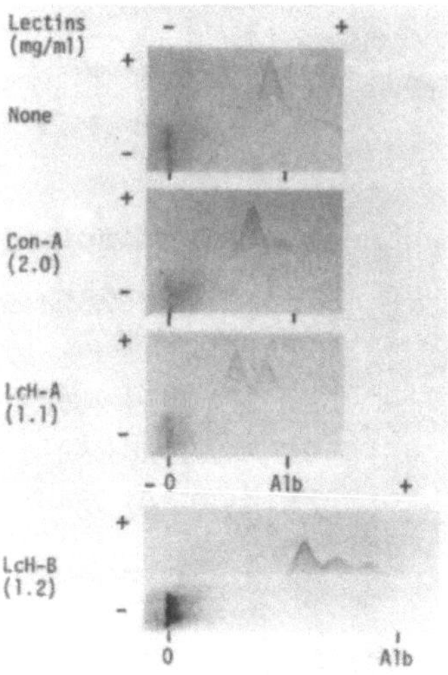

Fig. 1. CIAE patterns of human AFP with different lectins. "O" denotes the origin and "Alb" human albumin. Each slot received 50 ng of AFP and the second dimension gel contained 5 µl anti-human AFP immunoglobulin/ml.

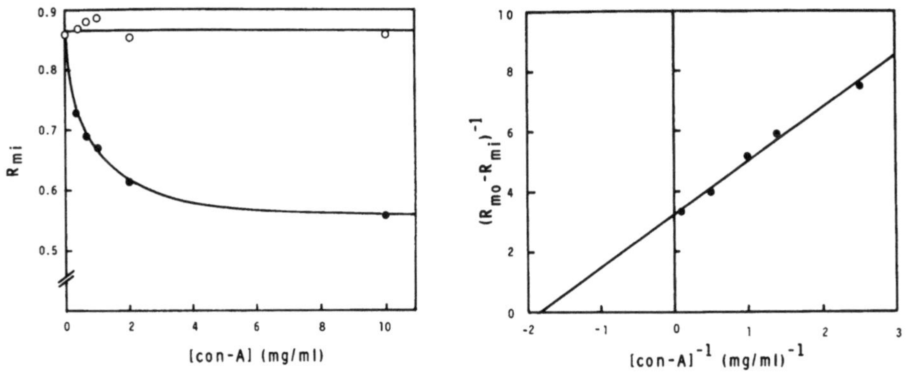

Fig. 2. Effect of the concentration of con-A on R_{mi} of ascites AFP in CIAE and its double reciprocal plot. ○, AFP_1(con-A); and ●, AFP_2(con-A).

Under the conditions described for CIAE, reproducible results were obtained with AFP below 50 ng (or 5 µl of ascites), indicating that the resolution of AFP's by CIAE was not interfered with by other glycoproteins present in the ascites.

AFP from a patient with primary hepatoma was resolved into two peaks by CIAE with con-A as reported (1)(Fig. 1). When the concentration of con-A was increased, the mobility of the anodic peak, AFP_1(con-A), was not altered, while the mobility of the cathodic peak, AFP_2(con-A), was reduced to a certain extent (Fig. 2). Those data were fitted to the formula of Bøg-Hansen and Takeo (4) for electrophoretic determination of dissociation constants:

$$\frac{1}{R_{mo} - R_{mi}} = \frac{K}{R_{mo} - R_{mc}} \times \frac{1}{C} + \frac{1}{R_{mo} - R_{mc}},$$

where K is the dissociation constant of the lectin-AFP complex, C is the concentration of lectin, R_{mo} is the mobility of AFP without lectin, R_{mi} is the mobility of AFP in presence of lectin and R_{mc} is the mobility of the lectin-AFP complex. The double reciprocal plot of the data gave a straight line (Fig. 2) with K and R_{mc} listed in Table 1. R_{mc} for AFP_1(con-A) was

Table 1. R_{mo}, R_{mc} and K values of different molecular species of AFP separated by CIAE with con-A, LcH-A or LcH-B and R_{mo} of the lectins employed

Lectins	Peak No. of AFP	R_{mo}	R_{mc}	K(mg/ml)
None	AFP	0.86		
Con-A		0.13		
	AFP_1(con-A)		0.86	infinite
	AFP_2(con-A)		0.55	0.55
LcH-A		-0.04		
	AFP_1(LcH-A)*		0.87*	infinite*
	AFP_2(LcH-A)		0.79	0.52
	AFP_3(LcH-A)		0.54	0.17
LcH-B		-0.14		
	AFP_1(LcH-B)		0.87	infinite
	AFP_2(LcH-B)		0.56	1.60
	AFP_3(LcH-B)		0.54	0.22

*, demonstrated with isolated con-A non-reactive species of AFP(the unbound fraction in Con A-Sepharose chromatography). Other values were determined with AFP in original ascites.

obtained by plotting R_{mi} against 1/[con-A] and extrapolating 1/[con-A] to zero. Since R_{mc} for AFP_1(con-A) was practically equal to R_{mo} for AFP, this should be taken as indicating that AFP_1(con-A) does not bind to con-A.

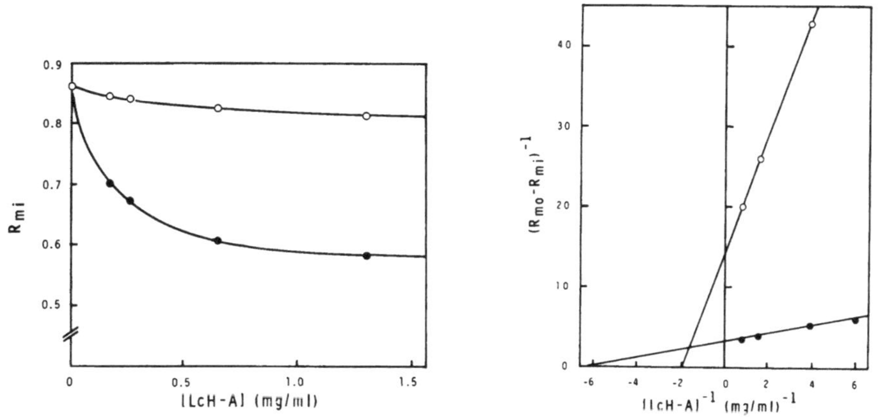

Fig. 3. Effect of the concentration of LcH-A on R_{mi} of ascites AFP's and their double reciprocal plots. ○, AFP_2(LcH-A); and ●, AFP_3(LcH-A).

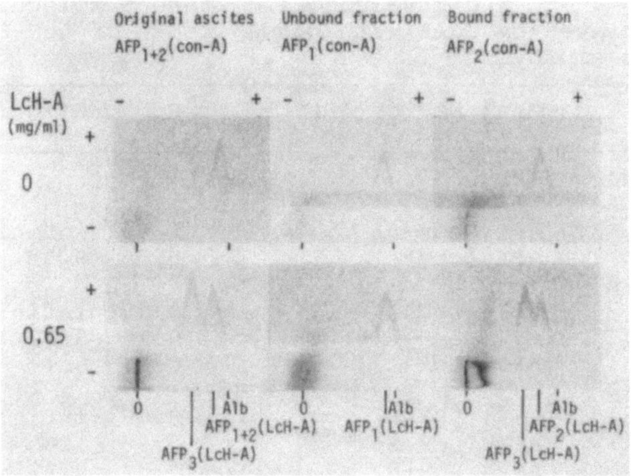

Fig. 4. Effect of LcH-A on CIAE pattern of AFP isolated with Con A-Sepharose chromatography. The second dimension gel contained 2.5 µl anti-human AFP immunoglobulin/ml and the amounts of AFP applied were reduced properly. AFP in the loosely bound fraction gave a CIAE pattern identical with that of the bound fraction.

Similar analysis with LcH-A gave two AFP peaks (Fig. 1), although the results were different from those with con-A; the anodic peak area with LcH-A being larger than that with con-A and its mobility being slightly reduced as the concentration of LcH was increased (Fig. 3). The results in Fig. 4 showed

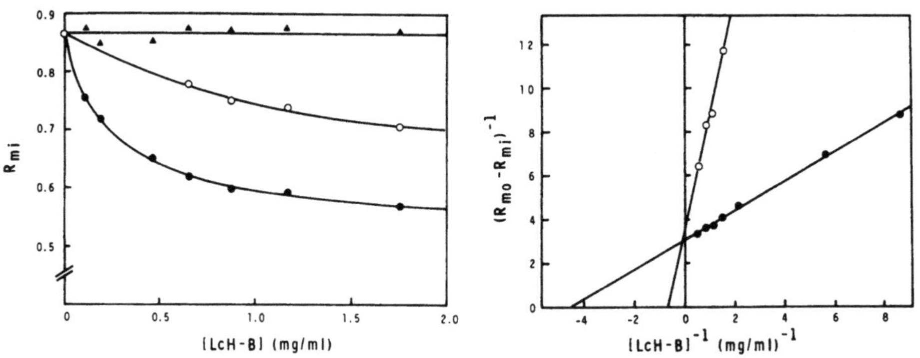

Fig. 5. Effect of the concentration of LcH-B on R_{mi} of ascites AFP's and their double reciprocal plots. ▲, AFP_1(LcH-B); ○, AFP_2(LcH-B); and ●, AFP_3(LcH-B).

that a fraction of ascites AFP bound to Con A-Sepharose, giving a single peak on CIAE with con-A (AFP_2(con-A)), revealed similar two peaks with a slightly smaller anodic one upon CIAE with LcH-A, while a fraction unbound to Con A-Sepharose, yielding also a single peak on CIAE with con-A (AFP_1(con-A)), gave an anodic peak with LcH-A, of which the mobility was not reduced by increasing LcH-A concentration. Thus, three species of AFP appear to be present with respect to the affinity to LcH-A; namely, AFP_1(LcH-A), AFP_2(LcH-A) and AFP_3(LcH-A), although AFP_1(LcH-A) and AFP_2(LcH-A) could not be resolved by CIAE with LcH-A alone even if the time of electrophoresis was prolonged,probably because R_{mc}'s of AFP_1(LcH-A) and AFP_2(LcH-A) were too close to each other (Table 1).

In contrast with LcH-A, LcH-B alone separated AFP into three peaks by prolonged running of first dimension electrophoresis (Fig. 1). The effect of LcH-B concentration on the mobilities of three molecular species is shown in Fig. 5 and their kinetic parameters in Table 1.

It may be noted from the R_{mc} and R_{mo} values summarized in Table 1 that the actual R_{mc} values of AFP-lectin complexes are, for all the AFP species except AFP_2(LcH-A), close to or slightly greater than the mean R_{mo} values of AFP and lectins. No suitable explanation is available at present for the limited reduction in R_{mc} of AFP_2(LcH-A).

CIAE with LcH, which is considered a mixture of LcH-A and LcH-B in varying proportions (3), gave results similar to those obtained with LcH-A. Agarose and disc electrophoretic patterns of LcH and LcH-A stained for proteins also resembled each other.

Molecular weights of lectins used in the present study, determined by polyacrylamide gradient gel electrophoresis, PAA 4/30 (Pharmacia Fine Chemicals), were 45,000 for con-A, 460,000 for LcH-A(also for LcH) and 42,000 for LcH-B. Subunit molecular weights were also determined by the gradient gel electrophoresis with sodium dodecyl sulfate. Con-A yielded a single major band with a molecular weight of 27,500. LcH-A also had a single major band with a molecular weight of 19,000, whereas LcH-B had two equally intense bands with molecular weights of 28,000 and 19,000. The subunit pattern of LcH was again similar to that of LcH-A, although a minor band with an approximate molecular weight of 12,000, common to all the LcH's, was somewhat pronounced in LcH.

Discussion

Beside the newly developed technique of a highly sensitive staining of immunoprecipitates with horseradish peroxidase-labeled protein A, the present study disclosed the presence of at least three distinct molecular species of AFP due to differential affinities to lectins. It is of some interest that a single molecular species of AFP with respect to the affinity to con-A can be further resolved into two molecular species by differential affinities to LcH-A in spite of the fact that both lectins have an identical sugar specificity. Tighter binding of LcH-A or LcH-B to AFP than con-A was not predicted from the results of other studies. Although the fundamental difference lying between LcH-A and LcH-B should be considered in separating AFP by CIAE with LcH, the variable molecular species in hepatoma cases shown by Miyazaki _et al._(2) appears to be AFP_2 (LcH-A) because of a strong affinity to con-A and weak binding to LcH-A. The identity of AFP_2(LcH-A) and AFP_2(LcH-B) remains to be solved in future studies.

The molecular weights determined in the present study reiterate the diversity and discrepancy present among the reported molecular weights for LcH-A and LcH-B (3,5). Thus, the K values were not expressed in terms of molar concentration in this report. However, it may be readily calculated that the lectins were present in the first dimension agarose gel at least ten times in excess in molar ratio to AFP, assuming that each subunit of lectin bind to AFP independently. Despite the existence of problems related to molecular structure of LcH's, the results of the present study are significant by themselves, as AFP can be readily separated into three distinct molecular species of known kinetic properties with specified molecules of lectins.

References

1. Kerckaert, J.-P., Bayard, B., Biserte, G.: Biochim. Biophys. Acta 576, 99-108 (1979)
2. Miyazaki, J., Endo, Y., Oda, T.: Acta Hepatol. Jap. 22, 1559-1568 (1981)
3 Howard, J.K., Sage, H.J., Stein, M.D., Young, N.M., Leon, M.A., Dyckes, D.F.: J. Biol. Chem. 246, 1590-1595 (1971)
4. Bøg-Hansen, T.C., Takeo, K.: Electrophoresis 1, 67-71 (1980)
5. Fligerová, O., Salvetová, A., Tichá, M., Kocourek, J.: Biochim. Biophys. Acta 351, 416-426 (1974)

ANALYSIS OF CENTRIFUGATION GRADIENTS BY FUSED ROCKET IMMUNO-ELECTROPHORESIS

Radu Portocala, Michel Lab, Katina Soteriadou, Niki Spyrou
Institut Pasteur Hellénique, Athènes, Grèce

Abstract

The localisation of the bands of a biological product along the gradients of centrifugation requires laborious techniques. In the case of viruses serological identifications or inoculation to cell cultures for each collected fraction are used.

In order to simplify the identification process and to reduce to a single step the number of steps required for the analysis of ultracentrifugation gradients, fused rocket immunoelectrophoresis has been applied. This technique allows us to test on a single electrophoretic plate 30-35 fractions collected from a centrifugation tube.

The results obtained are both satisfactory when either influenza virus or influenza hemagglutinin are located in a sucrose gradient. This technique was used in each gradient performed with influenza virus giving us very reproducible results, eventually allowing us to control the gradient profile in each run.

To test the sensitivity of this technique, the sucrose gradient of human serum was analyzed.

Monospecific serum enabled us to identify the fractions which contain IgM and IgG.

Electrophoresis '82

Material and Methods

Virus

The strain A/URSS/90/77 of the influenza virus (H_1N_1) is cultivated in the allantoic cavity of 11 days old embryonated eggs.

The allantoic fluid was collected 48 hours later and clarified by centrifugation at 2000 rpm in a SS34 Sorval rotor, at +5°, for 30 min.

Virus purification

The supernatant obtained above is centrifuged in an ultracentrifuge MSE at 100,000 g, +5°, for 60 min.

The pellet is suspended in TNE buffer (Tris-HCl 0.01M -PH=7.4- NaCl 0.1M, EDTA 1mM), placed on a lineaire sucrose gradient 20%-60% (W/W) prepared in the same buffer and ultracentrifuged at 100,000 g, +5° for 16 hours.

The band containing the virus has a density equivalent of 1.19-1.20 g/cm^3 and is withdrawn with the help of a peristaltic pump. Then is diluted in TNE buffer and ultracentrifuged again under the conditions indicated above. The sucrose gradient step is repeated once and the gradient content is fractioned in volumes of 23 drops per fraction. These fractions were used for the fused rocket immunoelectrophoretical analysis.

Virus dissociation

Triton X-100 (2%) was used as a dissociating agent.

The virus suspension is incubated at 37°for a period of one to 24 hours according to the desired dissociation degree.

Isolation and purification of the hemagglutinin (H_1) of the A/URSS virus

The purified virus is suspended in a TE buffer (Tris-HCl 0.1M -PH=7.2-, EDTA 1mM) and incubated at 37° for 12 hours in the presence of bromelain (1mg/ml) and 2-mercaptoethanon (0.05M) according to the method of Brand and Skehel (3). Hemagglutinin is isolated and separated from the viral particle by ultra-centrifugation in a sucrose gradient 5%-50% (W/W) at 100,000 g, +5°, for 16 hours. This technique allows the separation and isolation of the HA from the viral particles by a single centrifugation step. The position of the HAB band is identified by Fused Rocket Immunoelectrophoresis.

Antibodies

The antiserum against the viral A/URSS strain was prepared in rabbits using the Triton X-100 dissociated virus.

The immunisation schedule and the isolation of immunoglobulins was performed according to Harboe and Ingild (1975) (4).

Isolation of IgM from human serum

0.3 ml of serum is layered on a continuous sacharose gradient (4.5 ml) 12%-33% W/V) in PBS. After centrifugation at 150,000g, 18 h, +5°, the bottom of the tube is pierced and its content is fractioned in volumes of 5 drops per fraction.

Anti-human serum proteins

Rabbit antibodies (type 100 SF) against human serum proteins used in the immunoelectrophoresis to reveal the human serum fractions were obtained from commercial source (Dakopatts A/S, Copenhagen-Denmark).

Fused Rocket immunoelectrophoresis

The method used is that described by Svendsen (1) adapted for centrifugation.

1% (W/V) agarose Litex, Glostrup, Denmark (with low electro-endosmotic properties) disolved in Tris-Tricine buffer. Ca-lactate is added (5) to this buffer and 1% (W/V). Triton X-100. The non-ionic detergent is added in the gel in order to prevent the aggregation of membrane proteins by Demus and Mehl (6) and by Bjerum and Lundall (7) as well as by Portocala et coll. to prevent the aggregation of the Sendai virus polypeptides (8) of HBsAg (9) and influenza virus (2). The number of sample wells (ø 2.5 mm) punched in an antibody free agarose gel, is equal to the number of samples collected from the centrifugation tube. Samples of 5 µl volume from each fraction collected from the centrifugation experiment are transferred to the sample wells in the same order as they were collected, where they are allowed to diffuse at room temperature into the gel for 60 minutes. An agarose solution mixed with the antibody solution is poured onto the rest of the glass plate and allowed to congeal. The proteins are then forced by electrophoresis into the antibody containing agarose gel. Electrophoresis is carried out overnight with a potential gradient of 1.5 Vcm^{-1} at room temperature. The immunoprecipitates are colored with Coomasie brillant blue R-250.

Results and Discussion

After the second sacharose gradient centrifugation a net band is visible corresponding to the influenza virus. The density of this band determined by refractometrie is of 1.19 which is in accordance with the value found by others.

The content of the centrifugation tube is collected in fractions of 23 drops each. Figure 1 illustrates the immunochemical profile from a centrifugation gradient. A single peak is identified. Its precise position according to the sample wells determines exactly the wells, fractions, containing the virus. Moreover, it has to be remarked that the virus has conserved its physical and antigenical integrity as no viral fragments or subunits are observed in the gradient content.

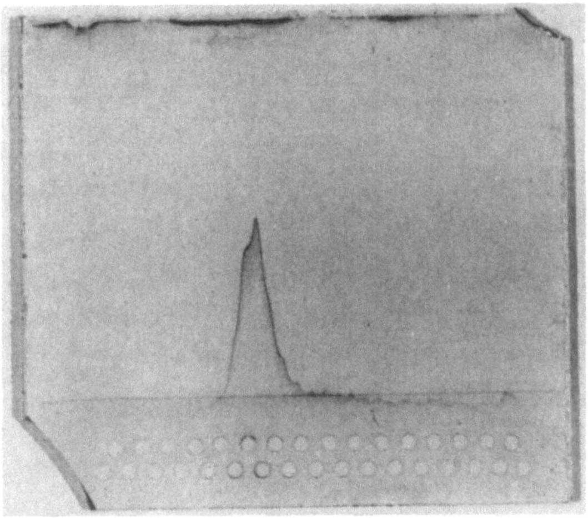

Fig. 1. Fused Rocket Immunoelectrophoresis
Immunochemical profile from a sucrose gradient centrifugation (20-60%) of the A/URSS influenza virus
Antigens: 5µl samples were transferred to the wells
Antiserum: 100 µl rabbit anti-influenza A/URSS purified serum
Antigenic fractions are located between fraction 10 and 16.

When influenza virus treated with bromelain under the conditions described by Brand and Skehel (3) hemagglutinin is released from the rest of the virus. According to this method several centrifugation gradients are required for isolating hemagglutinin. The last one permits to isolate hemagglutinin from a 5-20% lenear sucrose gradient. In this state of monomere, hemagglutinin loses the property of agglutinating erytrocytes and it is laborious to identify it along the gradient. Each fraction has to be analyzed by an immunoserological indirect method.

The method proposed allows to isolate hemagglutinin from the rest of the virus with a single centrifugation in a linear sucrose gradient (5-50% W/W). Fused Rocket immunoelectrophoretical analysis of the fractions collected from this gradient against a specific anti-influenza serum permits to identify and position with precision the sample wells containing hemagglutinin (Fig. 2).

Fig. 2. Fused Rocket Immunoelectrophoresis The antigenic peak on the right was identified as being hemagglutinin by crossed immunoelectrophoresis. The left hand complex peak represents the subviral particles.

(For more details see text)

It has to be emphasized that the technique proposed takes into consideration the physical state of the centrifugated virus. In fact, when the visible band of a sucrose gradient corresponding to the virus is collected, it is not a guarantee that the obtained virus is physically complete.

The analysis of a sucrose gradient centrifugation by fused rocket of an influenza A/URSS virus incorrectly conserved (congealed for four months at -10°) is illustrated in Fig. 3. After a centrigugation of 16 hours a single band is observed and this band is recognized by the anti-influenza A/URSS serum on the immunoelectrophoresis plate. On the right of the plate a second complex of peaks is observed. These antigenic fractions present at the top of the gradient were invisible (no band along the gradient) and correspond probably to different antigenic viral fragments. The virus may have spontaneously been desintegrated due to the incorrect conservation conditions (compare fig. 3 to fig. 1).

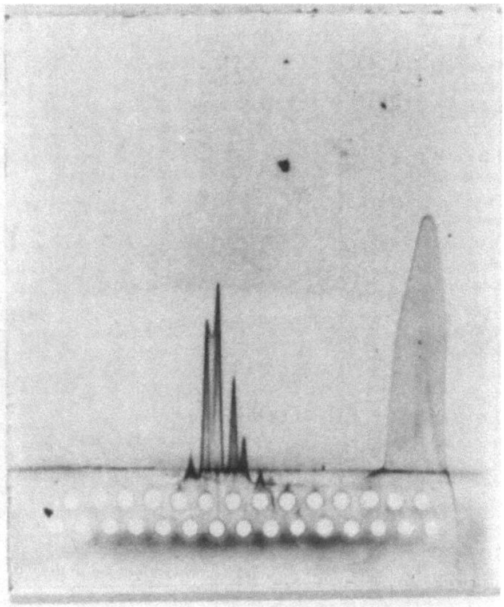

Fig. 3. Fused Rocket Immunoelectrophoresis
Immunochemical profile from a sucrose gradient centrifugation (20-60%) of the A/URSS virus spontaneously desintegrated.

(For more indications see text)

The presence of IgM in a serum is the result of a primary serological response of the organism to a viral infection. The research of IgM in a serum (taken as soon as clinical signs appear) of an individual suspected for a viral infection seems to be a means of a rapid diagnosis.

Moreover, the search of IgM in viral infections is very important in cases of congenital infections and infections of the newborn.

In the present study IgM were isolated by a sucrose gradient centrifugation 12-33% W/W (5 ml, PBS PH 7.2). Testing each fraction collected after gradient centrifugation by the classical serological methods (IHA, IF, ELISA)is quite long and laborious. All the fractions collected were analysed by Fused Rocket Immunoelectrophoresis against a reference anti-total human serum (Fig. 4), against a reference anti-human IgG serum (Fig. 5) and finally against a reference anti-human IgM serum (Fig. 6). On the plate the serum recognised the fractions containing IgM which are utilized for a serological diagnosis. This process of searching IgM was proved to be simple, rapid and reproducible.

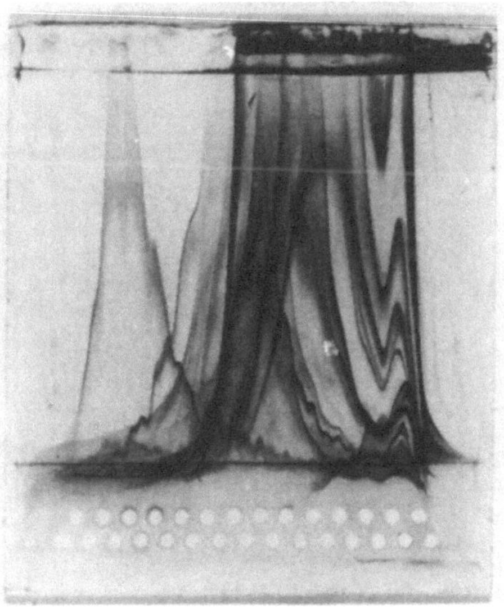

Fig. 4. Fused Rocket Immunoelectrophoresis
Immunochemical profile from a sucrose gradient centrifugation 12-33% of a total human serum
Antiserum: Reference anti-total human serum

Fig. 5. Fused Rocket Immunoelectrophoresis
Immunochemical profile from a sucrose gradient centrifugation 12-33% of a total human serum
Antiserum:Reference anti-human IgG serum

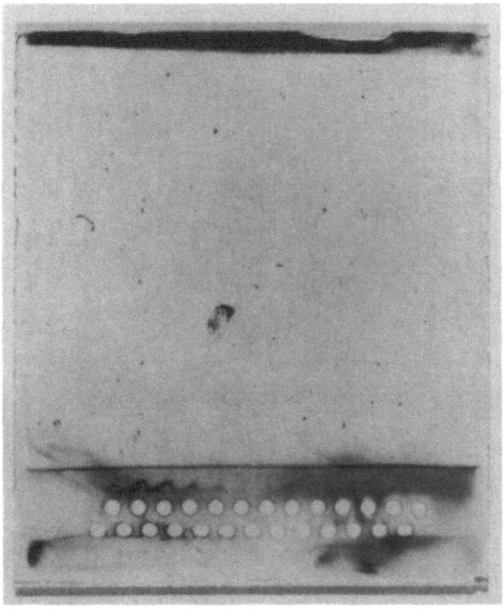

Fig. 6. Fused Rocket Immunoelectrophoresis
Immunochemical profile from a
sucrose gradient centrifugation
12-33% of a total human serum
Antiserum:Reference anti-human
IgM serum

A research of IgM was applied mainly in the case of rubella infections because this virus is pathogenic for the foetus especially during the first three months of pregnancy. A positive result must cause concern for the possible interruption of pregnancy.

In conclusion, the analysis of sucrose gradient centrifugations by Fused Rocket Immunoelectrophoresis has the following advantages:

a) Simplicity
b) Specificity and sensitivity
c) Small quantities of samples are required

References

1. Svendsen, P.J.: Scand. J. Immunol. 2, suppl. 1, 69 (1973).
2. Portocala, R., Soteriadou, K., Vincent, J., Sérié, Ch. Submitted.
3. Brand, C.M., Skehel, J.J.: Nature New Biology 238, 145 (1972).
4. Harboe, N., Inglind, A.: Scand. J. Immunol. 2, suppl. 1, 161 (1973).
5. Laurell, C.B.: Anal. Biochem. 10, 358 (1965).
6. Mehl, E., Demus, H.: Biochem. Biophys. Acta 211, 148 (1970)
7. Lundahl, P., Bjerrum, D.J.: Scand. J. Immunol. 2, suppl. 1, 139 (1973).
8. Portocala, R., Ghyka, Gr., Samuel, J.: Rev. roum. méd.-virol. 27, 261 (1976).
9. Portocala, R., Papevangelou, G., Kremastinou, J., Stathakos, D.: Rev. roum. méd. - virol. 29, 73 (1978).

QUANTITATIVE IMMUNOELECTROPHORESIS FOR COMPARATIVE ANALYSIS OF HYDROPHOBIC INTRINSIC MEMBRANE PROTEINS

H.-P. Schultheiß, M. Klingenberg

Institute of Physical Biochemistry, University of Munich

Dr.med. H.-P. Schultheiß
Institute of Physical Biochemistry, University of Munich
Goethestr. 33, 8000 Munich 2, Federal Republic of Germany

The development of procedures for the successful characterization of detergent-solubilized membrane proteins by means of quantitative immunoelectrophoresis has proved to be of great value for studying various properties and functions of membrane proteins (1).
Several methods like the use of the hydrophobic dye Sudan black B to colour detergent micelles (2) or the use of autoradiography after binding of a radiolabeled detergent to a protein (3) have been developed to characterize detergent binding proteins by immunoelectrophoretic methods. The detergent-induced shift in mobility (4) and its combination with electroimmunoprecipitation (5) provided a new and sensitive method for distinguishing between hydrophilic and amphiphilic proteins.
In this paper we descripe the detergent-induced charge shift as a prerequisite for the immunological characterization of an unusually hydrophobic and cationic protein - the ADP, ATP carrier.

Electrophoresis '82

The ADP, ATP carrier is an intrinsic membrane protein located in the inner mitochondrial membrane and it is the most active transport system of biomembranes in aerobic eukaryotic cells (6). It is the most abundunt protein in mitochondria and the major membrane protein in many eukaryotic cells. Calculated from the amount of carboxyatractylate (CAT)-binding - a specific ligand - the molecular weight of the ADP, ATP carrier corresponds to approximately 60.000. The noncovalent ligand protects the protein against denaturation maintaining it in a defined conformation state and gives so a quantitative assay for the intact protein share (7).
The charge shift of a typical intrinsic membrane protein like the ADP, ATP carrier offers a new methodical possibility for immunological studies of hydrophobic membrane proteins as we could also show for the phosphate carrier from chloroplasts or for the GDP-binding protein of brown adipose tissue mitochondria (unpublished data).

Materials and Methods

Methods for isolation of mitochondria from beef heart, loading of the mitochondria with the ligand carboxyatractylate (CAT) and solubilization and purification of the CAT-protein-complex were as described previously (8, 9).

Protein concentration was determined by the method of Lowry or by biuret.

Antibodies were raised in rabbits against the purified CAT-protein complex in a solution (pH 7.2) containing 5 mM MOPS, 100 mn Na_2SO_4 and 0,5% Triton X-100 (w/v). Initially the antigen (0,1 mg) was dissolved in Freud's complete adjuvant and the emulsion was injected subcutaneously at multiple sites in the back of rabbits. After three to four booster injections (0,2 mg) with Freud's incomplete adjuvant in four weeks intervals the blood was collected, allowed to clot at room temperature, centrifuged at 15.000 rpm and stored at -20 °C.

Charge shift crossed immunoelectrophoresis was performed with a standard buffer system of 0,1 M Glycine, 0.038 M Tris pH 8,7. 0,75% (w/v) agarose (type HSC, Litex, Glostrup, Denmark) was used in all cases. Electrophoreses were performed in water cooled immunoelectrophoresis apparatuses from LKB (Sweden). The following agarose-buffer systems were used: 0,08% (w/v) Triton X-100 and 0,06% (w/v) desoxycholate present in the agarose and buffer system during first dimension, 0,08% (w/v) Triton X-100 in the second dimension. The charge-shift of the proteins solubilized in 0,5% Triton X-100 where performed without pre-equilibration in the charged detergent during the first dimension. After application of the Triton-treated samples, first dimension was performed at 10 V/cm for 30 min. Then 1 cm wide strips were transferred to glass plates and the antibody-containing agarose with 0,08% Triton X-100 cast for second dimension (10). Immunoelectrophoresis was performed for 16 hrs at 1,5 V/cm. Immunoprecipitates were stained for protein with Coomassie Brilliant Blue R.

Results and discussion

The ADP, ATP carrier is characterized by its hydrophobicity and its high content of basic aminoacids such as lysine and arginine resulting in an isoelectric point of 10,2 as determined by isoelectric focusing. This property of the protein might be the cause for its tendency to form unspecific precipitates by an interaction with the agarose (fig.1). This unspecific precipitation was also seen by using an agarose with a very low content of charged groups, mainly anionic constituents like sulphate groups (fig.2). Furthermore, using different agarose gels of differing electroendosmotic properties and by varying the nature and pH of the buffer, it was not possible to hinder these unspecific precipitation and to induce the protein to migrate anodically. Therefore the charge-shift of this intrinsic membrane protein was used by us for mobilizing the carrier protein in the electrophoretic system. Using the combination of the ionic detergent desoxycholate (DOC) and the ionic detergent Triton X-100 in the first dimension, a condition could be established where a good migration velocity without unspecific interactions of the antigen with the agarose was reached (fig.3). Using such detergent mixtures the amphiphilic protein forms detergent-protein complexes - mixed micelles - containing both neutral and charged detergent molecules (4). In the case of Triton X-100 and desoxycholate this results in an anodal electrophoretic mobility. The final concentration of Triton X-100 and DOC was chosen in dependence on the migration velocity, the intactness of the protein and the undisturbed interaction of the antigen-antibody reaction.

The intactness of the ADP, ATP carrier protein in this detergent mixture was tested by the stability of the CAT-binding. After 24 hrs incubation of the CAT-protein complex under experimental conditions of electrophoresis (0,08% TX and 0,06% DOC) no dissociation of ^{3}H-CAT from the protein could be monitored by gelfiltration (G 25). This indicates that the protein was not denatured under the chosen condi-

tions as it can be assumed that the CAT-binding reflects a defined conformation state of the protein (7).
Using the cationic detergent cetyltrimethylammoniumbromide (CTAB) no experimental condition could be established to get a reproducible precipitation pattern without unspecific precipitation and an undisturbed CAT-protein complex simultaneously (fig.4). This indicates that the ADP, ATP carrier protein in its defined conformation state is extremely sensitive toward treatment with ionic detergents.
It was the aim of this work to provide a method for immunochemical analysis of hydrophobic intrinsic membrane proteins by crossed immunoelectrophoresis. Transferring the Triton-protein micelle into a mixed micelle of an ionic and non-ionic detergent hinders unspecific interaction beetween the protein and the agarose and leads to reproducible precipitation patterns. However the concentration of the ionic and nonionic detergent and their relative composition has to be elucidated such that the protein is not denaturated by the charged detergent and thus to assure an undisturbed antigen-antibody reaction in the second dimension.

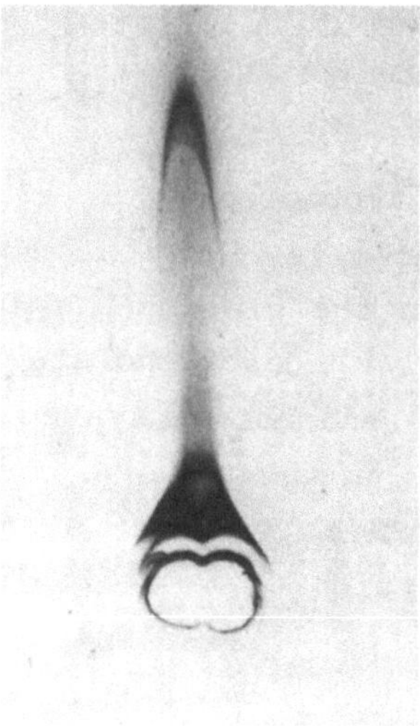

Fig.1: Electrophoresis of the ADP, ATP carrier protein (10 µg , protein, concentration 1 mg/ ml) in the presence of 0,5% (w/v) Triton X-100; agarose type HSA (Litex, Glostrup, Denmark). The precipitation is unspecific as no antibody was used.

Fig.2:

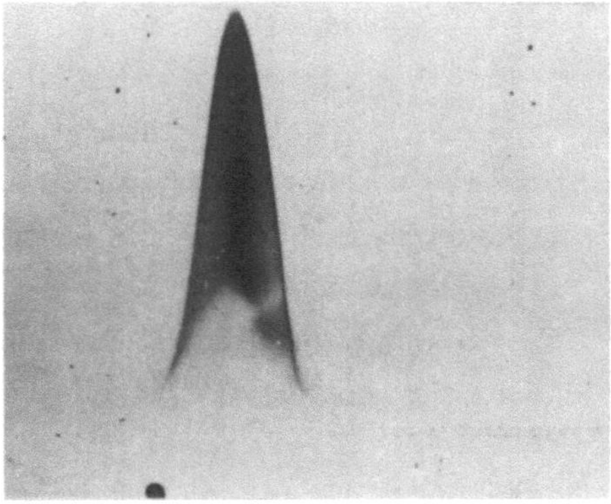

Fig.3:

Fig.2: Crossed immunoelectrophoresis of 10 µg ADP, ATP carrier protein (1 mg/ml). First dimension electrophoresis was performed in the presence of 0,5% (w/v) Triton X-100, agarose type HSC (Litex, Glostrup, Denmark). Second dimension was performed in the some agarose containing 0,5% (we/v) Triton X-100 and 12,5 µl/cm^2 of antiserum to the ADP, ATP carrier protein.

Fig.3: Charge-shift crossed immunoelectrophoresis of 10 µg ADP, ATP carrier protein (1 mg/ml). First dimension electrophoresis was performed in the presence of 0,08% (w/v) Triton X-100 and 0,06% (w/v) desoxycholate present in the agarose (HSC, Litex) and buffer system. Second dimension was performed in the same agarose containing 0,08% (w/v) Triton X-100 and 12,5 µl/cm^2 of antiserum to the ADP, ATP carrier protein.

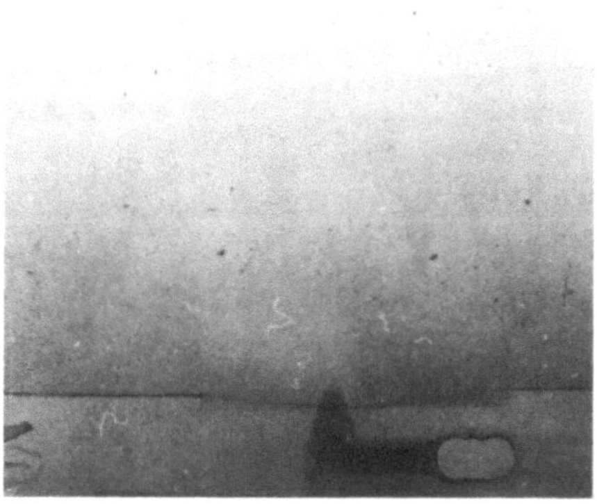

Fig.4: Charge-shift crossed immunoelectrophoresis of 10 µl ADP, ATP carrier protein (1 mg/ml). First dimension electrophoresis was performed in the presence of 1% (w/v) Triton X-100 and 0,0125 % (w/v) CTAB present in the agarose (HSC, Litex) and buffer system. Second dimension was performed in the same agarose containing 1% (w/v) Triton X-100 and 12,5 $\mu l/cm^2$ of antiserum to the ADP, ATP carrier protein.

References

1. Bjerrum, O.J., Bøg-Hansen, T.C.: In: Biochemical Analysis of Membranes (ed. Maddy, A.H.), Chapman and Hall, London, p. 378-425 (1976).

2. Bjerrum, O.J., Bhakdi, S., Rieneck, K.: J. Biochem. Biophys. Methods 3, 355-366 (1980).

3. Bjerrum, O.J., Bhakdi, S.: FEBS Lett. 81, 151-156 (1971).

4. Helenius, A., Simons, K.: Proc. Natl. Acad. Sci. USA 74, 529-532 (1977).

5. Bhakdi, S., Bhakdi-Lehnen, B., Bjerrum, O.J.: Biochim. et Biophys. Acta 470, 35-44 (1977).

6. Klingenberg, M.: In: The enzymes of biological membranes (ed. Martonosi, A.), pp. 383-438 (1976).

7. Klingenberg, M.: In: Atractyloside: chemistry, biochemistry and toxicology (ed. Santi, R., Luciani, S.), Piccin Medical books, p. 69-107 (1978).

8. Riccio, P., Aquila, H., Klingenberg, M.: FEBS Lett. 56, 129-132 (1975).

9. Riccio, P., Aquila, H., Klingenberg, M.: FEBS Lett. 56, 133-138 (1975).

10. Axelsen, N.H., Krøll, J., Weeke, B.: Scand. J. Immunol. 2, 2 (1973).

SENSITIVE MEASUREMENT OF 3H-GLUCOCORTICOID-RECEPTOR-COMPLEXES BY GEL - ELECTROPHORESIS

Karin Lenger
Institut für Biochemische Endokrinologie der Medizinischen Hochschule Lübeck
D - 2400 Lübeck (FRG)

Introduction

Steroid hormones are known to influence gene expression in their target organs. Studies of steroid hormone mechanism have been hindered by the use of crude receptor preparations and by the lack of a steroid hormone receptor assay being able to measure quantitatively the properties of the partially purified hormone receptors. In this study a gel-electrophoresis technique has been developed to measure quantitatively glucocorticoid receptors partially purified from cytosol of Morris Hepatoma 9121.

Results

Electrophoretic mobility of glucocorticoid-receptor-complexes.
After incubation of cytosol (8 mg protein/ml) from Morris Hepatoma 9121 with ^{3}H-glucocorticoids (1), ^{3}H-glucocorticoid-receptor-complexes could be separated from unbound steroid hormone by 7% disc polyacrylamide (2) (Fig.1) - and by 0.8% agarose (3) (Fig.2) gel-electrophoresis. Leaving the unbound hormone in the stacking gel, in polyacrylamide gels the ^{3}H-glucocorticoid-receptor-complexes moved 4-6 cm towards the anode (225 V;30-50 mA) (1) together with the tracking dye Bromphenol blue. After the run in agarose gels (1)(160 V; 30-50 mA), the unbound ^{3}H-steroid hormones, cortisol, dexa-

methasone and triamcinolone acetonide remained at the origin, whereas the ^{3}H-glucocorticoid-receptor-complexes moved 8-13 cm towards the anode (Fig.2).

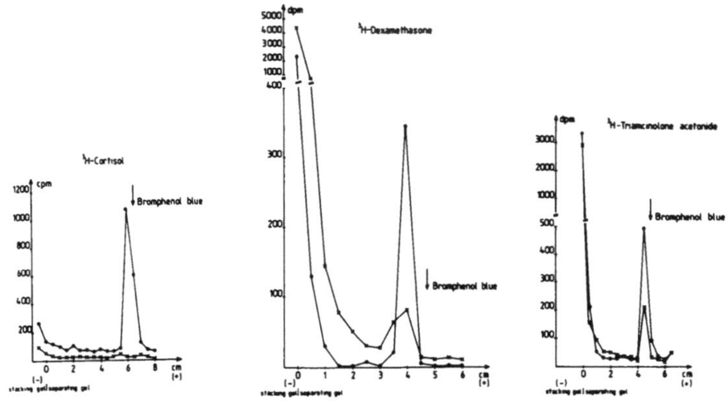

Fig.1: 7% disc polyacrylamide gel-electrophoresis (2): Electrophoretic mobility of the cytoplasmic ^{3}H-glucocorticoid-receptor-complexes incubated with ^{3}H-cortisol, ^{3}H-dexamethasone, and ^{3}H-triamcinolone acetonide (o - o), control values (x - x) were performed as described in (1).

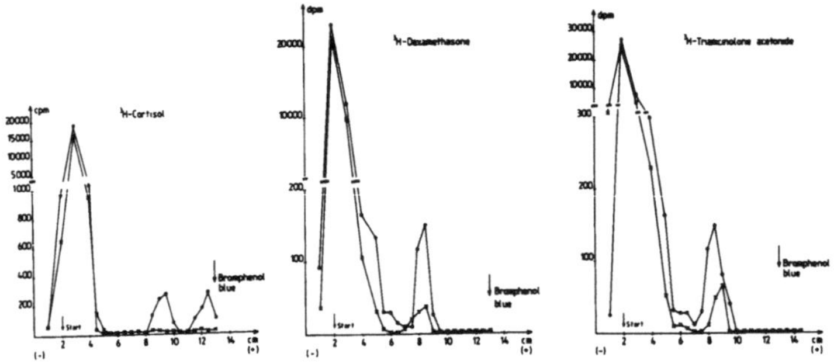

Fig.2: 0.8% agarose gel-electrophoresis (3) : Electrophoretic mobility of the cytoplasmic ^{3}H-glucocorticoid-receptor-complexes incubated with ^{3}H-cortisol, ^{3}H-dexamethasone and ^{3}H-triamcinolone acetonide (o - o), control values (x - x) were performed as described in (1).

Linearity. After incubation of increasing amounts of cytosol (8 mg protein/ml) with 8 x 10^{-8}M ^{3}H-cortisol, linearity for the ^{3}H-cortisol-receptor-complexes was achieved in polyacrylamide - and in agarose gels from 15-60 μg protein/50 μl assay (Fig.3). Linearity could also be proved from 2-10 μg protein/ 50 μl assay (Fig.4) using glucocorticoid-receptor solutions partially purified by removing the albumin fraction at pH 4.8 and then by 33% ammonium sulphate precipitation (1). Polyacrylamide gel-electrophoresis might be the more sensitive method. The values obtained by agarose gel-electrophoresis yielded only 50% of those obtained by polyacrylamide gel-electrophoresis.

Partial purification of the ^{3}H-glucocorticoid-receptor-complex

Partial purification of the ^{3}H-cortisol-receptor-complex was performed using 4 ml of the 33% ammonium sulphate fraction (0.096 mg protein/ml) incubated with 8 x 10^{-8}M ^{3}H-cortisol, which was subjected to a DEAE-cellulose column and eluted with a linear NaCl-gradient (0.0-1.0 M) (1). The ^{3}H-cortisol-receptor-complexes were eluted from 0.0-0.2 M and from 0.7-0.95 M, which had the same mobilities in both gel types as those from crude cytosol. A recovery of 50% could be measured when related to the radioactivity of the eluted fractions. (Tables 1;1a) Also in this case, the ^{3}H-cortisol-receptor-complex could be proved in the lower μg upper ng range of protein.

Measurement of K_d-values. After incubation of glucocorticoid-receptor solutions with increasing ^{3}H-cortisol, ^{3}H-dexamethasone and ^{3}H-triamcinolone acetonide, saturation curves could be obtained. The respective K_d-values were then calculated by Scatchard analysis (4) as 3.1 x 10^{-8}M for cortisol, as 4.4 x 10^{-9}M for triamcinolone acetonide, and as 2.8 x 10^{-9}M for dexamethasone.

Precision. Quantitative measurements of ^{3}H-glucocorticoid-receptors by gel-electrophoresis are associated with a coefficient of variation of 3.4 - 3.7%. Therefore, the final error expressed as the means of ± S.E.M. of the K_d-values calculated by Scatchard analysis (4) is in the range of 10-14%.

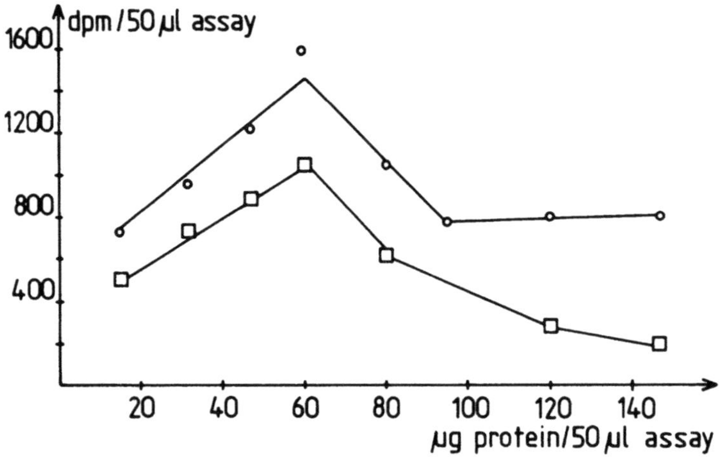

Fig.3: Increasing amounts of cytosol were incubated with 8 x 10^{-8}M cortisol and analyzed by 7% disc polyacrylamide (2) (o - o) - and 0.8% agarose (3) (-) gel-electrophoresis. Control values had been already substracted.

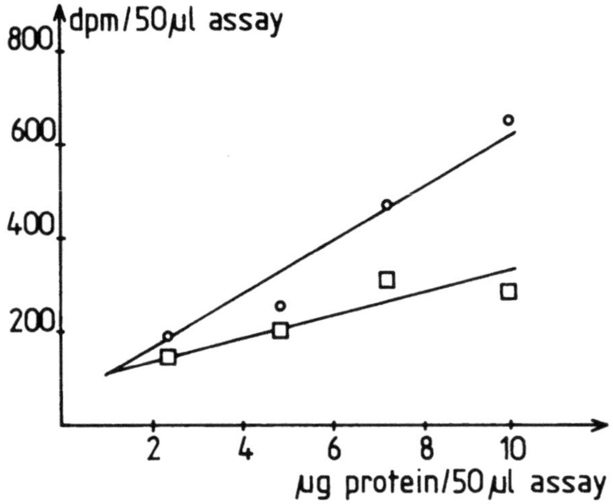

Fig.4: Increasing amounts of the 33% ammonium sulphate fraction were incubated with 8 x 10^{-8}M ^{3}H-cortisol and analyzed by 7% disc polyacrylamide (2) (o - o) - and 0.8% agarose (3) (-) gel-electrophoresis. Control values had been already substracted.

Table 1. Determination of bound ^{3}H-cortisol of the radioactive fractions obtained by different NaCl-elution concentrations after DEAE-cellulose chromatography of the 33% ammonium sulphate fraction incubated with 8×10^{-8} M ^{3}H-cortisol(1).

Elution concentr. [M] NaCl	Protein mg/ml	Vol. ml	Bound ^{3}H-cortisol dpm/50 µl	fmoles/50 µl	fmoles/mg protein
0.02	0.096	1.2	1 060	41.6	8 662
0.75-0.95	0.096	1.2	1 579	61.9	12 899

Table 1a. Gel analyses of the fractions with bound ^{3}H-cortisol obtained by DEAE-cellulose chromatography.

Elution concentr. [M] NaCl	Protein mg/ml	Vol. ml	Bound ^{3}H-cortisol fmoles/50 µl	fmoles/mg protein
0.02	0.096	1.2	16.6	3 472
0.75-0.95	0.096	1.2	28.4	5 920

Influence of bovine serum albumin. It could be shown that the addition of bovine serum albumin, mostly responsible for non-specific binding, did not influence receptor analyses by gel-electrophoresis (Table 3). The values obtained after incubation of serum albumine with ^{3}H-cortisol alone with the control values, which were performed by a) addition of 100-fold excess of unlabelled steroid hormone to the receptor solutions or by b) incubation of ^{3}H-cortisol with 10 mM MOPS-buffer pH 7.5 (1). Therefore, it is concluded that the proteins responsible for non-specific binding might be retained whereas the glucocorti-

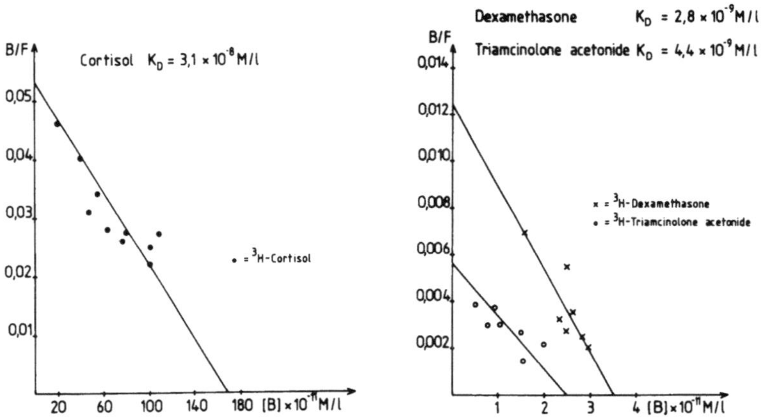

Fig.5 Scatchard analyses (4) of ^{3}H-glucocorticoid-receptor-complexes by gel-electrophoresis: the K_d-values were determined as 3.1×10^{-8} M/l for cortisol, as 4.4×10^{-9} M for triamcinolone acetonide and as 2.8×10^{-9} M for dexamethasone.

coid-receptor-complexes moved with fast mobility during electrophoresis.

Discussion

Sensitive gel-electrophoresis techniques were developed allowing quantitative measurements of cytosolic, partially purified glucocorticoid receptors even in the upper ng range of protein (1). In 7% polyacrylamide - and in 0.8% agarose gels, the ^{3}H-glucocorticoid-receptor-complexes moved with fast mobility; proteins with slow mobility as bovine serum albumin which are mostly responsible for non-specific binding might be then eliminated. The ^{3}H-glucocorticoid-receptor-complexes partially purified by removing the albumin fraction, by 33% ammonium sulphate precipitation and by DEAE-cellulose chromatography showed the same fast mobility in the gels as those from crude cytosol. Other properties (1) and the known high affini-

Table 3. Influence of bovine serum albumin. Cytosol of Morris Hepatoma 9121 (1) (65 µg/50 µl assa) was incubated with 8 x 10^{-8}M ^{3}H-cortisol. After addition of bovine serum albumin (17,5 µg/50 µl assay), aliquots of 50 µl were then analyzed by gel electrophoresis.

Fraction	fmoles bound ^{3}H-cortisol/50 µl assay 0.8% agarose gel	7% disc polyacrylamide gel
cytosol	174.39	231.5
cytosol + BSA	176.83	215.6
BSA	77.2	113.5
control ("non-specific" binding)	89.6	101.7

ty binding of the glucocorticoid-receptors (6,7,8) for cortisol, dexamethasone and for triamcinolone acetonide could also be shown by the new developed gel-electrophoresis technique. Competition experiments with other steroid hormones will be shown elsewhere. Wagner (3) and Miller et al.(5) reported fast and slow moving peaks in polyacrylamide - and agarose gels by performing quantitative measurements of receptor proteins. However,the slow moving peaks could not be found and a "diffusion" of free ^{3}H-steroid hormone towards the cathode (3) was not observed in these studies.Wagner (3) and Miller et al.(5) performed gel analyses after pretreatment with Dextran/charcoal with samples of crude cytosol or serum load up to 1.2 mg protein. However, analyses by direct gel electrophoresis without pretreatment with Dextran/charcoal lead to the detection limit of partially purified glucocorticoid receptors in the lower µg upper ng range of protein.

Acknowledgements

This work was supported by the Deutsche Forschungsgemeinschaft. The excellent assistance of Miss Regina Trunz is gratefully acknowledged.

References

1. Lenger, K.: Int.J.Biochem. in press (1982).
2. Maurer, H. R.: Disc-Electrophoresis, Walter de Gruyter, Berlin . New York (1971).
3. Wagner, R. K.: Acta Endocrin. Suppl. 218, 215-221 (1978).
4. Scatchard, G.: Ann. N. Y. Acad. Sci. 51, 660-672 (1949).
5. Miller, L. K., Diaz, S. C., Sherman, M. R.: Biochemistry 14, 4433-4443 (1975).
6. Goral, J. E., Wittliff, J. L.: Biochemistry 14, 2944-2951 (1975).
7. Snochowski, M., Dahlberg, E., Gustafsson, J.-A.: Eur. J. Biochem. 111, 603-616 (1980).
8. Failla, D., Tomkins, G. M., Santi, D. V.: Proc. Natl. Acad. Sci, USA 72, 3849-3852 (1975).

ELECTROPHORESIS AND ISOELECTRIC FOCUSING OF MAMMALIAN SEMEN PROTEINS

Georgi Georgiev
Institute of Biology and Immunology of Reproduction and Development of Organisms
1113 Sofia, Bulgaria

Introduction

The seminal plasma of mammals is an extracellular fluid which plays an important role in sperm metabolism, as well as in the preservation of sperm fertilizing ability.
In the last three decades intensive investigations were conducted into the bull seminal plasma proteins dealing with different aspects of protein composition and characteristics (1,2, 3,4).
The development of reproductive immunology also stimulated the research on seminal plasma proteins in relation to their antigenic properties (5,6).
The aim of this study was to separate and characterize the bull seminal plasma proteins. Seminal plasma proteins were separated with the aid of disc electrophoresis according to Hedrick and Smith (7) modified by Buzun (8) in different polyacrylamide gel concentration - 3.6%, 6.0%, 7.2%, 8.4%, 9.0%, 9.6% and 10.0%.
Isoelectric focusing was carried out in apparatus for disc electrophoresis according to Miner and Heston (9). With a view to producing the necessary gradient, 40% stock ampholine solutions (LKB, Sweden) of pH 4.0-6.0; pH 6.0-8.0; pH 8.0-10.0 were used.
The pH gradient was determined in 1 cm gel slices by means of Zeibold pH-meter.

Electrophoresis '82

Results

The application of disc electrophoresis(7,8) made it possible to analyse 6 of the most distinguishable and largest protein fractions of bull seminal plasma (Fig. 1 a,b).

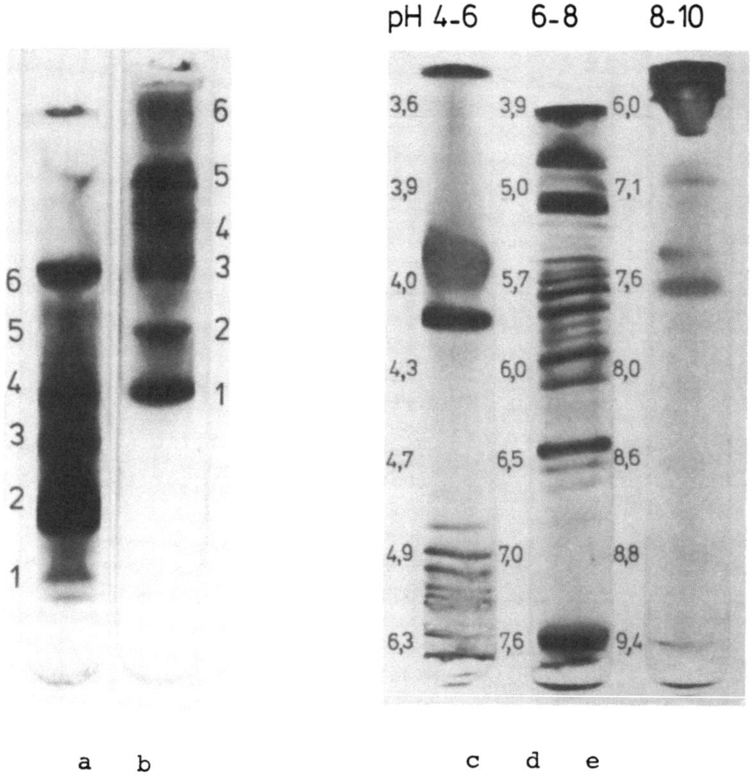

Fig. 1. a,b - the effect of different gel concentration (a-3.6%, b-9.6%) on the mobility of seminal plasma proteins; c,d,e - spreading of the protein fractions in the pH gradients after electrofocusing on polyacrylamide gel columns.

The results from the study on the relationship between the size of protein molecule, logarithm of its mobility 100(log Rm x 100) and gel concentration was expressed as linear regression for each protein fraction found in the separation on gels with different concentration are given in Fig. 2.

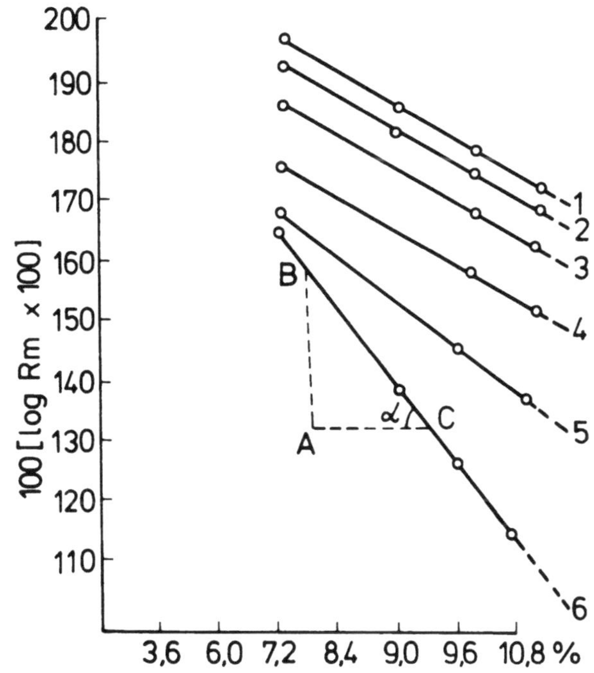

Fig. 2. Ferguson plots of bull seminal plasma proteins

Extrapolation based on these values revealed the following features. Protein fractions 1-5 were localized in parallel. In such cases Ferguson (10) and Hedrick and Smith (7) accept that these are protein molecules similar in size but different in net charge. Fraction 6 yielded non-parallel line showing a trend to cross the remaining lines in a direction to the 0% gel concentration which probably means that it had a higher molecular weight but equal electric charge.

The approximate molecular weight of fraction 1-5 was 30,000 and fraction 6-280,000 estimated by line slope (tg α) in the plot of Ferguson (10) and according to the data presented in Buzun's table (8).

The comparison of the results obtained from the experiments using different ampholines &pH 4.0-6.0; pH 6.0-8.0; pH 8.0-

10.0) revealed clearly expressed changes in the number and distribution of protein fractions (Fig. c,d,e).
When ampholine of pH 4.0-6.0 was applied the pH gradient varied between 3.6-6.3 and the proteins were separated into 12-16 fractions. An intensive grouping was recorded in the pH of 4.7-6.3 (Fig. 1c).
Figure 1d shows a model of protein distribution in the pH 6.0-8.0 the fluctuation of ampholine being from 3.9 to 7.6. Owing to this only the fractions observed in the pH range 6.3-7.6 were taken into consideration. In this pH interval there were 2-3 weak fractions and two considerably greater ones with isoelectric point about 7.6.
The application of ampholine pH 8.0-10.0 resulted also in a wider pH gradient 6.0-9.4. For this reason fractions in the pH range between 7.6-9.4 were recorded and 4.5 fractions were observed.
The results from this study indicate that proteins of acid nature which are mainly size or charge isomers dominate in bull seminal plasma.

References

1. Larson, S. L., Salisbury, G. W.: J. biol. Chem. 206, 741-749 (1951).
2. Salerno, G.: Zooprofilassi 14, 927-945 (1959).
3. Lavon, U.: J. Reprod. Fert. 31, 29-37 (1972).
4. Georgiev, G. H.: Int.J.Biochem. 9, 327-330 (1978).
5. Rao, S. S., Sadry, K. K.: J.J. Comp. Path. Therap. 70, 1-9 (1960).
6. Bratanov, K., Dikov, V.: Immunology of spermatozoa and Fertilization, Ed.: K. Bratanov,et al. Sofia, I, 41-45 (1969).
7. Hedrick, J. L., Smith, A. J.: Arch.Biochem.Biophys. 126 155-164 (1968).
8. Buzun, G. A.: Prikladnae biohimie i mikrobiologia, XII,

113-122 (1961).

9. Miner, G. D., Heston, L.L.: Analyt. Biochem. 50, 313-315 (1972).

10. Ferguson, K. A.: Met. Clin.Exp. 13, 985-1002 (1964).

POLYPEPTIDES SECRETED BY RESIDENT, INFLAMMATORY AND ACTIVATED MACROPHAGES.

Genevieve Lemaire, Anissa Bouchahda and Jean-François Petit
Institut de Biochimie (Bat 432) Université Paris-Sud
91405 ORSAY - FRANCE

Macrophages (MΦ) secrete a number of biologically active products into the extracellular medium and through this secretory function modulate the physiological state and the activity of surrounding cells and tissues (1). Among the proteins secreted are found 1) products involved in various host defense mechanisms : lysozyme, components of the complement..., 2) factors such as Interleukin I or Colony Stimulating Activity affecting other cell types (lymphocytes, fibroblasts) or MΦ and their precursors. 3) proteolytic enzymes (plasminogen activator, elastase...) which activate or degrade extracellular proteins.

Upon stimulation *in vivo* (or *in vitro*) MΦ display a series of functional alterations ; MΦ harvested from inflammatory exudates (induced by injection of a sterile irritant) present expanded secretory and endocytic capabilities but only activated macrophages (obtained from animals infected by intracellular parasites or treated by some immunostimulants) express enhanced antimicrobial resistance and antitumoral properties. (2)

We have used electrophoresis on polyacrylamide slab gels in the presence of SDS (SDS-PAGE) to analyse and compare the polypeptides secreted by 4 different MΦ populations (obtained from untreated animals, collected from inflammatory exudates or activated to become cytotoxic to tumor cells)- (Table I).

Electrophoresis '82

TABLE I : Characteristics of the 4 MΦ populations studied

eliciting agent (a)		Functional state		
		Secretion of Plasminogen Activator (b)	Cytostatic Activity (c)	
	R-MΦ	0	0 %	Resident
thioglycollate medium (1 ml)	TG-MΦ	4.5	0 %	Inflammatory
killed Group C Streptococci (d) (0.6 mg)	St-MΦ	26	75 %	Inflammatory and activated
trehalose dimycolate (e)	TDM-MΦ	∿0	92 %	Activated

a) injected i.p. 4 days before cell harvest

b) measured by a spectrophotometric assay using a chromogenic substrate of plasmin S2251 (3).
Activity is expressed as equivalents of urokinase CTA units secreted/mg of cellular protein/24 h.

c) the cytostatic activity corresponds to the capacity of macrophages to limit the incorporation of ^{3}H- thymidine in P815 mastocytoma cells (4) ; it is expressed as % of growth inhibition.

d) group C (Lancefield C74) Streptococci, grown in Todd Hewitt broth, treated by pepsin and kept at 4°C as a suspension in saline containing 0.5 % formaldehyde (5).

e) prepared from Mycobacterium tuberculosis strain Peurois ; suspended in water according to the method of Kato (6) it forms liposome-like structures ; the suspension is diluted in saline before use.

Methods

Macrophage cultures

Peritoneal MΦ were harvested from $B_6D_2F_1$ mice and plated in 35 mm petri dishes in 2 ml of medium (MEM + 5 % fetal calf serum (FCS) + antibiotics). After 4h at 37°C, non adherent cells were washed out and MΦ monolayers were incubated in medium supplemented with {^{14}C}amino acids (MEM + 0.1 % FCS + 0.05 % lactalbumin hydrolysate + antibiotics + {^{14}C} amino acids (1 or 2 μCi/ml)); medium was removed every 48 h and replaced with fresh medium (3 cycles). A low release (< 5 %) of lactate dehydrogenase into the medium was taken as an index of cell viability. Conditioned media were centrifuged, dialysed and lyophilised. The residue was dissolved in a denaturing mixture (62 mM Tris-HCl pH6.8, 3% SDS, 5% β-mercaptoethanol) (5 min. at 95° C). Cell monolayers were dissolved in the mixture. Aliquots (of conditioned media and cell lysates) were precipitated with 5 % TCA ; precipitates were recovered on GFC filters and counted. Protein and various cellular enzyme activities were measured on cells (from a second set of dishes) lysed with 0.1 % Triton X100.

Gel electrophoresis

Cell lysates and concentrated conditioned media were electrophoresed on 12.5 % and 7.5 % polyacrylamide slab gels according to the method of Laemmli (7). The gels were stained (0.13 % Coomassie blue in 50 % TCA), dehydrated and exposed to X-ray films (Kodak-No Screen). Appropriately exposed autoradiograms were scanned (with a Vernon PHI6 densitometer) and peak heights were measured for the major bands.

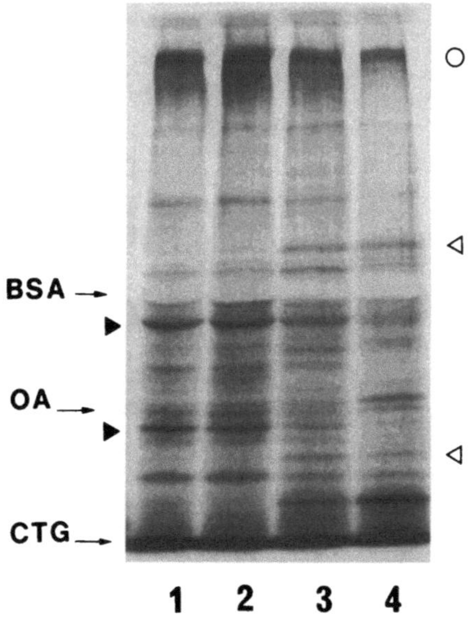

Fig.1 - Comparison of the polypeptides secreted by unstimulated and activated macrophages.

Unstimulated MΦ (R-MΦ) and MΦ elicited by injection of trehalose dimycolate (TDM-MΦ) were cultured <u>in vitro</u> in the presence of ^{14}C amino acids. Concentrated conditioned media were treated by SDS and β-mercaptoethanol and electrophoresed on a 7.5 % acrylamide slab gel (SDS-PAGE). An autoradiogram (exposed 5 days) is displayed.

Lane 1 TDM-MΦ first collection of conditioned medium (4h-52h)
2 TDM-MΦ second collection (52h-100h)
3 R-MΦ first collection of conditioned medium (4h-52h)
4 R-MΦ second collection (52h-100h)
◁ bands reduced in elicited MΦ
◀ bands increased in TDM-MΦ

M.W. markers : BSA (bovine serum albumin), OA (ovalbumin) and CTG (chymotrypsinogen).

RESULTS

1) Polypeptides secreted by unstimulated M Φ (R.MΦ)

The biosynthetic activity was approximately the same in the 4 MΦ populations but the % of radiolabeled proteins secreted into the extracellular medium (compared to total biosynthesis) was more important with unstimulated M Φ (40 %) than with inflammatory M Φ (15-20 %). In fact, analysis by SDS-PAGE revealed that the secretory activity of R.MΦ was important and diverse : on the 7.5 % acrylamide gel shown in fig 1 (lane 3) at least 18 polypeptides with an apparent MW >30 000 are resolved ; three smaller polypeptides (MW:28 000, 14 000 and 12 000) are visible on 12.5 % acrylamide gels.

2) Modifications in secretion patterns after stimulation in vivo

With elicited MΦ (compared to R.MΦ) the relative importance of several polypeptides was reduced : the reduction was especially pronounced, and was observed with the 3 eliciting agents used, for 2 peptides (indicated by ◁ in fig 1 ; MW : 78 000 and 35 000).

On the other hand, each eliciting agent stimulated the secretion of specific polypeptides : 1) the profiles of polypeptides secreted by TDM-MΦ(fig1, lane 1) were characterized by the presence of two very intense bands (indicated by ◀ - MW:59 000 and 39 000) ; 2) In the case of MΦ elicited by killed streptococci, a polypeptide of apparent MW 33 000 was prominent (data not shown) ; 3) Thioglycollate elicited M Φ essentially produced polypeptides of low MW (29 000 and 18 500) (data not shown).

The importance of the high MW component(s) (○ on fig.1) was always markedly reduced in the case of MΦ harvested from inflammatory exudates (TG-MΦ or St-MΦ). It is important to note that differences in secretion patterns were observed even when conditioned media were prepared from macrophages already cultured in vitro for 2 days (fig.1 lanes 2 and 4)

3) Modifications in cellular polypeptides after in vivo stimulation

The electrophoretic patterns of radiolabeled polypeptides, obtained when entire cell lysates were analysed, were basically similar for the 4 MΦ populations. However, some minor quantitative differences, characteristic for each macrophage population could be identified in lysates prepared from M Φ cultured 4h. The species presenting the most evident differences

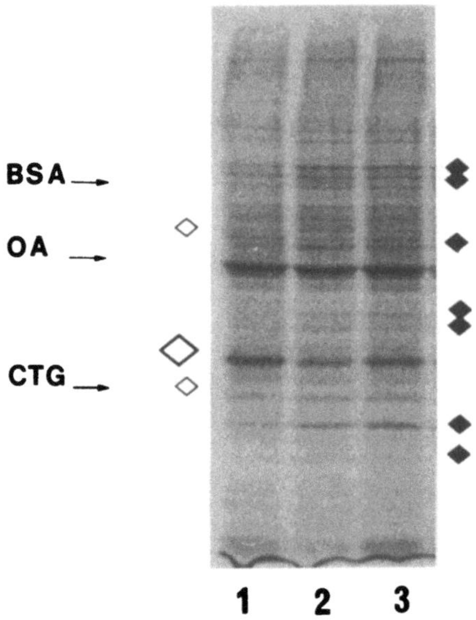

Fig. 2 - _In vitro_ modifications induced by LPS in electrophoretic patterns of macrophage lysates.

Cultures of thioglycollate elicited MΦ were washed 24h after plating and placed in the presence of LPS for 72h in a medium containing $\{^{14}C\}$ amino acids. Cell monolayers were lysed by 3% SDS and lysates electrophoresed on a 10 % acrylamide gel. An autoradiogram is displayed.

Lane 1 control - MΦ cultured in the absence of LPS.
2 250 ng/ml LPS.
3 25 ng/ml LPS.
◆ bands increased after _in vitro_ stimulation by LPS
◇ bands reduced.

(when compared to R-MΦ) was TG-MΦ : the relative importance of 9 polypeptides was altered by more then 30 % and the intensity of two bands (MW : 54 000 and 35 000) was increased by 100 %. As for secretion, differences between the various populations persisted after in vitro culture or were even amplified.

4) Modifications in secreted and cellular polypeptides after stimulation in vitro

The addition to the culture medium of LPS (lipopolysaccharide from the outer membrane of Salmonella enteritidis) (20 ng to 10μg/ml) introduced specific alterations in the biosynthetic and secretory activities of TG-MΦ . In patterns of polypeptides of cell lysates, there was an increase in intensity of at least 7 bands (◆ in fig.2) and the relative importance of 3 other bands (◇ in fig.2) was reduced (by 2 fold in the case of the polypeptide of MW : 27 000 ◇ in fig.2)

TABLE II : In vitro effects of LPS on the protein secretory activity of TG-MΦ

polypeptides apparent MW	peak heights (mm) 0	LPS 20 ng/ml	LPS 2 μg/ml
>100 000	25	77	88
59 000	33	67	57
39 000	62	72	84
29 000	77	40	45
22 000	<10	10	21
18 500	35	0	0
16 000	0	∿0	13
14 000	88	57	56

Cultures were washed 24h after plating and placed in medium containing {^{14}C} amino acids and LPS. 48h later conditioned media were collected and electrophoresed (12.5 % acrylamide gel) . Peak heights were measured on densitometer tracings of appropriately exposed (10 days) autoradiograms.

The secretion pattern of TG-MΦ was also markedly modified by LPS : the most evident alterations are summarized in Table II. In both cases(secreted proteins and intracellular polypeptides) two characteristics of the mode of action of LPS were revealed : 1) this bacterial immunomodulator exerts negative and positive controls on macrophage functions. 2) all the modifications do not require the same dose of LPS for induction.

In conclusion, a technique as simple as SDS-PAGE on one dimensional gels allows the detection of alterations induced in the expression of some cellular polypeptides after _in vivo_ or _in vitro_ stimulation of macrophages. It also permits an analysis of the polypeptides secreted by macrophages into the culture medium. The secretory function of MΦ is extensively modified upon stimulation _in vivo_ or _in vitro_ : various MΦ populations have clearly different patterns of secreted polypeptides. Thus these electrophoretic patterns may provide quantitative biochemical markers to identify the different functional states of MΦ.

References

1 - Gordon, S. : in : Mononuclear phagocytes, Van Furth (ed) p 1274, Martinus Nijhoff (1980).

2 - Cohn, Z.A. : J. Immunol. 121, 813-816 (1978).

3 - Drapier, J.C., Tenu, J.P., Lemaire, G., Petit, J.F. : Biochimie 61, 463-471 (1979).

4 - Tenu, J.P., Lederer, E., Petit, J.F. : Eur.J. Immunol. 10, 647-653 (1980) .

5 - Lemaire, G., Drapier, J.C., Tenu, J.P., Bouchahda, A., Lepoivre, M., Petit, J.F. : to be published in J.Reticulo-endothel. Soc.

6 - Kato, M., Am. Rev.Resp.Dis 95, 553 (1967).

7 - Laemmli, U.K. : Nature 227, 680-685 (1970).

DIRECT TISSUE ISOELECTRIC FOCUSING OF NERVOUS SYSTEM AND MUSCLE SECTIONS FOR DETECTION OF IgG PATTERNS

Tomas Olsson, Karl-Gösta Henriksson

Department of Neurology, University Hospital, Linköping

Hans Link

Department of Neurology, Karolinska Institutet, Huddinge Hospital, Huddinge, Sweden

Introduction

Protein analyses are mostly performed on body fluid samples. Qualitative and quantitative alterations in organ protein composition may, however, not be accurately reflected in body fluids. Previously, isoelectric focusing (IF) of saline extracts has been used for the characterization of tissue proteins under normal and pathological conditions. To circumvent time-consuming extraction procedures as well as any hypothetical denaturation and loss of proteins, Saravis et al. (1) applied frozen sections from a colonic adenocarcinoma directly on agarose gel for the qualitative characterization of tissue proteins by IF. This direct tissue isoelectric focusing (DTIF) technique enabled the detection of more proteins and higher numbers of protein bands compared to analysis after conventional tissue protein extraction. Potential application of DTIF should include analysis in biopsy and autopsy material. DTIF followed by antiserum immunofixation to identify and characterize proteins of interest would further extend the utility of the technique. In the present study, DTIF was applied on specimens from the central (CNS) and peripheral nervous system (PNS) and from skeletal muscle. Using monospecific antiserum for immunofixation, the IgG patterns were examined in normal and diseased tissues from these organs.

Electrophoresis '82

Materials and methods

The following tissues were examined: brain hemispheres from 3 patients with moderate to severe multiple sclerosis (MS); CNS and PNS from a patient with bilateral cerebral infarctions and with chronic polyneuropathy of unknown cause; "normal" brain and ischiadic nerves; muscle biopsies from 3 healthy volunteers and 5 patients with polymyositis. The tissues were frozen immediately after autopsy or biopsy. Blocks with one side measuring 8x8 mm were cut out from the tissues. 4-70 µm thick sections were cut on a cryostat microtome. The sections were collected on plastic film strips which were then put upside down on agarose gels used for IF. Thereafter IF, fixation and staining with Coomassie Blue was performed. The IgG pattern was characterized by subsequent immunofixation with antiserum against γ chains, alternatively with immunofixation using I 125-labelled antiserum against human IgG Fc fragments, followed by autoradiography.

Results

DTIF of 35 µm thick sections of CNS, PNS and muscle, followed by Coomassie Blue staining, revealed multiple well separated sharp protein bands. Increasing thickness as well as increasing section area resulted in more and stronger protein bands. However, no IgG pattern was seen using this procedure. However, when 35 µm thick PNS sections were used for DTIF and subsequent immunofixation with antiserum against γ chains, polyclonal IgG patterns similar to those of normal sera were identified in pH region 5.5 - 9.0. Using the more sensitive technique of radioimmunofixation after DTIF, IgG pattern could be visualized also from sections from CNS and muscle. This technique revealed oligoclonal IgG bands in all three MS brains, i.e. IgG of restricted heterogeneity not occurring when sections of normal brain or normal serum were examined in the same way. The oligoclonal IgG bands were mainly located in pH region 7.5 - 9.0 and they differed in number and mobility between the brains, but also between different regions whithin an individual brain.(Fig 1). No abnormal IgG patterns were seen when polymyositis muscle or PNS from the patient with polyneuropathy was subjected to DTIF.

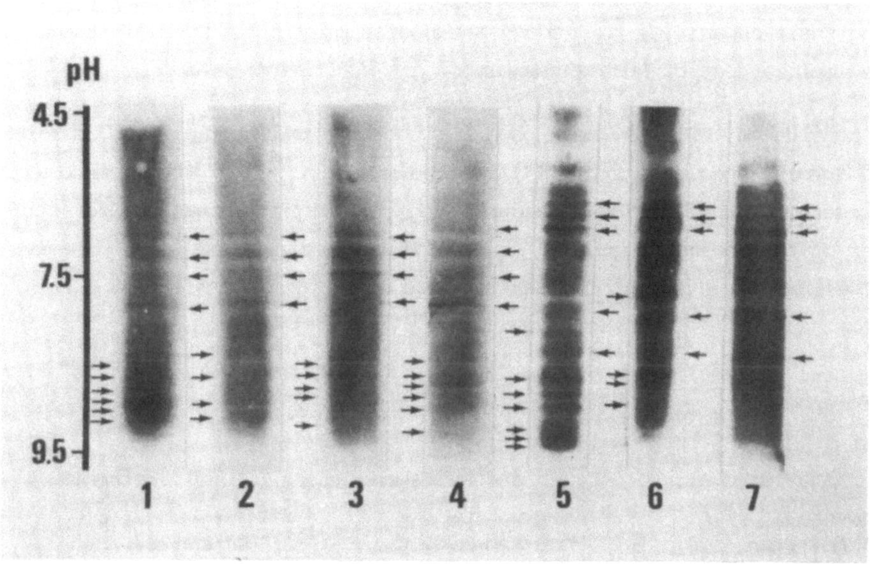

Fig 1. DTIF and subsequent radioimmunofixation with antihuman IgG Fc fragment antiserum performed on 35 µm thick sections of two different regions obtained from each of three different MS brains (1 and 2;3 and 4; 5 and 6;) and on a 35 µm thick section from a normal brain (7). 1-4 are obtained from one IF plate and 5-7 from another. Arrows from the left denote oligoclonal IgG bands, arrows from the right some of the polyclonal IgG bands. Note intraindividual variation in number and mobility of oligoclonal IgG bands, most evident when 5 and 6 are compared.

Discussion

The present study shows that DTIF is a simple and rapid method for separation of soluble tissue proteins from autopsy and biopsy specimens of CNS, PNS and skeletal muscle. Furthermore, particular proteins of interest can be identified by antiserum immunofixation after DTIF, and comparisons of patterns can be easily performed between normal and diseased tissues, as shown in this paper for IgG. The previous observation of different oligoclonal IgG band pattern in different regions of an individual MS brain after saline extraction and IF (2) was confirmed in the present study using DTIF on 3 MS brains. This phenomenon seems, therefore, not to be due to any protein extraction artifact. One possible reason for the oligoclonal IgG band pattern variation in MS brains may be a variation in distribution within the brain of plasma cell clones which vary regarding their capacities to produce antibodies of various specificities. The implications of such a phenomenon are not clear.

In conclusion, DTIF may be a powerful tool for protein analysis of biopsy and autopsy specimens from CNS, PNS and muscle tissues. If combined with antiserum immunofixation procedures, the technique is feasible for studies of tissue immunoglobulin abnormalities in immune-mediated diseases, as well as for studies of other types of protein abnormalities.

References

1. Saravis, C.A., M. O'Brien and N. Zamcheck; J. Immunol. Meth. 29, 97-100 (1979).

2. Mattsson, D.A., R.P. Roos and B.G. W. Arnason; Nature 287, 335-337 (1980).

DETERMINATION OF PREHbA$_{1C}$ BY USE OF ISOELECTRIC FOCUSING

Jan-Olof Jeppsson and Hans Tornqvist
Department of Clinical Chemistry, Malmö General Hospital, University of Lund
S-214 01 MALMÖ, Sweden

Introduction

Glycosylated hemoglobin, HbA$_{1C}$, has been increasingly used for monitoring long-term control of glucose metabolism in patients with diabetes mellitus (1,2). HbA$_{1C}$ is formed slowly and continously throughout the life span of the red cell and at a rate varying with the blood glucose concentration (3,4). The chemical steps in its synthesis start with a reversible Schiff base linkage of the aldehyde group of glucose with the N-terminal amino group of the ß-chains of hemoglobin. This labile intermediate, preHbA$_{1C}$, undergoes an Amadori rearrangement to the stable ketoamine HbA$_{1C}$ (5). The dissocation of pre-HbA$_{1C}$ to HbA is much faster than the HbA$_{1C}$ formation (6). Recent observations of fast day-to-day variations in HbA$_{1C}$ levels may mainly be explained by variations in preHbA$_{1C}$ which is codetermined in most methods (7,8). Preincubations of erythrocytes in glucose-free medium reduces the preHbA$_{1C}$ fraction and may diminish day-to-day variations of HbA$_{1C}$.

A number of methods have been employed in determinating HbA$_{1C}$ levels. So far, most laboratories have used different modifications of the original ion-exchange chromatography method of Trivelli (9). Rapid methods using commercial microcolumns estimate a combined fraction, HbA$_1$, in which HbA$_{1C}$ dominates among other minor hemoglobins (5). Other methods reported include electrofocusing (10), high pressure liquid

Electrophoresis '82

chromatography (11), colorimetric methods (12), a radioimmunologic method (13), an electrophoretic method on agar (14) and a fluorometric method (15). We have been using the electrofocusing procedure since it is specific and allows low-cost determinations in large series.

The aim of this report was first to identify the aldimide form of HbA_{1C}, and to investigate its rate of formation in the presence of glucose. We have by the use of electrofocusing localized this component close to the HbA_{1C} band. PreHbA_{1C} rapidly increases upon incubation of erythrocytes in 20 mM glucose at +37°C. The reaction is reversible if glucose is removed. PreHbA_{1C}, which interferes with HbA_{1C} determinations, can be abolished by preincubating the erythrocytes in 0.15 M NaCl at +37° for 6 hours. Thereafter electrofocusing offers a reliable and specific method for the analysis of HbA_{1C} on capillary blood samples. In children with insulin-dependent diabetes mellitus at different degree of metabolic control, levels of preHbA_{1C} varied between 0.2 and 3.7 % of total hemoglobin (median value 1.3 %, N = 25).

Materials and methods

Preparation of samples. Capillary blood samples (100 µl) were collected in heparinized tubes at 8 a.m. from insulin-dependent diabetic children. Erythrocytes and plasma were immediately separated by centrifugation. The erythrocytes were suspended in 1 ml 0.15 M NaCl. 0.5 ml of the suspension was recentrifuged and the supernatant replaced with 0.4 ml distilled water and two drops of CCl_4 followed by an another centrifugation. The hemolysate was bubbled with CO and stored at -70°C. The remaining 0.5 ml of erythrocyte suspension was incubated for 6 h at +37°C and then treated likewise.

HbA_{1C}. Electrofocusing of the preincubated hemolysates was employed to analyse HbA_{1C} content (10). Home made or ready made polyacrylamide gels (PAG-plates 1804-131 LKB-Produkter AB, S-161 25 Bromma, Sweden) were used, and the electrofocusing was performed in the LKB 2117 Multiphor apparatus with a constant power supply LKB no 2103. Water was circulated

5-10 l/min at 15°C through the cooling plate. Filter paper at the cathodal end was moistened with 0.1 M NaOH and for the anodal end with 0.04 M glutamic acid. The gels were prerun for 30 min at 2000 V, 20 W and 50 mA maximum, and then 20 µl of hemolysate was applied on 5 x 10 mm filter papers on the gels. These filter papers were removed after 40 min with the same electrical setting. After another 70 min the power was increased to 30 W for a further 10 min. The optimal gel temperature was 18-20°C.

Afterwards, the gel containing the HbA_{1C} fractions were cut out using a thin scalpel blade and transferred to 1 ml of elution buffer, 0.05 M Tris HCl, and 0.1 mM EDTA, pH 8.3. Previously, 20 µl of the same hemolysate had been transferred to 5 ml of elution buffer. The tubes were sealed and rocked at 4-10°C over night. HbA_{1C} was determined as the absorbance at 415 nm in percentage of the absorbance of total hemoglobin. Every 10th sample was a control from a pool of healthy donors stored at -70°C.

<u>PreHbA_{1C}</u>. The difference in triplicate HbA_{1C} determiantions before and after preincubation of the same erythrocyte sample for 6 h at +37°C represented the content of preHbA_{1C}, expressed as percentage of total hemoglobin.

Results

Figure 1 illustrates the time course in appearance of a band anodally to HbA_{1C} when erythrocytes from a non-diabetic donor were incubated with 20 mM glucose in 0.15 M NaCl at +37°C. Removal of glucose from the incubation medium at that time almost abolished this band within five hours. The HbA_{1C} band on the other hand was constant during the experiment. No other visible band showed these rapid glucose-dependent variation. Thus, this anodal band may be the aldimide form of HbA_{1C} or preHbA_{1C}. The preHbA_{1C} band can not in practise be sufficiently separated from HbA_{1C} for quantitative measurements. Instead preHbA_{1C} was estimated as the difference in HbA_{1C} before and after a 6 hours preincubation of erythrocytes at +37°C in 0.15 M NaCl .

During these experiments preHbA_{1C}, which was 0.6 % of total hemoglobin at start, reached a steady state value of 2.5 % after 6 hours of incubation in 20 mM glucose at +37° followed by a reduction to less than 0.1 % within 5 hours of further incubation in glucose free saline.

PreHbA_{1C} levels in blood samples from children with insulin dependent diabetes mellitus with different degrees of metabolic control varied between 0.2 % and 3.7 % with a median value of 1.3 % (N = 25). Median HbA_{1C} level was 11.4 % (range 7.5-13.5 %, N = 25).

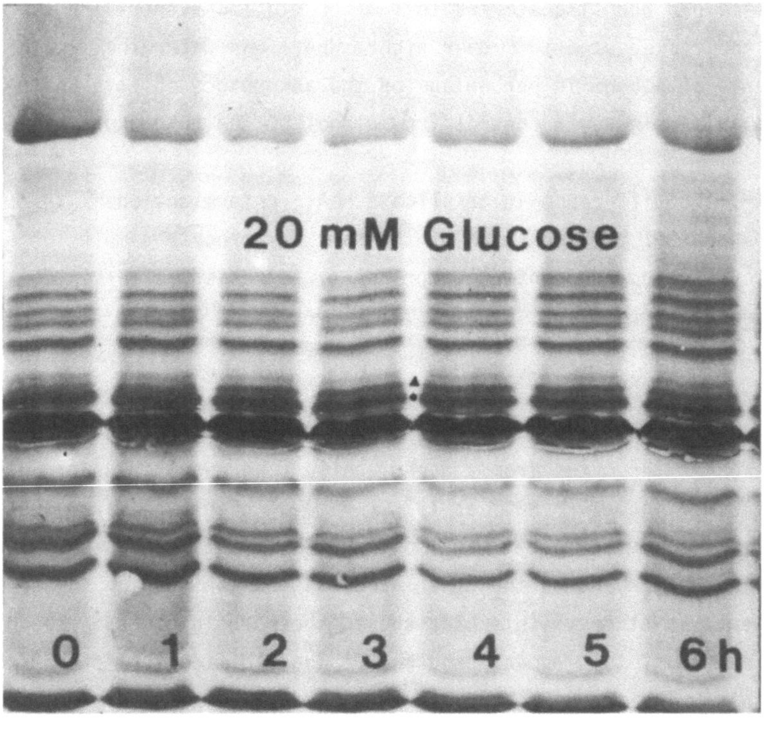

Figure 1. Changes in preHbA_{1C} illustrated om stained electrofocusing plates during incubation of fresh non-diabetic erythrocytes in 0.15 M NaCl containing 20 mM glucose (a) at +37°C for 0-6 hours. The triangel indicates the preHbA_{1C} and the dot the HbA_{1C} bands.

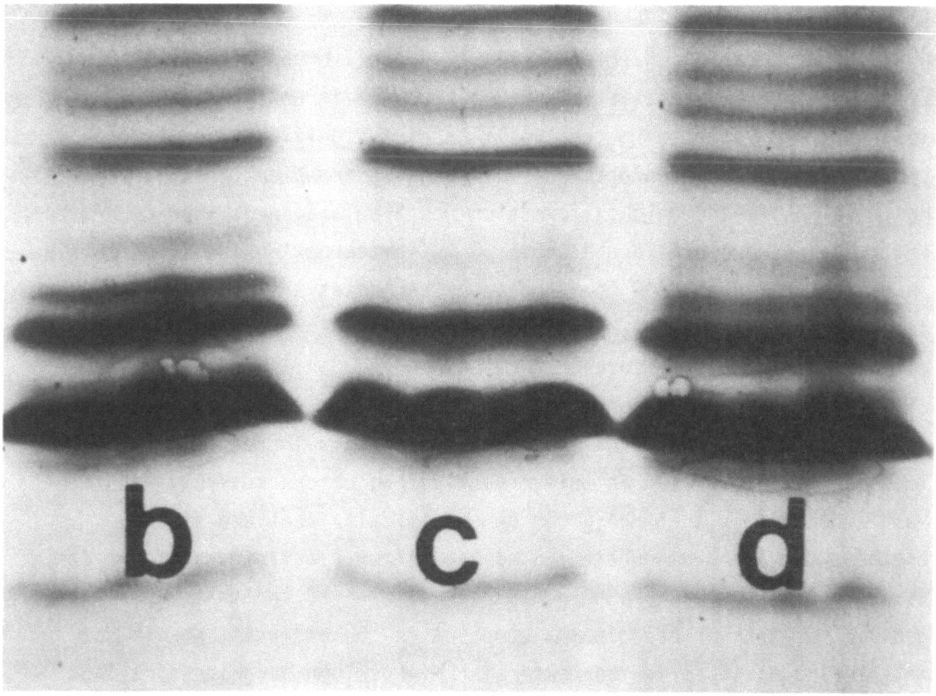

Figure 2. The effect on $preHbA_{1C}$ and HbA_{1C} of preincubation of erythrocytes in 0.15 M NaCl at +37°C for 6 hours (b). Before and (c) after incubation of erythrocytes from a diabetic and (d) from a non-diabetic before incubation.

Discussion

The appearance of a distinct isoelectric band barely separated from the HbA_{1C} band was shown to be dependent on the presence of glucose. The reaction seemed fast and reversible. Out data indicate that this band represents the intermediate aldimine form of HbA_{1C} or $preHbA_{1C}$, a designation proposed by Bunn (5). Welinder and Svendsen (8) could with electrofocusing demonstrate this component in diabetics with fasting blood glucose levels above 11 mM. We found that $preHbA_{1C}$ is always present in a concentration depending on the blood glucose level, and is thus also found in non-diabetics at a low concentration.

$PreHbA_{1C}$ was determined as the difference of HbA_{1C} levels before and after removal of $preHbA_{1C}$. This necessitated a method for complete removal of preHbA$_{1C}$ and a specific and accurate measurement of HbA$_{1C}$. Preincubation of erythrocytes in saline for 6 hours at 37°C removed preHbA$_{1C}$ quantitatively without change in HbA$_{1C}$, If, instead, hemolysates were treated likewise a significant methemoglobin formation occured, even after carbon monoxide treatment, giving false low HbA$_{1C}$ values. Methemoglobin formation was also prominent when erythrocytes were pre-incubated in 30 mM semicarbazide and 12 mM aniline, pH 5, for 30 min at 38°C as proposed by Nathan et al (16). The direct and accurate analysis of HbA$_{1C}$ has been difficult to perform mainly due to the small charge difference between HbA and HbA$_{1C}$ and the presence of several other minor components. This has promted the appearance of several different analytical methods and variants thereof. The electrofocusing technique used in this investigation offers several advantages. Most important is that HbA$_{1C}$ can be specifically analysed if the erythrocytes are preincubated in saline. After electrofocusing one can visually confirm that preHbA$_{1C}$ has been eliminated and that no genetic variants of hemoglobins complicates the analysis. HbF is not interfering as it is in the commercial microcolumn techniques, a fact that is important especially in pediatric practice. Furthermore, the electrofocusing technique allows determinations on small capillary samples in large series at a low lost.

Among children with insulin-dependent diabetes mellitus at varying degree of metabolic control we found that preHbA$_{1C}$ levels varied considerably. In some cases where the patients were clearly hyperglycemic preHbA$_{1C}$ reached a niveau of 25-30 % of the HbA$_{1C}$ value.
The present study clearly illustrates that preHbA$_{1C}$ levels mirror the very recent blood glucose levels since it correlates highly to urinary glucose excretion during the preceeding 12 hours period.

Acknowledgment
The skilful technical assistence of Mrs. Inger Solding is aknowledged. This investigation was supported by grant from the Swedish Medical. Research Council (Grant No B81-03X-05912-01A).

References

1 Koenig, R.J., Peterson, C.M., Jones, R.L., Sandele, Ch., Lehrman, M. and Cerami, A. The correlation of glucose regulation and hemoglobin A_{1C} in diabetes mellitus. N. Engl. J. Med. 295, 417-420 (1965).

2 Bunn, H.F., Gabbay, K.H. and Gallop, P.M. The glycosylation of hemoglobin: Relevance to diabetes mellitus. Science 200, 21-27 (1978).

3 Koenig, R.J. and Cerami, A. Synthesis of hemoglobin A_{1C} in normal and diabetic mice: Potential model of basement membrane thickening. Proc. Nat, Acad. Sci, USA, 72, 3687-3691 (1975).

4 Bunn, H.F., Haney, D.N., Kamin, S., Gabbay, K.H. and Gallop, P.M. The Biosynthesis of human hemoglobin A_{1C}: Slow glycosylation of hemoglobin in vivo (1976).

5 Bunn, H.F. Evaluation of glycosylated hemoglobin in diabetes mellitus. Diabetes, 30, 613-616 (1981).

6 Higgins, P.J. and Bunn, H.F. Kinetic analysis of the non-enzymatic glycosylation of hemoglobin. J. Biol. Chem. 256, 5204-5208 (1981).

7 Svendsen, P.Aa., Christiansen, J.S., Søegaard, U., Welinder, B.S. and Nerup, J. Rapid changes in chromatograpically determined hemoglobin A_{1C} induced by short-term changes in glucose concentration. Diabetologia 19, 130-136 (1980).

8 Welinder, B.S. and Svendsen, P.Aa. Heterogeneity of the hemoglobin A_{1C}-band in isoelectric focusing. Diabetologia 19, 465-467 (1980).

9 Trivelli, L.A., Ramneyh, H.M. and Lai, H.-T. Hemoglobin components in patients with diabetes mellitus. N. Engl. J. Med. 284, 353-357 (1971).

10 Jeppsson, J.-O., Franzén, B. and Gaal, A.B. Simplified determination of hemoglobin A_{1C} in diabetic patients by use of electrofocusing. Electrophoresis 79, p. 655, Editor J. Radola. Walter de Gruyters Co, Berlin & New York (1980).

11 Davis, J.E., McDonald, J.M. and Jarrett, L. A high-performance liquid chromatography method for hemoglobin A_{1C}. Diabetes 27, 102-107 (1978).

12 Flüchiger, R. and Winterhalter, K.H. In vitro synthesis of hemoglobin A_{1C}. FEBS letter 71, 356-360 (1976)

13 Javid, J., Pettis, P.K., Koernig, R.J. and Ceramin, A. Immunologic characterization and quantification of hemoglobin A_{1C}. Brit. J. Hemat. 38, 329-337 (1978).

14 Read, A., Tibi, L., and Smith, A.F. Assessment of a simple electrophoretic method for measuring HbA_1. Clin Chim Acta 108, 487-491 (1980).

15 Gallop, P.M., Flückiger, R., Hanneker, A., Minosohn, M.M. and Gabbay, H. Chemical quantitation of hemoglobin glycosylation: Fluorometric detection of formaldehyde release upon periodate oxidation of glycobin. Anal. Biochem. 117, 427-432 (1981).

16 Nathan, D.M., Avezzano, E.S. and Palmer, J.L. A rapid chemical means for removing labile glycohemoglobin. Diabetes 30, 700-701 (1981).

ELECTROPHORETIC ANALYSIS OF TISSUES IN DUCHENNE MUSCULAR DYSTROPHY

M. J. Dunn, A.H.M. Burghes, B.J. Thompson, H.E. Statham and V. Dubowitz
Jerry Lewis Muscle Research Centre, Department of Paediatrics and Neonatal Medicine, Hammersmith Hospital, London W12 OHS, U.K.

1. Introduction

The abnormality responsible for the progressive degeneration of skeletal muscle characteristic of Duchenne muscular dystrophy (DMD) has not been identified. In addition to the primary protein defect, genetic diseases can also result in specific secondary protein changes as the cell attempts to compensate for or react to the metabolic consequences of the primary defect (1). In studies of DMD only limited use has so far been made of electrophoretic techniques and therefore in our laboratory we have established a protein mapping programme using a battery of electrophoretic techniques to investigate changes in cellular protein profiles associated with DMD.

2. Direct Tissue Isoelectric Focusing of Muscle

Needle muscle biopsies performed for diagnostic purposes yield very small tissue samples which imposes severe constraints on the scope of biochemical analyses that can be performed. Saravis (2) described a method of direct isoelectric focusing (DTIF) in agarose which has the advantage of minimising deleterious effects due to proteolysis or denaturation. We have applied this procedure to human skeletal muscle (3) and now report further improvements to overcome some of the difficulties inherent in the use of agarose IEF.

Cryostat sections (20 µm) of needle muscle biopsy specimens were adhered to the hydrophilic side of Gel Bond (3). The gels contained 1% (w/v) Pharmacia Agarose IEF, 10% (w/v) sorbitol, 5% (w/v)

Electrophoresis '82

glycerol, 0.5% (w/v) Triton X-100, 1.25% (w/v) pH 3-10 Pharmalyte, 1.25% (w/v) pH 2-11 Servalyt and 0.5% (w/v) pH 5-8 Pharmalyte. Glycerol and sorbitol additives were used as these have been found to increase sharpness of bands and enhance resolution (4). Gels 0.5 mm thick were cast using the U-shaped gasket technique in a water-bath at 45°C. The anolyte was 0.1M aspartic acid (boiled 3 to 4 min. to get into solution) and 0.05M H_2SO_4. The catholyte was 0.25M lysine, 0.25M arginine, 1M NaOH, and 0.135M $Ca(OH)_2$ (spun to remove precipitate). The gel was blotted with Whatman No. 1 filter paper and care had to be exercised here due to decreased adherence of the gels in the presence of Triton X-100. Focusing was performed under a CO_2-free N_2 atmosphere at 10°C. The samples were applied and run in at 200V for 1 hr. The sample applicators were removed and the run continued at 8W, 1500V limiting for a further 2650 V.hr. The pH gradient was measured with a surface electrode and the gel refocused for 5 min at 3000V, giving a total of 3050 V.hr.

The gels were fixed in 20% (w/v) TCA containing 2M urea for 30 min. To remove Triton-ampholyte complexes with minimal loss of protein the gels were immersed in a solution containing 2M urea, 30% (v/v) methanol, 5% (w/v) TCA overnight. The gels were then washed for 60 min. in the same solution, followed by 2 x 30 min. washes in 5% (w/v) TCA, 30% (v/v) methanol and finally 2 x 15 min. in 30% (v/v) methanol, 10% (v/v) acetic acid. The gels were then dried, immersed for 5 min. in 30% (v/v) methanol, 10% (v/v) acetic acid and stained with Coomassie blue made up in the same solution.

A typical DTIF gel of normal muscle samples is shown in Fig. 1 and its pH profile in Fig. 2. As the majority of bands were in the pH 5-8 region we have attempted to spread the pH gradient in this region. LKB Ampholines for agarose IEF were found to be supplemented in the cathodic region and produced unwanted spreading of this region. Servalyt 2-11 gave good acidic and cathodic resolution whereas Pharmalyte 3-10 was superior in the mid-region (pH 5-8). The mixture of ampholytes which we finally adopted was found to be better than any of the ampholytes used alone.

We have compared a series of normal individuals by quantitation of densito-

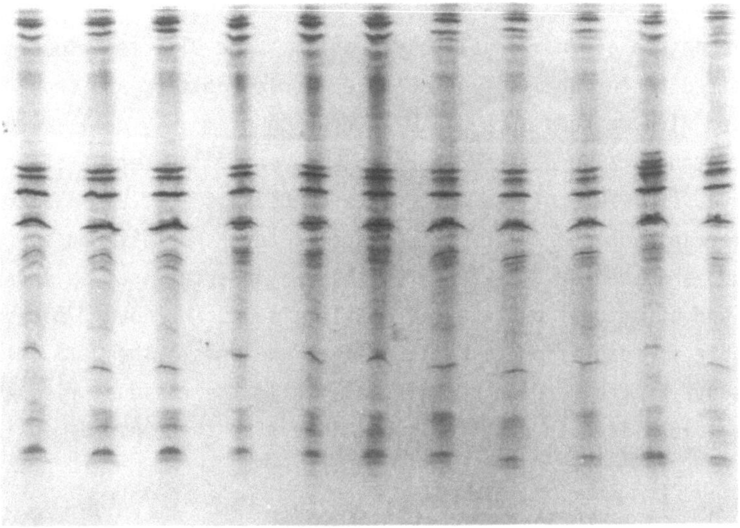

Fig. 1 : Typical DTIF gel of normal muscle samples.

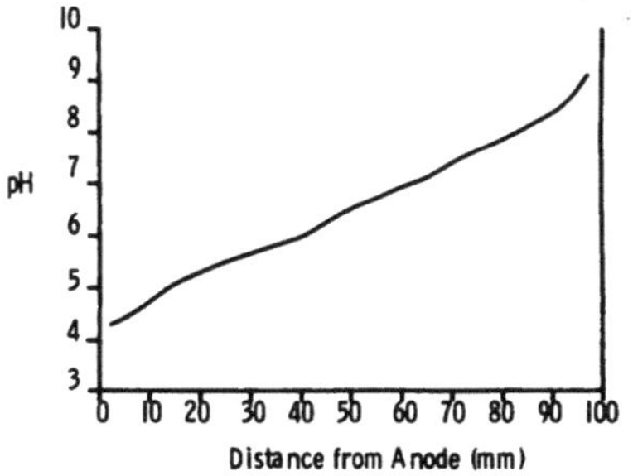

Fig. 2 : pH gradient of gel shown in Fig. 1 measured with a surface electrode.

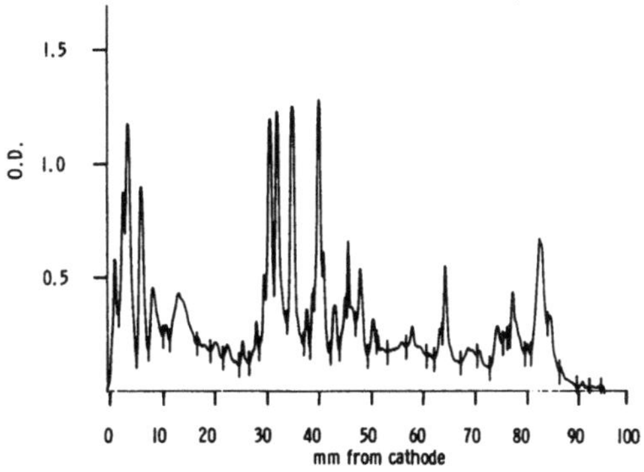

Fig. 3 : Densitometric scan of DTIF gel profile using a Joyce-Loebl Chromoscan 3 (50 µm slit width).

metric scans of stained gels (Fig. 3). Interestingly we have found a considerable degree of individual polymorphism. The DTIF technique is probably limited to the analysis of soluble muscle proteins and should be contrasted with methods such as 2D-PAGE which primarily deal with more abundant structural proteins, where an apparently low degree of polymorphism is observed (5). One of the advantages of DTIF is that residual non-soluble tissue remains adhered to the sample applicators and this can be solubilised in SDS and analysed by SDS-PAGE. In this way one can analyse separately the soluble and structural proteins of muscle tissue. Moreover, both techniques are capable of adaptation to 2-dimensional analysis. The DTIF gel would itself serve as the first dimension gel, while the residual tissue sample would be solubilised in a buffer compatible with the first dimension IEF.

3. Electrophoretic Analysis of Skin Fibroblast Proteins

As extensive muscle degeneration occurs in DMD many non-specific secondary changes in protein profiles might be expected. In an attempt to overcome some of the difficulties inherent in the study of diseased muscle we are using cultures of skin fibroblasts to compare protein profiles of cells from normal and DMD individuals.

3.1 SDS polyacrylamide gel electrophoresis

Total cell proteins of normal and DMD fibroblasts have been compared using gradient SDS-PAGE (6,7). The detection systems used were Coomassie blue staining, silver staining, and autoradiography of radioactively labelled samples. A typical autoradiograph of $[^{14}C]$-leucine labelled fibroblast proteins (200 000 cpm/sample) separated on a 0.7 mm thick 5 to 20% polyacrylamide gradient SDS gel is shown in Fig. 4. No qualitative differences were observed but quantitative analysis by gel slicing of samples labelled with $[^{35}S]$-methionine revealed differences associated with the disease state. Two methods of statistics were used to analyse the data (7) and regions of molecular weight 68-73 x 10^3 and 48 x 10^3 were found to be consistently decreased in DMD samples.

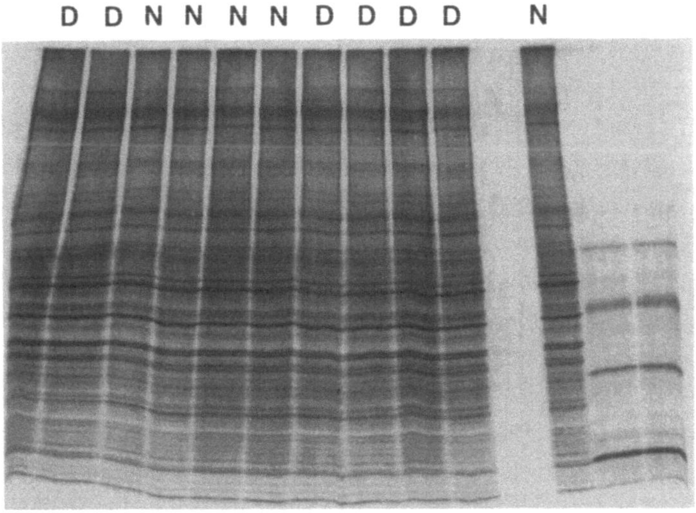

FIG. 4 Comparison of normal (N) and DMD (D) fibroblast proteins labelled with [^{14}C]-leucine and separated by gradient SDS-PAGE. The autoradiograph was exposed for 7 days using Kodak X-Omat H film with an intensifying screen.

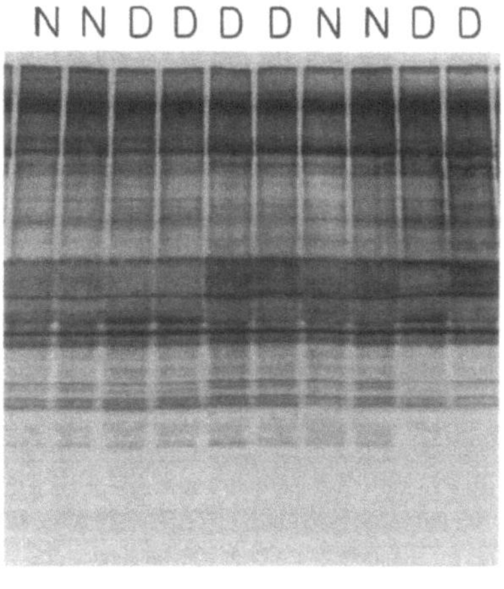

FIG. 5 [^{35}S]-methionine labelled proteins secreted by normal (N) and DMD (D) fibroblasts on 5 to 20% gradient SDS-PAGE. 10 day exposure using Kodak No-Screen film.

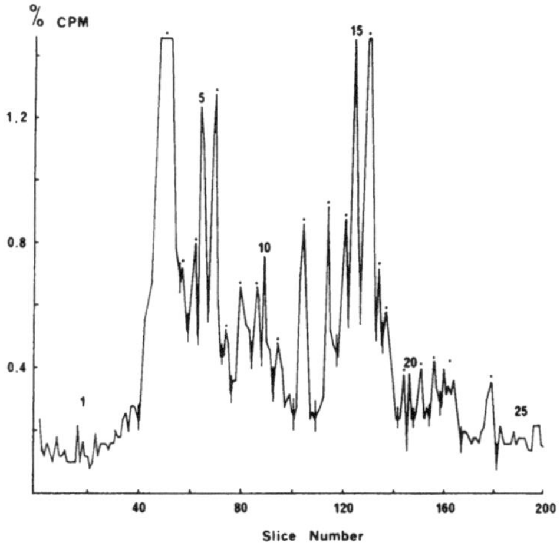

FIG. 6 Plot of [^{35}S]-methionine labelled proteins secreted by fibroblasts derived from a gel slicing experiment. The regions into which the gels were divided for quantitation are indicated.

We have compared the proteins labelled with $[^{35}S]$-methionine secreted into the medium by fibroblasts from 6 normal and 6 DMD individuals. Qualitative comparison of these profiles by autoradiography (Fig. 5) clearly demonstrates a variation between individuals in the abundance of a band of molecular weight 48 x 10^3. However it has not yet been established whether this change is associated with the disease state. A quantitative preliminary comparison of normal and DMD samples has been carried out by a gel slicing technique. The samples (200 000 cpm) were electrophoresed on rod-gels which were subsequently sliced into 0.5 mm segments, the proteins in each slice eluted and counted in a scintillation counter (7). A plot from such an experiment is shown in Fig. 6. No consistent significant differences between normal and DMD samples have been observed but there appears to be considerable variability between samples from different individuals.

3.2 Isoelectric focusing under dissociating conditions

Protein profiles of normal and DMD skin fibroblasts labelled with $[^{35}S]$-methionine have been compared by IEF on thin polyacrylamide gels containing 8M urea and 2% NP-40 cast on silanized glass or plastic supports (8). C4T4 gels were prepared containing 8M urea, 2% (w/v) NP-40 and 4% (w/v) Pharmalyte 3-10. The catholyte solution was 1M NaOH, 0.025M lysine, 0.025M arginine and 0.135M $Ca(OH)_2$ (spun to remove precipitate). The anolyte was 0.04M aspartic acid and 10 mM H_3PO_4 in 8M urea. A large number of sharp, well-resolved bands were observed, the majority of these being in the pH 5-8 region. Recently we have found that the use of a mixture of ampholytes (1.7% Servalyt 2-11, 1.7% Pharmalyte 3-10, 0.6% Pharmalyte 5-8) gave enhanced resolution (see Burghes et al, this volume). An autoradiograph of a typical C4T4 IEF gel is shown in Fig. 7 and no consistent qualitative differences between normal and DMD profiles were observed. A series of normal and DMD samples were quantitatively compared on gels cast on special plastic supports (8). A typical plot is illustrated in Fig. 8 and those regions found to differ significantly are summarised in Table 1.

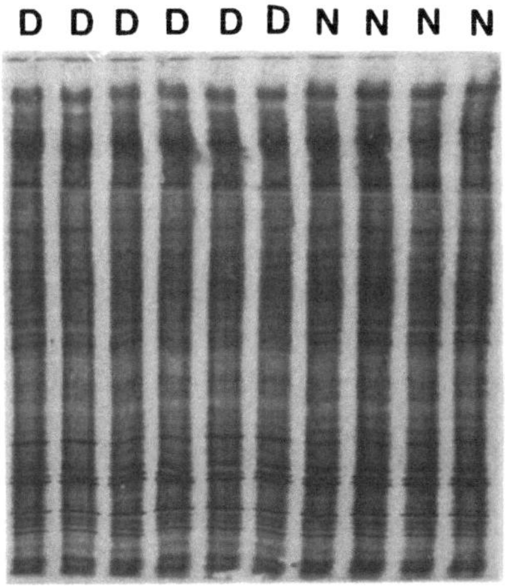

FIG. 7 C4T4 IEF gel (0.5 mm thick) using the ampholyte mixture described in the text of normal (N) and DMD (D) fibroblast proteins labelled with [^{35}S]-methionine. The gel was prefocused at 400V for 40 min. The samples (400 000 cpm) were applied and run at 15W, 1500V limiting for a total of 8500 V.hr. The pH gradient was measured with a surface electrode and the gel refocused at 3000V, giving a total of 8700 V.hr for the run. The autoradiograph was exposed for 1 day using Kodak X-Omat H film.

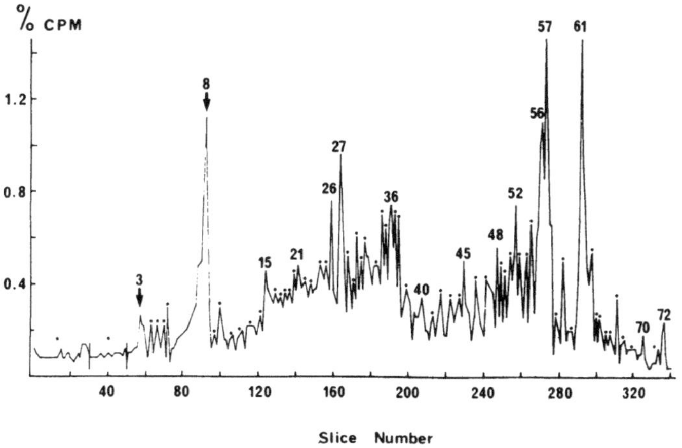

FIG. 8 Plot of IEF separation profile of [^{35}S]-methionine labelled fibroblast proteins derived from a gel slicing experiment. The areas into which the plots were divided for quantitation are indicated.

Table 1

Samples (400,000 cpm) were applied to an IEF gel cast on a plastic support. After electrophoresis each sample track was cut into 0.3 mm segments, the proteins eluted and counted in a scintillation counter. The proportion of radioactivity in each region (see Fig. 8) expressed as a percentage of the total counts recovered was calculated. The data were analysed by t-test.

Peak	pI*	Relative % total cpm				t value
		Normal (5)	[±SD]	DMD (5)	[±SD]	[df=17]
10	8.1	0.95	(0.25)	0.67	(0.26)	2.434†
18	7.6	0.79	(0.19)	1.24	(0.49)	2.577‡
26	7.3	1.70	(0.62)	2.31	(0.50)	2.380†
48	6.3	1.55	(0.54)	0.86	(0.25)	3.613§
52	6.2	1.39	(0.54)	0.93	(0.31)	2.304†
56	6.0	3.21	(1.26)	5.03	(1.33)	3.045§
59	5.9	1.26	(0.49)	0.83	(0.33)	2.247†

† $p < 0.05$, ‡ $p < 0.02$, § $p < 0.01$

* pI values uncorrected for presence of urea

4. Future Prospects

Using electrophoretic techniques we have been able to demonstrate that there are changes in the protein composition of skin fibroblasts which are associated with DMD, but the significance of these changes is unresolved. One method of improving the accuracy and sensitivity of comparisons between profiles of normal and DMD samples would be to use a dual labelling procedure (6). The increased resolution afforded by two-dimensional electrophoretic techniques is also ideal for such comparisons. We have already compared protein profiles by this technique (6) and our recent progress is presented separately at this meeting (see Burghes et al, this volume).

Acknowledgements

We would like to thank Ms. J. Carter and Ms. R. McDouall for their skilled technical assistance, Ms. L.J. White for her photographic expertise and Mrs. C. Trand for typing the manuscript. This work was supported by the British Medical Research Council, the Muscular Dystrophy Group of Great Britain and the Royal Society.

References

1. Merril, C.R.: in Electrophoresis '81 (ed. R.C. Allen and P. Arnaud), pp. 343-353, de Gruyter, Berlin (1981).
2. Saravis, C.A., O'Brien, M. and Zamchek, N.: J. Immunol. Methods 29, 97-100 (1979).
3. Thompson, B.J., Burghes, A.H.M., Dunn, M.J. and Dubowitz, V.: Electrophoresis 2, 251-258 (1981).
4. Coulson, S.E.: in Electrophoresis '81 (ed. R.C. Allen and P. Arnaud), pp. 229-236 (1981).
5. McConkey, E.H., Taylor, B.J. and Phan, D.: Proc. Natl. Acad. Sci. USA 76, 6500-6504 (1979).
6. Burghes, A.H.M., Dunn, M.J., Statham, H.E. and Dubowitz, V.: in Electrophoresis '81 (ed. R.C. Allen and P. Arnaud), pp. 295-308, de Gruyter, Berlin (1981).
7. Burghes, A.H.M., Dunn, M.J., Statham, H.E. and Dubowitz, V.: Electrophoresis 3, in press (1982).
8. Burghes, A.H.M., Dunn, M.J., Statham, H.E. and Dubowitz, V.: Electrophoresis 3, in press (1982).

CHANGES OF MYOSIN ISOENZYMES IN THE CARDIOMYOPATHIC SYRIAN HAMSTER AND IN THE HUMAN MYOCARDIUM.

V. Wiegand, H. Klein, E. Stroh, H. Hennekes, H. Kreuzer
Med. Klinik, Universität Göttingen, Germany

K. Lossnitzer
KH Sulzbach, Germany

Introduction

Changes in myocardial myosin isoenzyme pattern have been demonstrated in response to myocardial hypertrophy and dilatation (1). The present study was carried out to analyse if similar changes occur in the cardiomyopathic Syrian hamster and if this concept of myosin isoenzyme shifts is applicable to the human heart.

Methods

Non-dissociating pyrophosphate electrophoresis and Ca^{2+}-ATPase staining was performed according to (2). Left ventricular myosin pattern was analysed in 220 days old cardiomyopathic Syrian hamsters (n=6) and in control hamsters of the same age (n=6). Human myocardial tissue was obtained during open heart surgery by left ventricular biopsy in 5 patients with left ventricular hypertrophy (aortic valve disease) and 5 patients with normal left ventricles (atrial septal defect). Furthermore right ventricular biopsies were analysed in 7 patients with varying degrees of congestive cardiomyopathy.

Electrophoresis '82

Results

Similar to what has been shown in the rat heart (3), three different ventricular myosin isoenzymes (V_1, V_2, V_3 in order of decreasing mobility) are found in left ventricular hamster myocardium. The myosins exhibit different Ca^{2+}- activated ATPase activities (V_1:V_2:V_3=5.5:3:1.5). In the control group V_1 predominates (V_1 = 85 %, V_2 = 8 %, V_3 = 7 %) whereas in the cardiomyopathic hamster there is a shift to myosins of lower Ca^{2+}-ATPase activity (V_1 =33 %, V_2 = 30 %, V_3 = 37 %). In the human ventricular myocardium only one myosin comigrating with hamster or rat V_3 could be demonstrated;in overloaded gels a second minor component with slightly higher mobility was found.

Coelectrophoresis of left ventricular myosin

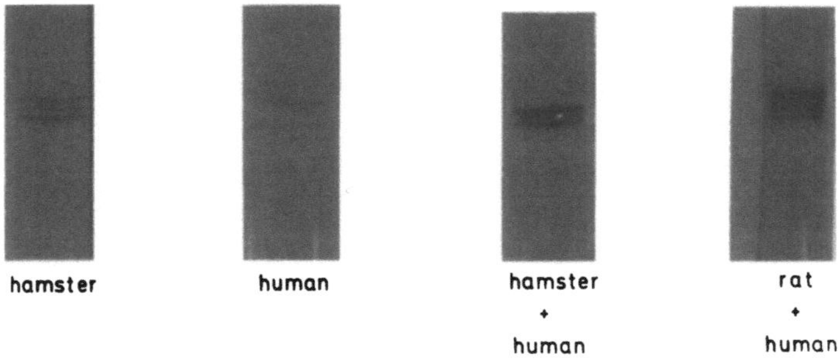

This closely resembled the pattern existing in the ventricular myocardium of the pig. Myosin from hypertrophied or dilated ventricle could not be differentiated from control left ventricular myosin on the basis of electrophoretic mobility; on coelectrophoresis both myosins comigrated.

Discussion

Myosin isoenzymes of the Syrian hamster ventricle show a pattern analogous to rat ventricular myocardium. In the cardiomyopathic hamster reduced myocardial Ca^{2+}-ATPase activity in the stage of insufficiency can be attributed to the redistribution of isoenzymes-a change, which obviously cannot be demonstrated using dissociating electrophoresis of myosin (4).
In contrast to rat and hamster isoenzyme distribution, human and pig ventricular myosins show a distinct pattern. We were unable to demonstrate an altered myosin pattern in the hypertrophied or dilated human ventricular myocardium. While this points out that strong species differences exist in ventricular myosins and results from animals with widely different myocardial contractilities (5) cannot be simply extrapolated to man, the possibility of a different myosin in the human hypertrophic myocardium cannot be excluded solely on grounds of electrophoretic mobility. Further work (e.g. peptide mapping) ist necessary to clarify this problem.

References

1. Mercadier J. J., Lompre A., Wisnewsky C., Samuel J., Bercovici J., Swynghedauw B., Schwartz K.: Circ. Res. 49, 525 - 532, (1981)
2. Hoh J.F., McGrath P.A., White R.: Biochem. J. 157, 87 - 95 (1976)
3. Hoh J.F., McGrath P.A., Hale P.: J. Mol. Cell. Cardiol. 10 1053 - 1076 (1977)

4. Venkatakrishman R., Rourke A.W.: Lab. Invest. 41, 198 - 205 (1979)

5. Henderson A.H., Craig R.J., Sonnenblick E.H., Urschel C.W: Proc. Soc. Exp. Biol., 134, 930 - 932 (1970)

CHROMATOFOCUSING OF HIGH DENSITY LIPOPROTEINS

A.C. Nestruck, P.D. Niedmann, H. Wieland, D. Seidel
Dept. of Clinical Chemistry, University of Göttingen,
University Hospital, Robert-Kochstr. 40, 3400 Göttingen, F.R.G.

Introduction

The structure and function of human plasma high density lipoproteins (HDL) is a subject of increasing interest, due in part to the size and metabolic heterogeneity of subclasses isolated by ultracentrifugation and/or various chromatographic or electrophoretic procedures (1-6), and that the metabolic role(s) of these subclasses remains to be elucidated -- i.e. their relationship to the enzyme lecithin:cholesterol acyltransferase(LCATase,EC 2.3.1.43); as accpetors of components released during hydrolysis of triglyceride-rich lipoproteins; or as "transporters" of cholesterol esters.

Using chromatofocusing(7,8), we report the isolation of four major subclasses of HDL, which presumably are focussed and elute according to their macromolecular isoelectric pH. This relatively new methodology offers the advantages of a high recovery of the starting material and for high density lipoproteins, appears to be a non-destructive and mild procedure based on the fact that the subclasses retain a typical electronmicroscopic morphology and retain α-mobility on agarose electrophoresis.

Electrophoresis '82

Materials and Methods

Blood was obtained from normal volunteers and collected in blood packs containing CPDA-1 as anticoagulant (Travenol, München,F.R.G.). To inhibit LCAT activity, dithio bis-(2-nitro benzoic acid)(DTNB) at 1.2mM was added to half of the plasma directly after plasma separation (within 30 min of blood drawing)(4). Sodium azide(0.02%) was added to all plasma and was included in all solutions and buffers. All procedures were carried out at 10°C and samples were kept at 4°C.

HDL was isolated by ultracentrifugation between non-protein solvent densities 1.063 - 1.21 $g.ml^{-1}$ by conventional methods (9), and dialyzed against either 0.025 M piperazine-HCl (pH 5.7) or Polybuffer 74, 1:15(pH 4.0)(Pharmacia, Uppsala, Sweden).

Chromatofocusing was performed essentially according to the Pharmacia Instruction booklet(10). Briefly, Polybuffer Exchanger 94(Pharmacia), equilibrated with 0.025M piperazine-HCl, pH 5.7, was packed into 1.6 x 30 cm columns and eluted until the pH of the eluant equalled that of the equilibrating buffer. Column packing was checked using bovine Cytochrome C. At a linear flow rate of 20 $cm.h^{-1}$, the HDL sample, preceeded by 5 ml Polybuffer 74, 1:15, pH 4.0 (the eluting buffer), was applied to the column. The samples usually in a volume of $<$ 9 ml, contained 45-55 mg protein and 11-14 mg cholesterol. The eluant was monitored at A 280 nm and for pH with an on-line flow through micro pH cell (Ingold, Frankfurt, F.R.G.). For estimation of sample recovery, effluent cholesterol was measured enzymatically(Ingomat, Boehringer Ingelheim,F.R.G.). A modification of the Lowry method for protein determination (11) was found to give satisfactory results in the presence of Polybuffer.

The Polybuffer was removed from the fractions by gel filtration on Sephadex G-75 (Pharmacia) with 0.15M NaCl, 0.001 M EDTA, 0.02% Na azide (pH 7.4) as eluant. When necessary the fractions were concentrated against 25% polyethylene glycol (Type 40,000) (Serva, Heidelberg, F.R.G.) in NaCl, EDTA, azide, as above.

The subfractions were subjected to agarose electrophoresis (Lipidophor, Immuno Diagnostica, Mannheim, F.R.G.), lipid analysis using High Performance Thin Layer Chromatography (HPTLC) with an anti-circular mode development (12), immunoelectrophoresis and immunoanalysis by double diffusion (13) against rabbit anti-human albumin (Behringwerke, Marburg, F.R.G.) and anti- apo A-I, -A-II, -B, -C, -D and -E (antisera produced in rabbits in our laboratory) and for apoprotein analysis to alkaline polyacrylamide gel electrophoresis (PAGE) on 10% gels (14), $NaDodSO_4$ PAGE (15) in the presence and absence of 2-mercaptoethanol and analytical isoelectric focussing (16).

For electron microscopy, the HDL subfractions were dialyzed against 5mM ammonium bicarbonate and at a final concentration of 125 ug lipoprotein . ml^{-1} in 1% phosphotungstate were placed on Formvar carbon-coated grids and examined with a Zeiss EM 10 electron microscope.

Results

A typical elution pattern of HDL is shown in Fig. 1 as monitored at A 280 nm and cholesterol determinations of the eluant. Over a volume of ~175 ml, 4 major fractions eluted with peak pH's of 5.05, 4.88, 4.67 and 4.50. This pattern was reproducible between subjects and was not changed by previous dialysis of the sample against the equilibrating buffer (pH 5.7) or the eluting buffer (pH 4.0), or by the addition of DTNB. An estimation of the relative proportions of the peaks

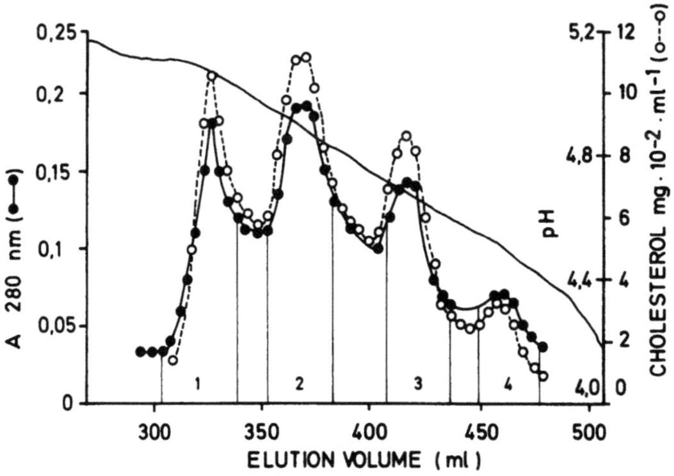

Fig. 1. Chromatofocusing of HDL on PBE 94 equilibrated with 25mM Piperazine-HCl (pH 5.7) and elution with Polybuffer 74, 1:15 (pH 4.0). A 280 nm (●—●); pH (——) and effluent cholesterol (o---o). Fractions 1-4 were pooled as indicated.

Table 1. Relative composition of HDL fractions

	% weight			
fraction	1	2	3	4
protein	49.2	52.9	56.5	59.2
UC	3.6	3.0	2.7	2.4
CE	20.6	19.6	19.3	16.8
PC + Sp	22.3	21.0	18.7	18.0
TG	4.0	3.3	3.4	3.2
	molar ratios			
UC:CE	.302	.264	.252	.240
TG:UC	.492	.485	.545	.605
TG:UC+CE	.112	.099	.105	.115

UC-unesterified cholesterol, CE-esterified cholesterol, PC-phosphatidylcholine, Sp-sphingomyelin, TG-triglyceride.

was attempted on the basis of eluant cholesterol and extrapolation of peak contours. About 85% of the starting cholesterol was recovered and appeared to be distributed in peaks 1 to 4 at ~23%, ~42%, ~25% and ~9% respectively. When DTNB was present, peak 1 cholesterol increased by about 3-4% at the expense of peak 2.

After separation from Polybuffer and concentration, all four subfractions (pooled as indicated on Fig.1) retained α-mobility on agarose electrophoresis (Fig.2) and all retained typical HDL morphology as seen in the electron microscope(17)(Fig.3).

Compositional analysis of the fractions revealed an increasing percentage weight of protein from fractions 1 through 4, with a concomittant decrease in all lipids (Table 1). On a molar basis however, unesterified cholesterol decreased most notably and triglyceride appeared to its replacement. The phospholipids quantitated were phosphatidylcholine and sphingomyelin and represented ≥95% of the phospholipids. Interestingly, in fractions 3 & 4, phosphatidyl ethanolamine was a minor but discreet component. Lysophosphatidylcholine was not consistently found and when present was estimated to represent < 1% of the phospholipids.

Apoprotein analysis also showed differences between the four fractions. After delipidation with tetramethylurea, the PAGE patterns showed increasing heterogeneity in the apoproteins from fraction 1 to 4 (Fig.4). This was also seen on analytical isoelectric focussing (Fig.5). Apoprotein A-I was identified in all fractions and in fraction 1, was the only apoprotein identified. Further, on $NaDodSO_4$ PAGE with 2- mercaptoethanol, no reduced lower molecular weight protein was seen, as would have been expected if A-II was present. Immunoelectrophoresis of fraction 1 (Fig.6) revealed a typical immunoprecipitation against anti-A-I and no reaction against anti-A-II.

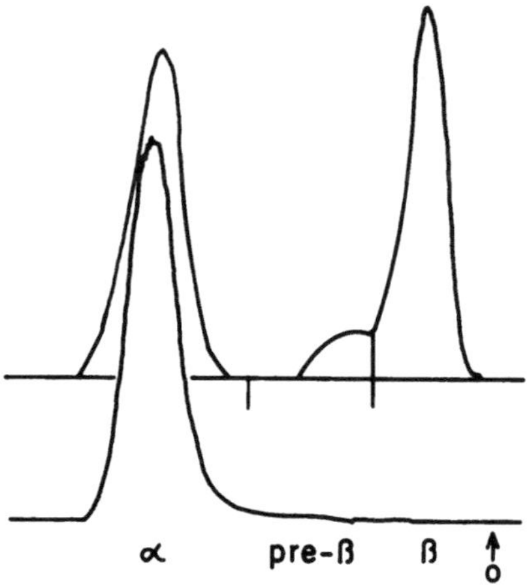

Fig. 2. Densitometric scans of poly-anion precipitated lipoproteins after electrophoresis in 1% agarose from Control serum (upper) and a representative scan of one of the HDL fractions (lower). o = origin.

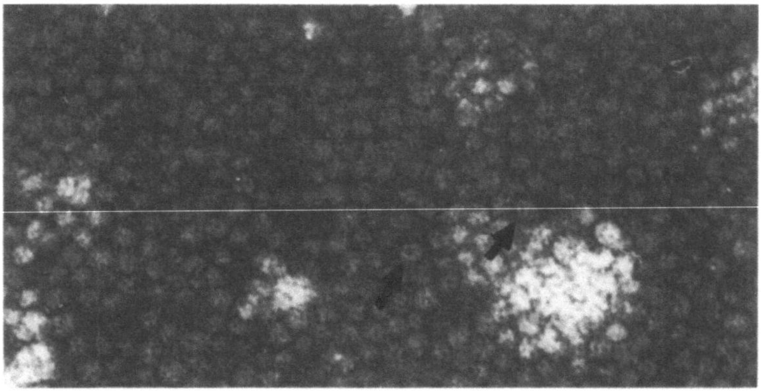

Fig. 3. A representative electron micrograph of a negatively stained HDL fraction. Particles are 70-80 Å. Arrows indicate particles with apparant subunit structure. Magnification x 500,000.

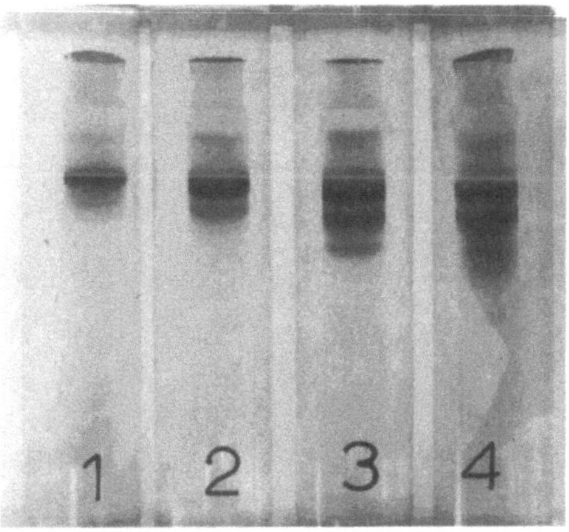

Fig. 4. Electrophoresis of fractions 1-4 after delipidation with tetramethylurea, on 10% polyacrylamide gels (pH 8.9, 7M urea). Gels are stained with Coomassie Blue - R 250).

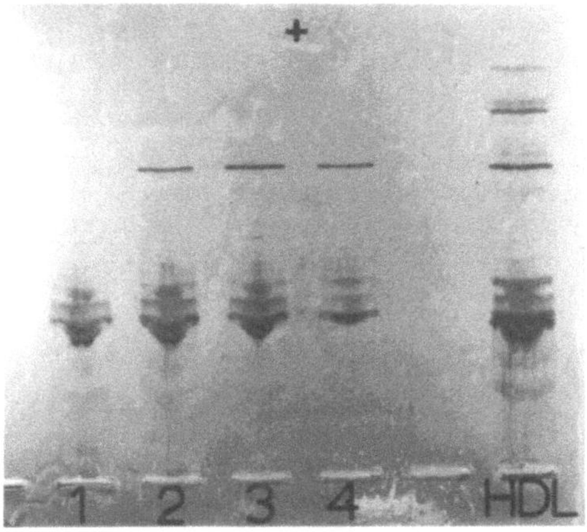

Fig. 5. Analytical isoelectric focussing of thin-layer polyacrylamide gel (pH gradient 4-6) of fractions 1-4 and the starting HDL after delipidation with tetramethylurea.

In addition to A-I, fractions 2-4 all contained A-II. Apo D was present in fractions 3 & 4 with some of the C apoproteins, and apo E seemed to be confined to fraction 4 (Figs.4-6).

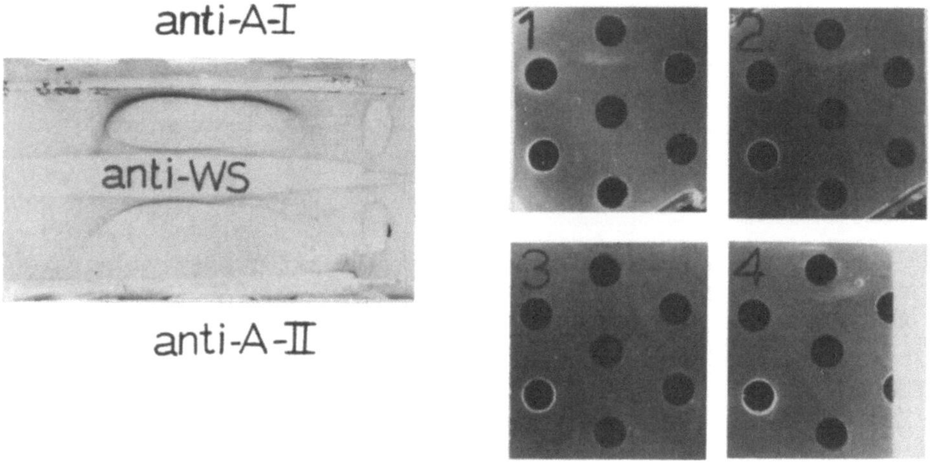

Fig. 6. Immunoelectrophoretic pattern of fraction 1 after staining with Amido Black, against anti-A-I(upper well), anti-human whole serum (middle well) and anti-A-II(lower well) (LEFT); and double diffusion immunoprecipitation patterns of fractions 1-4 against (from 12 o'clock, clockwise) anti-A-I, -A-II, -B, -C, -D and -E (RIGHT).

Discussion

Recent evidence suggests that high density lipoproteins are discreet populations of particles which can be separated from one another by electrophoretic, chromatographic or ultracentrifugal procedures(1-6). Chromatofocusing, a relatively new methodology, utilizes the isoelectric characteristics of molecules and allows a highly resolved separation of proteins with components of differing pI. In this study, high density lipoproteins, large molecular complexes of protein and lipid, were separated into 4 fractions within an interval of 0.5 pH units.

Although the relationship of these fractions to the subclasses isolated in the ultracentrifuge or by various electrophoretic or chromatographic procedures is unclear, the distinct apoprotein compositional differences between the fractions lead us to believe that these fractions may well represent metabolic entities. This relatively simple method is thought to be an important development for the further investigation of lipoprotein heterogeneity.

Acknowledgements

This work was supported by the Deutsche Forschungsgemeindschaft, SFB 89, Kardiologie Göttingen. We would like to thank the Zentrale Blutspendedienst, Med. Einrichtungen der Universität Göttingen, for supplying the plasma. Also, we would like to express our gratitude to Dr. P. Schwartz, Zentrum Anatomie, Universität Göttingen, for the electron microscopy.

References

1. Anderson, D.W.,Nichols,A.V.,Pan,S.S.,Lindgren,F.T. : Atherosclerosis 29,161-179 (1978)
2. Laggner,P.,Stabinger,H.,Kostner,G.M.:Prep. Biochem. 7, 33-43(1977).
3. Patsch,W.,Schonfeld,G.,Gotto,A.M.,Patsch,J.R.:J.Biol.Chem. 255, 3178-3185 (1980).
4. Jahani,M., Lacko,A.G.: J.Lipid Res. 22,1102-1110 (1980).
5. Marcel,Y.L., Vezina,C., Emond,D., Suzue,G.: Proc.Natl.Acad. Sci. USA 77, 2909-2973 (1980).
6. Blanche,P.J., Gong, E.L., Forte,T.M.,Nichols,A.V.:Biochim. Biophys. Acta 665, 408-419 (1981).
7. Sluyterman,L.A.AE., Elgersma,O.;J.Chromatogr.150,17-30(1978)
8. Sluyterman,L.A.AE., Wijdenes,J.:J.Chromatogr.150,31-44(1978)

9. Havel,R.J., Eder,H.A.,Bragdon,J.H.:J.Clin.Invest.34, 1345-1353 (1955).

10.Chromatofocusing with Polybuffer™ and PBE™, Pharmacia Fine Chemicals AB, S-75104, Uppsala 1, Sweden.

11. Bensadoun,A., Weinstein,D.: Anal.Biochem. 70,241-250(1976).

12. Niedmann, P.D. et al. (in preparation).

13. Ouchterolny,O., Nilsson,L.A.: in Handbook of Experimental Immunology (Weir, D.M. ed.) Blackwell, Oxford. pp. 1-44.

14. Kane, J.P.: Anal. Biochem. 53,350-364 (1973).

15. Weber,K., Osborne,M.: J.Biol.Chem. 244,4406-4412 (1969).

16. Marcel,Y.M., Bergseth,M., Nestruck,A.C.:Biochim.Biophys. Acta 573, 175-183 (1979).

17. Forte,T.,Nichols,A.V.: Adv.Lipid Res. 10, 1-41 (1972).

ISOELECTRIC FOCUSING OF PLASMA LIPOPROTEINS PRE-STAINED WITH TETRAZOLIUM BLUE: A PROMISING DIAGNOSTIC TOOL.

D. Stathakos

Laboratory for Enzyme Research, Department of Biology, Nuclear Research Center "Demokritos", Athens, Greece

G. Rekoumis, A. Avgerinos, N. Kalantzis and T. Kanaghinis

Research Unit, "Evangelismos" Hospital, Athens, Greece

Introduction

In recent years the generally acceptable classification and nomenclature of plasma lipoproteins was necessarily based on operational terms, mainly determined by the current methodology of fractionation. Thus, the prevailing trends encountered in the literature, classify lipoproteins:

a) according to their hydrated density (1, 2), during sequential ultracentrifugal steps, as: chylomicrons (chyl), very-low-density (VLDL), low-density (LDL) and high-density (HDL) lipoproteins;

b) according to the electrophoretic mobility of the charged molecule on various support media (i.e. paper, agarose, cellulose acetate, polyacrylamide gels, etc.) as: chylomicrons, pre-β, β and α lipoproteins, roughly corresponding in that order to the above four subgroups as derived by criteria of density (1, 2, 3).

Finally, in more elaborate studies of purified fractions lipoproteins can be characterized according to the apoprotein moiety (4, 5) and the lipid group attached to it.

Clearly, ultracentrifugal fractionation, although sensitive, is too time consuming to be routinely employed in current diagnosis. On the other hand, electrophoretic techniques, being relatively simple and fairly informative for diagnostic

Electrophoresis '82

purposes, have enjoyed widespread use in clinical laboratory practice (6, 7, 8, 9).
Over the years several efforts have been made to apply isoelectric focusing to the fractionation of lipoproteins (10, 11). The high resolving capacity of this method gave good reason to hope for more precise diagnostic techniques, once the inherent difficulties (staining, high molecular weights of lipoproteins, precipitation at their pIs, etc.) could be bypassed.
In the present report an experimental approach toward a reproducible isoelectric fractionation and eventual quantitation of plasma lipoproteins is described.

Materials and Methods

Reagents: Tetrazolium blue was purchased from Sigma; NADH (disodium salt) from Boehringer; phenazine methosulfate from BDH Chemicals; recrystallized acrylamide, N, N-methylenebisacrylamide (BIS) and N, N, N', N'-tetramethylethylenediamine (TEMED) from Serva Feinbiochemica; Ampholine carrier ampholytes from LKB Produkter AB.
Equipment: Centrifugation of the pre-stained samples was performed in an Eppendorf Microliter 3200 Centrifuge; electrofocusing in an M.R.A. vertical gel-rod apparatus with an LKB 2103 regulated power supply; scanning of the lipoprotein patterns in a Joyce-Loebl Chromoscan.
Samples: Venous blood was collected with EDTA as anticoagulant (0.6 mg EDTA/ml of blood) after a 14 h fast. After centrifugation the plasma was stored at 2°C and processed within 4 hours after collection.
Pre-staining: The procedure used by J. Green with nitroblue tetrazolium (12) was modified as follows.
To 1 ml of plasma were added:
a) 100 µl of 8×10^{-2}M NADH
b) 20 µl of 4×10^{-2}M phenazine methosulfate, and

c) 500 µl of 0.3% tetrazolium blue; this solution was prepared by heating for 10 min at 80°C.
After 60 min of incubation at 25°C the intensely-blue coloured mixture was centrifuged at 10.000 x g for 5 min. The supernatant was electrofocused as described below.
Isoelectric focusing: The vertical gel-rod technique was used, as described by Righetti and Drysdale (13). The anolyte and catholyte were 0.01 M H_3PO_4 and 0.02 M NaOH, respectively, with the catholyte in the upper chamber. In all experiments 3.8% polyacrylamide gels containing 3% Ampholines (pH 3.5-10) were used in 10 cm-long tubes, I.D. 3 mm. Prerunning time 30 min; duration of run: 6-8 hours at 500 V.
At the end of the run the gel was scanned and/or segmented with the use of a Hamilton syringe (13); the gel fractions were shaken in 500 µl aliquots of CO_2-free water in a nitrogen atmosphere; pHs were subsequently measured with an Orion 801 pH-meter.

Results and Discussion

In our pilot experiments we investigated the conditions which would ensure (a) uniform staining of the lipoprotein fractions and (b) their entrance into the gel matrix with a resolution of quantifiable level. These two problems have been frequently encountered and reported in several previous studies (12, 14, 15). Clearly, the apparatus of choice was the gel-rod type (13) which provides the possibility of multiple, simultaneous runs (up to 12) with varying compositions of gel and/or sample.
Pre-staining: Once an intense colour of lipoproteins is attained, pre-staining offers a time saving advantage of immediate visualization vs. long staining and destaining steps (12), especially when the classical techniques with Sudan Black B are used. One further advantage, first reported for the use

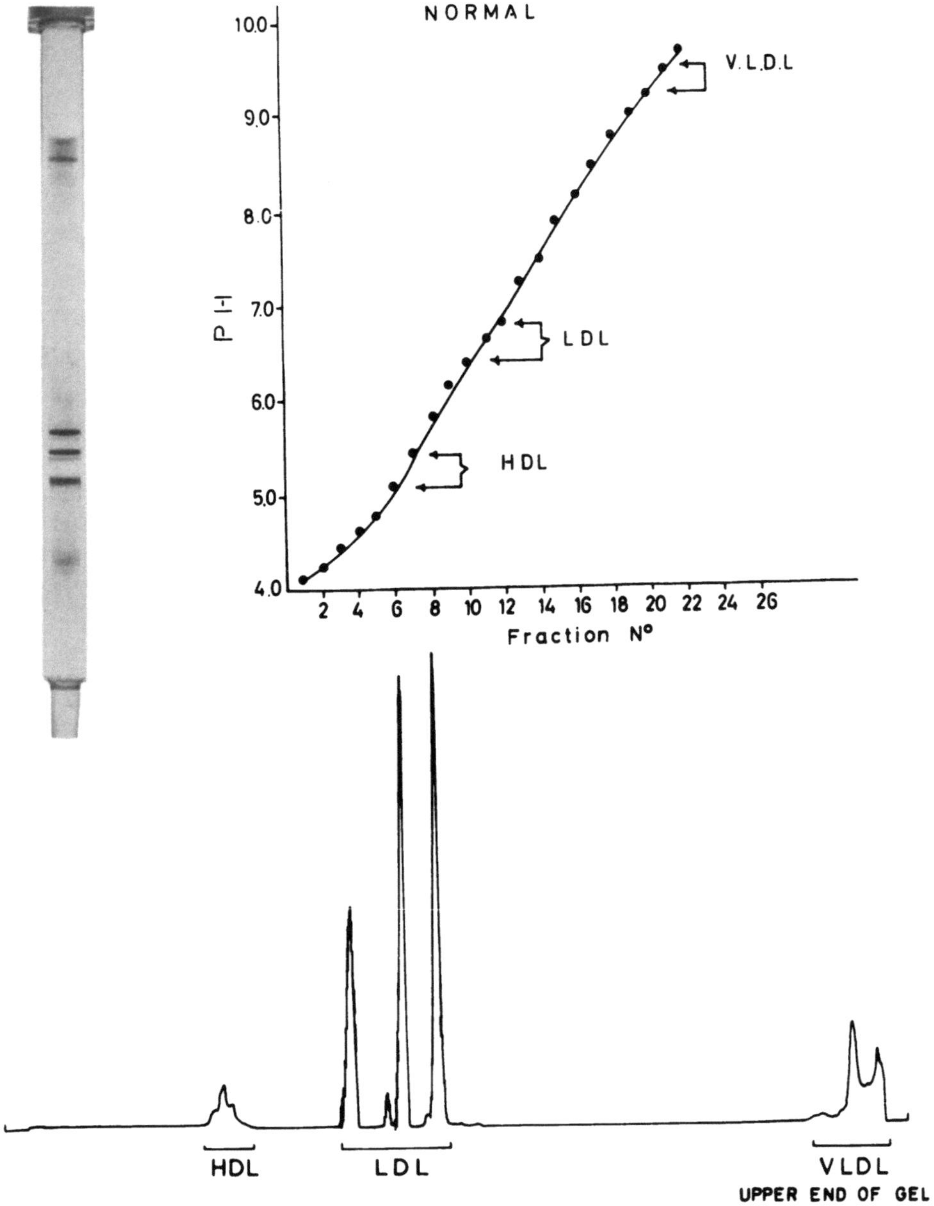

Fig. 1. Isoelectric focusing of normal plasma (for experimental details see text).

A. 20 µl of pre-stained plasma after focusing.
B. pH gradient.
C. Densitogram of the gel rod, 650 nm filter.

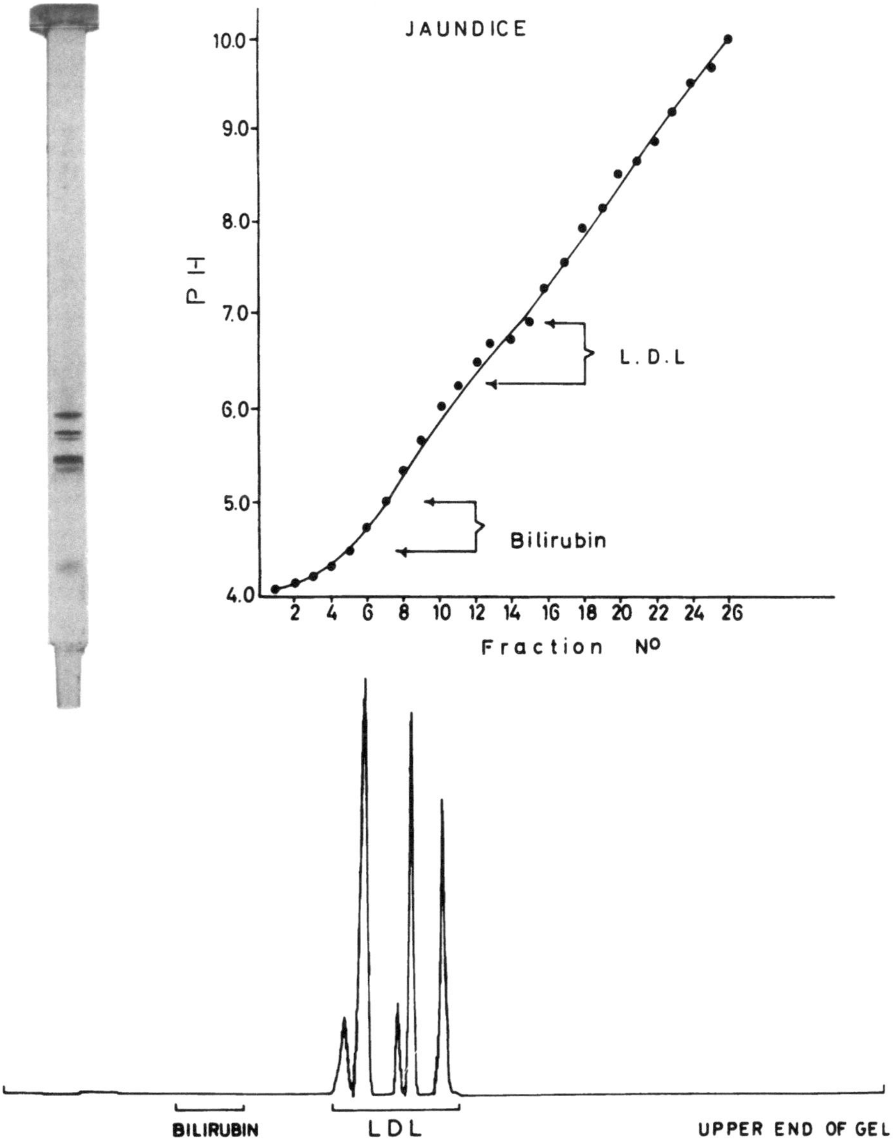

Fig. 2. Isoelectric focusing of jaundiced plasma.

Pancreatic carcinoma with long-standing (2 months) biliary obstruction. Experimental conditions as in Fig. 1. Note the pronounced (yellow) bilirubin band, marked by an arrow.

A. 20 µl of pre-stained plasma after focusing. B. pH gradient. C. Densitogram of the gel rod, 650 nm filter (not recording bilirubin).

of nitroblue tetrazolium (12, 15), is that pre-staining with this dye greatly facilitates the entrance of lipoproteins (particularly VLDL) into acrylamide gels, an observation we could also confirm. In our experiments this effect was even more pronounced with tetrazolium blue, which moreover gives a very vivid blue colour and intensely stained bands (Fig. 1 and 2). A crucial factor in the stoichiometry of the pre-staining reaction was phenazine methosulfate: it should always be added in freshly made solution, since its accurate weighing in solid form (15) lies beyond the capabilities of most laboratories. Yet, minor variations of its concentration can produce various metachromatic effects on the focused bands from light brown, through red to purple which make quantitation of the fractions virtually impossible.

Carrier-ampholytes: The concentration of Ampholine had to be raised to 3% in order to attain sharp focusing. Lower concentrations (1% to 2%) gave diffuse patterns or randomly non-reproducible ones, especially when abnormal plasma (e.g. in jaundice, hepatitis, etc.) was fractionated.

Chylomicrons: As expected, chylomicrons do not enter gels of such porosity and should as well be completely eliminated by centrifugation prior to pre-staining. Every attempt to focus lipoproteins in the presence of chylomicrons resulted in a dense aggregate at the application point followed by a pattern of faint atypical bands, always fewer than expected, probably due to retention of some fractions by the chylomicron-aggregate.

Fig. 1 shows the resolution of normal plasma into three main subgroups (Fig. 1A) and their respective pIs on the pH gradient formed (Fig. 1B). A further, quite sharp resolution into distinct bands can be observed within each subgroup which enabled us to obtain reasonably easy scannings of the gels (Fig. 1C).

In Fig. 2, representing a case of severe biliary obstruction, the VLDL and HDL fractions have disappeared, as ex-

pected (2, 3), and an altered profile of the LDL region with an additional band was observed. Clearly, this technique shows significant advantages over conventional electrophoresis by resolving each subgroup into several bands.
In this context it should be noted that the terminology "VLDL-LDL-HDL" is the one used throughout this text, though always with reference to the definitions mentioned above (pre-β, β, α, respectively; cf. introduction). In order to establish the identity of the three subgroups, normal plasma was electrophoretically fractionated on polyacrylamide gels (16) and each Lp fraction subsequently electrofocused. The three fractions focused at their anticipated pIs, respectively (Fig. 3).

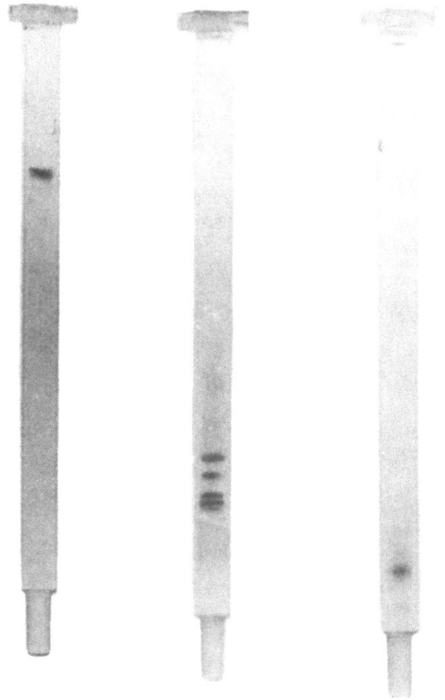

Fig. 3. Identification of focused bands in terms of conventional Lp-classification:

Normal plasma pre-stained with tetrazolium blue, as described above, was fractionated by electrophoresis on polyacrylamide gels (16). The gel segments corresponding to pre-β, β and α fractions, also identified immunoelectrophoretically, were cut out, positioned on top of the electrofocusing gels and the run was performed as usual (cf. Fig. 1).

When applied to current diagnosis, and namely in cases of hepatic disorders, this technique proved quite sensitive in providing reproducible results in good correlation with the

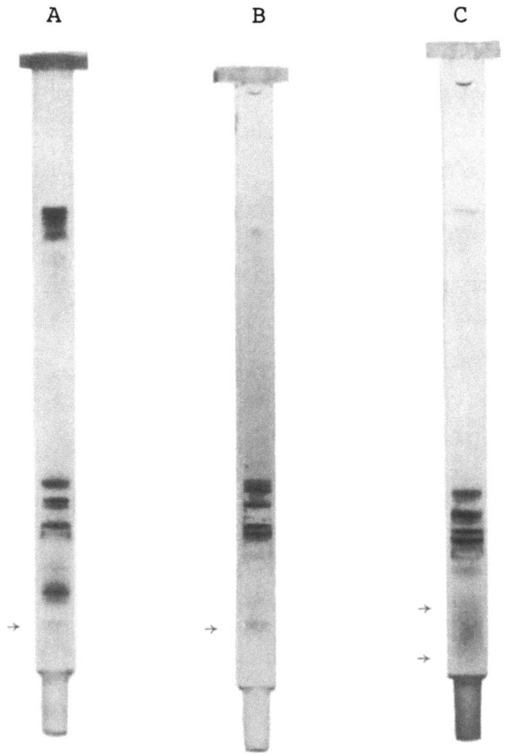

Fig. 4. Lipoprotein patterns in disease.

A. Hemolytic jaundice. Acute phase (4th day). Bilirubin band is indicated by an arrow.

B. Biliary obstruction disease. Acute phase (2nd week). Pronounced bilirubin band (indicated by arrow).

C. Acute B-viral hepatitis. (Clinical and biochemical evidence of cholestasis). Acute phase (3rd week). Broad bilirubin area (indicated by arrows).

clinical and biochemical findings: In Fig. 4 three characteristic patterns are shown, representing hemolytic jaundice (Fig. 4A), where - by definition - no hepatic lesion is present and the Lp-pattern remains intact; obstructive jaundice with considerable hepatocellular damage (Fig. 4B); and viral hepatitis - Type B (Fig. 4C). In the latter two cases the LDL region is heavily overpopulated at the expense of HDL and LDL, a transition process known to occur in liver disease (2, 3). In a series of experiments not shown here, we were also able to follow the day-by-day course of the disease, in several cases up to complete recovery. It is interesting to note in this respect, that the two "eliminated" subgroups (VLDL and HDL) reappear only after an interim phase of multiple intermediary bands on each side of the LDL region, until the pattern finally returns to normal as in Fig. 1 (to be published).

In conclusion this technique promises a rapid reproducible way to diagnose liver disorders, once enough systematic work is done on the specific Lp-pattern of each type of hepatic lesion and the individual bands are clearly identified. This also includes the localization and eventual quantification of Lp-X, the abnormal protein always present in obstructive jaundice (2). A limitation of this technique, i.e. the inadequate resolution of VLDL may be by-passed by the use of discontinuous gels (9, 12, 17), such as agarose for the pH region and polyacrylamide-gel layers of appropriate porosities below these pHs. These questions are currently under investigation.

Acknowledgement: We thank Mr. I. Lagouros for skillful technical assistance.

References

1. Hatch, F.T., Lees, R.S.: Adv. Lipid Res. 6, 1-68 (1968).
2. Day, R.C., Harry, D.S., McIntyre, N.: In "Liver and Biliary Disease", Eds. Wright, R., Alberti, K.G.M.M., Karran, S., Millward-Saddler, G.H., pp. 63-82, W.B. Saunders Co., London-Philadelphia-Toronto (1979).
3. Sabesin, S.M., Bertram, P.D., Freeman, M.R.: Adv. Int. Med. 25, 117-146 (1980).
4. Alaupovic, P.: Atherosclerosis 13, 141-146 (1971).
5. Zannis, V.I., Breslow, J.L.: In Electrophoresis '79, Ed. B.J. Radola, pp. 437-473, Walter de Gruyter, Berlin-New York (1980).
6. Frederickson, D.S., Levy, R.I., Lees, R.S.: N. Engl. J. Med. 276, 273-281 (1967).
7. Avgerinos, A., Scliros, P., Pyrovolakis, J., Stathakos, D.: J. Clin. Path. 30, 857-861 (1977).
8. Day, R.C., Harry, D.S., Owen, J.S., Foo, A.Y., McIntyre, N.: Clin. Sci. 56, 575-583 (1979).
9. Taketa, K., Ikeda, S., Watanabe, M.: In Electrophoresis '81, Eds. R.C. Allen, P. Arnaud, pp. 447-452, Walter de Gruyter, Berlin-New York (1981).

10. Vost, A., Pocock, D.M.E., Pleet, S.: Lipids 14, 864-871 (1979).

11. Emes, A.V., Latner, A.L., Rahbani-Nobar, M., Tan, B.H.A.: Clin. Chim. Acta 71, 293-302 (1976).

12. Green, J.: Clin. Chim. Acta 66, 295-301 (1976).

13. Righetti, P.G., Drysdale, J.W.: J. Chromatogr. 98, 271-321 (1974).

14. Godolphin, W.J., Stinson, R.A.: Clin. Chim. Acta 56, 97-103 (1974).

15. Segal, A.W., Putman, D., Levi, A.J., Lewis, B.: Clin. Chim. Acta 53, 361-367 (1974).

16. Masket, B.H., Levy, R.I., Frederickson, D.S.: J. Lab. Clin. Med. 81, 794-802 (1973).

17. Stathakos, D., Vellios, A., Koussoulakos, S.: In Electrophoresis '79, Ed. B.J. Radola, pp. 517-528, Walter de Gruyter, Berlin-New York (1980).

A NEW DETAILED LIPIDOGRAM : METHODS AND CLINICAL APPLICATIONS

Dominique Léglise, Gérard Lescoat, Jean François Menez,
Louis Georges Bardou

Laboratoire de Biochimie. Centre Hospitalier Universitaire Morvan, Brest, France.

Thierry Maudelonde

Service d'Endocrinologie, Centre Hospitalier Universitaire Morvan, Brest, France.

Introduction

The great number of studies dealing with lipid metabolism and their assessments, lead us to test new accurate methods. Lipoproteins (Lp) are now regarded as atherogenic Lp (preβ and βLp fractions) and antiatherogenic Lp (αLp fraction) (1,2) ; These α, preβ and βLp fractions are assimilated as HDL, VLDL, LDL obtained by ultracentrifugation (3). The lipoproteins separated electrophoretically can be stained whether by lipophylic dyes (e.g OIL RED O) (4) or by cholesterol (CH) enzymatic reagents (5). Using the latter technique we tried to stain Phospholipids (PL) and Triglycerides (TG). We investigated TG staining using two kinds of reagents. Consequently with these following scans (CH,PL,TG) and their serum concentration it is possible to outline a detailed lipidogram. We evaluated this procedure not only with a control but also with some patients affected by different lipoprotein diseases.

Material and Methods

1) Samples. Venous blood for control and patients was drawn after an overnight fast and serum were stored at + 4°C if not analyzed within 4 days.

2) Analytical methods. Standard equipments and materials : power supply (Titan III) electrophoresis chambers, scanning densitometer (Autoscan

fluor-vis Quant II), plates (Lipo III) and migrating proteins buffer were from Helena France.

3) Reagents. (CH, PL, TG) total : their measurements were performed with the following reagents using in each case the recommended method. (CH, PL, TG) electrophoresis reagents : the same reagents but concentrated according to a recently described method (6) : - Cholesterol : Worthington Diagnosics Ref. 27580, - Phospholipids B tests : Wako BIOLYON France Ref. 4451-148, - Triglycerides G tests : Wako BIOLYON France Ref. 4545-148 (TGT). These three methods were concluded with a Trinder's final reaction. The other Triglycerides reagents tested were from : Dow diagnostics, Ref. 47161 concluding with a formazan coloration (TGF). The procedure of these Trinder's methods are summarized in table (1).

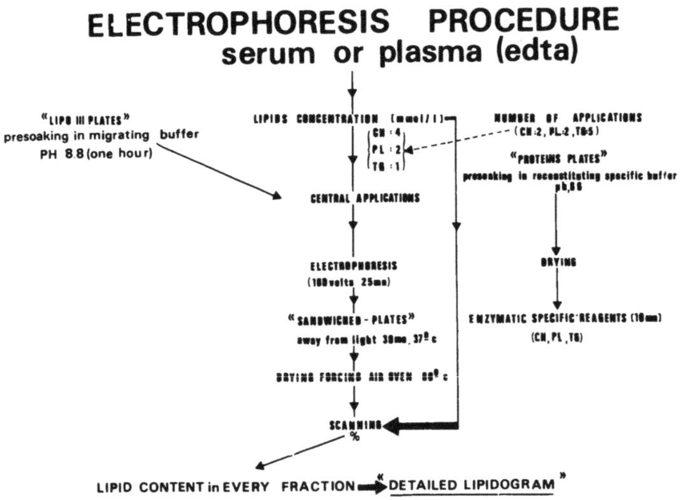

- table 1. Procedure for specific staining of CH, PL and TG (TGT)

The (TGF) method is based on the same methodology. However it requires cumbersome tasks like the use of disposable gloves and the manipulation away from light to avoid formazan's dye interference. The TG electrophoresis reagents were concentrated 4 times. At the end of the color development in a darkened dry oven (37°C) HCL (50 mmol/l) was spread over the separated strips to prevent self coloration and then dried in a forced air oven (60°C).

Results and Discussion

Two different sera were treated by both methods (TGT and TGF) and one illustration is given in figure 1. Then we compared (Student's paired-t test) the percentages of TG bound to αLp, preβLp and βLp with both procedures in 18 randomly choosen sera. αLp% ($\bar{m}$ ± SEM) = 9,8 ± 4,1 for TGT and 15 ± 3,8 for TGF ($p < 0{,}001$) ; preβLp = 44,4 ± 12,8 for TGT and 34,8 ± 13 for TGF ($p < 0{,}001$) ; βLp = 45,8 ± 12,3 for TGT and 50,2 ± 11,9 for TGF (NS). These two groups of tests were completed with CH and PL electrophoresis to establish detailed lipidograms. Then OIL RED O staining was performed with the same 18 sera. The difference between (CH + PL + TGT expresses in mmol/l) and OIL RED O lipidogram is not significant. A control group profile with (CH + PL + TGT) was found to have similar data in litterature (7) whilst (CH + PL +TGF) group data did not agree with the litterature.

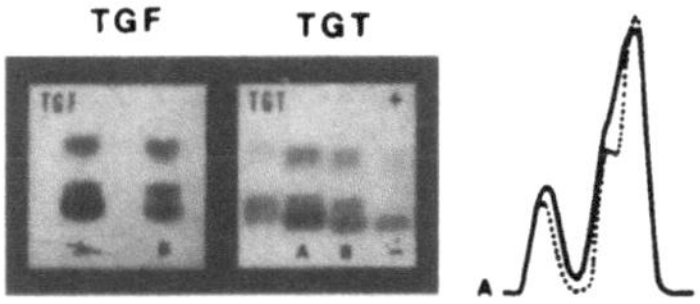

Fig 1 : TGF, TGT scans : Full line TGF, Dotted line TGT

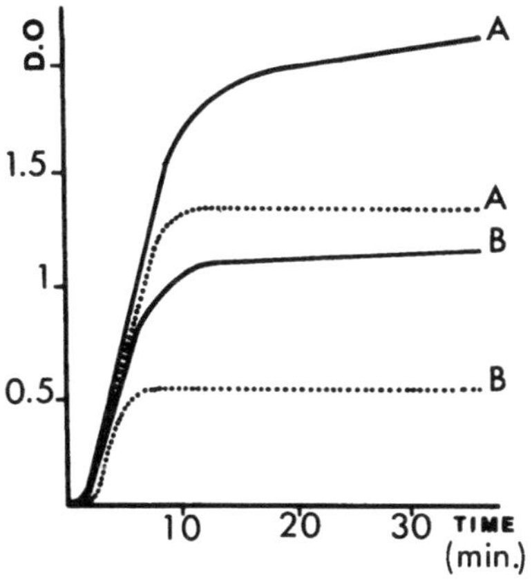

Fig 2 : Kinetics development of coloration for TGF and TGT

We considered the kinetics development of coloration with TGT and TGF reagents in a low and a high concentrated TG serum (cf fig. 2). We suggest that the least accurate results obtained with TGF methods for TG lipoprotein measurement could be explained by the kinetic development of coloration. In sandwiched plates it is not possible to add HCL (50 mmol/l) ten minutes after the beginning of the reaction as in total measurement. Even after 30 minutes the coloration doesn't seem to be completed, above all in rich fractions. In studying the kinetic graphs it is obvious that the reaction depends on the lipid content for each lipoprotein : the low concentrated fractions (e.g αLp) and the higher (e.g. preβ and βLp) could not give satisfactory results. The explanation of different kinetics in lipoproteins could be due to the enzyme substrate ratio which is never uniform on the plates. Consequently in this kind of procedure a constant peak level reaction is necessary. All these arguments plus an easier manipulation made us choose the TGT reagent. The CH, PL and TGT methodologies have been discussed (5), they allow us to establish a detailed lipidogram (DL) combined with the CH, PL and TG total values.

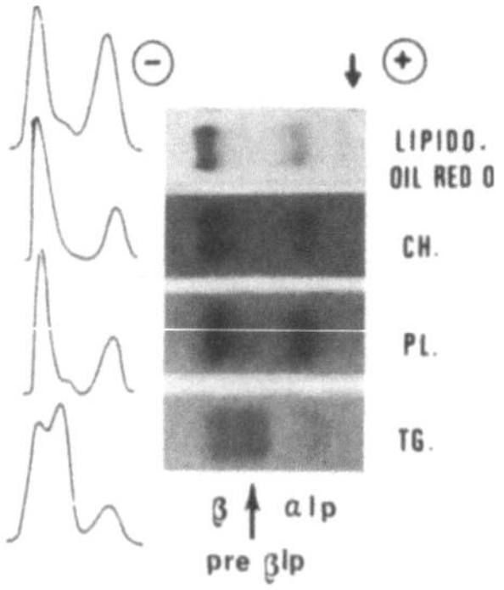

Fig. 3

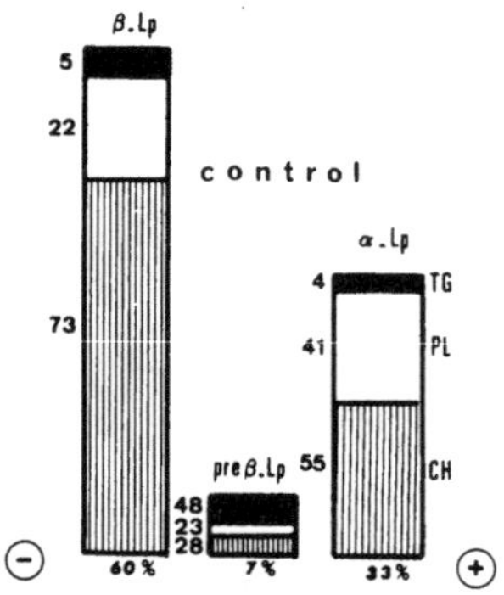

Fig. 3

An example of a young woman control is presented in figure 3. Different specific electrophoresis stainings are presented in figure 4 for some lipoprotein diseases.

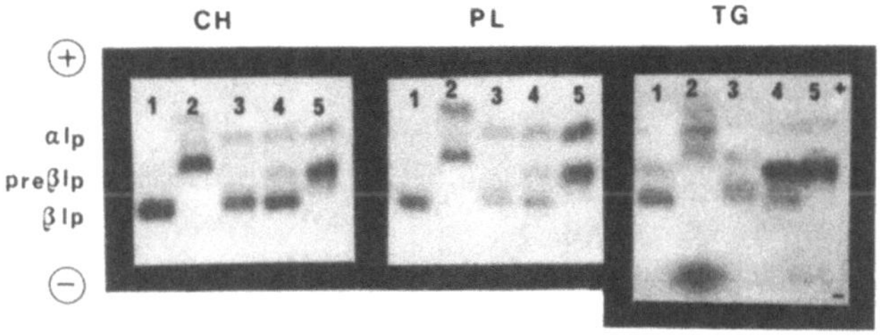

<u>Fig 4.</u> Specific staining of CH, PL and TG for patients with lipoprotein diseases

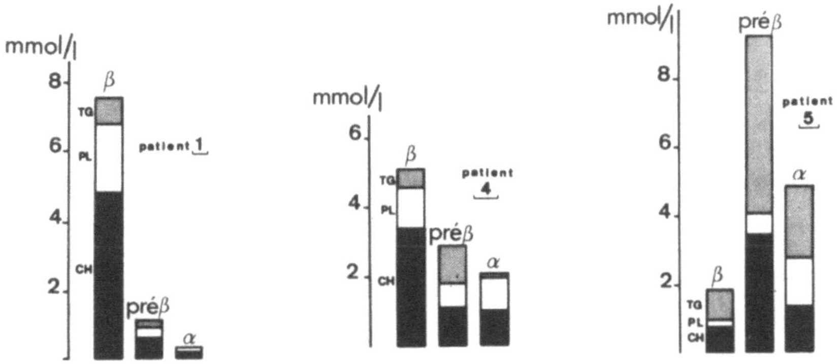

<u>Fig. 5</u> Detailed lipidogram established with patients 1,4,5

Patient N° 1. A 45 year old man (normal weight, non smoker, non alcohol drinker) suffered myocardial infractions (MI) and cerebral strokes. The lipid CH = 5,33 mmol/l, PL = 2,37 mmol/l, TG = 0,98 mmol/l were normal. DL (Cf. fig. 5) : faint αLp (CH and PL) and highly increased βLp.

Patient N°2. A 51 year old man had just suffered MI and was subcutaneously heparinized. Heparin interferes with both TG metabolism and electrophoresis migration. Free glycerol was present on application area and cannot be mistaken for chylomicrons. Consequently DL is to be avoided in heparinized patients.

Patient N°3. Was a normal lipoprotein electrophoresis used as control.

Patient N°4. A 60 year old man with a renal failure underwent continuous ambulatory peritoneal dialysis. Total TG = 1,83 mmol/l were increased. DL (Cf. Fig. 5) : this last abnormality was put forward by an increased preβ Lp fraction.

Patient N°5. An overweight alcoholic with normal CH (5,40 mmol/l) and PL 3,30 mmol/l but increased TG (7,10 mmol/l). DL (Cf. Fig.5) : TG electrophoresis showed the absence of chylomicrons, while in CH, PL, TG lipoproteins the preβLp were increased ; this abnormality could be classed as an endogenous hypertriglyceridemia induced by alcohol, whereas the DL profile could be compared with Frederickson's type IV ; in our case this type IV is improved by numerous lipoprotein data.

In conclusion the possibility of CH, PL and TG lipoprotein colorations after electrophoresis better the standard lipidogram by giving much more data and permits a more thorough investigation into lipoprotein diseases.

References

1. Goldstein, S.L., and Brown M.S. : Metabolism, 26, 11, 1257-1275 (1977)

2. Gordon, T., Castelli, W.P., Hjortland, M.C., Kannel, W.B., Dawber, T.R. : Amer. J. Med., 62, 707-713 (1977).

3. Charman, R.C. and Braevler, C. : Atherosclerosis, 17, 483-489 (1973).

4. Chin, H.P., Blankenhorn, D.H. : Clin. Chim. Acta, 20, 305-314 (1968).

5. Cobb, S.A. Sanders, J.L. : Clin. Chem., 24, 1116-1120 (1978)

6. Léglise, D., Menez, J.F., Person, B., Maudelonde, T., Bardou, L.G. : Clin. Chim. Acta, 118, 265-277. (1982).

7. Sezille, G., Fruchart, J.C., Jaillard, J., Dewailly, P. : Lipides et lipoproteines : Métabolisme et méthodes d'exploration, Crouan et Roques, Lille. France. (1976)

THE COURSE OF PROTEINURIA IN GENTAMICIN-TREATED RATS: IEF AND IMMUNOFLUORESCENCE MARKING OF PROTEINS ON THE PAG-PLATE.

Hans-Peter Brodersen, Klaus Sack

Klinik für Innere Medizin der Medizinischen Hochschule Lübeck, D-2400 Lübeck, Deutschland

Introduction

Since the introduction of aminoglycosides into antimicrobial therapy, much evidence of their causing renal tubular lesions has been gathered. Glomerular changes, although regularly caused by high dosages of aminoglycosides, have only seldom been reported (1,2,3). The 1st figure shows falciform swelling of the epithelium of Bowman's capsule, unspecific glomerulonephrosis with granular PAS-positive deposits and focal necroses of cappilar loops. These three types of lesions are often associated with one another. As far as we know, an analysis of the proteinuria, which always occured during therapy with high dosages of aminoglycosides, has not yet been performed, and was the aim of this study.

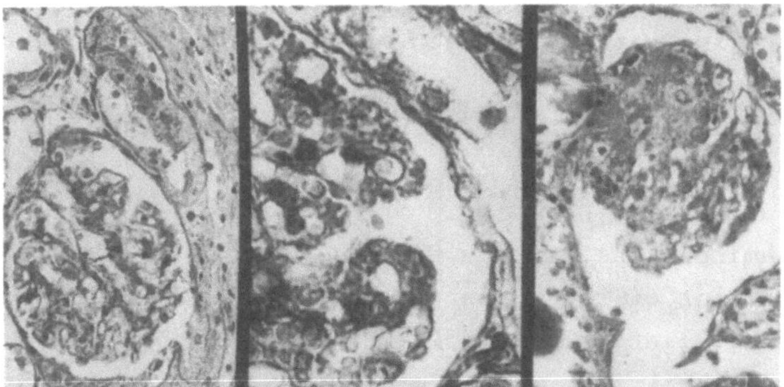

Fig. 1: Light microscopic photographs, PAS, obj. 40, 100 resp. Glomerular lesions caused by 5 days of treatment with 100 mg of aminoglycoside per kg bodyweight and day.

Electrophoresis '82

Methods

Female Wistar rats with an initial weight of 180 - 200 g were used. Groups of ten rats per gentamicin dosage (10 mg, 20 mg, and 100 mg/kg bodyweight and day) were kept in macrolon-cages, fed with altromin pellets and had free access to tap water. They were treated with nine single intramuscular injections of 5, 10 or 50 mg of gentamicin per kg at 12 hour intervals. Experimental design: An adaptation period (AP) is followed by a medication period (MP) of 5 days and a recovery period (RP) of 7 days. The medication was started on the last day of the AP at 19^h. The urinalyses were performed during the AP, on day 1, 3, 5, 8, 10 and 12 at 7^h. Twelve hours before the rats were put singulary into metabolic cages after being given 6 ml of tap water through a pharyngeal tube. Five rats of each group were sacrificed an day 5 and day 12 respectively.
The urinalyses consisted of:

1 biuret reaction
2 CA electrophoresis
3 IEF
4 immunreactions
 a) immunofixation
 b) immunofluorescence

The tests were begun immediately after collection. Tubule cells were counted in the Fuchs-Rosenthal-chamber. For further analysis the urine was centrifuged at 8000 g for 2 minutes and the supernatant stored at - 20° C. The protein content was measured photometrically by using the biuret reaction. The CA electrophoresis was performed in the Boskamp chamber after concentrating the urine 20 - 80-fold in the minicon CS 15 or 15 B produced by amicon. The IEF was done on the LKB Multiphor on commercial PAG-plates of the pH-range of 3.5 to 9.5. The stain was Coomassie brilliant blue R 250.

Results

Effect of the gentamicin dosage on the elimination of tubule cells: Even the lowest dosage of 10 mg/kg and day produced urinary cell counts exceeding the 95 % confidence limit (73 TC/min.) after day 5 and rising until the end of the experiment. Higher dosages caused an earlier and steeper increase of cells in the urine with an elimination peak on the 8th day for the highest dosage. The statistical distribution of the proteinuria of a control group of 117 rats is nearly log normal like that of tubule cell elimination and several tubule enzymes. The median is 0.8 mg/12 h, the confidence limit 1.85 mg/12 h. Gentamicin-induced proteinuria was dosagedependent like tubule cell elimination. Only the highest dosage caused a proteinuria exceeding the 95 % confidence limit of the control group. In the RP a wide scatter of protein excretion occurs.

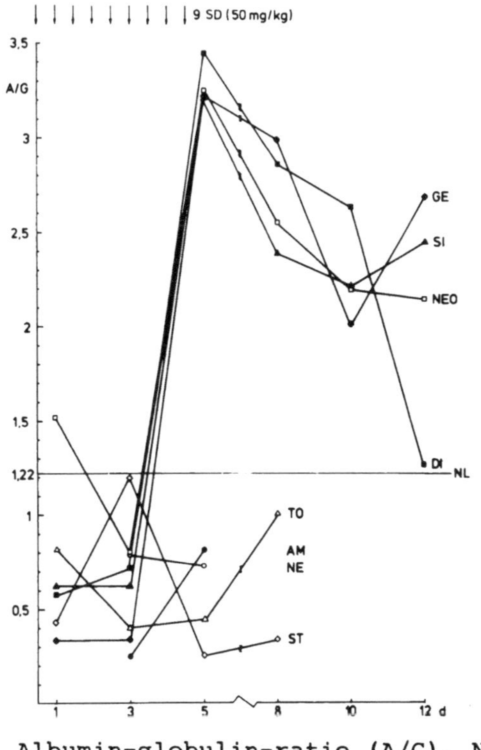

Fig. 2: Albumin-globulin-ratio (A/G). NL = norm limit. SD = single dosis. GE = gentamicin. SI = sisomicin. NEO = neomycin. DI = dibekacin. TO = tobramicin. AM = amicacin. NE = netilmicin. ST = streptomycin.

Figure 2: The analysis of proteinuria during therapy with 8 aminoglycosides shows that there were large differences. The albumin/globulin-ratio of gentamicin, neomycin, dibecacin and sisomicin rose steeply between the 3rd and 5th day, decreased without reaching normal values until the end of the experiment. An AG-ratio of more than 1.22 or an albumin portion of more than 55 % of the proteinuria are considered to be indicators of a glomerular damage. That means that high dosages of gentamicin were potentially glomerulotoxic.

To find more evidence for the assumed glomerular lesions we tried to demonstrate the excretion of immunoglobulins in the urine.

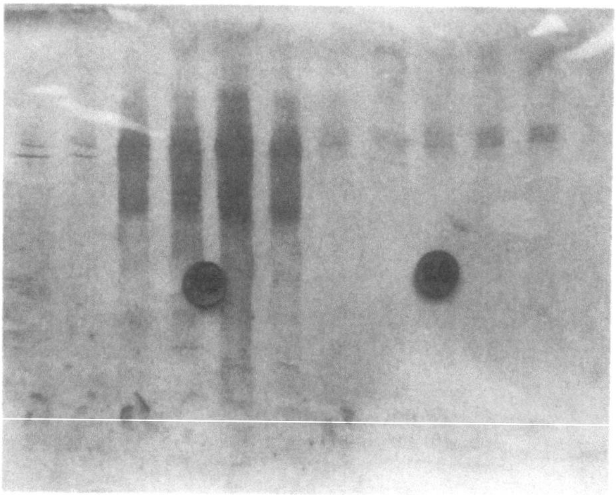

Fig. 3: Comparison of proteinuria caused by 2 different gentamicin dosages (20 and 100 mg/kg and day).The tracks have been loaded from the left to the right with urines of a rat receiving 100 mg/kg and day from day 1, 3, 5, 8, 10 and 12, and the urines (concentrated 10-fold) of an animal treated with 20 mg of gentamicin/kg and day from the same days of experiment.

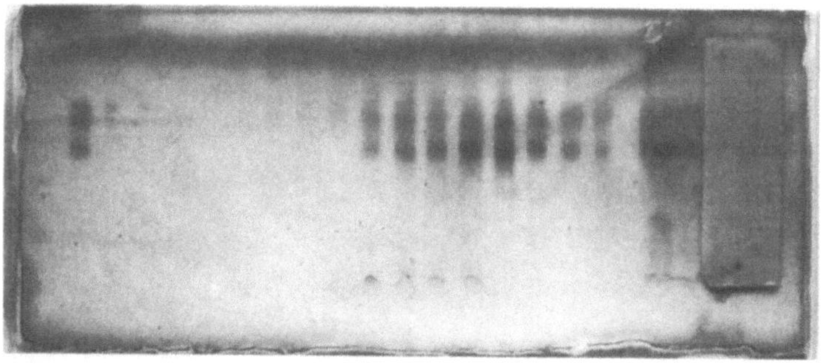

Fig. 4: The gel was loaded with rat albumin (5, 0.5, 0.05, and 0.005 mg/ml) and the urines of a rat which was treated with 100 mg of gentamicin/kg and day. In the last track urine from day 8 was loaded and an overlay performed containing antibodies against rat immunoglobulins.

The 4th figure reveals the proteinuria under therapy with 100 mg of gentamicin per kg bodyweight and day. From the fifth day on two bands appear that are identified by immunofixation as belonging to the immunglobulins.
To demonstrate the immunoglobulins more clearly we adapted the histochemical method of indirect fluorescence marking of proteins on the PAG-plate. Antibodies from rabbits against rat immunoglobulins and fluorescein conjugated antibodies from the swing against rabbit immunoglobulins were loaded at the anode and the cathode respectively. In between the urine-samples were loaded.

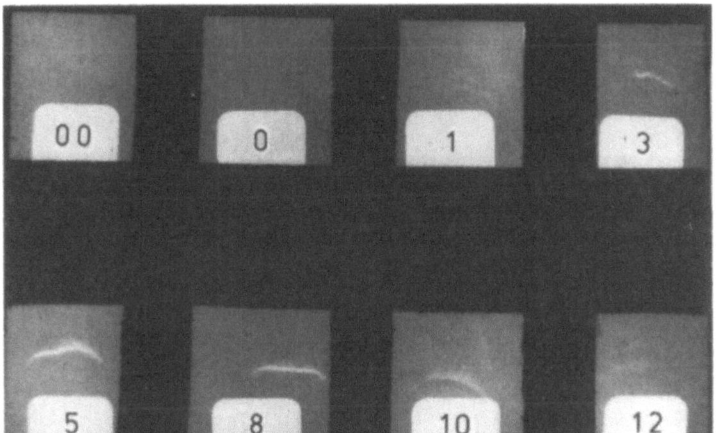

Fig. 5: Photographs through the immunofluorescence microscope track 00: only antibodies were loaded, track 0: swine of the AP, the following tracks: swines of the respective days of the experiment.

The 5th figure shows the result. In track 00 only the two antibodies were focused; in the other tracks urines from the days of AP, MP and RP were loaded. If the intensitiy of the immunofluorescence is related to the quantitiy of the immunoglobulins excreted, glomerulotoxicity could be shown by urine analysis.

Figure 6 shows the result of using direct immunofluorescence marking on the gel. Rabbit serum-samples of decreasing concentration were loaded at the cathod and the fluorescein conjugated antibodies from swine against rabbit immunoglobulins were loaded at the anode. Even a concentration of about 1.5 ng could be detected.

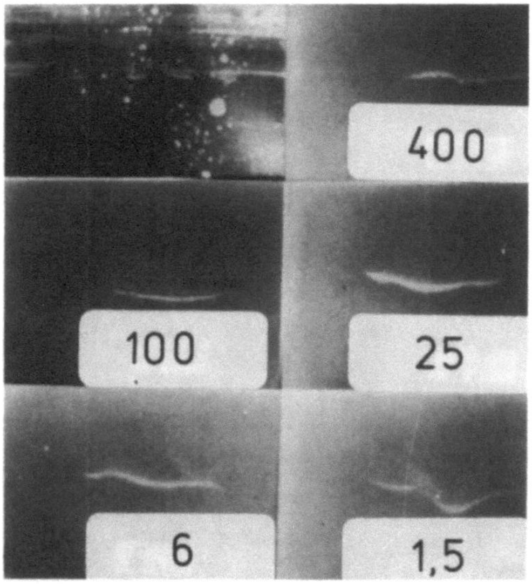

Fig. 6: Photograph of the unstained immunoprecipitates and photographs through the immunofluorescence microscope of the dilution series of rabbit immunoglobulins. The figures show the estimated load of immunoglubulins (ng) per track.

Summary

All aminoglycosides so far tested have been shown to be tubulotoxic. In sufficient dosage they also cause proteinuria. The relationship between these two effects is specific for each compound. Gentamicin-induced proteinuria war further analysed by CA electrophoresis, IEF and immunofluorescence marking on the PAG-plate. The increase of albumin and immunoglobulin excretion can be interpreted as evidence for glomerulotoxicity.

References:

1. Staemmler, M., Karhoff, B.: Virchows Arch. 382, 481 - 502 (1956)
2. Sack, K., Wilhelm, J.: Verh. dtsch. Ges. inn. Med. 79, 676 - 678 (1973)
3. Avasthi, P. S., Evan, A. P., Huser, J. W., Luft, F. C.: J. Lab. clin. Med. 98, 444 - 454 (1981)

AGAROSE ISOELECTRIC FOCUSING OF UNCONCENTRATED CSF AND RADIOIMMUNO-FIXATION FOR DETECTION OF OLIGOCLONAL BANDS IN PATIENTS WITH MULTIPLE SCLEROSIS AND OTHER NEUROLOGICAL DISEASES

Vasilios K. Kostulas

Department of Neurology, University Hospital, Linköping

Hans Link

Department of Neurology, Karolinska Institutet, Huddinge Hospital, Huddinge, Sweden

Introduction

Separation of immunoglobulin G (IgG) by electrophoresis or isoelectric focusing for the demonstration of oligoclonal bands has become an important part of the routine examination of cerebrospinal fluid. The procedure involves the study of the patient's serum taken simultaneously. Oligoclonal IgG bands present in CSF but not in serum are considered to represent the product of a restricted number of lymphocyte clones within the central nervous system and CSF. Oligoclonal bands have been reported in the CSF from most patients with multiple sclerosis (MS), in a substantial proportion of patients with optic neuritis and chronic myelopathy, i.e. disorders which may be related to MS, in about a third of patients with acute aseptic meningitis, meningo-encephalitis and Guillain Barré syndrome (1, 2). Occasionally, oligoclonal bands in CSF may be observed in diseases which are not considered to be related to infection or inflammatory reactions, such as in cerebrovascular diseases (3). One goal for many years has been to establish a method that makes it possible to examine CSF for oligoclonal bands without previous concentration of the fluid. Lasne, Chazot and co-workers (4, 5) recently reported the examination of unconcentrated CSF by agarose isoelectric focusing, followed by immunofixation with radiolabeled anti IgG, so-called radioimmunofixation, and autoradiography. They claimed that they found in normal CSF an

Electrophoresis '82

other neurological diseases (fig 1). The results obtained with agarose isoelectric focusing of concentrated CSF were quite similar.

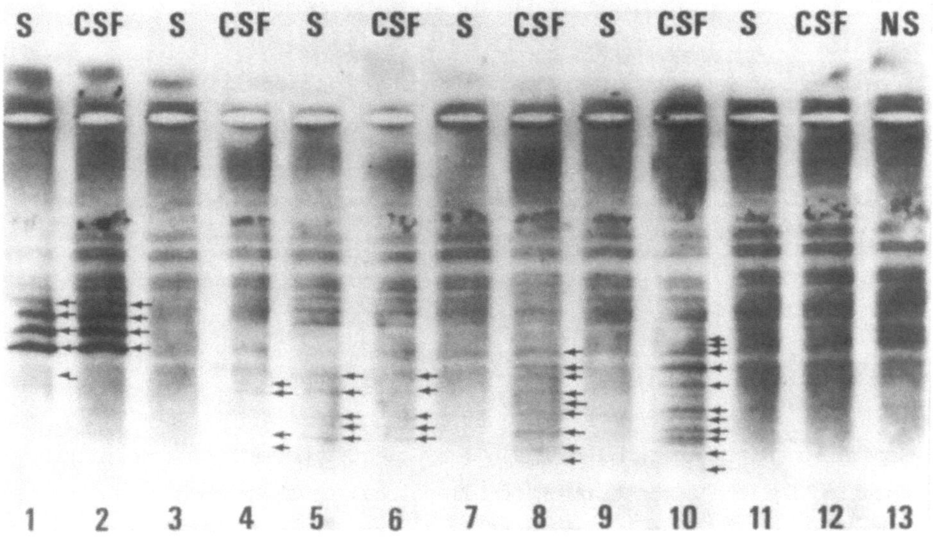

Fig 1. Agarose isoelectric focusing and radioimmunofixation of serum and unconcentrated CSF from one patient with benign monoclonal gammopathy (1, 2); one patient with paraesthesia (3, 4; bands in CSF only); Guillain Barré syndrome (5, 6; bands in CSF and serum); MS (7, 8; bands in CSF only); chronic aseptic meningitis (9, 10; bands in CSF only); psychoneurosis (11, 12; normal pattern) and pooled normal serum (13) run as control.

It has been debated since the introduction of techniques for the demonstration of oligoclonal bands in CSF, whether the procedure used for concentration of the fluid had any influence on the oligoclonal band pattern but, hitherto, no study has been reported to elucidate this question. Our data indicate that concentration of CSF by ultrafiltration has no influence on the oligoclonal IgG band pattern. CSF from 20 patients, 6 of them with MS, was first concentrated by ultrafiltration to about 4 g of IgG/l and then diluted to the original IgG concentration and examined by AIEF and radioimmunofixation together with the same patient's unconcentrated CSF and serum. The concentration procedure used had no influence on the number or intensity of oligoclonal IgG bands (fig 2).

oligoclonal IgG band pattern in the pH region 7 - 9, and they stated that the only difference for MS CSF was that this normal oligoclonal IgG band pattern was more pronounced in MS.

Materials and Methods

CSF and serum was obtained by lumbar puncture from 287 consecutive neurological patients, who have been examined by agarose isoelectric focusing of unconcentrated CSF and radioimmunofixation, for reference, agarose isoelectric focusing of concentrated CSF and of serum was carried out in parallel (6).
A modified version of the technique of Lasne et al. (5) enabling the demonstration of oligoclonal IgG bands in CSF without previous concentration of the CSF was used. Five µl of unconcentrated CSF with an IgG concentration of 20-30 mg l and serum diluted to the same IgG concentration were applied side by side. Carboxyhaemoglobin and pooled donor serum were run as references on every plate. AIEF was carried out as described above. Immediately after focusing the plates were fixed for 20 min in fixative solution and then rinsed 5 times over 24 h. For immunofixation cellulose acetate strips dipped in the radiolabeled antiserum were applied to the gel surface for 20 min. The strips were then removed and incubation continued in a moist chamber at room temperature over night. After rinsing over 24 h, and final rinsing in deionized water to avoid salt crystal precipitation, the gel plate was dried under warm air at 40° C. Antigen-antibody complexes precipitated on the agarose plates were visualized by autoradiography on the dried gel plate, using X-ray film for 6-12 h at room temperature.

Results

Two or more oligoclonal IgG bands were detected by AIF of unconcentrated CSF and radioimmunofixation in 98 % of the patients with clinically definite MS, 67 % of those with aseptic meningitis, and 21 % of those with

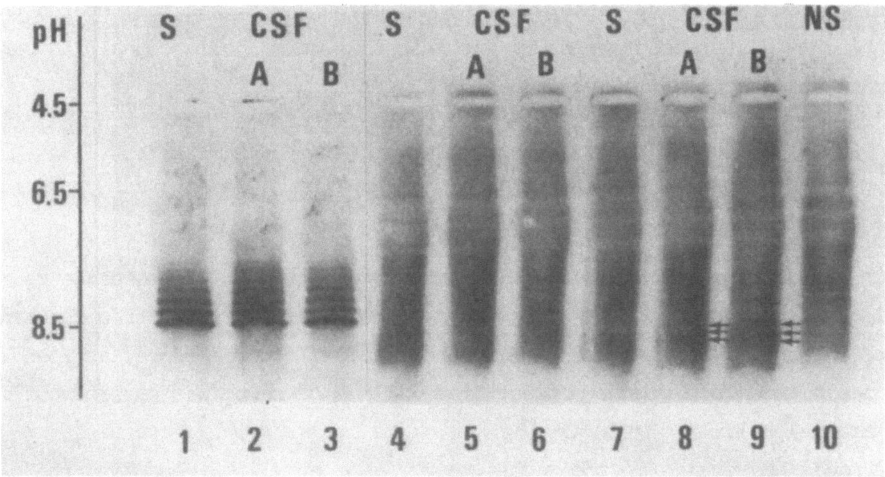

Fig 2. Serum (S), unconcentrated CSF (CSF A), and CSF that had been concentrated and diluted to the original IgG concentration (CSF B), from a patient with benign monoclonal gammopathy (1) (2) and (3): cerebral concussion (4) (5) and (6): and multiple sclerosis (7) (8) and (9). Normal serum (NS) was run as reference (10). The same numbers of IgG bands were seen in CSF, irrespective if unconcentrated CSF, or CSF that was first concentrated and then diluted to the original IgG level, were examined.

Discussion

With our modification of AIEF followed by radioimmunofixation of IgG, we were able to identify a pattern of IgG common to CSF and serum from patients with psychoneurosis who may be considered as "normal" controls. This pattern corresponds according to our opinion to polyclonal IgG. Bands of intrathecally produced oligoclonal IgG were found in patients with MS, infectious CNS diseases and in a minority of patients with other neurological disorders at frequencies similar to those obtained when concentrated CSF was examined by AIEF. AIEF of unconcentrated CSF seems therefore to be a new tool for the identification of locally produced oligoclonal IgG in neurological disorders. The main advantages of this technique - in addition to enable the examination of CSF without previous concentration - are that (1) oligoclonal IgG bands can be separated even when only small CSF quantities are available, such as in pediatric practice, and (2) the occurrence of artifacts such as the appearance of bands consisting of proteins other than IgG is minimized. The technique can also be adapted for the analysis of the IgG band pattern in e. g. AIEF-separated proteins from directly applied frozen nervous and muscular tissue sections (Olsson and Link manuscript in preparation) and for analysis of patterns of proteins other than that of IgG (Laurenzi and Link manuscript in preparation).

References

1. Link, H. Characteristics of the immune response within the CNS in neurological disorders. Acta neurol. scand., 57 (Suppl. 67): 177-190 (1978).

2. Trotter, J.L. and B.R. Brooks Pathophysiology of cerebrospinal fluid immunoglobulins. In: J.H. Wood (Ed.). Neurobiology of Cerebrospinal Fluid. Vol. 1. Plenum Press. New York, London, pp. 465-486 (1980).

3. Roström, B. and H. Link Oligoclonal immunoglobulins in cerebrospinal fluid in acute cerebrovascular diseases. Neurology (Minneap.). 31:590-596 (1981).

4. Chazot, G., Y. Lasne, O. Benzerara, C. Confavreux and B. Schott Radio-immunofixation - Une nouvelle technique de caractérisation des immunoglobulines dans le liquide céphalo-rachidien non concentré. Rev. neurol. (Paris), 136:783-786 (1980).

5. Lasne, Y., I. Benzerara, G. Chazot and R. Creyssel A sensitive method for characterization of oligoclonal immunoglobulins in unconcentrated cerebrospinal fluid. J. Neurochem., 36:1872-1874 (1981).

6. Saravis, C.A. and N. Zamcheck Isoelectric focusing in agarose. J. Immunol. Methods. 29:91-96 (1979).

PREPARATIVE ISOELECTRIC FOCUSING IN AGAROSE GELS AND ITS APPLICATION IN THE INVESTIGATION OF GAMMOPATHIES.

R. McLachlan & F.N. Cornell
Cancer Institute, 481 Lt. Lonsdale St., Melbourne, Australia 3000.

Sera from a small but significant proportion of patients presenting at our hospital contain two or more distinct monotypic immunoglobulins of the same isotype. Fig. 1. shows five such patients with spectrotypically distinct immunoglobulins of the same class, namely IgG.

We sought to determine if any clonal relationship existed between the cells producing these immunoglobulins. Three possibilities exist (1,2,3). The distinct immunoglobulins are being produced by:-

1) Separate, unrelated clones ie. bi/oligoclonal population of plasma cells secreting chemically distinct immunoglobulins.

2) Separate but related clones ie. bi/oligoclonal population of plasma cells secreting chemically related immunoglobulins, eg. somatic mutation.

3) A single clone ie. monoclonal population of plasma cells secreting chemically related immunoglobulins, eg. in vivo hybridoma.

Analysis of the H- and L-chains from the various monotypic immunoglobulins should determine if they contain any common chains and hence, by deduction if any relationship exists between the clones producing them.

To facilitate the interpretation of the H- and L-chain analysis, it is useful to eliminate the initial microheterogeneity of the proteins. Consequently it was decided to prepare spectrotypically

Electrophoresis '82

pure IgG by preparative isoelectric focusing.

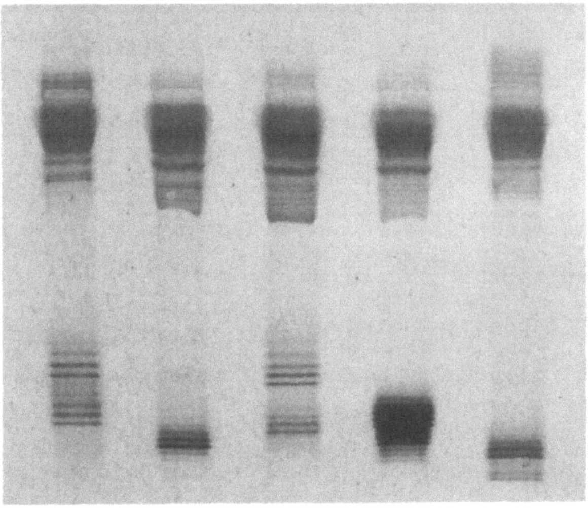

Fig. 1. Analytical IEF: Sera containing ≥2 monotypic IgGs.

METHODS

Preparative IEF method.

Gel composition.

1.0% IEF grade agarose

0.5% non-crosslinked polyacrylamide

2.0% carrier ampholytes, pH 3-10

Perspex templates, 11.0 cm. length, 0.5 cm. depth and 2.5-20.0 cm. variable width, are filled with gel to a depth of 0.4 cm. (Fig. 2).

Sample addition

Add sample to molten gel at 56°C then pour into a pre-cut sample slot. (Fig. 2).

Heat-labile samples may be added in Sephadex G-75 or Pevikon slurry and poured into a sample slot.

Dilute samples may be incorporated into gel constituents prior to pouring the gel. However, this method results in inferior resolution.

IEF conditions

Electrode contact is made through paper wicks (1-2 mm thick) moistened with 2% carrier ampholytes.

Focus: for 45.0 V.hr/cm gel length.
at 0.03 Watt/cm^2 gel surface.
with water cooling (14°C).

Duration of a typical run is 2½-3 hr.

Recovery of proteins

A paper imprint of the gel, stained for protein, is used as a guide for removal of appropriate gel sections. (Fig. 3).

Elution of proteins - the gel sections are frozen in a buffer of choice.
- the gel is removed by centrifugation (30-40,000 g).
- if necessary, the ampholytes are removed (dialysis etc.).

Analytical IEF is then performed on the recovered proteins to assess their purity. Isoelectrically homogeneous IgGs are reduced, alkylated with iodoacetamide and the H- and L-chains examined for spectrotype and M.W.

Analytical Methods

Analytical IEF: 1.0% IEF grade agarose, 0.5% non-crosslinked polyacrylamide, 2.0% carrier ampholytes pH 3-10. 1 Watt for 35 V.hr/cm gel length.

Urea agarose IEF: Analytical IEF gel containing 6.5 M urea. 1 Watt for 100 V.hr/cm gel length.

Urea PAGIF: 6.5 M urea, 4% T, 5% C, 2% carrier ampholytes pH 3-10. I Watt for 100 V.hr/cm gel length.

SDS-PAGE: Horizontal SDS-PAGE, 10% T, 2.7% C, 3 M urea, pH 8.8. Author's unpublished modification of the Laemmli method. (4).

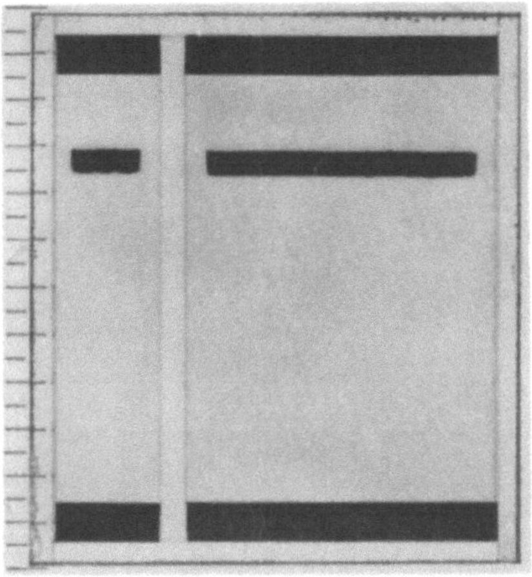

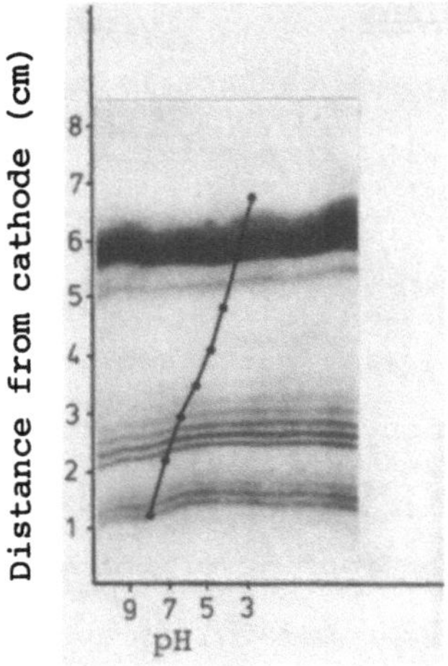

Fig. 2. Preparative IEF gels, separated by an adjustable spacer, prior to IEF.

Fig. 3. Paper imprint taken at completion of preparative IEF, with pH gradient superimposed.

Results

Sera from multiple myeloma patients containing typical monotypic IgGs are shown in Fig. 4. Protein A enriched IgGs from these sera were reduced and alkylated. The results (Fig. 5) revealed a major and a minor L-chain band and a group of H-chain bands.

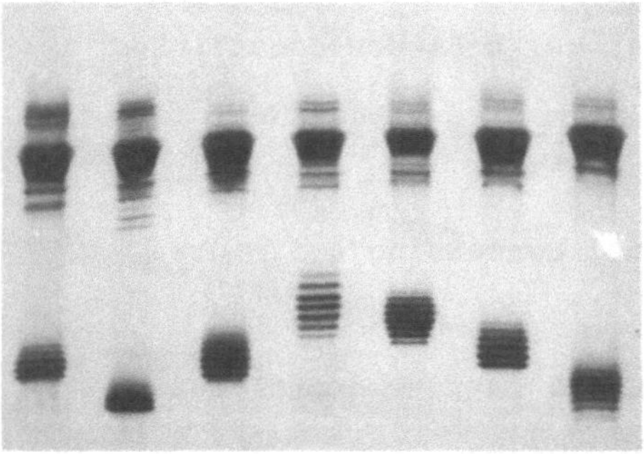

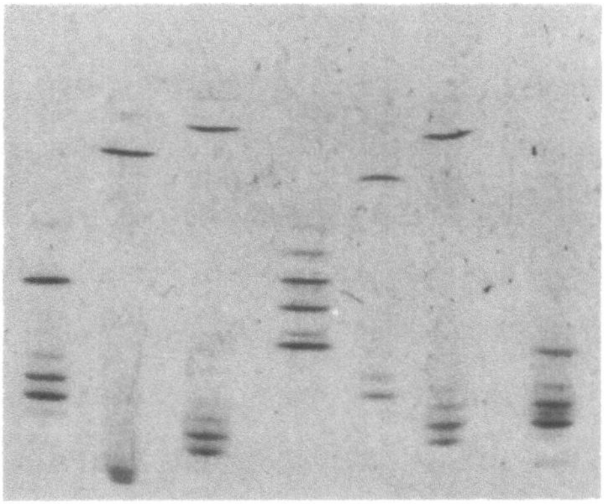

Fig. 4. Analytical IEF: Sera containing single monotypic IgGs.

Fig. 5. Urea PAGIF: Sera as in Fig. 4. reduced and alkylated.

The above monotypic IgGs were also purified by the preparative IEF method and an example is shown in Fig. 6. The spectrotypically pure bands were reduced and alkylated and each showed (Fig. 7) a major and a minor L-chain and a group of H-chain bands. The number of H-chain bands from the spectrotypically pure IgG is less than that obtained from the total IgG.

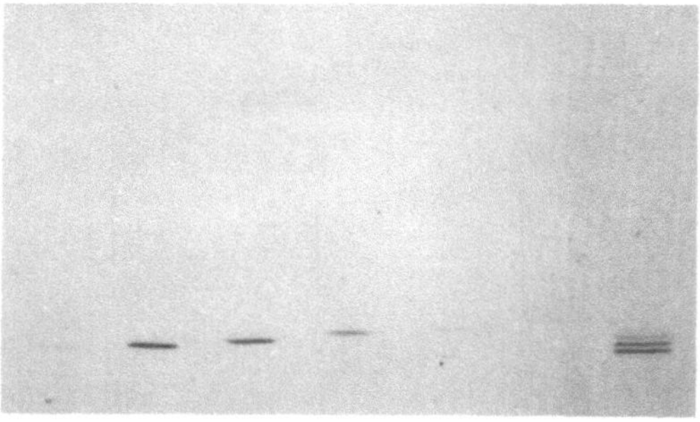

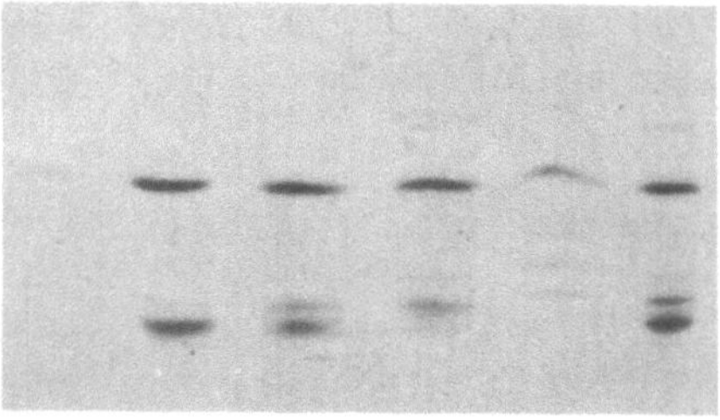

Fig. 6. Analytical IEF: A typical monotypic IgG purified by preparative IEF.

Fig. 7. Urea PAGIF: Purified IgG bands shown in Fig. 6. reduced and alkylated.

Two patients with more than one monotypic IgG in their sera have been investigated so far.

Fig. 8. shows isoelectrically homogeneous bands of IgG from a patient whose serum contained two monotypic $IgG_1(\lambda)$s. When reduced and alkylated, they showed different L-chain but similar H-chain spectrotypes.

One spectrotypically pure band from each monotypic IgG was reduced and alkylated with and without prior neuraminidase treatment (Fig. 9). Neuraminidase treated samples showed identical patterns to the untreated samples. Therefore, sialic acid differences did not contribute to the microheterogeneity of the H-chains. The most likely explanation of this microheterogeneity is post-reduction de-amidation.

SDS-PAGE analysis of the L- and H-chains showed only a single L-chain and a single H-chain M.W. (Fig. 10).

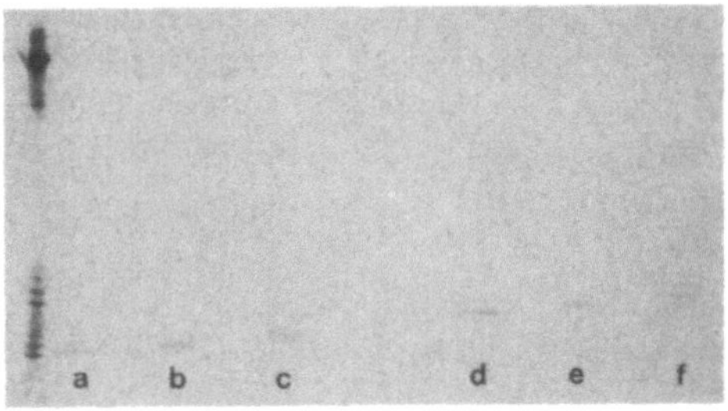

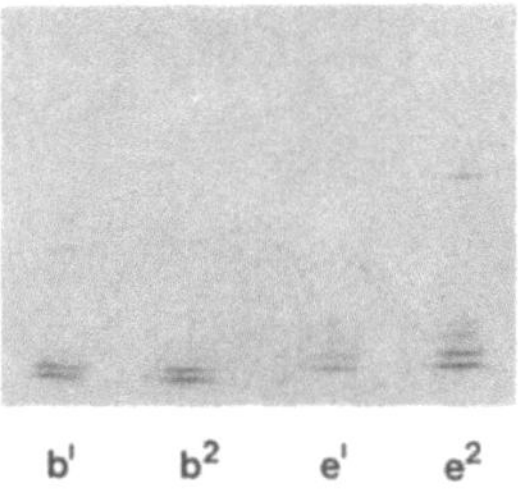

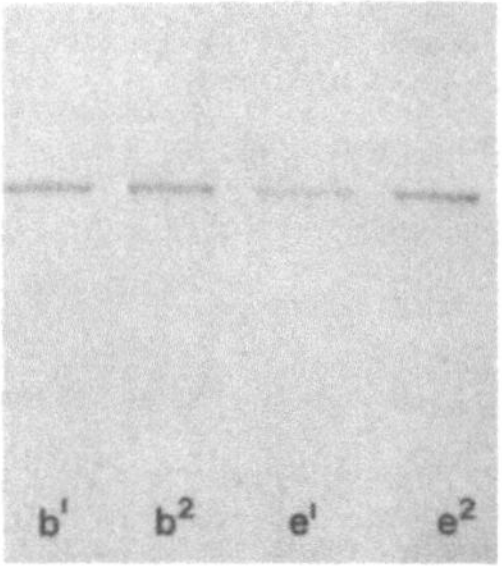

Fig. 8. Analytical IEF: Serum containing two monotypic $IgG_1(\lambda)$s purified by preparative IEF.

Fig. 9. Urea agarose IEF: Bands b & e shown in Fig. 8, reduced and aklylated without (b^1, e^1) and with (b^2, e^2) prior neuraminidase treatment.

Fig. 10. SDS-PAGE: Samples as in Fig. 9.

The second patient, whose serum contained two monotypic IgG(κ)s (Fig. 11) showed similar L-chain but different H-chain spectrotypes in each monotypic IgG (Fig. 12). Multiple H-chain bands of slightly different M.W. and a single L-chain band were observed on SDS-PAGE (Fig. 13).

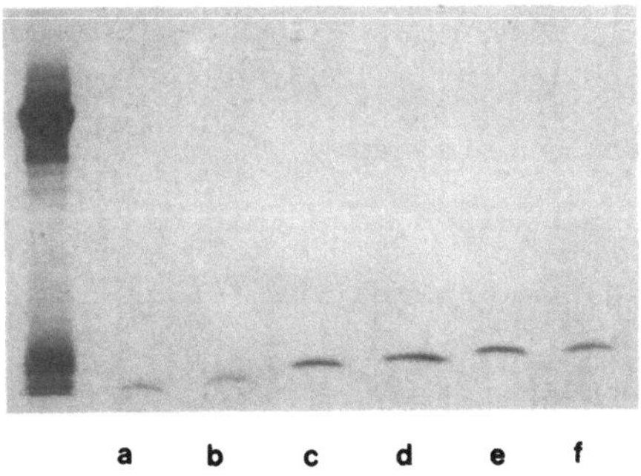

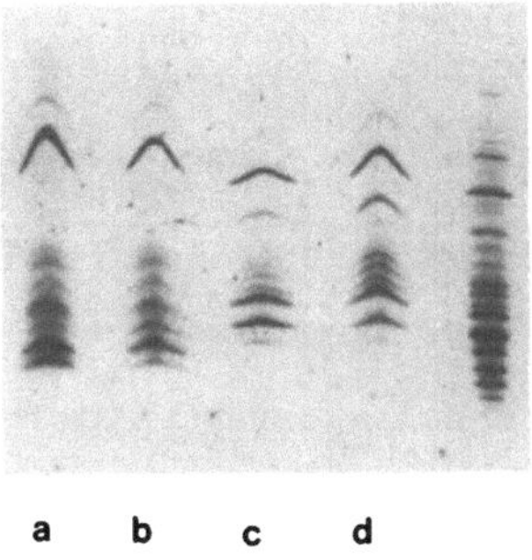

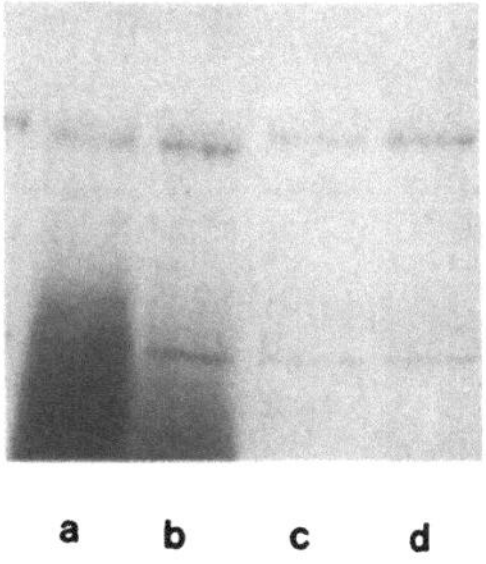

Fig. 11. Analytical IEF: Serum containing two monotypic IgG(κ)s purified by preparative IEF.

Fig. 12. Urea PAGIF: Two isoelectrically homogeneous bands from each monotypic IgG shown in Fig. 11., reduced and alkylated.

Fig. 13. SDS-PAGE: Samples as in Fig. 12.

Discussion

Preparative IEF Method

The use of agarose as a supporting medium for preparative IEF is a simple, convenient and high-resolution technique which allows large quantities of proteins to be recovered with negligible denaturation. (5,6). However even the highly purified commercial agarose preparations which are now available still exhibit appreciable electroendosmotic flow (6). The incorporation (7) of non-crosslinked polyacrylamide into the agarose gels:-

1) Significantly reduced the electroendosmosis by increasing the viscosity of the gel.

2) Significantly improves resolution by reducing the diffusion coefficients of the proteins. Bands which differ by as little as 0.05 pH units can easily be recovered.

3) Does not reduce the recovery of proteins. Recoveries of 80% or more are easily obtained from these composite gels.

Patient Results

Known monotypic IgGs were examined by the methods described above and an atlas of typical results was obtained.

So far, two patients, whose sera contained more than one monotypic IgG, have been examined. In one patient the results suggested that the two monotypic IgGs contained a common H-chain and different L-chains while in the other patient a common L-chain and different H-chains were observed.

The most likely explanation of these results is that the distinct monotypic immunoglobulins are not produced by separate clones but rather by single or related clones. Two possibilities are:- a) that a mutation has occurred in a parent clone leading to a sub-clone secreting a modified "parent" immunoglobulin (8) or

b) that hybridization has occurred between cells of two different clones leading to an *in vivo* hybridoma secreting two related immunoglobulins.

References:

1. van Camp, B.G.K., Shuit, H.R.E. Hijmans, W., Radl, J.: Clin. Immunol. Immunopath., 9, 111-119 (1978).
2. Kreuger, R.G., Hilton, P.M., Boehlecke, J.M. Kyle, R.A., Fair, D.S.: Cell Immunol., 54, 402-413 (1980).
3. Kyle, R.A., Robinson, R.A., Katzmann, J.A.: Am. J. Med., 71, 999-1008 (1981).
4. Laemmli, U.K.: Nature (London), 227, 680-685 (1970).
5. Ebers, G.C., Rice, G.P., Armstrong, H.: J. Immunol. Meth., 37, 315-323 (1980).
6. Chapuis-Cellier, C., Arnaud, P.: Anal. Biochem. 113, 325-331 (1981).
7. Johansson, B.G., Hjerten, S.: Anal. Biochem., 59, 200-213 (1974).
8. Wang, A.C., Fudenberg, H.H., Goldrosen, M.H., Freedman, M.H.: Immunochem., 9, 473-479 (1972).

ROCKET IMMUNOSELECTION USING PEROXIDASE LABELED ANTIBODIES IN THE DIAGNOSIS OF ALPHA HEAVY CHAIN DISEASE

Hocine Rabhi, Mohamed Ghaffor, Mohamed-Cherif Abbadi

Laboratory of Immunochemistry
Institut Pasteur d'Algérie, Alger, Algérie

Introduction

Alpha heavy chain disease is a human immunoproliferative disorder localized mainly to the gut-associated IgA secretory system (3, 13, 16). The proliferating lympho-plasmocytic cells produce and secrete an abnormal and incomplete alpha chain devoid of light chains, which is a constant marker of the disease (11, 14). Usually, such molecules can be easily detected in patients'sera by conventional immunoelectrophoresis with anti-human IgA antisera, particularly those containing antibody specificities against conformational determinants of Fab region (4, 6). However, it has became evident that in some circumstances, the diagnosis of the disease, because of the relative electrophoretic homogeneity and (or) low concentration of alpha chain protein, needs much more specific and sensitive laboratory testing. Different immunoselection techniques using either anti-Kappa and anti-Lambda antisera (5, 7, 10) or anti-Fab alpha antibodies (4, 6) have permitted to overcome such difficulties. We have tested convenient running conditions for a rocket-immunoelectrophoresis technique in which the selection of entire immunoglobulins and the demonstration of free alpha chains were obtained using successively anti-Kappa and anti-Lambda antiserum and peroxidase labeled anti-IgA antibodies.

Electrophoresis '82

Materials and methods

1) Patients'sera. 96 serum samples were obtained from patients with chronic diarrhoea and stored at -20°C until tested.

2) Antisera. Anti-human IgA antiserum was produced in goat by repeated subcutaneous injections of IgA myeloma protein (kindly provided by Dr C. Rivat), and subsequent absorption on polyclonal IgG Sepharose-4B columns (9). The specificity of the gammaglobulin fraction of this antiserum was controled by immunoelectrophoresis and Ouchterlony's immunodiffusion test.
Peroxidase labeled anti-IgA antibodies were purchased from Institut Pasteur Production, Paris, France.
Anti-Kappa and Lambda antiserum was obtained from goat hyperimmunized with 23,000 molecular weight gel filtration peak of reduced and alkylated polyclonal IgG.

3) Methods. Serum electrophoresis was done on cellulose acetate membranes in 0.05M pH 8.6 barbital buffer using Beckman Microzone chamber.
Immunoelectrophoresis was carried-out using 1.5 % agar gel in 0.05M pH 8.2 barbital buffer as described (12).
Immunoselection-Immunoelectrophoresis (5) was performed in the same agar gel containing 10 % of anti-Kappa and Lambda antiserum, using 10 microliters of pathological serum and 60 microliters of anti-IgA gammaglobulin fraction in the trough.
Rocket-immunoselection technique using 3 different gels (7) was achieved as represented in figure 1. The first gel containing 10 % of anti-Kappa and anti-Lambda antiserum consisted of mixture of 3 volumes of 1 % Agarose C (Pharmacia Fine Chemicals) and 1 volume of 1 % Indubiose A 37 (Industrie Biologique Française). The second gel was made of 1 % virgin agarose C.

Fig. 1. Diagram of rocket-immuno-selection plates.
GEL-1: 1% Agarose + 1% Indubiose containing anti-kappa and anti-lambda anti-serum.
GEL-2: 1% Agarose C.
GEL-3: 1% Agarose C containing labeled or unlabeled anti-IgA antibodies.

The third (anodic) gel was 1 % Agarose C containing 1.2 percent of peroxidase labeled anti-IgA antibodies. All gels were in 0.075M pH 8.6 acetate-barbital buffer. The electrophoresis was carried-out on cooled tank during 2 hours at 3.5mA/cm. The plates were extensively washed in buffered saline, dried and immersed during 20 mn in freshly prepared solution of 3,3'-diaminobenzidine tetrahydrochloride (BDH Ltd Chemicals) (8). Finally, the gels 1 and 2 were stained with 0.2 % Coomassie brilliant blue R250 solution (BioRad Laboratories).

Results and discussion

None of serum samples showed any abnormal broad band in cellulose acetate electrophoresis or any heterogeneous precipitin line in immunoelectrophoresis with anti-IgA antiserum. All sera tested by immunoselection-immunoelectrophoresis (5) were negative for abnormal heavy chains, whereas the rocket-immunoselection revealed alpha chain disease proteins in 8 cases. The alpha specificity of the upper rockets (fig. 2) was ascertained by negative reaction when anti-IgA antibodies were replaced by phosphate buffered saline, anti-IgM or anti-IgG antisera.

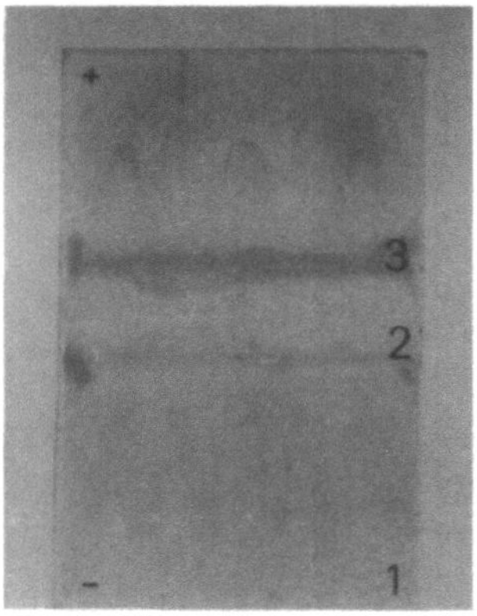

Fig. 2. Rocket-immunoselection plate revealing the presence of alpha heavy chain disease proteins in patient sera.
The alpha heavy chains were indetectable by immunoelectrophoresis or immunoselection-immunoelectrophoresis techniques.

The concentration of anti-light chain antiserum was sufficient to precipitate in the lower gel an IgA myeloma protein at 2 mg/ml.

The agarose C-Indubiose mixture gave a moderate electroendosmosis effect avoiding sample formylation (7) and limiting light-chain bearing immunoglobulin migration to the cathodic gel. It has been reported that the immunoselection combined with immunoelectrophoresis was proven to be specific and more sensitive (4) than other techniques (2, 7, 10). This was due to the availability of an anti-Fab alpha region antiserum, which precipitated both IgA myeloma proteins and increased concentrations of polyclonal IgA (4, 6). Our study demonstrated that an anti-Kappa and Lambda light chain antiserum may give satisfactory results concerning the specific detection of alpha chain disease proteins in low secreting cases (fig. 2).

Peroxidase labeled anti-IgA antibodies enhanced the sensitivity of the method and introduced a supplementary criterion of specificity.

Indeed, in two cases, abnormal alpha chains were detected by rocket-immunoselection technique two and six months before being demonstrated using conventional immunoelectrophoresis or

immunoselection-immunoelectrophoresis.
This seemed to correlate with an early diagnosis of the disease. This correlation could be established when histopathological and surgical data concerning the two patients will be obtained.

Acknowledgements

We are grateful to Dr C. Rivat (Unité de Recherches sur la Génétique des Protéines humaines, Bois-Guillaume, France) for kindly providing human IgA myeloma protein. We wish to thank Dr A. Benmansour (Institut Pasteur d'Algérie) for his pertinent advices, Mrs F. Matthys and Ms K. Hadj-Hamdi for secretarial assistance.

References

1. Axelsen, N. H., Harboe, M., Jonsson, V., Videbaek, A. A. : Scand. J. Haematol. 16, 218-225 (1976).
2. Dammacco, F., Antonaci, S., Miglietta, A. : Boll. Ist. Steroter. Milanese 54, 460-463 (1975).
3. Doe, W. F. : Br. J. Cancer 31, Suppl. II, 350-355 (1975).
4. Doe, W. F., Danon, F., Seligmann, M. : Clin. Exp. Immunol. 36, 189-197 (1979).
5. Doe, W. F., Henry, K., Hobbs, J. R., Avery-Jones, C. E., Dent, C. E., Booth, C. C. : Gut 13, 947-957 (1972).
6. Doe, W. F., Spiegelberg, H. L. : J. Immunol. 122, 19-23 (1979).
7. Gale, D. S. J., Versey, J. M. B., Hobbs, J. R. : Clin. Chem. 20, 1292-1294 (1974).
8. Graham, R. C., Karnovsky, M. J. : J. Histochem. Cytochem. 14, 291-293 (1966).
9. Marsh, S. C., Parikh, I., Cuatrecasas, P. : Anal. Biochem. 60, 149-152 (1974).
10. Radl, J. : Immunology 19, 137-149 (1970).

11. Rambaud, J. C., Bognel, C., Prost, A., Bernier, J. J., Le Quintrec, Y., Lambling, A., Danon, F., Hurez, D., Seligmann, M. : Digestion 1, 321-325 (1968).

12. Scheidegger, J. J. : Int. Arch. Allergy Appl. Immunol. 7, 103-105 (1955).

13. Seligmann, M. : Br. J. Cancer 31, Suppl. II, 356-361 (1975).

14. Seligmann, M., Danon, F., Hurez, D., Mihaesco, E., Preud' Homme, J. L. : Science 162, 1396-1397 (1968).

15. Seligmann, M., Mihaesco, E., Hurez, D., Mihaesco, C., Preud'Homme, J. L., Rambaud, J. C. : J. Clin. Invest. 48, 2374-2389 (1969).

16. WHO Meeting Report : Bull. WHO 54, 615-624 (1976).

CLASSICAL ELECTROPHORETIC TECHNIQUES VERSUS ULTRATHINLAYER FOCUSING AS TOOLS IN POPULATION GENETIC STUDIES.

Vibeke Simonsen

Institute of Ecology and Genetics, University of Aarhus
DK-8000 Aarhus C, Denmark

Introduction

Various biochemical methods have been used for determining genetic variation in natural populations.
In our department starch gel electrophoresis (SGE) as well as polyacrylamide electrophoresis (PAGE) has been used for studying the pattern of genetic variation of allozymes in natural populations of eelpout, Zoarces viviparus L., and other marine animals. The intention of these investigations has been to describe geographic variation and to discuss the action of evolutionary factors as natural selection, migration and genetic drift (1, 2).
The aim of this very preliminary study reported here is to demonstrate, whether the ultrathinlayer focusing method (UTIEF) can be as well applied for studying allozymes in natural populations and if the method reveals comparable amounts of genetic variation as SGE and PAGE.

Materials and methods

Extracts of kidney and pancreas from house mouse and of muscle from herring, stored at -18°C, were used. Extracts from whole individuals of fruit fly and of gammarus were tested, too.
The analysing method was UTIEF as described earlier (3). The procedures for visualization of the enzymes were similar to

those, which were used for revealing activity of enzymes after SGE or PAGE (4) with the exception, that for all staining solutions half the amount of the liquid contained 2% agar.

Results

The results of the enzymes and organisms tested are listed in Table 1, and Figure 1 shows the result of glucose-phosphate isomerase in house mouse obtained by SGE and UTIEF. If the staining solution is dried up, only amylase from pancreas of house mouse shows activity. Several prints of the same gel is possible for glucose-phosphate isomerase from kidney of house mouse.

Table 1. List of organisms and enzymes studied and a comparison of the analytic methods used.

Organism	Tissue	Enzyme	Result
Mouse	Pancreas	Amylase	PAGE better
	Kidney	Glucose-phosphate isomerase	See Figure 1
	Kidney	Lactate dehydrogenase	As SGE
	Kidney	Malate dehydrogenase	As SGE
	Kidney	Mannose-phosphate isomerase	As SGE
	Kidney	Phosphoglucomutase	As SGE
Herring	Muscle	Glucose-phosphate isomerase	As SGE
Fruit fly		Amylase	PAGE better
Gammarus		Glucose-phosphate isomerase	Enzyme unstable in UTIEF

Figure 1. Glucose-phosphate isomerase from kidney of house mouse. Ten individuals. Top: UTIEF, bottom: SGE.

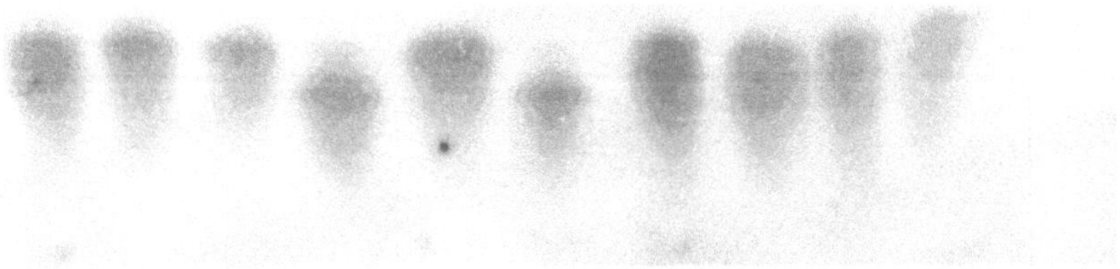

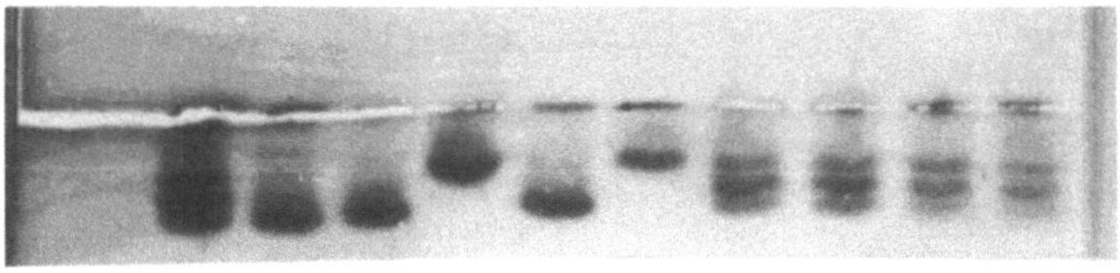

Conclusions

UTIEF gives at least for some of the enzymes tested results just as good as SGE, but with the advance, that the time used for the electrophoresis is reduced to one tenth and the amount of staining solution reduced to one fifth.

Acknowledgement

I want to express my sincere gratitude to Dr. B.J.Radola, Dr. A.Kinzkofer and M.Frey for introducing me to the UTIEF technique.

References

1. Frydenberg, O., Gyldenholm, A. O., Hjorth, J. P., Simonsen, V.: Hereditas 73, 233-238 (1973).
2. Christiansen, F. B., Frydenberg, O., Simonsen, V.: Hereditas 87, 129-150 (1977).
3. Kinzkofer, A., Radola, B. J.: Electrophoresis 2, 174-183

(1981).
4. Frydenberg, O., Simonsen, V.: Hereditas 75, 221-232 (1973).

C4 PHENOTYPES IN CAUCASIANS FROM THE SOUTHEASTERN UNITED STATES

Bruce Budowle,[1,3,4] Rodney C. P. Go,[2,3,4] Jeffrey M. Roseman,[2,4]
Bruce O. Barger,[2,4] Ronald T. Acton[1,2,4]

Departments of Microbiology,[1] Epidemiology,[2] Genetic Marker Laboratory,[3] and Diabetes Research and Training Center,[4] University of Alabama in Birmingham, Birmingham, Alabama, U.S.A.

Supported in part by PHS Grant No. CA09128 awarded by the National Cancer Institute, DHHS.

Introduction

The genetically-controlled polymorphic complement component C4 has gained much interest in the past couple of years. Using prolonged agarose gel electrophoresis, Rosenfeld et al. (1) were the first to observe polymorphic C4 patterns. However, they did not realize the extreme variability of C4 and, thus, were unable to offer a simple genetic model for this marker. Rittner et al. (2) found that C4 was linked to genes coding for the HLA antigens, probably lying between the B and DR loci (3). The genetics of C4 became evident when O'Neill et al. (4) demonstrated that two common blood group antigens--Rodgers and Chido--also known to be linked to HLA (5,6), represented two antigenic determinants on C4. Moreover, they showed that C4 was governed by two linked loci. By using neuraminidase-treated plasma samples, Awdeh and Alper (7) obtained relatively distinct electrophoretic separations of the C4 loci in a high pH agarose gel. They defined the C4 system so as to conform to the International System for Human Gene Nomenclature. The more acidic, faster-migrating, protein carrying the Rodgers antigen was named C4A; while the more basic, slower-migrating component containing

Electrophoresis '82

the Chido marker was called C4B. The variants (each producing a three-band pattern with the most anodal being the most intense) at each locus were arbitrarily numbered with integers according to their electrophoretic mobility from the cathode to the anode in agarose gel electrophoresis at pH 8.8. Seven alleles (six structural and one null) have been observed at the C4A locus, and three alleles (two structural and one null) have been found at the C4B locus by Awdeh and Alper (7) in their U.S. Caucasian population. Family data have shown that the alleles of the two linked loci are codominantly inherited. Moreover, since no crossovers have been found between these two loci, they are believed to be tightly linked.

This paper gives an account of the methodology for phenotyping C4, which is not clearly presented in the literature. Moreover, we compared the results of the C4 phenotypes found in Caucasians from the southeastern United States with a Caucasian population from the northeastern United States (7) to see if regional differences exist.

Materials and Methods

Plasma samples from 196 unrelated healthy random Caucasians from the southeastern United States were phenotyped for C4 according to a modified method of Alper and Awdeh (7).The plasma samples were desialated by incubating them at 4°C with Clostridium perfringens neuraminidase Type V (Sigma) at a concentration of 10 mU of enzyme/ml of plasma, while being dialyzed (8) overnight against 0.1 M phosphate buffer (pH 6.8) containing 0.005 M disodium EDTA.

The neuraminidase-treated plasma samples were then subjected to electrophoresis in 0.8% agarose gels (Seakem ME) using à continuous Tris/Glycine/Barbital buffer system. The electrode buffer consisted of a solution containing 180.0 g tris, 224.8 g glycine, 52.4 g sodium barbital and 8.3 g bartital in a total of 8 ℓ of H_2O at a pH of 8.8. The gel buffer was prepared by mixing 35 ml of electrode buffer with 5 ml of 0.2 M EDTA, pH 7.2 and 60 ml H_2O. Ten µl of each sample were applied to the gels (10x20x0.1 cm) a half centimeter from the cathodal wick for 20 minutes via an LKB application mask (excess material was blotted). Electrophoresis was carried out at a constant current of 65 mA and a temperature of 4°C for two hours and subsequently immunofixed with a 1:3 dilution of antisera to C4 (Atlantic Antibodies). After a one-hour incubation at room temperature, the gels were pressed with absorbent paper under a one kg weight for one minute and then set in phosphate-buffered saline overnight at room temperature. The following day, the gels were dried, stained in a solution of 2.5g Coomassie brilliant blue R250, 450 ml ethanol, 450 ml H_2O and 100 ml Glacial Acetic Acid and destained in a solution of 450 ml ethanol, 450 ml H_2O and 100 ml Glacial Acetic Acid.

Those heterozygous for the null phenotype were determined by comparing the stain intensity between the A and B loci. The stained pattern, which was half as intense as the other, carried a null phenotype.

Results

The electrophoretic patterns of the various C4 alleles found in our Caucasian population are shown in Figures 1a and 1b. Seven alleles (six structural and one null) were found at the C4A locus, and five alleles (four structural and one null) were detected at the C4B locus. The alleles of these two linked loci have 420 possible genotypes, of which 386 are distinguishable. The other 34 genotypes are heterozygous for the null phenotype at both loci, which cannot be differentiated from homozygosity at both loci. Forty-one different C4 phenotypes were observed in 196 Caucasians from the southeastern United States (Table 1).

Discussion

This report confirms the findings of Awdeh and Alper (7) of the highly polymorphic nature of C4 in Caucasians. We observed the same variants at the C4A locus and two more variants (a total of five variants) at the C4B locus. Out of a possible 386 phenotypes, we detected 41 different C4 phenotypes in 196 individuals. Therefore, it is evident that C4 on chromosome 6 is one of the most complex genetic systems within a single human population, second only to HLA.

This is the first example of the variant C4B5 (named by Alper and Awdeh, personal communication) in healthy Caucasians. In addition, this is the first known report of the arbitrarily named rare variant C4B9. C4B9 is a slow-migrating B locus phenotype; the most anodal band migrates slower than the most cathodal band of the C4B1 phenotype (Figures 1a and 1b).

For the most part, the phenotypic expression of the C4 variants in our population is similar to that reported by Awdeh and Alper (7) (Tables 2a and 2b). The greatest differences between the two populations were for the AO, B1 and BO phenotypes. AO is increased (29.6% vs. 21.0%), while B1 and BO are decreased (87.8% vs. 96.0% and 14.3% vs. 25.0%, respectively) in our southeastern U.S. Caucasian population compared with the Caucasian population of Awdeh and Alper (7).

It is interesting that the null phenotypes at both C4 loci are relatively common in the Caucasian population, unlike the null phenotypes of the other C3 convertases, C2 and properdin factor B, which are tightly linked to C4 and probably arose through tandem duplication. The C4AO and C4BO phenotypes were detected in 29.6% and 14.3% of the population, respectively. It is likely that the observed phenotypic frequency of the null phenotypes is low. The method for detecting an individual carrying a null allele at one of the C4 loci is by crossed immunoelectrophoresis (9), or more easily by comparing the intensity of the banding patterns of each loci. If the pattern of C4A is twice as intense as C4B, it is likely that the individual carries a C4B0 phenotype. At present, it is not possible to detect an individual from the general population carrying a null allele at both loci. For example, unless family data is used, there is no known way to discern between the A33B11 and A30B10 phenotypes. We are presently investigating families in the Caucasian and black American populations to obtain C4 gene frequencies and haplotype data by direct counting methods rather than maximum likelihood methods.

The highly polymorphic nature of C4 can be very beneficial as a preliminary screening marker for Major Histocompatibility Complex disease association studies. It is becoming increasingly evident that of the HLA loci DR is primarily associated with many diseases. Due to the fact that C4 is closely linked to and presently more polymorphic than DR, that there is no need for viable cells, and the relative ease and low cost of obtaining a phenotype, this complement component may be as good as or even a better marker for disease association studies than DR. We are presently investigating the strength of C4 as a risk predictor for several HLA-associated diseases, such as insulin-dependent diabetes mellitus and acute lymphocytic leukemia.

Moreover, the features of C4 (polymorphic nature, ease of assay, sample material, and low cost) make this genetic marker excellent for paternity and forensic applications. We are compiling family data and performing persistence (i.e., stability) tests on C4 for such utility.

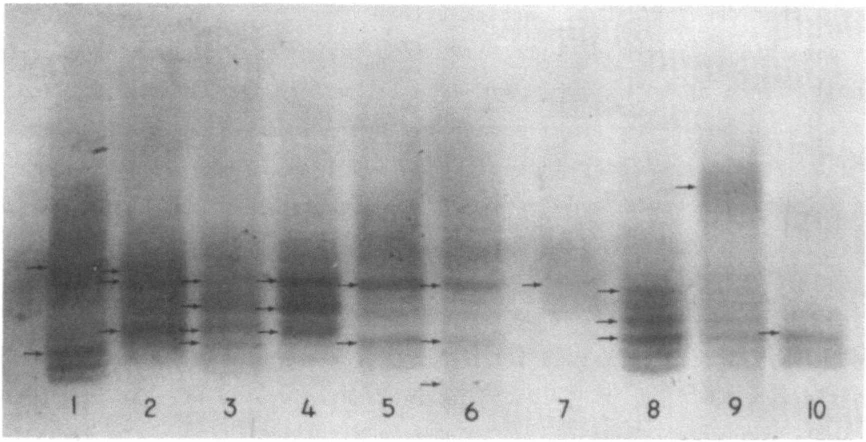

Figure 1a. Agarose gel electrophoretic patterns of C4 from plasma of selected individuals showing the C4 variants in the Caucasian sample. The arrows indicate the most anodal band of the variant. The phenotypes are 1) A6B1; 2) A54B22; 3) A41B21; 4) A40B52; 5) A3B1; 6) A33B91; 7) A3B00; 8) A20B21; 9) C4 degradation product; 10) A00B1.

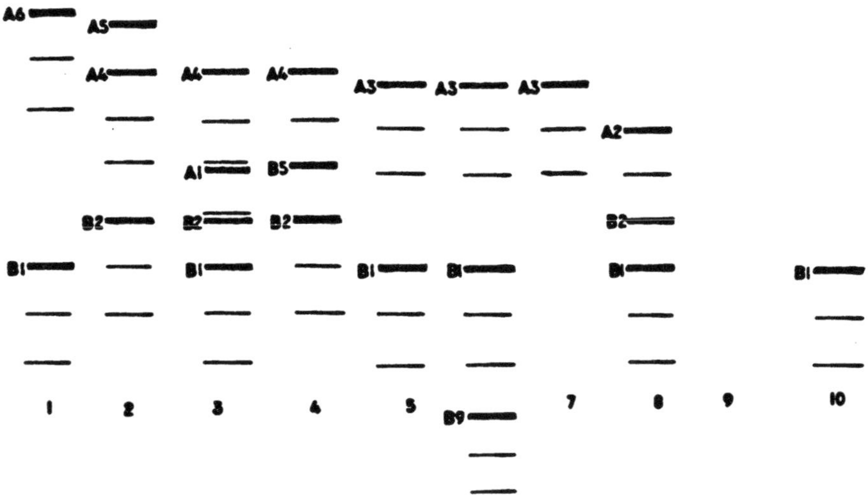

Figure 1b. Diagram of Figure 1a. DP = C4 degradation product.

Table 1. Observed C4 phenotypes in 196 random unrelated Caucasians.

Phenotype	No. Observed	Phenotype	No. Observed
A6B1	1	A40B21	5
A63B11	10	A40B52	2
A63B21	1	A3B1	51
A62B11	1	A3B2	7
A60B11	4	A3B00	6
A60B21	1	A33B10	14
A54B22	1	A33B21	9
A50B11	1	A33B20	1
A4B1	1	A33B91	1
A4B2	1	A32B11	10
A44B10	1	A32B10	2
A44B20	1	A32B22	2
A44B21	1	A32B21	2
A43B11	2	A30B11	30
A43B10	1	A30B22	1
A43B20	1	A30B21	5
A43B21	7	A20B11	3
A42B11	1	A20B21	1
A42B10	1	A00B1	3
A41B21	1	A00B2	1
A40B11	1		

Table 2a. Phenotypic expression (%) of the C4A variants in Caucasians

	Pop. No.	A6	A5	A4	A3	A2	A1	A0
Budowle et al.	196	9.2	1.0	14.3	83.2	11.7	0.5	29.6
Awdeh/Alper (7)	100	11.0	1.0	11.0	89.0	15.0	0	21.0

Table 2b. Phenotypic expression (%) of the C4B variants in Caucasians

	Pop. No.	B9	B5	B2	B1	B0
Budowle et al.	196	0.5	1.0	26.0	87.8	14.3
Awdeh/Alper (7)	100	0	0	21.0	96.0	25.0

References

1. Rosenfeld, S., Ruddy, S., Austen, K. F.: J. Clin. Invest. 48, 2283-2293 (1969).
2. Rittner, C., Hauptmann, G., Grosse-Wilde, H., Grosshan, E., Tongio, M. M., Mayer, S.: Histocompatibility Testing 1975, Munksgaard, Copenhagen, pp. 945-954 (1975).
3. Bruun-Petersen, G., Lamm, L. U., Sorensen, I. J., Buskjaer, L., Mortensen, J. P.: Hum. Genet. 58, 260-267 (1981).
4. O'Neill, G. J., Yang, S., Tegoli, J., Berger, R., DuPont, B.: Nature 273, 668-670 (1978).
5. Middleton, J., Crookston, M. C., Falk, J. A., Robson, E. B., Cook, P., Batchelor, J. R., Bodmer, J., Ferrara, G. B., Festentein, H., Harris, R., Kissmeyer-Nielsen, F., Lawler, S. D., Sachs, J. A., Wolf, E.: Tissue Antigens 4, 366-373 (1974).
6. Giles, C. M., Gedde-Dahl, T., Robson, E. B., Thosby, F., Olaisen, B., Arnason, A., Kissmeyer-Nielsen, F., Schreuder, I.: Tissue Antigens 8, 143-149 (1976).
7. Awdeh, Z. L., Alper, C. A.: Proc. Natl. Acad. Sci. U.S.A. 77, 3576-3580 (1980).
8. Awdeh, Z. L.: Anal. Biochem. 71, 601-603 (1976).
9. Awdeh, Z. L., Raum, D., Alper, C. A.: Nature 282, 205-207 (1979).

ISOELECTRIC FOCUSING OF SEED PROTEINS FROM NATURAL GREEK POPULATIONS OF TWO *AGROPYRON* SPECIES

Michael Moustakas, Elli Coucoli and Ioannes Tsekos
Botanical Institute, University of Thessaloniki, Greece

Introduction

Seed protein electrophoresis has been increasingly utilized in the last decade as an additional approach for species identification and as a useful tool for tracing back the evolution of various groups. A paper summarizing the main features of the seed protein profile has been published by Ladizinsky and Hymowitz (1).

Seed proteins are mainly storage proteins not likely to undergo changes in dry mature seeds. Furthermore their composition is considered to be highly constant not reflecting variation in nutritional status and environmental factors (2, 3, 4 and others).

The additive properties of the seed protein profile have successfully been utilized in studies aiming at elucidating the origin and the evolution of polyploid plants (*Triticum* and *Secale* (5), *Gossypium* (6), *Triticum* and *Aegilops* (7), *Solanum* (8) and others).

The genus *Agropyron*, in particular its group *Elytrigia*, though extensively studied by certain workers appeared rather questionable in taxonomy and nomenclature and consequently is still poorly understood.

According to Cauderon (9), information about relationships within *Elytrigia* group does not allow precise formulation of

Electrophoresis '82

the genomes for each species. The need for further research in this group was also pointed out by Heneen and Runemark (10) and Dewey (11). Cauderon (9), summarizing the present knowledge, recognized in *Elytrigia* group two polyploid complexes and four aggregate species. It is in *Agropyron junceum* (L.) Beauv. complex that the diploid *A. striatulum* and the hexaploid *A. junceum* subsp. *mediterraneum* are included. Their genomic formulas as proposed by Cauderon (9,12) are respectively J_1J_1 and $J_1J_1J_2J_2E_3E_3$.

Unpublished cytological data of a research program being currently in progress in our laboratory support the view that the polyploid *A. junceum* subsp. *mediterraneum* is comprised of differentiated J genomes ($J_1J_2J_3$) while the participation of another genome different in origin and identity proved rather improbable.

The purpose of the present work is to document the genomic constitution of both diploid and hexaploid by using the seed protein profiles obtained by isoelectric focusing.

Materials and Methods

Plant material

Plants were recently collected in several North-Greek littoral zones. Precisely, 23 plants of *A. striatulum* representing five populations from four biotopes and 24 plants of *A. junceum* subsp. *mediterraneum* representing five populations from five biotopes were examined.

Protein extraction

Seed proteins were extracted from individual plants by grinding 100 mg of mature seeds with 1.0 ml cold (4°C) distilled water over an ice bath. The resulting mixture was centrifuged at 10.000 g for 15 min. The clear supernatant was lyophilized, then it was redissolved in 100 µl of cold (4°C) distilled water and subsequently used as the protein sample.

Isoelectric focusing

Isoelectric focusing was conducted in a LKB Multiphor apparatus 2117 connected to an LKB 2103 power supply. Polyacrylamide gels of 5% total acrylamide concentration and 3% cross-linkage with 2.2% LKB Ampholines in pH ranges 3.5-9.5 and 4.0-6.5 were used. Electrode paper strips were soaked in 1M sodium hydroxide (cathode) and 1M o-phosphoric acid (anode) for 3.5-9.5 gels and in 0.1 M b-alanine (cathode) and 0.1M glutamic acid in 0.5 M o-phosphoric acid (anode) for 4.0-6.5 gels.

Protein samples of 20 µl were applied to the focusing gel on pieces of Whatman 3 MM paper (1.0X0.5 cm) 1.0 cm from the cathode of the isoelectric focusing gel. Focusing was across the width, with a constant power 30 W for 1.5 h for 3.5-9.5 gels and 25 W for 2.5 h for 4.0-6.5 gels. During the focusing, the gel plate was cooled at 4°C by connection to a thermostat.

Fixation-staining-destaining

The gels were fixed immediately at room temperature after the current was switched off in a solution containing 30 gr trichloroacetic acid, 7 gr sulfosalicylic acid, 75 ml methanol and 175 ml distilled water. After 30 min. the gels were rinsed in the destaining solution for 5 min. and then stained for 15 min. at 60°C with 0.10% w/v Serva Blau R added in the destaining solution. The gels were subsequently destained for a couple of days with ethanol-water-glacial acetic acid 25:65:10 v/v.

Results and Discussion

Although variability in the protein pattern was detected within (more) and between (less) the populations of *A. striatulum* and *A. junceum* subsp. *mediterraneum*, all the examined plants possessed the same basic seed protein profile which was composed of 34 bands (in gels pH 4.o -6.5).

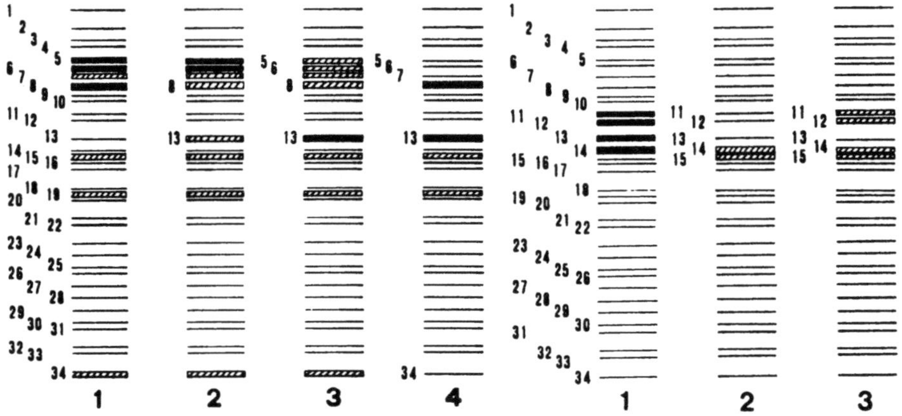

A. striatulum

A. junceum subsp. mediterraneum

No qualitative protein phenotypic differences were found and all the observed variation concerned the intensities of some bands.

The most conspicuous intensity differences involve bands numbered 5,6,7,8,11,12,13 and 14. Protein bands 5,6,7 and 8 can be characterized as the diagnostic ones for the profile of *A. striatulum*, while bands 11,12,13 and 14 identify *A. junceum* subsp. *mediterraneum* protein phenotype.

The most frequent protein profiles of the diploid are two, precisely number 1 (observed in four out of five populations) and number 2 (detected in three populations). The profile number 3, common enough, showed mutually reverse intensities in bands 5,6 and 13 comparatively with the profile 2. The pattern number 4 was detected only once (in the most abundant population).

The profile number 1 of the hexaploid species was characteristically detected in all populations. The patterns number 2 and 3 were observed only in one population and should be emphasized that in the same population frequent chromosomal structural rearrangements were also found (unpublished data).

The relatively higher variation recognized in the diploid comparatively to the hexaploid indicates that the polyploid species tends to comprise one or a few types rather uniform with respect to protein profiles, whereas diploid species represent a more heterogeneous situation as it is referred in (13). Similar results, as regards the variation of diploid and polyploid wheats, were given by Jaaska (14) who suggested that polyploids of the wheat group are neopolyploids of relatively recent origin entailing limited time possibilities to develop intraspecific genetic differentiation. Instead they developed highly balanced and stabile genomes capable of maintaining a low level of spontaneous changes.

Analogous situation must have been of the hexaploid *A. junceum* subsp. *mediterraneum* which appeared more uniform protein profile than the diploid, thus representing a more balanced genomic state. After all, it has been repeatedly pointed out that the protein profile of polyploids was found much more uniform in comparison with their diploid progenitors (for review see 1).

The observed differences in the darkness of bands suggest that the formation of certain bands in seed protein profiles are probably under control of quantitative gene systems as reported by Ladizinsky and Hymowitz (1).

The abovementioned slight variation in band intensities should be a consequence of a specific structural arrangement modified in activity by gene regulation (15).

Regulatory genes may therefore be of greater importance for

the variation of fitness or quantitative traits than structural genes (16).

Genome constitutions judged from morphological and caryological data are fully corroborated by the protein profiles, which indicate that the two species are closely related and possess a common genome, designated J_1. Thus it is suggested that *A. striatulum* and *A. junceum* subsp. *mediterraneum* have the genome formulas J_1J_1 and $J_1J_1J_2J_2J_3J_3$ respectively.

The data obtained confirm the stability of the seed protein profile and its interspecific similarity. Therefore seed protein electrophoresis can be appreciated as a powerful tool in studies concerning the origin of polyploid plants.

Concluding we think that the method is of promising value in biosystematic investigation of the genus *Agropyron*.

References

1. Ladizinsky, G., Hymowitz, T.: Theor. Appl. Genet. 54, 145-151 (1979).
2. Dunnill, P.M., Fowden, L.: Phytochem. 4, 933-944 (1965).
3. Lee, J.W., Ronalds, J.A.: Nature 213, 844-846 (1967).
4. Gray, J.R., Fairbrothers, D.E., Quinn, J.A.: Bot. Gaz. 134, 166-173 (1973).
5. Johnson, B.L., Hall, O.: Am. J. Bot. 52, 506-513 (1965).
6. Cherry, J.P., Katterman, F.R.H., Endrizzi, J.E.: Can. J. Genet. Cytol. 13, 155-158 (1971).
7. Johnson, B.L.: Am. J. Bot. 59, 952-960 (1972).
8. Edmonds, J.M., Glidewell, S.M.: Plant Syst. Evol. 127, 277-291 (1977).
9. Cauderon, Y.: Proc. Conf. Broadening Genet. Base Crops, Wageningen, 1978. Pudoc, Wageningen (1979).
10. Heneen, W.K., Runemark, H.: Hereditas 70, 155-164 (1972).
11. Dewey, D.R.: Am. J. Bot. 68, 216-225 (1981).
12. Cauderon, Y.: Proc. 2nd Int. Wheat Genetics Symp., Lund 1963, Hereditas, suppl. Vol. 2. p. 218-234 (1966).

13. Ladizinsky, G., Johnson, B.L.: Can. J. Genet. Cytol. 14, 875-888 (1972).

14. Jaaska, V.: Pl. Syst. Evol. 137, 259-273 (1981).

15. Kreft, I., Javornik, B., Milkovic, N.: Proc. 3rd Int. Barley Genet. Symp. Garching 1975. Barley Genetics III. p. 30-35 (1976).

16. Hedrick, P.W., McDonald, J.F.: Heredity 45, 83-97 (1980).

PLACENTAL ENZYME POLYMORPHISMS IN THE HUMAN POPULATION OF NORTHERN GREECE

Anastasia Kouvatsi, Costas Triantaphyllidis
Laboratory of General Biology, University of Thessaloniki, Thessaloniki, Greece

INTRODUCTION

We have initiated a program for the study of the genetic structure of the population residing in Northern Greece (20, 21). These studies were based on starch gel electrophoresis of haemolysates or serum from adults. From 1980 we started working on cord blood haemoglobins as well as on placental extracts using the same technique. Furthermore, since thin layer isoelectric focusing in polyacrylamide gels has recently been reported as the method of choice for cord blood haemoglobin screening (17) and as a much more powerful separation technique for isozymes (22), this program was undertaken in an attempt to reinvestigate: a) the incidence of the occurence of the haemoglobin variants, and b) the gene frequencies of five enzyme systems by a more sensitive method.
In this work, we want to report on the gene frequencies of 13 gene markers and to compare them with the frequencies obtained in other Greek and European populations.

MATERIAL and METHODS

Placental speciments and umbilical cord blood samples were obtained at delivery from different obstetric clinics of the cities Thessaloniki and Kastoria (Northern Greece) and kept fro-

Electrophoresis '82

zen (-30° C) until they were analyzed (within about thirteen days from collection). The placental extracts were prepared according to Beckman et al. (8). Haemolysates were prepared by repeated freezing and thawing washed red cells, and stored at -30° C for later analysis. The following systems were studied: (a) from red cells: fetal haemoglobins and (b) from placental extracts: alkaline phosphatase (Aph), acid phosphatase (Acph), glucose-6-phosphate dehydrogenase (G-6PD), phosphogluconate dehydrogenase (6PGD), lactate dehydrogenase (LDH), phosphoglucomoutase (PGM_1), esterase D (EsD), adenylate kinase (AK_1), glutamate-oxaloacetate transaminase (GOT), and glucosephosphate isomerase (GPI). Starch gel electrophoresis and staining procedures were performed according to previously published methods (8, 18). LDH, EsD and Aph were analyzed also using polyacrylamide gels [4.5% and 7.5% (W/V) acrylamide and 5% (W/v) N,N, methylene bisacrylamide] using the buffer system described by Ashton and Braden (2). Fetal haemoglobins were further studied by isoelectric focusing in polyacrylamide gels (4). For the LDH thermostability experiments, a standard amount of the placenta extract was placed in small glass tubes. The tubes were incubated in a constant temperature water bath at 50^{0} C, 55° C, 60° C, 65° C and 70° C for 20 min. The next day the samples were run in starch and polyacrylamide gels. In each gel unheated samples of the normal and variant LDH phenotypes were also included.

RESULTS and DISCUSSION

The phenotype and gene frequencies found for the Aph, PGM_1, EsD, Ak_1 and 6PGD systems are shown in Table 1. The observed numbers and those expected from the gene frequencies calculated for Aph and EsD on the assumption of Hardy-Weinberg equilibrium were in good agreement. This was not the case for the PGM_1 system. In this system, a significant deficiency of hete-

Table 1. The phenotype and gene frequencies of five gene markers in samples of placental extracts from Northern Greece

Loci	Phenotypes	Frequencies Observed	Frequencies Expected	Gene Frequencies
Aph	F_1	16	15.3	F_1 = 0.2689
	F_1S_1	71	76.1	S_1 = 0.6674
	S_1	98	94.4	I_1 = 0.0566
	S_1I_1	13	16.0	S_2 = 0.0047
	F_1I_1	11	6.5	9 = 0.0024
	Rare	3	3.7	$x^2 = 4.4$
	Total	212	212	$P > 0.100$
PGM_1	1	138	127.4	PGM^1 = 0.7698
	1 - 2	54	75.4	PGM^2 = 0.2279
	2	22	12.1	PGM^6 = 0.0023
	1 - 6	1	0.1	$x^2 = 16.6$
	Total	215	215	$P < 0.100$
EsD	1	143	140.4	EsD^1 = 0.8237
	1 - 2	55	60.2	EsD^2 = 0.1763
	2	9	6.4	$x^2 = 1.50$
	Total	207	207	$P > 0.05$
AK_1	1	205		AK^1 = 0.9744
	1 - 2	9		AK^2 = 0.0256
	2	1		
	Total	215		
6PGD	A	192		PGD^A = 0.9638
	AC	15		PGD^C = 0.0362
	Total	207		

rozygotes was observed. A similar excess of PGM_1 homozygotes has also been reported in placentae from diverse populations (1, 12). Ananthakrishnan *et al.* (1) have discussed the possible reasons for this significant deviation from the Hardy - Weinberg equilibrium. None of these reasons was sufficient to explain the deviation. Perhaps the study of this system with isoelectric focusing, now in progress in our laboratory, will give us an answer as to whether this deviation is actually real or artifactual.

We would also like to report the following: As far as the data on the AK_1, 6PGD and PGM_1 polymorphisms in adult samples from Macedonia (20, 24) are concerned, we found almost exactly the same frequencies for the AK_1^1 (X^2=0.8, p>0.10) and PGD^A (X^2=0.9, p>0.10) alleles, but remarkable differences in the PGM_1^1 and PGM_1^2 alleles. The significant differences (X^2=14.9,p<0.05) between our PGM_1 data and those found by Stamatoyannopoulos _et al_. (24) are probably due to the small size of the two samples. Thus, it is difficult at present to draw any conclusions.

The Aph gene frequency distribution in this Northern Greek sample showed no significant difference (X^2=1.2, p>0.10) when compared with that of a Southern Greek sample (6). Comparison between Greek and Italian (Rome) populations (9) showed no significant difference conserning the Aph gene. Therefore, the Aph allele frequencies in Greece are similar to those of other European populations (3, 6, 9), with the exception of the I_1 allele. It is worthwhile to mention that four European populations (Brithish, Greek, Italian and German) are characterized by an excess of Aph homozygotes (3, 6, 9, 14 and this investigation), while the Canadian population has no such excess (13). The reason for such a difference is not clear at present.

The frequencies of the alleles of EsD in our sample showed considerably higher EsD^2 and lower EsD^1 values as compared with populations from Northern and Central Europe (15). The EsD gene frequency distribution in this Greek sample, on the other hand, showed no significant difference (X^2=1.12, p>0.05) when compared with an Italian sample of newborns (10). The results show that the EsD^2 allele seems to be more frequent in Greek and Italian than in Central or Northern European populations ; however, this finding should be confirmend by further research.

Besides the enzyme systems given in Table 1, the following were also analysed:

GPI: The total of 215 placental samples showed the phenotype

GPI 1 (gene frequency GPI^1=1.00); GPI variants were not dete - cted in the Greek population (23 and this investigation).

GOT: 215 placental samples could be typed for GOT.One of them showed the phenotype $GOTm^1/GOTm^2$. Thus, the frequency of the $GOTm^1$ allele in our sample is 0.998. On the other hand,two samples were heterozygotes ($GOTs^{1-2}$) for the GOTs gene. The gene frequency in the sample is therefore: $GOTs^2$=0.005. Thus, the Greeks do not differ in this enzymic system from other European populations hitherto investigated (18).

Acph: The investigation of 215 placental samples did not show any Acph variants. This means that all of them had the common Acph phenotype (5). The absence of Acph variants in our Greek sample is in accordance with observations on other European populations in which Acph variants are also absent or very rare (7, 18).

G6PD: 110 placental samples could be typed for G6PD. One of them had the phenotype Gd^- and another one the phenotype Gd^{106}. Thus, the frequency of the Gd^B allele is 0.98.

Hb: Among 215 cord blood samples, two (about 1%) were found to have Hb Bart's. This percentage is about the same with that reported among 500 infants from Southern Greece (16). The incidence of about 10% of Hb Bart's in the cord blood of the two infants is an indication that they are probably a-thalassaemia heterozygotes (aa^o/aa). Follow-up examinations on the two babies perhaps will varify if they are carriers of the a-thalassaemia-1 or the a-thalassaemia-2 type.

No γ-chain variants were seen in the sample of 215 cord bloods so far tested. This observation is in accord with the finding that the frequency of γ-chain variants appears to be very low in a variety of ethnic groups (25).

LDH: In our sample of 215 unrelated subjects an unusual electrophoretic pattern was observed (Fig. 1). According to a generally accepted rule, a new LDH variant is called by the name of the city where it is discovered. We have therefore, tentatively desingated this variant found in Kastoria as LDH Kastoria-1. As shown in Fig. 1 the isozymes LDH-2, LDH-3, and LDH-4

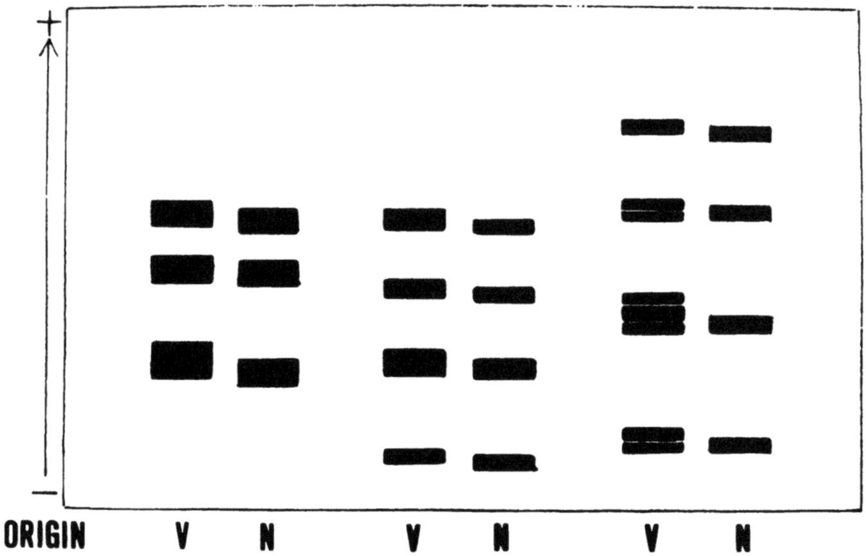

Figure 1.Diagram illustrating the LDH pattern in extracts from placentae. (N) Normal and (V) variant phenotype.The separation was obtained by 10% starch gel electrophoresis (Left) and 7.5% (Middle) and 4.5% (Right) polyacrylamide gel electrophoresis.

of this variant are resolved into two,three and two components, respectively. Furthermore, LDH-1, LDH-2, LDH-3 and LDH-4 isoenzymes in the variant phenotype exhibit an anodal displacement in comparison with the LDH isoenzyme pattern from normal individuals. In the LDH-2 and LDH-4 positions the two components have a roughly equal intensity. The slower one has the same mobility as the normal LDH-2 and LDH-4 components, respectively. In the LDH-3 position the slower component has the same mobility as normal LDH-3, and the middle component shows significantly greater intensity than either of the others.

The thermal stability experiments are summarized in Fig.2. The results suggest that: a) LDH-2, LDH-3 and LDH-4 isozymes from the KAS-1 variant were more heat sensitive than those of the normal LDH phenotype.b) In the variant LDH phenotype,the LDH-2 components are more susceptible to denaturation than the LDH-3 components. The opposite situation is true for the normal LDH phenotype and c) There is no LDH activity after incubation of placental extracts at 70° C for 20 min.

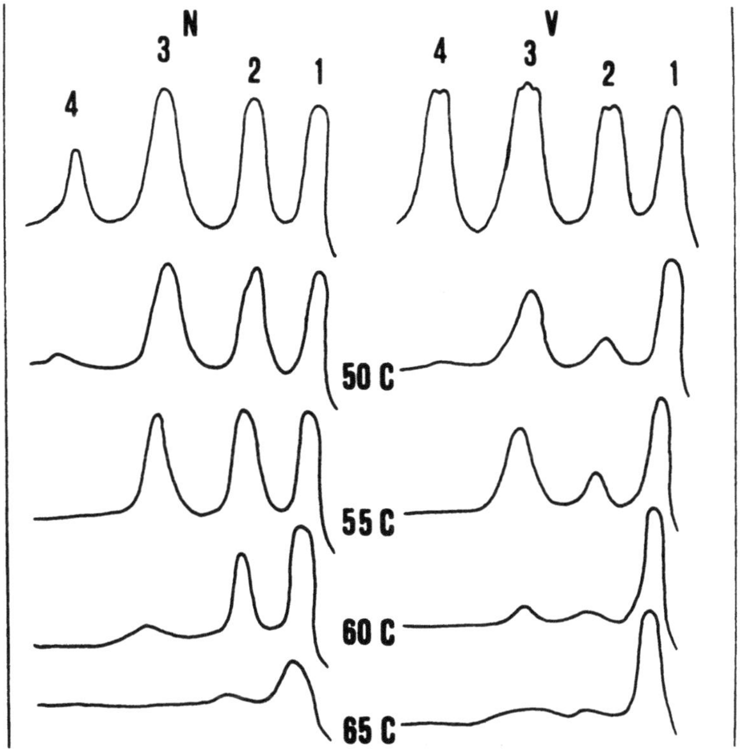

Figure 2. Densitometric evaluation of the LDH 4.5% polyacrylamide gel electrophoresis of the normal (N) and the Kastoria- 1 variant (V) LDH phenotypes: *Upper line*: Densitometric diagram of an unheated placental extract of the N and V phenotypes. *Other lines*: Densitometric diagram of the N and V phenotypes after incubation of the extracts at 50° C, 55° C, 60° C and 65° C for 20 min.

LDH-Kastoria-1 seems to be similar to the LDH Calcutta-1 variant. This variant has been thought to be a faster than normal A variant albeit "an unexplecable anodal displacement" in the LDH-1 zone was consistently observed (11). Herbschleb-Voogt and Khan (19), on the other hand, demonstrate that the CAL-1 variant is not a variant of the LDH-A but of the LDH-B locus. Thus, direct comparison is needed between the KAS-1 and CAL-1 variants. Further experimentation with appropriate approaches is also needed to resolve if the Kastoria-1 variant is due to a mutation at the A or B locus of the human LDH system.

Acknowledgements: We are grateful to Dr C.D. Kastritsis for his valuable discussions in the organization of the project and his continuing encouragement. Thanks are also due to Mrs E. Kara - mitrou-Kouskounelou for typing the manuscript.

REFERENCES

1. Ananthakrishnan, R., Beck, W., Walter. H.:Humangenetic 19, 247-252 (1973).
2. Ashton, G.C., Braden, A.W.H.: Aust. J. Biol. Sci. 14, 248-254 (1961).
3. Beck, W., Ananthakrishnan, R.: Humangenetic 25,127-132 (1974).
4. Basset, P., Beuzard, Y., Garel, M.C., Rosa, J.: Blood 51, 971-982 (1978).
5. Beckman, L., Beckman, G.: Biochem. Genet. 1, 145-153 (1967).
6. Beckman, L., Beckman, G., Christodoulou, C., Ifekwunigwe , A.: Acta genet. Stat. med. 17, 406-412 (1967).
7. Beckman, G., Beckman, L., Tärnvik, A.: Human Heredity 20, 81-85 (1970).
8. Beckman, L., Björling, G., Christodoulou, G.: Acta genet . Basel 16, 59-73 (1966).
9. Bottini, E., Lucarelli, P., Palmarino, R., Spennati, G.F., Reynaud, G.: Hum. Genetic 11, 62-65 (1970).
10. Corbo, R.M., Spennati, G.F., Scasshi, R., Palmarino, R. , Penna, M.R.D., Lucarelli, P.: Human Heredity 31,167-171 (1981)
11. Das, S.R., Mukherjee, B.N., Das, S.K., Ananthakrishnan, R., Blake, N.M., Kirk, R.L.: Hum. Genet. 9, 107-109 (1970).
12. Donald, L.J.: Hum. Hered. 27, 280-284 (1977).
13. Donald, L.J.: Human Hered. 28, 19-25 (1978).
14. Donald, L.J., Robson, E.B.: Ann. Hum. Genet. 38, 7-18 (1974).
15. Ebeli-Strùijk,A.C., Wurzer-Figurelli, E.M.,Ajmar, F., Khan, P.M.: Hum. Genet. 34, 299-306 (1976).
16. Fessas, P.: Proc. 7th Congr Europ. Soc. Haemat. London 2, 1043 (1960).
17. Galacteros, F., Kleman, K., Caburi-Martin, J., Beuzard, Y., Rosa, J., Lubin, B.: Blood 56, 1068-1071 (1980).
18. Harris, H., Hopkinson, D.A.:Handbook of enzyme electrophoresis in human genetics, North-Holland, Amsterdam 1976.
19. Herbschleb-Voogt, E., Khan, P.M.: Hum.Genet. 57, 290-295 (1981).
20. Kaplanoglou,L.,Triantàphyllidis C.: Human Hered. (in press).

21. Parisi, H.A., Lanara, E.E., Triantaphyllidis, C.D.: Human Heredity 30, 181-184 (1980).

22. Scherz, R., Pflugshaupt, R., Bütler, R.: Human Heredity 31, 187-190 (1981).

23. Schneider, P., Ananthakrishnan, R., Walter, H.,Xirotiris, N., Abele, R.: Humangenetik 27, 217-222 (1975).

24. Stamatoyannopoulos, G., Thomakos, A., Giblett, E.R.: Humangenetik 27, 23-30 (1975).

25. Schroeder, W.A.: Hemoglobin 1, 513-525 (1977).

ELASTASE INHIBITION BY SUBTYPES OF THE M GENETIC VARIANT OF α_1-ANTITRYPSIN

Kathleen P. Brooks and Richard M. Iammarino

Departments of Biochemistry and Pathology, West Virginia University Medical School, Morgantown, WV 26506

Alpha$_1$-antitrypsin (α_1AT) is the major serine protease inhibitor in human serum (1,2) and in the human lower respiratory tract (3). The protease inhibitor (Pi) system contains more than 40 alleles, which are named by letters according to their relative positions in isoelectric focusing (4). The M-allele is the most common, with an allele frequency of 0.86 to 0.99 depending on the population studied (5). PiS and PiZ are the next most-frequent alleles. The PiZZ individual has a deficiency of serum α_1AT and runs a high risk of developing emphysema, presumably because the leukocyte elastases in the lung are unopposed in their action (6). It has been hypothesized that the PiMZ phenotype, associated with an intermediate protease inhibitor deficiency, is also a risk factor for lung disease (7) but this hypothesis has not been accepted without reservation; (8) moreover, it is well known that most clinical cases of obstructive lung disease are associated with individuals showing a PiM phenotype. It is now generally accepted that the PiM phenotype contains at least three alleles and six phenotypes. This has prompted a search to determine the PiM subtypes in a group of adults with obstructive lung disease (9).

The association between Pi subtypes and diseases such as obstructive lung disease have to this point concluded that serum concentration of α_1AT is a major determinant factor in assessing individual risk. Put in other terms, no functional differences have been associated with major Pi variants that could not be explained by differences in serum concentration _per se_.

We report a study which measures the elastase-inhibiting properties of serum showing functional differences between major homozygous subgroups

Electrophoresis '82

of M (M1, M2 and M3). These differences are more marked at high dilutions (low concentration) of serum.

Material and Methods

Specimens used in these studies were obtained from volunteer blood donors under a protocol approved by the Committee for the Protection of Human Subjects, and consisted of serum or fresh-frozen plasma containing the anticoagulant CPDK (citrate phosphate dextrose adenine solution). They were stored at -20° or -80°C until used. α_1AT phenotypes were determined by isoelectric focusing of fresh or frozen-and-thawed serum using polyacrylamide gel plates at pH 4-5 (Ampholine PAG-plates, LKB), a prerun with power supply set at 2000 V, 50 mA, 20 W for 1 hr, followed by a separating run at 2000 V, 13.5 mA, 30 W for 1 h, and a final focusing at 2000 V, 60 W for 1 min. Reference sera (M1M2) was used at frequent intervals along each gel. Gels were fixed in 12.5% trichloroacetic acid and stained for 5 min at 55°C in 0.2% Coomassie Blue R250. The homozygous M1, M2 and M3 sera were well resolved under these conditions.

α_1AT concentration was determined by rate immunonephelometry (Beckman Human α_1AT - Immunochemistry System) according to the manufacturer's instructions.

Inhibition of porcine pancreatic elastase (Sigma Type III) by serum or fresh-frozen plasma was determined by incubating a fixed amount of elastase (usually 15ug) with 15ul aliquots of varying dilutions of serum or plasma in a total volume of 300ul of 0.05M sodium phosphate buffer, pH 6.5 for 10 min, and assaying 200ul of the incubation mixture by the method of Visser and Blout (10) using the synthetic substrate N-tert-butoxycarbonyl-L-alanine p-nitrophenyl ester (Sigma). Results were expressed as per cent inhibition (%I) in which %I = 100 x (Vo-Vi)/Vo where V_o is the initial rate of hydrolysis of the synthetic substrate in the presence of the uninhibited control and V_i is the rate in the presence of serum or plasma. Data were plotted as %I vs concentration of α_1AT except in cases where there was insufficient serum to determine inhibitor concentration by rate immunonephelometry, in which cases, data were plotted as %I vs fractional concentration of α_1AT with undiluted serum taken as 1.0.

Results

The distribution of α_1AT phenotypes in 189 blood donors is shown in Table 1. Allele frequencies (Table 2) were similar to those reported for other populations. From this population we were able to identify two M3 and seven M2 specimens as well as 83 M1 homozygotes. From these samples and others sent to us for clinical evaluation, the concentrations of the various phenotypes of α_1AT by the method we use are presented in Table 3.

TABLE 1. Distribution of Pi phenotypes in a sample of 189 blood donors in northern West Virginia.

Pi Phenotypes	No.
M1	83
M2	7
M3	2
M1M2	46
M1M3	21
M2M3	9
M1S	8
M2S	3
M3S	1
M1Z	2
M2Z	2
M3Z	0
SZ	3
FM1	2
	189

TABLE 2. Alpha$_1$-antitrypsin allele frequencies of West Virginia University Hospital blood donors. Data from a study by Klasen (11) is included for comparison.

Pi Allele	WVU Blood Donors	Klasen Study
M1	.648	.703
M2	.196	.170
M3	.093	.080
S	.040	.026
Z	.018	.013
F	.005	.007

TABLE 3. Alpha$_1$-antitrypsin concentration and phenotype of selected blood donors. All available sera homozygous for M-subtype, together with a random selection of sera of other available phenotypes, were quantitated by rate immunonephelometry.

Phenotype	α_1AT (mg/ml) mean ± S.D.	No.
M1	1.85 ± .32	11
M2	1.67 ± .18	5
M3	1.74 ± .16	3
M1M2	1.91 ± .27	3
M1M3	1.80 ± .46	4
M1S	1.42	1
M2Z	1.73	1
M3Z	0.74 ± .26	2
SZ	0.64	1

Variation in elastase inhibitory capacity with dilution of serum over a wide range is shown in Figure 1 for the various homozygous subytpes of M studied. We noted with the synthetic substrate used in this assay only 85% of total elastase activity was inhibited. This maximum level of inhibition was similar for all subtypes of M studied under the assay conditions we used.

Figure 1. Elastase inhibitory activity of PiMM subtypes as a function of specimen dilution. Diluted serum or plasma was incubated with elastase and aliquots assayed as described in Methods.

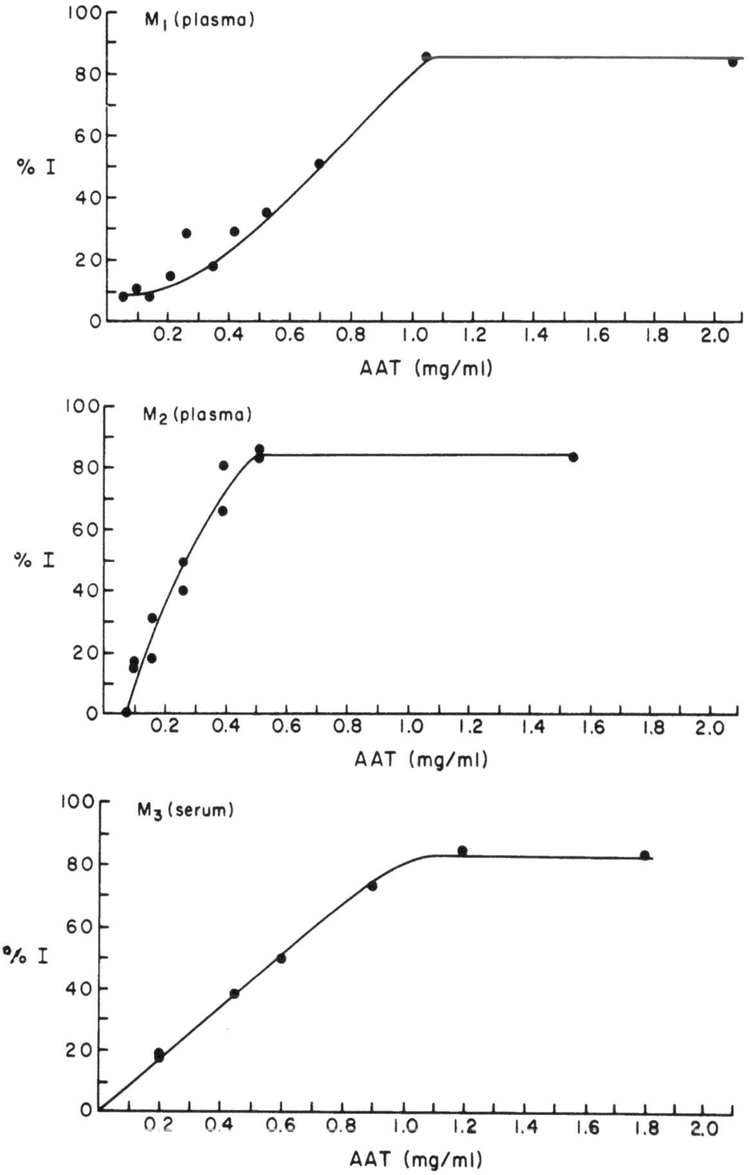

(A)
M1 plasma

(B)
M2 plasma

(C)
M3 serum

The differences between the three major homozygous subtypes show that M1 has a roughly sigmoidal curve of inhibition with measurable inhibitory capacity noted at the lowest levels of concentration used, which contained 0.7 ug α_1AT incubated with 15 ug elastase. With the PiM2 subtype the elastase inhibiting property dropped off radically as lower serum concentrations were studied. No elastase inhibiting properties could be identified at the lowest concentration used. In the homozygous PiM3 subtype the inhibitory curve was essentially linear but similar to M2 in that at the lowest concentration no residual elastase inhibiting property was noted. Patterns similar to that of Figure 1A were obtained for seven different serum samples. The M2 pattern was the same for 2 different serum samples as well as the plasma sample shown in Figure 1B. The M3 pattern seen in serum (Figure 1C) was duplicated in one additional plasma sample as well.

Discussion and Conclusions

We started this study to determine the specific elastase inhibitory capacity of the three major M subtypes of α_1AT. It was not possible to develop a simple normalizing function for comparing the elastase inhibitory capacity of these three genetic and electrophoretic variants because the rate of decline of elastase inhibition with dilution was different for each, although maximum inhibition of 15 ug of elastase was the same for all 3 PiM subtypes. At the lowered concentrations of α_1AT that might occur in dilute body fluids such as alveolar fluid or inflammatory exudates (5 ug/ml of incubation mixture), PiM1 showed a significantly increased capacity for elastase inhibition when contrasted to M2. M3 was intermediate between M2 and M3 in its ability to inhibit elastase at this concentration.

We see the need for caution before concluding that these _in vitro_ differences in inhibition of porcine pancreatic elastase have biologic significance, since individual differences in leucocyte elastase, as suggested by Janoff (12) and supported by Kramps _et al_. (13), add another variable to this protease-protease inhibitor system _in vivo_. Nevertheless, we suggest that the general approach of examining PiM subtypes individually and varying the α_1AT concentration may prove to be valuable for studies of α_1AT function.

References

1. Eriksson, S.: Acta Med. Scand. 177, Suppl. 432, 5-85 (1965).
2. Morse, J.O.: N. Engl. J. Med. 299, 1099-1105 (1978).
3. Gadek, J.E., Fells, G., Zimmerman, R., and Crystal, R.: Am. Rev. Respir. Dis. 121, (Part 2), 341 (1980).
4. Cox, D.W., Johnson, A.M. and Fagerhol, M.K.: Hum. Gen. 53, 429-433 (1980).
5. Fagerhol, M.K. and Cox, D.W.: Adv. Hum. Gen. 11, 1-62 (1981).
6. Larsson, C.: Acta. Med. Scand. 240, 345-351 (1978).
7. Cox, D.W., Hoeppner, V.H. and Levinson, H.: Am. Rev. Respir. Dis. 113, 601-606 (1976).
8. Madison, R., Mittman, C., Afifi, A., and Zelman, R.: Am. Rev. Respir. Dis. 124, 149-153 (1981).
9. Bencze, K., Sabatke, L., and Fruhmann, G.: Chest 77, 761-763 (1980).
10. Visser, L., and Blout, E.R.: Biochim. Biophys. Acta. 268. 257-260 (1972).
11. Klasen, E.C.: Hum. Hered. 31, 1980-182 (1981).
12. Janoff, A.: in Pulmonary Emphysema and Proteolysis (Mittman, C., editor) Academic Press, New York, 205-224 (1972).
13. Kramps, J.A., Bakker, W. and Dijkman, J.H.: Am. Rev. Resp. Dis. 121, 253-261 (1980).

ISOELECTRIC POINTS AND TITRATION CURVES OF THE THREE HAPTOGLOBIN PEPTIDE CHAINS α, α_2 AND β.

K. Suzuki*, J. Constans**, K. Shibata*, M. Viau** and H. Matsumoto*

* Department of Legal Medicine, Osaka medical school, Takatsuki city, Osaka, Japan 569.
** Centre d'Hémotypologie - CNRS - CHU Purpan, 31052 Toulouse Cedex, France

Introduction

Haptoglobin, a serum α_2 glycoprotein, is one of the first protein investigated for its polymorphism based on the existence of three electrophoretic phenotypes Hp 1-1, Hp 2-1 and Hp 2-2 (1). Biochemical studies then developed, demonstrated the existence of three different polypeptide chains α, α_2 and β (2). The α chain is produced by the Hp^1 allele. Two isoforms of the α chain, αS and αF were described corresponding to the existence of two subtypes Hp 1S and Hp 1F (3). The α_2 chain was the result of a gene duplication associated to an unequal crossing-over between two α genes (4). The postulated existence of four α_2 mutants α_2SS, α_2FF, α_2FS and α_2SF was then confirmed by the detection of the α_2SS and α_2FF peptides in population samples, but up to now it was not possible to separate and to identify the two peptides α_2FS and α_2SF. Other rare mutants of the α and α_2 chains were also identified in some families only. Our interest was the study of the distribution of the Hp subtypes among human populations, but the traditional techniques used were, despite some improvements (5, 6) poorly resolutive, long time consuming and not adapted to the study of samples on large series. Recently we developed the isoelectric focusing method to the separation of the different Hp peptides (7). In this paper we discuss the problem related to the focusing conditions of the peptides. In addition, we present the titration curves of the three Hp chains. On a methodological point of view, we discussed the interpretations of these titration curves and the isoelectric points of the peptides by comparison to their structures.

Material and Methods

The detailed description of the techniques used are under publication (7,8). Briefly we adapted the partial Hp purification by DEAE cellulose chromatography to a micro method. (9,10). The Hp recovered was controlled after PAGE and immuno electrophoresis. Some minor contaminants such as albumin were detected. The following sera were used as standards : Hp 1F-1S ; Hp 2SS-1F ; Hp 2SS-1S ; Hp 2FS-2FS ; Hp 2FF-1F ; and Hp 2FF-1S.

Isoelectric focusing was performed as described (7) on a 1 mm thick gel polyacrylamide (5 %). The original point was the preparation of an ampholyte mixture made of 0,8 % pH range 5-7 ; 0,8 % pH range 3,5-10 and 0,4 % pH range 6-8. Samples were applied using a rubber punched by 50 holes, each of them receiving 10 µl of the Hp peptide solution. After migrations (2 h under 10 W., at 10 °C) the gel was stained by a H_2SO_4-Coomassie-Brillant Blue G 250 solution (11). The titration curves of the peptides were developed according to the methods described by Rosengren (12) and Righetti (13).

Electrophoresis '82

Results

The first difficulty encountered in the Hp subtyping by IEF was the location of the Hp peptides. In a first step we used the double one dimension electrophoresis described by Atland (14) (first dimension PAGE urea, second one IEF) (Fig. 1-A). For further steps only IEF was performed and from the pattern of migration the identification of the peptides was directly obtained (Fig. 1-B). After the migrations, the pH curve was plotted with the pH values obtained by direct measurements on the surface of the gel with an Ingold surface electrode. The values obtained were controled by comparison with the mobilities of isoelectric point reference molecules (Pharmacia) (Fig. 1-C). The titration curves of the α, α_2 and of the desialylated form of the β chain are presented (Fig. 3 and 4) and discussed.

Discussion

The Hp α and α_2 chains : After IEF (Fig. 1 A and B) the α and α_2 chains were detected as a single band pattern in absence of urea in the gel. The αS peptide is more anodic than the α F peptide which is in agreement with their mobilities after PAGE urea pH 3.7. The distance of migration between the two peptides is large enough to avoid any mistyping of these chains. This result constitutes an additional example of the resolvent power of the IEF procedure developed. On the same patterns, the α_2FS peptide was identified with an intermediate mobility between the αS and αF peptide but very close to the αF chain.

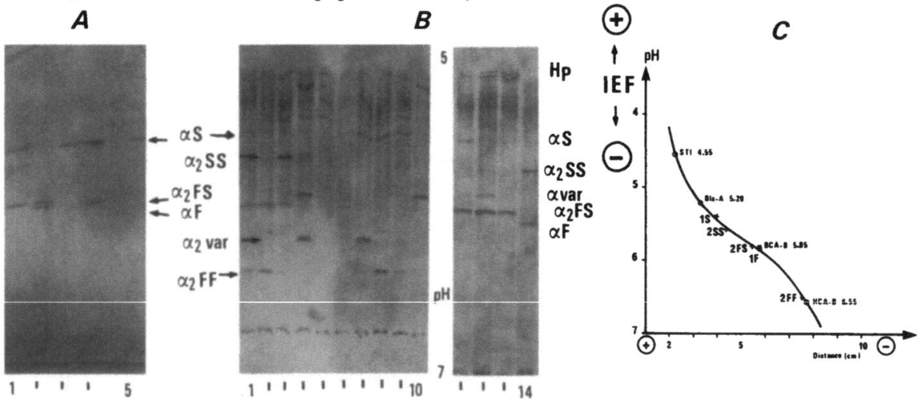

Figure 1 : IEF pattern of the Hp α and α_2 peptides :

A - Two one dimension electrophoresis (PAGE-IEF) patterns. IEF conditions were 5 % polyacrylamide gel and ampholytes pH range 3.5-10:1 ml; 5-7: 0.5 ml ; and 6-8 : 0.5 ml. Samples 1,5 : Hp 2FS-1S ; sample 2 : Hp 2FS-1F ; sample 3 : Hp 1S ; sample 4 : Hp 1F-1S.

B - IEF pattern. Samples 1,2 : Hp 2FF-1F, sample 4 : Hp 2FS-2 var., samples 3,14 : Hp 2SS-1F ; sample 7 : Hp 2 var.-1S ; samples 8,9 : Hp 2FF-1S ; sample 10 : Hp 2FS-2FS ; sample 11 : Hp 2FS-1S.

C - pH gradient curve

On the same patterns we were able to identify the α_2SS and α_2FF peptides which are well separated from the α_2FS chain. This result shows that the distinction of the three

α_2 mutants cannot be now so difficult and hazardous as before. The consequences of the new method proposed are :

- an important difference of frequencies of the α_2 alleles after Hp subtyping in a sample already studied (15)

	Hp 1S	Hp 1F	Hp 1 var.	Hp 2SS	Hp 2FS	Hp 2FF	Hp 2var.
Blood donor sample studied by PAGE urea gel pH 3.7	0.175	0.200	-	-	0.625	-	-
This study	0.139	0.245	0.005	0.045	0.547	0.012	0.007

- the detection of two new α and α_2 mutants not detectable after PAGE as we verified. The pI of the Hp α mutant is 5.65 and 6.20 for the α_2 mutant.

The pI variations and the peptide structures : The IEF method is not only a convenient method to obtain the separation of the Hp peptides in good conditions but ought to it we were able to determine the pI of these peptides (Fig. 1C). The values obtained are :
αS : 5.39 αF : 5.84 α_2 FS : 5.77 α_2SS : 5.55 and α_2FF : 6.52
Between the peptides the pI variations calculated are :
pI αF - αS : 0.45 pI α_2FS - α_2SS : 0.27 pI α_2FF - α_2FS : 0.75 and pI α_2FF - α_2SS : 1

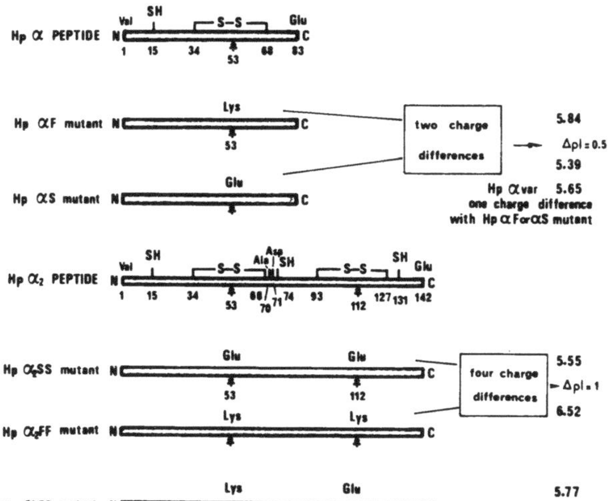

Figure 2 : Structure and isoelectric points of the Hp α and α_2 peptides.

From these data we noticed that the two charge differences between the αF and αS peptides were associated with a Δ pI equivalent to almost 0.5 pH unit and in a similar way a Δ pI equal to 1 unit pH between the α_2FF and α_2SS chains (four charge differences). In other words the lysin (αF,53) substituted by a glutamic acid (αS) was followed by aΔpI of 0.45 and for amino acid substitutions in position 53 and 112 by lysin or glutamic acids in the α_2 SS and FF peptides, theΔpI obtained is of 1 pH unit (Fig. 2). According to these results, it was possible to consider that the pI between the Hp peptides was correlated to their difference of structure. In these conditions, the expected pI for the α_2FS chain would differ of 0.5 pH unit from the pI of the α_2SS or α_2FF peptides. The observed value for α_2FS is 5.77 which is 0.23 pH unit below the expected value (6.05). This difference might be related to the structure of the α_2 FS chain. (8)

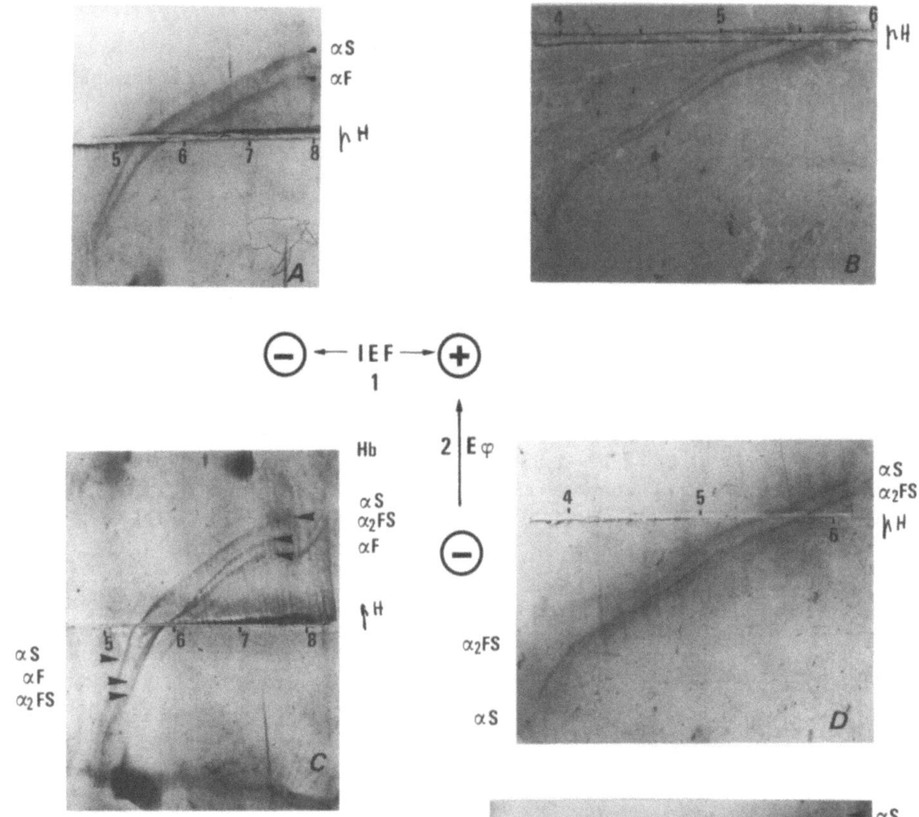

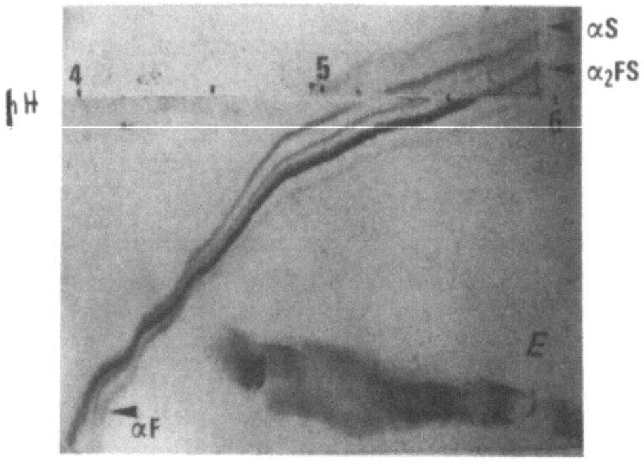

Figure 3 : Titration curves of the Hp α and α_2 peptides chains.
The conditions of the electrophoresis were in the first dimension, 2 h at 10 W constant ; in the second dimension, 40 mn at 800 V constant. In both dimensions the electrolytes were 1M H_3PO_4 at the anode and 1M NaOH at the cathode. The gel was cooled at 4°C with a thermostat. The amount of protein loaded in all cases was 100 µg in 100 µl. The second electrophoresis was started as soon as the sample was applied in the trench. The gel contained 5 % T and 3 % CBis, when 7 and 9 % T gels were prepared the 3 % C Bis was not modified. Two ampholyte mixtures were used : mixture 1 made of ampholytes pH range 3.5-10 : 1 ml ; 4-6 : 0.5 ml and 5-7 : 0.5 ml. Mixture 2 made of ampholytes pH range 3.5-10 : 0.5 ml ; 2.5-4 : 0.5 ml ; 3.5-5 : 0.3 ml ; 4-6 : 0.3 ml and 5-7 : 0.3 ml.

A - Curves of the Hp 1F and 1S peptides. 5 % T. Ampholyte mixture 1
B - Curves of the Hp 1F and 1S peptides. 5 % T. Ampholyte mixture 2
C - Curves of the Hp 2FS, 1S and 1F peptides. 5 % T. Ampholyte mixture 1.
D - Curves of the Hp 2FS and 1 S peptides. 7 % T. Ampholyte mixture 2
E - Curves of the Hp 2FS, 1S and 1F peptides. Similar electrophoretic conditions as in D.

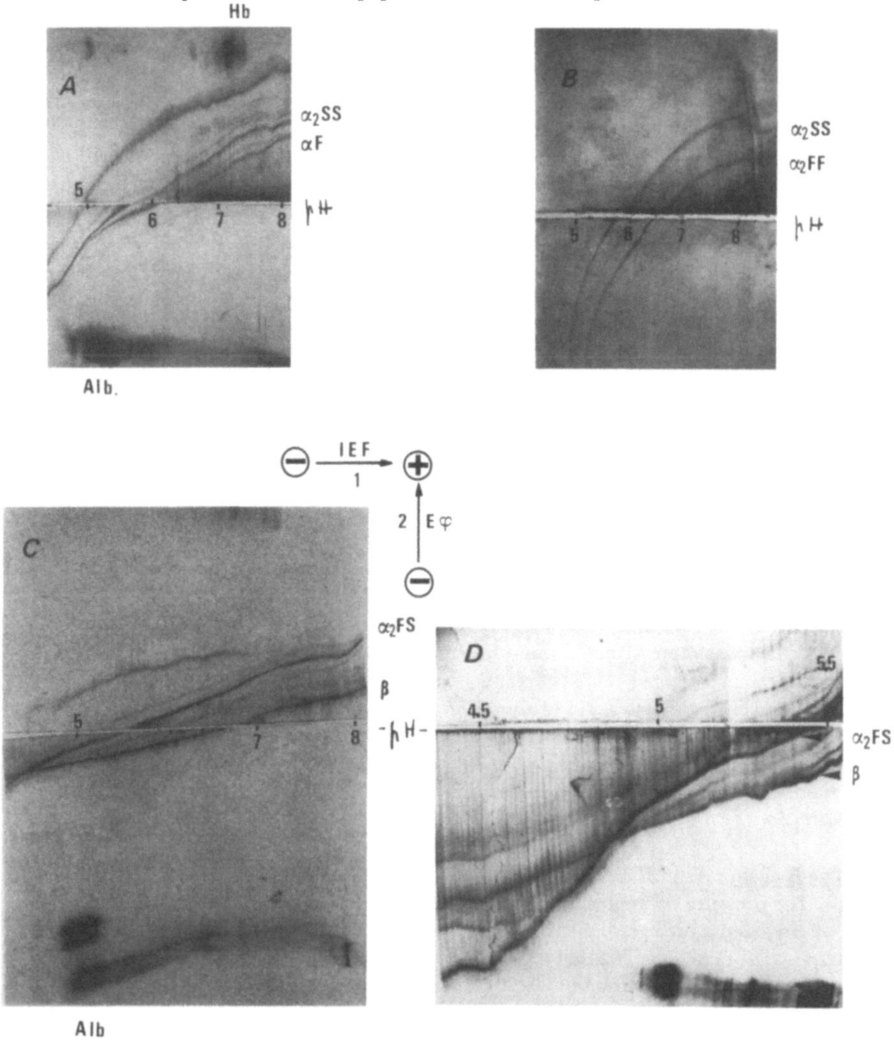

Figure 4 : Titration curves of the Hp α_2 mutants and of the desialylated form of the β chain :
A - Curves of the Hp α_2SS and Hp αF peptides. 5 % T. Ampholyte mixture 1
B - Curves of the Hp α_2SS and Hp α_2FF peptides. 5 % T. Ampholyte mixture 1
C - Curve of the desialylated β chain. 6M urea IEF. 9 % T. Ampholyte mixture 2 %:pH 3.5-10 1.5 ml and 5-7 0.5 ml
D - Curve of the desialylated β chain. 6M urea IEF. 5 % T Ampholyte mixture 2

Titration curves : After the description by Rosengren, the method received several developements and applications, such as the titration curves of the haemoglobin mutants (13), the enzyme-substrat reaction (16), the protein-lectin affinities or the ligand-carrier protein dissociations (17). If the technical difficulties can be solved by some improvements, the most delicate question is the correct intepretation of these curves due to the influence of several factors which may interferre with the establishment of these curves. The titration curves of the Hp peptides are now proposed as an interesting application of this method and they would give us the opportunity to discuss their interest and limits. With the Hp 1F - 1S genotype (Fig. 3 A and B) the curves obtained are characterized by a fast variation of the peptide mobility in the pH zone below 5 followed by a reduced mobility when the pH is higher than 6. In this second pH zone (from 6 to 8) the curves are almost parallel. Their crossing point with the trench confirms the zero mobility of the peptides and their pI. For pH below the two curves converged but they never intercrossed as we observed when narrow pH ampholyte mixture was used (Fig. 3 B). It is to notice that these curves behave in a similar way as the Hb A and C titration curves established by Righetti (13) and Krishnamoorthy. The Hb mutants are also characterized by a lysin-glutamic substitution. We performed then the titration curves of the α_2FS peptide in the presence of the α F and αS chains. Two distinct curves were obtained for α_2FS and αS peptides while the curves of the α_2FS and αF peptides were very similar. For pH 5.8, the curves were superimposed ; in the pH zone 6 to 8, the α_2FS curve is slightly more anodal than the αF one and in the pH zone 5.5 to 4 the αF curve is more anodal than the α_2FS one (Fig. 3 C). When the second ampholyte mixture is used (narrow pH range) the two curves were more easily differenciated and the presence of the two intersectionswere confirmed. A 7 % polyacrylamide gel produced a wider separation of the α_2FS and αF titration curves. The α_2FS curve was more anodal than the αS and αF curves in the acidic zone, the αS curve being in an intermidiate position between the α_2FS and αF curves as it sould be expected according to the PAGE urea pH 3.7 patterns (18). In this experiment it was possible to confirm that the αF curve intercrossed with the α_2FS curve for pH 5.80 and again for pH 4.10. This last experiment also shows that when the reticulation of the gel increased (5 to 7 %) the mobility of the α_2 chain was more affected than the mobility of the α peptide.

The titration curves of the α_2SS and of the αF peptides were observed with only one crossing point for pH 4.5. The α_2SS and α_2FF curves were observed as almost parallel and they never joined in the acialic pH zone which is a pattern very similar to the one obtained with the αS and αF chains. This observation is in agreement with the structure of the α and α_2 chains. (Fig. 4 B)(19, 20)

The β chain : After IEF, the β chain was undetectable by comparison to the PAGE urea procedure. The presence of urea was considered as indispensable to its focusing. For this purpose we prepared a urea gradient IEF gel from 0 to 8M urea (Fig. 5 A). A progressive focusing pattern was obtained constituted by eight bands (6 major and 2 minor bands) from 4 to 8M urea. After 48 h of the neuraminidase treatment the pattern was cleared out and only two bands were present. The most cathodal one was the desialylated form of the β chain (pI 6.70). The presence of the minor band (pI 6.50) was explained by the difficult removal of the last sialic acid residue present. From the successive steps of the cathodal drift of the 8 isoproteins we calculated a total of 33 sialic acids differently distributed among the 8 isoproteins (7)(21) (Fig. 5 - B).

The sialic acid residues distribution among the different isoproteins of the β chain is the following one :

Isoprotein	1	7 sialic acid residues
	2	6 sialic acid residues
	3	5 sialic acid residues
	4	5 sialic acid residues
	5	4 sialic acid residues
	6	3 sialic acid residues
	7	2 sialic acid residues
	8	1 sialic acid residue

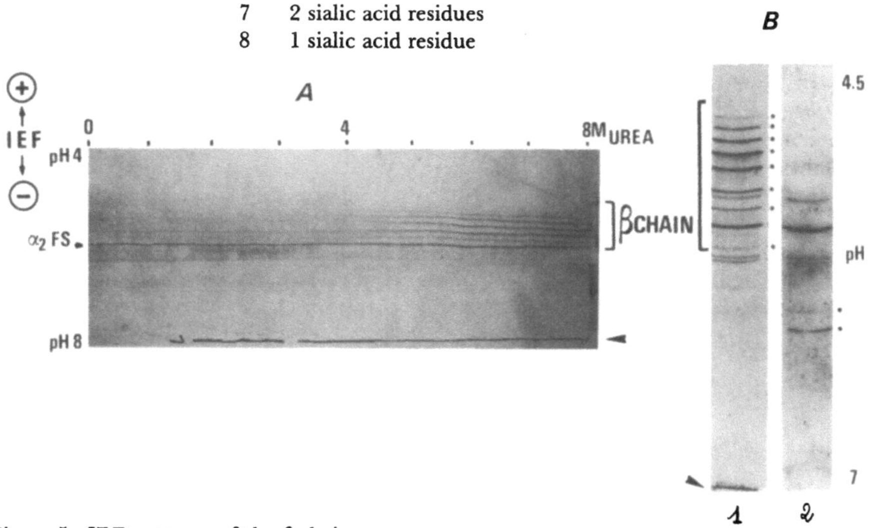

Figure 5 : IEF patterns of the β chain

A - 0 to 8M urea gradient IEF. A progressive focusing pattern of the β chain was observed for 4M urea concentration.

B - 6M urea IEF pattern of the β chain (sample 1). Eight bands were observed ranging from pH 4.90 to 5.95. After the neuraminidase treatment during 48 h (sample 2) only two bands (pI 6.50 and 6.70)

Titration curve : The sample used was belonging to the Hp 2FS - 2FS genotype. The titration curve was performed on the desialylated form of the β chain (DS-β)chain) after 48 h of the neuraminidase treatment. This curve was characterized by a regular line between pH 4.5 to 8 with a weak slope (Fig. 4 - C) corresponding to a limited charge variation towards the pH gradient. The zero mobility of the DS-β chain was confirmed (pI 6.70). The titration curve of the non desialylated β chain would be of course different (Fig. 5 B). These results show that the heterogeneity of the β chain was not an artifact of focusing. The interesting point is that the β chain only focused when the urea concentration was over 4M. This might be due to the insolubility of the peptide or to interactions with the ampholytes.

In conclusion the titration curves can be considered as a useful tool in order to collect some analytical data on the molecules but their interpretations have to be done cautiously when molecules of different size and shape were involved as in the case studied.

In addition from these experiments, we were able to select some influential parameters to be considered in the interpretation of the titration curves :

1 - the electrophoretic conditions such as power, voltage, time of the two electrophoresis may significantly modified the shape of the curves.

2 - the reticulation of the gel : this influence was commented by Righetti (13). We added that it may affect differently the proteins but also the ampholytes according to their size and shape. This effect was observed in this work when the titration curves of the α and α_2 peptides were compared.

3 - the composition and the nature of the ampholytes is also an important parameter mostly towards the interactions between the charged amino acids of the peptides or proteins and the active residues of the carrier ampholytes (electrostatic ligands, hydrogen bindings).

4 – the temperature of the gel is of course essential for the titration of the amino acids of the molecules but also in the kinetic of the reaction such as ligand dissociation, protein-protein binding etc.

5 - finally the time for the equilibrium of the sample in the trench with the pH gradient before beginning the second electrophoresis. If the sample is left in the trench during increasing period of time, the results obtained are not similar after the second electrophoresis (usual run 20 mn).

Our experiments confirm that the Hp polymorphism can be now easily performed. Original data are presented on the peptides such as their isoelectric points and titration curves.

References

1 - Smithies, O. and Walker, N.F. (1955). Nature 176. 1265-1266
2 - Connell, G.E., Dixon, G.H., and Smithies, O. (1962). Nature 193, 505-506
3 - Smithies, O., Connell, G.E. and Dixon, G.H. (1962). Am. J. Hum. Genet. 14, 14-21
4 - Smithies, O., Connell, G.E. and Dixon, G.H. (1962). Nature 196, 232-236
5 - Pastewka, J.V., Reed, R.A., Nees, A.T. and Peacock, A.C. (1973). Analyt. Biochem. 51, 152-162
6 - Thymann, M. (1977). VII Internationale Tagung der Gesellschaft für forensische Blutgruppenkinde. Sonderdruck Ed. 187-192
7 - Constans, J., Shibata, K.. Viau, M., Suzuki, K. and Matsumoto, H. (1982). Biochem. Genet. to be published.
8 - Shibata, K.. Constans, J., Viau, M. and Matsumoto, H. (1982). Hum. Genet. to be pubished.
9 - Christopherson, R.I., Jones, M.E. and Finch, L.R. (1979). Analyt. Biochem. 100, 184-187
10 - Fry, D.W., Courtland White, J. and Goldman, I.D. (1978). Analyt. Biochem. 90, 809-815
11 - Blakesly, R. and Boezi, J.A. (1977). Anal. Biochem. 82, 580-582
12 - Rosengren, A., Bjellqvist, B. and Gasparic, V. (1977). in Electrofocusing and Isotachophoresis (Radola B.J. and Graesslin, eds) 165-171, de Gruyter, Berlin
13 - Righetti, P.G., Krishnamoorthy, R., Gianazza, E. and Labie, D. (1978). J. Chromat. 166, 455-460
14 - Atland, R., Schmidt, S.R., Kaiser, G., and Knoche (1975). Hum. Genet. 28, 207-211
15 - Constans, J., Richard, P. and Viau, M. (1978). Hum. Hered. 28, 328-324
16 - Krishnamoorthy, R., Rosisio, A.B., Labie, D. and Righetti, P.G. (1978). FEBS Lett. 94, 319-323
17 - Constans, J.. Viau, M., Gouaillard, C., Bouissou, C. and Clerc, A. (1980) in Electrophoresis 79 (Radola, B.J.) 701-709, de Gruyter, Berlin
18 - Constans, J. and Viau. M. (1977). Am. J. Hum. Genet. 29, 280-286

19 - Kurosky, A., Barnett, D.R., Lee, T.H., Touchstone, B., Hay, R.E., Arnott, M.S., Bowman, B.H. and Fitch, W.M. (1980). Proc. Natl. Acad. Sci. USA, 77, 3388-3392
20 - Kurosky, A. (1980). in Protides of the Biological fluids (Peeters, H. Ed.) 99-102, Pergamon Press, New York
21 - Yang, H.I. and Przyloylska, M. (1973). Can. J. Biochem. 51, 597-605

ANTITHROMBIN III: A NEW POLYMORPHISM REVEALED BY ISOELECTROFOCUSING AND IMMUNOFIXATION

Sebastian Weidinger, Friedrich Schwarzfischer, Hartwig Cleve

Institut für Anthropologie und Humangenetik der Universität München,
D-8000 München 2, Germany

Introduction

Antithrombin III (AT III) is the most prominent physiological protease inhibitor of the coagulation system (1). It is a glycoprotein with a molecular weight of 65 000 D and is synthesized in the liver. AT III irreversibly inactivates thrombin (Factor II a), activated Factor X, as well as Factors VII a, IX a, XI a, and XII a (2,3,4). It furthermore inhibits serin proteases of other systems, such as plasmin (5) and kallikrein (6,7). The inactivation of these proteases by AT III takes place in a relatively slow reaction (8). In the presence of heparin this reaction is considerably accelerated. AT III can be regarded as a heparin cofactor.

In normal persons, AT III concentration in plasma vary between 14-20 mg/dl. The range of its concentration is normally rather narrow (9). Already small decreases below the normal level results in increased risk to develop thrombosis.

Electrophoresis '82

Materials and Methods

Gels (245 x 110 x 0.5 mm) were prepared from 2 g saccharose, 2.5 ml acrylamide solution (28 %), 2.5 ml N, N'-methylenbisacrylamide solution (2%), 0.9 ml carrier ampholytes of pH range 4-6, 5 µl N, N, N', N'-tetramethylethylendiamine (TEMED), and 10 ml aqua. dest. After degasing for 5 min. gels were polymerized by 0.5 ml ammonium persulfate solution (1 %).

Human sera (8 µl) were applied on small filter paper pieces 2 cm from the cathodal edge of the gel. Separation was carried out at maximum values of 1800 V, 25 mA, and 18 W in a LKB-Multiphor chamber for 210 min. without prefocusing. The cooling temperature was + 11 °C. At the cathode 0.1 M ß-alanin was used, at the anode a mixture of 0.025 M aspartic acid and 0.025 M glutamic acid was employed.

For identification of AT III types gels were analyzed by immunofixation with a monospecific anti-human-AT III antiserum purchased from Atlantic Antibodies (Merz + Dade). A dilution of 1 : 3 was used. Soaked and dried cellulose acetate strips were placed on the gel surface for 2 min. and washed overnight. The strips were stained with Coomassie Blue R 250.

Results and Discussion

Figure 1 shows the different antithrombin III phenotypes which are obtained after isoelectric focusing on polyacrylamide gels (PAGIF) at pH range 4-6 and subsequent immunofixation with a monospecific anti-AT III antiserum. The most common phenotype AT III 1-1 has in sera a three band pattern in the region of pH 5.0. The rare phenotypes AT III 2-1, 3-1, and 4-1 have a four or five band pattern. In plasma, a two band pattern of AT III 1-1 is observed. In comparison to sera, the anodal band is missing.

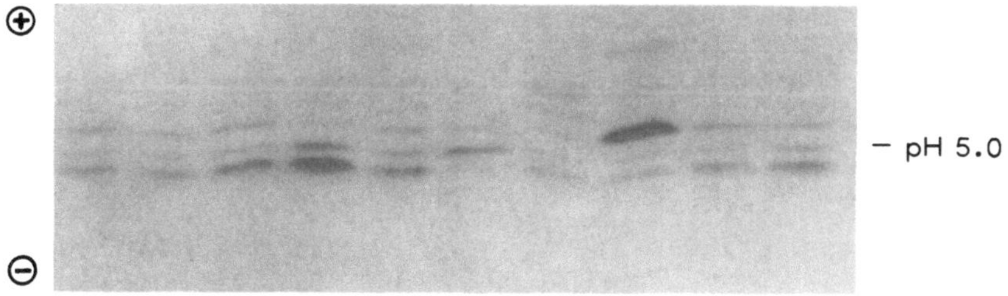

Fig. 1. Immunofixation patterns of antithrombin III in sera samples following PAGIF. Phenotypes from left to right: 1-1, 1-1, 1-1, 1-1 (plasma), 1-1, 3-1, 2-1, 4-1, 1-1 and 1-1

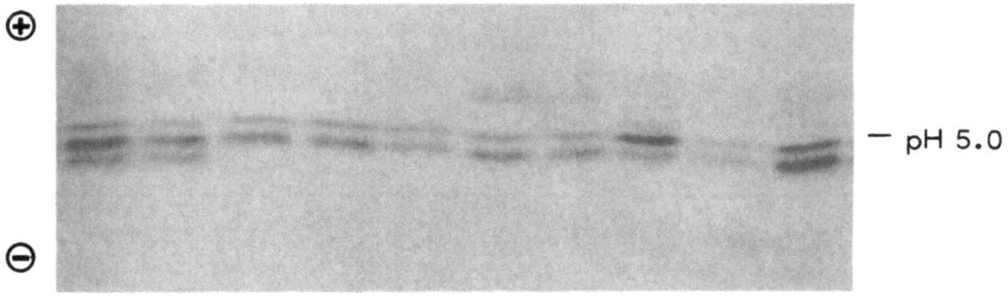

Fig. 2. Immunofixation patterns of antithrombin III in plasma from patients with coagulation disorders. Phenotypes from left to right: 5-1 (mother), 5-1 (child), 1-1, 1-1, 1-1, 1-1, 1-1, 1-1, 1-1 and 1-1

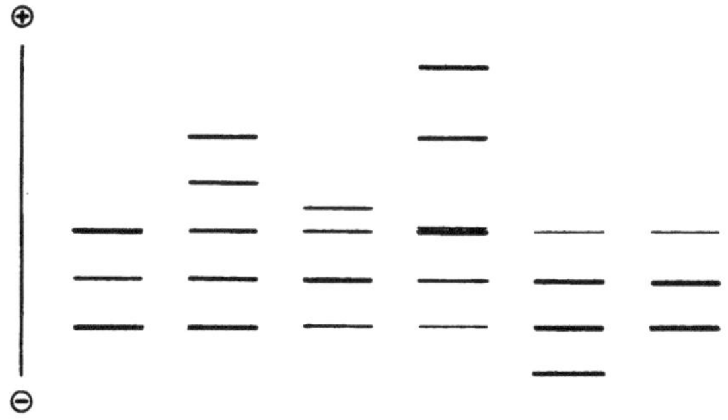

Fig. 3. Schematic diagram of antithrombin III phenotypes following PAGIF (pH range 4-6) and immunofixation. Phenotypes from left to right: 1-1, 2-1, 3-1, 4-1, 5-1 and 1-1 (plasma)

Table 1. Distribution of AT III phenotypes and alleles in a sample from Southern Germany

Phenotypes	Observed n	Observed %	Expected n	Expected %	Allele frequencies
AT III 1 - 1	205	96.70	205.06	96.76	AT III^1 = 0.9835
2 - 1	5	2.36	4.96	2.34	AT III^2 = 0.0119
3 - 1	1	0.47	0.96	0.45	AT III^3 = 0.0023
4 - 1	1	0.47	0.96	0.45	AT III^4 = 0.0023
Total	212	100.00	211.94	100.00	

$\sum \chi^2 = 0.0036$; df = 1; $P > 0.20$

Table 2. Distribution of AT III phenotypes in 41 families with a total of 42 children

Matings		Children 1 - 1	2 - 1	3 - 1	4 - 1
1 - 1 x 1 - 1	35	36 (36)	-	-	-
1 - 1 x 2 - 1	4	2 (2)	2 (2)	-	-
1 - 1 x 3 - 1	1	1 (0.5)	-	0 (0.5)	-
1 - 1 x 4 - 1	1	1 (0.5)	-	-	0 (0.5)
Total	41	40	2	0	0

Expected values are given in brackets

Figure 2 demonstrates AT III phenotypes in plasma from patients with coagulation disorders. In two patients, AT III deficiencies were present with decreased protein concentrations as well as functional defects. In two other patients, a mother-child pair, a functional defect was associated with normal protein concentrations as determined immunochemically. This type was provisionally classified as AT III 5-1. In plasma a third band was noted in a cathodal position.

Figure 3 shows schematically the AT III phenotypes observed thus far. The proposed nomenclature may be regarded as provisional.

Observed and expected phenotype numbers in a sample of 212 unrelated healthy persons from Southern Germany are presented in Table 1. The frequency of alleles in this sample was AT III^1= 0.9835, AT III^2= 0.0119, AT III^3= 0.0023, and AT III^4= 0.0023. The distribution of phenotypes fitted the Hardy-Weinberg equilibrium.

Table 2 shows the distribution of AT III phenotypes in 41 families. The results are in agreement with the assumption of an autosomal codominant mode of inheritance.

Analysis of antithrombin III appears to be useful for the classification of AT III deficiencies with coagulation disorders, particularly thrombosis.

References

1. Rosenberg, R.D., Damus, P.S.: J. Biol. Chem. 248, 6490 (1973)
2. Yin, E.T., Wessler, S., Stoll, P.J.: J. Biol. Chem. 246, 3694 (1971)
3. Damus, P.S., Hicks, M., Rosenberg, R.D.: Nature 246, 355 (1973)
4. Kurachi, K., Fujikawa, K., Schmer, G., Davie, E.W.: Biochem. 15, 373 (1976)
5. Highsmith, R.F., Rosenberg, R.D.: J. Biol. Chem. 249, 4335 (1974)
6. Burrowes, C.E., Habal, F.M., Movat, H.Z.: Thromb. Res. 7, 175 (1975)
7. Venneröd, A.M., Laake, K.: Thromb. Res. 7, 223 (1975)
8. Rosenberg, R.D.: New Engl. J. Med. 292, 146 (1975)
9. Heimburger, N., Karges, H.E.: Behringwerke Lab. 28, 65 (1978)

COAGULATION FACTOR XIII B (FXIIIB) ALLOTYPING BY IMMUNOFIXATION ELECTROPHORESIS : CONFIRMATION OF THE THREE-ALLELE-MODEL

P.Kühnl,P.Kreckel and W.Spielmann

Institute of Immunohematology,University Frankfurt/M.
Sandhofstr.1,D-6000 Frankfurt/M.,FRG

A genetically determined polymorphism of the two subunits A and B of the human coagulation factor XIII (syn.:FXIIIA, FXIIIB,or FSF= fibrin-stabilizing factor) was first described by Board in the years 1979 (FXIIIA) and 1980 (FXIIIB).In functional assays or immunofixation agarose gel electrophoresis(IAGE) procedures, using monospecific anti-human-FXIIIA/B antisera three FXIIIA phenotypes A1,2-1 and 2 were identified in plasma or platelet lysates of Australian Whites.The two corresponding alleles, which coded for the isozymes of FXIIIA(fibrinoligase) were called FXIIIA*1(=0.79),and FXIIIA*2(=0.21).The description of unstable A subunits(FXIIIA*3),a silent gene (FXIIIA*Q0), and a fourth structural gene FXIIIA*4, which occurs in the Pacific region (Fiji) at non-polymorphic frequencies completed the present formal genetic model (Board and Coggan,1981;Castle et al.,1981).
The heterogeneity of the B subunit, which serves as a carrier molecule for A in plasma(A_2B_2 complex, linked together by non-covalent bonds) was also detectable by IAGE of plasma or serum.Five phenotypes FXIIIB 1,2-1,3-1, 3-2 and 3 were observed among 245 Australian blood donors; the sixth phenotype,B2, which would have proven the three-allele-model (FXIIIB*1=0.7469 ; FXIIIB*2=0.0836; FXIIIB*3= C.1693),had not yet been detected in this population sample(Board,1980).The technical problems of resolving the

Electrophoresis '82

different FXIIIB allotypes recently led to a controversial discussion of the genetic model, when Kera et al.(1981) proposed a two-allele hypothesis with only three common phenotypes S,FS and F.The corresponding allele frequencies, calculated from 290 Japanese were FXIIIB*F= 0.336 and FXIIIB*S=0.664.Family studies of 16 matings with 27 children and 38 matings with monozygous twin pairs confirmed the assumed pattern of inheritance; the employed separation technique was agarose gel isoelectric focusing in the pH range 4-6.5 of desialylated samples and subsequent immunofixation with B antiserum.

The objective of our study was the investigation of the FXIIIB polymorphism in the German population, especially to settle the question of the existence of the B2 homozygote, which had not been found in Caucasians so far.In addition, the use of native plasma, concentrated and/or desialylated material for optimal resolution of B allotypes after IAGE had to be tested.

Materials and Methods

Blood samples of 236 random blood donors and families from Hessen,Germany were drawn into glass tubes with 1 ml ACD solution as anticoagulant and centrifuged after 1 to 6h.Serum was obtained from venous blood by 5 min centrifugation at 900 g. Plasma or serum were used immediately or stored at -20°C up to 3 months. Concentration of the samples was performed with Centriflo CF 25 or 50 ultrafiltration membrane cones(Amicon Co.) which yielded a 95% retention of molecules with a molecular mass above 25,000 or 50,000 dalton, respectively.A fivefold concentration was obtained by 30 min centrifugation at 750 g and 20°C(1 ml→0.2ml). For neuraminidase treatment 15 μl of a neuraminidase (Type V, 10 U/ml, from clostridium perfringens(=CPN), Sigma Co.) solution were incubated with 30 μl of plasma or serum for 3 h at 37°C. IAGE was performed on a thin-layer(1.5x200x200mm)agarose (HGT,Seakem,FMC Co.) gel, 3.5 h separation at 30V/cm in a discontinuous pH 8.0 LiOH/boric acid (bridge buffer), citric acid/tris(gel buffer) system,according to Board's modification(1980) of the buffer system described by Ashton and Braden(1961).The agarose concentration was 1%, the cooling temperature 10°C and the start of the 15 μl sample 3 cm from the cathodal edge of the gel. For immunofixation 0.2 ml of a monospecific anti-human FXIIIB antiserum(Clotimmun Factor XIII S,Behringwerke Co) and 0.3 ml aqua dest. were applied to a gel area 2-8 cm from the start. After a 1.5 h incubation at 37°C in a moist chamber the gel was rinsed briefly, pressed under filter paper, washed in saline overnight, dried and stained with Coomassie Brillant Blue R 250 (0.2% in 250 ml ethanol,75 ml acetic acid, aqua dest. ad 1000 ml;Kreckel and Kühnl,1982).

Results and Discussion

Plasma or serum samples yielded a characteristic IAGE-proteinogram after a fivefold concentration.Homozygotes for one of the three B alleles were identified by an intense double-banded pattern(major bands) and two faint,anodal and equidistant minor bands.Whereas B3-1 heterozygotes had four bands of similar intensity(Fig.1), B2-1 carriers revealed an partial overlap of the fast

"2-component" and the slow "1-component", resulting in a triplet pattern. This was also observed in B3-2 individuals in a similar manner. Identification problems arose with B3 homozygotes, whose minor bands had an almost identical position as the major bands of B1, thus imitating B3-1 heterozygotes. For routine

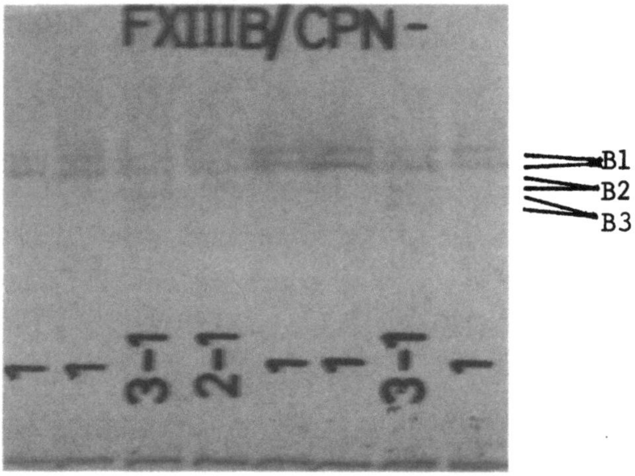

Fig.1a FXIIIB phenotypes from concentrated plasma samples, which were not treated with neuraminidase (CPN-) after IAGE

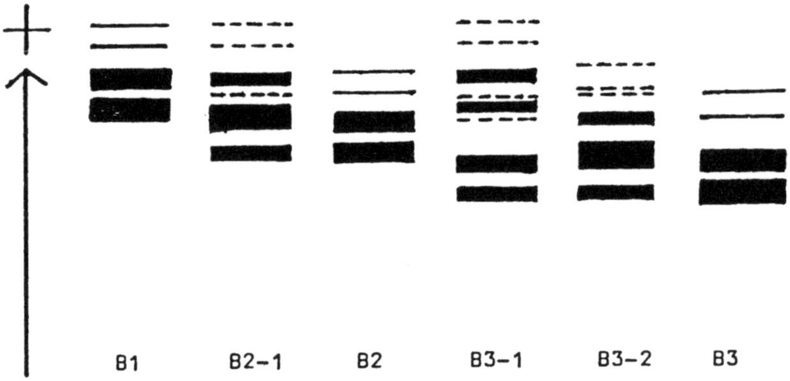

Fig.1b Diagram of FXIIIB phenotypes from concentrated CPN- plasma samples after IAGE. B2 is found symmetrically between B1 and B3 (from Kreckel and Kühnl, 1982)

analysis we therefore used CPN+ , concentrated plasma samples. Desialylation of FXIIIB molecules shifted the zones of activity towards the cathode and "reduced" the problematic B3-2 and B2-1 triplet patterns - especially in aged samples- into double-banded patterns with only very faint anodal components. B1 homozygotes were accordingly characterized by one intense band (Fig. 1c, channels 8,9). With this modification for the first time two homo-

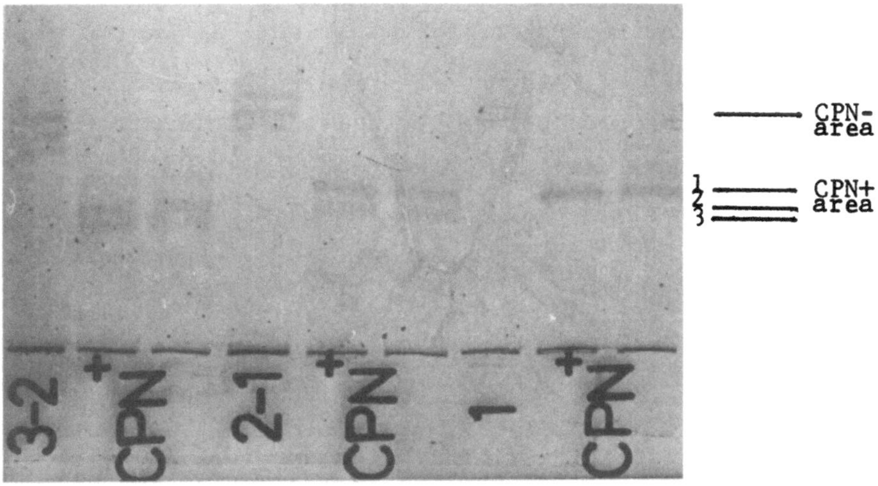

Fig. 1c FXIIIB 3-2,2-1 and 1 phenotypes before and after treatment with neuraminidase(CPN+) and separation by IAGE

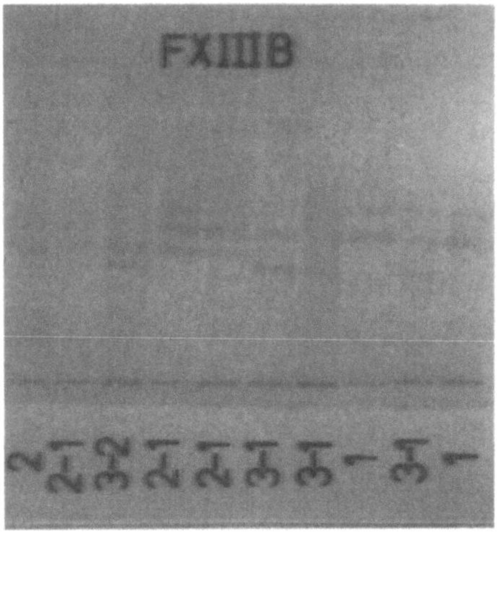

Fig. 2 FXIIIB phenotypes(CPN+) after IAGE. The new allotype B2 in channel 1. Note the different width of B 2-1 vs. B 3-2 corridors, indicating an asymmetrical position of the B2 molecule between B1 and B3

zygous carriers of the *B2 allele were observed .The B2 position was asymmetrically between B1 and B3(closer to B3) in desialylated samples, whereas it was located symmetrically between B1 and B3 in CPN- samples.A possible explanation for the different behaviour of the B2 asialo-molecule is the existence of an additional sialic acid residue, compared with B1 and B3.CPN treatment causes a stronger cathodal shift of the polypeptide chain in this case, thus indicating that the molecular heterogeneity within the FXIIIB system depends not only on a different amino acid composition (primary structure) but also a different carbohydrate moiety(sialic acid content).

The existence of six B phenotypes, including the hitherto missing"new" B2 allotype proves the three-allele-model of Board(1980) correct.In our population sample the distribution of observed and expected phenotypes is in accordance with Hardy-Weinberg equilibrium conditions($\Sigma\chi^2 = 0.6$; $0.7 < p < 0.8$ at df =2) (Table 1);compared with a previous investigation in Hessen, which was based on a smaller sample size and CPN- material(Kühnl et al.1981) and the data from Australia(Board,1980), the allele frequencies differ only slightly(B*1 = 0.725 , B*2 = 0.095 , B*3 = 0.180; Table 2).Family data (n=15,22 children)further confirm the assumed way of autosomal - codominant inheritance of three common alleles at the B locus.

The finding of only three common phenotypes (F,FS and S) in Japanese by agarose isofocusing (Kera et al.1981) might possibly be due the absence of a third common allele in Orientals; an inverse ratio of fast/slow migrating components(*S>*F ; but B*1>B*2+B*3!) then would be a remarkable difference between these two

Table 1. FXIIIB phenotypes and allele frequencies in a German population (Hessen)

	FXIIIB phenotypes						
	1	2-1	2	3-1	3-2	3	total
n (obs)	126	32	2	58	9	9	236
n (exp)	(123.91)	(32.59)	(2.14)	(61.60)	(8.10)	(7.65)	(235.99)

FXIIIB*1 = 0.725 ; FXIIIB*2 = 0.095 ; FXIIIB*3 = 0.180 . $\Sigma\chi^2 = 0.6$; $0.7 < p < 0.8$; df= 2

Table 2. FXIIIB allele frequencies in various populations

Authors	Population	Method/Material	n	FXIIIB*1	FXIIIB*2	FXIIIB*3	$\Sigma\chi^2$
Board (1980)	Australian Whites	IAGE / plasma	245	0.7469	0.0836	0.1693	4.377
Kühnl et al. (1981)	Germany (Hessen)	IAGE / plasma	110	0.75	0.07	0.18	1.4
This paper (1982)	Germany (Hessen)	IAGE / concentrated desialylated plasma or serum	236	0.725	0.095	0.180	0.6
				FXIIIB*S	FXIIIB*F		$\Sigma\chi^2$
Kera et al. (1981)	Japan	Agarose gel isofocusing (AGIF)/ desialylated plasma	290	0.664	0.336		0.004

Table 3. FXIIIB phenotypes and alleles : 2- vs. 3-allele model

FXIIIB phenotypes

Board (1980);Kühnl et al.(1981) This paper(1982)	Kera et al.(1981)	
1	F3	F
2-1	F1 or F2	F
3-1	FS	
2	S3	S
3-2	S2	S
3	S1	S

FXIIIB alleles

FXIIIB*1	FXIIIB*F
FXIIIB*2	FXIIIB*F, FXIIIB*S
FXIIIB*3	FXIIIB*S

major races.If we reclassify our phenotypes according to the nomenclature of this Japanese group(Table 3), the Hardy-Weinberg equilibrium is still given in our population sample.We encountered, however, one mating in which the two double-heterozygous parents(B3-2x2-1) had two children(B2-1 and B3-1); one phenotype was incompatible with the two-allele model:a B2-1(="F") child is an "impossible" offspring from a FxS mating. A B*QO was excluded in this case with respect to the heterozygozity of all four family members,an illegiti-

macy of the propositus by investigation of other genetic markers(Kreckel and Kühnl,1982).
In view of an application of the FXIIIB system in blood group genetics, its high informative value should render it a useful marker.In affiliation cases the single exclusion chance for non-fathers reaches 23%(subunit A polymorphism= 13.5%).Decreased levels of B activity were reported in rare cases of congenital FXIIIA deficiency, suggesting a regulation of B subunit synthesis by an equilibrium between free B subunits in plasma and those bound to A subunits.

References

Ashton,G.C.,Braden,A.W.H.:Serum β-globulin polymorphism in mice.Aust.J.Biol.Sci.14,248(1961)
Board,P.G.:Genetic polymorphism of the A subunit of human coagulation factor XIII.Am.J.Hum.Genet.31,116(1979)
Board,P.G.:Genetic polymorphism of the B subunit of human coagulation factor XIII.Am.J.Hum.Genet.32,348(1980)
Board,P.G.,Coggan,M.:Polymorphism of the A subunit of coagulation factor XIII in the Pacific region.Description of new phenotypes.Hum.Genet.59,135(1981)
Castle,S.,Board,P.G.,Anderson,R.A.M.:Genetic heterogeneity of factor XIII deficiency.First description of unstable A subunits.Brit.J.Haemat.48,337(1981)
Kera,Y.,Nishimukai,H.,Yamasawa,K.:Genetic polymorphism of the B subunit of human coagulation factor XIII: another classification.Hum.Genet.59,360(1981)
Kreckel,P.,Kühnl,P.,Spielmann,W.:Human coagulation factor XIIIA (FXIIIA) phenotyping by immunofixation agarose gel electrophoresis(IAGE).Blut(1982,in press)
Kreckel,P.,Kühnl,P.:Improved coagulation factor XIII B (FXIIIB) phenotyping after neuraminidase treatment of plasma and first description of the FXIIIB 2 phenotype(submitted for publication)
Kühnl,P.,Kreckel,P.,Spielmann,W.:Faktor XIII-Polymorphismen:Subunits FXIIIA und FXIIIB.Abstr.Vol.9th Int. Congr.Soc.Forens.Haemogenetics,Berne 1981

SEPARATION OF LEUKEMIC B CELLS FROM PATIENTS WITH CHRONIC LYMPHOCYTIC LEUKEMIA BY DENSITY GRADIENT ELECTROPHORESIS. FUNCTIONAL AND PHENOTYPIC HETEROGENEITY OF THE SEPARATED CELL FRACTIONS.

Chris D. Platsoucas
Memorial Sloan-Kettering Cancer Center, New York, NY 10021

Introduction

Patients with B cell chronic lymphocytic leukemia (CLL) possess in the peripheral blood leukemic cells of B cell origin, as well as non-malignant B and T lymphocytes (1-3). T lymphocytes from patients with CLL have a number of abnormal phenotypic and functional characteristics (4-8). Leukemic B cells in CLL possess Fc receptors for IgM, IgG and IgA, have complement receptors (6,9-11), and form spontaneous rosettes with mouse erythrocytes (12). These cells exhibit in the cell surface immunoglobulin of monoclonal origin, present in lower density than in normal B lymphocytes (9,13). Leukemic B cells from these patients can be induced to differentiate *in vitro* to plasma cells, in the presence of allogeneic T cells from normal donors, in the pokeweed mitogen (PWM)-induced differentiation system. These plasma cells synthesize and secrete monoclonal immunoglobulin (14-17). Synthesis and secretion of monoclonal immunoglobulin by CLL leukemic cells is significantly accelerated by phorbol esters (18).

We applied the density gradient electrophoresis method for the separation and analysis of leukemic B cells from patients with CLL and observed significant heterogeneity of these cells. Leukemic B cells able to differentiate into plasma cells in the presence of T cells and PWM were found to be only in the low mobility fractions and were highly enriched by density gradient electrophoresis.

Electrophoresis '82

Methods

Untreated patients with B cell CLL were employed in this study. Monocyte-depleted peripheral blood lymphocytes from patients with CLL were depleted of T cells before electrophoresis by rosetting with sheep erythrocytes (7). E-rosette forming cells from normal donors were prepared by the same method. Density gradient electrophoresis was performed in 2.5-6.25% ficoll, inverse 6.25-5.725% sucrose linear density gradient (19-21). Cell surface markers were determined by methods described in detail elsewhere (21). *In vitro* treatment with ubiquitin (100 ng/ml) was carried out for 14 hr at 37°C in RPMI 1640 supplemented with 10% fetal calf serum. The ability of B cells from patients with CLL to differentiate into immunoglobulin-synthesizing and secreting plasma cells in the presence of T cells from normal donors and PWM was determined by a modification of the method of Kneightley *et al*. (22).

Results and Discussion

Two representative experiments of the separation of purified E-rosette negative (largely leukemic) cells from patients with CLL by density gradient electrophoresis is shown in figures 1 and 2. The separated cells were pooled in seven fractions on the basis of their relative position in the electrophoretic distribution (23) and their phenotypic and functional properties were investigated. The results of the analysis by a panel of surface markers (Fc receptors for IgG and IgM, complement receptors I and II) are shown in figures 1 and 2. The ability of the separated cell fractions to differentiate into immunoglobulin-synthesizing and secreting plasma cells, in the PMW induced differentiation system and in the presence of T cells from normal donors is shown in figures 1 and 2. The results are expressed as plasma cells

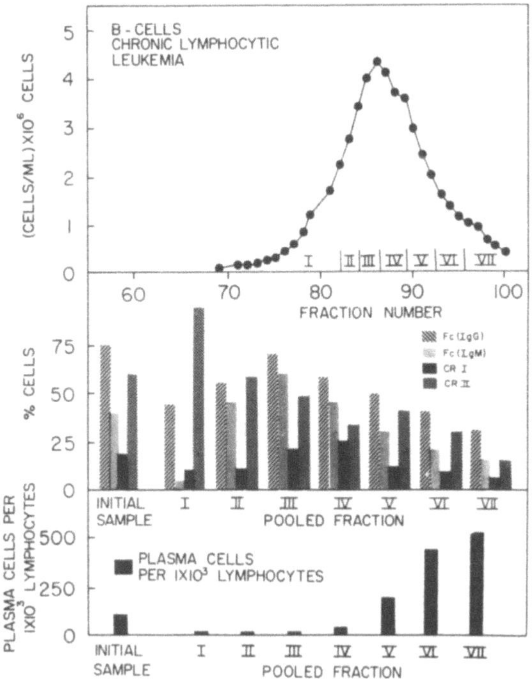

Figure 1. Separation of E-rosette negative (largely leukemic) cells from an untreated patient with CLL (WBC 25,000, Stage I) by density gradient electrophoresis. Upper panel: electrophoretic distribution profile. Middle panel: analysis of the separated cell fractions by surface markers. Lower panel: differentiation of leukemic B cells (μκ) to plasma cells (κ-positive, λ-negative) upon provision of allogeneic T cells in the presence of PWM. Results are expressed as κ-positive plasma cells per 1 x 10^3 viable cells. Lambda positive plasma cells were less than 1%.

per 1 x 10^3 viable lymphocytes. Approximately 82% to 92% viable cells were recovered from these cultures. Lymphocytes able to differentiate into plasma cells were highly enriched in the low mobility fractions V, VI, and VII. These plasma cells were of leukemic B cell origin because they contained only the light chain (kappa) that was present on the cell surface of the leukemic cells before stimulation. Plasma cells containing lamba light chain were 1% or less in the separated cell fractions or the initial sample. Results of the analysis of immunoglobulin-positive cells

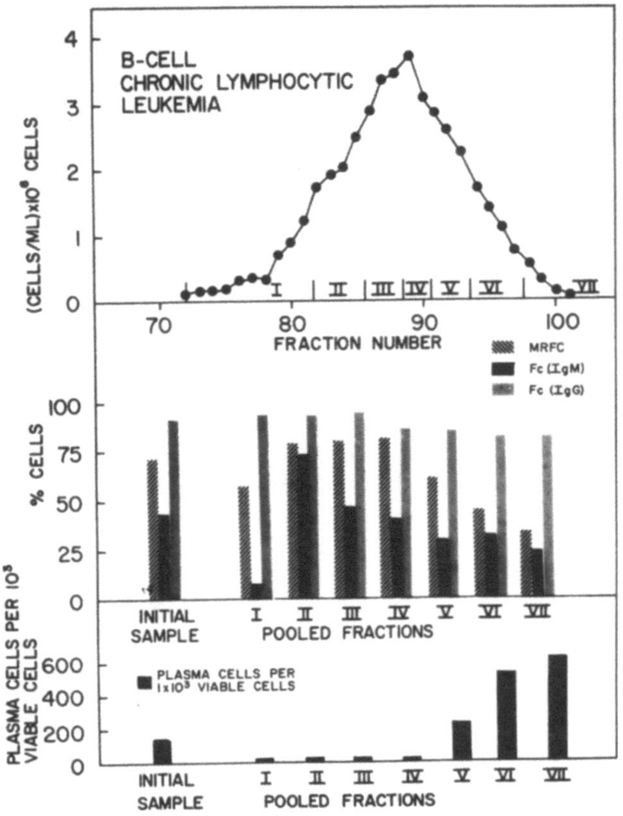

Figure 2. Separation of E-rosette negative cells from another untreated patient with CLL (WBC 74,000; stage I) by density gradient electrophoresis.

using the fluorescence activated cell sorter are shown in figure 3. These cells were present in large proportions (higher than 80%) in all the fractions; however, significant differences were observed in the density of cell surface immunoglobulin. In vitro treatment with ubiquitin resulted in a highly significant increase of cell surface immunoglobulin expression in the low mobility fractions VI and VII (figure 3).

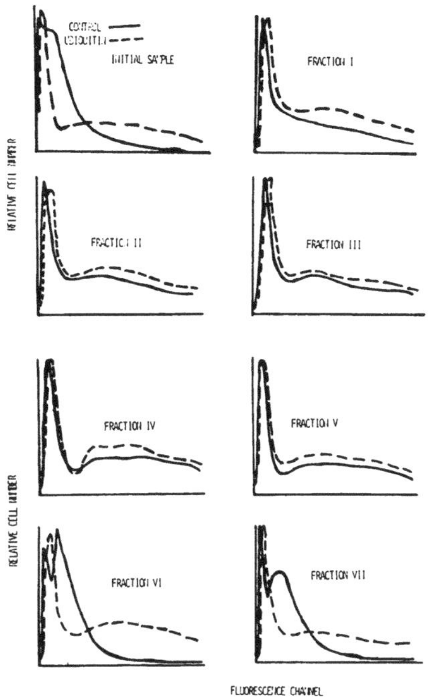

Figure 3. Analysis of immunoglobulin positive cells in the electrophoretically separated cell fractions, using the fluorescence activated cell sorter. In vitro treatment with ubiquitin (100 ng/ml for 14 hrs at 37°C) resulted in significant increase in the density of cell surface immunogloublin expression in the low mobility fractions VI and VII.
Control: ———
Ubiquitin: ----

These results demonstrate significant phenotypic and functional heterogeneity of leukemic B cells from patients with CLL. It appears that subpopulations of leukemic B cells in different differentiation or maturation stages are present in these patients. Leukemic B cells of low electrophoretic mobility were able to differentiate into plasma cells in the presence of allogeneic T cells from normal donors in the PWM-induced system. These cells were highly enriched by density gradient electrophoresis.

Acknowledgment

Supported in part by grant CH-151 from the American Cancer Society, grant PCM-8119178 from NSF, CA-32070 and CA-8748 from the National Institutes of Health and a grant from The Richard Molin Memorial Foundation for Cancer Research.

References

1. Preud'homme, J.L. and Seligmann, M. Blood 40, 777 (1972).
2. Aisenberg, A.C. and Block, K.J. New Engl. J. Med. 287, 272 (1972).
3. Hersh, E.M., Guinn, G.A., Rosen, R., Wallace, S., Rose, S. and Freireich, E.S. Proceedings of the 4th Leukocyte Culture Conference, O.R. McIntype, ed., Appleton-Century-Crofts, New York, p. 273.
4. Fu, S.M., Chiorazzi, N., Kunkel, H.G. and Harris, H.S. J. Exp. Med. 140, 1570 (1978).
5. Kay, N.E., Johnson, J.D., Stannek, R. and Douglas, S.D. Blood 54, 540 (1979).
6. Platsoucas, C.D., Kempin, S., Karanas, A., Clarkson, A., Good, R.A. and Gupta, S. Clin. Exp. Immunol 40, 256 (1980).
7. Platsoucas, C.D., Fernandes, G., Gupta, S.L., Kempin, S., Clarkson, B., Good, R.A. and Gupta, S. J. Immunol. 125, 1216 (1980).
8. Platsoucas, C.D., Galinski, M., Kempin, S., Clarkson, B. and Good, R.A. J. Immunol. (in press)
9. Dickler, H.B., Siegal, F.P., Bentwick, Z.M. and Kunkel, H.G. Clin. Exp. Immunol 14, 97 (1973).
10. Pickler, W.J. and Knapp, W. Scand. J. Immunol. 1, 105 (1978).
11. Gupta, S., Platsoucas, C.D. and Good, R.A. PNAS, 76, 4025 (1979).
12. Stathopoulos, G. and Elliott, E.V. Lancet 1, 600 (1974)
13. Foulis, A.K., Cochran, A.J. and Anderson, J.R. Clin. Exp. Immunol. 14, 481 (1973).
14. Fu, S.M., Chiorazzi, N., Kunkel, H.G. and Harris, H.S. J. Exp. Med. 140, 1570 (1978).
15. Fu, S.M., Chiorazzi, N. and Kunkel, H.G. Immunol. Rev. 48, 24 (1979).
16. Robert, K.-H. Immunol. Rev. 48, 123 (1979).
17. Saiki, O., Kishimoto, T., Kuritani, T., Muraguchi, A. and Kamamura, Y. J. Immunol. 124, 2609 (1980).
18. Totterman, T.H., Nilsson, K. and Sundstrom, C. Nature 288, 176 (1980).
19. Boltz, R.C., Todd, P., Streibel, M.J. and Louie, M.K. Prep. Biochem. 3, 383 (1973).
20. Griffith, A.L., Catsimpoolas, N. and Wortis, H.H. Life Sci. 16, 1693 (1975).
21. Platsoucas, C.D., Good, R.A. and Gupta, S. Cell. Immunol. 51, 238 (1980).
22. Keightley, R.G., Cooper, M.D. and Lawton, A.R. J. Immunol. 117, 1538 (1976).
23. Platsoucas, C.D., Griffith, A.L. and Catsimpoolas, N.J. Immunol. Meth. 13, 145 91976).

ELECTROPHORETIC MOBILITY TEST BY STIMULATED LYMPHOCYTE SUPERNATANTS

Nobuya Hashimoto, Masaki Ageshio, Shunichi Nose,
Masakazu Horita, Toshiko Kobayashi and Masakazu Abe

Department of Internal Medicine,
Jikei University School of Medicine,
Minato-ku, Tokyo

Introduction

The supernatants of human lymphocytes stimulated with antigen and non specific mitogen influence electrophoretic mobility of sheep erythrocytes as the indicator cells of lymphocytes sensitization.
This method was originated by Field and Caspary as macrophage electrophoretic mobility test (1), and then modified by Porzsolt and Ax (2), Jenssen and Shenton (3) as tanned sheep erythrocyte electrophoretic mobility test.
This test system has been used for cancer diagnosis and for application to clinical immunology by many investigators (4) (5).
We have previously reported that electrophoretic mobility of sheep erythrocytes was reduced by supernatant from lymphocytes with tuberculin test positive subjects in incubation with PPD (6), and was affected by supernatants of lymphocytes stimulated with the various immunopotentiators (7).
The purpose of the present study is to measure alteration of surface charge of the sheep erythrocytes added with supernatants of lymphocytes from normal subjects and systemic lypus erythematosus (SLE) with mitogens and antigen.

Electrophoresis '82

Materials and Methods

Lymphocytes were isolated from human peripheral blood and adjusted in 10^6 cells in 0.15 ml. Viability of the lymphocytes was more than 90 %. The whole lymphocytes were separated into T cell and non T cell by the ordinary E-rosetting technique. T cells were confirmed as 80 - 85 % of E-rosett forming cells, and non T cell as 90 % of surface Ig positive cells. The whole lymphocytes or separated T and non T cell were stimulated with mitogens and antigen in 5 % CO_2 at 37°C. As stimulating agents, PHA (Phytohaemagglutinin, HA16, Wellcome) was used at 30 µg / 15 µl, Con A (Concanavalin A, EY Lab.,) at 5 µg or 10 µg / 0.1 ml, and calf thymus DNA (Sigma) at 100 µg / 0.1 ml.
Then, the supernatants of the stimulated culture mixtures were obtained and added with indicator cells at 37°C, 45 min. As indicator cells, sheep erythrocytes, tanned and stabilized with sulphosalicylic acid were used as a lyophilized preparation from Behringwerke, West Germany (Courtesy of Japan Hoechst Co., Tokyo), and adjusted to 10^7 cells in 0.1 ml of electrophoretic buffer. As a control, supernatant from another test sample to which stimulating agents were not added was prepared.
Measurement of erythrocyte electrophoretic mobility was performed in an analytical cell microelectrophoretic apparatus, manufactured by Sugiura Laboratory Co., Tokyo.
Electrophoretic buffer was triethanolamine solution devised by Prof. Hannig (8). All measurements were carried out at a potential difference of 40 V, 200 uA and at temperature of 25°C.
The time for each of 30 cells from each sample to cross one square of the microscope grid was determined, each cell being timed in both directions of the field (1 grid square = 34.3 µ), and recorded by an electric digital stop-watch with an automatic printer (reading to 0.01 sec). Any

measurements in which the difference in timing for the two directions of the field was greater than 10 % were discarded. The cells in focus in the stationary layer were measured. The mean actual migration time was measured. The percentage slowing in electrophoretic mobility was expressed by the formula :

$$\% \text{ slowing} = \left(\frac{T_1 - T_0}{T_0} \right) \times 100$$

T_1 : mean electrophoretic time of the sample

T_0 : mean electrophoretic time of the control

Results and Discussion

Percentage slowing of electrophoretic mobility of sheep erythrocytes by supernatant of the whole normal lymphocytes stimulated with PHA was first investigated. In one hour of incubation time with PHA and the lymphocytes, mean percentage slowing was 19.00 ± 3.65 in fifteen cases, and in forty eight hours of incubation time it was 26.24 ± 5.02. When incubation time with PHA and the lymphocytes was longer, percentage slowing was more increased. In the supernatant from the normal lymphocytes stimulated with PHA, percentage slowing of sheep erythrocytes electrophoretic mobility was 20.81 ± 5.04 in total.

In the supernatants from lymphocyte subpopulation stimulated with PHA, electrophoretic mobilities of sheep erythrocytes were measured. Mean percentage slowing of sheep erythrocytes was 16.62 ± 12.80 in the supernatant from T cell culture, and 8.14 ± 5.35 in the supernatant from non T cell culture. This implies that electrophoretic mobility of sheep erythrocyte was more delayed by effect of the supernatant from T cell culture mixed with PHA.

To exclude direct effect of PHA on electrophoretic mobility, electrophoresis was carried out in addition with PHA to

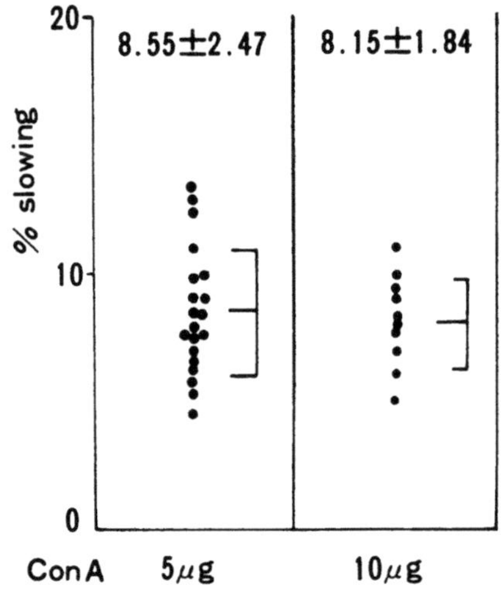

Fig 1. Effect of concentration of Con A which was added to lymphocyte on electrophoretic mobility of sheep erythrocyte

sheep erythrocyte. As a result, it was confirmed that no direct effect of PHA on electrophoretic mobility of sheep erythrocyte was observed.

In the supernatant from the whole lymphocyte from normal subjects stimulated with Con A, mean percentage slowing of electrophoretic mobility of sheep erythrocyte was 8.55 ± 2.47. This is lower than a result of percentage slowing by supernatant of lymphocyte stimulated with PHA. To investigate effect of concentration of Con A which was added to lymphocyte on electrophoretic mobility of sheep erythrocyte, 5 μg and 10 μg of Con A were used. In supernatant from lymphocyte to which 5 ug of Con A was added,

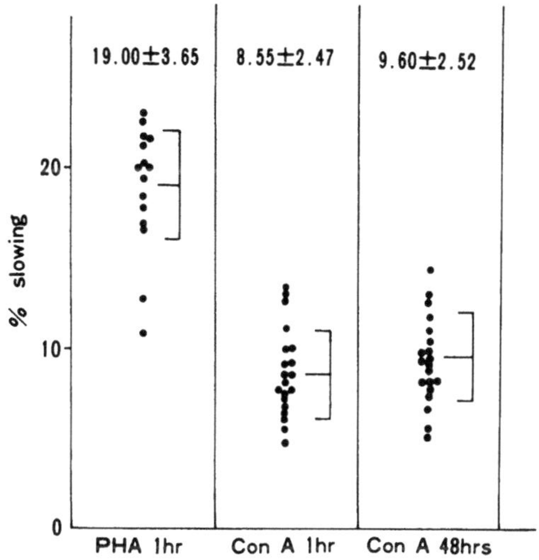

Fig 2. Effect of incubation time of Con A and lymphocyte on electrophoretic mobility of sheep erythrocyte (concentration of Con A used ; 5 µg)

mean percentage slowing of electrophoretic mobility of sheep erythrocyte was 8.55 ± 2.47, and in 10 µg of Con A it was 8.15 ± 1.84 (Fig 1). There is no significant difference in 5 µg and 10 µg of Con A.

Incubation time for stimulation of lymphocyte with Con A was investigated. In one hour of incubation time of lymphocyte culture stimulated with Con A, percentage slowing of electrophoretic mobility of sheep erythrocyte was 8.55 ± 2.47, and in fourty eight hours of incubation time of culture of lymphocyte and Con A, percentage slowing was 9.60 ± 2.52. It was not so altered by duration of incubation time with Con A as stimulation with PHA (Fig 2).

Table 1. % slowing of electrophoretic mobility of sheep erythrocyte in normal T-and non T cell stimulated with Con A

subjects No.	control (sec)	T-cell		non T-cell	
		Ep time	% slowing	Ep time	% slowing
1	3.34	3.76	12.57	3.45	3.29
2	2.81	3.08	9.60	3.06	8.89
3	2.84	3.32	16.90	3.06	7.75
4	4.24	4.59	8.25	4.26	0.47
5	3.27	3.57	9.17	3.30	0.91
6	3.69	4.19	13.55	3.88	5.14
7	3.56	4.10	15.17	3.70	3.93
8	3.55	4.02	13.23	3.74	5.35
9	3.62	4.09	12.98	3.76	3.86
10	3.26	3.60	10. 2	3.49	7.05
mean % slowing ± SD		12.18±2.63		4.66±2.62	

For separated T cell and non T cell, electrophoretic mobility test was performed in supernatant of lymphocyte stimulated with Con A. Electrophoretic time and percentage slowing of each case are shown in table 1. Mean percentage slowing of T cell was 12.18 ± 2.63, and in non T cell it was 4.66 ± 2.62. It is suggested that much electrophoretic slowing factors may be concentrated in the supernatant of T cell stimulated with Con A.

Subsequently, sheep erythrocyte electrophoretic mobility test was performed for supernatant of the whole lymphocytes from SLE patients stimulated with PHA. The SLE patients diagnozed by the criteria of ARA were admitted to our

Table 2. % slowing of electrophoretic mobility of sheep erythrocyte in T cell and non T cell from SLE patients stimulated with Con A

SLE patients	control (sec)	T-cell		non T-cell	
		Ep time	% slowing	Ep time	% slowing
24 ♀	3.51	3.52	0.28	3.82	8.83
33 ♀	3.81	3.80	0.26	4.06	6.56
17 ♀	3.66	3.75	2.46	3.86	5.46
25 ♀	4.02	4.03	0.25	4.23	5.22
29 ♀	3.71	3.80	2.42	4.01	8.09
mean % slowing		1.13±1.07		6.83±1.42	

University Hospital. Mean percentage slowing in five cases was 16.03 ± 8.47, slightly lower than that in normal lymphocyte stimulated with PHA, but not significant.
In supernatant of the whole lymphocytes from SLE patients stimulated with Con A, mean percentage slowing of electrophoretic mobility in five cases was 11.58 ± 1.20, slightly higher than that in normal lymphocyte stimulated with Con A, but not significant.
In subpopulation of SLE patient lymphocytes stimulated with Con A, mean percentage slowing of T cell was 1.13 ± 1.07 and in non T cell it was 6.83 ± 1.42 (Table 2). It is interesting that percentage slowing of electrophoretic mobility of sheep erythrocyte was higher in supernatant from non T cell-Con A culture than that of T cell-Con A culture.

Table 3. Influence of DNA on electrophoretic mobility of sheep erythrocytes

electrophoretic mobility time (sec)		% slowing
TSRBC	TSRBC+DNA	
4.37	3.26	−25.40
4.37	3.34	−23.56
4.37	3.32	−24.03

Table 4. Electrophoretic mobility of sheep erythrocyte in supernatants of normal lymphocyte treated with DNA

age	sex	electrophoretic mobility time (sec)		% slowing
		TSRBC+lymph	TSRBC+lymph+DNA	
20	♀	4.92	3.58	−27.23
49	♂	4.92	3.65	−25.80
28	♂	4.89	3.96	−19.01

It may be speculated that T cells from SLE patients are unable to release an electrophoretic slowing factor even by stimulation with Con A.

For SLE patient lymphocytes, calf thymus DNA as antigen was used in place of mitogen such as PHA and Con A. Influence of DNA on electrophoretic mobility of sheep erythrocyte was first investigated. Electrophoretic mobility of sheep erythrocyte to which DNA was added was faster than that of sheep erythrocyte alone (Table 3).

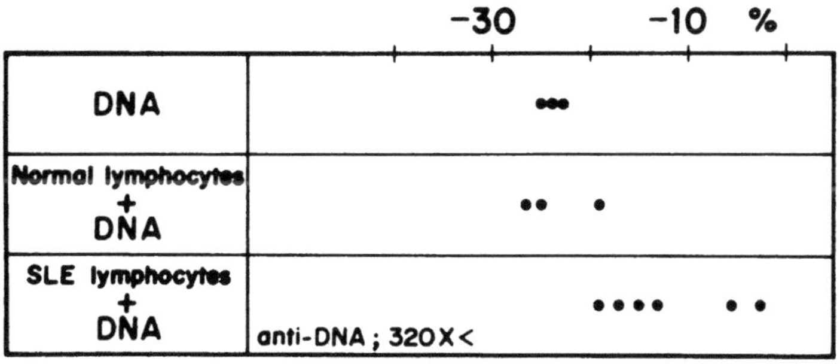

Fig 3. % slowing of electrophoretic mobility of sheep erythrocyte in SLE lymphocyte stimulated with DNA

Similarly, electrophoretic mobility time of sheep erythrocyte was also shortened in the supernatant of the normal lymphocyte treated with DNA (Table 4).
This implies that surface negative charge of sheep erythrocytes was increased by addition with DNA.
Then, to the whole lymphocytes from SLE patients DNA was added, and for the obtained culture supernatant sheep erythrocyte electrophoresis was carried out. The result is shown in Fig 3. In the supernatant of SLE patients lymphocyte stimulated with DNA, electrophoretic mobility of sheep erythrocytes was delayed, as compared with two other controls.
This implies that increase of the electrophoretic mobility by DNA was impaired in the supernatant from lymphocytes of SLE patients.
Subpopulation of SLE patients lymphocytes stimulated with DNA was investigated in electrophoretic mobility test.
Mean percentage slowing in T cell from six cases of SLE was -31.31 ± 3.60, and in non T cell -37.24 ± 3.57.

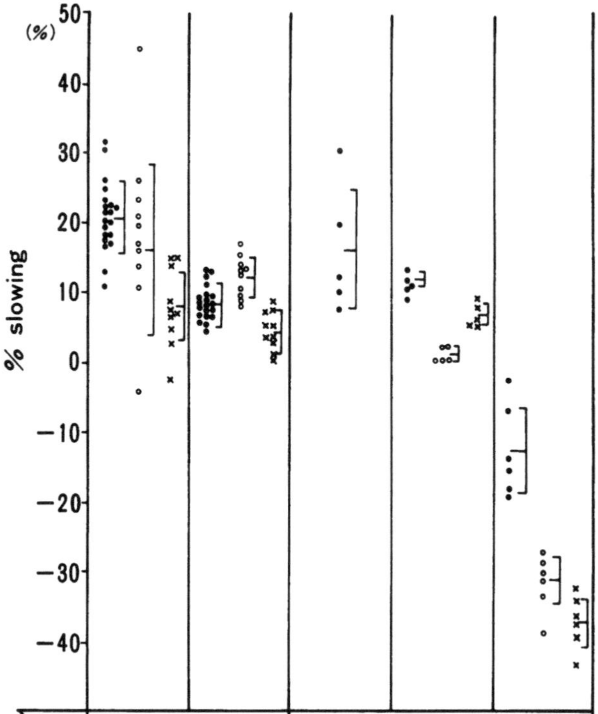

mean % slowing ±SD	normal lymphocytes PHA	normal lymphocytes ConA	SLE patients lymphocytes PHA	SLE patients lymphocytes ConA	SLE patients lymphocytes DNA
● whole lymph.	20.81± 5.04	8.55± 2.47	16.03± 8.47	11.58± 1.20	-12.70± 6.10
○ T-cell	16.62±12.80	12.18± 2.63	—	1.13± 1.07	-31.31± 3.60
× non T-cell	8.14± 5.35	4.66± 2.62	—	6.83± 1.42	-37.24± 3.57

Fig 4. % slowing of electrophoretic mobility of sheep erythrocyte stimulated with PHA, Con A and DNA

Electrophoretic mobility of sheep erythrocytes was more delayed in the supernatant from T cell of SLE patients stimulated with DNA than that in non T cell.

Conclusion

The above mentioned results are summarized in Fig 4. In the supernatants of normal whole lymphocytes stimulated with PHA and con A, electrophoretic mobilities of sheep erythrocyte were delayed in both cases. Percentage slowing of electrophoretic mobility was higher in PHA than in Con A. On subpopulation of normal lymphocytes stimulated with PHA and Con A, percentage slowing was higher in T cell stimulated with any mitogen. However, in the SLE patients lymphocytes stimulated with Con A, percentage slowing was higher in non T cell than that of T cell. In stimulating with DNA, electrophoretic mobility was delayed in T cell from SLE patient.

In the present study, we showed influence of supernatants of lymphocytes stimulated with mitogens (PHA and Con A) and antigen (DNA) on electrophoretic mobility of sheep erythrocytes. Electrophoretic mobility of sheep erythrocytes varied dependent on subpopulation of lymphocytes stimulated with mitogen and antigen.

References

(1) Field, E. J. and Caspary, E. A., : Lancet ii, 1337, (1970)

(2) Porzsolt, F. et al., : Behring Inst. Mitt., 57, 128, (1975)

(3) Jenssen, H. I. and Shenton, B. K., : Acta biol. med. germ., 34, 29, (1975)

(4) Preece, A.W. and Sabolovic, D. (Eds.) : Cell Electrophoresis, Clinical Application and Methodology, North-Holland, Amsterdam (1979)

(5) Preece, A. W. and Light, P. A. (Eds.) : Cell Electrophoresis in Cancer and Other Clinical Research, North-Holland, Amsterdam (1981)

(6) Hashimoto, N., Physico-Chemical Biology, 23, 209, (1979)

(7) Hashimoto, N., et al., In Electrophoresis '81 (ed. by Allen and Arnaud) 983, Walter de Gruyter, Berlin (1981)

(8) Hannig, K., Methods in Microbiology, : (ed. by Norris and Ribbons) 5B, 513 (1979)

SECTION V ROUND-TABLE DISCUSSIONS

ROUNDTABLE DISCUSSION ON ACCURACY AND PRECISION IN 2-DIMENSIONAL MACROMOLECULAR MAPPING.*

Chairman: A. Chrambach

Endocrinology and Reproduction Research Branch, NICHD, NIH, Bethesda MD 20205, USA

Panel Members: **

R. C. Allen

Dept. of Laboratory Animal Medicine, Medical University of South Carolina, Charleston SC 29401, USA

J. Klose

Institut fuer Toxikologie und Embryonalpharmakologie, Freie Universitaet Berlin, 1000 Berlin 33, Germany

C. R. Merril

Laboratory of General and Comparative Biochemistry, NIMH, NIH, Bethesda MD 20205, USA

P. G. Righetti

Istituto di Chimica Organica ed Analitica, Universita di Milano, 20133 Milano, Italy

Audience Participant:

H. Kronberg

Abteilung Neurochemie, Max-Planck-Institut fuer Experimentelle Medizin, D-3400 Goettingen, BRD

* This report summarizes the questions with which the panel members were confronted, and the outcome of a round-table discussion on April 24, 1982. Neither the questions themselves, nor the tentative answers given, reflect necessarily the full consensus of every panel member in every detail.

** N. Leigh Anderson

Division of Biological and Medical Research, Argonne National Laboratory, Argonne IL 60439, USA, participated in the panel but has dissociated himself from the report.

Electrophoresis '82

Summary

While 2-dimensional macromolecular mapping appears to be capable of resolving systems of more than 1,000 components, the question was asked whether some of these components might be generated artifactually, and to what degree spots can be identified objectively and in a reproducible fashion. Six conceivable sources of artifact and misidentification were considered: 1) Detergent inadequacy in dissolving and dissociating the protein. 2) Transient state electrofocusing patterns. 3) Ionic, hydrophobic and H-bond interactions. 4) Failure to solubilize, saturate and stack the isoelectric proteins derivatized with SDS. 5) Spot identification problems arising from pattern irreproducibility. 6) Overshadowing of minor components by major ones.

Introduction

The dream underlying biochemistry is to be able to account for life in terms of chemical compounds and their reactions. Classically, the experimental approach toward this goal has been to first simplify the system under investigation as much as possible by tissue and cell fractionations yielding defined biological starting material (specific organs, cell types, centrifugal fractions etc.), then to dissolve that material as completely as possible and to subfractionate it until the fractions could be resolved into distinct, relatively few components. Attempts to directly resolve complex biological systems by such methods as chromatography linked to mass spectrometry largely failed until O'Farrell demonstrated the possibility of exhibiting on a single electrophoretic 2-D macromolecular map 15% of the estimated 10,000 protein components of the cell simultaneously (1). The hope engendered by that demonstration led hundreds of scientists to apply the fractionation conditions developed by O'Farrell for the analysis of *E. coli* to many other systems, and in medicine, to the notion of a "human protein index", i.e. a catalogue of macromolecules and their distributions in health and disease (2).

The macromolecular maps, with thousands of spots allegedly representing the building stones of nature, necessarily also invited critical questions. Could the spots, at least in part, be artifacts generated by either electrofocusing (EF) or polyacrylamide gel electrophoresis in SDS-containing buffers (SDS-PAGE)? And if so, does the presence of artifacts matter when purely comparative data are sought? Would isolated and well characterized standard proteins run as single spots on the 2-D-maps? And if this were the case - could one identify such a protein reproducibly in terms of its isoelectric point (pI) - molecular weight (MW) coordinates? Can one identify a protein in a mixture reproducibly by reference to internal markers? And particularly - are the numerous "secondary spots" of identical molecular weight but different apparent net charge relative to a given major "primary" spot in the pattern real or artifactual? These questions to date have not been the subject of direct methodological investigations. It seemed important, therefore, to attempt obtaining answers to these questions to the degree that they are buried in reports dealing with the biological applications of 2-D macromolecular mapping [e.g. (3)].
The present discussion represents an attempt to confront practitioners of 2-D macromolecular mapping with these questions, to unearth evidence in favor of the accuracy and precision of spots on the maps and, as much as possible, to dispose of relevant doubts.
The round table discussion of 110 minutes on which this report is based was limited to the problems related to spot accuracy and precision. Other areas of questions of great importance for 2-D macromolecular mapping exist that should be discussed at some future occasion. We'll indicate a few:
a) Do methods exist, or can methods be developed, to replace trial-and-error in the establishment of EF conditions which give rise to a uniform zone distribution in the first dimension, and of pore gradient engineering capable of providing an even zone distribution in the second dimension, and thus making full use of the resolving capacity of the method? Such uniform zone distribution in both dimensions is the geometric basis for covering the entire area of a map with spots (1).
b) Is 2-D mapping of native macromolecules of importance to biology and medicine? And if yes, what are the most appropriate conditions of separations based on net charge differences (charge fractionation) and based on size differences (size fractionation) for native macromolecules?

c) What distinctions exist between the various automated spot identification and quantitation methods (4-8)? What are the possible errors in automated pattern recognition? In the use of photographic images? In data processing?

d) How do the various spot detection methods compare with regard to spot quantitation? Does isotope analysis, silvering (9-11), and protein staining with Coomassie Blue R 250 (G 250?) give identical quantitation? Within what quantitative limits?

e) Finally, to what degree is the theoretical geometric advantage of zone spreading over two dimensions being realized in practice? How does the resolution of complex systems compare to that, for instance, of 1-stage EF gels at 400-800 V/cm (12), or to that obtained by 2-stage or 3-stage gel electrophoretic separations (13)?

Prior to considering the potential sources of inaccuracy and irreproducibility of 2-D macromolecular mapping, it seems important to point out that at least inaccuracy, i.e. the formation of artifactual spots by the electrophoretic system, is not necessarily detrimental to the purposes of a particular study. An example is the comparative pattern analysis as a function of disease states or genetic situations in large numbers of samples. What matters here exclusively is the recognition of a particular disease state or genetic defect in a population on the basis of a particular pattern different from that in the rest of the population. If the patterns allow for that recognition, it may be immaterial whether or not artifactual spots contribute to the pattern.

1) Detergent problems.

Problems in the solubilization, maintenance of the dissolved state during EF, and of complexing within detergent micelles may arise in the 2-D macromolecular mapping of water insoluble proteins.

a) Solubilization. The 4 chemical categories of detergents, with generic structures of ROH, RCOOH, $R(NH_2)COOH$ and RSO_3H, differ specifically in solubilizing and disaggregating power and in their protein inactivating and denaturating effects (14-16). Furthermore, RCOOH and RSO_3H alter the net charge of proteins so as to confer upon them an equal charge density, or at least a very high one, thus abolishing charge differences and the possibility of separation by such "charge fractionation" methods as EF. These differences in detergent properties with regard to solubility are exemplified for a membrane protein system (17) as shown in Table 1. Moreover, certain detergents or mixtures of detergents disaggregate the solubilized protein more efficiently than others (15). These facts seem to support the notion that a specific detergent optimization was desirable both in the preparation of the sample and the 1st dimension of 2-D macromolecular mapping. Moreover, detergents with low CMC such as Triton X-100 (alias NP-40) seem particularly unsuitable since their mechanism of solubilization of hydrophobic proteins involves in general the incorporation of one or several copies of the protein into a detergent micelle or complex, together with other proteins, lipid, phospholipid etc., frequently giving rise to particle weights of the order of 10^6 or more (16) (see section c).

The problem of poor solubilizing power of non-ionic detergents with polyoxyethylene backbone (inclusive of Triton X-100) in a membrane protein system such as that of Table 1 appears to be resolved in many cases in the presence of high concentrations of urea. The solubilizing agent introduced by O'Farrell (1), 2% Triton X-100 in 9.5 M urea, is being considered by many practitioners of 2-D macromolecular mapping to be as effective with regard to protein solubilization as SDS and therefore to be generally applicable. Urea raises the critical micelle concentration (CMC) of the non-ionic detergent [Fig.8 of (14)] and therefore tends to disrupt the solubilizing micellar organization of the detergent. Since Triton monomers (or other monomers of detergents with low CMC) are not solubilizers for proteins, this destruction of micellar organization by urea cannot be total - otherwise proteins would be insoluble in this medium. Evidently, the effect of urea in abolishing micellar organization is balanced by the low CMC value for Triton - two orders of magnitude less than that of SDS (14), and by the relatively high concentration of Triton used (2%), giving rise to Triton-protein complexes of reduced complexity and probably weight than obtained in the absence of urea. Other factors that affect micellar organization - temperature, lipid and phospholipid concentrations, presence of other hydrophobic proteins, detergent concentration, adducts like glycerol - will also affect protein solubility in this solvent. Large molecular weight complexes may exceed the exclusion limit of the EF gel (see below). Furthermore, if the dissolved species is separated from the membrane by sedimentation, the high density (d=1.14) of urea may prevent the membrane from sedimenting. Thus, protein solubility in 2% Triton - 9 M urea appears sensitive to many factors which vary between experiments and laboratories.

The degree of protein inactivation by the detergent is irrelevant for the purpose of this discussion, as long as we are not concerned with the macromolecular mapping of native proteins (see above).

The modification of net charge through the use of charged detergents is, of course, not a problem with the uncharged solubilizing agent, 2% Triton X-100 - 9 M urea, as long as the urea is freshly prepared from a highly purified reagent and thus free of isocyanate. It may present a problem, however, when SDS is admixed to that solubilizing agent as in the procedure of Anderson and Anderson (18). To the degree that the SDS remains bound to protein under the conditions of EF, it must perturb the EF pattern and does so in several reported cases (19-21); however, at least in a single case specifically tested the SDS appears to be removed from the protein quantitatively under the conditions of EF used [data of A. Burghes shown in Fig.3, Chapter on Two-Dimensional Electrophoresis of (22)]. The same data, showing pattern identity in presence and absence of SDS, also indicate that no significant mobilization of micellar neutral detergent in the EF gel occurred by reaction with SDS (see below).

b) Maintenance of the solubilized state during EF. Maintenance of the solubilized state during EF depends on the maintenance of non-ionic detergent concentration in the EF gel and that of the solubilizing micellar or complexing structure comprising protein and detergent. The concentration of non-ionic detergent in an EF gel can dramatically decrease through participation of charged detergents in the micelle (forming a mixed micelle) and transport of the mixed micelle in the electric field [Fig.12 of (14); Fig. 3 of (23)]. Triton mobilization through SDS and resulting protein precipitation or pattern change due to addition of SDS to non-ionic detergent has not been a problem in the experiments of Dr. Burghes (see above), but should be critically tested for in other applications where the steady-state concentration of non-ionic detergent as influenced by the various parameters alluded to above and the dependence of solubility on detergent concentration may differ from the experimental conditions of Dr. Burghes. As a rule, it seems advisable to eliminate charged hydrophobic compounds from the solubilization mixture as much as possible. The second mechanism leading to protein insolubility in detergent-PAGE or -EF in general is the disruption of the solubilizing micellar complex during electrophoresis. This disruption may be due to reaction of the neutral detergents with charged compounds as discussed above, or it may be due to an alteration during electrophoresis of any of the factors influencing micellar structure which have been alluded to above. A low temperature of EF will have the effect of lowering the micellar weight [Fig.10 of (14)]. At high temperature, such as in "hot spots" along the pH gradient (see below), the detergent will aggregate to the point of precipitation (the "cloud point"). Urea will tend to disrupt the micellar structure, thus augmenting the effect any of the other factors may have during EF. It appears therefore advisable to avoid the use of neutral detergents with low CMC such as Triton X-100 and Lubrol in favor of detergents with high CMC such as β-octylglucoside, digitonin or CHAPS (24) which are capable of reacting as monomers with proteins in effecting solubilization.

The consensus among the practitioners of 2-D macromolecular mapping appeared to be that in general the presence of 2% nonionic detergent and 9 M urea in an EF system effectively prevents protein precipitation during EF. However, in view of the sensitivity of this system (see above), this generalization may not apply in a particular application, and the reader may find it useful to recall some of the mechanisms which promote protein precipitation in EF, and ways of recognizing it. Protein precipitation during EF may be detected by following the position of a protein zone on the gel with time. Since the pH gradient is continuously changing in EF [Fig.2 of (25)], a constant zone position indicates protein precipitation which is non-isoelectric and irreversible [see Fig.7 of (26)]. Precipitation would also give rise to vertical streaking in the 2nd dimension (due to the gradual solubilization of a protein precipitate by SDS).

The tendency of proteins to precipitate during EF is due to 2 mechanisms in addition to that of detergent withdrawal discussed above. The first relates to isoelectric precipitation which is favored by the low ionic strength of EF systems and occurs over an extended pH range which is the more extended the lower the ionic strength [Fig.4 of (16)]. Isoelectric precipitation of proteins is not a problem for EF in gels (as opposed to fluids where precipitates sediment), as long as it remains reversible and the precipitated zones remain mobile, migrating with the ever-changing pH gradient.(This migration proceeds presumably through a reversible process of re-solubilization and re-precipitation). The second mechanism of protein precipitation in EF is thermal precipitation at loci of high voltage due either to relative deficiency of carrier ampholytes in the zone (see below) or to high protein concentration within the band. The latter effect is evidenced by the behavior of protein kinase which precipitates in pH gradients under those conditions which give rise to sharp activity peaks (Sephadex G-75), while it remains native under those conditions where its distribution curve is relatively broad (Ultrogel) (27) (Table 2). The closer the asymptotic approach of the protein toward its pI-position, and the longer the protein is exposed to a near-isolelectric pH, the more irreversible the precipitation becomes. As a consequence, the changing pH gradient leaves behind an artifactual protein zone due to precipitate which is not resolubilized in the electric field [Fig.2 of (27)].

c) Complexing. The nature of the detergent solubilized protein may be a protein-detergent-lipid-phospholipid complex comprising one or several copies of one protein and one or several protein species (16). The nature of the complex depends on the detergent/protein ratio [Fig.10 of (16)]. It is this complex which migrates into the EF gel, and which produces the zones of the 2nd dimensional gel. The retardation of the large complexes in the EF gel, and the charge density of the complex as opposed to the protein of interest, could give rise to artifactual zone positions in EF. Since solubility of the protein in such a case depends on the maintenance of the complex, the resulting artifactual EF zones cannot be prevented by disrupting the complex. The remedy might consist of replacing the large micellar structure of a low CMC detergent such as Triton X-100 (micellar weight 90,000) which may contain many heterogeneous components by a high CMC detergent capable of reacting as the monomer with the protein and exhibiting a small micellar weight (e.g. 2,000 for detergents with a cholic acid type backbone) even under experimental conditions where micelles form (14). The uncharged detergent with cholic acid-type hydrophobic backbone, CHAPS, may be a leading candidate in this respect (24).

2) Transient state EF patterns.

Two-dimensional patterns frequently exhibit multiple "secondary" spots aligned horizontally in the 2nd dimension with a particular primary spot. Although we will survey a number of mechanisms by which artifactual zones in EF may arise, there are clearly cases in which these secondary spots have been shown to be in vivo products [e.g. (28)-(33), and for EF alone (34)]. Thus, the accuracy of secondary spots needs to be evaluated in each particular case.

The steady-state of EF is defined as one in which all species contained in the system are aligned in order of their constituent mobilities (35) or isoelectric points (36). Accordingly, transient states obtain whenever a) proteins of identical pI but differing in aggregation state (size) approach their common isoelectric position at different rates; b) when a continuous ligand loss or gain occurs during EF, prior to the complete removal of, or saturation with, the ligand; and c) when voltage minima along the migration path prevent the attainment of isoelectric positions during the finite lifetime of the pH gradient. For the purpose of macromolecular mapping, the transient states of EF diminish necessarily pattern reproducibility since the reproducibility of kinetic phenomena depends much more stringently on a meticulous control of time and all factors affecting migration rates (temperature, ionic strength, gel concentration etc.) (37) than a steady-state separation, i.e. reproducibility of the latter is less perturbed by variations in experimental conditions. Thus, "non-equilibrium focusing" (38) while not incompatible with the 2-D technique certainly adds to its reproducibility problems (see section 5); it may also give rise to artifactual zones as discussed below.

a) Protein migration rates. Each protein approaches its steady-state pH at a particular rate governed by its pH-mobility profile (titration curve) and %T-mobility profile (Ferguson plot), as well as by the continuously changing pH gradient [Fig.2 of (25)]. Thus, the steady-state of a complex protein system can only be recognized in terms of pattern constancy. "Pattern constancy" implies that after some time of EF the zone pattern assumes a recognizable "gestalt" or constancy which persists while the pH gradient drift progressively compresses the "constant pattern" toward the cathodic terminus [see Fig.7 of (39)]. Without pattern constancy with EF time, the number of zones in EF is arbitrary and necessarily both inaccurate and irreproducible. The number of zones approaching their isoelectric positions may be either smaller [Fig.7 of (39)] or larger [Fig.1 of (40)] than those at the steady-state. The panel therefore agreed that a reasonable degree of pattern constancy with EF time was a prerequisite for the application of an EF gel to the 1st dimension. Thus, transient state EF artifacts in 2-D macromolecular mapping may cease to be a problem to the degree that constant patterns with EF time are obtained through the use of immobilized carrier constituents (41) if it should become possible to extend "immobilized pH gradients" to a wide pH range without generating electroendosmotic problems, and if such systems do not exhibit uneven conductivity profiles.

A particular source of artifactual zone formation in EF appears to be protein aggregation, when EF is applied in relatively restrictive gels. Protein aggregation is furthered at the high protein concentrations within zones, the low ionic strength of EF, and a pH at or near the isoelectric point. To the degree that aggregation, like complexing, is abolished by the presence of 9 M urea - 2% non-ionic detergent, aggregation is not a problem in 2-D macromolecular mapping. To the degree that it is not under the specific experimental conditions, it needs to be considered here. Gel restrictiveness leads to multiple zones per protein if the protein aggregates, since each aggregation state approaches the common isoelectric position at a different rate, which decreases with increasing molecular size, unless the EF time is sufficient for the coalescence of all oligomeric states of a protein in a single zone [Fig.1 of (40)]. The instability of pH gradients with time makes this frequently difficult. This mechanism of artifactual zone formation in EF can be detected by observing a decrease of the number of zones with increasing pore size of the gel and with increasing time of EF.

Aggregates of a single protein should not persist after reaction with SDS in the second dimensional gel. If due to the inadequacy of reaction conditions for the derivatization with SDS (see section 4) they do persist, they are recognizable in the 2-dimensional technique as vertical alignments of spots in the second dimension. However, this is clearly not obtained unless all aggregation states attain to their common isoelectric position, i.e. in a "non-restrictive" gel.

The practical remedy for this problem is to select a disaggregating milieu for EF and to select a gel with "non-restrictive pore size" [1]. The most "non-restrictive" gels are those polyacrylamide gels crosslinked more than 10% which are presumably supercoiled (45). Their mechanical stability decreases, however, as %C is increased from 10 to 50%. For work at 0°C, a more mechanically stable highly DATD crosslinked gel "nonrestrictive" to molecules of less than 0.5×10^6 MW has been recommended (42), but is controversial, since some authors consider its use counterindicated by the chain-terminating property of DATD and its asymmetric incorporation into the polymer (46,47). Other candidates for "non-restrictive" gels with even larger pore sizes are agaroses (48). Their degree of non-restrictiveness also, as in polyacrylamide, appears to depend on their capacity for super-coiling (in the decreasing order of IsoGel to SeaPlaque to SeaPrep 15/45, Marine Colloids) (49). However, the adherence to (agarose coated) glass walls of these gels is less strong [2] than that of such polyacrylamide gels as 4 %T, 4 $\%C_{Bis}$ or 5 %T, 15 $\%C_{DATD}$ at 0°C, and their mechanical stability under conditions of EF appears less, so that agaroses do not seem to be the universal answer for a "non-restrictive" pore size in EF, even if one neglects residual electroendosmosis (51).

In summary, as long as protein aggregation persists under conditions of EF, due to the ineffectiveness of the disaggregating solvent, or as long as gel restrictiveness prevents the various aggregation states of a protein from merging at their common pI-position within the lifetime of the pH gradient, an artifactually high number of horizontally displaced spots is produced on the 2-dimensional map. Failure of disaggregation under the conditions of SDS-PAGE used in 2-dimensional mapping may in addition displace these artifactual spots vertically (see section 4).

b) Ligands. Continuous ligand loss, e.g. of SDS, or ligand gain, e.g. of isocyanate derived from urea, may prevent the steady-state and the onset of pattern constancy. EF gels then become unsuitable for 2-D mapping for the reasons outlined above, and experimental remedies such as urea purification or replacement of SDS by an effectively neutral detergent like CHAPS (24) are needed.

c) Voltage minima. The conductance across the migration path in EF is uneven for at least 4 reasons: i) Unevenness of carrier ampholyte concentrations; ii) increased specific conductance of acidic and basic carrier ampholyte species as compared to neutral ones; iii) the conductance contribution of protons and hydroxyl ions to the acidic and basic regions of a pH gradient. iv) In addition, there may exist a "microscopic" variation in conductance along the migration path between the distribution maxima of carrier ampholyte zones and the minima between them.

The concentrations of carrier ampholytes are uneven across the migration path in EF: Acidic and basic species exist at much higher concentrations than neutral ones [Fig.2 of (25)]. In general, this is probably true for synthetic carrier ampholyte mixtures (SCAMs) as well as simple buffer mixtures (52). Data showing even carrier ampholyte distributions [e.g. Fig.2-4 of (53)] are misleading in this regard since they reflect an arbitrary single early time point in EF. At an early time of EF, acidic and basic species appear as salt-complexes across the entire migration path. These complexes separate in the field with time. This results in a progressive increase of voltage at neutrality, and a corresponding decrease of voltage in the acidic and basic regions (54). This has the effect that proteins are progressively decelerated in their migration through the acidic and basic regions of the pH gradient towards their pI-position.

The relatively high conductance per mole of acidic and basic carrier ampholytes as compared to neutral ones [Fig.2 of (55)] and the contribution of protons and hydroxyl ions to conductance will also serve to promote the low mobility of zones across the terminal segments of the pH gradient. This low mobility delays the condensation of uniformly, or both cathodically and anodically applied proteins and of the various oligomeric forms of a protein into a single focused zone. Such delay may give rise to multiple zones, either of doublets of the same protein which approximate one another, or of oligomers of the same protein (section a), if the time allotted for the separation is insufficient. Correspondingly, regions of high wattage can give rise to irreversible thermal protein precipitation (see section 1b). Carrier ampholytes do not distribute continuously across the migration path but rather form discrete zones detectable by Schlieren optics (56,57), by the naked eye as waves of gel on the surface of horizontal gel slabs, by "caramelization" (58), or by staining (59). Protein and carrier ampholyte zones are strikingly coincident [Figs. 2-VI, 3, 4, 5 and 6 of (57); p. 151 of (58)]. The coincidence of protein and carrier ampholyte patterns may indicate a slowing down of protein migration rates at or between the loci of maximal carrier ampholyte concentration along the pH gradient where the voltage is low. Or it may reflect binding of carrier ampholytes by the protein (see below). The idea was also entertained during the discussion that the apparent carrier ampholyte patterns obtained in the presence of protein could be due to the osmotically swollen protein zones themselves. This notion would obviously dispense with the "coincidence phenomenon", but seems counterindicated by the specificity of such methods as caramelization for carrier ampholytes and by the identity of patterns in presence and absence of protein. When bands are slowed down in their migration by either of these mechanisms to the degree that they can be considered "arrested" within the time frame of an EF separation, they pile up forming artifactual heterogeneous zones in EF. These pileups may precipitate thermally, due to the high voltage across a highly concentrated protein zone. Even homogeneous bands under these conditions remain in a transient state and are subject to its augmented irreproducibility relative to the isoelectric steady-state, which again leads to greater irreproducibility of the 2-D map.

3) Ionic, hydrophobic and H-bond interactions.

Ionic interactions during EF appear favored through the very low ionic strength of the system. However, the dominant opinion of the panel was that the energies of ionic interactions of proteins were insufficient to withstand separation in the electric field under the conditions of EF. Ionic protein-protein interactions can be detected for a short time when isoelectric carrier ampholytes are used as buffers in the second dimension of pH-mobility profiles. Such biologically important interactions have been revealed for the first time between methemoglobin and cytochrome b_5

(60). As a rule, however, it is to be expected that during the transient state of an EF fractionation ionic complexes will have ample time to dissociate.

Other evidence for interactions exists that must at least in part have ionic character. The addition of a basic carrier constituent to a mixture of carrier constituents affects the acidic side of the pH gradient, presumably through salt formation (52). In the development of pH gradients with time, acidic and basic constituents separate gradually (25) at rates that clearly are less than the migration time needed to traverse the gel length. The EF of nucleic acids (61) provides a good example for an artifactual charge interaction between the basic, protonated oligoamino backbone of carrier ampholytes and an acidic species; another such example is the well documented and quantitated interaction between heparin and basic carrier ampholytes (62). It is very likely that similar interactions do take place between basic proteins and acidic carrier ampholytes. Such interactions have been observed with polyarginine, but not with polylysine or histone (63). No clear evidence for binding of neutral proteins to carrier ampholytes exists in the literature (63,64). These polar interactions should not be affected by the addition of either hydrophobic bond breaking agents (detergents) or H-bond breaking agents (urea) to the EF system.
Non-polar, hydrophobic interactions have been demonstrated between Triton X-100 and carrier ampholytes [Fig.4 of (65)]. Alkaline carrier ampholytes, being the most non-polar, were found to bind the detergent preferentially. Cooperative hydrophobic and polar interactions with carrier ampholytes have been demonstrated by dyes (60). Protein-protein binding under the conditions of EF probably falls into the same cooperative category. It was the consensus of the panel that 2-D macromolecular mapping under the conventional conditions is not perturbed by the hydrophobic and H-bond interactions of proteins since they are overwhemingly disrupted by 9 M urea, 2% non-ionic detergent. It has been clearly demonstrated that dissociating agents such as 8 M urea, formamide or DMSO are effective in the case of dye- and polyanion-binding to carrier ampholytes (60,62). Electrofocusing in simple buffers (MW 200) (52) can provide a control for artifacts caused by binding of proteins to SCAMs (MW 700) as well as an alternative mode of EF that should decrease the incidence of binding artifacts.

4) Failure to saturate the isoelectric proteins with SDS.

a) Solubilization and mobilization by SDS: The solubilization and mobilization of the isoelectric zones obtained in the 1st dimension proceeds by reaction with SDS either in a "equilibration buffer" (1) and/or by electrophoretic passage of SDS through the EF gel. In either case, 2% non-ionic detergent-9 M urea are substituted for the conventional application of heat to fully denature the protein and to unravel the polypeptide chains for full saturation with SDS. In the latter case, no net reducing conditions are applied to hydrolyze disulfide bonds for the same purpose [although net reducing conditions could be obtained by means of charged sulfhydryl compounds such as thioglycolate (66)]. Solubilization depends on the physical state of the protein in the EF zone and the degree to which the isoelectric precipitation of proteins in the 1st dimension remains reversible (irreversible protein precipitation progresses with time after the attainment of the isoelectric endpoint as pointed out in section 3). Solubilization also appears to depend on the matrix: It has been found much less efficient in polyacrylamide EF gels polymerized into a polyacrylamide stacking gel than upon embedding the 1st dimensional polyacrylamide gel in an agarose stacking gel (67). This is apparently due to covalent binding between protein and gel matrix under the conditions of acrylamide polymerization and should apply also to those situations where a 1-D gel is polymerized into a 2-D Resolving Gel.

Failure to solubilize the EF zone quantitatively yields an erroneously low number or low concentration of spots. Obvious other causes of the same problem would be a loss of protein into the equilibration buffer or failure to efficiently fix and stain SDS-proteins[3].

b) Saturation with SDS: While solubilization and mobilization of the isoelectric proteins would succeed within wide quantitative limits of the reaction between protein and SDS, saturation is required for obtaining valid MW estimates in SDS-PAGE (the 2nd dimension). The validity of deriving MW estimates from SDS-PAGE at a single gel concentration derives from the saturation of unconjugated proteins with 1.4 g SDS/g of protein, and the corresponding common y-intercept of the protein derivatives on their Ferguson plots (73). To saturate a protein with SDS, one usually reacts at 100°C for 5 min in the presence of reducing agent. The high temperature unravels the protein chains to random coils, while the reducing agent splits disulfide linked chains, to allow for a maximal exposure of the polypeptide chains to SDS. In some cases, conformational stability is so high that much higher boiling times [e.g. 30 min, Fig. 1 of (72)] are required to saturate the protein with the detergent. Saturation with SDS requires more stringently than solubilization that 2% Triton-9 M urea be capable of fully replacing the action of heat and reducing agent in opening up the entire protein surface for reaction with SDS. This capability should be viewed critically and tested (e.g. by obtaining the identical MW at different gel concentrations).

There may also exist a specific impediment for the reaction of isoelectric proteins with SDS which is presumably related to isoelectric aggregation and precipitation of proteins. The evidence for this is still tentative and indirect: Firstly, very basic proteins migrate abnormally in mildly alkaline SDS-PAGE. These proteins, e.g. RNAse, should bind more SDS electrostatically than neutral proteins - but they bind less, resulting in "abnormally low" mobilities in SDS-PAGE. This is presumably due to the fact that the usual slightly basic pH of reaction with SDS is also near-isoelectric for basic proteins. Secondly, SDS-proteins in all buffers at pH's above 2 or 3 should exhibit the same mobilities, since the net charge of SDS-proteins is nearly entirely due to the sulfate charges with pK near 0, i.e. the sulfonate groups are fully ionized. Yet, mobilities of SDS-proteins between pH 7 and 4 (where most proteins are isoelectric) were found to be much less than those in alkaline buffers. The evidence for this comes from stacking experiments (unpublished), which show that at neutrality or below, a trailing ion mobility (relative to Na^+) of 0.200 cannot be attained, while in mildly alkaline buffers all SDS-proteins exhibit mobilities above that value (i.e. are stacked).

Finally, failure to saturate proteins with SDS may be due to their being conjugated with non-protein (lipo-, glyco- or nucleo-) moieties. To the degree that proteins fail to take up 1.4 g SDS/g, migration distance is no longer proportional to MW independently of gel concentration. In other words, different "MWs" are derived from a standard curve at 2 different gel concentrations, and evidently these "MWs" are in error. This is an important consideration for 2-D electrophoresis since one of its main applications is to membrane and other conjugated proteins.

Whatever the mechanisms leading to non-saturation of the protein with SDS may be in a particular application, it must lead to erroneously high MW estimates and therefore misidentification. It should be noted, however, that a correct identification even of SDS-saturated SDS-proteins on the basis of their size also depends on avoiding 2 further sources of possible error: Nonlogarithmic interpolation between the migration distances of the standards, and failure to take into account the sigmoidal, rather than linear, nature of the relation between log MW and migration distance [Fig.3 of (73)]. Thus, it is important to ascertain that the selected gel concentration lies within the central, approximately linear, segment of the sigmoidal standard curve.

c) Stacking: Most users of 2-D macromolecular mapping do not apply a specific stacking gel, although most employ a discontinuous buffer system, i.e. moving boundary electrophoresis, for the 2nd dimensional gel. It should be obvious that under those latter circumstances a stack, (a system of sequential moving boundaries) forms in the sample phase, i.e. the buffer in which the 1-D gel is suspended, independently of the presence or absence of a gel matrix. Thus, for the purpose of the following discussion, a "stacking gel" may be a buffer solution.
Failure to saturate proteins with SDS may cause a failure to stack under the conditions of SDS-PAGE. This is the case when trailing ion mobilities in SDS-PAGE are tailored to the relatively high mobilities of SDS-proteins (with mobilities relative to Na^+ of about 0.3) as opposed to the underivatized proteins (with representative relative mobilities of 0.05) (74). Stacking of the solubilized SDS-proteins appears important to provide a uniform starting zone for the simultaneous 2nd dimensional analysis of all proteins derived from the zones of the EF gel. It provides independence, to a large degree, of the relative amounts of proteins in the EF zones, if sufficient stacking gel length is provided to concentrate even the zones with the largest amount of protein into a stack. Given enough stacking gel volume, the stacked protein zones, visualized by a stacked tracking dye, assume a uniform migration distance prior to their arrival at the resolving gel surface. This fact provides, to some degree, a uniform starting zone independently from a perfectly horizontal resolving gel sur-

face as required by a 2-D gel in a continuous buffer for forming uniform starting zones at the resolving gel surface. Stacking also helps to overcome a geometric impediment to an immediate entry of the EF zones into the 2nd dimension. The 1st dimensional zone is formed at a right angle to the 2nd dimensional one, and the quantitative migration of protein from the former into the latter is slow relative to the rate at which the moving boundary sweeps the stacking gel. Since there is no way to predict the rate of solubilization and stacking of the 1st dimensional zones, that stacking gel length sufficient for the concentration of all proteins into sharp starting zones with equal migration distances for simultaneous entrance into the resolving gel needs to ascertained by systematic experiment (in which, of course, resolving gels need not be cast). In such stacking experiments, note should be taken of the spatially extended nature of the stack in SDS-PAGE [Fig.1 of (74)] and the need, therefore, to define with suitable tracking dyes or by observing refractile zones both the leading and the trailing edges of the stack (Fig.4, ibid.). The width of the stack depends on the amount of micellar SDS within the stack; it increases in proportion to migration distance, since SDS enters the stack continuously from the cathodic buffer (74).

5) Spot identification problems arising from pattern irreproducibility. Conventionally, spots are identified by visual comparison between two gel maps obtained within a single experiment, juxtaposing a control or standard pattern with that of the unknown run in parallel. This approach avoids the interexperimental error sources (differences in gel or buffer composition, temperature etc.). It fails, of course, to answer the question whether the same identification would be obtained if the 2-D macromolecular map was obtained 2, 3 or 25 times. This failure to reproduce the identification renders it of questionable value.

The identification of spots on a 2-D macromolecular map rests on their positions within the pI-MW coordinate system. The reproducibility of spot position in such a fixed coordinate system is, however, not sufficient to allow one to characterize each spot in terms of migration distances, or in terms of two numbers representing pH and MW. This has been attempted within the statistical confidence limits of these two numbers on a complex protein map, with the result that only the very strongest spots could be reproducibly identified in that fashion (75). There appears to be a general consensus that "unfortunately, neither (isoelectric point, molecular weight) can be measured directly from the gels with an accuracy close to the resolving power of the system unless standards are also incorporated" [p.316 of (76) and (77)].

More specifically, identification by pI and MW of a single spot, considered independent of any other spot, has a 95% probability of being accurate if both parameters are defined within ± 2 standard deviations, calculated from repetitive separations. When, as is necessary in practice, one spot is to be distinguished from another one, the distribution overlap between different spots taken as individual independent entities has to be evaluated. The 95% probability of correct distinction between the spots is then evidenced by the non-overlap of the ellipsoidal "95% confidence envelopes" in a pI vs. MW plot [Fig.3 of (75)]. In the single study, in which spot distinction by these criteria has been tested on a 2-D macromolecular map (75), it has not been possible to identify any component within 95% confidence limits by a non-overlapping MW - pI ellipse. This result may, however, change when a shifting coordinate system, and relative pI - MW values, are applied to the analysis. It is also possible that the data base in the single reported case was significantly inferior to that in other laboratories.

Hence, if pI and MW are used for identification, it appears that inter-experimental variations are too large to allow for a statistically valid spot identification in a rigid coordinate system. Data of Dr. Anderson, however, show that the migration distances between adjacent spots are reproducible "within 0.2 mm". Therefore, several authors (77-79) have introduced a locally shifting coordinate system, based on the relative position in terms of pI and MW of a spot in relation to the major, dominant spots of the protein system. This is analogous to band identification in a single dimension by relative rather than absolute mobility. The identification of a spot by relative pI - relative MW pairs of values has been statistically validated in some of the best evaluated 2-D analyses [e.g. (79)].

Significance of spot identification in 2-D patterns has frequently been claimed on the basis of biological and particularly genetic evidence. This of course raises the danger of circuitous reasoning: The spot is used to support a genetic phenomenon, and the genetic phenomenon is used to support the pattern change. The validity of using such extraneous sources of evidence to support spot identification does, of course, increase when several independent biochemical and genetic techniques are used to identify the spot of interest. An example is a study aiming at the identification of galactokinase (80). This protein was identified by specific antibody [see Fig.2 of (80)] and by electrophoresis of isogenic strains in which the kinase gene was either present as an amber mutant (Fig.5, ibid.) or in which multiple copies of the kinase gene were expressed (Fig.1, ibid.).

Pattern irreproducibility appears predominantly due to the EF problems discussed in sections 1-3. Therefore, it is possible that they will disappear once the first dimension of separation can be carried out in an immobilized pH gradient (41) with wide pH range. To date, it has not been possible to set up such gradients.

6) Overshadowing of minor components by major ones.

Protein concentrations in biological fluids and extracts may vary in concentration by several orders of magnitude. This poses the dilemma that at low enough sample concentrations, one obtains a "map" of only the major components, e.g. the 30 protein zones of serum in PAGE [Fig.8 of (81)], while at concentrations sufficient to detect the minor components, the major zones overshadow the minor ones [Figs.1,2 of (82)]. Thus, to avoid the overshadowing of the minor components by the major ones in the analysis of systems containing dominant proteins, such as serum [Figs.1,2 of (82)], the sample must be freed of its major components prior to 2-D macromolecular mapping. But such removal brings with it the danger of also removing, inadvertently, other chemically similar or adsorbed minor components. Critical testing for the simultaneous removal of minor components is needed, e.g. by testing the specificity of the antibody or the affinity matrix used to adsorb the major component from the system. This is not a routine task - the methodology of removing major components selectively remains in fact quite underdeveloped.

References

1) O'Farrell, P. H. (1975) J. Biol. Chem. 250, 4007-4021.

2) Anderson, L. and Anderson, N. G. (1977) Proc. Natl. Acad. Sci. USA 74, 5421-5425.

3) "Clinical Applications and Developments in Two-Dimensional Electrophoresis" (1982) Clin. Chem. 28, 737-1092.

4) Vo, K.-P., Miller, M. J., Geiduschek, E. P., Nielsen, C., Olson, A. and Xuong, N. H. (1981) Anal. Biochem. 112, 258-271.

5) Merril, C.R., Goldman, D. and Ebert, M. (1981) Proc. Natl. Acad. Sci. USA 78, 6471-6475.

6) Kronberg, H., Zimmer, H. G. and Neuhoff, V. (1980) Electrophoresis 1, 27-31.

7) Klose, J., Nowak, J. and Kade, W. (1980) in Electrophoresis '79 (Radola, B. J., ed.), W. de Gruyter, Berlin - New York NY, pp.297-312.

8) Capel, M., Redman, B. and Bourque, D. P. (1979) Anal. Biochem. 97, 210-228.

9) Merril, C. R., Goldman, D. and Van Keuren, M. L. (1982) Electrophoresis 3, 17-23.

10) Poehling, H. M. and Neuhoff, V. (1981) Electrophoresis 2, 141-147.

11) Guevara, J. Jr., Johnston, D. A., Ramagali, L. S., Martin, B. A., Capetillo, S and Rodriguez, L. V. (1982) Electrophoresis 3, 197-205.

12) Allen, R. C. (1980) Electrophoresis 1, 32-37.

13) Altland, K., Rauh, S. and Hackler, R. (1981) Electrophoresis 2, 148-154.

14) Hjelmeland, L. M., Nebert D. W. and Chrambach, A. (1979) Anal. Biochem. 95, 201-208.

15) Newby, A. C., Chrambach, A. and Bailyes, E. M. (1982) in Techniques in Lipid and Membrane Biochemistry - Part I (Hesketh, T. R., Kornberg, H. L., Metcalfe, J. C., Northcote, D. H., Pogson, C. I. and Tipton, K. F., eds.) Vol. B4/I, No. B409, Elsevier/North Holland Biomedical Press, Amsterdam, The Netherlands, pp.1-22.

16) Hjelmeland, L. M. and Chrambach, A. (1981) Electrophoresis 2, 1-11.

17) Gonenne, A. and Ernst, R. (1978) Anal. Biochem. 87, 28-38.

18) Anderson, N. G. and Anderson, N. L. (1978) Anal. Biochem. 85, 331-340.

19) Klose, J. and Feller, H. (1981) Electrophoresis 2, 12-24.

20) Horst, M. N., Basha, S.M.M., Baumbach, G.A., Mansfield E.H. and Roberts, R.M. (1980) Anal. Biochem. 102, 399-408.

21) Wilson, D. L., Hall, M. E., Shone, G. C. and Rubin, R. W. (1977) Anal. Biochem. 83, 33-34.

22) PAGE and PAGIF (Allen, R. C. and Maurer, H. R., eds.), 2nd ed., W. de Gruyter, Berlin - New York NY, 1983, in press.

23) Newby, A. C. and Chrambach, A. (1978) Biochem. J. 177, 623-630.

24) Hjelmeland, L. M. (1980) Proc. Natl. Acad. Sci. USA 77, 6368-6370.

25) Baumann, G. and Chrambach, A. (1975) in Progress in Isoelectric Focusing and Isotachophoresis (Righetti, P. G., ed.) Elsevier, Amsterdam -New York NY, pp. 13-23.

26) Ben-Or, S., Sullivan, J. V. and Chrambach, A. (1982) in Electrophoresis '82 (Stathakos, D., ed.), W. de Gruyter, Berlin-New York NY, in press.

27) Salokangas, A., Eppenberger, U. and Chrambach, A. (1981) Prep. Biochem. 12: 299-320.

28) Garrels, J. I. and Hunter, T. (1979) Biochim. Biophys. Acta 564, 517-525.

29) Izant, J. G. and Lazarides, E. (1977) Proc. Natl. Acad. Sci. USA 74, 1450-1454.

30) O'Farrell, P. Z. and Goodman, H. M. (1976) Cell 9, 289-298.

31) Latner, A. L., Marshall, T. and Gambie, M. (1980) Electrophoresis 1, 82-89.

32) Parker, J., Pollard, J. W., Friesen, J. D. and Stanners, C. P. (1978) Proc. Natl. Acad. Sci. USA 75, 1091-1095.

33) Anderson, N. L. and Anderson, N. G. (1979) Biochem. Biophys. Res. Commun. 88, 258-265 (1979).

34) Klose, J. and Spielmann, H. (1975) Biochemical Genetics 13, 707-720.

35) Hjelmeland, L. M. and Chrambach, A. (1982) Electrophoresis, in press.

36) Catsimpoolas, N. (1975) in Progress in isoelectric focusing and isotachophoresis (Righetti, P. G., ed.) Elsevier, Amsterdam-New York NY, pp.76-92.

37) Allan, B. J., White, T. T., Kirk, J. and Schilling, J. J. (1981) Anal. Biochem. 113, 1-25.

38) Bravo, R., Small, J. V., Fey, S. J.Larsen, P. M. and Celis, J. E. (1982) J. Mol. Biol. 154, 121-143.

39) Finlayson, R. and Chrambach, A. (1971) Anal. Biochem. 40, 292-311.

40) Nguyen, N. Y. and Chrambach, A. (1980) Electrophoresis 1, 14-22.

41) Bjellqvist, B., Ek, K., Righetti, P. G., Gianazza, E., Goerg, A., Postel, W. and Westermeier, R. (1982) J. Biochem. Biophys. Methods 6, 317-339.

42) Baumann, G. and A. Chrambach. (1976) Anal. Biochem. 70, 32-38.

43) Chrambach, A. (1980) J. Molec. Cell. Biochem. 29, 23-46.

44) Yadley, R. A., Rodbard, D. and Chrambach, A. (1973) Endocrinology 93, 866-873.

45) Rodbard, D., Levitov, C. and Chrambach, A. (1972) Sep. Sci. 7, 705-723.

46) Bianchi Bosisio, A., Loeherlein, C., Snyder, R. S. and Righetti, P. G. (1980) J. Chromatogr. 189, 317-330.

47) Gelfi, C. and Righetti, P. G. (1981) Electrophoresis 2, 213-219.

48) Buzas, Zs. and Chrambach, A. (1982) Electrophoresis 3, 121-129.

49) Buzas, Zs. and Chrambach, A. (1982) Electrophoresis 3, 130-134.

50) Nguyen, N. Y., McCormick, A. G. and Chrambach, A. (1978) Anal. Biochem. 88, 186-195.

51) Serwer, P. and Hayes, S. J. (1982) Electrophoresis 3, 80-84.

52) Cuono, C. B. and Chapo, G. A. (1982) Electrophoresis 3, 65-75.

53) Vesterberg, O. (1973) Ann. N. Y. Acad. Sci. 209, 23-33.

54) Jackiw, B. A., Chidakel, B. E., Chrambach, A. and Brown, R. K. (1980) Electrophoresis 1, 102-106.

55) Gelsema, W. J., De Ligny, C. L. and Van der Veen, N. G. (1979) J. Chromatogr. 173, 33-41.

56) Righetti, P. G., Pagani, M. and Gianazza, E. (1975) J. Chromatography 109. 341-356.

57) Bonitati, J. (1981) J. Biochem. Biophys. Methods 4, 49-70.

58) Felgenhauer, K. and Pak, S. J. (1973) Ann. N. Y. Acad. Sci. 209, 147-153.

59) Frater, R. (1970) Anal. Biochem. 38, 536-538.

60) Righetti, P. G., Gacon, G., Gianazza, E., Lostanlen, D. and Kaplan, J. C (1978) Biochem. Biophys. Res. Commun. 85, 1575-1581.

61) Righetti, P. G. and Drysdale, J. W. (1973) Ann. N. Y. Acad. Sci. 209, 163-186.

62) Righetti, P. G. and Gianazza, E. (1978) Biochim. Biophys. Acta 532, 137-146.

63) Righetti, P. G., Gianazza, E. and Bianchi Bosisio, A. (1979) Recent Developments in Chromatography and Electrophoresis (Frigerio, A. and McCamish, M., eds.) Elsevier, Scientific Publ. Co., Amsterdam, pp. 89-117.

64) Baumann, G. and Chrambach, A. (1975) Anal. Biochem. 64 530-536.

65) Gianazza, E., Astorri, C. and Righetti, P. G. (1979) J. Chromatogr. 171, 161-169.

66) Dirksen, M. L. and Chrambach, A. (1972) Sep. Sci. 7, 747-772.

67) Poehling, H. M. and Neuhoff, V. (1980) Electrophoresis 1, 90-102.

68) Chrambach, A., Reisfeld, R. A., Wyckoff, M. and Zaccari, J. (1967) Anal. Biochem. 20, 150-154.

69) Diezel, W., Kopperschlaeger, G. and Hoffman, E. (1972) Anal. Biochem. 48, 617-620.

70) An der Lan, B. and Chrambach, A. (1982) this volume

71) Newby, A. C. and Chrambach, A. (1978) Biochem. J. 177 623-630.

72) An der Lan, B., Cowell, J. L., Manclark, C. R. and Chrambach, A. (1982) Arch. Biochem. Biophys., submitted.

73) Rodbard, D. (1976) in Methods of Protein Separation (Catsimpoolas, N., ed.) vol.2., Plenum Press, New York NY, pp. 145-218.

74) Wyckoff, M., Rodbard, D. and Chrambach, A. (1977) Anal. Biochem. 78, 459-482.

75) Hurley, P. M., Catsimpoolas, N. and Wogan, G. N. (1978) in Electrophoresis '78 (Catsimpoolas, N., ed.) Elsevier, Amsterdam-New York NY, pp. 283-296.

76) Anderson, N. L., Edwards, J. J., Giometti, C. S., Willard, K. E., Tollaksen, S. L., Nance, S. L., Hickman, B. J., Taylor, J., Coulter, B., Scandora, A. and Anderson, N. G. in Electrophoresis '79 (Radola, B. J., ed.), W. de Gruyter, Berlin-New York NY, pp. 313-328.

77) Taylor, J., Anderson, N. L. and Anderson, N. G. (1981) in Electrophoresis '81 (Allen, R. C. and Arnaud, P., eds.) W. de Gruyter, Berlin-New York NY, pp. 383-400.

78) Kronberg, H., Zimmer, H. G. and Neuhoff, V. (1981) in Electrophoresis '81 (Allen, R. C. and Arnaud, P., eds.) W. de Gruyter, Berlin-New York, pp.413-423.

79) Merril, C. R., Goldman, D. and Ebert, M. in Electrophoresis '81 (Allen, R. C. and Arnaud, P., eds.) W. de Gruyter, Berlin-New York NY, pp. 343-353.

80) Merril, C. R., Gottesman, M. E. and Adhya, S. L. (1981) J. Bact. 147, 875-887.

81) Davis, B. J. (1964) Ann. N. Y. Acad. Sci. 121, 404-427.

82) Felgenhauer, K. (1979) J. Chromatogr. 173, 299-311

Footnotes

[1] The criterion for "non-restrictiveness" has been Steady-State Stacking of a protein at pH 10.5 with a trailing ion (relative) mobility of 0.064 (42). This is obviously arbitrary.

The determination of "non-restrictiveness" of a gel for a particular protein cannot be made in simple or unequivocal fashion. The sloped segment of the Ferguson plot of a protein comprises the range of gel concentrations which are unequivocally restrictive. However, its horizontal segment, i.e. the $R_f = 0$ region of the Ferguson plot, does not necessarily define the range of "non-restrictive" gel concentrations. This fact is due to the dependence of the stacking of a protein on the trailing ion mobility of the moving boundary (stack) in question (43). The larger that mobility, the more reliably is stacking of the protein ($R_f = 1.0$) a measure of non-restrictiveness. To achieve high protein mobilities, the test for non-restrictiveness should be carried out at very high or very low pH. At neutral pHs, the degree of restrictiveness, estimated by the criterion of stacking at an extreme of pH, appears frequently too low, in view of protein aggregation at and near the pI. The molecular shape change concurrent with titration also has to be considered [e.g. Table 3 of (44)] but appears frequently negligible in the context of "non-restrictiveness". Thus, within the constraints of protein aggregation, the estimates of non-restrictiveness made by the criterion of stacking at high pH and low trailing ion mobility appear applicable to EF.

[2] Our experience of differences in wall adherence properties refers necessarily to gel tubes since they cannot be observed in horizontal gel slabs. In gel tubes, so-called wall adherence manifests itself in two ways: Mechanical separation of the gel from the tube, and, in its absence, a dependence of pH in the gel on the pH of the electrolyte in which it is suspended (50). Appropriate coating of the tube overcomes both forms of wall separation (50). The second form of wall separation is presumably a capillary effect that should be equally important for gel slabs as for tubes, although it cannot be measured in the former.

[3] Staining of SDS-proteins: SDS-proteins are highly soluble and therefore resist fixation under the standard conditions of precipitation in protein chemistry; they are also highly negatively charged and resist staining by the anionic dyes commonly applied to protein staining in gel electrophoresis. Both of these problems can lead to spot disappearance from the map. By contrast, the SDS-derivatives of large molecular weight SCAMs may be stained with protein stains (e.g. Coomassie Violet) (59) and produce artifactual zones in the 2nd dimension. Blank gels should be stained to detect this type of artifact. This seems particularly important when the classical protein stains are replaced with the more sensitive, but also less specific, silver stains (9-11).

Commonly, 2-dimensional gel patterns are obtained by fixing in dilute acetic acid solutions containing one of the lower alcohols. The objection against that procedure is that it does not provide a reliable SDS-protein or even protein fixative. An example of zone disappearance in such a medium is given in Fig.5 of (14). Using a reliable protein fixative such as TCA, under the standard conditions of no-background protein staining (68,69), the most efficient way to stain 2-dimensional patterns with minimal risk of protein solubilization is one recently devised for SDS-PAGE (70): Removal of SDS (and other detergents) from the gel by rapid diffusion of the detergent monomer from the gel, using 12.5% TCA for secure protein fixation, enough (usually 45%) methanol to increase the CMC of the detergent and thereby prevent micelle formation, and a mechanical device for effective magnetic stirring (70). This procedure, followed by staining according to Diezel et al. (69), allows for pattern development on the day of electrophoresis. If the presence of TCA-precipitable detergents present in the system [i.e. derived from the 1st dimension and carried into the second one as in Fig.3 of (71)] presents a problem, protein fixation by crosslinking with glutaraldehyde can be applied (14). A suitable procedure consists of 2-3 h staining in 2% glutaraldehyde (40 ml/2 ml gel) containing 1 ml 0.25% Coomassie Brilliant Blue G-250 per 40 ml, followed by storage in 5% acetic acid (10 ml/gel) (72).

Table 1

Detergent[a]	% Protein[b] Solubilized[c]
1% SDS, ME, 100°C	100
1.2% SB_{14}	65
2% NP-40	46
None	40

[a] Detergent/protein = 30-40/1

[b] 3T6 fibroblast cells

[c] Supernatant 100,000g, 20 min.

Data of (17)

Table 2

Recovery[a] (%) from IEF in Triton X-100	Matrix	Zone Width	Carrier Ampholytes
65	Ultrogel ACA 54	broad	buffers
0-5	Sephadex G-75	sharp	buffers
0-8	Polyacrylamide,	18mm	Ampholine
40-50	"	6mm	"
10	" + ethyleneglycol	6mm	"

[a] of cAMP-dependent protein kinase

Data of (27)

ROUND-TABLE DISCUSSION: A SILVER STAIN WORKSHOP*

Chairman: C.R. Merril

Laboratory of General and Comparative Biochemistry, NIMH, NIH, Bethesda MD 20205, USA

Panel Members:

R.C. Allen

Dept. of Laboratory Animal Medicine, Medical University of South Carolina, Charleston SC 29401, USA

W. Ansorge

European Molecular Biology Laboratory, Postfach 10.2209, D-6900 Heidelberg, FRG

V. Neuhoff

Max-Planck-Institut für experimentelle Medizin, Forschungsstelle Neurochemie, D-3400 Göttingen, FRG

D. Sammons

The Upjohn Co., Diabetes Research, Kalamazoo MI 49001, USA

Audience Participants:

A.H.M. Burghes

Jerry Lewis Muscle Res. Center, Department of Paediatrics and Neonatal Medicine, Hammersmith Hospital, London W12 OHS, U.K.

Ch. Cuono

Yale University, School of Medicine, 333 Cedar St., New Haven CT 06510, USA

M. Lasky

Pharmacia Fine Chemicals, 800 Centennial Avenue, Piscataway NJ 08854, USA

* This round-table discussion was held on April 24, 1982.

Electrophoresis '82

A SILVER STAIN WORKSHOP

Chairman: Dr. C.R. Merril

Panel Members: Dr. R. Allen
Dr. W. Ansorge
Dr. V. Neuhoff
Dr. D. Sammons

Audience Participants: Dr. A. Burghes
Dr. Ch. Cuono
Dr. M. Lasky

Dr. Merril: Following the introduction of silver staining to the electrophoretic community a number of staining variations of the method have been developed. Most of these stains have similar problems. I made a list of these problems: 1) protein fixation, 2) loss of sensitivity, 3) background staining and surface and other artifacts. If anybody in the audience wants to bring up other problems, that will be fine too. I thought we'd go through each of these one by one. Also we will have some of the individuals on the panel present their techniques for minimizing these problems. We will begin with fixation. Let me just state my position: in general we have found that 50% methanol and 10% acetic acid, is a fairly good fixing solution. At least for gels that are 1 mm thick. For gels 0.5 mm or less we generally use 20% TCA. Protein loss becomes fairly dramatic for gels thinner than 0.5 mm, if methanol and acetic acid are used to fix them. I'd like to call on some of the other panel members for their comments:

Dr. Neuhoff: I have heard just this morning that there would be a silver stain session. So I am not very well prepared and I have no material with me. But if you will allow me to make a general comment, it is my feeling that silver staining is not a real staining procedure. It's not chemistry, it's just alchemy. I do not like procedures where we have to use techniques very carefully just to have a clear background. I very much hope that in the future development of silver stains they will end up as simple as imido-black or Coomassie blue staining techniques. So that we will have not all the problems we have now.

Dr. Merril: As for silver staining being a form of alchemy, the major problem with the alchemists was that they couldn't fulfill their promises. They really couldn't convert base metals into gold. With the silver stain we really do have a stain with sensitivity 100 fold or greater then Coomassie blue. As for the underlying chemical mechanism of the silver stains, the stain techniques have been developed from adaptions of histological silver stains and photographic processes. In all the applications: histological stain, production of silver photographic images and silver gel stains, the basic reaction involves the reduction of ionic silver to metallic silver. The unknown for the gel protein silver stains is: why this reduction is differentially affected by the proteins or other polymers in the gel. There is no question that it is important to try to understand how the stain works, such understanding will help in developing simpler more reliable stains, however, even without a full understanding, the stains will serve useful purposes, just as the current histological silver stains do. As long as we make no false claims or promises than there should be no fear of alchemy.

In my prepared presentation yesterday, I tried to review some of the experiments in my laboratory which give some insight on how silver staining works. In this regard I believe the experiment in which, simple reduction of a gel soaked in a solution of silver nitrate, produces negative protein spot images, strongly suggests that the proteins on the gel are physically altering the local conditions and it is this physical effect which causes the differential silver staining. Perhaps someone else on the panel or in the audience might have some further insights on the mechanism of silver staining.

Dr. Allen: I'd make one more comment on fixation, Dr. Schwabe and I at the Medical University of South Carolina, have been working with the protein relaxin with a molecular weight of about 2,500. It turns out that TCA is a very interesting fixative if you buy commercial grade, the cheapest material, the relaxin will precipitate in 10% TCA. If you buy the purified grade, relaxin stays in solution. The ability of relaxin to stay in solution in 10% purified TCA is now being used to purify relaxin. One should perhaps use the cheapest material possible and go to 20% TCA, particularly for the ultra thin gels, otherwise you are going to lose protein. You are not fixing it, and as Dr. Merril said you do not have that much success with methanol and acetic acid with ultra thin gels.

Dr. Sammons: I would like to make a comment about fixation that relates to color. With the color silver stain Gelcode (trademark) it is essential for one to use ethanol-acetic acid as a fixative. There is also a difference in fixation between methanol and ethanol, with ethanol being the better fixative when used according to Gelcode (trademark) instructions.

Audience question: What about the use of glutaraldehyde fixation?

Dr. Allen: I think the people that try to fix with aldehydes usually use them as a secondary step following fixation with TCA, or methanol-acetic acid. They certainly do increase sensitivity. When we discussed the diamine procedures with the various glutaraldehydes, paraformaldehydes, and other reagents that have been used we'll try to answer that question. I wouldn't suggest that anybody try to fix protein with aldehydes alone.

Dr. Neuhoff: I'd just like to add a comment, I would like to recommend that you use a fixation time long enough to wash out all the glycine, otherwise you will have black gels. The glycine will reduce the silver.

Dr. Sammons: I second that motion. We recently made some changes in our washing procedures after fixation. We just wash in water after the fixation to remove the glycine and ampholytes. We have found after fixation if one changes the water and this is for 1.5 mm gels, four times (one hr each change) one will get out everything that is going to wash away.

Dr. Merril: One of the problems that exists is that many of us are staining gels of different thicknesses. Colleagues who are using thicker gels, have noted that the thicker the gel the more important it is to lengthen the wash time prior to staining, I think some agitation during the wash is also very helpful.

Participant: I tried to fix 1.0 mm thick isoelectric focusing gels with foil on one side with ethanol or methanol and acetic acid. It doesn't work, I don't know the reason for that, but it might be that when there's a foil on one side, the fixatant solution can penetrate only from one direction. Another point concerns the presence of the ampholytes, they may in some way interfere or compete, perhaps in fixation between the protein and the ampholytes, the ampholytes are in excess; I don't know the reason but you have to use TCA if you want to silver stain isoelectric focusing gels.

Dr. Cuono: Have any of the panelists have any comments on other fixatives which may be important to use with sialic acid containing glyco-proteins which are notoriously soluble in just about everything. Some of the ones with which I'm working are only precipitable by mixtures of sulfosalicylic acid and TCA or alternatively by phosphotungstic acid which I do know precipitates ampholytes. Are there any comments on that, as far as whether sulfosalicylic or phosphotungstic interfere with the silver stain?

Dr. Merril: I haven't had any experience with those fixatives.

Dr. Allen: What percentage TCA have you been using? I have not found any problem with glycoproteins using 20% TCA to fix them. We've been working with proteins such as alpha-one-anti-trypsin and orosomucoid. In fact we have so much protein that we have to dilute our samples.

Dr. Merril: At this time I would like to have some more discussion on the mechanism of the silver stain. As I mentioned in my presentation yesterday it's my personal impression that the silver staining of proteins in gels works primarily by some physical effect, perhaps with a mechanism similar to the Donnan equilibrium effect. As I mentioned previously, if a gel is soaked in silver nitrate, in the dark, and then developed (by placing it in a reducing agent) a negative image is obtained. Not a positive image. It is as if the silver ions have been excluded or perhaps in a state in which reduction is differentially inhibited where there are charged polymers such as proteins. Other charged polymers such as RNA, DNA and lipopolysaccharide also stain with silver. The other piece of information that fits with a "physical effect" is the fact that by adding other ions which might help neutralize charge effects, the stain effect can be eliminated. I added, for example, ammonium nitrate, (which I chose because it doesn't precipitate the silver and it also doesn't change the pH of the silver nitrate solution). Ammonium nitrate at 1M concentration totally cancels out the silver staining effect. Experiments with radioactive silver (silver, 110^{M} a gamma emitter) also gave some insight into the stain mechanism. Gels that had proteins separated on them were placed in this radioactive silver solution. The silver had a very high specific activity so we used a 10^{-9}M solution. After 20 minutes in the silver solution the gel was covered with plastic and overlayed with x-ray film. When the film was developed the radioactivity was primarily located in the areas in which there are proteins. Proteins do in fact have groups on them, such as sulfur, that will react with silver. On the other hand, when nonradioactive silver nitrate was added to the solution so that the silver concentration was 10^{-1}M silver, (a concentration similar to that used in the stain) an even distribution of the radioactive silver is observed on the autoradiogram. So at the concentration of silver used in the stain experiments the physical distribution of the ions is uniform. The ability to form a negative image by reduction must indicate that the silver near the protein is in a different form. Many of the steps involved in the histological stains are there for an image reversal purpose so that a positive stain is achieved. Image reversal, in itself is a very complicated process. Are there any comments or does anybody on the panel want to make any suggestions about the mechanism?

Dr. M. Lasky: I've done some investigation into the mechanism and as far as the protein is concerned I think that possibly both adelhyde and sulfur groups might be involved in the mechanism. I've done some studies with proteins that have a high serine content. These proteins give an intense color with the silver stain. Serine is the only amino acid which has a primary hydroxyl group. Generally in the silver staining mechanism there's an oxidation step which would oxidize primary hydroxyl groups to aldehydes, at which point the silver diamine complex, which is also known as Tollens reagent, could complex to the aldehydes. Later the formaldehyde is used to form a silver mirror. The same thing happens with silver and sulfhydryl groups. You obtain silver mercaptides which also are known to give a color. I think possibly these groups may be involved somehow in the mechanism.

Dr. Merril: It may be that different silver stains rely more or less on physical effects or chemical effects. In this respect the current silver stain used in my laboratory (which consists of dichromate solution followed by sodium carbonate formaldehyde solution) probably relies more heavily on physical effects than the silver diamine stains. In a set of proteins from a crude *E. coli* lysate about one out of two hundred spots will not show up. I'm not talking about minor spots, some are major spots. If I then take the gel and just put it back into silver nitrate without additional prior treatment with dichromate, the image reversal reagent, then those spots will appear following reduction (in the carbonate-formaldehyde solution). I think what's happening is that those particular proteins, which do not appear at first, are reacting chemically with the silver. For these spots you probably don't want an image reversal. Hopefully someone will design some clever experiments to give more insight. Are there any other comments?

Dr. A Burghes: Relating to the aldehyde that was talked about. Do you think that possibly glutaraldehyde reacts with the amines on the proteins and this reaction actually increases the sensitivity in the histological types of silver staining?

Dr. Merril: Possibly. But I think that given the probability that you have two different processes here, a physical one and a chemical one, there's a balance thats got to be met. You might decrease the apparent sensitivity for some proteins with glutaraldehyde if it enhanced the reactivity and thereby decreased image reversal for that protein, at least in the none silver diamine stains. It would be interesting to do a densitometry survey of proteins stained with and without glutaraldehyde pretreatment, to see if some intensities were increased and others decreased.

Dr. Ansorge: Is this actually an oxidation step with dichromate or is it the chromium that's reacting?

Dr. Merril: I do not think it is a specific reaction for chromium. There is actually a whole series of compounds that can be used for carrying out image reversal. Image reversal reagents range all the way from oxidizing agents to reducing agents. Prior to using dichromate we used potassium ferricyanide. Just recently there was a paper in Analytical Biochemistry where Morrissey replaced the dichromate with dithiothreitol (DTT).

Participants: I've compared your method of using dichromate and Morrissey's using DTT. I stained hyaluronic acid which has only carboxide groups, and the DTT or Morrisseys method produced strong red band. With the dichromate it was slightly green, but with Oakley's method there was no stain at all. This adds another compound, hyaluronic acid glycosaminoglycans to the list that you can stain with silver, but, not with all stain procedures apparently.

Dr. Ansorge: To the points of the oxidation or not oxidation. In our technique, that I presented at this conference, we use potassium permanganate which normally, of course, is known as an oxidant. But the concentrations, (about 0.01%) that we are working at, they are very low. We believe that it is not primarily the action of oxidant that we're using there. We think it is a metal amplification of the image. We first bind the copper on the protein, then manganate (or its oxide or hydroxide), one could verify each of these steps by following the kinetics of the binding. At the end we bind silver on the protein, but there do not seem to be too many binding places (as e.g. the SH groups) available. The silver could bind more on the metals that are already available there. When we tried this with the dichromate it didn't work, the background was very high. With the permanganate it seems to be mainly a physical adhesion to the proteins. We tried about twenty, other metals, but none of them worked as well as the copper. The procedures have been polished for one year, with different biological groups all over Europe involved.

Dr. Neuhoff: Dr. Ansorge, a simple question, can you increase the sensitivity by your treatment?

Dr. Ansorge: The obtained sensitivity was 0.2 ng per band, or 0.01-0.02 ng/mm^2, as calculated in the way used by Dr. Merril. The figure depends upon protein type (bands of different colors), but is highly reproducible.

Dr. Merril: The sensitivity achieved by Dr. Ansorge's stain is similar to that which we observed in our stains. The fact that DTT, a reductant as Morrissey demonstrated did work points out clearly that you don't need an oxidizing agent. I think that reagents such as dichromate, permanganate and DTT are involved in the process of image reversal. In fact, I found that silver nitrate with just a very intense light will also give a positive stain (however it is not a very sensitive stain).

Dr. Sammons: With the color stain I think we may be dealing with a slightly different mechanism since it is totally independent of light. We don't know what the mechanism is. One point I want to make concerning Gelcode (trademark) is that sensitivity and background is also a function of the time one leaves the gel in the sodium hydroxide. If one leaves the gel in this step too long before the sodium carbonate one doesn't get the sensitivity. Thus, it's important to time the steps so that one puts the gel into the sodium carbonate at a time when the reaction will be maximized. Although we don't understand the chemistry of the process we do know that if one runs thinner gels than 1.5 mm one only needs to shorten the time in the sodium hydroxide and formaldehyde. All other reagents and times can remain the same.

Dr. Merril: To achieve image reversal, you don't actually need light. In fact when you photograph with color slide film, a positive image is developed for you, directly, by using chemical processes very similar to the ones we use. In fact I got the idea to use ferricyanide, and dichromate from the photo-chemical literature. I'd like Dr. Allen to present his variation on the stain method.

Dr. Allen: A basic point about stain is that it was developed back in the 50's for degenerated nerve tissue and it was, in retrospect, tailored to look mainly at the glycoproteins. Without correct adjustment of the ammonium/sodium ratio, there was nonspecificity of the stain. Of course, for our work we wanted nonspecificity. If one looks at the literature where the diammine stain has been used one finds that the ratio of ammonium to sodium varies from 29 parts of ammonium to 1 part sodium down to about 4 to 1. We have looked at this point and found that one does not achieve specificity or sensitivity if you have too low a sodium hydroxide concentration. Another point is that everyone stresses that the NH_4OH must be absolutely fresh. One must not forget that sodium hydroxide will absorb CO_2 from the air and become sodium carbonate if it is allowed to stand. This is more pronounced in a dilute solution and continual loss of sensitivity occurs after about a week of storage. So one must use fresh reagents throughout. With our particular stain procedure modification we can complete staining in 1/2 hr or less on ultrathin-layer gels. However, 18-25 hr wash times are required to wash our glutaraldehyde in 1.5 mm gels for an absolutely clear background. These latter were counterstained SDS gels that had been stained previously with Coomassie blue and thus were grossly overloaded for silver stains. The trick here with the silver step is, as Dr. Neuhoff pointed out in his excellent paper with Dr. Poehling, that this procedure is highly temperature sensitive. We ran about 1,000 gels with the original diammine method slightly modified to increase the sodium level with absolute reproducibility. After the meeting in Charleston last year the air-conditioning in our laboratory didn't work and the ambient temperature in the room varied between 4 and 5 degrees higher than usual. We suddenly got heavy background staining. This led to the next step of purposely using higher temperatures throughout. We had been using 55°C, then went to 60°C for the first steps and the silver diammine at 37°C because everybody normally has a waterbath. Simply, if you look at the literature and try to repeat various methods and are forced to do it in different parts of the world, one is going to run into ambient temperature problems. A 3° increase in room temperature can change the silver step kinetics. If you stain at 55°C or

60°C it doesn't make much difference whether you alter the temperature by 3°C to 5°C because one is no longer appreciably increasing the reaction rate. With these techniques in ultrathin gels one can stain with high sensitivity and detect about 100 picograms per mm of gel in a 20 min total procedure time. With free gels such as 1.5 mm SDS gels there is no real background problem as long as the glutaraldehyde is completely removed. Even at 65°C this takes an over-night wash. On gels backed on "Gel-Bond" you have very little problem although one would expect a problem with this material because "Gel-Bond" contains agar coat on it. We have treated it with hot glycerol prior to casting and while agar stains normally with silver, following this procedure, we got no background problem. If one uses silane to bond the gel to glass or polyester film, as Dr. Ansorge pointed out yesterday, the silane layer may be stained within minutes. This is a critical point in the timing of the diammine reaction step. Too long and the silane layer is reacted. As far as reproducibility goes we have found that we can use a number of different commercial glutaraldehydes as long as they are purified or electron-microscope grade. It's far better to dilute the glutaraldehyde 1:2, otherwise one has a much greater difficulty in washing it out of denser or thicker gels. As for quantification we think with silver that microdensitometry is appropriate, to compare similar proteins between samples, although not all proteins within samples. In regard to storage, we have gels that we stained over a year ago and they are just the same as when first stained. They haven't been stored under any specific conditions. However, they may sometimes leak losing the color. I think this occurs with use of older reagents. The procedure we have is as follows; slightly modified to employ only one temperature throughout.

A brief summary of my procedure for rapid silver diammine stain for ultrathin-layer gels is given:

Reagents

1. 10% Alcohol

2. Glutaraldehyde
   3.36 gm Sodium Cacodylate
   25.0 ml Glutaraldehyde (Electronmicroscopic grade)
   375 ml Double Glass Distilled Water

3. Diammine Silver
   0.5 ml .5% Cupric Nitrate
   12 ml 0.175 N NaOH (.69 gm NAOH/100 ml) Fresh weekly
   5.0 ml 95% Alcohol
   25 ml $AgNO_3$ (1.25 gm $AgNO_3$ in 25 ml Double Dist. Water)
   3.5 ml 25% NH_4OH

   Add reagents in order given

4. Reducing Agent
   100 ml 95% Alcohol
   894 ml Double Glass Distilled Water
   6.0 ml 1% Citric Acid
   1.0 ml 37% Commercial Formalin or Formaldehyde

5. Stop Bath
   10% Acetic Acid

1. Place gel into glutaraldehyde solution 1.5 min at 50°C with agitation.
2. Wash gel in 10% alcohol 4 x 3 min at 50°C with agitation.
3. Place gel in diammine silver for 3-4 min at 50°C.
4. Wash (2 times) 30 seconds in 10% alcohol at 50°C.
5. Place in reducing solution at 50°C for 1-2 min and observe intensity. If still developing, leave in 1 min more. However, if background begins to appear, stop immediately in 10% Acetic acid. then if you wish to, place back in Diammine Silver 2 min at 50°C. wash again in 10% alcohol and develop bands to desired intensity in reducer. Stop in Acetic acid and wash in H_2O. Wiping any excess silver from surface with a Q-tip dipped in water.

Dr. Merril: We're beginning to run a little bit short on time so I'd like to go fairly briefly through some points now. One of them concerns the background. Dr. Ansorge, yesterday, when he gave his talk, presented a good demonstration of the way the background staining and the protein staining really don't follow the same kinetics. It's important, whichever stain one chooses, to try to determine the kinetics for your stain and then to optimize it so that the maximum sensitivity can be achieved with the least background. There's a problem I've noticed with almost every stain that I've tried, that is, when you stain a gel, a 2-D gel in particular, bands at about 67,000 MW and 60,000 MW are often observed. I honestly don't know what those bands are and I was hoping that somebody here might have some idea. The bands are generally not visible unless you stain the gel for maximum sensitivity, although it varies from time to time. When I first saw those bands I thought it was due to the agarose used to cement the 1st dimension to the 2nd. I've actually run gels without putting agarose on top of them and I can still see those bands so I wonder if there are any suggestions? Yes!

Dr. Ansorge: In our experience we obtained bands in two cases. First, when we use mercaptoethanol we always see some two or three additional bands. Secondly I know that a group from Strasbourg had a certain agarose from which they were getting these bands as well.

Dr. Sammons: We also believe that mercaptoethanol is involved. Whether the mercaptoethanol itself is the band or whether mercaptoethanol causes the increased appearance of the bands is unclear. We have made several changes that reduce the problem. We now are running our gels with 2% mercaptoethanol and importantly we filter the stock the mercaptoethanol solution through .22 micron filters prior to use. If one looks in bottom of their mercaptoethanol bottle one will see little black precipitates. We have run several different lots of mercaptoethanol with similar results and are now using a lot of Eastman Kodak mercaptoethanol that works very well. The other thing that we do routinely which may contribute reducing the band is that we filter through .22 micron filters all of our reagent solutions to take out any particles that might be in there that might be aggregating.

Dr. Merril: I've tried some filtering and that hasn't removed the bands. It might be important to make a few comments on sensitivity at this point. There are many different variants of silver stains right now in the literature. Each person who has developed a stain usually optimizes it for the gels that they are running. If you want to try one of these stains it's very important to be very careful as to what the thickness was of the gels for the technique that you're following, and realize that you probably are going to have to modify that particular method if you want to use it for a thicker or thinner gel. It's not just a matter of adjusting the prewashes it's also a matter of adjusting the concentration of the components. When you put the gel into one solution and then you switch it to another solution the diffusion of chemicals into the gel and out of the gel depends n the gel characteristics and thickness. These variables will change the speed of this diffussion. On top of that, you've got the various chemical reactions that are going on, so it's not a simple problem.

Dr. Ansorge: I don't know exactly how the other processes work, but in my process I did notice this in fact the silver staining is mainly a business of the surfaces. Professor Neuhoff and his group noted this sometime ago, and we found that the process that we had been using is, in fact, thickness independent. The only thing which one has to change is the first fixation wash time and (in our technique) the second wash after the permanganate. Those seem to be the only two critical times for the process. When we increased, for instance, gel thickness from 1 mm to 1.5 mm we added 5 min for each of the above steps, and similar extensions seem to be working at least up to 3 mm thick gels.

Dr. Sammons: With the color silver stain method it really is a matter of adjusting just one step. One doesn't need to change the concentrations of any reagents, or change the solution volume to gel volume ratio. The time however does change. If one is staining a 1.5 mm gel, the reaction time in sodium hydroxide-formaldehyde is 10 min, or 9.5 to 10 minutes depending on how much orange background you want. If one goes to a .75 mm thick gel then one must cut the time back to 2 min, or 2 to 3 minutes. If one goes to a 200 micrometer thick gel then one must cut the time back to 25 seconds.

Dr. Merril: We should discuss quantitation. We've generally found that the silver stain is quantitative over about a forty fold range but the quantitation varies from one protein to the next. That is, different proteins have different slopes of staining.

We used eight different purified proteins and then we've also performed quantitation on spots from lymphocyte lysates and *E. coli* lysates with fairly good results.

Dr. Ansorge: We found that modifications of ionic strength or PH value in certain washes can change the color of bands as well as their relative intensities. It is a very interesting problem, and it is clear that when you are using silver staining you have to stick, for comparative purposes, to one technique. One should try several techniques and choose the one which works best in his problem.

Dr. Merril: For quantitation on a 2D gel, a number of spots are measured with a microdensitometer. We then analyse the ratio between the spots and find out which spots have constant ratios, comparing these from one gel to the next. We then define those spots with constant ratios as operationally constitutive and use them to normalize the densities from one gel to the next. We have found that we have problems with quantitation if we modify our techniques to enhance color effects. Enhancement of color often produced spots with negative staining (less density than the background), or spots with "punched out centers". I wonder if Dr. Sammons would comment on this?

Dr. Sammons: Yes, we don't get holes in our peaks or spots, but sometimes we get spots in our spots. The reason for this phenomenon is overloading for a given protein. For instance, occasionally yellow spots may have brown centers. Sometimes a brown protein may have a red center. This can be eliminated in all cases simply by diluting your loading sample to a lower concentration. Due to the wide dynamic range of protein concentration within a cell, we will have to run multiple gels of a sample at different concentrations.

Dr. Merril: It may be possible to extend the dynamic range of the stain by photographing the image during staining. In this way spots that would saturate can be detected and the density measured prior to saturation, from a photograph. Calibrated density wedges should be used, of course, next to the gel, so that variations due to the photographic process or gel scanning can be corrected.

Dr.Allen:It should be noted we all are still developing the silver stains.

Dr. Merril: Our time has run out. In closing - I would like to thank the panel and the audience.

ROUND-TABLE DISCUSSION ON: PEPTIDE FOCUSING*

Chairman: Pier Giorgio Righetti

Department of Biochemistry, University of Milano, Via Celoria 2, Milano 20133, Italy

Participants:

N.L. Anderson

Molecular Anatomy Program, Division of Biological and Medical Research, Argonne National Laboratory, Argonne IL 60439, USA

C. Cuono

Yale University, School of Medicine, 333 Cedar St., New Haven CT 06510, USA

E. Gianazza

Department of Biochemistry, University of Milano, Via Celoria 2, Milano 20133, Italy

V. Neuhoff

Max-Planck-Institut für experimentelle Medizin, Forschungsstelle Neurochemie, D-3400 Göttingen, FRG

* This round-table discussion was held on April 24, 1982.

Electrophoresis '82

ROUND TABLE DISCUSSION ON: PEPTIDE ISOELECTRIC FOCUSING

Chairman: Pier Giorgio Righetti

Department of Biochemistry, University of Milano, Via Celoria 2, Milano 20133, Italy

P.G.R.: Ten years have elapsed since my first experiments with Dr. F. Chillemi on peptide IEF (1). The success of our separations of synthetic hGH fragments was to be ascribed to the presence of a DNP-His residue along the peptide chain, which allowed gel scanning at 350 nm or fluorescence detection. The next year, together with A. Kopwillem, we started the first extensive investigation on separation of peptides by IEF and ITP (2).

There are two main problems with peptide IEF: analytical and preparative. In the analytical scale, peptides cannot be stained, after IEF, with the common stains for primary amino groups (e.g. ninhydrin, fluorescamine) since the carrier ampholytes in the background react very strongly too. Alternatively, if one tries to first fix the peptides in TCA and wash out the buffering ampholytes, there will be a concomitant, size-dependent, heavy loss of peptides from the gel matrix. My group has developed methodologies, over the years, to overcome these problems.

In the preparative scale, it appears that separation between peptides and commercial carrier ampholytes will never be achieved with present-day IEF systems. However, the newly developed IEF technique in immobilized pH gradients (3) might be a valid answer to these problems. I should like now to ask Dr. E. Gianazza to give us a brief introduction to this topic.

E.G.: let us first describe the problems connected with the staining of short peptides after IEF. If the gel is fixed in TCA, only peptides longer than 25 amino acids are fixed (this is true even when using as a fixing agent 100% TCA). Shorter peptides are lost, to different extents, in the washing solution. If, however, the gel is bathed directly in a colloidal dispersion of 0.1% Coomassie Brilliant Blue G-250 in 12.5% TCA (4), even peptides as short as 12 amino acids in length are fixed and stained. Righetti and Chillemi (5) have hypothesized that the mechanism for this fixation is

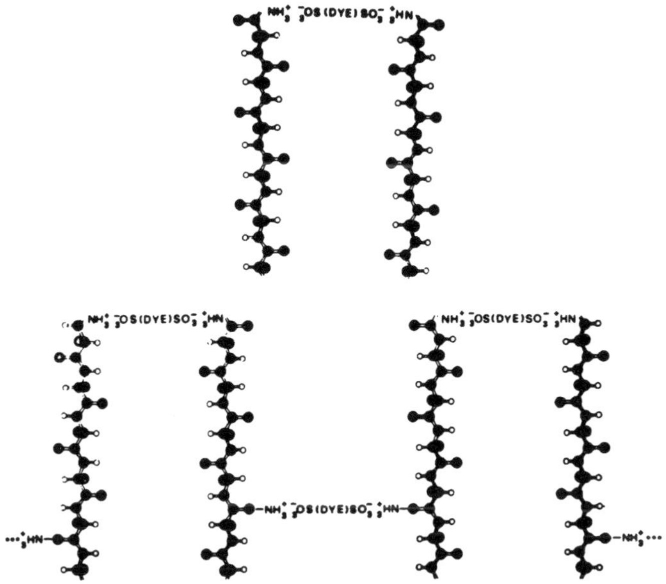

Fig. 1. Proposed dye-peptide binding model. A peptide with no basic residues (upper part) will only bind the dye via its $-NH_2$ terminus, thus forming a dimeric complex that readily splits because it is extremely unstable in H_2O. The lower part shows a peptide with an additional binding site in the chain (an ε-NH_2 group of Lys). In this case, a macromolecular aggregate is formed, which is trapped in the gel matrix and is insoluble in water. This model should not be seen as a linear, as depicted, but as a three-dimensional array of chains. The aggregate is further stabilized by stacking among aromatic rings of different dye molecules, made possible by the ionic interactions of the two sulphate groups of the dye with the peptide amino groups (from Righetti and Chillemi; see ref. 5).

due to the dye molecules acting as bridges over different peptide chains, via salt links between their sulphate groups and peptide amino groups(Fig. 1). This implies that, in addition to the terminal free $-NH_2$ groups, the peptides contain at least another amino group in the side chains available for binding. The dye-peptide interaction will also be favored by hydrophobic residues in the peptide chain, able to bind via hydrophobic interaction with the bulky aromatic rings in the CBB molecule. When using this direct CBB staining technique, it is also important to perform IEF in thin gels (0.5 mm or thinner) of relatively high %T (8-10%) so as to minimize diffusion phenomena during the fixation and staining step.

Unfortunately,by this method, shorter peptides (in the tri- to dodeca- peptide range) are lost. We have tried to cover this gap by using a combina-

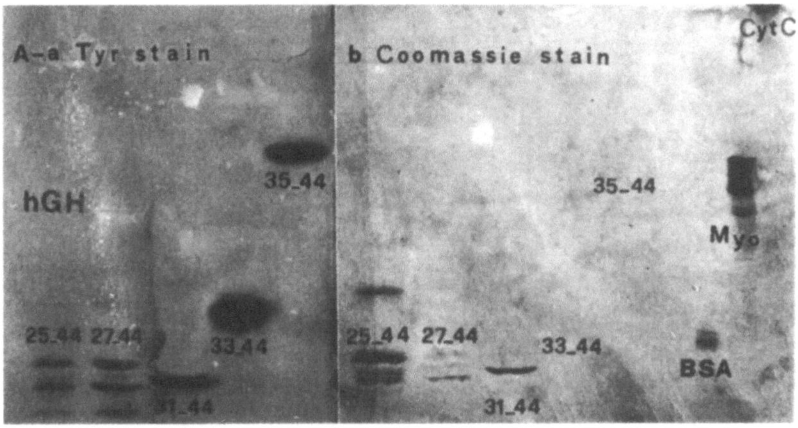

Fig. 2. Detection of oligopeptides. IEF separation in a 7%T gel, 0.3 mm thin, containing 2% Ampholine in the pH 3.5-10 range. Samples: 100 µg each of the human growth hormone sequences 25-44, 27-44, 31-44, 33-44 and 35-44. A: specific stain for Tyr (dark red bands on a pale yellow background); B: Coomassie Brilliant Blue G-250 stain. Notice the total loss of the two peptides 33-44 and 35-44 in the CBB stain. pH markers: bovine serum albumin (BSA), myoglobin (Myo) and cytochrome C (Cyt C) (from Gianazza *et al.*; see ref.7).

tion of ultrathin gels (6) (0.2 to 0.3 mm thick) and specific spray stains for individual amino acids. To this purpose, after the IEF step, the ultrathin gel is pasted to a sheet of Whatmann 3MM paper (slightly larger, in size, than the IEF gel) and immediately dried in the oven at 110°C. The paper bound gel can now be sprayed with any of the specific stains for the following amino acids: Arg, Met, Cys, Tyr, Trp and His. In this way, 30% of all the possible amino acids in an oligopeotide can be detected, with a sensitivity level of only a few µg of peptide per IEF zone (7) (Fig. 2). It is important that the gel is bound to an absorbing surface, like paper, and not to glass or plastic, since, once dried, its surface will be rather impermeable to the bulky, aromatic stains commonly used in specific amino acid spray cocktails. A method for direct His staining of focused peptides, in a wet, instead of dried gel, has also been independently described by Faupel and Von Arx (8), but it is not generally applicable to other specific stains (PA gels curl and shrink severely in the organic solvents usually employed in these staining solutions). A major drawback of this technique is that it is not fully compatible with 8M urea in the gel: urea decreases the specific stain sensitivity, moreover, upon gel drying, it forms a crust on the gel

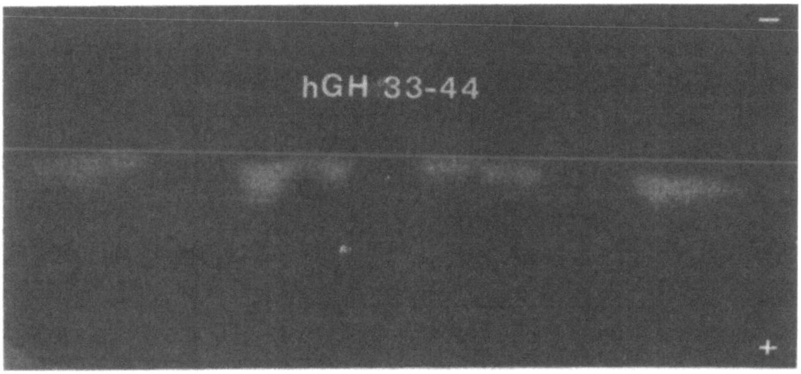

Fig. 3. Small-scale preparative IEF of hGH 33-44 fragment (150 μg/sample) on a polyacrylamide gel slab. The spots were visualized by a short (a few seconds) exposure to iodine. The gel was then left in an air stream until the background faded, the bands were cut, eluted in 80% acetic acid, and the eluate was then hydrolized and run on an automatic amino acid analyzer (from Gianazza *et al.*; see ref. 10).

surface. Alternative peptide solvents, instead of urea, could be: tetramethyl urea, dimethyl sulphoxide and dimethyl formamide, since they are fully compatible with the IEF process (9). At present, we are exploring other solvents with higher solubilizing properties, compatible with the gel matrix and fully inert toward peptides.

As an alternative to the two methods described above (general CBB stain and specific amino acid stains) we have tried to develop a general stain, independent from peptide length (10). When a focused and dried ultrathin layer gel is exposed to iodine vapors in a jar, the carrier ampholytes adsorb the iodine molecules quickly, giving a uniformly brown stain in the gel-paper sheet. By a mechanism still not completely understood, the focused peptides inhibit this adsorption, thus being detected by negative staining, as white zones over a brown background. The staining is fully reversible, since no chemical reaction occurs, and is developed within a few feconds to a few minutes of exposure to iodine. In order to obtain a uniform brown color, 4 to 6% carrier ampholytes are used in the focusing gel, instead of the usual 2% levels, so that the gaps in between the "stripes" of focused Ampholine are smoothed. A drawback of the present method is its low sensitivity, a minimum of 30 μg peptide/zone being required for detection. However this technique can be exploited for small scale preparative applications: after iodine sublimation, the peptides can be eluted from the gel in 80% acetic acid with recoveries up to 85% (Fig. 3). As an alternative to all these staining procedures, Steck *et al.* (11) have tried, after IEF of peptides, to cross-link them in 5% formaldehyde and then detect them in common Coomassie stains in acid-alcohol. This method, however, does not seem to be satisfactory because also carrier ampholytes are fixed and stained. IEF of peptides is now of widespread use, especially in the field of hormone research (12-16).

Notwithstanding the success of analytical IEF of peptides, its preparative counterpart is still not quite feasible. This is mainly due to the fact that carrier ampholytes and peptides have quite similar physico-chemical properties, both in terms of net charge and mass, so that their separation, after IEF, by gel chromatography, dialysis or ion-exchange is an extremely complicated proposition. Notwithstanding the fact that carrier ampholytes generally have a molecular mass ($\overline{M}_r$) not exceeding 1000, their separation even from higher $\overline{M}_r$ peptides is complicated by the fact that they tend to aggregate in solution (17). Recently, Gelsema *et al.* (18) have studied the feasibility of separating carrier ampholytes from peptides by hydrophobic interaction chromatography on Octyl Sepharose. However, one cannot expect this method to be generally applicable, because most peptides below 5000 in molecular mass are not adsorbed to amphiphilic gels (19), which is in agreement with the author's conclusion: "the separation of peptides from carrier ampholytes by hydrophobic interaction chromatography in the presence of salt is in general not feasible for peptides with $\overline{M}r$'s in the most interesting range (say 2000-8000)". As an alternative to this method, one could try to focus against a background of small $\overline{M}_r$ carrier ampholytes (Ampholines have an average $\overline{M}_r$ = 750). This approach has been used by Gasparic and Rosengren (20) who have synthesized small $\overline{M}_r$ Ampholine ($\overline{M}_r$ = 300) but these compounds have never been tried for actual separation of peptides and are not available on the market. Righetti and Hjertén (21) have tried to solve this problem at the other end of the $\overline{M}_r$ scale: instead of using smaller and smaller ampholytes, they have synthesized extremely large carrier ampholytes. A giant polyethylene imine ($\overline{M}_r$ 40,000-60,000) is mixed in a linear gradient of acrylic acid in a flow-through system and let to react at 80°C for 70 hours (Fig. 4). Giant carrier ampholytes ($\overline{M}_r$ range 50,000-90,000) are thus obtained. Albeit these compounds interact strongly among themselves, especially in acidic pH ranges, giving rise to aggregates and precipitates, they interact only wea-

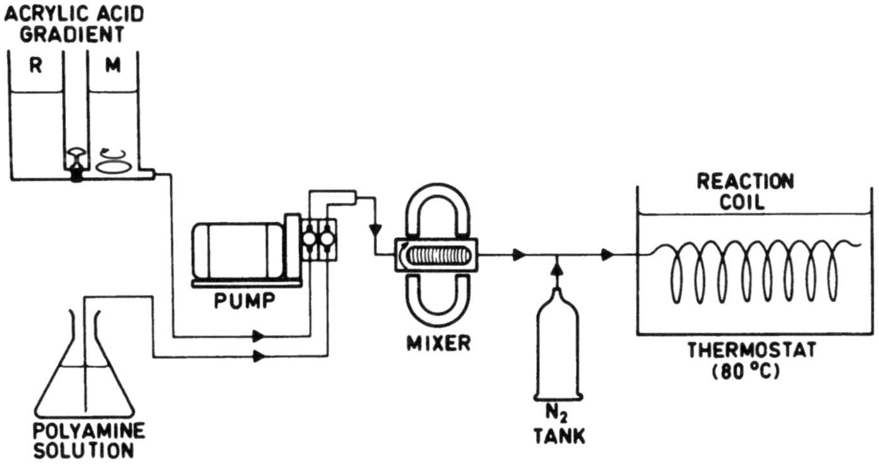

Fig. 4. Synthesis of giant carrier ampholytes. A 10% polyethylenimine solution is mixed with a linear 2-18% (v/v) acrylic acid gradient with the aid of an ultrograd mixing chamber (in this chamber the magnet bar, immersed in the magnetic field, rotates over its long axis). The flowing gradient is segmented, every 1 ml, with a small nitrogen bubble. Once the reaction coil is filled, the reaction is let to proceed for 70 hours at 80°C (from Righetti and Hjertén; see ref. 21).

kly and reversibly with proteins and no interactions are apparent with model dipeptides (His-Ser, His-Met, His-Phe and His-Lys) which could be recovered with ca. 85% yields from preparative runs. All these data indicate that, although with some difficulties, analytical and preparative separations of peptides are feasible. Even these last difficulties might be totally eliminated with the use of the new technique of immobilized pH gradients.

P.G.R.: Thank you Dr. Gianazza. Are there any specific questions or comments on peptide purification?

Dr. Neuhoff: A few years ago we have published a short paper describing dansylation of proteins before electrophoresis (Hoppe Seyler's Z.Physiol. Chem. 361, 77-78 (1980)). It appears that dansyl groups bound to proteins do not alter their electrophoretic mobility. Perhaps this pre-staining method could be excellent for revealing peptides after IEF.

P.G.R.: But in theory you could expect to change the isoelectric point of a protein or a peptide because you block a primary amino group.

Dr. Neuhoff: Not quite, since you are substituting a primary amino group in proteins with the amino group of dansyl chloride. Apparently, the pK's of the two groups are quite similar, so that only in a few cases there is a change in mobility of proteins.

P.G.R.: Perhaps dansylation might be a good technique for peptide detection in IEF but I still think that, in the case of proteins, where many groups are available for reaction, you might introduce a microheterogeneity due to conformational transitions and trace labelling, so that a pure protein will split into multicomponents in narrow pH gradients in IEF. But definitely this seems to be worthwhile trying in the case of peptides.

Dr. Cuono: You have described how it is impossible, from a preparative point of view, to separate a focused peptide from commercially available carrier ampholytes. But this should not be a problem any longer with our multicomponent mixture (22) now available from Polysciences (Poly/Sep 47).

P.G.R.: Yes, possibly, but your 47 component mixture has only been made available now, so I am afraid no one has tried it as yet for preparative peptide fractionation. By the way, what is the highest $\overline{M}_r$ of your compounds?

Dr. Cuono: around 200 daltons.

P.G.R.: This is quite small indeed, as compared with carrier ampholytes. I think one should be able to separate them even from relatively small peptides. Are there any other questions?

Dr. Anderson: Cleveland *et al.* (23) have devised an ingenious method for peptide mapping by digesting with proteases a protein, still entrapped in an electrophoretic gel slice, and then running it in an SDS-gel of high %T. Do you see this as a generally applicable method?

P.G.R.: I think that in the Cleveland technique (23) there are in general no staining problems because he performs only a partial digestion, leading to relatively large fragments. In the published pictures I have seen, the peptide pattern is in general not quite complex. This is also true in the case of the fragmentation technique by chemical cleavage of aspartyl-prolyl bonds of Sonderegger *et al.* (24). But upon extensive digestion, I doubt one could generate a good peptide map in an SDS gel, even at high %T, since discrimination by size in the low mass range would be difficult to achieve. As an alternative to this method, or complementary to it, I suggest generating peptide maps in the single IEF dimension. We have done that with globin chain (25) and casein (26) digests with an excellent resolution.

Well, if there are no other questions, I will give back the microphone to the meeting organizer, Dr. Stathakos. Thank you, Dimitri, it was an excellent meeting; I hope to see you folks next year in Japan.

Abbreviations: P.G.R.: Pier Giorgio Righetti; E.G.: Elisabetta Gianazza; IEF: isoelectric focusing; SDS: sodium dodecyl sulphate; TCA: trichloroacetic acid; CBB: Coomassie Brilliant Blue; PA: polyacrylamide; %T: (grams acrylamide + grams cross-linker)/100 ml solution; hGH: human growth hormone; DNP: dinitro phenyl-; ITP: isotachophoresis.

References

1. Righetti, P.G., Drysdale, J.W.: Ann. N. Y. Acad. Sci. 209, 163-186 (1973)
2. Kopwillem, A., Chillemi, F., Righetti, A.B.B., Righetti, P.G.: Prot. Biol. Fluids 21, 657-665 (1973)
3. Bjellqvist, B., Ek, K., Righetti, P.G., Gianazza, E., Görg, A., Postel, W., Westermeier, R.: J. Biochem. Biophys. Methods 6 (1982) in press
4. Blakesley, R.W., Boezi, J.A.: Anal. Biochem. 82, 580-582 (1977)
5. Righetti, P.G., Chillemi, F.: J. Chromatogr. 157, 243-251 (1978)
6. Görg, A., Postel, W., Westermeier, R.: Anal. Biochem. 89, 60-70 (1978)
7. Gianazza, E., Chillemi, F., Gelfi, C., Righetti, P.G.: J. Biochem. Biophys. Methods 1, 237-251 (1979)
8. Faupel, M., Von Arx, E.: J. Chromatogr. 157, 253-257 (1978)
9. Righetti, P.G., Gianazza, E., Brenna, O., Galante, E.: J. Chromatogr. 137, 171-181 (1977)
10. Gianazza, E., Chillemi, F., Righetti, P.G.: J. Biochem. Biophys. Methods 3, 135-141 (1980)
11. Steck, G., Leuthard, P., Bürk, R.R.: Anal. Biochem. 107, 21-24 (1980)
12. Noble, R.L., Yamashiro, D., Li, C.H.: Int. J. Peptide Protein Res. 10, 385-393 (1977)
13. Biscoglio de Jimenez Bonino, M.J., Cascone, O., Arnao de Nué, A.I.,

Santomé, J.A., Sanchez, D., Oré, R., Villavincencio, M.: Int. J. Peptide Protein Res. 17, 374-379 (1981)

14. Schwartz, G.P., Burke, G.T., Katsoyannis, P.G.: Int. J. Peptide Protein Res. 17, 243-255 (1981)

15. Kohnert, K.D., Schmid, E., Zuhlke, H.,Fiedler, H.: J. Chromatogr. 76, 76, 263-267 (1973)

16. Bibring, T., Baxandall, J.: Anal. Biochem. 85, 1-14 (1978)

17. Bianchi Bosisio, A., Snyder, R.S., Righetti, P.G.: J. Chromatogr. 209, 265-272 (1981)

18. Gelsema, J.W., De Ligny, C.L., Blanken, W.M., Hamer, R.J., Rozen, A.M. P., Bakker, J.A.: J. Chromatogr. 196, 51-58 (1980)

19. Hjertén, S.: J. Chromatogr. 101, 281-288 (1974)

20. Gasparic, V., Rosengren, A.: in: Isoelectric Focusing (Arbuthnott, J.P., Beeley, J.A., eds.) Butterworths, London, pp. 178-181 (1975)

21. Righetti, P.G., Hjertén, S.: J. Biochem. Biophys. Methods 5, 259-272 (1981)

22. Cuono, C.B., Chapo, G.A.: Electrophoresis 3, 65-75 (1982)

23. Cleveland, D.W., Fischer, S.G., Kirschner, M.W., Laemmli, U.K.: J. Biol. Chem. 252, 1102-1106 (1977)

24. Sonderegger, P., Jaussi, R., Gehring, H., Brunschweiler, K., Christen, P.: Anal. Biochem. 122, 298-301 (1982)

25. Gianazza, E., Righetti, P.G.: Prot. Biol. Fluids 27, 715-718 (1979)

26. Righetti, P.G., Muneroni, P., Todesco, R., Carini, S.: Electrophoresis 1, 37-42 (1980)

AUTHOR INDEX

SUBJECT INDEX